Stem Cell Engineering

Gerhard M. Artmann · Stephen Minger · Jürgen Hescheler
Editors

Stem Cell Engineering

Principles and Applications

Springer

Editors
Prof. Dr. Gerhard M. Artmann
Institute of Bioengineering and Centre of Competence in Bioengineering
Aachen University of Applied Sciences
Heinrich-Mußmann-Straße 1
52428 Jülich
Germany
artmann@fh-aachen.de

Prof. Dr. Jürgen Hescheler
Universitätsklinikum Köln
Inst. Neurophysiologie
Robert-Koch-Str. 39
50931 Köln
Germany
j.hescheler@uni-koeln.de

Dr. Stephen Minger
King's College London
Wolfson Centre for Age-Related Diseases
GKT School of Biomedical Sciences
London
United Kingdom SE1 1UL
stephen.minger@kcl.ac.uk

ISBN 978-3-642-11864-7 e-ISBN 978-3-642-11865-4
DOI 10.1007/978-3-642-11865-4
Springer Heidelberg Dordrecht London New York

Library of Congress Control Number: 2010927175

"We, the editors edited to the best of our knowledge. However we would like to emphasize, that the chapter authors are fully responsible for the contents of their articles, citations etc."

Cover design: WMXDesign, Heidelberg

Printed on acid-free paper

Springer is part of Springer Science+Business Media (www.springer.com)

About the Authors

Prof. Arti Ahluwalia, PhD She obtained her B.Sc. in physics and Ph.D. in bioengineering; Arti Ahluwalia is associate professor of bioengineering at the Faculty of Engineering, University of Pisa, and vice director of the Interdepartmental Research Center "E. Piaggio." She is also affiliated with National Council of Research Institute of Clinical Physiology (CNR-IFC) and carries out part of her research activities within the laboratories of the Institute. Her main scientific interests are focused on the interaction between biological systems and man-made devices or structures. In this context, her research activities span from biomolecular films, bioreactors, surface engineering, and biosensing to microfabrication and biomaterials for tissue engineering.

Omonigho Aisagbonhi, MD, PhD She is an M.D./Ph.D. graduate student in the Medical Scientist Training Program at Vanderbilt University Nashville, TN, USA. She obtained her B.Sc. in biochemistry from California State University, Long Beach, in 2004.

Prof. Dr. habil. Gerhard M. Artmann He studied physics at the Technical University of Dresden (1970–1974). He obliged to leave East Germany in 1985 and became a Ph.D. student at RWTH Aachen (1986–1988). Since 1989 he is a professor at Aachen University of Applied Sciences where he was a dean of the Faculty of Applied Physics (2002–2004). He habilitated in 1999 at TU Ilmenau and is chair of the Center of Competence in Bioengineering since 2003 as well as managing director of the Institute of Bioengineering.

Kuppusamy Aravindalochanan He has completed his bachelor of engineering in Madras University, India. He acquired his masters in biomedical engineering from University of Applied Sciences, Albstadt-Sigmaringen University, Germany. Since 2004, he is working as a scientist in Laboratory for Sensors, Department of Microsystems Engineering, Freiburg, Germany. His fields of interest are multi-physics simulation, MEMS, and electro-chemical sensor development for clinical and diagnostic applications.

Prof. Andrew H. Baker He is a non-clinical professor in molecular medicine and the British Heart Foundation Glasgow Cardiovascular Research Centre at the University of Glasgow, Glasgow, UK. He specialized in vascular biology, gene therapy and, more recently, stem cell biology and the application of genetic engineering to manipulate human pluripotent stem cells.

Prof. Christoph Becker, PhD He studied biology at the RWTH Aachen University, Aachen, Germany. In his Ph.D. studies he investigated regulation mechanisms of subcellular trafficking of glucose transporters in cardiomyocytes at both the Department of Physiology, Medical Faculty, RWTH Aachen University and the Department of Biochemistry and Molecular Biology, Biological Faculty, University of Barcelona. As a postdoc he joined the Interdisciplinary Center of Clinical Research "BIOMAT" at the University Hospital Aachen and the National Cancer Institute at the National Institutes of Health in Bethesda, MD, USA. Since 2003 CB is head of the laboratory of urology at the University Hospital Aachen. In several projects he investigates the myogenic differentiation potential of bone marrow stromal cell and develops collagenous biomaterials for the tissue engineering of urinary bladder and ureter, respectively.

Dr. Anne Bernhardt Anne Bernhardt studied biochemistry at Martin-Luther University Halle-Wittenberg. She worked at the Max Planck Research Unit for Enzymology of Protein Folding in Halle and received Ph.D. in 1997. She is a member of the group "Tissue Engineering and Biomineralization" at the Max Bergmann Center for Biomaterials/Institute of Materials Science of Technical University, Dresden, Germany, since 2001.

Paul R. Bidez III, PhD He earned his B.S. from West Chester University in 1997, his MBA from Pennsylvania State University in 2002, and his Ph.D. from Drexel University in 2006. He is also a Certified Six Sigma Black Belt. With funding from the National Aeronautics Space Administration (NASA) he completed his MBA working on novel pathways of stem cell differentiation. Upon completion of his Ph.D., he joined the Research Planning and Integration subdivision of Merck Research Laboratories, a division of Merck & Co, North Wales, PA, USA.
Dr. Bidez currently applies Six Sigma and Change Management Methodologies to a variety of opportunities in the drug discovery and development processes.

Thomas Brevig, MD, PhD He is director of Global Research in Labware & Speciality Plastics in Thermo Fisher Scientific, Roskilde, Denmark. He graduated in medicine in 1998 and received his Ph.D. in neuroscience in 2001, both from University of Southern Denmark. He has made scholarly contributions in the fields of transplantation immunology, brain repair, biomaterials, and cell culture and analysis.

Melissa Brown, BSc, PhD She received a B.S. in bioengineering from Syracuse University in 2004 and a PhD in biomedical engineering from Duke University, Durham, NC USA in 2009. Her Ph.D. dissertation focused on cord blood-derived endothelial progenitor cells and their use in prevention of restenosis in vessels that have undergone various atherosclerosis treatment methods. She received fellowships from an Institutional Training Grant on Biomolecular and Tissue Engineering and the American Heart Association.

Cindy Cheng, BSc She is a biomedical engineering graduate student studying under Dr. George A. Truskey at Duke University, Durham, NC, USA. Cindy received a B.S. in Biomedical Engineering from Duke University in 2008. She is a Fellow in the Center for Biomolecular and Tissue Engineering at Duke and is supported by an Institutional Training Grant from the National Institutes of Health.

Matthew Dalby, PhD He is a lecturer in cell Engineering at the University of Glasgow, UK. He started the post in the Faculty of Biomedical and Life Sciences after completing a 5-year BBSRC David Phillips research fellowship. Prior to this he was a research assistant with Profs. Adam Curtis and Chris Wilkinson and Dr. Mathis Riehle and a Ph.D. student at the IRC in Biomedical Materials, Queen Mary College, University of London. His interests are bone cell engineering, nanotopography, and cell mechanotransduction; and he works within an interdisciplinary environment comprising biologists, orthopedic surgeons, and engineers. He has published around 60 papers and reviews.

Bernd Denecke, PhD He studied biology at the Ruhr University Bochum and received his Ph.D. at the Institute of Cell Biology (Tumor Research), University of Essen Medical School. After his postdoctoral fellowship at the Institute of Medical Microbiology, Hannover Medical School (MHH), he started his work at the Interdisciplinary Center for Clinical Research on Biomaterials and Material–Tissue Interaction of Implants "BIOMAT" of the RWTH Aachen, Germany. Since 2005 he is head of the Chip-Facility which he established at the Interdisciplinary Center for Clinical Research "BIOMAT".

Lucia Denk, PTA Lucia Denk finished her apprenticeship as pharmaceutical technical assistant in 2000. Afterward she worked at a pharmacy. Since 2002 she has a position as technical assistant at the Institute of Cellular and Molecular Anatomy, University of Regensburg, Regensburg, Germany.

Prof. Chris Denning He is an associate professor and reader in the Wolfson Centre for Stem Cells, Tissue Engineering and Modelling at the University of Nottingham, Nottingham, UK. His interests are in improving culture, scalability, cardiac differentiation, and transgenesis in human embryonic and induced pluripotency stem cells with a view to drug screening and in vitro modeling of genetic disease.

Paolo Di Nardo, MD He is a cardiologist, directs the Laboratory of Cellular and Molecular Cardiology at the University of Rome Tor Vergata and is member of the Board of the National Institute for Cardiovascular Research (INRC), Italy. Scientific advisor of major international organizations and member of international scientific societies, he is author of several papers published in peer-reviewed international scientific journals. In 1994, he organized the First International Congress in which the possibility of heart regeneration in mammalians was analyzed. Since then, his major interests have been in stem cell and tissue engineering technology.

Ilya Digel, PhD He was born in 1973 in Zhambyl (Kazakhstan). In 1995 graduated from Kazakh National University (biology and chemistry) in Almaty with a Diploma of Honors. He received his Ph.D. in microbiology from the same University (1998). Since 2001, he is research member in the Laboratory of Cellular Biophysics at Aachen University of Applied Sciences. He is leading a group working on molecular biophysics, microbiology, and cell biology.

Rita R. Fiñones, PhD Rita Fiñones, Ph.D. candidate in Materials Science and Engineering, University of California at San Diego, CA, USA, is a graduate student in the laboratory of Dr. Martin Haas. Her dissertation is focused on developing a stem-cell-inspired model of prostate cancer progression. She is interested in the translation of stem cell concepts to the study of cancer and to the development of regenerative therapies.

Lesley Forrester, PhD She received her Ph.D. in the clonal analysis of hematopoietic stem cells from the University of Edinburgh, Edinburgh, in 1986 and carried out postdoctoral training in Toronto, ON, Canada. She established her own group at Edinburgh University in 1994 where her work is focused on the production of hematopoietic cells from embryonic stem cells.

Jane Frimodig, PhD Jane Frimodig is a Ph.D. candidate in materials science and engineering, currently working in the Department of Bioengineering at the University of California, San Diego, CA, USA. She received a master of science degree with distinction in materials engineering from California State University Northridge. In 2007 and 2008 Ms. Frimodig received the Chancellor's Collaboratories Grant, awarded to promote interdisciplinary collaborations.

Andreas Geerts, Dipl. Biol. He received diploma in biology (1993) at the University of Cologne (genetics, biochemistry, pharmacology). Scientific work in Institut für Enzymtechnologie der Universität Düsseldorf, Max Planck Institut für Züchtungsforschung Cologne, and Centro National de Biotechnologia Universidad Autonoma de Madrid. Scientist at institute for target discovery at Bayer Schering Pharma AG, Wuppertal, Germany.

Michael Gelinsky, PhD He studied chemistry at the Albert Ludwigs University in Freiburg (Breisgau) and did his Ph.D. in the field of bioinorganic chemistry. Since 1999 he is working at the Institute of Materials Science of Technical University in Dresden, Germany. Since 2002 he is head of the research group "Tissue Engineering and Biomineralization" which is located in the newly founded Max Bergmann Center of Biomaterials (Institute of Materials Science, Technical University, Dresden). He and his group focus on the development of new biomaterials, especially for bone and cartilage reconstruction, scaffold-based tissue engineering, novel methods for scaffold fabrication, and bone biomineralization.

Stefan Golz, PhD He received Ph.D. in molecular biology (1999) at the University of Cologne (Institute for Genetics) about protein–protein interactions of T4 endonuclease VII. He is senior scientist at the Institute for Target Discovery, Bayer Schering Pharma AG, Wuppertal, Germany.

Sabrina Gordon-Keylock, PhD She completed her Ph.D. in 2007 in Dr. Lesley Forrester's laboratory on the hematopoietic differentiation of murine embryonic stem cells and is currently carrying out her postdoctoral training in the Institute for Stem Cell Research at the University of Edinburgh, Edinburgh.

David A. Gough, PhD He received his Ph.D. from the University of Utah in 1974 and was a postdoctoral fellow at the Joslin Clinic of the Harvard Medical School. He is a founding fellow of the American Institute of Medical and Biological Engineering and a recipient of the M. J. Kugel Award presented by the Juvenile Diabetes Foundation and the Jacobs School's Teacher of the Year Award in 1996. He is a former chair of the Bioengineering Department at UCSD, CA, USA.

Marijke Grau, PhD She received master of science from the University of Bremen (Germany), in 2006. She recently completed her PhD at the RWTH Aachen University (Germany). She now continues with her postdoctoral research studies at the Institute for Molecular and Cellular Sports Medicine (German Sports University Cologne, Germany). She investigates the regulatory mechanisms of the red blood cell nitric oxide synthase and the physiological and pathophysiological relevance of the enzymatic synthesized nitric oxide.

Rylie Green, PhD She completed her Ph.D. in neural interfaces in the Graduate School of Biomedical Engineering, at the University of New South Wales, Australia, where she is now undertaking postdoctoral research studies. Her expertise is focused on the development of biosynthetic systems for improving the interactions of tissues at device interfaces. Dr. Green is a member of the Institute of Engineers, Australia, and the Australasian Society for Biomaterials and Tissue Engineering (ASBTE). Dr. Green has received a Fresh Science Award which recognises upcoming scientists throughout Australia. Rylie has also received the Ruldolf Cimdens Award from the European Society of Biomaterials and number of a international travel grants from ASBTE and UNSW.

Martin Haas, PhD He received his B.Sc. degree in electrical engineering from the Technicon, Israel Institute of Technology and his Ph.D. in biophysics from the University of California, Berkeley, CA, USA. During his postdoctoral research at the Salk Institute, he worked on the DNA tumor viruses under the direction of Drs. Renato Dulbecco and Marguerite Vogt. As senior scientist at the Weizmann Institute and then as professor at UCSD, Dr. Haas studied radiation-induced leukemia in mice, followed by work in the tumor suppressor field. Based on his extensive experience in the areas of virology, cell biology and the use of gene transfer vectors, Dr. Haas recently entered the field of stem cell research.

Prof. Antonis Hatzopoulos, PhD He is an associate professor of Medicine and Cell and Developmental Biology at Vanderbilt University, Nashville, TN, USA. He obtained his B.Sc. in chemistry from Aristotelian University of Thessaloniki, Greece, and his Ph.D. degree from Northwestern University, Evanston, IL, USA.

Prof. Dr. med. Dr. h.c. Jürgen Hescheler He is Chairman and Director of the Institute of Neurophysiology at the University of Cologne. He has been working with embryonic stem cells of the mouse for over 20 years. Beginning with studies on cellular signal transduction, he has defined many important basic aspects both of fundamental research and of clinical applications. He was the first scientist worldwide to perform electrophysiological experiments on stem cells thus pioneering the establishment of stem cell research for application in transplantation medicine. In 2002 he was the first scientist in Germany to obtain permission to work with human hmbryonic stem cells. In March 2004 he was appointed coordinator of the European Consortium FunGenES (Functional Genomics of Engineered Embryonic Stem Cells) followed by CRYSTAL (Cryobanking of Stem Cells for human therapeutic application) in 2005 and ESNATS (Embryonic stem cell-based novel alternative testing strategies) in 2007. He is also coordinator of the BMBF consortium "iPS and adult bone marrow cells for cardiac repair", which started work in March 2009. In 2005 he founded the German Society for Stem Cell Research.

Isgard S. Hueck, MSc Born in Germany, she studied biology at the Westfaelische Wilhelms University of Munster and received her license as cytopathologist in Cologne, Germany, in 1987. After many years of clinical work experience in hematology, cancer diagnosis, and therapy, she received a master of science degree in biomedical engineering at the Aachen University of Applied Sciences, Germany, in conjunction with the University of California, San Diego, CA, USA, in 1998. Based on her extensive expertise in the areas of cellular engineering, cytohistopathology, and cancer research, Isgard Hueck is a key connection between interdisciplinary and international cooperation in biomedical research. Her main research interests are in the studies of cellular cancer diagnosis, angiogenesis, and stem cells.

Prof. Gerhard Jakse, MD He studied medicine at the Medical Faculty of Graz. In several temporary employments at the New York State Hospital in Utica and Rochester and at Urology Departments in Klagenfurt, Linz, Innsbruck, Mainz, and Iowa City, he underwent professional training in both general medicine and urology. Following his venia legendi he became associate professor of urology at the University of Innsbruck and continued his career as full professor of urology at the Technical University of Munich and the RWTH Aachen University, Aachen, Germany. GJ is member of several national and international Urological Societies and collaborates in various Urological Journals.

Nicole M. Kane, PhD She is a postdoctoral research associate at the British Heart Foundation Glasgow Cardiovascular Research Centre at the University of Glasgow. Her research is focused on the genetic manipulation of human embryonic and induced pluripotent stem cells to further delineate pluripotency and differentiation commitments, in particular to a cardiovascular lineage.

Prof. Malte Kelm, MD He is university professor of medicine and chairmen of the Department of Cardiology, Pneumology, and Vascular Medicine University Hospital Düsseldorf (Germany). Beside his interests including interventional catheterization and cardiac imaging techniques, he has developed a profound expertise in the underlying mechanisms and therapy of coronary micro- and macroangiopathy with special focus on molecular mechanisms of endothelial dysfunction and nitric oxide pathways.

Stefanie Keymel, MD She studied at Heinrich-Heine University Düsseldorf and received M.D. for the investigation of EPC and endothelial dysfunction with aging. Medical and scientific education was obtained at the University Hospital Aachen (Germany) and since 2009 in the Department of Cardiology, Pneumology, and Vascular Medicine University Hospital Düsseldorf (Germany) and since 2009 in the Department of Cardiology, Pneumology, and Vascular Medicine University Hospital Düsseldorf (Germany). Research was focussed in the field of NO and microvascular biology.

Jochen Kieninger studied physics and micro system technology at the University of Freiburg. He received his diploma in 2003 ("Dipl.-Ing. Mikrosystem-technik."). Afterwards he worked at the Laboratory for Sensors (Prof. Gerald A. Urban) in the field of electrochemical sensors, micro fluidics, and thin-film technology. Since 2008 he is head of the group "Sensors for cell culture monitoring & micro technology". 2009 he joined the School of Soft Matter Research of the Freiburg Institute for Advanced Studies (FRIAS) in the group of Prof. Andreas Manz.

John Paul Kirton, PhD He received his B.Sc. in biochemistry in 2001 and a Ph.D. in molecular biology (2006) following studies at the Wellcome Trust Centre for Cell-Matrix Research, The University of Manchester.

He then attained a postdoctorate position in vascular biology within the same department. His current postdoctorate research role is held at King's College London, UK, studying vascular and stem cell biology.

Petra Kleinbongard, PhD She studied biology and received the master of science 1999 from the Ruhr University Bochum (Germany). She has done her Ph.D. in biology in 2003 in cooperation between the Division of Cardiology, Heinrich-Heine-University Düsseldorf (Germany) and the Department of Biology, Ruhr University Bochum (Germany). Since 2003 she is postdoctoral fellow (Institute of Pathophysiology, University of Essen Medical School, Essen, Germany) and did research in the field of circulating blood, blood cell functions, and interactions with vascular cells.

Frank Kloss He attended both Dental School at the University of Würzburg and Medical School at the University of Erlangen-Nürnberg, where he also started his medical career at the Department for Maxillofacial and Oral Surgery. From there, he moved on to the Medical University Innsbruck, Austria, to finish specialty training. He now holds a residency in oral-, maxillofacial surgery and guides various research projects focused on bone implantology and cancer. He is winner of the Scientific-Hans-Pichler-Award 2008 granted by the Austrian Society for Oral-, Maxillofacial Surgery. He is a research committee member of the International Bone Research Association (IBRA).

Prof. Karl-Heinz Krause, MD He is professor of medicine at the University of Geneva. He has two major research interests: (i) oxidative stress and NOX family NADPH oxidases and (ii) neuronal differentiation and brain development, using pluripotent stem cells as models and tools. He is elected member of the Swiss Academy of Medical Sciences, as well as the American Society for Clinical Investigation.

Prof. J. Yasha Kresh, PhD He is currently professor of cardiovascular medicine and surgery, research director of Cardiothoracic Surgery and Cardiovascular Biophysics Laboratory at the Drexel University College of Medicine, Philadelphia, PA, USA. He received his Ph.D. in biomedical (cardiovascular) engineering from Rutgers.
He works at the interface of medicine and engineering, dedicated to the discovery and development of innovative methodologies for the treatment of system disorders. He is fellow of the American College of Cardiology, American Heart Association, American Institute for Medical and Biological Engineering, Biomedical Engineering Society and senior member of many biological, medical, and engineering organizations.
He resides in Pennsylvania with his number one wife (Myrna) and their two black Labradors (Gidget and Melody) which do not retrieve.

Prof. Peter I. Lelkes, PhD He is the Calhoun Chair Professor of cellular tissue engineering at Drexel University's School of Biomedical Engineering, Science and Health Systems (BIOMED) and honorary professor at the Changchun Institute of Polymer Chemistry and Physics, at Chinese Academy of Sciences, and at the Aachen University of Applied Sciences, Germany.
Dr. Lelkes earned his Ph.D. in 1977 from the Technical University (RWTH) in Aachen, Germany. He joined Drexel University in 2000 after working for 5 years at the Weizmann Institute of Science (Rehovot, Israel), 5 years at the National Institute of Health (Bethesda, MD), and 12 years at the University Wisconsin.

Prof. Günter Lepperdinger He was principal investigator and section head at the Institute for Biomedical Aging Research of the Austrian Academy of Sciences in Innsbruck, Austria, since 2002. He is an adjunct professor of biochemistry at the University of Salzburg and an adjunct professor of developmental biology at the University of Innsbruck. He is a recipient of the Christian-Doppler-Award, the Best of Biotechnology Award, and the Werner-Welzig-Award; he is an APART fellow of the Austrian Academy of Sciences, Burgen Scholar of the Academia Europeae, and Elected Life-Time Member of the International Society for Hyaluronan Sciences. He served as an associate editor for "Experimental Gerontology – Elsevier" and he currently is section editor for Regenerative Medicine and Technology of Karger's international journal *Gerontology*.

Dr. Anja Lode She studied biology (University Potsdam) and did Ph.D. in genetics (Technical University Dresden, 2001). She is a member of the group "Tissue Engineering and Biomineralization" at the Max Bergmann Center for Biomaterials/Institute of Materials Science of Technical University Dresden, Germany, since 2002.

Penny Martens, PhD She is a lecturer in the Graduate School of Biomedical Engineering at the University of New South Wales, Australia. She obtained her PhD in Chemical Engineering from the University of Colorado in 2003. Her main research activities involve the development of biosynthetic polymeric hydrogel composites for use in biomaterial applications and the evaluation of their material and biological properties. She is currently serving as treasurer of the Australasian Society for Biomaterials and Tissue Engineering and is on the board of the Federation of Australian Scientific and Technological Societies. Penny has also been awarded a 2009 NSW Young Tall Poppy Science Award.

Dr. Stephen Minger He was appointed the Global Director for Research and Development for Cell Technologies at GE Healthcare in September 2009. Dr. Minger received his PhD in Pathology (Neurosciences) in 1992 from the Albert Einstein College of Medicine. After post-doctoral work in CNS gene therapy, neural transplantation and neural stem cell biology, he moved to the UK in 1996 and was appointed a Lecturer in Biomolecular Sciences at King's College London in 1998. Over the past 18 years, his research group has worked with a wide range of tissue-derived stem cell populations, as well as mouse and human embryonic stem (ES) cells. In 2002, his research team was awarded one of the first two licenses granted by the UK Human Fertilisation and Embryology Authority for the derivation of human ES cells and his group was the first to deposit a human ES cell line into the UK Stem Cell Bank. Minger was also one of the first two groups in the UK to be granted a research license by the HFEA in 2008 to pursue Somatic Cell Nuclear Transfer (SCNT) to generate "hybrid human embryos" for fundamental research into genetic forms of neurodegenerative conditions. Stephen is the Stem Cell Expert and a Member of the UK Gene Therapy Advisory Committee (GTAC) at the Department of Health and was until recently the Focal Point for Regenerative Medicine, Drug Discovery and Modernisation of Traditional Chinese Medicine for the UK Department of Business, Innovation and Skills in China.

Marilena Minieri, PhD She is an experimental pathophysiologist, works at the Laboratory of Cellular and Molecular Cardiology, University of Rome Tor Vergata, and is senior scientist of the National Institute for Cardiovascular Research (INRC). She is author of several papers published in peer-reviewed international scientific journals. Her major scientific interests span from intracellular signaling to stem cell and tissue engineering technology.

Prof. Will W. Minuth He is a professor of anatomy at the University of Regensburg in Germany. He is teaching medical students in gross anatomy, histology, and neuroanatomy. The research interest is related to the field of renal epithelial barrier function including the action of steroidal hormones. The driving force in his experimental work is the unsolved question, if stem/progenitor cells can be applied in future to treat acute or chronic renal failure. For that reason the actual research of Dr. Minuth is focusing to renal stem/progenitor cells developing into a functional epithelium under the tubulogenic action of aldosterone. Sophisticated perfusion culture including an artificial interstitium plays an essential role in this hot spot of research.

Katrin Montzka, PhD She studied bioengineering at the University of Applied Sciences in Aachen, Germany. She joined the Department of Genetics, Cell Biology and Anatomy, Munroe-Meyer Institute for Genetics and Rehabilitation, University of Nebraska Medical Center, USA, as a visiting fellow. In her Ph.D. studies she investigated the potential therapeutic application of human bone marrow-derived mesenchymal stromal cells in experimental spinal cord injury at the Department of Neurology, RWTH Aachen University. Employment followed as a Research Associate at the Department of Urology at the RWTH Aachen University, Germany.

Guido Nikkhah, PhD He studied medicine in Giessen, Germany (1984–1990) and received his Ph.D. degree in neurobiology from the University of Lund, Sweden (1994). He trained in neurosurgery in Hanover, Germany, where he also habilitated (1997). From 2001–2006, he was the vice-chairman of the clinical department of Stereotactic and Functional Neurosurgery and head of the research group "Molecular Neurosurgery" at the University of Freiburg, Germany. In 2007 he became the chairman of this department. He has also been the president and board member of NECTAR (Network of European CNS Transplantation and Restoration) and the treasurer of the Academia Eurasiana Neurochirurgica.

Robert Nordon, PhD He obtained his medical degree and Ph.D. at the University of New South Wales (UNSW), Australia. Robert is currently working in the Graduate School of Biomedical Engineering at UNSW. His major research interest is the development of bioreactor technologies for the study and delivery of cell-based therapies. He is the inventor of hollow fiber cell separation and expansion device that has been commercialized for clinical application. His more recent work focuses on progenitor cell fate mapping and mathematical modeling of stem cell development.

Prof. Richard Oreffo, PhD He holds the chair of musculoskeletal science and is co-founder of the Centre for Human Development, Stem Cells and Regeneration. He is associate dean for innovation and enterprise within the Faculty of Medicine, Health and Life Sciences, University of Southampton. Richard graduated from Liverpool and Oxford with degrees in biochemistry and a D.Phil. in bone biology. He leads a research group focused on understanding bone development and developing strategies to regenerate bone and cartilage using stem cell technology and innovative scaffolds for orthopedic application (see www.skeletalstemcells.org). Much of the work is undertaken in multidisciplinary programs in collaboration with clinicians, bioengineers, modelers, and bone biologists. He has published over 120 peer-reviewed papers and 20 contributed reviews/book chapters.

Burcin Özüyaman, MD He received his M.D. in the field of nitric oxide research with a special regard to its role in platelet function. He continued his research career at the Institute of Cardiovascular Physiology, Heinrich-Heine-University, Düsseldorf (Germany). Afterward, he started with a residency at the Department of Cardiology, University Hospital RWTH Aachen (Germany). In parallel he focused in the field of EPC research and RBC function always with a special refer to nitric oxide.

Prof. Laura Poole-Warren, PhD She is a professor of biomedical engineering at the University of New South Wales (UNSW). She is a leader in the biomaterials and tissue engineering field with specific emphasis on development of bioactive polymers and understanding cell and tissue interactions with these materials. She is the International Liaison Delegate for the IUS-BSE representing Australia and New Zealand and is a council member of the Asian Biomaterials Federation.

Olivier Preynat-Seauve, PhD He is a pharmacist, initially involved in the study of anti-tumor immunity. Actually, he is a biologist interested in the field of embryonic stem cell differentiation. He is working on neuronal differentiation of stem cells focusing on three main application: tissue engineering, cell therapy of neurodegenerative disorders, and understanding of early events of neural differentiation.

Jan Pruszak, PhD He obtained his medical degree from Hannover Medical School (MHH) in Germany in 2004. He then pursued postdoctoral training at Harvard Medical School, Boston, MA, USA, with Dr. Ole Isacson at the Center for Neuroregeneration Research. Since 2007, he holds an academic appointment as an instructor at Harvard Medical School and has been an affiliated faculty member of the Harvard Stem Cell Institute. His research interests include the regulation of growth and lineage specification of stem cells, in the context of developing novel therapeutic strategies for neurological disease.

Anne Roessger, PhD She studied biology and did her Ph.D. in bacterial resistance development at the University Hospital of Regensburg. In 2008 she started working as a lecturer at the Institute of Cellular and Molecular Anatomy, University of Regensburg, Regensburg, Germany.

Prof. Dr. med. Angela Rösen-Wolff Professor She is head of the Clinical Research Unit of the Department of Pediatrics at the University Clinic *Carl Gustav Carus* in Dresden since 1995. Since then she has been investigating the underlying molecular mechanisms of human phagocyte defects and bone tissue engineering. She has collaborated in several research initiatives focusing on tissue engineering of bone and development of new biomimetic bone substitute materials. She developed an in vitro model of hydroxyapatite remodeling in a co-culture system of osteoblasts together with osteoclasts. She is author of more than 80 peer-reviewed international publications in the field of molecular virology, pediatric immunology, biotechnology, biomaterials, and tissue engineering.

Silke Schwengberg, PhD After her studies in biology at the Ruhr University Bochum, she received her Ph.D. in cell biology and immunology at the University of Hannover. From 1999 to 2001 she worked as research scientist at the Janssen Research Foundation in Neuss (Germany) and Beerse (Belgium). In 2001 she joined the biotechnology startup Axiogenesis AG in Cologne, Germany, as senior scientist and is now head of the laboratory.

Sarvpreet Singh student at the Medical University of Innsbruck, is involved in a clinical study to investigate recurrent or refractory high-grade glioma and anaplastic astrocytoma at the Neuro-Oncology Innsbruck since 2004. Since 2006 his research is focused on the irradiation sensitivity of mesenchymal stem cells and thus affiliated to the Department of Cranio-Maxillofacial and Oral Surgery at the Medical University of Innsbruck in collaborative conjunction with the Institute for Biomedical Aging Research of the Austrian Academy of Sciences.

Prof. Hwal (Matthew) Suh, PhD He received D.D.S. and M.S.D. degrees in 1978 and 1984 from Yonsei University, Seoul, Korea. He attended at the Columbia Institute of Materials Science and Technology in New York, USA, from 1985 till 1987 and moved to Japan to receive Ph.D. in 1992 from Osaka University. As his major is nanobiomaterials for the regenerative medicine, he had served as the director of National Brain Korea 21 Project Team of Nanobiomaterials for the Cell-Based Implants, the chair of Department of Medical Engineering in College of Medicine, and at present, he is chair professor of the Program of Nano Science and Technology in the Graduate School at Yonsei University. He is also the founder of the Korean Society for Biomaterials in 1993.

Prof. Edda Tobiasch, PhD She is professor of Genetic Engineering and Cell Culture, has published several articles in international journals and in books. She holds several international patents and is referee member for various scientific committees at home and abroad. She has studied at the University of Kaiserlautern and made her Ph.D. at the German Cancer Research Center followed by postdoc positions at the Ruprecht-Karls University, Heidelberg, and the Forschungszentrum Karlsruhe. She was instructor at the Beth Israel Deaconess Medical Center in the Harvard Medical School, Boston, MA, USA.

Prof. George Truskey, PhD is professor and chair of the Department of Biomedical Engineering at Duke University, Durham, NC, USA. His research interests are in cardiovascular tissue engineering and the effect of physical forces on cells. He is a fellow of the American Institute of Medical and Biological Engineering, the American Heart Association, and the Biomedical Engineering Society (BMES). He is currently president of BMES.

Tsung-Neng Tsai, PhD He received clinical medical training and awarded the position of attending in Cardiovascular Division, Tri-service, General Hospital in Taiwan (1997–2007). He won a national scholarship to allow his current studies toward a Ph.D. in stem cell research in King's College London, London.

Prof. Gerald A. Urban, PhD He received the diploma (Dipl.-Ing.) for technical physics at Technical University Vienna. In 1985 he received the Ph.D. in electrical engineering at the TU Vienna. He was co-founder of the company OSC in Cleveland and Vienna. In 1994 he received the Venia Legendi for Sensor Technology. In 1997 he became full professor of sensors at the Institute for Microsystem Technology at the Albert Ludwig University Freiburg/Germany. He joined the Freiburg Institute of Advanced Studies (FRIAS) in 2008. His main interest focuses on research and development of microsensor applications including microthermistors, flow sensors and chemosensors and biosensors including oxygen. Recently nano- and microsystem technology is the main field of interest including the development of microarrays for proteomics and cell-based microassays.

Jean Villard, MD, PhD He is head of the Immunology and Transplant Unit and chargé de cours at the University Geneva Hospital and medical school. His main research of interests are the immune response after tissue and solid organ transplantation, with a special focus on the natural killer cells immunity at the genetic and functional level.

Prof. Yen Wei, PhD He received B.S. and M.S. in chemistry at Peking University in China in 1979 and 1981, respectively. He then earned his Ph.D. from the City University of New York in 1986. After postdoctoral work at MIT, he joined Drexel University, Philadelphia, PA, USA, in 1987, where he is currently the Herman B. Wagner Professor of Chemistry and Director of the Center for Advanced Polymers and Materials Chemistry. His awards include the ACS Philadelphia Section Award, Drexel Research Achievement Award, and DuPont Young Faculty Award. Recently, he was honored with an endowed Chang Jiang Lecture-Chair Professorship at Shanghai Jiao Tong University in China, a Nanochemistry Chair Professorship at Chung-Yuan Christian University, and an Honorary Chair Professorship at National I-Lan University in Taiwan.

Robin Wesselschmidt She is president, Primogenix, Inc., a company specializing in the derivation, characterization, and distribution of mouse embryonic stem cells. She has been actively working with and deriving mouse embryonic stem cells since 1992. She has derived and commercialized several widely used lines, including the RW4 and GSI-1 lines. She is focused on developing, streamlining, and translating research techniques into robust operations that can be utilized in core laboratories or developed into viable business operations. Since 2004, she has been a co-director and instructor for two of the seven NIH-sponsored human embryonic stem cell training courses held biannually in Southern California. She has written and co-edited papers and book chapters, as well as a laboratory manual in the area of pluripotent stem cell technologies.

Andreas Wilmen, PhD He received his Ph.D. in biology in 1994. He received his postdoc at the Clinical Research Group for Gastrointestinal Endocrinology, University of Marburg, Germany. Employment followed at the Institute for Human Gene Therapy, University of Pennsylvania, Philadelphia, USA. He is presently a senior scientist at the Institute for Target Discovery, Bayer Schering Pharma AG, Wuppertal, Germany.

Prof. Qingbo Xu, MD PhD He is professor and BHF John Parker Chair of Cardiovascular Sciences, Cardiovascular Division at King's College London, and studied in China, USA, and Austria before moving to London. He received several awards nationally and internationally. His research interests focus on stem cells and vascular diseases.

Prof. Masayuki Yamato, PhD He is a professor of the Institute of Advanced Biomedical Engineering and Science at Tokyo Women's Medical University. He was originally trained with a background in cell biology and biochemistry, but over the past decade, his research interests have been focused on the regeneration of various tissues and organs, such as the cornea, using cell sheets, instead of traditional tissue engineering approaches using cells seeded into biodegradable scaffolds. In particular, his work with both corneal and oral mucosal epithelial cell sheets has already been applied to human patients suffering from ocular surface dysfunctions. He is currently engaged in research collaborations with physicians and surgeons from various medical departments, with the aim of taking regenerative medicine using cell sheets from the level of basic laboratory science to clinical applications.

Contents

Contributors

Arti Ahluwalia Interdepartmental Research Centre "E. Piaggio", University of Pisa, Pisa, Italy, arti.ahulwalia@centropiaggio.unipi.it

Omonigho A. Aisagbonhi Division of Cardiovascular Medicine, Department of Cell and Developmental Biology, Department of Medicine, Vanderbilt University Medical Center, Nashville, TN, USA

Kuppusamy Aravindalochanan Laboratory for Sensors, Department of Microsystems Engineering, University of Freiburg, Freiburg, Germany

Aysegül Temiz Artmann Laboratory of Medical and Molecular Biology, Institute of Bioengineering, Aachen University of Applied Sciences, Jülich, Germany

Andrew H. Baker Faculty of Medicine, British Heart Foundation Glasgow Cardiovascular Research Centre, University of Glasgow, Glasgow, UK, ab11f@clinmed.gla.ac.uk

Christoph Becker Department of Urology, University Hospital and Medical Faculty, RWTH Aachen University, Aachen, Germany, cbecker@dgu.de

Anne Bernhardt The Max Bergmann Center of Biomaterials, Institute of Materials Science, Technische Universität Dresden, Dresden, Germany

Paul R. Bidez, III School of Biomedical Engineering, Science, and Health Systems, Drexel University, Philadelphia, PA, USA; Merck Research Laboratories, Merck & Co., North Wales, PA, USA, PBidez@aol.com

Thomas Brevig Research and Development, Thermo Fisher Scientific, Roskilde, Denmark, thomas.brevig@thermofisher.com

Melissa A. Brown Department of Biomedical Engineering, Duke University, Durham, NC, USA

Cindy S. Cheng Department of Biomedical Engineering, Duke University, Durham, NC, USA

Matthew J. Dalby Centre for Cell Engineering, Joseph Black Building, Institute of Biomedical and Life Sciences, University of Glasgow, Glasgow, G12 8QQ, Scotland, UK

Bernd Denecke IZKF "BIOMAT.", RWTH Aachen, Germany, bernd.denecke@rwth-aachen.de

Lucia. Denk Department of Molecular and Cellular Anatomy, University of Regensburg, Regensburg, Germany

Chris Denning Wolfson Centre for Stem Cells, Tissue Engineering & Modelling (STEM), Centre for Biomolecular Sciences, University of Nottingham, Nottingham, UK

Paolo Di Nardo Laboratory of Molecular and Cellular Cardiology, Department of Internal Medicine, University of Rome Tor Vergata and National Institute for Cardiovascular Research (INRC), Bologna, Italy

Ilya Digel Laboratory of Cell- and Microbiology, Institute of Bioengineering, Aachen University of Applied Sciences, Jülich, Germany, digel@fh-aachen.de

Máté Döbrössy Laboratory of Molecular Neurosurgery, Department of Stereotactic and Functional Neurosurgery, Albert Ludwigs University Freiburg, Freiburg, Germany

Rita Finones Department of Mechanical and Aerospace Engineering, Division of Biology and Material Science, University of California San Diego, La Jolla, CA, USA

Lesley Forrester Centre for Regenerative Medicine, University of Edinburgh, Queens Medical Research Institute, 47 Little France Crescent, Edinburgh EH16 4YJ

Jane Frimodig Department of Bioengineering, University of California San Diego, La Jolla, CA, USA

Andreas Geerts Bayer Schering Pharma AG, Target Discovery, Wuppertal, Germany

Michael Gelinsky The Max Bergmann Center of Biomaterials, Institute for Materials Science, Technische Universität Dresden, Dresden, Germany; DFG Research Center and Cluster of Excellence for Regenerative Therapies Dresden (CRTD), Technische Universität Dresden, Dresden, Germany, michael.gelinsky@tu-dresden.de

Stefan Golz Bayer Schering Pharma AG, Target Discovery, Wuppertal, Germany, stefar.Golz@Bayerhealthcare.com

Sabrina Gordon-Keylock Centre for Regenerative Medicine, University of Edinburg, Rodger Land Building, Kings Buildings, West main road, Edinburg E49 3JQ

David A. Gough Department of Bioengineering, University of California San Diego, La Jolla, CA, USA

Marijke Grau Department of Medicine, Medical Clinic I, University Hospital RWTH Aachen, Aachen, Germany

Rylie A. Green Graduate School of Biomedical Engineering, University of New South Wales, Sydney, NSW, Australia

Martin Haas Department of Mechanical and Aerospace Engineering, Division of Biology and Material Science, and School of Medicine, Cancer Center, University of California San Diego, La Jolla, CA, USA

Antonis K. Hatzopoulos Division of Cardiovascular Medicine, Department of Cell and Developmental Biology, Department of Medicine, Vanderbilt University Medical Center, Nashville, TN, USA, antonis.hatzopoulos@vanderbilt.edu

Isgard S. Hueck Department of Bioengineering and School of Medicine, Moores Cancer Centre, University of California San Diego, La Jolla, CA, USA

Gerhard Jakse Department of Urology, University Hospital and Medical Faculty, RWTH Aachen University, Aachen, Germany

Nicole M. Kane Faculty of Medicine, British Heart Foundation Glasgow Cardiovascular Research Centre, University of Glasgow, Glasgow, UK

Malte Kelm Department of Medicine, Medical Clinic I, University Hospital RWTH Aachen, Aachen, Germany

Stefanie Keymel Department of Medicine, Medical Clinic I, University Hospital RWTH Aachen, Aachen, Germany

Jochen Kieninger Laboratory for Sensors, Department of Microsystems Engineering, University of Freiburg, Freiburg, Germany

John Paul Kirton Cardiovascular Division, King's College London BHF Centre, London, UK

Petra Kleinbongard Institute of Pathophysiology, University of Essen Medical School, Essen, Germany, petra.kleinbongard@uk-essen.de

Frank R. Kloss Department of Cranio-Maxillofacial and Oral Surgery, University Hospital Innsbruck, Innsbruck, Austria

Karl-Heinz Krause Laboratory of Experimental Cell Therapy, Department of Genetic and Laboratory Medicine, Geneva University Hospital, Geneva, Switzerland

J. Yasha Kresh Department of Cardiothoracic Surgery, Drexel University College of Medicine, Philadelphia, PA, USA, JYasha.Kresh@DrexelMed.edu

Peter I. Lelkes Laboratory of Cellular Tissue Engineering, School of Biomedical Engineering, Science, and Health Systems, Drexel University, Philadelphia, PA, USA, pilelkes@drexel.edu

Günter Lepperdinger Institute for Biomedical Aging Research, Austrian Academy of Sciences, Innsbruck, Austria, guenter.lepperdinger@oeaw.ac.at

Anja Lode The Max Bergmann Center of Biomaterials, Institute of Materials Science, Technische Universität Dresden, Dresden, Germany

Penny J. Martens Graduate School of Biomedical Engineering, University of New South Wales, Sydney, NSW, Australia

Marilena Minieri Laboratory of Molecular and Cellular Cardiology, Department of Internal Medicine, University of Rome Tor Vergata and National Institute for Cardiovascular Research (INRC), Bologna, Italy

Will.W. Minuth Department of Molecular and Cellular Anatomy, University of Regensburg, Regensburg, Germany, will.minuth@vkl.uni-regensburg.de

Katrin Montzka Department of Urology, University Hospital and Medical Faculty, RWTH Aachen University, Aachen, Germany

Guido Nikkhah Laboratory of Molecular Neurosurgery, Department of Stereotactic and Functional Neurosurgery, Albert Ludwigs University Freiburg, Freiburg, Germany

Robert Nordon Graduate School of Biomedical Engineering, University of New South Wales, Sydney, NSW, Australia

Richard O.C. Oreffo Bone and Joint Research Group, Developmental Origins of Health and Disease, University of Southampton, Southampton S016 6YD, UK

Burcin Özüyaman Department of Medicine, Medical Clinic I, University Hospital RWTH Aachen, Aachen, Germany

Laura A. Poole-Warren Graduate School of Biomedical Engineering, University of New South Wales, Sydney, NSW, Australia

Olivier Preynat-Seauve Laboratory of Experimental Cell Therapy, Department of Genetic and Laboratory Medicine, Geneva University Hospital, Geneva, Switzerland

Jan Pruszak Laboratory of Molecular Neurosurgery, Department of Stereotactic and Functional Neurosurgery, Albert Ludwigs University Freiburg, Freiburg, Germany; Center for Neuroregeneration Research, Harvard Medical School, McLean Hospital, Belmont, MA, USA

Anne. Roessger Department of Molecular and Cellular Anatomy, University of Regensburg, Regensburg, Germany

Angela Rösen-Wolff DFG Research Center and Cluster of Excellence for Regenerative Therapies Dresden (CRTD), Technische Universität Dresden, Dresden, Germany; Department of Paediatrics, University Hospital Carl Gustav Carus, Technische Universität Dresden, Dresden, Germany

Silke Schwengberg Axiogenesis AG, Köln, Germany

Sarvpreet Singh Department of Cranio-Maxillofacial and Oral Surgery, University Hospital Innsbruck, Innsbruck, Austria; Institute for Biomedical Aging Research, Austrian Academy of Sciences, Innsbruck, Austria

Hwal (Matthew) Suh Laboratory of Tissue Regenerative Medicine, Department of Medical Engineering, College of Medicine, Yonsei University, Seoul, Republic of Korea, hwal@yuhs.ac

Edda Tobiasch University of Applied Sciences Bonn-Rhein-Sieg, Rheinbach, Germany, EddaTobiasch@h-brs.de

George A. Truskey Department of Biomedical Engineering, Duke University, Durham, NC, USA, gtruskey@duke.edu

Tsung-Neng Tsai Cardiovascular Division, King's College London BHF Centre, London, UK

Gerald A. Urban Laboratory for Sensors, Department of Microsystems Engineering, University of Freiburg, Freiburg, Germany

Jean Villard Immunology and Transplant Unit, Division of Immunology and Allergology and Division of Laboratory Medicine, Geneva University Hospital and Medical School, Geneva, Switzerland

Yen Wei Department of Chemistry, Drexel University, Philadelphia, PA, USA, weiyen@drexel.edu

Robin Wesselschmidt Primogenix, Inc., Laurie, MI, USA

Andreas Wilmen Bayer Schering Pharma AG, Target Discovery, Wuppertal, Germany

Qingbo Xu Cardiovascular Division, King's College London BHF Centre, London, UK, qingbo.xu@kcl.ac.uk

Masayuki Yamato Institute of Advanced Biomedical Engineering and Science, Tokyo Women's Medical University, Tokyo, Japan

Editorial Board

Part I
Instead of an Introduction – The Emperor's New Legs

The Emperor's New Body: Seeking for a Blueprint of Limb Regeneration in Humans

Ilya Digel and Aysegül Temiz Artmann

Abstract Aspiring to comprehend and control regeneration – the ability to recreate lost or damaged cells, tissues, organs or even limbs – has been the mind-boggling challenge for over 250 years. Regeneration is a common feature in many animal species, whereas its capacity in mammals is notoriously limited. The partial or complete loss of digits or limbs and the deformation of facial injuries profoundly affect the quality of life of the wounded and present a set of challenges for the medical community. This chapter is devoted to some of the problems and prospects of human limb regeneration. It briefly reviews the appearance of regenerative abilities in different tribes across the animal kingdom as well as the genes and cellular signaling pathways involved. Special emphasis is placed upon blastema and scar formation as well as on morphogenetic pattering.

The analysis of evolutionary manifestations of regeneration ability, the apparent relation to the embryonic development mechanisms, as well as consideration of some existing clinical approaches suggest the possibility to "awaken" the regeneration ability in mammals and even in human beings. The ultimate goal of further research will be to identify ways for enhancing the capacity for wound healing and tissue restoration in humans. This would open new exciting prospects for future regenerative medicine.

Keywords Limb regeneration · Regenerative medicine · Blastema formation · Scar formation morphogenesis · Wnt/β-catenin pathway · Developmental patterning · BMP signaling · Stem cells

Abbreviations

AEC – apical epithelial cap
AER – apical ectodermal ridge

I. Digel (✉)
Laboratory of Cell- and Microbiology, Institute of Bioengineering, Aachen University of Applied Sciences, Jülich, Germany
e-mail: digel@fh-aachen.de

G.M. Artmann et al. (eds.), *Stem Cell Engineering*,
DOI 10.1007/978-3-642-11865-4_1, © Springer-Verlag Berlin Heidelberg 2011

BMP – bone morphogenetic protein
FGF – fibroblast growth factor
NEJ – neuroepidermal junction
PDGF – platelet-derived growth factor
Shh – sonic hedgehog
TGF – transforming growth factor
ZPA – zone of polarizing activity

... our wise Sardaukar commander had Idaho's corpse preserved for the axolotl tanks.
Why not? This corpse held the flesh and nerves of one of the finest swordsmen in history.
(Frank Herbert, "Dune Messiah", 1969)

1 Introduction: A Dream of Regeneration

In this short review we will define regeneration as a complex biological process by which living organisms restore shape, structure, and function of body parts lost due to injury or experimental amputation. Strictly speaking, this process should be defined as "reparative regeneration" in contrast to the "physiological" one, though both kinds are closely related to each other. Physiological regeneration involves a constant renewal of structures on the cellular (blood cells, epidermis) and the intracellular (renewal of cellular organelles) level, providing the basis for whole-body functioning. Reparative regeneration is the elimination of structural damage produced by trauma or pathology. Since reparative regeneration represents an important universal component of adaptation to environmental threats, its significance cannot be overestimated.

The term "regeneration" was suggested in 1712 by the French scientist René Antoine Réaumur, who was among the first in studying limb regeneration in crayfish (*Astacus astacus*) [1]. His pioneering studies were followed over the past 250 years by brilliant works of Abraham Trembley [2], Charles Bonnet, Lazzaro Spallanzani, Thomas Hunt Morgan, and many others. Still we know surprisingly little about the biological significance and molecular mechanisms underpinning this remarkable phenomenon. While our skin and hair cells constantly renew themselves, a human who loses a leg has no chance to grow a new one.

The potential of the regeneration is recognized in the attention currently given to tissue engineering and stem cell approaches. Regeneration studies are now helping doctors treating wounds to minimize scarring. Without a doubt, controlling regeneration will be an ideal solution for many clinical problems. If a patient with renal failure gets only as little as 10% of renal cells revived, she or he would not be bound to a dialysis apparatus any more. If a possibility of reinvigorating insulin-producing β-cells in the pancreas is found, millions of diabetics would be relieved from everyday injections. Replenishing *substantia nigra* neurons would help patients with Parkinson's disease. Moreover, stimulating regenerative capability might circumvent some of the notorious tissue deteriorations that accompany old age. Yet, regeneration as one of the most dramatic and important of the healing processes remains an unrealizable dream of mankind. This is perhaps because our attitude toward regenerative ability is inevitably weighted toward biomedical considerations, whereas other (evolutionary, physiological, and embryological) perspectives often remain neglected.

Biological insights might be especially important as researchers will attempt to reconstruct entire organs rather than just one type of cell. Studies on the regenerative magicians of the animal world will perhaps reveal the rules that guide reformation of complex structures, findings that could eventually allow diseased human hearts or other organs to rebuild themselves. Why are some animals able to regenerate complex structures following transection, tissue removal, or amputation, and others apparently not? Is regeneration a basic, primordial attribute of metazoans or a mechanism which has evolved independently in a variety of contexts? What are the exact mechanisms of regeneration? Why in phylogenetic groups where regeneration occurs, closely related species are observed which do not possess this ability? Is there any way to evoke or activate regeneration pathways for use in regenerative therapy? These and other questions remained unanswered for very long time and only today we begin to understand the complexity of the issue.

The regeneration process is both the morphological and the functional reconstruction of a part of a living organism, which has previously been destroyed. Among most vertebrates, the capacity to regenerate is limited to some tissues. It is however possible to observe the regeneration of appendages (limb, tail, fin, jaw, etc.) among several amphibians and fish. The essence of true regenerative process is the appearance of a mass of primitive, presumably totipotent cells called the *blastema* at the site of injury. After reaching a critical size, this cellular mass begins to grow and redifferentiate to produce the multi-cellular, multi-tissue, complex missing structure. This regeneration leads to reforming of the amputated part with a complete restoration of its shape, segmentation, and function. For example, in salamanders, the responding cells are genetically programed to become the cell types of the lost structure. As a result of this, full limb growth is completed within 2 months.

Humans cannot perform the same trick of regrowing a severed limb like salamanders or newts can because complete regeneration of complex tissues and organs is usually precluded by fibrotic reactions leading to scar formation. In humans, true regeneration is limited to the healing of fractures of the long bones; many other processes commonly called regenerative (i.e., skin and peripheral nerve fiber) are simply increased rates of cellular multiplication or growth. However, humans already have a certain capacity for regeneration. For instance, the pool of liver cells and red blood cells can self-renew. During embryonic development, mammals and birds can regenerate such diverse tissues and structures as their skin and spinal cord.

The limb is composed of a diverse number of distinct cell types making up the various tissues of the digit (e.g., epidermis, nails, muscles, dermis, adipose, and bone), and in response to amputation, a perfect replica of the limb including all those structures must appear. The prospect of enhanced limb regeneration in humans is encouraged also by the observation that many of the tissues of the adult limb have an inherent capacity to undergo a limited regenerative response to injury.

Mammalian tendons and ligaments undergo limited regenerative healing [3]. Other tissues of the limb, such as the skin, are less successful in their response to injury and heal imperfectly. However, there is indication that significant advances in inducing scar-free healing of adult skin wounds are on the horizon [4]. The fact that individual injured tissues of the human limb have the capacity to undergo a regenerative response indicates the existence of rudimentary repair processes which might be enhanced to initiate a limb regeneration response.

The aspiration to arouse regenerative ability in obviously nonregenerating animals and human beings has been stimulating many interesting physiological studies in the past centuries. Oren E. Frazee [5], for example, studied the effect of passing electrical current through the water in which salamanders were kept. In 1909, he demonstrated that this appeared to increase the rate of limb regeneration in these animals. In their long series of investigations extending from the 1920s throughout the 1940s, both Burr [6] and Lund [7] reported growth effects of applied electrical currents on a variety of plants and animals. Their observations were later on confirmed and extended by Becker and Spadaro [8]. By the mid-1950s, Singer published a series of papers in which he demonstrated the dependence of limb regeneration in the salamander on the presence of a threshold amount of nerve tissue in the amputation stump [9]. He was also able to produce a small amount of limb regeneration in the adult frog by transplanting additional functioning nerve tissue into an amputation stump. The evidence suggested that the nerves somehow controlled normal growth; in their absence, normal growth was inhibited and abnormal growth was facilitated.

Another successful attempt was reported by Rose in 1944 when he produced a small measure of regeneration of the amputated foreleg of the frog (a species that, despite folklore, is a nonregenerator) by dipping the extremity daily in hypertonic saline. Six years later, a similar result was reported by Polezhaev [10] in the same type of animal, by repeatedly needling the stump daily. Person et al. [11] described spontaneous regenerative responses following subtotal forelimb amputation in the young white rat. In one group of animals, the forelimb was amputated through the lower humerus and the skin sutured. In a second group, adjacent muscle tissue still attached to bone at its origin was interposed between the cut surface of the humerus and the skin. Among animals of the first group (skin closure only), bone growth and limb regenerative responses were generally not observed. Animals of the second group displayed significant elaborations of cartilage and bone at the limb terminus. The appearance and the subsequent modification of these tissues suggest that some capacity for limb regeneration exists innately in the young rat and can be more readily evoked than has been recognized heretofore [11].

While it was the intent of all these investigators to delay the overgrowth of skin over the end of the amputation stump, the procedure used in each experiment was obviously repeatedly traumatic. A similar approach is being applied nowadays in Ukraine, where Boris Bolotov successfully stimulates growth of new fingers by applying a compress with solid sodium chloride, which prevents normal scarring processes (http://boris-bolotov.org.ua/). His method looks rather medieval but the number of cured patients impresses.

In spite of plenty of studies describing the facilitation of limb regeneration by electrical and other forms of stimulation, little is known of innate regenerative capacity in the mammalian limb. Indeed, why is the amputation of limbs not followed by regeneration in mammals and birds? Is it the absence of stem cells, the absence of recruitment signals for these cells, or rather the absence of signal receptivity? Getting to a regeneration phase in mammalian limbs would be a major breakthrough which would be based upon extensive understanding of limb development mechanisms. That is why a large part of the future research will involve examining the particular cellular and molecular systems that allow certain animals to completely regenerate lost tissue and organs.

According to their biological context, regeneration processes are classified either as *tissue renewal* or as *re-growth of appendages or body parts*. The first process does not require a de novo formation of new structures and will not be considered in this work; in contrast, the second type of regeneration relies on morphogenetic processes that can be regarded as developmental events taking place during adult life. The regeneration process can be divided into a wound-closure, dedifferentiational phase and a redevelopmental phase and usually involves three interlinked mechanisms – an acute wounding response, the generation of an appropriate population of precursors, and morphogenesis.

For successful regeneration, several processes have to be managed:

- Cell dedifferentiation or alternatively stem cell recruitment with further proliferation
- Spatial organization of morphogenesis using regulatory gradients
- Differentiation: muscle, nerve, bone, cartilage, tendons, blood vessels, and skin

The sub-tasks of critical importance are the following:

- Preventing infection
- Preventing tumor formation
- Preventing autoimmune response

This review constitutes a short report on the current understanding of the basis of the regeneration of appendages in tetrapods and in fish with an emphasis on its feasibility and its implications to regenerative medicine.

2 Biological Aspects of Regeneration

A big experimental material on animal regeneration accumulated since the eighteenth century was summarized by Albert von Kölliker, Charles Darwin, and Heinrich Wilhelm Waldeyer in the so-called three rules of regeneration [10].

The first rule: The regenerative ability is decreasing with an increase of complexity and level of organization of an animal. Many invertebrates are capable of renewing great parts of their body, while high-organized animals and humans can regenerate only some organs or their parts.

The second rule: The regenerative ability is decreasing with age. The embryo has the capacity to undergo regulative growth when cells or tissues are removed or rearranged. In the adult, regeneration can only partially replace missing parts by growth and remodeling of somatic tissues. For example, while most of the mammals cannot regenerate limbs, many (including young children) can regenerate the ends of their fingers.

The third rule: During regeneration, every tissue arises from the same or similar tissue.

Further studies have shown that all these three rules are not absolute but rather reflect some existing trends. Indeed, some primitive animals like sponges, hydropolyps, flatworms, tapeworms, and annelid worms, pearlweeds, echinoderms, and tunicates are "regeneration champions" as they can regenerate the whole organism from a small fragment of the body (Fig. 1). Especially notable is the capability for regeneration in sponges. If a body of adult sponge is pressed through a sieve so that all cells will be separated from each other and when these separated cells are placed into water, they start to reunite and produce an entire sponge again. The tapeworm is capable of renewing whole individuals from every single piece of its body. Theoretically it is possible to cut such a worm into 200,000 pieces and obtain the same number of new worms as a result of regeneration.

Mollusks, arthropods, and vertebrates are not able to regenerate whole individuals from one fragment, but most of them are able to restore the lost organ. Some species use *autotomy* in case of emergency. From one amputated arm of a starfish, a whole animal can be grown. Among vertebrates, urodele amphibians (e.g., axolotls, salamander, and newts) have the unique ability to perfectly regenerate complex body parts after amputation, whereas mammalians and birds tend to regenerate poorly. Still, birds can regrow feathers and some parts of a beak. Mammals are capable of renewing claws and partly the liver. Also deer can regrow antlers after being shed.

However, the belief is wrong that the regeneration ability absolutely always negatively correlates with the level of organization. For every good regenerator, there are closely related species that have lost the ability to regenerate, suggesting that complexity alone does not correlate with the talent. It would also be wrong to state that the regeneration ability always gradually decreases with age. In general, the older organisms have reduced regeneration ability, but in ontogenesis this capability can increase [12].

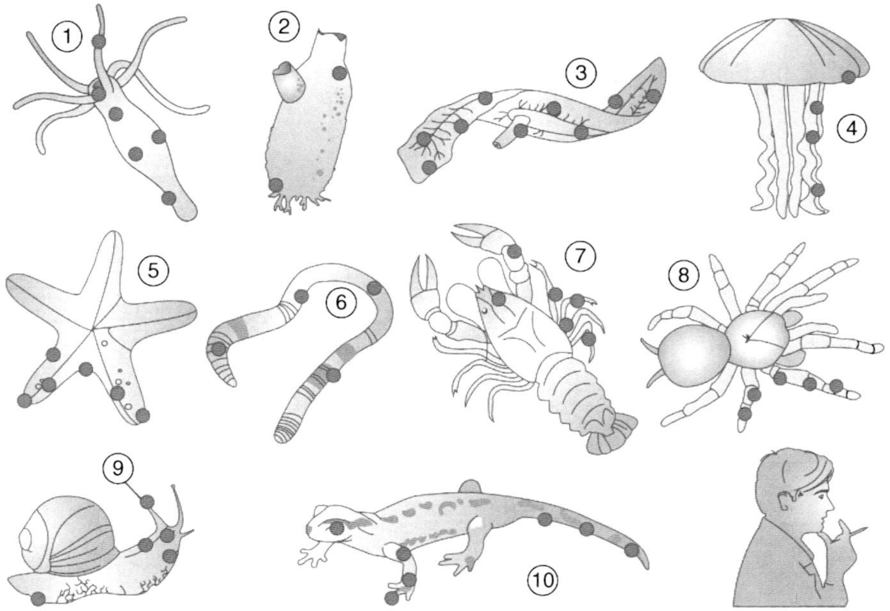

Fig. 1 The number of species that are capable of some form of regeneration is immense, and the spectrum of structures that can be generated is vast. Hydras (*Hydrozoa*) (*1*), sponges (*Porifera*) (*2*), and flatworms (*Planaria*) (*3*) engage in dazzling feats of renewal being able to regenerate an entire body from a small fragment. Each of the two fragments resulting from the amputations in *Cubozoa* and *Scyphozoa* (*4*), *Ctenophora*, *Echinodermata* (*5*), and *Annelida* (*6*) is capable of regenerating an entire organism (bidirectional regeneration). Many other animals can restore some lost body parts or their fragments [limbs, antennas, eyes (*Crustacea, Arachnea, Insecta*)] (*7, 8*), mantia, foot, and tentacles (*Mollusca*) (*9*). Among the vertebrates, some fish (*Danio rerio*) and urodeles (*Urodela=Caudata*) (*10*) retain a significant limb regenerative ability (including limbs, tail, eyes, and jaws) during adulthood. Nowadays, we are beginning to understand the molecular basis for these body-building tricks, hoping to decipher how humans might perform similar processes

There are excellent reviews by Richard Goss, Alvardo Sanchez [13, 14], and others devoted to evolutionary aspects of regeneration and its appearance in different groups. Comparison of regeneration abilities of different species (Fig. 2) shows that this feature appeared and disappeared in different groups as a result of action of many ecological, genetic, and morphological factors. The body of evidence is increasing and indicates that in almost all invertebrate groups, there are many species showing weak or no regenerative abilities. Such "primitive" animals as ctenophores (known as comb jellies) and rotifers are practically not able to regenerate, whereas more complex crustaceans and amphibians have this ability. Many other examples are known as well. Some closely related species are very different in their regenerative capacity. For example, an earthworm can regenerate itself and form a new individual from a single body piece, whereas a leech is not even able to

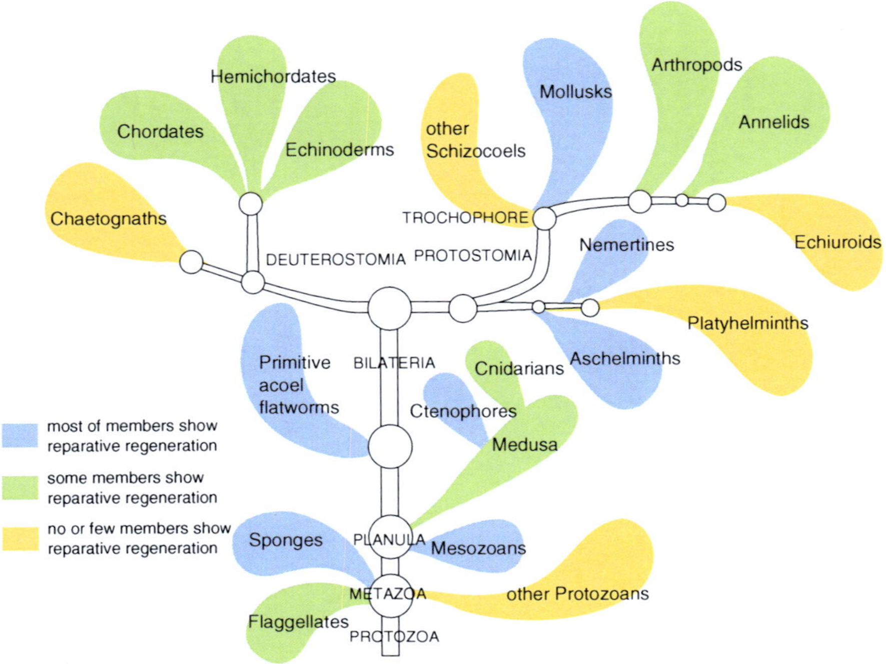

Fig. 2 Regeneration processes are observed in a large number of animal species. Examples of adult regeneration are found in most phyla although not in all. A cursory look at regenerative ability across phylogenetic groups leads to the conclusion that regenerative potential declines with the evolution of complexity [14]. Is regenerative ability the result of a gain of function in some groups or a loss of function in other groups?

regrow a lost organ. After amputation, a new extremity forms in Urodela (salamanders, newts, and axolotls), but in frogs and toads (Anura), a stump is usually simply healed.

3 Regeneration Mechanisms: From the Hydra to a Human Being

For a given organism, the regeneration process unfolds simultaneously on different levels of organization – cellular and intercellular, tissue, organ, and whole-body levels. In his monograph *Regeneration* and his paper (both appeared in 1901) published in *Science*, T.H. Morgan recognized two modes of regeneration according to cellular criteria: *morphallaxis*, which involves a redeployment of existing cells in the absence of division and depends upon repatterning of tissues, and *epimorphosis*, which involves division.

3.1 Morphallactic Regeneration

The morphallactic mode of regeneration is realized as the recreation of the missing body parts in the absence of proliferation. This mode mostly applies to regeneration in *Hydra* and other *diploblasts* (i.e., the animals having just two germ layers). For example, regeneration of the swimming muscle of the medusa *Polyorchis* involves cell migration in the absence of proliferation.

The adult Hydra polyp is composed of three distinct cell lineages: the ectodermal epithelial cell lineage, the endodermal epithelial cell lineage, and the interstitial cell lineage. The first two lineages form the epithelial layers of the polyp. Epithelial cells are in a differentiated state allowing them to form these two layers and, intriguingly, they can be considered as stem cells that give rise to new body column tissue. The interstitial cell lineage contains multipotent stem cells that give rise to nerve cells, gametes, nematocytes (the stinger cells used to capture prey), and mucus and gland cells. These three lineages do not interconvert in the adult polyp.

At the oral end of the polyp is the mouth opening and a ring of tentacles (that together constitute the head). At the aboral end, there is a disk of adhesive cells that constitute the foot. The epithelial cells in the body column are mitotic, whereas the epithelial cells of the tentacles and the foot are arrested in G_2 of the cell cycle (Fig. 3). As new cells are produced by mitosis in the body column, cells are displaced into asexual buds and sloughed from the ends of the tentacles and the center of the foot.

Regeneration in Hydra is polarized and does not depend on growth. This means that heavily irradiated Hydra unable to undergo cell division will still regenerate.

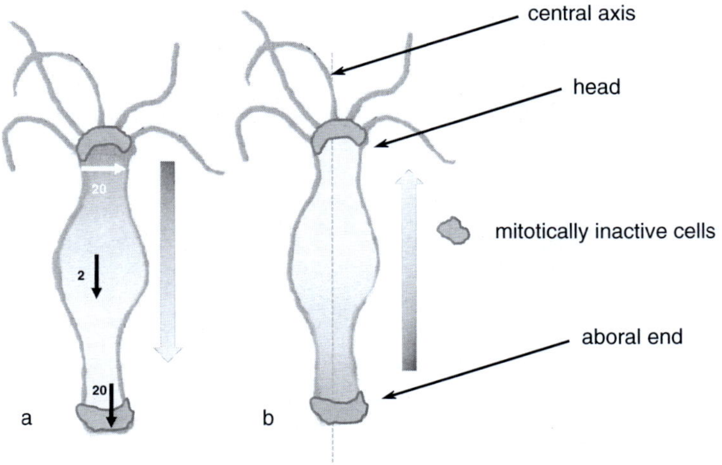

Fig. 3 Tissue dynamics and morphogenetic gradients in Hydra. (**a**) Head activation and head inhibition. (**b**) Foot activation and foot inhibition. The *arrows* indicate some directions of cellular movements. The number of days required for the cellular displacement is indicated next to the *arrow*

When cut in two, the lower piece will develop a head and the upper will develop a foot. A piece excised from the Hydra body will regenerate both a head and a basal disc in the same polarity. A small fragment will produce a small Hydra that will grow after feeding. The only process that takes place, at least during the first day, is the differentiation of the pre-existing cells in the region of the injury. Immediately after amputation, epithelial cells of the stumps will migrate and stretch to cover the wound (wound-healing phase) [15].

In fact, the Hydra undergoes continuous growth and pattern formation and cells are lost at the tentacle tips and from the basal disc. The cells continually change their position and form new structures as they move up and down the body column. When budding occurs at two-thirds down the body axis, a head develops there, which then detaches as a small new Hydra. Hicklin and Wolpert [16] have linked the two potentials for the formation of head or foot structures with the term of positional value, implying that diffusible extracellular signals would account for axis patterning in Hydra. Coupled to theoretical considerations [17], these data were interpreted in terms of developmental potentials; the formation of head and foot is supposed to result from two pairs of antiparallel morphogenetic activation and inhibition gradients of extracellular signals [18]. This means that head regeneration in Hydra can be accounted for in terms of two gradients: (1) a head inhibitor gradient and (2) positional information gradient (along the body axis). Diacylglycerol, a potent second messenger (i.e., phosphatidylinositol signaling), causes ectopic head formation, while lithium induces ectopic feet. Later, however, sophisticated grafted experiments have illustrated the fact that pattern formation can occur in Hydra in the absence of developmental gradients [19]. Homologues of *Hox* genes and fork-head transcription factors acting in the organizing regions of Hydra are the subject of intensive study now.

3.2 Epimorphic Regeneration

Regeneration in Hydra is a classical example of morphallaxis, whereas other familiar regeneration examples are epimorphic. This mode of regeneration is characteristic of triploblasts (i.e., animals having three germ layers) and requires the formation of a local growth zone of undifferentiated cells (*blastema*). The cells for epimorphic processes may be recruited either by mobilization of a reserve population or by reversal differentiation in cells at the site of injury or tissue removal [20]. Mixed contributions of both types (sources) in some cases should not be excluded as well.

The blastema is made up of two compartments: the superficial sheet of cells of epithelial origin covering the entire stump and the underlying layer of cells of mesenchymal origin. The formation of the blastema itself represents a transition phase in which limb cells respond to injury by dedifferentiating to become embryonic limb progenitor cells that can undergo redevelopment. During this phase, rapid wound closure is followed by the dedifferentiation of limb cells to form the blastema. The interactions that will take place between these two layers of the blastema will lead to the differentiation of the missing structures. This type of regeneration is common to planarians, gastropods, echinoderms, urochordates, and limb and tail regeneration in vertebrates [14].

Thus, vertebrate limb regeneration involves cell dedifferentiation and growth. Studies on various regeneration models describe a highly dynamic process in which the wound-healing response that results in the formation of a regeneration blastema interfaces with a redevelopment process that ultimately regenerates the limb [21, 22]. Following amputation of the limb, the regeneration process involves a number of stages, including the closure of the wound by the epidermis, the dedifferentiation of stump tissue to release cells, and the transformation of the wound epithelium into a specialized structure called the *apical epithelial cap* (AEC). The latter is required for limb outgrowth, the migration of dedifferentiated cells to the center of the wound where a blastema forms, and the proliferation of blastema cells [23, 24]. Fish, salamanders, and larval anurans are among the few vertebrates capable of regenerating lost appendages. This process seems to recapitulate ontogenic development of the structure in most respects.

Regeneration begins as a local response of the cells of the limb stump and results in a perfect replacement regardless of the level of amputation. The earliest events of the urodeles' injury response distinguish the pathway leading to regeneration from a response leading to wound healing and scar formation in mammals. The first involves the epidermis, which is able to close the amputation wound in a matter of hours. As early as 1962, Rose and Becker had called attention to the importance of a peculiar relationship between the epidermis and the nerves in the epimorphic limb regeneration process [25]. The first event in such regeneration is the overgrowth of the epidermis alone (not the dermis) over the cut end of the amputation stump. Following this, the cut ends of the nerves remaining in the amputation stump begin to grow into this epidermal "cap" where they form peculiar "synapse-like" junctions with the epidermal cells. This "neuroepidermal junction" (NEJ) is apparently the primary structure in the regenerative process, since following its formation the blastema appears. If the formation of the NEJ is prevented by interference with either the nerve or the epidermis, or by simply interposing a layer of the dermis under the epidermis, blastema formation does not occur and regeneration is absent. In experiments in which limb regeneration was stimulated by electrical means, no NEJ formed and it was postulated that its function had been taken over by the applied electrical currents. Therefore, the NEJ could be postulated to be the single structure that produced the "regeneration-type" potentials, neither the nerve nor the epidermis acting alone.

The cells of the blastema arise from beneath the wound epidermis, dedifferentiate, and start to divide. Over weeks, these cells become cartilage, muscle, and connective tissue. This is dramatically different by comparison to mammalian wounds that take multiple days to close. Once the wound is closed, the wound epithelium undergoes changes in morphology and gene expression to form the *apical epithelial cap* (AEC), which directs the outgrowth of limb mesenchymal cells during morphogenesis. The AEC is considered to be homologous to the *apical*

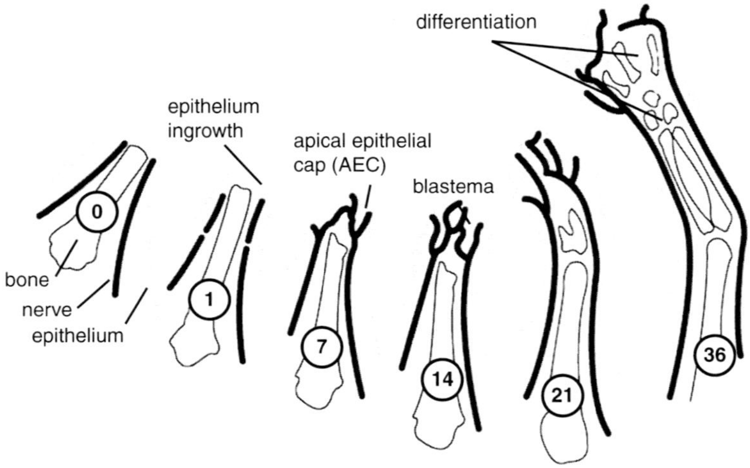

Fig. 4 Limb regeneration stages in Urodela. When a salamander loses a limb, the wound sends out molecular signals that prompt surrounding tissue to begin production of new progenitor cells, also referred to as precursor cells. These progenitor cells continue to divide and form a large pool of cells at the wound site, called a blastema, that will later specialize and mature to help form the bone, muscle, cartilage, nerves, and skin of the regenerated limb. The *numbers* in the *circles* represent the days passed after amputation. Models such as the salamander might advance practical efforts to spur regeneration in human

ectodermal ridge (AER) of the developing limb bud [26]. Like the limb bud, outgrowth of the blastema involves interactions between the blastemal mesenchymal cells and the AEC (Fig. 4).

Skeletal muscle regeneration is perhaps the best studied example, since muscle represents a highly differentiated cell type. When amphibian myotubes are exposed to medium containing high concentrations of serum from various sources, they respond by re-entry to the S phase [27]. It was proposed that prothrombin activation on wounding or injury leads to generation of a signal which is found in most or all vertebrates but which selectively activates newt myotubes to re-enter the cell cycle. Interestingly, the mammalian p16 (SDK inhibitor) tumor repressor protein and X-irradiation are very effective at blocking the serum-induced S-phase re-entry by the newt myotubules but they cannot prevent dedifferentiation if the myotubes are implanted into a blastema [28].

It is also well known that normal-appearing limbs that lack muscle tissue are formed during development or regeneration when muscle precursor cells are prevented from participation in limb outgrowth [29, 30]. These studies indicate that the contribution of myoblasts in limb development or tissue-derived muscle blastema cells in limb regeneration is not required for limb outgrowth per se, but they are required for forming the limb musculature. The demonstration that myofibers can dedifferentiate provides a striking example of cellular plasticity in limb regeneration. There is also an evidence that mammalian myotubes can undergo a similar dedifferentiation response in vitro [31]. In mammals, however, muscle tissue regeneration occurs routinely but is not associated with a dedifferentiation response;

instead, myogenic mononucleate stem cells called satellite cells provide a cellular source for the regeneration response [32, 33].

3.3 Regeneration by Induction

In addition to morphallaxis and epimorphosis, Polezhaev [34] introduced the third method of regeneration: *regeneration by induction*. This method was demonstrated by investigations on the restoration of lost regenerative power of the cranial bones in adult dogs and in some other cases. Antler regeneration in deer has traditionally been viewed as an epimorphic process closely resembling limb regeneration in urodele amphibians, and the terminology of the latter process has also been applied to antler regeneration [13]. More recent studies, however, showed that, unlike urodele limb regeneration, antler regeneration does not involve cell dedifferentiation and the formation of a blastema from these dedifferentiated cells. These studies rather suggest that antler regeneration is a stem cell-based process that depends on the periodic activation of, presumably neural crest derived, periosteal stem cells of the distal pedicle. A new concept of antler regeneration as a process of stem cell-based epimorphic regeneration is proposed that does not involve cell dedifferentiation or transdifferentiation. Antler regeneration illustrates that extensive appendage regeneration in a postnatal mammal can be achieved by a developmental process that differs in several fundamental aspects from limb regeneration in urodeles [35]. During regeneration by induction, an inducing agent, a reactive or a competent material, and the right (and mostly unknown!) conditions for induction must be present.

The methods of regeneration differ not only in their formal and secondary features but also in the essence of their fundamental processes: true reorganization, growth, and induction. Each method of regeneration can be found not only in the pure form but also in combinations of varying degree and type. For example, regeneration of embryonic mouse digit tips and anuran amphibian (*Xenopus*) limbs shows intermediate regenerative responses between the two extremes. Thus, adult mammals (least regenerative) and urodele amphibians (more regenerative) provide a range of models to study the various abilities of limbs to regenerate [36]. Besides regeneration, hypertrophy is also very important for the partial restoration of parenchymatous internal organs in mammals, but it is not identical with complete regeneration of original morphology and function [34].

4 Regeneration in Nonregenerators

The overall process of limb regeneration is complex and not fully understood, but it is clear that it is the early events of wound healing and dedifferentiation that distinguish the amphibian from the mammalian response. In spite of known limitations of the regenerative ability in mammals, they are characterized by four types of regenerative processes: regeneration of parenchymatous organs, regeneration of parts of organs, intraorganic regeneration, and intracellular regeneration. The restoration following the damage caused by pathological factors is an extremely widespread process, and it should be paid more attention when analyzing the regenerative ability in mammals.

Cells that specialize to a high degree must retrace many steps to return to a form that can rekindle regeneration. Thus, some animals might have sacrificed renewal proficiency for specialized cells and tissues with quicker, more efficient proliferation of differentiated cells versus a lengthy, stepwise, and less controllable dedifferentiation of cells. Evidence occurs that the absence of regenerative ability in mammals is a loss of function. This is perhaps a consequence of physiological adaptations important for survival following injury. Humans might also have lost the ability to regenerate because maintaining such large numbers of dividing cells would increase the risk of cancer. Although planaria are practically immortal, they maintain exquisite control of cell division – an attribute that might have waned as human tissues took on specialized tasks.

Although mammals are not endowed with great ability to regenerate limbs and other complex structures, they have the potential to regenerate a surprisingly large array of injured tissues [37–40]. The latter includes cases of antler replacement in deer as well as ingrowth from the margins of holes cut in bat wing membranes and in the external ears of rabbits, pikas, cats, and echolocating bats [41]. The digit tip in mammals, including humans, is also regeneration competent and offers an interesting mammalian model for regeneration [42]. The observation that fingertips of children have regenerative capacity indicates that the ability for a modest level of regeneration was retained in a context-specific format even in higher mammals [43].

Unlike that in urodele amphibians, in which regeneration, once started, typically results in the formation of an almost perfect replica of the structure that was lost, mammalian regeneration proceeds with varying degrees of success [33]. One of the major challenges in the scientific study of regeneration in mammals and its clinical application in humans is to understand why regeneration proceeds very well under some circumstances and very poorly under others.

It is suggested that the relative inadequacy of regeneration in warm-blooded vertebrates may be attributed to the precocity with which they tend to form dermal scars in healing wounds. Scars are believed to preclude blastema production. Wound healing around the margins of rabbit ear holes is uniquely characterized by the development of prominent epidermal downgrowths adjacent to the severed sheets of dermis in the integument on either side of the ear. These downgrowths act as epidermal blockades that prevent scar formation in favor of allowing blastema cells to accumulate. A logical approach to the experimental induction of regeneration in normally nonregenerating mammalian appendages would involve manipulation of the mechanisms by which epidermis heals amputation stumps [13]. The particular role of scar formation will be addressed in more detail later in this chapter.

The necessity of injury, nerves, and wound epidermis for urodele limb regeneration is well accepted. In urodeles, the wound epidermis has recently been shown to have the role of maintaining dedifferentiated cells of the amputated limb stump in the cell cycle. The result of this wound epidermal stimulus is a sufficient number of cell divisions such that blastema formation occurs [36]. In contrast to that, in amputated limbs of higher vertebrates, the wound epidermis is nonfunctional. Dedifferentiated or undifferentiated cells are not maintained in the cell cycle and blastema formation therefore does not occur. Instead, tissue regeneration occurs precociously due to the lack of a cycling stimulus. Thus, the scar tissue which forms at the limb tips of nonregenerating vertebrates can be regarded as the result of a nonfunctional wound epidermis [44].

An important further aim will be to prove that mammals can form the required progenitor cells for regeneration. A certain type of mouse, known as the Murphy Roths Large (MRL) mouse, having enhanced regenerative capabilities is a very promising object in this respect. The MRL mouse was first shown to regenerate a portion of the ear but the range of its regenerative skills is much broader and includes toes, tail, liver, kidneys, and, importantly, the heart muscle. This animal has an unusual ability to show both cell proliferation and lack of scarring, which are two key features of successful regeneration.

The MRL mouse was generated through the interbreeding of the LG mouse (75%; H-2d/f), the AKR mouse (12.6%; H-2k), the C3H mouse (12.1%; h-2k), and the C57B1/6 mouse (0.3%; H-2b) [45]. The MRL *(lpr/lpr)* mouse was selected originally for its large size and was found to have a major defect in immune regulation. It was found that with age, lymphocytes in the lymph nodes and spleen showed unregulated proliferation. This phenotype was found to be due to a retrotransposon insertion into the second intron of the *fas* gene [46] and led to an absence of cell death. Further works have shown that other genes, *mmp9* and *mmp2*, may be implicated too.

Young adult autoimmune MRL *lpr/lpr* as well as normal nonautoimmune MRL *lpr/-* (MRL/MpJ) mice were found to be capable of completely closing 2 mm through-and-through surgical ear holes within 30 days, whereas all other mice have residual open holes, usually stable over the lifetime of the animal [47]. The healing seen in these MRL mice displays normal gross- and microanatomy reminiscent of regeneration seen in amphibians [48, 22], whereas the normal outcome in mammals is scarring [49]. In fact, the type of healing seen in MRL mice is mechanistically different from the "normal" one.

Heber-Katz's team has carried out an experiment where they injected fetal liver stem cells from an MRL mouse into the tail vein of a nonhealing mouse [49]. One month later they produced a small injury to the heart of the nonhealing mouse and no scar formation was seen. Instead, islands of newly formed cardiomyocytes were observed in the damaged regions. Interestingly, if fetal liver stem cells from the nonhealing mice were transferred into the MRL mice, a lot of scar tissue and little cardiomyocyte island formation were detected. This experiment implies the existence of certain cell factors causing inhibition of scar formation, which can be used to enhance the regenerative ability of mammals in general.

By studying urodeles and MRL mice, it might be possible to identify the specific types of cells, molecular signals, genes, and cellular scaffolding required for regenerative cell growth. In future, approaches should be found to orchestrate the formation of a blastema in response to an injury at the site where nature would normally direct the accumulation of scar tissue. The next step would be to prove that following blastema formation, a functionally normal limb or digit develops. The last would mean that achieving full restoration of limb structure and function is feasible. Transferring this ability to people will be a challenge taking into account the fact that at least 20 different genes are involved in the regeneration ability of the MRL mouse [50]. The genes involved in heart regeneration seem to overlap with the ear hole closing ones, but it requires more or some different genes. The good news is that the initial studies have found that the genes responsible for the MRL mouse's predilection for autoimmune diseases are different than the ones involved in fast healing. An important work of mapping genes shall be done now to shed more light upon the genetic combinations that may be at the root of the mouse's regenerative process.

There may indeed be plausible reasons of why some animals are normally incapable of regenerative growth. First, the initial phase may fail to produce a blastema because of either an inadequate signal or an inability of the cells to respond to an adequate signal by dedifferentiation. If an adequate blastema was formed, the second-phase informational signal might be missing or inadequate to produce the subsequent patterning, redifferentiation, and growth. These aspects will be discussed in more detail in the following sections.

5 Regeneration as Redevelopment

The key for regeneration is, in fact, the capacity to rekindle mechanisms responsible for an organism's embryonic development. A hydra, for instance, can be seen as a kind of permanent embryo (Fig. 5). Even before an injury occurs, it has the cells it needs to step in and reconstruct the organism. Similarly, planaria (flatworms) maintain a cadre of stem cells – perhaps as many as 20% of its total number – ready to attend to a wound. As a matter of fact, the planarian invests a huge part of its energy in maintaining adult tissues and can be regarded as an animal that has taken stem cells to an extreme [51]. Urodele amphibians take a more complicated path. When they require a new limb or organ, they deprogram specialized cells rather than recruit stem cells that already exist. Regeneration comes down to how you convert an adult differentiated cell back into an embryonic underdifferentiated one [43].

Like the salamander, the zebrafish (*Danio rerio*) engages specialized cells in deprograming and forming a blastema. These cells turn on *msx* gene which apparently keeps cells in a malleable state by dampening programs needed for specialization [52]. Like other champion regenerators, zebrafish readily retreat to an embryonic state – at least at the cellular level. When gauged against heart cells from mammals, for example, those from zebrafish seem more like fetal cells than adult ones.

Deeper understanding of the cellular and molecular mechanisms involved in dedifferentiation is important for thinking about cell reprogramming necessary for

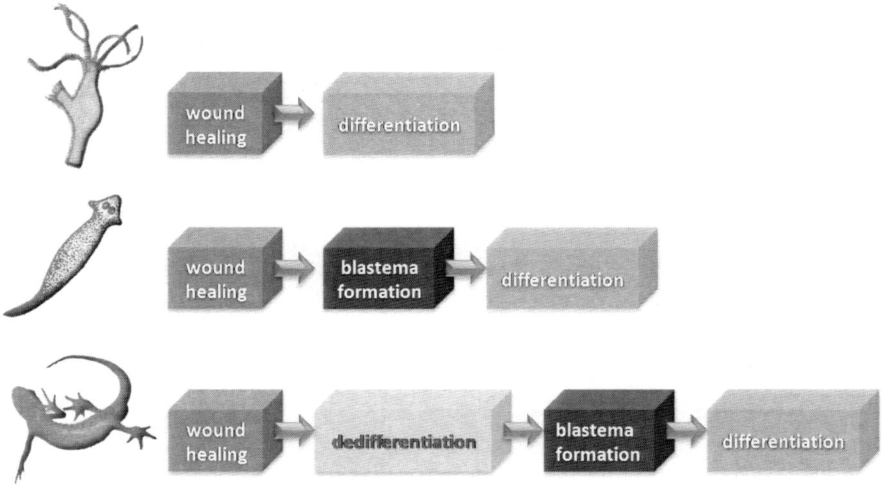

Fig. 5 The steps required before regeneration are fully achieved in various model systems. In hydra, the head organizer activity is established within several hours after cutting; in planaria, the blastema appears after 1 day; in urodeles, the two first phases last about 2 weeks. Completion of regeneration and wound healing takes a long time in regenerative and nonregenerative limbs, respectively. However, it is the early steps that are critical for determining the extent of regenerative response after limb amputation, ranging from wound healing with scar formation, scar-free wound healing, and partial limb regeneration to complete restoration

inducing a regeneration response [53, 20]. Getting to a redevelopment phase in mammals would be a major breakthrough, given our extensive understanding of limb development. The acute response to tissue injury or removal generally includes the events of wound healing, hemostasis, and tissue repair. One critical role played by tissue injury is likely to be the initiation of regeneration. A regenerative system has to sense the removal or damage of tissue, and it is almost a logical requirement that some aspect of the injury response should signal for regeneration to occur. There are many potential signals and a variety of candidates have been suggested, some of them we are going to discuss below.

Prominent in urodele amphibian limb regeneration is the formation of a blastema of undifferentiated cells that goes on to reform the limb. After limb/fin amputation, the axolotl and zebrafish epidermal cells migrate to cover the wound surface. For regeneration to occur, this event needs to be preceded by the formation of the apical ectodermal cap that covers the wound surface after amputation [54, 55]. During the last decade, and in part with the help of the knowledge gathered during the embryogenesis of the vertebrate limb, some of the molecular and cellular processes involved in AEC and blastema formation have been unveiled [56, 54].

The blastema shares many properties with the developing limb bud; thus, the outgrowth phase of regeneration can be thought of as cells going through development again, i.e., redevelopment. Regeneration of body parts shares many other similarities with their embryonic development and it is unlikely that there are different programs for morphogenesis during embryonic development and adult regenerative processes. These similarities can be traced up to gene regulation level since both the processes are regulated by the same mean signaling cascades: Wnt/β-catenin and BMP. Specifically, members of the Wnt and bone morphogenetic protein (BMP) signaling pathways have been shown to be required in vertebrates for the formation of the apical ectodermal ridge (AER). In embryogenesis, the AER forms at the junction between the dorsal and the ventral ectoderm and becomes morphologically visible as a ridge of epithelium at the distal tip of the limb bud. Kawakami and co-workers have reviewed remarkable similarities in the molecular (Wnt and BMP pathways) and cellular (i.e., AEC and AER formation) processes involved in limb regeneration and embryogenesis [57].

Although the majority of work in the field to date has focused on β-catenin-dependent, or canonical, Wnt signaling (Fig. 6), examples continue to accumulate

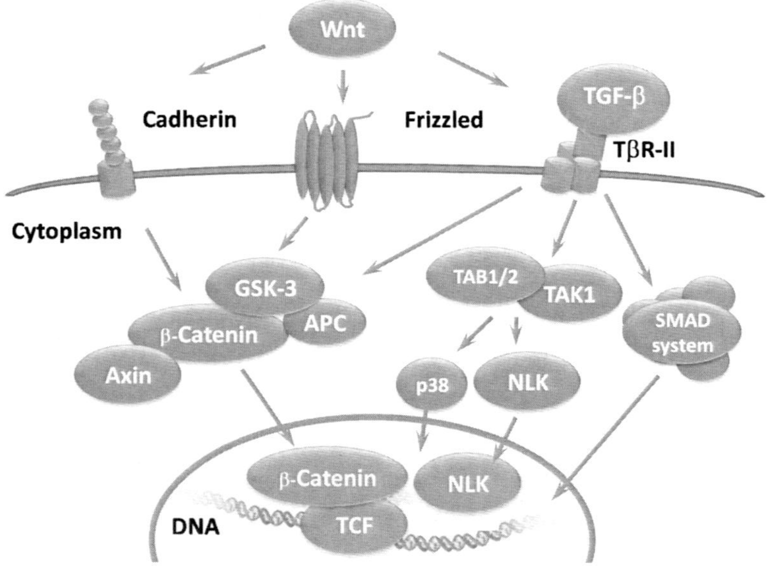

Fig. 6 Main components of the Wnt/β-catenin signaling pathway. The Wnt's comprise a large family of highly conserved growth factors that are responsible for important developmental and homeostatic processes throughout the animal kingdom. The defining event in canonical Wnt signaling is the cytoplasmic accumulation of β-catenin and its subsequent nuclear translocation and activity. Under unstimulated conditions, a β-catenin destruction complex formed by proteins that include axin, adenomatosis polyposis coli (APC), and glycogen synthase kinase (GSK-3) keeps cytoplasmic levels of β-catenin low through phosphorylation by GSK-3. Phosphorylated β-catenin becomes ubiquitylated and is targeted for degradation by the proteasome. Following Wnt binding to a receptor complex composed of members of the Frizzled (Fz) family of seven transmembrane, serpentine receptors, and low-density lipoprotein receptor-related protein (LRP), the Axin/APC/GSK-3 complex is inhibited, leading to a block in β-catenin phosphorylation by GSK-3. Hypophosphorylated β-catenin accumulates in the cytoplasm and is translocated to the nucleus where it regulates target gene expression

in which Wnt's and/or other key components of the canonical signaling cascade participate in β-catenin-independent processes [58].

Bone morphogenetic proteins (BMPs) as well as their antagonists are involved in controlling a large number of biological functions including cell proliferation, differentiation, cell fate decision, and apoptosis in many different types of cells and tissues during embryonic development and postnatal life. BMPs exert their biological effects via using BMP–Smad and BMP–MAPK intracellular pathways (Fig. 7). The magnitude and the specificity of BMP signaling are regulated by a large number of modulators operating at several levels (extracellular, cytoplasmic, and nuclear). In developing and postnatal skin, BMPs, their receptors, and BMP antagonists show stringent spatiotemporal expressions patterns to achieve proper regulation of cell proliferation and differentiation in the epidermis and in the hair follicle. Genetic studies assert an essential role for BMP signaling in the control of cell differentiation and apoptosis in developing epidermis, as well as in the regulation of key steps of regeneration (initiation, cell fate decision, and cell lineage differentiation) [59].

Three steps are involved in AER formation: induction of the AER precursors, migration of the precursors to the distal tip, and compaction of these cells to form the tall morphological ridge. The key signals that mediate AER function are the fibroblast growth factors (FGFs). *Fgf8* is expressed throughout the AER, whereas

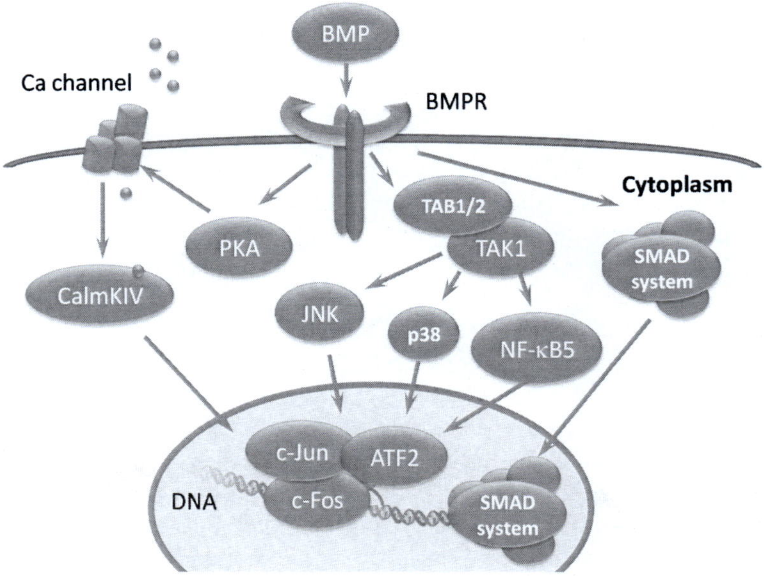

Fig. 7 Main components of BMP signaling pathway. Bone morphogenetic proteins (BMPs) are multifunctional growth factors that form a major part of the transforming growth factor beta (TGF-β) superfamily. These proteins play a pivotal role in embryonic development and cellular functions and are involved in nearly all processes associated with skeletal morphogenesis. BMP signals are transduced from the MBPR plasma membrane receptors to the nucleus through both Smad-dependent and Smad-independent (MAPK, JNK, etc.) pathways

Fgf4, *Fgf9*, and *Fgf17* are restricted to the posterior and distal AER. Many other genes are expressed throughout the AER, including *Msx1* and *Msx2*, *Dlx5* and *Dlx6*, and *Bmp2*, *Bmp4*, and *Bmp7*, although only BMPs have so far been implicated in AER formation and function [60, 57].

Later, Kawakami and colleagues have shown that reduction in Wnt and BMP signaling during limb regeneration in axolotls, *Xenopus laevis,* and zebrafish induces alterations in the formation of the AEC that prevent normal fin/limb regeneration [61]. In tissues with blocked Wnt pathway, the early phase of epidermal migration appeared to be unaffected, but the subsequent process that led to the formation of the AEC was altered so that epithelium lacked the cuboidal shape and smooth stratification characteristic of blastema formation and regeneration [62, 57]. More importantly, by performing gain-of-function experiments of the Wnt/β-catenin pathway during appendage regeneration, it has been demonstrated that this pathway promotes *Xenopus* and zebrafish limb/fin regeneration.

To gain further insights into the molecular mechanisms by which downregulation of Wnt signaling might alter the formation of the AEC, the expression of several genes known to be involved in AEC formation during *Xenopus* limb regeneration was analyzed [61]. Of particular interest were the changes observed in p63 expression. p63 is required in higher vertebrates for the stratification and maintenance of the AER [63], a pseudostratified epithelium that, like the regenerating AEC, is required for the proliferation of underlying mesenchymal cells, and therefore for normal limb development [56].

While obviously not identical processes, the similarities encountered in the molecular and cellular processes involved during limb embryogenesis and limb regeneration suggest a mechanism whereby variations in the concentration and/or spatiotemporal distribution of developmental regulators may allow regeneration to occur. By means of switching these cascades on and off, one can not only inhibit regeneration in otherwise regenerationable animals but also provoke regeneration in animals that have lost this ability.

6 Blastema Formation Versus Scarring

During any regeneration event, the final outcome involves the differentiation of blastema cells into various tissue types originally present in the amputated structure. The cycle of dedifferentiation and redifferentiation raises the possibility that cells may be able to transdifferentiate into cell types that differ from their tissue of origin. There is evidence that this type of plasticity does occur during limb and tail regeneration [64, 65]. Many regeneration studies have focused on transdifferentiation as a measure of cell plasticity and these studies assume that in order to transdifferentiate, a cell must first dedifferentiate. A critical signal for the activation of cell cycle re-entry and regeneration in salamander appears to be linked to coagulation and the local activation of thrombin from its zymogen prothrombin [66, 67, 53]. For instance, re-entry to the cell cycle by striated muscle in both the newt [20] and the medusa of the jellyfish *Podocoryne* [68] is apparently initiated by proteolytic activation. These insights are valuable in part because they place the regenerative response in a wider context.

In contrast to urodele regeneration, the wound-healing response following limb amputation in adult mammals has not been studied in great detail. Key events of nonamputation wound healing include the formation of a fibrin clot, a relatively slow re-epithelialization, an inflammatory response that helps to populate the wound site, the formation of granulation tissue, the differentiation of fibroblasts to become myofibroblasts that contract the wound, and the deposition of parallel collagen bundles that form scar tissue [69].

The role of immune "interference" in regeneration has been reviewed recently [70] and will not be considered in detail here. It is often suggested that the ongoing events of wound healing may be incompatible with regeneration, and immune modulation is widely recognized as an important factor [4]. The role of the immune system in inflammation, fibrosis, and scar formation in mammals is highlighted by the occurrence of scar-free wound healing in embryos at stages which precede the development of some immune cell types [69].

It has been pointed out that there is a correlation in phylogeny between the development of adaptive immunity and the progressive loss of regenerative ability [71]. It is possible that there may be contexts where regeneration is directly linked to an aspect of the immune response, and this would be of great interest. Mammals have highly developed adaptive immunity and relatively poor capacity to regenerate. Among amphibians the urodeles are better regenerators but relatively immunodeficient compared with anurans such as *Xenopus* as they have an IgM-based humoral immune response with slow allograft rejection [72–74]. In *Xenopus* the development of adaptive immunity occurs in the late larval stages approaching metamorphosis, and this is the same period when regenerative ability in the hind limb is progressively lost. In a comparison of genes expressed in blastemas at a regeneration-competent and an older regeneration-incompetent stage, many were noted to be regulators of adaptive immunity in mammals [70].

As a matter of fact, human injury response might impede renewal. One might speculate that evolutionary pressures have been exerted on intermediate-sized, widespread, dirty wounds with considerable tissue damage, e.g., bites, bruises, and contusions. "Modern" wounds (e.g., resulting from trauma or surgery) caused by sharp objects and healing in a clean or sterile environment with close tissue apposition are new occurrences, not previously encountered in nature and to which the evolutionary-selected wound-healing responses are somewhat inappropriate. Both repair with scarring and regeneration can occur within the same animal, including man, and indeed within the same tissue, thereby suggesting that they share similar mechanisms and regulators. Big wounds really are life threatening. To solve that problem, humans possibly abandoned regeneration and went to immediate wound closure. The fact that animals that regenerate do not display scarring supports this idea. In addition to the presence of dividing cells, the lack of scarring in the MRL mouse is what distinguishes it from others. Scar formation implies how rushed and disorganized the mammalian injury response is.

The lack of a scar after injury is not totally foreign to mammals. This is typically found in early development where it has been reported that fetal animals heal without scar tissue [75, 76]. The wound heals and a scar is formed after injury in adult mammals such as mice and humans, while wound healing is completed without scarring in an embryonic mouse. The fetal wound area develops hair follicles and has normal tissue architecture and normal collagen superstructure. These are elements that would be ascribed to regeneration in adults. In contrast, adults usually heal with scar formation in which there are no newly developing hair follicles, there is abnormal collagen structure, and abnormal tissue architecture and function. For instance, skin wound scars are weaker and heal poorly upon rewounding. There is an age-related developmentally directed component to this transition as well.

Experiments done with marsupials show that the transition point between no scarring and scarring occurs at pouch day 9. This implies that immune response is involved in the initiation of scar formation [4]. In fetal mice, the transition point is found to be embryonic day 16, again coincident with the appearance of inflammatory responses to wounds in mouse ontogeny [77].

The cellular and molecular differences between scar-free healing in embryonic wounds and scar-forming healing in adult wounds were reviewed by Ferguson and O'Kane [4]. They pointed out that important differences include the inflammatory response, which in embryonic wounds consists of lower numbers of less differentiated inflammatory cells. This, together with high levels of morphogenetic molecules involved in skin growth and morphogenesis, means that the growth factor profile in a healing embryonic wound is very different from that in an adult wound. The cells of the wound epithelium are known to come from the epidermis and they go on to differentiate into the regenerated epidermis; so the evidence indicates that the epidermis dedifferentiates to form the AEC. The formation of the mesenchymal component of the blastema is more complex as multiple tissue types provide cells that contribute to the blastema. Nevertheless, it is clear that the relative contribution from different tissues in the stump is not equivalent and that some tissues, such as the dermis, overcontribute, while other tissues, such as skeleton and muscle, undercontribute [78, 79, 64].

Among the factors that block regeneration is the formation of scar tissue and the secretion of a basement membrane around reforming tissues in the wound site. The problem of scar formation as a pathway alternative to regeneration was addressed in some reviews by Polezhaev [10, 34, 80]. On the one hand, it might be a good solution in the short term as it represents a rapid response to injury that could prevent infection and further damage. On the other hand, it blocks the reorganization and the more plastic responses that are needed in order to regenerate.

Perhaps one of the most interesting aspects of ear hole closure in the MRL mouse is seen in the ability of this mouse to break down its own basement membrane that forms during and after epidermal coverage of the wound. With the usual repair mechanisms generally associated with mammalian wound healing, including the formation of a provisional matrix and a remodeling response, a basement membrane forms and separates the epidermis from the dermal layer. However, in the regenerative response seen in amphibians, the basement membrane either never forms or forms and disappears so quickly that it has not been reported to be present. The growth of the blastema soon follows [20, 81, 22]. The importance of the basement membrane and its presence or absence has been demonstrated by the induction of a basement membrane in the amphibian limb stump. This event leads to the cessation of the regenerative response and to the formation of a scar.

An important group of molecules shown to be involved in both wound repair and regeneration, and specifically in a remodeling response involving collagen as well as other extracellular matrix molecules, is the family of metalloproteinases (MMPs) [82–84]. This is especially true for two family members MMP-2 and MMP-9 that are produced and secreted by multiple cell types, including neutrophils and macrophages.

Analysis of the genes and gene products affected in the MRL mouse shows that MMP-2 and MMP-9 are upregulated, whereas inhibitors of the metalloproteinases, TIMP-2, and TIMP-3 are downregulated. This gene activation pattern results in breaking down the basement membrane, thus keeping tissue labile in the wound site. In addition to this, the MRL mice demonstrate the improved ability to regenerate from spinal cord injury, which can be, again, related to the inhibition of scar tissue formation. Also, consequently, when the MRL mice are treated with a metalloproteinase inhibitor (such as minocycline), the rate of regeneration is greatly slowed [85].

Recent studies on MRL mice suggest that processes of regenerative growth and patterning for the formation of new structures such as hair follicles may also involve modulation of the inflammatory response to the injury in a way that reduces fibrosis and formation of scar tissue. This modulation includes changes in cytokine signaling and may involve properties of the extracellular matrix mediated by factors that include hyaluronic acid and "anti-adhesive substrates" such as tenascin-C. Changing properties of the immune system may also underlie the declining regenerative potential in this system. Recent studies in comparative and developmental immunology raise the possibility that phylogenetic changes in regenerative capacity may be the result of evolutionary changes in cellular activities of the immune system [86].

The differences between fetal and adult skin wound healing appear to reflect processes intrinsic to fetal tissue (such as the unique fetal fibroblasts) a more rapid and ordered deposition and turnover of tissue components and, particularly, a markedly reduced inflammatory infiltrate and cytokine profile. Scarless fetal wounds are relatively deficient in the inflammatory cytokine, transforming growth factor beta (TGF-β). In contrast, the fibrosis characteristic of adult wound repair may be associated with TGF-β excess. Existing experimental studies suggest that specific anti-TGF-β therapeutic strategies can ameliorate scar formation in adult wound repair and fibrotic diseases [87].

The fact that embryonic wounds that heal without a scar have low levels of TGF-β1 and TGF-β2, low levels of platelet-derived growth factor, and high levels of TGF-β3 offers promise in manipulating healing adult wounds in mice, rats, and pigs to mimic the scar-free embryonic profile. Notably neutralizing PDGF, neutralizing TGF-β1 and TGF-β2, or adding exogenous TGF-β3 resulted in scar-free wound healing in the adult. Exogenously applied FGF-10 results in the reinduction of the expression of the genes involved in regeneration (*shh*, *msx-1*, *fgf-10*, and others) and could be a key molecule in possible regeneration of nonregenerative limbs in higher vertebrates. Thus, by subtly altering the ratio of growth factors present during adult wound healing, it is possible to induce adult wounds to heal perfectly with no scars, with accelerated healing, and with no adverse effects (e.g., on wound strength or wound infection rates).

A new concept for the regulation of muscle regeneration by the TWEAK/Fn14 pathway as a regulator of mesenchymal progenitor cells was suggested by Burkly et al. [88]. According the model, cardiotoxin-induced injury triggers release of soluble factors that recruit early inflammatory cells and induce Fn14 expression on muscle precursor cells. Secreted by infiltrating inflammatory cells, TWEAK engages Fn14 and promotes the proliferation and activation of myoblasts, which in turn can contribute to a robust inflammatory response by producing additional chemotactant. Following this logic, one can speculate that mammalian myotubes have lost their responsiveness by, for example, the receptor mutation [89].

Summarizing, the existing observations suggest that many mammalian organs harbor a latent capacity to regenerate but that our rapid healing process blocks renewal. This hypothesis for loss of regenerative ability represents an interesting opportunity to identify the responsible ligands and their receptors and to manipulate regenerative processes pharmacologically.

7 Spatial Patterning of Morphogenesis

Nematode *Caenorhabditis elegans* is a perfect model object for developmental biology and for its introduction S. Brenner was awarded with the Nobel Prize in 2002. Bischoff et al. [90] reported that in the *C. elegans* embryo, the descendants of P_1, the posterior blastomere of the two-cell stage, constitute a polarizing centre that orients the cell divisions of most of the embryo. This polarization depends on an MOM-2/Wnt signal which is similar to the known cell polarity pathways in other organisms. These findings may provide a paradigm of how polarity can be organized in a morphogenetic field of cells in other organisms or their regenerating parts.

In Hydra, the early studies on developmental patterning initially suggested that nerve cells are the primary source of signals during regeneration and limb budding [15]. However, nerve cells-depleted animals are still able to regenerate, probably due to morphogenetic substances produced by the epithelial cells [91, 15]. Peptides (e.g., pedibin and Hym-346), produced by nerve cells as head activator or by epithelial cells [92, 91], were shown to accelerate apical or basal regeneration, respectively. The other signals that control patterning in Hydra remain largely unknown.

The head region of Hydra apparently acts as an organizing region and as an inhibitor of inappropriate head formation. The hypostome and the basal discs act as organizing centers to give polarity and act to induce head and tail formation. Grafts of the hypostome to the gastric region will induce a second head (and eventually a new body). Grafts of the region next to the head to the gastric region will not generate a new head unless the original head is removed but will generate a new head in the foot region.

Using PCR, six genes of the HOM/HOX class have been identified in hydra [93]. One of them, *Cnox-2*, has been sequenced and resembles the so-called deformed gene of *Drosophila*. The expression pattern of *Cnox-2* suggested that it may be involved in axial patterning as it was strongly expressed in the epithelial cells of the body column and foot but only weakly expressed in the head. In addition, expression of *Cnox-2* decreased sharply in body column tissue as it was converted into head tissue, thereby providing further support for a role in Hydra pattern formation.

In 1952, Marsh and Beams [94] reported on an interesting series of experiments on *Planaria*, a species of relatively simple flatworm with a primitive nervous system and simple head-to-tail axis of organization. Since electrical measurements had indicated a simple head–tail dipole field, they postulated that the original head–tail electrical vector persisted in the cut segments and that it provided the morphological information for the regenerate. If this was so, then reversal of the electrical gradient by exposing the cut surface to an external current source of proper orientation should produce some reversal of the head–tail gradient in the regenerate. While performing the experiment they found that as the current levels were increased, the first response was to form a head at each end of the regenerating segment. With still further increases in the current, the expected reversal of the head–tail gradient did occur, indicating that the electrical gradient which naturally existed in these animals was capable of transmitting morphological information.

Interesting phenomena were observed following foreleg amputations in salamanders as compared to the same amputation in frogs. While the immediate postamputation electrical potentials were positive in polarity and of about the same magnitude in both species, the frog's potential slowly returned to the original slightly negative potential as simple healing by scarification and epithelialization took place. In the salamander, the positive potential very quickly (3 days) returned to the original baseline but then became increasingly negative in polarity, coinciding with blastema formation and declining thereafter as regeneration occurred.

After demonstration of the presence of an organizer region in the *Xenopus* embryo, the graded distribution of differentiated cells along the developmental axis was observed [95], and the graded distribution of biological activities was demonstrated by grafting experiments. At gastrula stage, perpendicular activity gradients of Wnt's and bone morphogenetic proteins (BMPs) regulate head-to-tail and dorsal–ventral patterning. The formation of head, trunk, and tail requires increasing Wnt activity.

Since limb development itself is dependent on epithelial–mesenchymal interactions, both the timing and the spatial pattern of dedifferentiation in the various tissues must be coordinated for a successful regenerative response. The AEC is necessary for proximal–distal limb outgrowth because the AEC provides a permissive signal to enable the realization of proximal–distal pattern. Surgical removal of the AEC results in limb truncation. Removal of the AEC at progressively later stages results in progressively more distal truncations. The recent studies with chickens and mice propose that the AER controls the initial size of the limb bud, cell survival, and proliferation and that it generates sufficient numbers of mesenchymal cells to form appropriate size condensations [26]. In urodeles and fish, regeneration always proceeds in a direction distal to the cut surface. Grafting a distal blastema to a proximal stump will often induce the stump to generate a normal limb and the distal blastema forms the wrist and hand.

Crawford and Stocum [96] showed that the position of the regeneration blastema cells can be changed by the introduction of retinoic acid. Retinoic acid is present in developing vertebrate limbs in a distinct pattern and is higher in concentration in more distal blastemas. Wounded epidermis is also a source of retinoic acid. Exposure to retinoic acid changes the positional value of a blastema to more proximal ones such that elements proximal to the cut as well as those distal are generated. Therefore, retinoic acid acts as a morphogen to establish the positional values on the cell surfaces of the developing limb and the limb blastema cells. This may have important consequences for the interpretation of retinoic acid as a ZPA (zone of polarizing activity) morphogen. Later, Bryant and her colleagues speculated that retinoic acid does not provide a gradient of positional information but rather is capable of transforming anterior limb bud tissue into posterior limb bud tissue [55]. The host limbs would now contain "posterior" tissue (the graft of retinoic acid-treated anterior cells) next to anterior tissue (the host limb bud). The result would be intercalary regeneration to restore the positions between the two normally nonapposed tissues. In this interpretation, retinoic acid is an agent that can modify positional values within the limb field.

Studies from K. Muneoka's laboratory established an initial link between the production of fibroblast growth factors (FGFs) by the AER and the signaling ability of the zone of polarizing activity. Signaling by the ZPA is now known to be mediated by a secreted factor called sonic hedgehog (Shh) and it is now well established that FGFs modulate Shh production [97, 98]. *Hedgehog* (*Hhh*) genes encode a family of signaling molecules involved in a variety of developmental processes in both vertebrates and invertebrates [99]. In *Drosophila*, *Hhh* patterns the body segments, the wing, leg, eye imaginal discs, and regions of the fly brain, either directly or through the recruitment of other signaling molecules such as decapentaplegic (*Dpp*), and *Wingless (Wg)*. In contrast to the single *Hhh* family member in the fly, there are three *Hedgehog* members in mammals: sonic *Hhedgehog (Shh)*, Indian *Hhedgehog (Ihh)*, and desert *Hhedgehog (Dhh),* this latter one likely representing the most ancestral *Hhh* gene in vertebrates [100]. Historically, the ZPA is identified and characterized based on its ability to induce supernumerary digits and more recently, there is evidence that the production of sonic Hhedgehog proteins by ZPA cells is responsible for this induction.

Following dedifferentiation, regenerating cells are classified as mesenchymal and undifferentiated (i.e., appearing embryonic); they display embryonic characteristics such as migratory behavior and differential adhesion. In addition, they readily undergo cell division and re-express many developmentally regulated genes, including members of the *Hoxa* and *Hoxd* gene clusters, *Msx1*, *Msx2*, and *Shh* [43]. Their migratory and proliferative behaviors play a key role in the accumulation of the blastema at the center of the wound [78], and their differential adhesive character

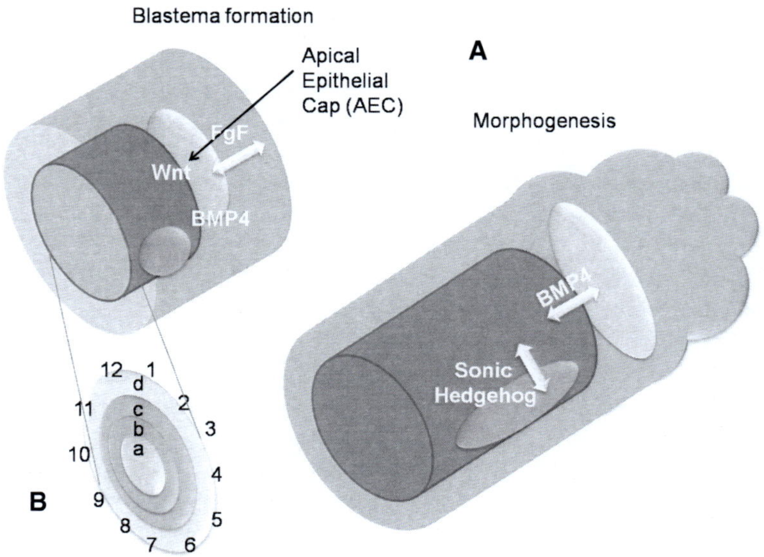

Fig. 8 Molecular signals in the spatial patterning of a regenerating limb. (**a**) The regenerating limb is formed by a series of cellular and molecular interactions that occur between a specialized group of ectodermal cells at the tip of the limb bud that form a structure called the apical epithelial cap (AEC) and the mesenchymal cells that underlie the AEC. The apical ectoderm produces factors that are necessary for distal outgrowth by the mesenchyme. Mesenchymal cells interact with one another and with the AEC to establish spatially distinct patterns of gene expression followed by the differentiation of specific structures. (**b**) Polar coordinate model for the specification of positional information. Each cell has a circumferential value (0–12) specifying the anterior–posterior axis and a radial value (a–d) specifying the proximal (a) to distal (d) axis

plays a role in distinguishing their proximal–distal position in the limb (Fig. 8a) [96]. The re-expression of developmental genes warrants the classification of later stages of regeneration as redevelopment [55].

FGFs are secreted factors that interact with specific cell surface receptors to elicit a response. One of the features of the FGF family is that they have an affinity for heparan sulfate moieties present in the extracellular matrix (ECM). Remarkably, stripping the ECM preparation of bound factors enhances signaling levels, suggesting that this activity involved the inactivation of inhibitory activity in the limb bud. These studies are truly unique in that they represent the first experimental evidence that ECM components are playing a key role in modulating cell–cell signaling during limb development.

Reginelli et al. [101] have discovered that FGF-4 modulates cell movements during digit morphogenesis. They found that FGF-4 induces expansion of the digit tip leading to digit bifurcation. FGFs are known to regulate the expression of the *Msx* homeobox-containing genes (*Msx1* and *Msx2*) during limb outgrowth. Mutant mice lacking the *Msx1* gene are defective in their ability to regenerate the digit tip. Their studies also show that during digit formation, the expression of the *Bmp4* gene is dependent on *Msx1* and *Msx2* expression and that adding BMP4 to amputated *Msx1* mutant digits rescues the regeneration response. Inhibition of the activity of BMPs in wild-type mice results in the inhibition of digit tip regeneration. Later, it had been shown that FGFs can induce a regenerative response from the nonregenerating chick limb bud and that FGF-4 acts as a chemotactic factor produced by the AEC to regulate cell movements important for limb morphogenesis [102]. The authors have suggested that cell motility and differential cell adhesion play a critical role in limb morphogenesis, leading to outgrowth.

Research on developing and regenerating limb buds has also shown another level of pattern formation. When axolotl limb buds were transferred to regenerating axolotl blastema stumps, in a way that maintained the original polarity with respect to the stump, normal limbs developed. However, when anterior and posterior cells from regeneration blastema or normally nonadjacent tissues of limb buds are juxtaposed, the results are dramatic and unusual. When a newly emerged limb is severed at its base and rotated 180° on its stump so that the anterior–posterior axes are reversed between the host and the graft, the developing limb contains three areas of outgrowth, producing supernumerary structures, often with two or three sets of digits [103, 104]. Similarly, grafting of left hind limb blastema onto the right hind limb stump produces three sets of distal regions on the limb [104].

By comparing such results in regenerating vertebrate limbs, insect limbs, and insect imaginal discs, French and his colleagues [105] have proposed a series of empirical rules that predict the outcome of a wide variety of experimental perturbations with regenerating appendages. Starting from Wolpert's [106] premise that pattern arises from a cell's recognition of its relative positions in a developing population, they speculate that a cell assesses its physical location in a system of polar coordinates. In this system, each cell has a circumferential value (from 0 to 12) as well as a radial value (from a to e). In regenerating limbs, the outer circle represents the proximal (shoulder) boundary of the limb field and the innermost circle represents the most distal regions (Fig. 8b).

The suggested polar coordinate model has been extremely useful in predicting the extent of duplicated structures and several empirical rules on regenerative patterning were developed. The *shortest intercalation rule* states that when two normally nonadjacent cells are juxtaposed, growth occurs at the junction until the cells between these two points have all the positional values between the original points. The circular sequence, like a clock, is continuous, 0 being equal to 12 and having no intrinsic value. Being circular, however, means that there are two paths by which intercalation can occur between any two points. For example, when cells having the values 4 and 7 are placed next to each other, there are two possible routes between them: 4, 5, 6, 7 and 4, 3, 2, 1, 12, 11, 10, 9, 8, 7. According to this model, the shortest route is taken. The exception, of course, is when the cells have values that fall exactly opposite to each other in the coordinate system so that there is no one shortest route. In this case, all values are formed between the two opposites.

The other rule is the *complete circle rule for distal transformation*. Once the complete circle of positional values has been established on the wound surface, the cells proliferate and produce the more distal structures. The predictive value of these rules can be seen when a transplant is made between regeneration blastemas, a transplant in which the anterior and posterior axes are reversed. The result is a limb with three distal portions [104]. This outcome can be explained by viewing the anterior–posterior axis on the grid as having two opposite numbers, say 3 and 9. In juxtaposing the values of 3 and 9, one generates a complete circle of values at each of the extreme sites and a smaller intercalating series at all other sites. The result is three complete circles, which, by the law of distal transformation, will generate three complete limbs from that point on.

In insects, when tissues of vastly different positional value are placed in conjunction, intercalary growth occurs to replace the missing values. For instance, grafting of amputated cockroach legs demonstrate intercalation. A distal cut tibia grafted onto a proximal cut will grow to intercalate the missing pieces. However, a proximally cut tibia, grafted onto a distally cut host, will also grow by intercalation. In the latter case, the regenerated portion is in the reverse orientation (by bristle direction). Circumferential values can also be regenerated by intercalation [107].

The polar coordinate model also predicts the effects of regulation when a portion of the tissue is lost. The newly formed cells would have positional values intermediate between those of the remaining cells and would reconstruct the appropriate part of the tissue. Because the basis of regeneration appears to be the recognition of differences between adjacent tissues, it is probable that epimorphic pattern formation during regeneration and normal pattern formation during embryonic limb development are the result of the proximate interactions between adjacent cells rather than the result of long-range gradients [104].

In their brilliant review, Han et al. [43] addressed an important question of the feasibility of a human regeneration and the fact that it may be limited by the physical size of the human limb. As the authors lawfully noted, the examples of successful regeneration of tails in some large reptiles (such as crocodile) imply the minor role that size plays in regeneration.

8 Limb Tissue Differentiation

In the conclusion of the chapter, we would like to discuss briefly the finalizing steps of the regeneration process, where the leading role is played by the signaling pathways accomplishing post-regenerational tissue redifferentiation.

An important player in cell fate determination and differentiation is *Notch* signaling. The biological function of this pathway is critically dependent on context-specific interactions with other signaling pathways. In many mammalian systems, over the last few years the role of the Notch signaling pathway in keratinocyte growth/differentiation control and tumor development has been intensively studied [108]. Notch signaling enhances stem cell potential and suppresses differentiation, while in others, notably keratinocytes, it exerts an opposite function [109]. Dotto et al. have proposed a hypothesis that there are three main determinants of specificity: (a) a cross talk of Notch signaling with cell type-specific regulatory molecules; (b) cell type-specific cross talk of Notch with other regulatory pathways, which are themselves not cell type specific; (c) cell type- and tissue-specific global organization, to which Notch contributes and in the context of which it functions. Also, important interconnections between Notch signaling and other pathways were established [110, 111].

Another important agent in cellular differentiation and survival is p63 protein, belonging to the p53 tumor suppressor family which also includes the p73 protein. While Notch signaling promotes commitment of keratinocytes to differentiation, a transcription factor p63 has been implicated in the establishment of the keratinocyte cell fate and/or the maintenance of epithelial self-renewal. These proteins are thought to act as transcription factors that regulate the progression of the cell through its cell cycle and cell death (apoptosis) in response to environmental stimuli, such as DNA damage and hypoxia. The p63 protein is expressed in the nucleus of basal cells in many types of epithelium [112]. It is expressed in proliferating keratinocytes of the basal layer of the epidermis and hair follicles, and in the basal layers of the mammary gland and the prostate, while it is strongly downmodulated with differentiation. p63 exerts a dual function in suppressing Notch signaling in epidermal cells with high self-renewal potential, while synergizing with other aspects of Notch function in early stages of differentiation.

An inverse gradient of p63 expression versus Notch activity exists in the lower versus upper epidermal layers which results from their reciprocal negative regulation. Nguyen et al. [113] found that p63 expression is under Notch control in both mouse and human keratinocytes. Specifically, p63 expression is negatively regulated by Notch activation through a mechanism independent of cell cycle withdrawal and involving modulation of genes in the interferon response pathway. In turn, p63 counteracts the ability of Notch1 to promote the irreversible versus reversible commitment. Notch1 and p21WAF1/Cip1, a "canonical" Notch target in keratinocytes, suppress Wnt ligand expression and signaling, and function as negative regulator of stem cell potential and tumorigenesis. In UV light exposure studies, it was found that the *Notch1* gene is a p53 target with a key role in human keratinocyte tumor suppression [111].

Little is known of pathways involved in upstream control of *Notch1* gene expression and activity. Kolev et al. [114] identified epidermal growth factor receptor (EGFR) as a key negative regulator of *Notch1* gene expression in primary human keratinocytes, intact epidermis, and skin squamous cell carcinomas (SCCs). EGFR signaling functions in the stem cell compartment of epithelial tissues, as a "built-in" mechanism to maintain self-renewal and, at the same time, suppress differentiation. These findings point to a novel role of EGFR within this context, of negative control of the *Notch1* gene, through a mechanism involving transcriptional downregulation of the *p53* gene by the EGFR effector c-Jun.

Summarizing, existing studies suggest that signaling and regulatory cascades of embryonic development served as a raw material for regeneration ability. In other words, body part regeneration is performed on the basis of the same program which controls embryonic development. This program can in principle be activated under certain conditions. Thus, even in the animals that lost the regeneration ability, this program can be awakened by activating specific embryonic development pathways which, in turn, open the prospect of activation of blocked regeneration abilities in human beings. Inducing a regenerative response in humans was once an unrealistic endeavor but is now becoming a very reasonable goal. Progress in our understanding of the molecular basis of the regenerative response is advancing rapidly, and the goal of eventually inducing this response in humans may be attained in the not so distant future.

Acknowledgement We express our gratitude to Prof. Dr. Gerhard M. Artmann (Laboratory of Cellular Biophysics; AcUAS) for his continuous support and valuable comments. We are also deeply indebted to Thomas Wulf (M.Sc.) and Marina Digel (M.A.) for their great help in preparation of this manuscript.

References

1. Dinsmore CE. Conceptual foundations of metamorphosis and regeneration: from historical links to common mechanisms. Wound Repair Regen. 1998; 6:291–301.
2. Trembley A. Mémoires, Pour Servir à l'Histoire d'un Genre de Polypes d'Eau Douce, à Bras en Forme de Cornes. Leiden: Jean and Herman Verbeek; 1744.
3. Molloy T, Wang Y, Murrell G. The roles of growth factors in tendon and ligament healing. Sports Med. 2003; 33:381–394.
4. Ferguson MW, O'Kane S. Scar-free healing: from embryonic mechanisms to adult therapeutic intervention. Philos Trans R Soc Lond B Biol Sci. 2004; 359:839–850.
5. Frazee O. The effect of electrical stimulation upon the rate of regeneration in *Rana pipiens* and *Ambystoma jeffersonianum*. J Exp Zool. 1909; 7:457–475.
6. Burr HS. The meaning of bioelectric potentials. Yale J Biol Med. 1944; 16:353–360.
7. Lund EJ. Bioelectric fields and growth. Austin: University of Texas Press; 1947.
8. Becker RO, Spadaro JA. Electrical stimulation of partial limb regeneration in mammals. Bull NY Acad Med. 1972; 48:627–641.
9. Singer M. Induction of regeneration in the forelimb of post metamorphic frog by augmentation of the nerve supply. J Exp Zool. 1954; 126:419–426.
10. Polezhaev LV. Regeneration of organs in animals. Usp Sovrem Biol. 1950; 30:258–270.
11. Person P, Libbin RM, Shah D, Papierman S. Partial regeneration of the above-elbow amputated rat forelimb. I. Innate responses. J Morphol. 1979; 159:427–438.
12. Godwin JW, Brockes JP. Regeneration, tissue injury and the immune response. J Anat. 2006; 209:423–432.
13. Goss RJ. Prospects of regeneration in man. Clin Orthop Relat Res. 1980; 151:270–282.
14. Sanchez AA. Regeneration and the need for simpler model organisms. Philos Trans R Soc Lond B Biol Sci. 2004; 359:759–763.
15. Schaller HC, Hermans-Borgmeyer I, Hoffmeister SA. Neuronal control of development in hydra. Int J Dev Biol. 1996; 40:339–344.
16. Hicklin J, Hornbruch A, Wolpert L. Inhibition of hypostome formation and polarity reversal in Hydra. Nature. 1969; 221:1268–1271.
17. Meinhardt H, Gierer A. Applications of a theory of biological pattern formation based on lateral inhibition. J Cell Sci. 1974; 15:321–346.
18. MacWilliams HK. Numerical simulations of hydra head regeneration using a proportion-regulating version of the Gierer-Meinhardt model. J Theor Biol. 1982; 99:681–703.
19. Ando H, Sawada Y, Shimizu H, Sugiyama T. Pattern formation in hydra tissue without developmental gradients. Dev Biol. 1989; 133:405–414.

20. Brockes JP, Kumar A. Plasticity and reprogramming of differentiated cells in amphibian regeneration. Nat Rev Mol Cell Biol. 2002; 3:566–574.
21. Straube WL, Tanaka EM. Reversibility of the differentiated state: regeneration in amphibians. Artif Organs. 2006; 30:743–755.
22. Stocum DL. Amphibian regeneration and stem cells. Curr Top Microbiol Immunol. 2004; 280:1–70.
23. Tsonis PA. Bridging knowledge gaps on the long road to regeneration: classical models meet stem cell manipulation and bioengineering. Mol Interv. 2007; 7:249–250.
24. Nye HL, Cameron JA, Chernoff EA, Stocum DL. Regeneration of the urodele limb: a review. Dev Dyn. 2003; 226:280–294.
25. Becker RO. The bioelectric factors in amphibian-limb regeneration. J Bone Joint Surg Am. 1961; 43-A:643–656.
26. Christensen RN, Tassava RA. Apical epithelial cap morphology and fibronectin gene expression in regenerating axolotl limbs. Dev Dyn. 2000; 217:216–224.
27. Tanaka EM, Gann AA, Gates PB, Brockes JP. Newt myotubes reenter the cell cycle by phosphorylation of the retinoblastoma protein. J Cell Biol. 1997; 136:155–165.
28. Velloso CP, Kumar A, Tanaka EM, Brockes JP. Generation of mononucleate cells from post-mitotic myotubes proceeds in the absence of cell cycle progression. Differentiation. 2000; 66:239–246.
29. Holder N. Organization of connective tissue patterns by dermal fibroblasts in the regenerating axolotl limb. Development. 1989; 105:585–593.
30. Kieny M, Chevallier A. Autonomy of tendon development in the embryonic chick wing. J Embryol Exp Morphol. 1979; 49:153–165.
31. Odelberg SJ. Cellular plasticity in vertebrate regeneration. Anat Rec B New Anat. 2005; 287:25–35.
32. Le GF, Rudnicki M. Satellite and stem cells in muscle growth and repair. Development. 2007; 134:3953–3957.
33. Carlson BM. Some principles of regeneration in mammalian systems. Anat Rec B New Anat. 2005; 287:4–13.
34. Polezhaev LV. Methods of regeneration. Sov J Dev Biol. 1975; 5:134–139.
35. Kierdorf U, Kierdorf H, Szuwart T. Deer antler regeneration: cells, concepts, and controversies. J Morphol. 2007; 268:726–738.
36. Yokoyama H. Initiation of limb regeneration: the critical steps for regenerative capacity. Dev Growth Differ. 2008; 50:13–22.
37. Colitti M, Allen SP, Price JS. Programmed cell death in the regenerating deer antler. J Anat. 2005; 207:339–351.
38. Allen SP, Maden M, Price JS. A role for retinoic acid in regulating the regeneration of deer antlers. Dev Biol. 2002; 251:409–423.
39. Goss AN. Intra-uterine healing of fetal rat oral mucosal, skin and cartilage wounds. J Oral Pathol. 1977; 6:35–43.
40. Goss RJ. The evolution of regeneration: adaptive or inherent? J Theor Biol. 1992; 159:241–260.
41. Goss RJ, Grimes LN. Epidermal downgrowths in regenerating rabbit ear holes. J Morphol. 1975; 146:533–542.
42. Borgens RB. Mice regrow the tips of their foretoes. Science. 1982; 217:747–750.
43. Han M, Yang X, Taylor G, Burdsal CA, Anderson RA, Muneoka K. Limb regeneration in higher vertebrates: developing a roadmap. Anat Rec B New Anat. 2005; 287:14–24.
44. Tassava RA, Olsen CL. Higher vertebrates do not regenerate digits and legs because the wound epidermis is not functional. A hypothesis. Differentiation. 1982; 22: 151–155.
45. Murphy ED, Roths JB. A Y chromosome associated factor in strain BXSB producing accelerated autoimmunity and lymphoproliferation. Arthritis Rheum. 1979; 22:1188–1194.

46. Fukuyama H, Adachi M, Suematsu S, Miwa K, Suda T, Yoshida N, Nagata S. Transgenic expression of Fas in T cells blocks lymphoproliferation but not autoimmune disease in MRL-lpr mice. J Immunol. 1998; 160:3805–3811.
47. Heber-Katz E. The regenerating mouse ear. Semin Cell Dev Biol. 1999 Aug;10(4): 415–419.
48. Brockes JP, Kumar A. Comparative aspects of animal regeneration. Annu Rev Cell Dev Biol. 2008; 24:525–549.
49. Heber-Katz E, Leferovich J, Bedelbaeva K, Gourevitch D, Clark L. The scarless heart and the MRL mouse. Philos Trans R Soc Lond B Biol Sci. 2004a; 359:785–793.
50. Heber-Katz E, Leferovich JM, Bedelbaeva K, Gourevitch D. Spallanzani's mouse: a model of restoration and regeneration. Curr Top Microbiol Immunol. 2004b; 280:165–189.
51. Cebria F, Guo T, Jopek J, Newmark PA. Regeneration and maintenance of the planarian midline is regulated by a slit orthologue. Dev Biol. 2007; 307:394–406.
52. Keating MT. Genetic approaches to disease and regeneration. Philos Trans R Soc Lond B Biol Sci. 2004; 359:795–798.
53. Tanaka EM. Cell differentiation and cell fate during urodele tail and limb regeneration. Curr Opin Genet Dev. 2003; 13:497–501.
54. Poss KD, Keating MT, Nechiporuk A. Tales of regeneration in zebrafish. Dev Dyn. 2003; 226:202–210.
55. Bryant SV, Endo T, Gardiner DM. Vertebrate limb regeneration and the origin of limb stem cells. Int J Dev Biol. 2002; 46:887–896.
56. Capdevila J, Izpisua Belmonte JC. Patterning mechanisms controlling vertebrate limb development. Annu Rev Cell Dev Biol. 2001; 17:87–132.
57. Nakatani Y, Kawakami A, Kudo A. Cellular and molecular processes of regeneration, with special emphasis on fish fins. Dev Growth Differ. 2007; 49:145–154.
58. Gordon MD, Nusse R. Wnt signaling: multiple pathways, multiple receptors, and multiple transcription factors. J Biol Chem. 2006; 281:22429–22433.
59. Botchkarev VA, Sharov AA. BMP signaling in the control of skin development and hair follicle growth. Differentiation. 2004; 72:512–526.
60. Slack JM, Lin G, Chen Y. The *Xenopus* tadpole: a new model for regeneration research. Cell Mol Life Sci. 2008; 65:54–63.
61. Kawakami Y, Rodriguez EC, Raya M, Kawakami H, Marti M, Dubova I, Izpisua Belmonte JC. Wnt/beta-catenin signaling regulates vertebrate limb regeneration. Genes Dev. 2006; 20:3232–3237.
62. Bayliss PE, Bellavance KL, Whitehead GG, Abrams JM, Aegerter S, Robbins HS, Cowan DB, Keating MT, O'Reilly T, Wood JM, Roberts TM, Chan J. Chemical modulation of receptor signaling inhibits regenerative angiogenesis in adult zebrafish. Nat Chem Biol. 2006; 2:265–273.
63. Mills AA. p63: oncogene or tumor suppressor? Curr Opin Genet Dev. 2006; 16:38–44.
64. Echeverri K, Tanaka EM. Ectoderm to mesoderm lineage switching during axolotl tail regeneration. Science. 2002; 298:1993–1996.
65. Casimir CM, Gates PB, Patient RK, Brockes JP. Evidence for dedifferentiation and metaplasia in amphibian limb regeneration from inheritance of DNA methylation. Development. 1988; 104:657–668.
66. Imokawa Y, Brockes JP. Selective activation of thrombin is a critical determinant for vertebrate lens regeneration. Curr Biol. 2003; 13:877–881.
67. Imokawa Y, Simon A, Brockes JP. A critical role for thrombin in vertebrate lens regeneration. Philos Trans R Soc Lond B Biol Sci. 2004; 359:765–776.
68. Lin YC, Grigoriev NG, Spencer AN. Wound healing in jellyfish striated muscle involves rapid switching between two modes of cell motility and a change in the source of regulatory calcium. Dev Biol. 2000; 225:87–100.
69. Nodder S, Martin P. Wound healing in embryos: a review. Anat Embryol (Berl). 1997; 195:215–228.

70. Mescher AL, Neff AW. Regenerative capacity and the developing immune system. Adv Biochem Eng Biotechnol. 2005; 93:39–66.
71. Flajnik MF, Kasahara M. Comparative genomics of the MHC: glimpses into the evolution of the adaptive immune system. Immunity. 2001; 15:351–362.
72. Fahmy GH, Sicard RE. A role for effectors of cellular immunity in epimorphic regeneration of amphibian limbs. In Vivo. 2002; 16:179–184.
73. Tsonis PA, Del Rio-Tsonis K, Wallace JL, Burns JC, Hofmann MC, Millan JL, Washabaugh CH. Can insights into urodele limb regeneration be achieved with cell cultures and retroviruses? Int J Dev Biol. 1996; 40:813–816.
74. Plytycz B, Seljelid R. MHC molecules and lymphocytes: evolutionary perspective. Arch Immunol Ther Exp (Warsz). 1998; 46:137–142.
75. Hantash BM, Zhao L, Knowles JA, Lorenz HP. Adult and fetal wound healing. Front Biosci. 2008; 13:51–61.
76. Colwell AS, Longaker MT, Lorenz HP. Fetal wound healing. Front Biosci. 2003; 8:s1240–s1248.
77. Hopkinson-Woolley J, Hughes D, Gordon S, Martin P. Macrophage recruitment during limb development and wound healing in the embryonic and foetal mouse. J Cell Sci. 1994; 107(Pt 5):1159–1167.
78. Muneoka K, Sassoon D. Molecular aspects of regeneration in developing vertebrate limbs. Dev Biol. 1992; 152:37–49.
79. Holder N, Bryant SV, Tank PW. Interactions between irradiated and unirradiated tissues during supernumerary limb formation in the newt. J Exp Zool. 1979; 208:303–310.
80. Polezhaev LV. Regeneration of the fingers in children. Khirurgiia (Mosk). 1980; 12:76–77.
81. Globus M, Alles P. A search for immunoreactive substance P and other neural peptides in the limb regenerate of the newt *Notophthalmus viridescens*. J Exp Zool. 1990; 254:165–176.
82. Page-McCaw A. Remodeling the model organism: matrix metalloproteinase functions in invertebrates. Semin Cell Dev Biol. 2008; 19:14–23.
83. Ghahary A, Ghaffari A. Role of keratinocyte–fibroblast cross-talk in development of hypertrophic scar. Wound Repair Regen. 2007; 15(Suppl 1):S46–S53.
84. Akerman B, Haegerstam G, Pring BG, Sandberg R. Penetration enhancers and other factors governing percutaneous local anaesthesia with lidocaine. Acta Pharmacol Toxicol (Copenh). 1979; 45:58–65.
85. Peled ZM, Phelps ED, Updike DL, Chang J, Krummel TM, Howard EW, Longaker MT. Matrix metalloproteinases and the ontogeny of scarless repair: the other side of the wound healing balance. Plast Reconstr Surg. 2002; 110:801–811.
86. Harty M, Neff AW, King MW, Mescher AL. Regeneration or scarring: an immunologic perspective. Dev Dyn. 2003; 226:268–279.
87. Adzick NS, Lorenz HP. Cells, matrix, growth factors, and the surgeon. The biology of scarless fetal wound repair. Ann Surg. 1994; 220:10–18.
88. Burkly LC, Michaelson JS, Hahm K, Jakubowski A, Zheng TS. TWEAKing tissue remodeling by a multifunctional cytokine: role of TWEAK/Fn14 pathway in health and disease. Cytokine. 2007; 40:1–16.
89. Brockes JP, Kumar A, Velloso CP. Regeneration as an evolutionary variable. J Anat. 2001; 199:3–11.
90. Bischoff M, Schnabel R. A posterior centre establishes and maintains polarity of the *Caenorhabditis elegans* embryo by a Wnt-dependent relay mechanism. PLoS Biol. 2006; 4:e396.
91. Schaller HC, Hoffmeister SA, Dubel S. Role of the neuropeptide head activator for growth and development in hydra and mammals. Development. 1989; 107(Suppl):99–107.
92. MacWilliams HK. Hydra transplantation phenomena and the mechanism of Hydra head regeneration. II. Properties of the head activation. Dev Biol. 1983; 96:239–257.
93. Shenk MA, Gee L, Steele RE, Bode HR. Expression of Cnox-2, a HOM/HOX gene, is suppressed during head formation in hydra. Dev Biol. 1993; 160:108–118.

94. Marsh HG, Beams HW. Electrical control of morphogenesis in regenerating *Dugesia tigrina*. I. Relation of axial polarity to field strength. J Cell Physiol. 1952; 39:191–213.
95. Slack JM. Inducing factors in *Xenopus* early embryos. Curr Biol. 1994; 4:116–126.
96. Stocum DL, Crawford K. Use of retinoids to analyze the cellular basis of positional memory in regenerating amphibian limbs. Biochem Cell Biol. 1987; 65:750–761.
97. Thesleff I, Jarvinen E, Suomalainen M. Affecting tooth morphology and renewal by fine-tuning the signals mediating cell and tissue interactions. Novartis Found Symp. 2007; 284:142–153.
98. Li S, Muneoka K. Cell migration and chick limb development: chemotactic action of FGF-4 and the AER. Dev Biol. 1999; 211:335–347.
99. Hammerschmidt M, Brook A, McMahon AP. The world according to hedgehog. Trends Genet. 1997; 13:14–21.
100. Zardoya R, Abouheif E, Meyer A. Evolutionary analyses of hedgehog and Hoxd-10 genes in fish species closely related to the zebrafish. Proc Natl Acad Sci U S A. 1996; 93:13036–13041.
101. Reginelli AD, Wang YQ, Sassoon D, Muneoka K. Digit tip regeneration correlates with regions of Msx1 (Hox 7) expression in fetal and newborn mice. Development. 1995; 121:1065–1076.
102. Wanek N, Gardiner DM, Muneoka K, Bryant SV. Conversion by retinoic acid of anterior cells into ZPA cells in the chick wing bud. Nature. 1991; 350:81–83.
103. Javois LC, Iten LE, Murphy DJ. Formation of supernumerary structures by the embryonic chick wing depends on the position and orientation of a graft in a host limb bud. Dev Biol. 1981; 82:343–349.
104. Bryant SV, Iten LE. Intercalary and supernumerary regeneration in regenerating the mature limbs of *Notophthalmus viridescens*. J Exp Zool. 1977; 202:1–16.
105. French V, Rowlands TF. Regeneration in the anterior–posterior axis of the insect thoracic segment. J Embryol Exp Morphol. 1986; 98:137–165.
106. Wolpert L. Positional information and pattern formation. Curr Top Dev Biol. 1971; 6:183–224.
107. Ventegodt S, Hermansen TD, Flensborg-Madsen T, Nielsen ML, Merrick J. Human development VI: supracellular morphogenesis. The origin of biological and cellular order. ScientificWorldJournal. 2006; 6:1424–1433.
108. Dotto GP. Notch tumor suppressor function. Oncogene. 2008; 27:5115–5123.
109. Lefort K, Dotto GP. Notch signaling in the integrated control of keratinocyte growth/differentiation and tumor suppression. Semin Cancer Biol. 2004; 14:374–386.
110. Devgan V, Mammucari C, Millar SE, Brisken C, Dotto GP. p21WAF1/Cip1 is a negative transcriptional regulator of Wnt4 expression downstream of Notch1 activation. Genes Dev. 2005; 19:1485–1495.
111. Lefort K, Mandinova A, Ostano P, Kolev V, Calpini V, Kolfschoten I, Devgan V, Lieb J, Raffoul W, Hohl D, Neel V, Garlick J, Chiorino G, Dotto GP. Notch1 is a p53 target gene involved in human keratinocyte tumor suppression through negative regulation of ROCK1/2 and MRCKalpha kinases. Genes Dev. 2007; 21:562–577.
112. Petitjean A, Ruptier C, Tribollet V, Hautefeuille A, Chardon F, Cavard C, Puisieux A, Hainaut P, Caron de FC. Properties of the six isoforms of p63: p53-like regulation in response to genotoxic stress and cross talk with DeltaNp73. Carcinogenesis. 2008; 29:273–281.
113. Nguyen BC, Lefort K, Mandinova A, Antonini D, Devgan V, Della GG, Koster MI, Zhang Z, Wang J, Tommasi diVA, Kitajewski J, Chiorino G, Roop DR, Missero C, Dotto GP. Cross-regulation between Notch and p63 in keratinocyte commitment to differentiation. Genes Dev. 2006; 20:1028–1042.
114. Kolev V, Mandinova A, Guinea-Viniegra J, Hu B, Lefort K, Lambertini C, Neel V, Dummer R, Wagner EF, Dotto GP. EGFR signalling as a negative regulator of Notch1 gene transcription and function in proliferating keratinocytes and cancer. Nat Cell Biol. 2008; 10:902–911.

Part II
Basics and Basic Research

Engineering the Stem Cell Niche and the Differentiative Micro- and Macroenvironment: Technologies and Tools for Applying Biochemical, Physical and Structural Stimuli and Their Effects on Stem Cells

Paolo Di Nardo, Marilena Minieri, and Arti Ahluwalia

Abstract In recent years there has been an explosion of interest in stem cell research, given their promising medical applications in cell-based tissue regeneration, drug testing and of course basic research. A decade of restless experimental and clinical research has demonstrated that the routine use of stem cells to repair solid organs is not at hand in spite of recent excessively enthusiastic announcements in the press and even serious scientific journals. Indeed, biologists only partially comprehend cell-differentiating mechanisms and have mapped only a few of the extrinsic and intrinsic factors involved. Even less is understood the complex qualitative, quantitative and temporal orchestration of these factors in the different steps featuring the whole differentiating process.

Most of the current research is centred on the identification of soluble ligands which regulate and control signalling pathways, and our knowledge on the role of the physical and structural microenvironment is still scarce. In this chapter, we focus only on cues which can be controlled externally using mechanical and structural parameters, and so can be easily defined using appropriate engineering and design. Firstly, the influence of the single parameters on cell behaviour is described, and then we discuss how technological tools such as biomaterials, scaffolds and bioreactors, as well as well-constructed and defined multiscale classification models can be best employed to engineer artificial biomimetic in vitro systems.

Keywords Engineering · Microenvironment · Stem cells · Stimuli

"Da mihi ubi consistam,. . .et terram caelumque movebo"
Give me where to stand. . . and I will move the earth
(Archimedes, 287 b.c.).

A. Ahluwalia (✉)
Interdepartmental Research Centre "E. Piaggio", University of Pisa, Pisa, Italy
e-mail: arti.ahluwalia@centropiaggio.unipi.it

G.M. Artmann et al. (eds.), *Stem Cell Engineering*,
DOI 10.1007/978-3-642-11865-4_2, © Springer-Verlag Berlin Heidelberg 2011

1 Introduction

Stem cells are inherently unstable and their fate in vivo is modulated by the so-called niche, i.e. an anatomic and functional unit in which the unstable stem cells are confined in a semi-quiescent state. Upon specific requirements to preserve tissue integrity and function, stem cells asymmetrically divide and the daughter cells leave the niche towards a new decoy environment directing natural stem cell instability in favour of the acceptance of the required phenotype. It is generally accepted that it is the "symmetry" or "featurelessness" of the niche microenvironment which maintains an undifferentiated status in stem cells and perturbations are either self-generated due to instabilities or arise from external factors. This is certainly true of the embryo in the first stages of development. Some external signals or non-linear perturbations must lead to non-symmetry of the system to pull it out of quiescent stemness [1, 2]. Therefore, stem cells are in a state of metastable equilibrium within and without their niches, and this has led to a huge number of papers fervently reporting on the new factors which drive stem cell differentiation. The past few years have been spent in a futile attempt to identify "the soluble factor" (growth factor, cytokine, chemical, etc.) able to spark gene programmes leading stem cells to adopt different phenotypes. The injection of one or a mixture of soluble factors into the diseased organ in order to recruit, expand and differentiate the rare stem cells available to substitute damaged cells has even been suggested [3]. Most of this fervour is driven by the promise of tissue regeneration by stem cells. In fact the field of tissue engineering has undergone a definite shift towards the explicit use of stem cells, although few recognize that it is these cells that are largely responsible for the success of engineered skin, blood vessels and the few others we may care to count on one hand. Figure 1 shows how the fraction of papers explicitly dealing with stem cells in the journal *Tissue Engineering* has changed since it was first established in 1995. Over 35% of papers now focus on the factors that guide stem cell differentiation and the numbers are certain to increase. The general impression that one gains from the current literature is that anything will make mesenchymal stem cells (MSCs) or bone marrow stem cells (BMSCs) as well as embryonic stem cells (ESCs) or embryoid bodies differentiate, and it is highly likely that the very act of isolating them and plating in vitro causes changes in their phenotype.

The attempt to solve extremely complex problems with oversimplistic solutions generated inconsistent results that, however, were able to further fuel hopes in patients in which conventional treatments were no longer able to counteract their diseases. A mindful survey of the current literature on cell differentiation processes ought to have suggested a more careful and articulated approach to stem cell manipulation and clinical application.

Indeed, the territory of stem cell biology and engineering is still riven with pitfalls and obstacles – and more importantly blind spots. We face a huge number of technological tasks before being able to gain a mechanistic understanding of stem cell destiny in vitro. Without a basic comprehension of the mechanisms, trying to control stem cell destiny will get us nowhere. And the number of papers published

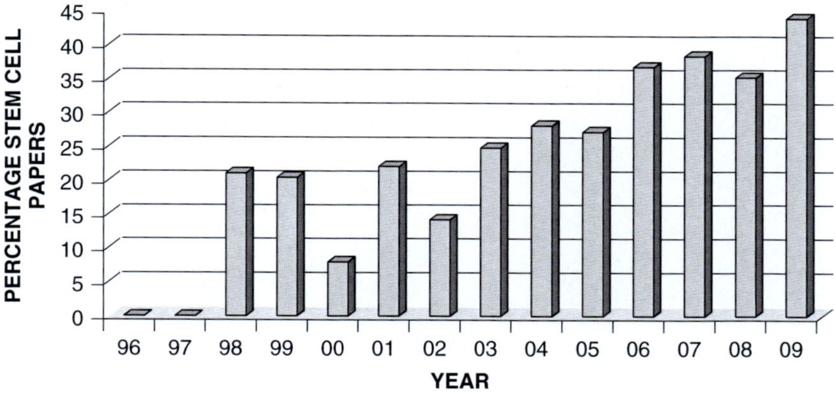

Fig. 1 Percentage of papers dedicated to stem cells in the journal *Tissue Engineering* since its inception in 1995. By the end of this decade, 50% of all papers in regenerative medicine will probably focus on the use of stem cells

on the subject, often presenting conflicting results, is a strong evidence that so far the research has been leading up blind alleys.

We need to find solutions to innumerable biological and technical problems and the majority of these are related to three main issues:

- *Stem cell source*. There is a scarcity of stem cells in the adult body. Often they are difficult to identify, locate and isolate.
- *Control of stemness in vitro*. The cells should ideally be amplified in vitro, but only in such a way so as to maintain symmetry: that is, each daughter is an identical copy of the mother.
- *Control of differentiative status*. The cells need to be differentiated in a controlled manner without becoming tumoural.

Can we surmount these? The question of stem cell source is intricately linked with the question of aging and the role of stem cells in tissue homeostasis and repair as well as with the natural limits of reparative potential of these cells and the human body in general. While blood cells, skin and epithelial tissue in general have a high turnover and therefore an abundant supply of stem cells, in other tissues such as the liver and pancreas, homeostasis is maintained by adult cells which can proliferate to a certain extent and a small number of resident stem cells are called into action in injury. In those tissues with low cell turnover and low regenerative potential, such as the heart and brain, adult or autologous sources are probably out of the question, since as illustrated in Fig. 2, they have very few if any stem cells [4]. Here, the reparative capacity is limited and homeostasis is maintained by differentiated cells. As far as the other two issues mentioned are concerned, keeping in mind the capacity and limits of intrinsic self-renewal amongst specific tissues, there is little doubt that the key lies in developing a mechanistic understanding of the influence of the microenvironment on cell behaviour and function at multiple levels [5].

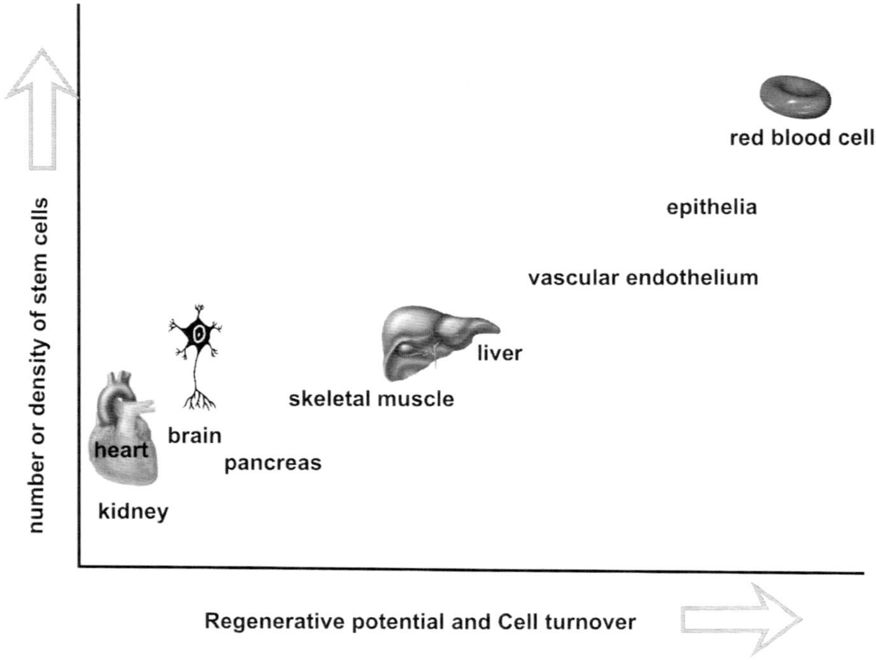

Fig. 2 The possibility of isolating stem cells from adult tissue depends on their density, which in turn depends on the regenerative potential of the tissue in question

2 The Micro- and Macroenvironment

Three principal extrinsic cues are known to guide cell function and fate: the biochemical microenvironment, including ligands, signalling molecules and other cells, the physico-chemical environment which comprises gradient-dependent factors such as surface properties, oxygen tension, pH and temperature, and finally the mechano-structural environment. The mechano-structural environment is the architecture in two and three dimensions as well as mechanical forces such as stress and strain, all of which act in a non-linear but fairly constant manner. Together the three classes of stimuli represent what we define as the tripartite axes of cues. Superimposed upon this is the one and only linear factor which acts on cells and which we have no control over: Time, with a capital T. Figure 3 illustrates this concept, and as shown the number of possible variables is almost infinite.

Cells in vivo are surrounded by all these cues together in the microenvironment in the form of the prestressed extracellular matrix, prestressed neighbouring cells, endocrine and paracrine signals, blood flow, body movement and forces and nutrient diffusion. Furthermore, cells themselves will remodel and modulate their own habitat, an aspect which is often ignored in in vitro experiments, as are most of the mechano-structural stimuli.

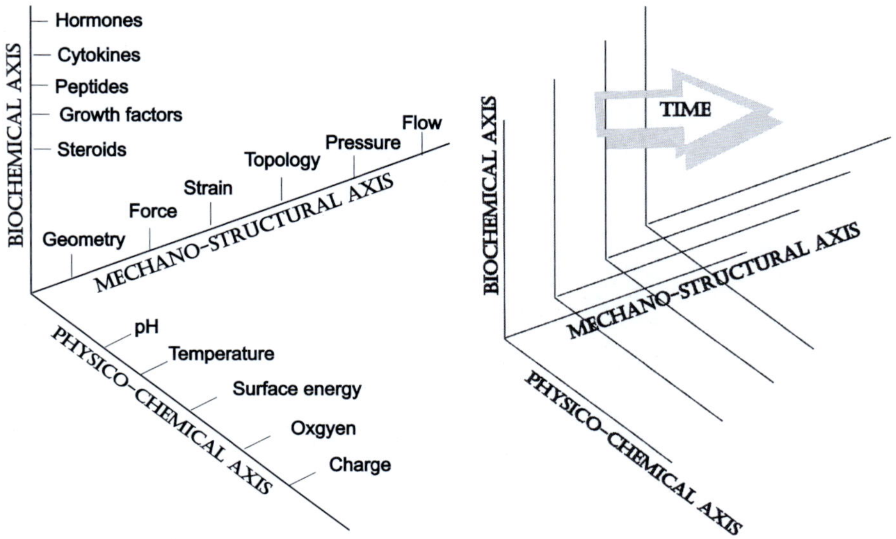

Fig. 3 The tripartite axes of cues which act on cells. (**a**) The three principal axes must be referred to a fourth axis Time, which moves inexorably forward in a linear manner. Although we have no control over Time, "the biological clock", it must be taken into consideration. (**b**) Each of the principal axes is composed of a subset of factors, and their "intensity" influences cell function in a non-linear manner

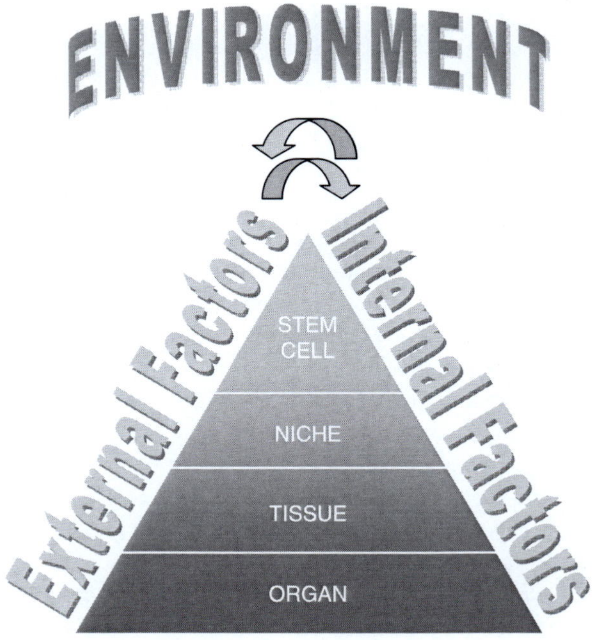

Fig. 4 Both extrinsic and intrinsic factors influence the orchestration of cellular function. Note that there is an interdependency between the two

In Fig. 4 we schematize the hierarchic organization of the stem cell habitat. All cells coordinate their behaviour and function in response to signals from the macro- to the microscale, within and without their tissues. The signals are not necessarily soluble ligands but can be external or internal.

3 Time

Time is probably one of the most important but least considered variables which determine stem cell fate. Even stem cells are equipped with a biological clock which ticks constantly. Therefore, unless the tick frequency is decreased by techniques such as freezing or hypoxic conditions, the functional ability of stem cells decreases with their age [6]. There is however some question as to whether the aging is due to intrinsic aging of the stem cells or due to the aging microenvironment of the organism itself, since aged skeletal progenitor stem cells are as effective as young ones when implanted in young animals. According to Rando [4] the extent to which aging deteriorates the stem cell or its microenvironment depends on the regenerative potential of stem cells (see Fig. 2). The implication is that stem cells with a high density and regenerative potential age intrinsically, due to the Hayflick limit or "replicative senescence", whereas those resident in tissues with low cellular turnover are influenced by senescence of the microenvironment.

4 Biochemical Microenvironment

In the human body, biochemical cues are generally provided by soluble ligands which may be secreted by paracrinal cells or supplied by a capillary network. Insoluble ligands are also present; these are adhesion proteins or molecules such as collagen, laminin and carbohydrates. They are considered insoluble because they have very small diffusion constants. Other insoluble cues arise from neighbouring cells themselves, through adhesive and cohesive junctions. The recognition that cells themselves remodulate their extracellular milieu is of crucial importance in all aspects of cell biology, and particularly in regenerative medicine and engineering. Much of the work on stem cell biology to date has focused on the use of a handful of soluble factors in vitro to differentiate cells. In most cases the signalling pathways are still not clearly identified and methods used to differentiate cells are determined empirically on the basis of trial and error. For example, human or mouse embryonic stem cells can be coerced to differentiate into neuron-like, pancreatic β-cell-like, adipose-like, hepatocyte-like and osteoblast-like cells by the addition of substances such as dexamethasone, retinoic acid and dimethyl sulphoxide [7] or by gene transfection with transcription factors such as osterix [8]. Clearly, none of these routes are clinically acceptable or in anyway controllable. Likewise, bone marrow- or adipose tissue-derived mesenchymal stem cells can be differentiated into osteocytes, chondrocytes or adipocytes by the addition of dexamethasone, ascorbic acid,

TGF-β and indomethacin [9]. Differentiation into cells such as neurons, hepatocytes and cardiomyocytes has also been reported [10, 11], although several investigators also provide more recent data, in which mesenchymal stem cell plasticity is disproved [12, 13]. The type of trial-and-error studies conducted in the past 20 years should have induced investigators to probe the spectrum of biochemical stimuli in the stem cell microenvironment more profoundly but so far has only led to great confusion. The story of cardiac stem cells is a case in point; investigators have passively accepted the concept that stem cell fate is mostly governed by soluble factors (growth factors/cytokines, epigenetic factors, etc.). As a consequence, the in vitro replication of niche conditions is focused on the identification of a mixture of soluble factors. Indeed, attempts at activating both in vitro and in vivo stem cell lineage specification have been based on the assumption that a few soluble factors, sometimes injected into either the bloodstream or the damaged tissue, can govern the whole process leading stem cells to full differentiation, i.e. stem cell specification and determination, sarcomere assembly, and cardiomyocyte integration into the myocardial architecture [14, 15]. To this end, only a relatively small number of soluble factors have been investigated to verify their actual potential in inducing cardiac stem cell differentiation into cardiomyocytes. These factors pertain to two distinct categories: biochemical and epigenetic factors. Biochemical factors, such as IGF-1, HGF, oxytocin, TGF-β1 and FGF-2, added in different concentrations in stem cell culture media have induced the expression of early and late cardiac markers, but the demonstration of a fully controlled differentiation into cardiomyocytes is still lacking [16]. Similar results have also been obtained with epigenetic modulators regulating chromatin stability and gene expression. Chromatin modifications are important for several nuclear processes, including DNA repair, replication, transcription and recombination, biological processes that probably change with Time or age.

Given the complexity of the biochemical microenvironment, in our opinion the logical way to define a niche or a differentiative microenvironment is by the delivery of tissue-specific stimuli through an in vitro system that recapitulates the in vivo functions of a natural niche. In this way, cells are provided a biomimetic habitat in which stem cell fate, e.g. self-renewal versus differentiation, can be controlled in a defined and, to a large extent, predictable manner through the provision of cell–cell and soluble-mediator cross talk through induction by tissue-specific cells [17]. In fact, distinct cell types (e.g. stromal cells and endothelial cells) and the extracellular matrix [18] are major players in the regulation of stem cells within their niche. Recapitulating this environment requires innovative and infinitely flexible cell culture systems in which the presence of feeder cells, tissues or appropriate ligands in specific space- and time-dependent configurations can be combined in a tightly controlled manner to provide a wide array of multi-parametric signals which can be tailored to guide stem cell fate, leading to new paradigms in stem cell culture. One of the authors has already proposed and patented such a system, the MCB (multicompartmental bioreactor), which is schematized in Fig. 5 with specific reference to the differentiation of stem cells for cardiac regeneration [19].

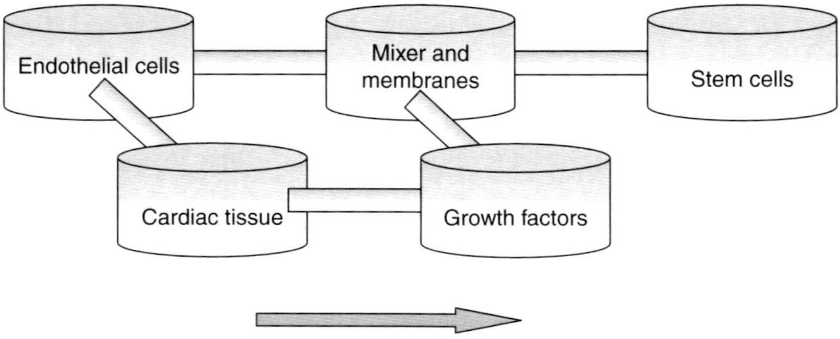

Fig. 5 Concept of the MCB (multicompartmental bioreactor) for providing tissue-specific biochemical and physical cues for cardiac tissue regeneration in vitro. Here, the biochemical cues are provided by the tissues and cells which condition cardiac habitat in vivo. The biochemical cues are vehicled to the mixer and can be sorted and conveyed to the stem cells through membranes with custom pore size. Physical and structural cues can be supplied by three-dimensional scaffolds, flow and hydrodynamic or compressive forces

5 Physico-chemical and Mechano-structural Axes – The Macroenvironment

Several studies have suggested the potential effects of other non-soluble stimuli, such as those induced by physical factors, on cell differentiation processes. So far at least 10 different non-biochemical cues which influence cell phenotype have been identified and new ones are constantly being added. The stimuli known to influence stem cell behaviour and fate are illustrated in Fig. 6. Most of these are the external factors noted in Fig. 4, whilst others are internal, but modulated somehow by the external environment. For example, the oxygen concentration will depend firstly

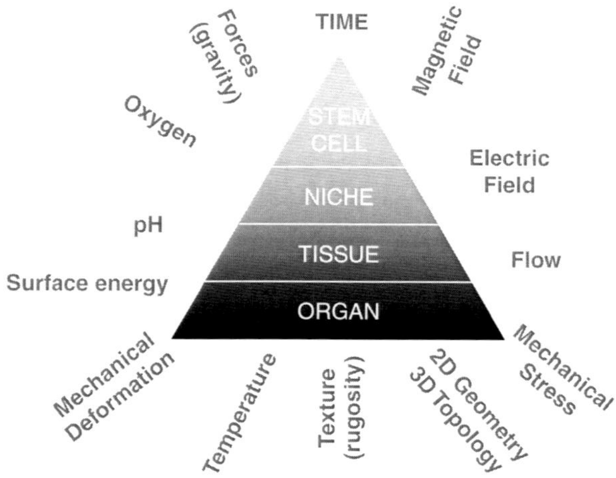

Fig. 6 The multitude of physico-chemical and mechano-structural cues known to influence stem cell fate. From the figure we can distinguish the physico-chemical cues (pH, oxygen, surface energy, temperature and electromagnetic fields) from the mechano-structural factors (topology, gravity, etc.) and Time which stands on its own and is unidirectional

on atmospheric partial pressure and then on the distance between cells and oxygen source (capillaries).

Numerous recent papers have sprouted showing how even minor experimental modifications can change cell phenotype. Indeed, stem cells are so very sensitive and unstable that even cell seeding density and seeding protocol have been observed to influence cell shape and gene expression [20]. Note, however, that seeding density is well known to influence shape and spreading of even adult cells and many cells only express specific genes and proteins when at confluence. Similarly, wettability has long been known to play a role in cell adhesion, but only in the short term until cells themselves modulate their extracellular environment [21], and recent reports show that the same effects are observed on stem cells [22]. In fact, Liu et al. report that in a foetal osteoblastic cell line, spreading and initiation of the differentiating cascade depends on surface energy, but over long timescales, the cells do converge to the same phenotype independent of the initial surface energy.

6 Physico-chemical Factors

Physico-chemical factors typically influence the cell microenvironment in a concentration-, field- or gradient-dependent manner. This includes chemical gradients, temperature gradients and time-varying electric or magnetic fields.

Probably one of the best known physico-chemical cues is oxygen concentration. In incubators, the acid–base balance is strictly regulated through the use of buffers, which ensure that the pH and CO_2 tension remain similar to those in vivo. On the other hand, typical oxygen concentrations in vivo vary from 12.5 to 5%, whilst in cell culture incubators, the oxygen concentration is the same as that in air, or 20%. It is important to realize that while the atmospheric oxygen tension is relatively straightforward to control in hypoxic incubators, the actual oxygen tension at the cell surface will vary greatly from one experiment to the other because of the low solubility of oxygen. It will depend particularly on the height of the medium and the thickness of any construct used. Several reports show that lowered oxygen concentrations (5%) increase stem cell proliferation [23–25]. Grayson et al. [26] have shown that even lower oxygen concentrations of about 2% increase MSC proliferation whilst maintaining an undifferentiated state, thus suggesting that hypoxic conditions are characteristic of the niche environment. Some authors have observed an induction of adipose-like phenotype in MSCs in severe hypoxia (1%) [27], whilst others note that adipogenesis is suppressed at 6% oxygen with respect to 20% oxygen [28]. Lennon et al. report that rat MSCs exposed to 5% oxygen during amplification show enhanced osteogenesis after implantation, compared with cells amplified in 20%, and this is probably due to increased proliferation as suggested above [29]. More sophisticated oxygen sensing at the cell surface is mandatory for ensuring adequate control of this critical parameter and micro-pH sensors should be integrated into any bioreactor system for stem cell engineering. It is also useful to model oxygen consumption and gradients in stem cell culture systems using finite-element modelling tools.

Temperature is also known to play a role in modulating stem cell behaviour. In particular, at lower temperature (32°C), MSCs from young rats display reduced apoptosis and proteasome activity as well as increased expression of heat-shock proteins [30]. The same authors also observed that at reduced temperatures, the stemness of MSCs is better maintained as they may be inducing a quiescent state in the cells and protecting them against spontaneous differentiation [31]. Note that both reduced temperature and hypoxia also slow down the biological clock.

Surface tension, wettability and contact angle all are expressions of the interfacial or surface energy of a solid–liquid or solid–air interface. The surface energy can be changed by modulating the chemistry of a substrate. Studies on the interactions between cells and surfaces and proteins and surfaces date back to the 1970s, and it has been widely observed that cell adhesion and function is better on hydrophilic surfaces than on hydrophobic ones [32]. It is also known that cell adhesion is mediated by proteins. Furthermore, there is a critical limit of contact angle, 20°, below which no adhesion will occur because the surface is too hydrophilic and a superficial layer of water prevents protein adhesion. Curran et al. [33] show that stem cell differentiation is guided by surface chemistry and energy, independent of inductive media. Although all the surfaces tested maintained cell viability, silanized hydrophobic surfaces with CH_3 end groups (so low surface energy) maintain MSC phenotype, while increasing the surface energy by adding NH_2- or SH-terminal groups promotes osteogenesis. Further increase of surface energy by the addition of OH or COOH moieties promotes chondrogenesis.

Electric, magnetic and ultrasound fields are routinely used for therapeutic purposes, particularly for bone, cartilage and muscle injury, even though the mechanisms of wound healing and regeneration under these fields are not clearly understood. In this context, the cardiomyocyte and myocyte models are probably the most well known, since electrical stimulation (1 Hz, 5 V/cm, 2 ms pulses) induces contraction coupling and structural organization of cardiomyocytes and smooth muscle cells [34]. Almost 20 years ago, Sauer et al. [35] demonstrated that a single 1990s burst of electromagnetic stimulation (5 V/cm) of ESCs led to differentiation towards a cardiomyocyte phenotype. The same authors show that treatment of ESC-derived embryoid bodies with field strengths ranging from 2.5 to 7.5 V/cm, applied only for the 1960s, dose-dependently increased differentiation towards the endothelial phenotype [36]. These findings highlight the acute sensitivity of embryonic stem cells as well as the need for more rigorous and standardized experimental protocols. Endothelial progenitor cells and muscle precursor cells can also be stimulated by electromagnetic fields to promote myocyte differentiation [37, 38]. Interestingly, electrical stimulation (10–40 V, 5 ms, 0.5 Hz pulses) of human embryonic fibroblasts was shown to cause loss of cell proliferation and cell number but also led to differentiation of fibroblasts into multinucleated myotube-like structures [39].

Weak magnetic pulses with higher frequency (30–120 Hz, 1.25 mT) have been used to differentiate preosteoblastic cells [40], and the authors found that the fields affected only undifferentiated cells, not differentiated osteoblasts, indicating that the former were more sensitive. The increased sensitivity also leads to cell damage at high frequencies; for example, bone marrow-derived stem cells under high-frequency (50 Hz) and low-intensity magnetic (0.8 mT) fields showed a reduction in the proliferation and differentiation of the granulocyte-macrophage progenitor (CFU-GM) compared to non-exposed bone marrow cells [41], while ESC which did not possess tumour suppressor genes showed adverse effects under magnetic fields of intensity 2.3 mT at a frequency of 50 Hz [42]. Ultrasound has also been shown to induce differentiation. Finally in low-intensity ultrasound field studies, MSCs differentiate towards a chondrocytic phenotype [43, 44].

7 Mechano-structural Microenvironment

As represented in Fig. 3, the mechano-structural axis comprises those cues which condition the static nature of the cell habitat. Here we include 3D architecture, surface physical features such as roughness, bulk properties such as elastic modulus, as well as stress, strain and force. The most striking and repeatable differences are found between two- and three-dimensional (3D) environments. In fact, there is a dramatic change in habitat when cells are removed from an in vivo context to a Petri dish, and in particular, stem cells find themselves in a highly non-symmetrical context. Therefore, several reports have shown that encapsulated cells or spheroids are capable of maintaining stemness, because the microenvironment does not condition the cells to express differentiated proteins for coupling to an extracellular environment. Although there are few systematic studies on the comparative effects of 2D and 3D structures, there are clear indications that cells respond quite differently to the 3D environment which allows greater cell–cell interaction and maintenance of spherical morphologies. In this context, hydrogels such as those derived from alginate, collagen and hyaluronic acid have been shown to be quite promising – they provide a homogeneous, structureless soft 3D environment which is probably ideal for stem cell proliferation and maintenance, as well as for differentiation into softer tissues such as neural or hepatic [45, 46].

Three-dimensional environments can also provide more controlled spatial information to cells if they are architectured using techniques such as microfabrication based on rapid prototyping. In this case, the 3D system is usually nominated as a "scaffold" and provides a rigid and porous framework for cells to adhere on and spread in three dimensions. One of the first reports on architectured scaffolds employed for stem cell engineering describes a random pore (250–500 μm) salt-leached synthetic polymer scaffold [a blend of polylactide (PL) and polylactide-*co*-glycolide (PLGA)] seeded with ESCs [47]. The cells generated complex capillary-like organized features which cannot be formed in a 2D environment. Moreover, the porous scaffold permitted the organization and orientation of ESCs, whereas a homogeneous isotropic soft gel did not, suggesting that both stiffness and a structured topology are important features to which the cells respond.

Further studies by Liu and Roy [48, 49] also confirmed that a porous rigid scaffold (tantalum in this case) promotes differentiation of ESCs into hematopoietic cells with respect to classical 2D cultures. Here, the cells were observed to interact with the scaffolds forming smaller aggregates, and thus increasing cell–substrate rather than cell–cell contact, probably led to increased expression of ECM and adhesion proteins. Similar results for porous PLGA scaffolds have been observed for ESC osteogenesis [16].

An important aspect of the mechano-structural environment is the chemical nature of the substrate, that is, the nature of the biomaterial used as an interface for cell adhesion and a structural support for tissues. Recent work has demonstrated that biomaterials, i.e. matrices, scaffolds and culture substrates, can present key regulatory signals to create artificial surrogate microenvironments that control stem cell fate. Although little is known about the influence of specific biomaterial features, factors, such as ligand density, and material mechanical properties have also been shown to play a role in determining phenotype [50]. We would argue that the biomaterial cue belongs to the mechano-structural axis because it is intimately bound with the elastic modulus; although if the biomaterial is degradable, it can be dramatically modified by the cells themselves. A number of reviews listing the different biomaterials used in stem cell engineering are available [51–53]. However, very little comparative information can be found regarding the performance and effect of materials on stem cell differentiation, since most approaches use a single material and then test various inducing media to assess the differentiative stimulus provided by the material.

Probably the best studied mechano-structural feature is the elastic modulus or the stiffness of a substrate or a scaffold and its effect on the lineage specification. Engler et al. [50] have pioneered studies on matrix elasticity and its effect on MSCs. Through the use of a well-defined elastically tunable polyacrylamide gel (nominally

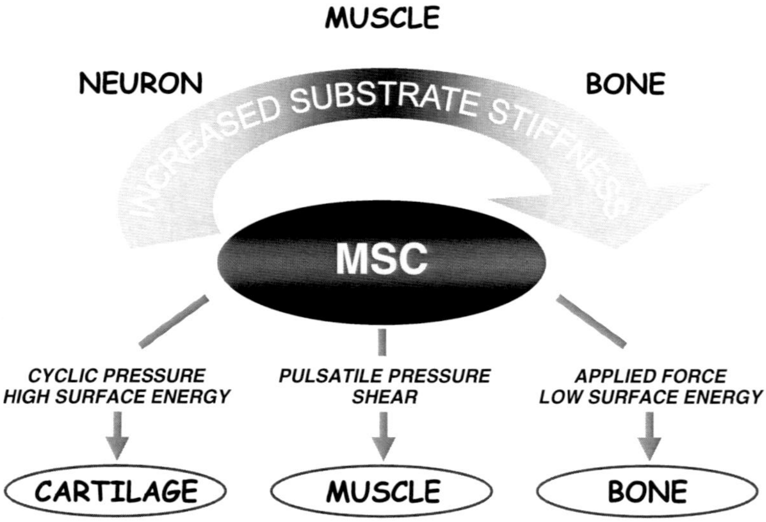

Fig. 7 Some of the factors which influence mesenchymal stem cell differentiation. Very few systematic studies exist, but it is clear that cell substrate interactions, which depend on substrate surface energy and elastic modulus, play a key role in determining stem cell destiny

2D since cells did not penetrate the cell but adhered on the surface), they observed lineage specification independent of inductive biochemical factors. Softer gels (0.1–1 kPa) were neurogenic, the hardest (24–40 kPa) were osteogenic, while gels with intermediate elastic moduli (8–17 kPa) were myogenic. In all three cases, the elastic modulus matches that of the corresponding native tissue. Figure 7 summarizes the physico-chemical and mechano-structural factors which are known to influence MSC lineage specification and commitment.

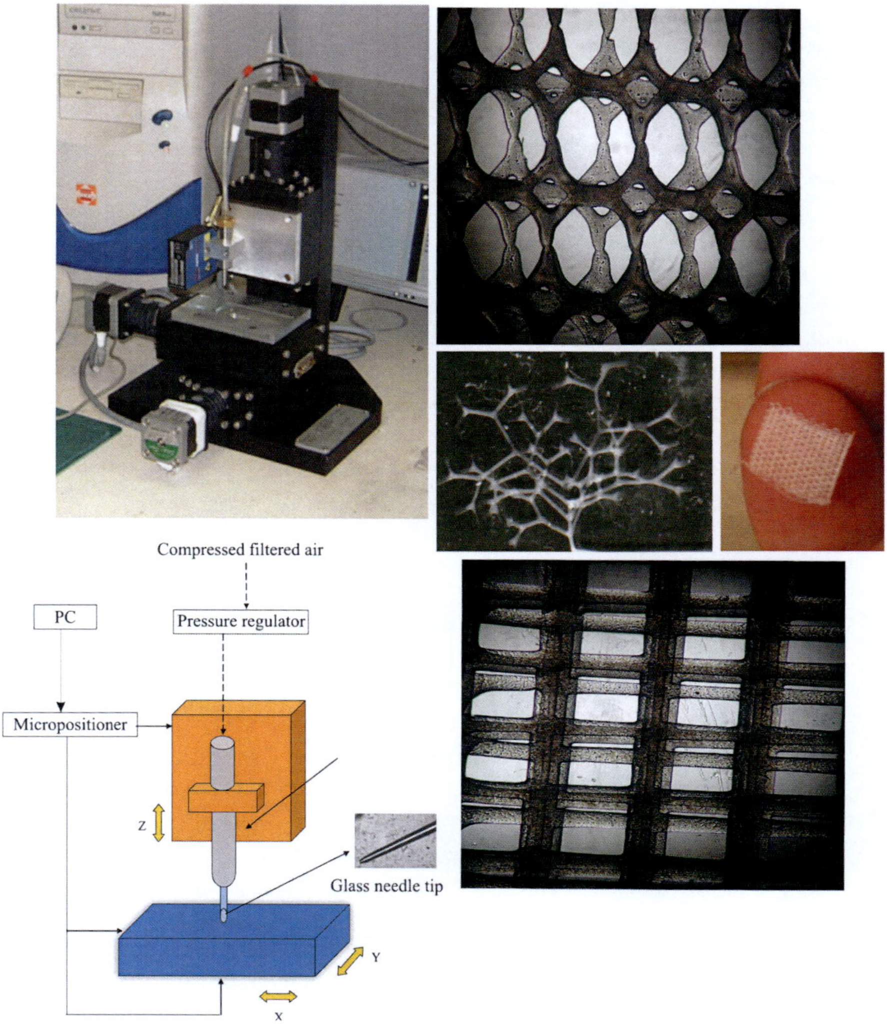

Fig. 8 PAM (pressure-assisted microsyringe) system, its working principle and sample 2D and 3D scaffolds (polylactide and polylactide-*co*-glycolide). PAM is a CAD/CAM microfabrication method which enables the production of biomaterial scaffolds with tunable geometries and mechanical properties [55]

The potency of scaffold stiffness and topology in driving cardiac stem cell differentiation in a three-dimensional culture context was confirmed by Forte et al. [54]. Cardiac stem cells adopted the cardiomyocytic phenotype only when cultured in strictly controlled conditions characterized by a critical combination of chemical, biochemical and physical factors, and emulation of the inner myocardial environment. In these studies, the emulation of myocardial environment was achieved by fine-tuning the array of growth factors dissolved in the culture medium and above all the chemistry, topology and stiffness of three-dimensional supports on which stem cells were seeded. The absence of one or more appropriate growth factors or the presence of a polymeric scaffold with stiffness higher than that passively expressed by myocardium did not trigger the differentiating cascade leading to the cardiomyocytic phenotype. Here scaffold stiffness was modulated by changing the topology of the structure using a rapid prototyping technique PAM, illustrated in Fig. 8. The optimal stiffness to induce cardiomyocyte differentiation was in the range 300 kPa on the scaffolds with square pores of about 150 μm.

Shear stress induced by flow is also an important physical stimulus, particularly for driving cells towards vascular phenotypes. Flow at physiological wall shear stresses typical of blood vessels (0.1–1.5 Pa) induces differentiation towards endothelial-like characteristics [56–58]. Adamo et al. [59] have recently demonstrated that shear stresses of the order of 0.5 Pa increase haematopoietic colony-forming potential and expression of haematopoietic markers in mouse embryos. Therefore, fluid shear stresses also influence the differentiation pathways of cells indirectly associated with the vascular system. Like scaffold stiffness, wall shear stress is a parameter which can be easily controlled and modulated using engineering design tools and technology.

8 Putting It All Together – Making Space for Engineers in Biology

Many of the factors discussed here have already been studied and identified in adult differentiated cell systems, although several questions are still open to discussion even in this sphere, for example, as concerns cell adhesion in vivo, in vitro and after implantation [60].

Providing these cues in a controlled and coordinated manner in vitro is an enormously challenging task, not currently available in traditional Petri dishes or microwells and cell culture incubators. Indeed, the most commonly cited tools for engineering stem cells are bioreactors [53, 61] and biomaterials [51, 52]. A limiting feature of most of the bioreactor systems described is their narrow range or field of application. Each bioreactor is different and specifically tailored for a single type of tissue and can usually manage only a handful of stimuli.

On the other hand, modelling tools are not considered as belonging to the gantry of methods applied to the study of stem cell fate in vitro. The inherent complexity of the stem cell niche and the differentiative microenvironment and the infinite range of cell responses to different stimuli suggest that stem cell biology and control are

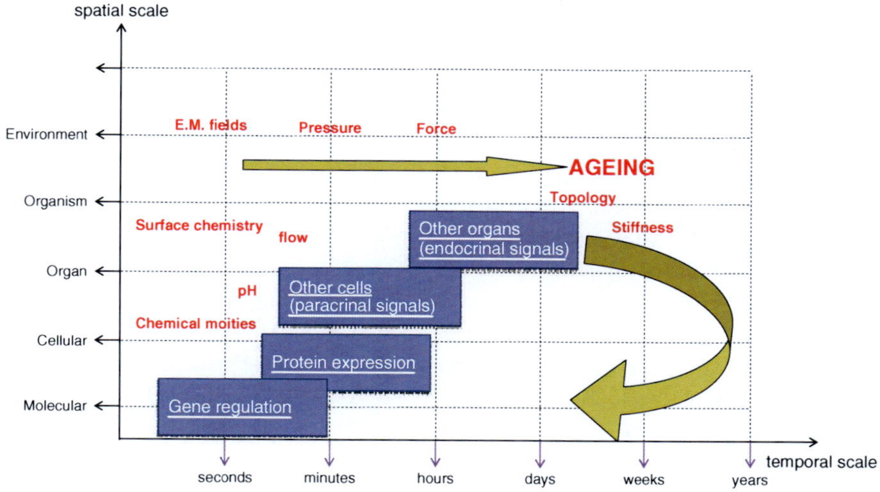

Fig. 9 A multiscale systems biology approach to study the effect of the micro- and macroenvironment of stem cells should take account of the multiple factors which play a role in the scenario and map these using PCA and phase diagrams

ideal candidates for assessment and simulation using a multiscale systems biology approach. The approach consists in coupling network models, be they deterministic or stochastic, across large ranges of length- and timescales to describe complex systems, as described in Fig. 9 [62]. Firstly, we need to single out and identify each parameter and study it individually, in the absence of other factors, or at the very least the other factors should be controlled. Cell response then should be studied at multiple hierarchical levels, genomic, proteomic and metabolomic, followed by a higher system and network level. Large-scale external factors such as flow and pressure can then be linked with molecular factors such as gene expression and biochemical inducers to give a complete picture of the micro- and macroenvironmental framework which orchestrates cell and tissue function.

The cues then need to be combined and appropriate phase diagrams constructed. Alternatively, advanced statistical methods such as three-way PCA (principal component analysis) enable phase maps to be generated from experiments in which multiple cues are present. Since many of the cues or stimuli overlap (such as elastic modulus and strain, surface energy and chemistry) or interact synergically (such as porosity and oxygen supply), this type of analysis could help pinpoint the most crucial factors which drive phenotypic expression in stem cells and their derivatives.

9 Conclusion

We are slowly building up a knowledge base on stem cell biology, even though several issues still need to be resolved. Amongst the many obstacles to the clinical application of stem cell technology is the availability of suitable procedures to expand the very few multipotent cells present in a tiny tissue sample and to address their fate to create functional portions of tissue.

These goals can be achieved by supplying stem cells with a complex array of biochemical signals and environmental stimuli by means of culture microenvironment with specific and well-controlled physical and chemical factors as well as a suitable three-dimensional environment providing physical support to stem cell adhesion, proliferation and differentiation and geometrical guidance in tissue organization. Together they form what we call here the triaxis of cues: a trio of biochemical, biophysical and biomechanical signalling systems which interact synergically to support both form and function in all living tissues. Providing these cues in a controlled and coordinated manner in vitro is a challenging task, particularly in the case of stem cells, which are known to be inherently unstable within and without their niches. Only a systematic and well-informed approach will bring us closer to understanding the roles of these cues in regulating stem cell fate, providing us with the tools to intervene and control these cells for therapeutic purposes.

References

1. Ho A. Kinetics and symmetry of divisions of hematopoietic stem cells. Exp Hematol. 2005; 33(1):1–8.
2. Schiffmann Y. Symmetry breaking and convergent extension in early chordate development. Prog Biophys Mol Biol. 2006; 92:209–231.
3. Vandervelde S, van Luyn MJ, Tio RA, Harmsen MC. Signaling factors in stem cell-mediated repair of infarcted myocardium. J Mol Cell Cardiol. 2005; 39:363–376.
4. Rando TA. Stem cells, ageing and the quest for immortality. Nature. 2006; 441:1080.
5. Segers VFM, Lee RT. Stem cell therapy for cardiac disease. Nature. 2008; 451:937.
6. Ho AD, Wagner W, Mahlknecht U. Stem cells and ageing: the potential of stem cells to overcome age-related deteriorations of the body in regenerative medicine. EMBO Rep. 2005; 6:s36–s38.
7. Trounson A. The production and directed differentiation of human embryonic stem cells. Endocr Rev. 2006; 27(2):208–219.
8. Kurata H, Guillot PV, Chan J, Nicholas M. Osterix induces osteogenic gene expression but not differentiation in primary human fetal mesenchymal stem cells. Tissue Eng. 2007; 13(7):1512–1523.
9. Bobis S, Jarocha D, Majka M. Mesenchymal stem cells: characteristics and clinical applications. Folia Histochem Cytobiol. 2006; 44(4):215–230.
10. Makino S, Fukuda K, Miyoshi S, Konishi F, Kodama H, Pan J, Sano M, Takahashi T, Hori S, Abe H, Hata J, Umezawa A, Ogawa S. Cardiomyocytes can be generated from marrow stromal cells in vitro. J Clin Invest. 1999 Mar; 103(5):697–705.
11. Jackson KA, Majka SA, Wang H, Pocius J, Craig JH, Majesky MW, Entman ML, Michael LK, Hirschi KK, Goodell MA. Regeneration of ischaemic cardiac muscle and vascular endothelium by adult stem cells. J Clin Invest. 2001; 107:1395–1402.
12. Murry CE, Soonpaa MH, Reinecke H, Nakajima H, Nakajima HO, Rubart M, Pasumarthi KB, Virag JI, Bartelmez SH, Poppa V, Bradford G, Dowell JD, Williams DA, Field LJ. Haematopoietic stem cells do not transdifferentiate into cardiac myocytes in myocardial infarction. Nature. 2004; 428:664–673.
13. Ratajczak MZ, Kucia M, Reca R, Majka M, Janowska-Wieczorek A, Ratajczak Shinji J. Stem cell plasticity revisited: CXCR4-positive cells expressing mRNA for early muscle, liver and neural cells 'hide out' in the bone marrow. Leukemia. 2004; 18:29–40.

14. Dawn B, Stein AB, Urbanek K, Rota M, Whang B, Rastaldo R, Torella D, Tang XL, Rezazadeh A, Kajstura J, Leri A, Hunt G, Varma J, Prabhu SD, Anversa P, Bolli R. Cardiac stem cells delivered intravascularly traverse the vessel barrier, regenerate infarcted myocardium, and improve cardiac function. Proc Natl Acad Sci USA. 2005; 102:3766–3771.
15. Zampetaki A, Kirton JP, Xu Q. Vascular repair by endothelial progenitor cells. Cardiovasc Res. 2008; 78:413–421.
16. Tian XF, Heng BC, Ge Z, Lu K, Rufaihah AJ, Fan VT, Yeo JF, Cao T. Comparison of osteogenesis of human embryonic stem cells within 2D and 3D culture systems. Scand J Clin Lab Invest. 2008; 68(1):58–67.
17. Park J, Setter V, Wixler V, Schneider H. Umbilical cord blood stem cells: induction of differentiation into mesenchymal lineages by cell–cell contacts with various mesenchymal Cells. Tissue Eng Part A. 2008 15(2):397–406.
18. Ho AD, Wagner W. The beauty of asymmetry: asymmetric divisions and self-renewal in the haematopoietic system. Curr Opin Hematol. 2007; 14(4):330.
19. Mazzei D, Vozzi F, Cisternino A, Vozzi G, Ahluwalia A. A high-throughput bioreactor system for simulating physiological environments. IEEE Trans Ind Electron. 2008; 55:9.
20. McBride SH, Knothe Tate ML. Modulation of stem cell shape and fate A: the role of density and seeding protocol on nucleus shape and gene expression. Tissue Eng Part A. 2008 Sep; 14(9):1561–1572.
21. Bianchi F, Vozzi G, Pescia C, Domenici C, Ahluwalia A. A comparative study of chemical derivatisation methods for spatially differentiated cell adhesion on 2-dimensional microfabricated polymeric matrices. J Biomater Sci Polymer Edn. 2003; 14:1077–1096.
22. Liu X, Lim JY, Donahue HJ, Dhurjati R, Mastro AM, Vogler EM. Influence of substratum surface chemistry/energy and topography on the human fetal osteoblastic cell line hFOB 1.19: phenotypic and genotypic responses observed in vitro. Biomaterials. 2007; 28: 4535–4550.
23. Csete M. Oxygen in the cultivation of stem cells. Ann NY Acad Sci. 2005; 1049:1–8.
24. Studet L, Csete M, Lee SH. Enhanced proliferation, survival, and dopaminergic differentiation of CNS precursors in lowered oxygen. J Neurosci. 2000; 20:7377–7383.
25. Moussavi-Harami F, Duwayri Y, Martin JA, Moussavi-Harami F, Buckwalter JA. Oxygen effects on senescence in chondrocytes and mesenchymal stem cells: consequences for tissue engineering. Iowa Orthop J. 2004; 24:p15–p20.
26. Grayson WL, Zhao F, Izadpanah R, Bunnell B, Ma T. Effects of hypoxia on human mesenchymal stem cell expansion and plasticity in 3D constructs. J Cell Physiol. 2006 May; 207(2):331–339.
27. Fink T, Abildtrup L, Fogd K, Abdallah BM, Kassem M, Ebbesen P, Zachara V. Induction of adipocyte-like phenotype in human mesenchymal stem cells by hypoxia. Stem Cells. 2004; 22:1346–1355.
28. Csete M, Walkikonis J, Slawany N, et al. Oxygen-mediated regulation of skeletal muscle satellite cell proliferation and adipogenesis in culture. J Cell Physiol. 2001; 189:189–196.
29. Lennon DP, Edminson JM, Caplan AI. Cultivation of rat marrow-derived mesenchymal stem cells in reduced oxygen tension: effects on in vitro and in vivo osteochondrogenesis. J Cell Physiol. 2001; 187:345–355.
30. Stolzing A, Sethe S, Scutt A. Stressed stem cells: temperature response in aged mesenchymal stem cells. Stem Cells Dev. 2006; 15:478–487.
31. Stolzing A, Scutt A. Effect of reduced culture temperature on antioxidant defences of mesenchymal stem cells. Free Radic Biol Med. 2006; 41:326–338.
32. Webb K, Haldy V, Tresco PA. Relationships among cell attachment, spreading, cytoskeletal organization, and migration rate for anchorage-dependent cells on model surfaces. J Biomed Mater Res. 2000; 49:362.
33. Curran JM, Chen R, Hunt JA. The guidance of human mesenchymal stem cell differentiation in vitro by controlled modifications to the cell substrate. Biomaterials. 2006 Sep; 27(27):4783–4793.

34. Radisic M, Park H, Shing H, Consi T, Schoen FJ, Langer R, Freed LE, Vunjak-Novakovic G. Functional assembly of engineered myocardium by electrical stimulation of cardiac myocytes cultured on scaffolds. Proc Natl Acad Sci USA. 2004 Dec 28; 101(52):18129–18134.
35. Sauer H, Rahimi G, Hescheler J, Wartenberg M. Effects of electrical fields on cardiomyocyte differentiation of embryonic stem cells. J Cell Biochem. 1999 Dec 15; 75(4):710–723.
36. Sauer H, Bekhite MM, Hescheler J, Wartenberg M. Redox control of angiogenic factors and CD31-positive vessel-like structures in mouse embryonic stem cells after direct current electrical field stimulation. Exp Cell Res. 2005 Apr 1; 304(2):380–390.
37. Abilez O, Benharash P, Miyamoto E, Gale A, Xu C, Zarins CK. P19 progenitor cells progress to organized contracting myocytes after chemical and electrical stimulation: implications for vascular tissue engineering. Endovasc Ther. 2006 June; 13(3):377–388.
38. Serena E, Flaibani M, Carnio S, Boldrin L, Vitiello L, De Coppi P, Elvassore N. Electrophysiologic stimulation improves myogenic potential of muscle precursor cells grown in a 3D collagen scaffold. Neurol Res. 2008 Mar; 30(2):207–214.
39. Genovese JA, Spadaccio C, Langer J, Habe J, Jackson J, Patel AN. Electrostimulation induces cardiomyocyte predifferentiation of fibroblasts. Biochem Biophys Res Commun. 2008 Jun 6; 370(3):450–455.
40. Yamaguchi DT, Huang J, Ma D, Wang PK. Inhibition of gap junction intercellular communication by extremely low-frequency electromagnetic fields in osteoblast-like models is dependent on cell differentiation. J Cell Physiol. 2002 Feb; 190(2):180–188.
41. Van Den Heuvel R, Leppens H, Nêmethova G, Verschaeve L. Haematopoietic cell proliferation in murine bone marrow cells exposed to extreme low frequency (ELF) electromagnetic fields. Toxicol In Vitro. 2001 Aug–Oct; 15(4–5):351–355.
42. Czyz J, Nikolova T, Schuderer J, Kuster N, Wobus AM. Non-thermal effects of power-line magnetic fields (50 Hz) on gene expression levels of pluripotent embryonic stem cells—the role of tumour suppressor p53. Mutat Res. 2004 Jan 10; 557(1):63–74.
43. Ebisawa K, Hata K, Okada K, Kimata K, Ueda M, Torii S, Watanabe H. Ultrasound enhances transforming growth factor beta-mediated chondrocyte differentiation of human mesenchymal stem cells. Tissue Eng. 2004 May–Jun; 10(5–6):921–929.
44. Cui JH, Park SR, Park K, Choi BH, Min BH. Preconditioning of mesenchymal stem cells with low-intensity ultrasound for cartilage formation in vivo. Tissue Eng. 2007 Feb; 13(2): 351–360.
45. Baharvand H, Hashemi SM, Ashtiani SM, Farrokhi A. Differentiation of human embryonic stem cells into hepatocytes in 2D and 3D culture systems in vitro. Int J Dev Biol. 2006; 50:645–652.
46. Brännvall K, Bergman K, Wallenquist U, Svahn S, Bowden T, Hilborn J, Forsberg-Nilsson K. Enhanced neuronal differentiation in a three-dimensional collagen–hyaluronan matrix. J Neurosci Res. 2007 Aug 1; 85(10):2138–2146.
47. Levenberg S, Huang NF, Lavik E, Rogers AB, Itskovitz-Eldor J, Langer R. Differentiation of human embryonic stem cells on three-dimensional polymer scaffolds. PNAS. 2003; 100:12741–12746.
48. Liu H, Roy K. Biomimetic three-dimensional cultures significantly increase hematopoietic differentiation efficacy of embryonic stem cells. Tissue Eng. 2005 Jan; 11(1–2):319–330.
49. Liu H, Lin J, Roy K. Effect of 3D scaffold and dynamic culture condition on the global gene expression profile of mouse embryonic stem cells. Biomaterials. 2006; 27:5978–5989.
50. Engler AJ, Sen S, Sweeney HL, Disher DE. Matrix elasticity directs stem cell lineage specification. Cell. 2006; 126:677–689.
51. Saha K, Pollock JF, Schaffer DV, Healy KE. Designing synthetic materials to control stem cell phenotype. Curr Opin Chem Biol. 2007; 11:381–387.
52. Hwang NS, Varghese S, Elisseeff J. Controlled differentiation of stem cells. Adv Drug Deliv Rev. 2008; 60:199–214.
53. Burdick JA, Vunjak-Novakovic G. Review: engineered microenvironments for controlled stem cell differentiation. Tissue Eng A. 2008; 14:1809–1820.

54. Forte G, Carotenuto F, Pagliari F, Pagliari S, Cossa P, Fiaccavento R, Ahluwalia A, Vozzi G, Vinci B, Serafino A, Rinaldi A, Traversa E, Carosella L, Minieri M, Di Nardo P. Criticality of the biological and physical stimuli array inducing resident cardiac stem cell determination. Stem Cells. 2008 May; 26(8):2093–2103.
55. Mariani M, Rosatini F, Vozzi G, Previti A, Ahluwalia A. Characterisation of tissue engineering scaffolds microfabricated with PAM. Tissue Eng. 2006; 12(3):547–558.
56. Wang H, Riha GM, Yan S, Li M, Chai H, Yang H, Yao Q, Chen C. Shear stress induces endothelial differentiation from a murine embryonic mesenchymal progenitor cell line. Arterioscler Thromb Vasc Biol. 2005; 25:1817–1823.
57. Metallo CM, Vodyanik MA, de Pablo JJ, Slukvin II, Palecek SP. The response of human embryonic stem cell-derived endothelial cells to shear stress. Biotechnol Bioeng. 2008 July 1; 100(4):830–837.
58. Wu CC, Chao YC, Chen CN, Chien S, Chen YC, Chien CC, Chiu JJ, Linju Yen B. Synergism of biochemical and mechanical stimuli in the differentiation of human placenta-derived multipotent cells into endothelial cells. J Biomech. 2008; 41(4):813–821.
59. Adamo L, Naveiras O, Wenzel PL, McKinney-Freeman S, Mack PJ, Gracia-Sancho J, Suchy-Dicey A, Yoshimoto M, Lensch MW, Yoder MC, García-Cardeñ G, Daley GQ. Biomechanical forces promote embryonic haematopoiesis. Nature. 2009; doi:10.1038/nature08073.
60. Wilson CJ, Clegg RE, Leavesley DI, Pearcy MJ. Mediation of biomaterial–cell interactions by adsorbed proteins: a review. Tissue Eng. 2005 Jan–Feb; 11(1–2):1–18.
61. Godara P, McFarland CD, Nordon RE. Design of bioreactors for mesenchymal stem cell tissue engineering. J Chem Technol Biotechnol. 2008; 83:408–420.
62. Evans DJW, Lawford PV, Gunn J, et al. The application of multiscale modelling to the process of development and prevention of stenosis in a stented coronary artery. Philos Trans R Soc A. 2008; 366:3343–3360.

Differentiation Potential of Adult Human Mesenchymal Stem Cells

Edda Tobiasch

Abstract Mesenchymal stem cells (MSCs) can be found in various tissues of the adult organism. These cells display a multilineage potential and can be in vitro differentiated toward the osteogenic, adipogenic, chondrogenic, and myogenic lineage, which makes them promising candidates for future tissue replacement strategies.

Keywords Mesenchymal stem cells · Multilineage potential · Lipoaspirate · Dental follicle · Adipose tissue · Diabetes · Osteogenic differentiation · Scaffolds · Regenerative Medicine

Abbreviations

CD	cluster of determinants
CSD	critical size bony defects
DF	dental follicle
DMEM	Dulbecco's modified Eagle medium
MSCs	mesenchymal stem cells
FCS	fetal calf serum
IBMX	isobutylmethylxanthine
MTT	3-(4,5-dimethylthiazol-2-yl)-2,5-diphenyltetrazoliumbromide
SCE	sister chromatid exchange

For every complex problem, there is a solution that is simple, neat, and wrong.

E. Tobiasch (✉)
University of Applied Sciences Bonn-Rhein-Sieg, Rheinbach, Germany
e-mail: Edda.Tobiasch@h-brs.de

G.M. Artmann et al. (eds.), *Stem Cell Engineering*,
DOI 10.1007/978-3-642-11865-4_3, © Springer-Verlag Berlin Heidelberg 2011

1 Introduction

We all benefit from the huge progresses in medicine of the last century by enjoying a statistically significant prolonged lifespan. This generally positive effect unfortunately also gave rise to novel major health problems. An increased probability to fall ill with cancer or to develop type II diabetes as well as to suffer from age-related degenerative diseases such as loss of teeth or cartilage, muscle or bone degeneration must be expected.

In adolescents and young adults, an altered recreational behavior with accident-prone sports leads to an increased number of cartilage, tendon, and critical bone defects, too. Taken together, tissue replacement strategies must be developed to meet these therapeutic challenges.

Interestingly, a lot of these medical problems could be theoretically solved by means of adult mesenchymal stem cell-derived substitution approaches. These cells can differentiate into the needed lineages [1] and they pose no major ethical issues. MSCs can be isolated from the patient himself, which, in addition, will decrease immunological rejection problems. Or, they can be isolated in large amounts from plastic surgery lipoaspirates of healthy donors. Or, with respect to dental implants, the isolation from the defect side itself is feasible, thus causing no additional clinical interventions for the patient. It seems that these cells provide a perfect solution for each of the mentioned problems.

But stem cell differentiation is not fully understood yet and obviously cells are not the same as tissues or organs, lacking the three-dimensional structure and the highly ordered arrangement of various cell types found within an organ. Even for simple tissues this problem is far from being solved. Thus, for stem cell-directed replacement strategies such as critical size bone repair to be clinically successful, a scaffold must be identified and optimized to support cellular adherence, cell recruitment, osteoinduction, osteoconduction, and angiogenesis. In addition, the material must be biocompatible, degradable, and resilient to shear forces and pressure. At the moment the questions and difficulties which come with the above-listed requirements have only begun to be addressed.

MSCs might also be useful to achieve new insights into basic biomedical sciences and thereby provide the fundamentals for a better understanding of the illness development, even for multifactorial syndromes such as type II diabetes.

In this disease the gaining of overweight is a key factor, because the differentiating (much more than the differentiated) adipocytes are metabolically highly active [2]. They release inflammatory cytokines and other mediators which can increase the risk for cardiovascular diseases, the major cause of death for these patients. Investigation of the gene expression profile of the differentiating adipocyte which can be derived from MSCs and its interaction with endothelial cells might therefore increase the knowledge of diabetes-related medical problems and lead to new treatment strategies in the future.

2 The Source of Mesenchymal Stem Cells

Mesenchymal stem cells are ultimately derived from the mesenchyme, the embryonic connective tissue which develops mainly from the mesodermal germ layer. But it also contains cells from the endoderm and to some extent wandering cells from the neuronal crest [3]. If neuronal crest cells are involved, the tissue is often termed ectomesenchyme and plays a critical role in the development of cranial hard and soft tissues such as bone and muscles. In embryogenesis, the star-formed mesenchymal cells give rise to differentiated cells and tissues such as bone, blood cells, vessels, tendons, and connective tissue [4].

In the adult organism, connective tissue cells meet several conditions. In bone marrow they form the supportive stromal niche for hematopoietic stem cells and are thought to play a role in their hierarchical differentiation processes [5], whereas the stroma within the fat tissue has very different functions. Cells isolated out of the stroma can proliferate in vitro and differentiate into several lineages (Fig. 1), which indicates that they are multipotent. In addition, these cells are self-renewable and

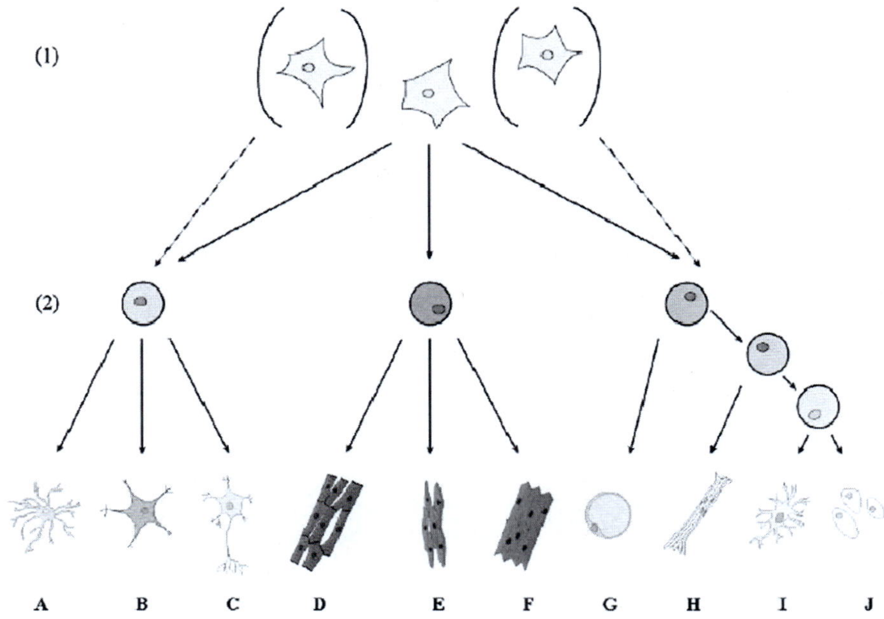

Fig. 1 Derivation of tissue cells from mesenchymal stem cells. The mesenchymal stem cells (*1*), which are self-renewing, give rise to various cells. It is suggested that isolated MSCs are a mixture of cells some of which are already committed with a greater differentiation potential toward related tissue types. These "pre-committed" stem cells are indicated in *brackets* with *dashed lines*. Upon activation of specific signal cascades, stem cells are thought to differentiate into committed progenitor cells (*2*) and then further develop into fully differentiated cells (*A–J*): *A*, astrocyte; *B*, oligodendrocyte; *C*, neuron; *D*, cardiac muscle; *E*, smooth muscle; *F*, skeletal muscle; *G*, adipocyte; *H*, tenocyte; *I*, osteocyte; *J*, chondrocyte

possess long-term viability, all attributes of stem cells. Accordingly, they are named mesenchymal stem cells.

3 The Isolation of Mesenchymal Stem Cells

In the beginning, MSCs have been isolated as by-product during the isolation of hematopoietic stem cells and they have been investigated due to their on-site and developmental connections. Hematopoietic cells are of major interest since they can be used, next to basic research, for leukemia patient treatment. The first successful bone marrow transplantation and later Nobel laureate-awarded work, conducted by E. Donnall Thomas in 1990, is therefore also the first successful stem cell-based therapy.

Over the years, scientists got more and more interested in the MSCs as well, mainly due to their self-renewal capacity and multipotency. In the meantime, these cells can be isolated from various tissues with bone marrow, umbilical cord blood, and fat (Fig. 2a, b) being the major sources [6, 7, 1].

The isolation procedure of MSCs is rather simple, because they attach to standard cell culture plastic material [8] and can therefore be easily separated from cells

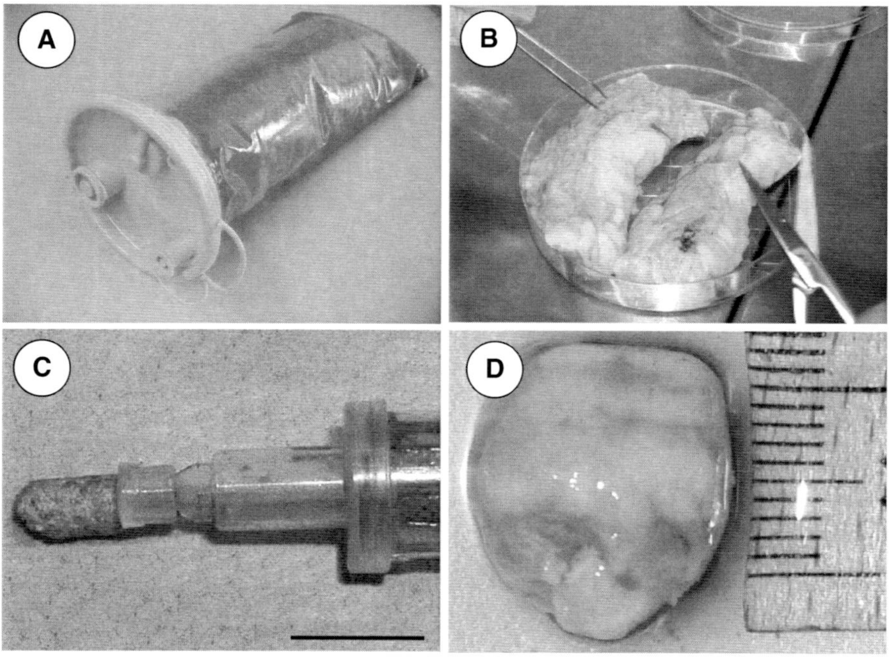

Fig. 2 Sources of mesenchymal stem cells and precursor cells. (**a**) Human liposuction material in a typical collection bag (AescoLOGIC, volume 3 l); (**b**) two parts of a human abdominal plastic (50 g, each) in a 10-cm Petri dish; (**c**) syringe with collected bone chips from a human mandible (bar = 1 cm); (**d**) human dental follicle

which do not attach, such as hematopoietic cells, or from cells which need coating of the culture material with collagen, gelatine, poly-L-lysine, or other suitable substances.

All mesenchymal stem cell populations exhibit a characteristic and consistent fibroblast-like phenotype and are similar in cell size and granularity in monolayer cultures (Fig. 3b) [9, 10].

In the following, the procedure of isolating human mesenchymal stem cells is explained in detail to demonstrate the course of action using liposuction material derived from plastic surgery as source for the stem cells.

Procedure: Determine the amount of received fat and add PBS to the fat tissue in a ratio of 1:2. Then shake it well and let it stand at room temperature (RT) for 30 min to separate the fatty and the aqueous phase (see Fig. 3a). Discard the bottom layer (fluid phase) since it yields only relatively few stem cells. Add to the upper layer, PBS (ratio 1:1) and collagenase (0.15 U/ml), incubate at 37°C for 60 min in the shaking water bath, and additionally shake thoroughly every 15 min. Centrifuge at $200 \times g$ at RT for 10 min. Combine the pellets and wash with 10 ml PBS followed by centrifugation at $200 \times g$ at RT for 10 min. Discard the supernatant, resuspend the pellet in 10 ml erolysis buffer [9], and incubate at RT for 10 min. Centrifuge again at $200 \times g$ at RT for 10 min. Discard the supernatant and resuspend the pellet

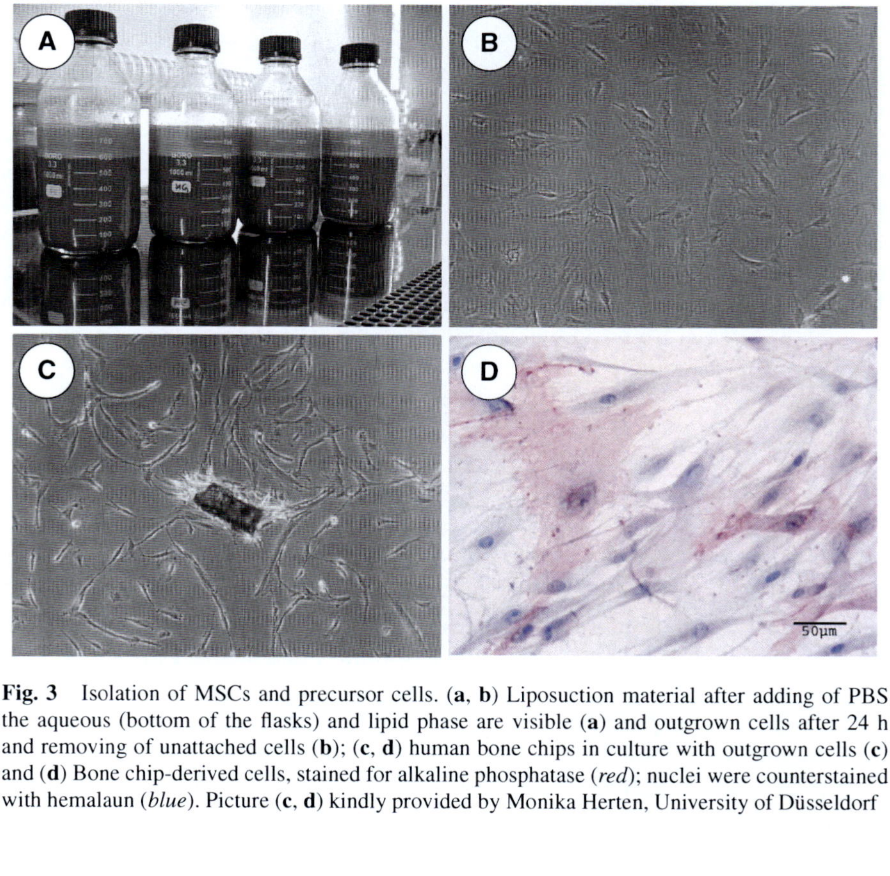

Fig. 3 Isolation of MSCs and precursor cells. (**a**, **b**) Liposuction material after adding of PBS the aqueous (bottom of the flasks) and lipid phase are visible (**a**) and outgrown cells after 24 h and removing of unattached cells (**b**); (**c**, **d**) human bone chips in culture with outgrown cells (**c**) and (**d**) Bone chip-derived cells, stained for alkaline phosphatase (*red*); nuclei were counterstained with hemalaun (*blue*). Picture (**c**, **d**) kindly provided by Monika Herten, University of Düsseldorf

in an adequate volume of medium for seeding. Change medium after 24 h to remove non-adherent cells (Fig. 3b).

Another interesting source of mesenchymal stem cells is the bone marrow. The bone marrow is embedded in the spongy tissue (cancellous bone, *substantia spongiosa ossium*) of the inner cavity of long bones, which is surrounded by the more dense corticalis (*substantia compacta*). Although the hip is the major site for the isolation of bone marrow-derived MSCs, depending on the treatment area, where the isolated cells might be of use, other bones such as the mandible for dental implant surgery might be of interest as cell source as well.

One of the first steps for the later insertion of a dental implant is the drilling of a hole in the mandible or the maxilla, depending on where the implant is needed. During this process, bone chips can be accumulated as a waste product in a bone collector and isolated (Fig. 2c). This isolation of bone chips as putative stem and precursor cell source (Fig. 3c, d) causes no further pain for the patients and the material can later be used for alveolar bone augmentation [11].

Ectomesenchymal stem cells, which differ in potency and commitment from mesenchymal stem cells, can be isolated from the human dental follicle (Fig. 2d) since it is a developmental precursor for essential periodontal tissues such as root development and periodontal ligament. The dental follicle can be gained from wisdom teeth in adolescents if they are extracted prior to tooth eruption [12]. The detailed isolation procedure is given in the next paragraph.

Procedure: Transfer the dental follicle (DF) from the transport medium into a small culture dish. Remove transport medium and wash the DF three times with 1× PBS. After the last washing step, remove PBS and determine the weight of the DF. To prevent desiccation of the DF, cover each sample again with 1× PBS until the cutting process starts. Cut the DF into very small pieces. Cover the tissue pieces with 3 ml collagenase/Dispase (1 mg/ml) and incubate them for 2 h at 37°C with 5% CO_2. Place a 100-μm filter onto a 50-ml tube and filter the cell suspension with additional 3 ml culture medium (DMEM). Centrifuge the filtrate for 5 min at 200×*g*. Discard the supernatant, resuspend the pellet in 5 ml culture medium, and culture the cells at 37°C with 5% CO_2. Change the medium after approx. 24 h of incubation and wash cells carefully with 1× PBS.

4 The Differentiation of Mesenchymal Stem Cells

In 2006, the "Mesenchymal and Tissue Stem Cell Committee" of the International Society for Cellular Therapy defined the minimal criteria to determine an MSC [13]. This got necessary, because the sources for MSCs increased and each scientific group working with MSC derived from a specific tissue defined their own more or less specific marker set of differentiation antigens (cluster of determinants, CDs) located on the surface of those cells.

Nevertheless a lot of information could be deduced from this extensive surface marker expression profiles which are very similar but not identical, for example, to the populations derived from bone marrow or adipose tissue [7, 14, 8, 15]. Both

MSC populations express CD9, CD10, CD13, CD29, CD34, CD44, $CD49_d$, $CD49_e$, CD54, CD55, CD59, CD105, CD106, CD146, and CD166 [15]. Furthermore, both populations are uniformly negative for the expression of CD3, CD4, CD11c, CD14, CD15, CD16, CD19, CD31, CD3, CD34, CD38, CD45, CD56, CD61, CD62P, CD104, and CD144. However, distinctions were observed for the following three markers: STRO-1, CD49d, and CD106. The fat tissue-derived MSCs are positive for CD49d and negative for STRO-1 antigen and CD106 expression, whereas the opposite is true for bone marrow-derived MSCs.

The question arises if these dissimilarities could be expected to count for divergences in the differentiation potential of the stem cells. STRO-1 is an antigen which was first identified by Simmons and Torok-Storb in 1991 and is used to identify bone marrow-derived MSCs [16, 36]. It is also commonly used to select dental pulp stem cells which can be differentiated toward hard tissues, such as cementoblasts for periodontal tissue repair experiments (Fig. 7). CD49d (ITGA4, VLA4, and VLAA4) which interacts with VCAM-1 is an antigen which is found in a variety of different blood cell lineages [17]. CD106 (alpha-4-beta-1 ligand, ICAM-110, and VCAM-1) is mainly expressed on endothelial and follicular dendritic cells [18]. The presence of CD49d and CD106 can therefore be expected on bone marrow-derived cells, whereas the absence of their expression in fat-derived MSCs is consistent with the localization of these cells in a non-hematopoietic tissue.

To reduce the necessary antigen stain to a feasible minimum, the International Society for Cellular Therapy suggested that CD73, CD90, and CD105 must be positive in at least 95% of the cells and that CD34, CD45 CD14 (or CD11b), CD79α (or CD19), and HLA-DR must be negative in at least 98% of the isolated cells. This criterion, together with the ability of the cells to adhere under standard culture conditions and with the specific stain after differentiation toward chondroblasts, adipocytes, and osteoblasts in vitro (Fig. 4), is now the accepted standard to define a mesenchymal stem cell [13].

The differentiation of MSC toward the myogenic lineage is not included in that list. The reason for this is simple. There is no specific dye for myoblasts. Of course, several specific immunohistological staining could be performed, but this would counteract the idea of a simple and fast testing, and it is unnecessary as the differentiation toward three distinct directions should be enough to underline the multipotentiality of these cells. In addition and in contrast to several publications [8, 7], the differentiation in the myogenic lineage is not trouble free. The stroma contains cells which exhibit myogenic markers and the isolated cells, depending on the tissue source, can be easily contaminated with smooth muscle cells from vessels.

The differentiation toward the other three lineages is surprisingly simple. For the chondrogenic lineage, a well-grounded cell culture technique is necessary since micromass cell culture must be applied, but the differentiation protocol itself is easy to perform [19]. Even simpler is the differentiation toward adipocytes and osteoblasts. A simple cocktail of hormones and other signaling molecules must be given to the cells in a passage number preferably between 4 and 8 for 3–4 weeks followed by the specific staining with oil red O for lipids in adipocytes and alizarin red for calcium phosphate depositions of the osteoblasts [20].

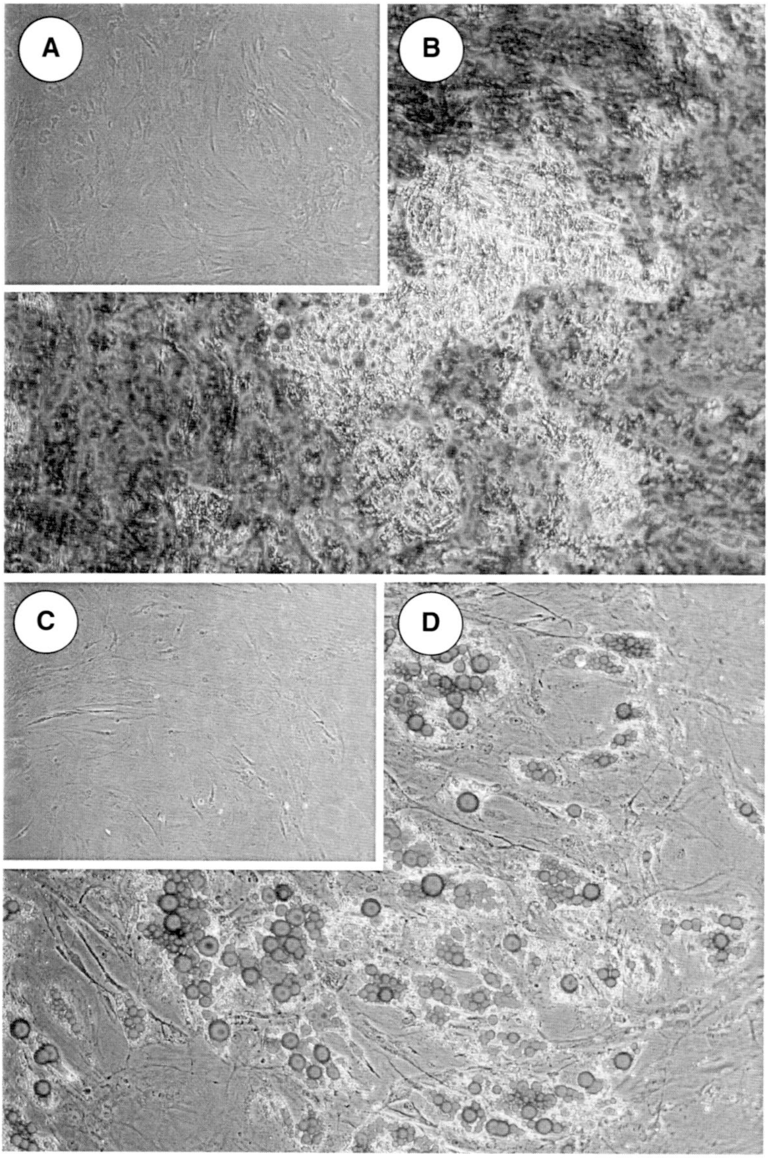

Fig. 4 Differentiation of MSCs. (**a**, **b**) Undifferentiated (control) (a); and specific differentiated adult MSCs toward the adipogenic lineage, stained with *oil red* O (**b**); (**c**, **d**) Undifferentiated (control) (c); and specific differentiated adult HMSCs toward the osteogenic lineage, stained with *alizarin red* (**d**). All magnifications 38×

Procedure: Adipogenic differentiation. MSC with a confluency of 80–100% are incubated in differentiation medium [10 μM insulin, 1 μM dexamethasone, 0.5 mM isobutylmethylxanthine (IBMX), 200 μM indomethacin in DMEM with 10% FCS and antibiotics/antimycotics]. The medium must be exchanged every 3–4 days. The cells are stained with oil red O after 3 weeks (see Fig. 4a, b).

Procedure: Osteogenic differentiation. MSC with a confluency of 60–70% are incubated in differentiation medium (0.01 μM 1,25-dihydroxyvitamin D3, 10 mM β-glycerophosphate, 0.05 mM ascorbic acid 2-phosphate in DMEM with 10% FCS and antibiotics/antimycotics). The medium must be exchanged every 3–4 days. The cells are stained with alizarin red after 4 weeks (see Fig. 4c, d).

During the differentiation process, the adipocyte profile of excreted signaling molecules and their impact on endothelial cells can be investigated for a better understanding of the developmental steps leading to the metabolic syndrome (syndrome X, Reaven's syndrome, CHAOS) which is also named "deadly quartet" due to its role as major risk factor for cardiovascular disease and diabetes [21, 22].

Of course it is very tempting not only to use the differentiated cells to increase the basic knowledge in science but also to use them for the treatment of patients with degenerative diseases. Aside from the adipogenic lineage, all lineages in which MSCs can differentiate would be of interest for this. But a chondroblast is not a cartilage, a myoblast is not a muscle fiber, and an osteoblast is not a bone. Techniques must be developed to force the cells, which grow in culture only as a monolayer, into a three-dimensional structure. To do so scaffolds have been developed; however, this is still an ongoing process due to the multifactorial requirements to these materials.

5 Biomaterials for Three-Dimensional Scaffolds

Scaffolds can be manufactured from a huge variety of materials. These materials can be divided into three main classes: synthetic polymers, biobased synthetic polymers (such as polyester or poly(ester)urethane), natural polymers (such as cellulose or polypeptides), and materials obtained from a donor organ [23, 24].

Materials obtained from a donor organ have several advantages. They have the perfect structure with respect to form and pore size, they exhibit the best flexibility and rigidity necessary for the corresponding tissue, and they are biocompatible. Disadvantages are possible rejections caused by the recipient's immune system and a danger for the transmission of viral and prion-based diseases.

An example for such an approach is the transplantation of specially treated donor heart valves. The valves are depleted of the donor cells and the leftover is a collagen scaffold which can now be seeded with cells from the patient. This new technique has already been used in young children. It is expected that this specific cardiac valve will grow with the young patient, thus reducing the need for further surgical interventions [25]. Before, the defect valve could only be replaced with artificial valves or valves derived from pigs. Next to their inability to grow with the patients, both had additional disadvantages. The artificial valve necessitated the lifelong use of anticoagulant drugs and the pig-derived valve had only a very limited stability. Both had to be replaced after some years causing additional risk, stress, and pain for the young patients.

Scaffold materials are used for bone replacement as well. Especially bone lesions above a critical size become scarred rather than regenerated leading to non-union (Fig. 6d). These critical size bone defects (CSDs) caused by trauma, tumors, and endoprosthesis revision surgery present an unsolved therapeutic problem in orthopedic and trauma surgery [26]. Due to missing alternatives, autologous bone transplantation is still the gold standard for treatment of osseous defects. But the resources of autografts are limited and associated with several problems including infection risks and donor site morbidity. In contrast, successful transplantation of vital bone from donors depends on an immunosuppressive therapy due to the immunological barrier between host and donor, which is associated with severe drug-related side effects.

Periodontal diseases are widespread in industrial countries and a frequent cause of tooth loss. In addition, they are associated with coronary artery disease [27]. The periodontium is a complex organ consisting of epithelium, soft, and mineralized connective tissues. To achieve regeneration, the formation of these tissues (root cementum, connective tissue fibers, and bone) is required [28]. To date, considerable efforts have been made to identify potential stem cells within dental tissues, but the availability of dental stem cells is problematic, as it is restricted to specific time points. Several therapies, such as implantation of autografts, allografts, alloplastic materials, or guided tissue regeneration, show divergent results and are mostly unpredictable [29].

Taken together, there is an increasing need for new scaffold materials which should be optimized in various aspects for the respective tissue. Synthetic polymers might be an alternative in some of the cases. There would be no limit in supply and they could be theoretically modified to meet the needs of the particular organ and patient. However, there is a difficulty which is not found in natural biomaterials. These materials are often not biocompatible and therefore must be tested thoroughly before they can be used in humans.

6 Biocompatibility of Scaffold Materials

The first step to investigate the biocompatibility of a synthetic material is the testing for cytotoxicity in vitro to meet the international standards compiled as ISO 10993. In this standard, there is a list of test methods with which acute adverse biological effects of extractables from scaffold materials can be evaluated. They are based on the testing of mammalian cells, which are exposed to the new material directly or indirectly by means of fluid extracts [30–32].

The easiest and quickest testing is the microscopic examination of the cells to investigate if they can attach to the material and keep their usual structure (Fig. 5a) or

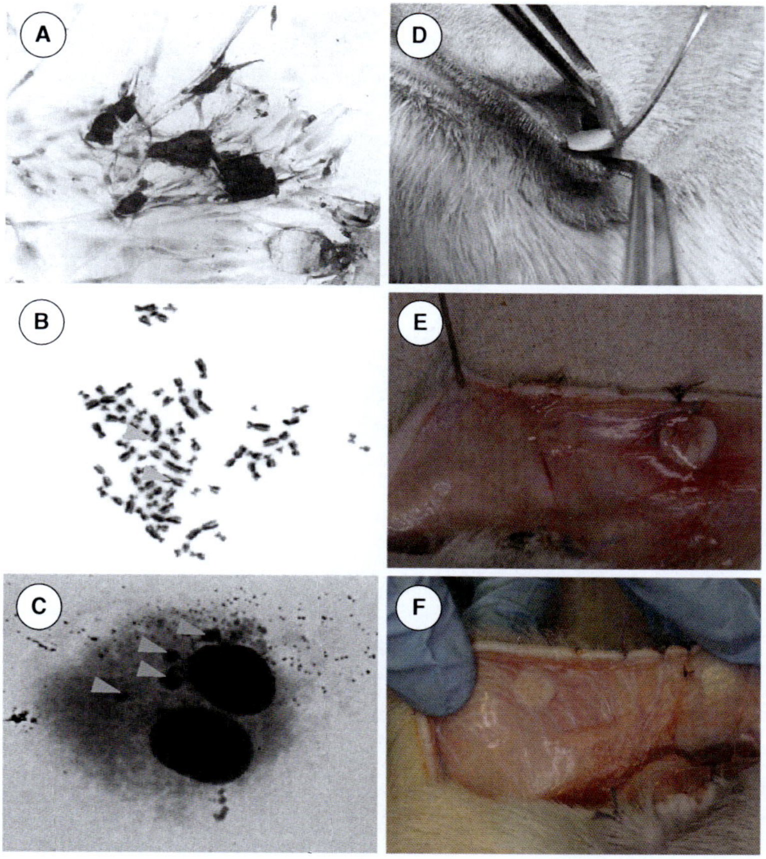

Fig. 5 Biocompatibility testing of scaffold materials. (**a–c**) In vitro biocompatibility testing. (**a**) Attachment of cells to Navigraft® (Tutogen Medical GmbH), a substance composed of bovine hydroxyapatite, stained with *toluidine blue*. (**b**) Detection of single-strand breaks with sister chromatid exchanges (SCEs) in metaphase chromosomes of HeLa cells after incubating with a polymer for scaffolds. SCEs are indicated by *arrowheads*. Magnification 1,000×. (**c**) Detection of double-strand breaks via several micronuclei emerging in a cell exposed to a polymer for scaffolds. Micronuclei are indicated by *arrowheads*. Magnification 1,000×. (**d**) In vivo biocompatibility testing via subcutaneous implantation of scaffold material in rats; (**e**, **f**) removal of implant after 8 weeks, with (**e**) and without (**f**) visible signs for inflammation. Picture (**d**, **e**, **f**) kindly provided by Monika Herten, University of Düsseldorf

if a change of appearance, for example, a more rounded cell shape, can be monitored which will be a first hint to a cytotoxic feature of the substance.

The most commonly used cytotoxicity test is the MTT test. This colorimetric assay measures the enzyme activity responsible for the reduction of 3-(4,5-dimethylthiazol-2-yl)-2,5-diphenyltetrazoliumbromide (MTT) to formazan. By doing so, the yellow basic product will turn purple in living cells, thus indicating the function (or if not, the dysfunction) of the mitochondria [33].

A negative result in such a test is one indication that a substance is free of harmful by-products or has insufficient quantities of them to cause acute toxicity in cultured cells. However, that does not mean that a material can be considered biocompatible.

Long-term effects might be based on accumulation even when the harmful effect of the agent is below the threshold for causing acute cytotoxicity. It is therefore sensible to test direct effects on the human DNA as well, even if this is not mandatory by law.

Single-nucleotide exchanges or other small size mutations cannot be tested on a regular basis since they require techniques such as sequencing the whole genome, which would be too time consuming and too expensive at the moment. Nevertheless, this might be a future possibility, given that it has been done already to identify all point mutations causing a specific cancer in humans [34, 37].

Larger size DNA mutations such as strand breaks leading to sister chromatid exchange events or double-strand breaks causing the loss of DNA parts in chromosomes can be investigated in cell culture (Fig. 5b, c). To examine strand breaks, a technique is applied that uses the semi-conservative replication of the DNA to stain both chromatids differently, which makes the exchange visible. Loss of chromosome parts can be visualized by a method in which the resulting micronuclei are stained. In the next paragraph, the protocols for both methods are given.

Procedure: Sister chromatid exchange. HeLa cells are synchronized by serum starvation for 24–48 h. Cells are treated for 6 h with the scaffold material or a solution in which the scaffold material was incubated for 15 min at 70°C or for 24 h at 37°C. 5-Bromo-2-deoxyuridine (BrdU) is added for two cell cycles (48 h). The cells are arrested in metaphase by colcemid and stained with Hoechst followed by Giemsa (Fig. 5b). The amount of metaphases from each sample, which has to be evaluated for the ascertainment of valid statistical data, depends on the occurrence of SCEs and the calculated number of their differences within the samples.

Procedure: Micronucleus assay. HeLa cells are synchronized by serum starvation for 24–48 h. Parts of the cells are incubated for 6 h with the scaffold material or a solution in which the scaffold material was incubated for 15 min at 70°C or for 24 h at 37°C. Untreated cells serve as negative controls. Another part of the cells is subjected for 5 s to UV irradiation and used as positive control. Cytokinesis is inhibited by the addition of cytochalasin B. All cells are stained with Giemsa. One hundred binucleated cells of each sample have to be evaluated for the presence of micronuclei (Fig. 5c) to permit valid statistical data.

A new substance for a medical device which passes the cytotoxicity test in vitro must then be tested in vivo to investigate mechanisms such as the reaction of the immune system and degradation processes, both of which cannot be evaluated in cell culture. For this, a rodent model where the scaffold material is implanted subcutaneously is the most common choice (Fig. 5d, e). After several weeks of monitoring, the implant will be removed (Fig. 5e, f). The side of implantation will be investigated for inflammation (Fig. 5e) and the structure of the material for changes caused by degradation.

The rat model can also be used for more sophisticated experiments as a further needful evaluation step toward a future treatment of human tissue defects by applying the new scaffold material [35]. In Fig. 6 an example for such an experiment investigating different bone replacement strategies is shown.

A critical size cranial bone defect was investigated for several independent parameters. The regeneration quality and quantity of the CSD was compared for three different approaches: without fill-in material, filled with autologous bone chips

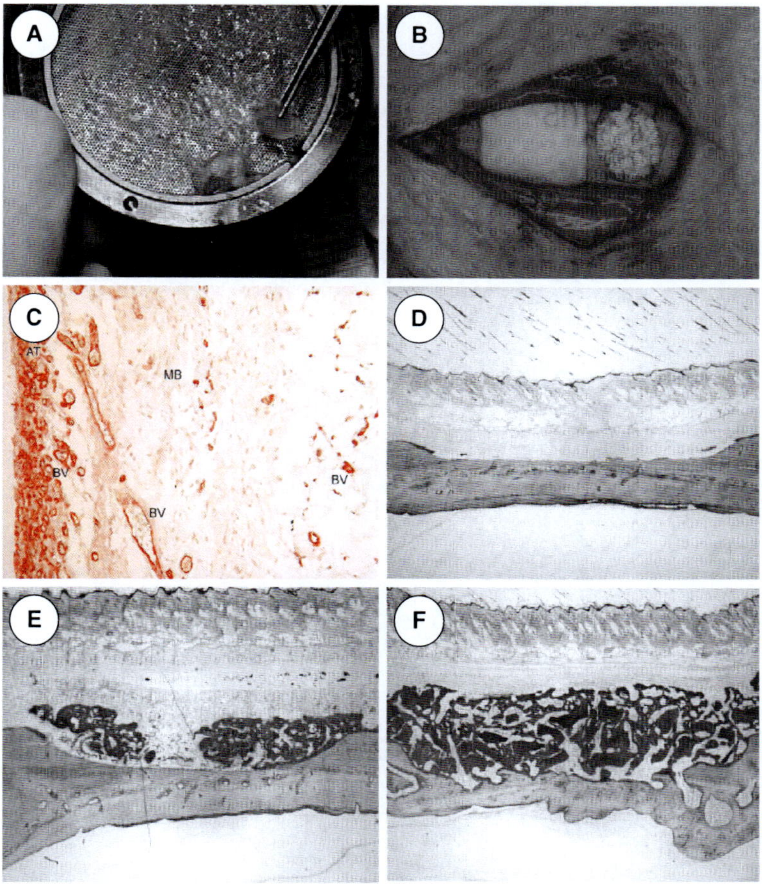

Fig. 6 Regeneration of a critical size cranial bone defects in rat. (**a**) Isolation of bone chips. (**b**) Insertion of bone scaffold material with and without additional Bio-Gide® membrane (Geistlich Biomaterials) cover. (**c**) Ingrowth of blood vessels into the biomaterial. Transmembraneous angiogenesis using Bio-Gide®, 2 weeks after subcutaneous implantation in rats. Specific immunohistochemical staining of endothelial cells with transglutaminase II. AT, Adjacent connecting tissue; BV, blood vessels; MB, membrane bodies (magnification 40×). (**d–f**) Result of the regeneration of the critical size defect. Staining with *toluidine blue*: (**d**) without fill-in material, (**e**) with autologous bone chips, (**f**) with bovine hydroxyapatite. Pictures kindly provided by Monika Herten, University of Düsseldorf

(Fig. 6b), and filled with bovine-derived hydroxyapatite (Fig. 6d, e, f). Another experiment provided data for the insertion of bone scaffold material with or without an additional membrane cover which might inhibit the ingrowth of unwanted cells into the cavity of the defect (Fig. 6a). An experiment investigating the wanted ingrowth of blood vessels into the biomaterial to support transport of nutrients, regeneration of the bone, and degeneration of the scaffold material is shown in Fig. 6c.

Despite the large amount of data which can be collected by using a rat model for such investigations, a further step is necessary before clinical trials can be initiated. Several major differences between this mammal and humans are the reason for this. A rodent is very resistant to all kinds of treatments. It can easily survive toxic drugs to an extent which would be deadly for humans, if the same amount per bodyweight would be applied. The life expectancy of a rat is rather short, so long-term effects cannot be investigated to the full extent. For critical size bone defects or dental regeneration, the exposure to strong forces is a major

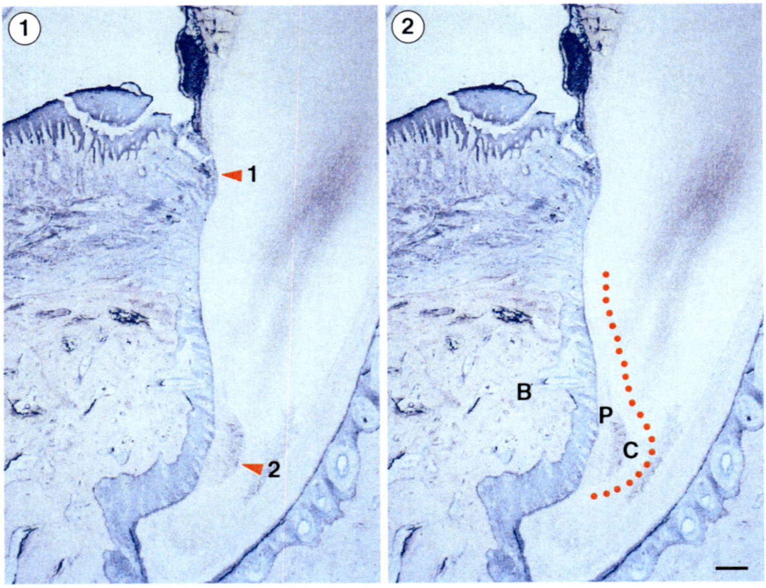

Fig. 7 Periodontal regeneration with autologous progenitor cells. Interdental defect at a molar of a minipig (*1* and *2*, indicated by arrow *heads*) treated with cultured periodontal ligament progenitors after 90 days. The periodontium has been restored only in part. Cementum deposition can be seen on the apical half of the root surface (see *C*, and *dotted line*), and new bone fills (*B*) on the apical half of the defect. Parts of the coronal root surface show connective tissue attachment without new cementum formation. Connective tissue attachment on the apical part of the root surface is visible in which the fiber bundles are attached to the newly formed cementum and bone. *1*, Coronal mark; *2*, apical mark filled with new cementum. *B*, New alveolar bone; *P*, newly formed periodontal ligament with oriented fiber bundles; *C*, apical mark filled with new cementum. Stain is *toluidine blue*. The *scale line* indicates 800 μm. Picture kindly provided by Hermann Lang, University of Rostock

parameter for the implant quality and this cannot be addressed in such a small-sized animal. A large animal model is required and a sheep for critical size bone defects or a minipig for periodontal regeneration experiments is a suitable choice (Fig. 7).

7 Future Aspects

So far considerable efforts were made to identify adult stem cells in various tissues. But most organs have only progenitor cells with a very limited potential, with skin cells being a characteristic example. If the potential of the progenitor cells is higher, the availability is often restricted to specific time points as for dental stem cells or MSCs from the umbilical cord.

Nevertheless, the multipotential of human mesenchymal stem cells makes them ideal candidates for tissue replacement approaches. These cells can be derived in especially large abundance from fat and their isolation and application causes no major ethical problems. They can be derived from plastic surgery waste material of healthy donors so that allograft and autograft strategies could generally be practicable. Although bone marrow transplants are evaluated in first clinical trials, there is no standardized MSC–scaffold composite protocol available at the moment for the treatment of bone defects or other defects where a three-dimensional structure is needed.

The existing biomaterials for three-dimensional structures are not optimized to control cell differentiation or to achieve angiogenesis to the full. Neither can they be adequately injected into the body nor are they biodegradable in the needed time frame. The degradation processes which should be distributed equally within the scaffold material occur mainly on the surface. There is an imperious need to design highly porous biodegradable tissue-engineered scaffolds that offer an improved level of functionality than those currently available. These materials should be apt to be functionalized and to have a direct manageable influence on cell growth and differentiation for a patient-tailored approach.

An easy applicable “ready-to-go” method for MSC-based patient treatment is far from being available for most replacement strategies. The devil is as usual in the details which have to be investigated and understood first. Major problems must be solved, such as the risk caused by stem cell tumorigenicity and the construction of the correct three-dimensional structure for the tissue which is supposed to be repaired. For some tissues such as cartilage and bone, replacement strategies might be achieved soon, whereas for others, it will take more time to develop approaches which are not only “ready to go”, but also safe.

Acknowledgments My cooperation partners Dr. Monika Herten, University of Düsseldorf, and Prof. Dr. Hermann Lang, University of Rostock, provided beautiful pictures for this work. Without their highly appreciated input, this chapter would have been way less vivid. I thank Juliane Czeczor for drawing the scheme. This work was supported by BMBF-AIF, AdiPaD; FKZ: 1720X06.

References

1. Zuk PA, Zhu M, Mizuno H, Huang J, Futrell JW, Katz AJ, Benhaim P, Lorenz HP, Hedrick MH. Multilineage cells from human adipose tissue: implications for cell-based therapies. Tissue Eng. 2001; 2:211–228.
2. Eeg-Olofsson K, Cederholm J, Nilsson PM, Zethelius B, Nunez L, Gudbjörnsdóttir S, Eliasson B. Risk of cardiovascular disease and mortality in overweight and obese patients with type 2 diabetes: an observational study in 13,087 patients. Diabetologia. 2009 Jan:52 (1): 65–73. Epub 2008 Nov 5.
3. Weston JA, Yoshida H, Robinson V, Nishikawa S, Fraser ST, Nishikawa S. Neural crest and the origin of ectomesenchyme: neural fold heterogeneity suggests an alternative hypothesis. Dev Dyn. 2004; 229(1):118–130.
4. Morrison SJ, White PM, Zock C, Anderson DJ. Prospective identification, isolation by flow cytometry, and in vivo self-renewal of multipotent mammalian neural crest stem cells. Cell. 1999 Mar 5; 96(5):737–749.
5. Phillips RL, Ernst RE, Brunk B, Ivanova N, Mahan MA, Deanehan JK, Moore KA, Overton GC, Lemischka IR. The genetic program of hematopoietic stem cells. Science. 2000 Jun 2; 288(5471):1635–1640.
6. Hung SC, Chen NJ, Hsieh SL, Li H, Ma HL, Lo WH. Isolation and characterization of size-sieved stem cells from human bone marrow. Stem Cells. 2002; 20(3):249–258.
7. Lee OK, Kuo TK, Chen WM, Lee KD, Hsieh SL, Chen TH. Isolation of multipotent mesenchymal stem cells from umbilical cord blood. Blood. 2004 Mar 1; 103(5):1669–1675.
8. Zuk PA, Zhu M, Ashjian P, De Ugarte DA, Huang JI, Mizuno H, Alfonso ZC, Fraser JK, Benhaim P, Hedrick MH. Human adipose tissue is a source of multipotent stem cells. Mol Biol Cell. 2002; 12:4279–4295.
9. Winter A, Breit S, Parsch D, Benz K, Steck E, Hauner H, Weber RM, Ewerbeck V, Richter W. Cartilage-like gene expression in differentiated human stem cell spheroids: a comparison of bone marrow-derived and adipose tissue-derived stromal cells. Arthritis Rheum. 2003 Feb; 48(2):418–429.
10. De Ugarte DA, Morizono K, Elbarbary A, Alfonso Z, Zuk PA, Zhu M, Dragoo JL, Ashjian P, Thomas B, Benhaim P, Chen I, Fraser J, Hedrick MH. Comparison of multi-lineage cells from human adipose tissue and bone marrow. Cells Tissues Organs. 2003a; 174:101–109.
11. Zaffe D, D'Avenia F. A novel bone scraper for intraoral harvesting: a device for filling small bone defects. Clin Oral Implants Res. 2007 Aug; 18(4):525–533.
12. Haddouti E-M, Skroch M, Zippel N, Müller C, Birova B, Pansky A, Kleinfeld C, Winter M, Tobiasch E. Human dental follicle precursor cells of wisdom teeth: isolation and differentiation towards osteoblasts for implants with and without scaffolds. Mater Sci Eng Technol. 2009 40(10): 732–737.
13. Dominici M, Le Blanc K, Mueller I, Slaper-Cortenbach I, Marini F, Krause D, Deans R, Keating A, Prockop Dj, Horwitz E. Minimal criteria for defining multipotent mesenchymal stromal cells. The International Society for Cellular Therapy position statement. Cytotherapy. 2006; 8(4):315–317.
14. De Ugarte DA, Alfonso Z, Zuk PA, Elbarbary A, Zhu M, Ashjian P, Benhaim P, Hedrick MH, Fraser JK. Differential expression of stem cell mobilization-associated molecules on multi-lineage cells from adipose tissue and bone marrow. Immunol Lett. 2003b; 89:267–270.
15. Gronthos S, Franklin DM, Leddy HA, Robey PG, Storms RW, Gimble JM. Surface protein characterization of human adipose tissue-derived stromal cells. J Cell Physiol. 2001 Oct; 189(1):54–63.
16. Gronthos S, Simmons PJ. The biology and application of human bone marrow stromal cell precursors. J Hematother. 1996 Feb; 5(1):15–23.
17. Jin H, Aiyer A, Su J, Borgstrom P, Stupack D, Friedlander M, Varner J. A homing mechanism for bone marrow-derived progenitor cell recruitment to the neovasculature. J Clin Invest. 2006 Mar; 116(3):652–662.

18. Lévesque JP, Takamatsu Y, Nilsson SK, Haylock DN, Simmons PJ. Vascular cell adhesion molecule-1 (CD106) is cleaved by neutrophil proteases in the bone marrow following hematopoietic progenitor cell mobilization by granulocyte colony-stimulating factor. Blood. 2001 Sep 1; 98(5):1289–1297.
19. Barry F, Boynton RE, Liu B, Murphy JM. Chondrogenic differentiation of mesenchymal stem cells from bone marrow: differentiation-dependent gene expression of matrix components. Exp Cell Res. 2001; 268:189–200.
20. Pansky A, Roitzheim B, Tobiasch E. Differentiation potential of adult of human mesenchymal stem cells. Clin Lab. 2007; 53:81–84.
21. Grundy SM, Brewer HB, Cleeman JI, Smith SC, Lenfant D. For the conference participants. Definition of metabolic syndrome: report of the National Heart, Lung, and Blood Institute/American Heart Association conference on scientific issues related to definition. Circulation. 2004; 109:433–438.
22. Kahn R, Buse J, Ferrannini E, Stern M. The metabolic syndrome: time for a critical appraisal. Joint statement from the American Diabetes Association and the European Association for the Study of Diabetes. Diabetes Care. 2005; 28:2289–2304.
23. Williams DF, ed. Medical and dental materials. In: Cahn RW, Haasen P, Kramer EJ, eds. Materials science and technology. Vol. 14. Weinheim: Wiley; 2001.
24. Nicholson JW, ed. The chemistry of medical and dental materials, RSC materials monographs. London: Royal Society of Chemistry; 2002.
25. Lichtenberg A, Cebotari S, Tudorache I, Hilfiker A, Haverich A. Biological scaffolds for heart valve tissue engineering. Methods Mol Med. 2007; 140:309–317.
26. Herten M, Rothamel D, Schwarz F, Friesen K, Koegler G, Becker J. Surface- and nonsurface-dependent in vitro effects of bone substitutes on cell viability. Clin Oral Investig. 2008 Aug 8; 13(2):149–155.
27. Geerts SO, Legrand V, Charpentier J, Albert A, Rompen EH. Further evidence of the association between periodontal conditions and coronary artery disease. J Periodontol. 2004 Sep; 75(9):1274–1280.
28. Lang H, Schüler N, Nolden R. Attachment formation following replantation of cultured cells into periodontal defects—a study in minipigs. J Dent Res. 1998 Feb; 77(2):393–405.
29. Bartold PM, McCulloch CA, Narayanan AS, Pitaru S. Tissue engineering: a new paradigm for periodontal regeneration based on molecular and cell biology. Periodontology. 2000 Oct; 24:253–269.
30. Braybrook JH, ed. Biocompatibility: assessment of medical devices and materials. New York: Wiley; 1997.
31. Williams DF. Definitions of biocompatibility. Amsterdam: Elsevier; 1987.
32. Williams DF, ed. Systemic aspects of biocompatibility. Vols. I and II. Boca Raton: CRC Press; 1981.
33. Berridge MV, Tan AS. Characterization of the cellular reduction of 3-(4,5-dimethylthiazol-2-yl)-2,5-diphenyltetrazolium bromide (MTT): subcellular localization, substrate dependence, and involvement of mitochondrial electron transport in MTT reduction. Arch Biochem Biophys. 1993; 303:474–482.
34. Timothy JR. J R Army Combat casualty care. Med Corps. 2008 Sep; 154(3):214–5.
35. Bruder SP, Kraus KH, Goldberg VM, Kadiyala S. The effect of implants loaded with autologous mesenchymal stem cells on the healing of canine segmental bone defects. J Bone Joint Surg. 1998; 80:985–996.
36. Gronthos S, Graves SE, Ohta S, Simmons PJ. The STRO-1 + fraction of adult human bone marrow contains the osteogenic precursors. Blood. 1994 Dec 15; 84(12):4164–4173.
37. Ley TJ, Mardis ER, Ding L, Fulton B, McLellan MD, Chen K, Dooling D, Dunford-Shore BH, McGrath S, Hickenbotham M, et al. DNA sequencing of a cytogenetically normal acute myeloid leukaemia genome. Nature. 2008; 456:66–72.

The Potential of Selectively Cultured *Adult* Stem Cells Re-implanted in Tissues

Isgard S. Hueck, Martin Haas, Rita Finones, Jane Frimodig, and David A. Gough

Abstract For stem cell research the focus in the past has been on embryonic stem cells, cells from which humans are initially constructed, but which are limited in availability. A new direction in stem cell research is looking at a different kind of pluripotent cells – *adult* stem cells, which are responsible for maintenance and repair of tissue. To retrieve and grow stem cells remains an important challenge in medicine, but carries the hope to cure many diseases. Pluripotent *adult* stem cells have been found in many organs and also in cancer tumors. Adult stem cells can both self-renew and differentiate to replace compromised tissue in the organ where they reside. If single adult tumor stem cells can re-grow into a new tumor, just as adult stem cells can rebuild a physiological organ, the potential of *adult* stem cells would be significant in new cancer therapy and organ engineering approaches.

This chapter introduces the methods behind isolation of adult stem cells, in vitro culturing, characterization, genetic manipulation for imaging purposes, and the promising results of re-implantation into healthy tissues. Adult stem cells derived from human early-stage prostate tumors are used in a newly developed, novel research model. Since the environmental niche is an important factor for stem cell growth ex vivo, stem cell isolation is achieved via culturing in a characterized environment that mimics the stem cell niche for these types of adult stem cells. The human prostate tumor stem cell niche is described and prostate tumor stem cells (Pr*Tu*SCs) are used to show the so-called stem cell center (SCC) growth in cell culture. It is also demonstrated how Pr*Tu*SCs are epigenetically altered with eGFP for tracking and imaging purposes. Cultured Pr*Tu*SCs are characterized as *adult* stem cells by stem cell marker analysis and expression of stem cell-associated transcription factors. To further demonstrate the pluripotent potential of Pr*Tu*SC, the effects of re-implantation into healthy tissues in vivo are presented in the orthotopic xenografting and tissue recombination methods. When implanted into immune-suppressed SCID mice, results show that cultured Pr*Tu*SC not only

I.S. Hueck (✉)
Department of Bioengineering and School of Medicine, Moores Cancer Center, University of California San Diego, La Jolla, CA, USA

G.M. Artmann et al. (eds.), *Stem Cell Engineering*,
DOI 10.1007/978-3-642-11865-4_4, © Springer-Verlag Berlin Heidelberg 2011

can re-grow a cancerous prostate tumor, but, depending on the implantation site and its microenvironment, also have the ability to generate normal benign human prostate glandular structures in vivo. The dorsal mouse-skinfold window chamber method is introduced as an experimental model that allows direct observation of implanted stem cells and their behavioral characteristics, e.g., promotion of healthy or tumor-associated angiogenesis, which is critical in tissue renewal and cancer metastasis.

Clinical applications in the near future might include short-term SCC-colony counts as an in vitro surrogate method that shows capability of predicting the probability of tumor re-occurrence and progression. Finally, it is discussed that methodologies similar to this may be used to derive human tumor stem cells from biopsies or from surgical specimens of other human epithelial tumor sources, e.g., breast, colorectal, liver, and others. Using *adult* stem cell models will provide many future applications in medicine covering new strategies in cancer diagnosis and treatment as well as promising regenerative tissue and organ reconstruction approaches.

Keywords Adult stem cells · Stem cell niche · Microenvironment · Tumor stem cells · Stem cell centers · Tissue recombination method · Angiogenesis · Window chamber

1 Introduction

In the early 1960s, Korean anatomist Bong Han Kim proposed the existence of stem cells, called the Bonghan microcells, for the first time [1]. Since then, stem cells have been discussed as the originating cell for the growth of many organs and the cure for many diseases. But until the recent decade, stem cells remained a mythical cell, which everyone talked about, but which no one had reliably identified. Today, stem cells are again becoming the primary hope of curing disease in the new millennium due to modern technologies that allow isolation and characterization of stem cells. But new knowledge brings changes in the old theory: the mystical stem cell of the past is found to be an entire population of different cells. Until recent years, the main focus had been on embryonic stem cells, cells from which humans are initially constructed, but which are limited in availability for research. A new direction of stem cell research is looking at other kinds of pluripotent stem cells, *adult* stem cells, which are responsible for maintenance and repair of tissue. Just like embryonic stem cells, pluripotent *adult* stem cells can self-renew, but can also differentiate into organ-specific tissue of the organ they reside in. Adult stem cells have been characterized in many organs like the bone marrow [2], human skin [3], the cornea of the eye [4], mammary glands [5], the esophagus [6], the heart [7], the kidneys [8, 9], the lungs [10, 11], the pancreas [12], the nervous system [13, 14], the prostate [15], as well as in benign tumor organs and cancer tumors [16, 17]. If single adult tumor stem cells can re-grow into a new tumor, just as adult stem cells can rebuild a physiological organ, then the potential of *adult* stem cells in cancer therapy and in organ engineering approaches would be significant. Furthermore, the isolation of adult stem cells from early-stage tumors makes them readily available for research to study their capability of tissue renewal.

In this chapter, a newly developed novel research model using adult stem cells derived from human early-stage prostate tumors is introduced as an example for the isolation of *adult* stem cells from tumor tissues. Since the environmental niche has become one of the most important factors that alter stem cell behavior and differentiation [3] and is the key prerequisite information for stem cell culturing ex vivo [18], technologies will be introduced that explore imitation of the stem cell niche for these types of cells in vitro via a specialized growth medium as well as in vivo via re-implantation in different tissues with varying environments and components (tissue recombination methods). The differences in growth and differentiation of the implanted stem cells in two different implantation sites is surprising and supports the significance of the interaction with the environmental niche [19, 20]. Furthermore, these first results allow the hope of the possibility to manipulate *adult* stem cells via the environmental niche to create healthy tissues in regenerative tissue and organ reconstruction approaches. Finally, this technology is not limited to early-stage prostate tumor stem cells, but, in the future, might be applied in a similar way to other human epithelial tumor sources from biopsies or surgically dissected specimens.

2 *Adult* Stem Cells in Cancer

The majority of human cancers are carcinomas, composed of epithelial cells. Carcinomas arise in normal epithelial tissues by a stepwise progression from early benign lesions, which may develop into a final, often-lethal form of metastasis. Indeed, metastatic cancers are the major cause of death from cancer [21], as pre-metastatic lesions can be often cured or slowed down by surgery, hormonal intervention, chemotherapy, or radiation. Metastases of most human carcinomas are largely incurable. The tumor progression process, in which early adenomas develop in stepwise fashion into fully developed metastatic carcinomas, has been studied extensively for a century and especially since the inception of molecular and genetic techniques of modern biology [22]. Current molecular models endeavor to explain selected features of the complex tumor progression process. For instance, unlimited proliferation – a hallmark of malignant cancers but not necessarily of early-stage tumors, whether benign or potentially invasive – can be attributed to *cancer stem cells* [23]. It is thought that, following models developed by Vogelstein and colleagues for the initiation and the progression of colon carcinomas [24, 25], a sequential buildup of genetic alterations in tumor suppressor genes and oncogenes is responsible for the development of malignant metastases.

Recently, the breakdown of epithelial cell homeostasis has been linked to cancer [26]. Cancer progression has been linked to the loss of epithelial characteristics and the concomitant acquisition of an invasive phenotype. The loss by a cell of the epithelial phenotype and the gain of mesenchymal characteristics has been coined as the epithelial to mesenchymal transition (EMT). EMT is judged to be a central event in the development of malignancy [26–28]. Epithelial–mesenchymal transition is potentially reversible, and therefore cannot be explained solely by irreversible, cell-autonomous genetic alterations [29]. This implies the idea of the potential reversibility of metastases and the existence of a powerful dynamic component to human tumor progression [30, 31]. Specifically, the tumor environment has been demonstrated to be of significant regulatory role in this process [30, 32]. Individual tumors show distinct subareas of proliferation and cell-cycle arrest, epithelial differentiation and EMT, and cell adhesion and dissemination [31]. Clearly, tumor cells do not act as independent proliferation units, but tumors are heterogeneous in their morphological and functional aspects, communicating with their environment.

The experimental models introduced in this chapter demonstrate the role of the environment by isolation of human adult prostate tumor stem cells (Pr*Tu*SCs) through selective culturing in vitro and even more in the outcome of xeno-transplantation of small numbers of prostate *tumor* stem cells in different re-implantation sites in vivo [33, 34]. In this chapter, novel methods are described that selectively culture human carcinoma-derived *tumor stem cells* that fully maintain their tumor-initiating phenotype. Since the field of tumor-initiating cells (TICs) has generated many reviews in the past decade, several aspects of cell characteristics and the relationship between *adult* stem cells (ASC), tumor-initiating cells (TICs), and cancer stem cells (CSCs) need to be discussed at this point.

Tumor initiation model (Chart 1): The hallmarks of most cancer cells include (i) self-sufficiency for growth signals; (ii) insensitivity to growth-inhibitory signals; (iii) evasion of programmed cell death; (iv) unlimited replicative potential; and (v) tissue invasion and metastasis [35]. Those cell abilities such as the possibility to grow in alternative microenvironments have been appointed to cancer stem cell (CSC) characteristics and are thought to be tumor initiating [35]. Interestingly, CSCs could *only* be found in *metastases* of solid cancers and have been

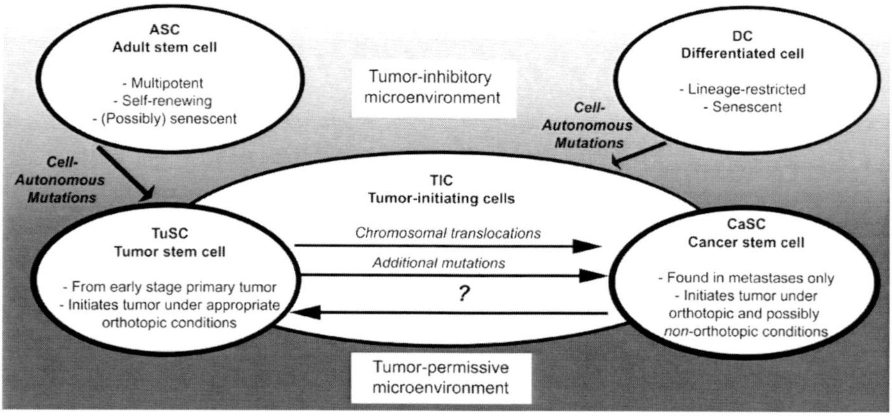

Chart 1 Model of the relationship between *adult* stem cells (ASC), tumor-initiating cells (TICs), and cancer stem cells (CSCs)

successfully isolated *only* from human *metastatic* cancer sources, namely nervous system neoplasia, colorectal cancers, prostate or hepatocellular carcinomas, and mammary cancers [35]. Whereas the frequency of CSCs was initially thought to be a fraction of 1% of a typical tumor's cells, recent evidence suggests that the frequency of CSCs may be much higher and depends on the level of immune rejection of the host [36–39]. These results indicate an important role of the microenvironment on the implanted tumor-initiating cells (TICs) and suggest that additional factors are at work to turn TICs into CSCs [35, 37]. TICs are a subpopulation of cells in a tumor, capable of self-renewal and multilineage differentiation, but need to be distinguished into two kinds of cells: (a) xeno-transplantation-capable *cancer* stem cells (CSCs), derived from metastases *only*, and (b) *tumor* stem cells (*Tu*SCs), derived from human early-stage primary solid *tumor* tissues, which are not necessarily invasive. Whereas CSCs may possess unlimited replicative potential and are "immortal" after having discarded their senescent phenotype, *Tu*SCs have a senescent phenotype [33, 34] and probably are capable of asymmetric self-renewal only. This could indicate that the senescent *Tu*SC is a precursor to the immortal CSC. *Tu*SCs would undergo continued mutation to acquire a full CSC phenotype, a view consistent with the relatively long latency between the incidence of an early-stage tumor and the appearance of a fully metastatic cancer. One example would be a breast tumor nodule prior to the involvement of draining lymph nodes and metastases. In this and similar cases, there would appear to be a need to strictly distinguish between a *Tu*SC in the local breast nodule and a CSC in a metastatic site. There also is a need to rigorously culture, isolate, and study the alterations that are associated with both these types of TICs. A working model of the hypothesis of the relationship between *adult* stem cells (ASC), tumor-initiating cells (TICs), and cancer stem cells (CSCs) are demonstrated in detail in Chart 1. *Adult* stem cells and differentiating cells (DC) maintain homeostasis within a tumor-inhibitory microenvironment of a stem cell niche. Changes in the niche interact with cell-autonomous mutations of ASC and DC to generate a tumor-permissive microenvironment that supports the growth of TICs. Results suggest that ASC in tumor-permissive conditions first undergo an identity change into localized growths of tumor stem cells (*Tu*SCs) [33, 34]. *Tu*SCs can undergo additional oncogenic insults, such as chromosomal translocations and other mutations. Compromised *Tu*SCs result in the adoption of cancer stem cell (CSC) characteristics, such as the ability to grow in alternative environments that progress the disease. Whereas cell-autonomous mutations are permanent, the reversibility of the transition from *Tu*SC to CSC remains an open question (Chart 1).

There is an apparent need to isolate, culture, and grow human carcinoma-initiating cells in a biologically active form, whether to study sequential cell-autonomous genetic alterations in the development of human carcinomas, to study genes that control EMT in the development of human malignancy, or to study the role of the microenvironment in the conduct of human tumor and cancer cells. The methods illustrated here, using human early-stage prostate cancers to isolate *adult* prostate tumor stem cells (Pr*Tu*SCs) via selection through the environmental components of a culture, can be readily applicable in modified form to the isolation and culturing of human early-stage tumors from other organs that develop carcinomas such as the mammary gland, liver, and lungs.

3 Isolation and Characterization Methods of *Adult* Stem Cells

The methodologies in this chapter are taken from the original work performed to culture adult *Tu*SCs from specimens of human *early-stage* prostate tumors obtained from radical prostatectomies [33]. In theory, specimens that are derived from metastases at the time of presentation should also be amenable to the methods described, but the isolation of human *cancer* stem cells (CSCs), per se, has been recently described by Roberts et al. [40] and will not be discussed in this context. The focus of this chapter is to explore the potential of *adult tumor* stem cells (*Tu*SCs) in cancer development as well as in promising approaches to reconstruct regenerative tissue and possibly healthy organs from *adult* stem cells that have not acquired irreversible characteristics of cancer stem cells as described earlier.

Primary early-stage human prostate tumor samples were obtained from radical prostatectomies, with care to be collected prior to fixation in formalin, as surgical pathologists follow a rigid routine of fixing surgical tissue in formalin [41]. Since live cells are needed for cell culturing, the time between tissue removal at surgery and enzymatic digestion of the tissue should be kept to a minimum, ideally within less than 2 h of prostatectomy. Prostate tissue weighing 500–1,500 mg was washed several times in sterile PBS and samples were aseptically cut into small pieces of approximately 1 mm^3. The tissue was digested overnight in a 0.2 μm filter-sterilized collagenase I (150 units/ml) solution (Sigma-Aldrich) diluted in growth medium at 37°C. Continuous stirring of 100 rev/min was applied in a 10% CO_2 environment (incubator). After 12 h of dispersion in collagenase the tissue pieces had digested into a homogeneous cell suspension and the cells were pelleted in a clinical centrifuge for 3 min at 300×*g*. The cells were frozen at ~10^7 cells/ml in 90% fetal bovine serum plus 10% DMSO (Sigma-Aldrich) and stored in liquid nitrogen until use. Routinely, each frozen vial contained the cell equivalent of 50 mg of dissected tissue.

Isolation of *adult* Pr*Tu*SCs was obtained by selective culturing of the suspended tissue. Tissue culture-treated well plates (Corning) were coated with 10 μg/ml laminin (Sigma-Aldrich) in PBS for 1 h at 37°C and washed twice with PBS to remove unattached laminin. Live-frozen prostate tumor samples were quickly thawed and pelleted in 10 ml of medium, then cultured in growth medium. The growth medium contained keratinocyte serum-free medium (Gibco) with 40 mM L-glutamine (Gibco), 12.5 μg/ml gentamycin, and 2.5 μg/ml amphotericin B, supplemented with 10 ng/ml basic fibroblast growth factor (FGF from R&D), 40 ng/ml epidermal growth factor (EGF from R&D), 58 μg/ml bovine pituitary extract (BPE from Gibco), 1 mM $CaCl_2$, and 0.025 wt% bovine serum albumin (BSA from Sigma-Aldrich). Cultures were incubated at 37°C in a 10% CO_2 and 5% O_2 environment and the medium was changed daily with care not to lose floating cells. Routinely, cells representing 50 mg of tissue were plated in one to three 6-well cell clusters. By adjusting cell plating in a number of 6-well tissue culture clusters, the number of *Tu*SC colonies could be adjusted and a pre-designed number of *Tu*SC colonies could be obtained. For example, a well with only one colony or any predetermined number of colonies can be trypsinized and used for analysis in vitro or for xeno-transplantations in vivo. Fast-growing epithelial colonies developed from plated tumor samples after approximately 1 week and were subcultured at a ratio of 1:3 upon reaching ~90% confluence.

Early Pr*Tu*SC colonies at a stage of only a few hundred cells can be infected with GFP-encoding viruses (titer ~2–5 × 10^6), if green fluorescent protein (GFP) labeling of *Tu*SCs is required, e.g., in xeno-transplantation experiments. Cell cultures used for in vivo orthotopic xenograft assays were retrovirally infected with GFP retrovirus. Monolayer cultures were infected with viral supernatant and 10 μg/ml polybrene for 45 min at 900 RCF (Beckman TJ-6, TH-4 rotor) and incubated at 37°C for 20 min, followed by aspiration of excess viruses and replenishment with growth medium. Infection with serum-containing retrovirus supernatants needs to be limited to a short time, e.g., 45 min, as the stem cells readily differentiate and cease to proliferate in the presence of even 2% serum for a few hours. Once the Pr*Tu*SCs are differentiated by serum, the differentiation step cannot be reversed and the differentiated cells *cannot* be further grown. Infected cells were *not* selected by antibiotic resistance, to maintain integrity of primary cultures. Efficiency of GFP infection was ~50% by visual assessment of fluorescent cells under the microscope.

*Stem cell centers (*SCCs*)*: Single colonies of epithelial cells are first observed about 3 days after plating into growth medium, and at 7 days the number of colonies was stable and did not increase further. Seven-day colonies have approximately 20–200 cells each. Significantly, out of 50 early-stage prostate tumor samples, *only* samples that were pathologically classified as carcinomas gave rise to epithelial stem cell colonies. All cultures from samples classified as adenocarcinomas generated between a few (2–5) and many (200–500) single colonies per 50 mg of surgically dissected prostate tissue. There was no apparent correlation between the number of stem cell colonies generated after 7 days of culturing and the differentiation grade (Gleason Score) of the individual carcinoma cases. On the other hand, all cultures from samples classified as benign prostatic hyperplasia ($n = 32$) generated zero stem cell colonies under the specific culture conditions described. Therefore, it can be concluded that physiological prostate stem cells, as are present in benign hyperplasia cases, do not yield epithelial stem cell colonies under these conditions. With no exception, the results suggest that the stem cell colonies grown from cultures of prostate cancer samples are specific for *tumor* stem cells (*Tu*SCs) and the culture conditions used are highly selective for the prostate *tumor* stem cell niche.

Several independent examples of individual prostate tumor stem cell (Pr*Tu*SC) colonies are shown in Fig. 1. Pr*Tu*SC colonies appear as a tight honeycomb of prostate epithelial cells. It can be suggested that early-stage carcinomas of other human tumor organs may produce very similar types of stem cell colonies. The tight epithelial colonies maintained close cell–cell contact through E-cadherin bridges. Only cells that were part of the tight epithelial colonies continued growing in the cultures. Single cells that attached to the laminin-coated culture dish without contact to neighboring epithelial cells did not grow, but underwent apoptosis and were lost from the cultures. The E-cadherin bridge-mediated cell–cell contact is a sine qua non in all colonies, without exception. Each epithelial stem cell colony had a prominent feature that appeared in the presence of a small, compact group of

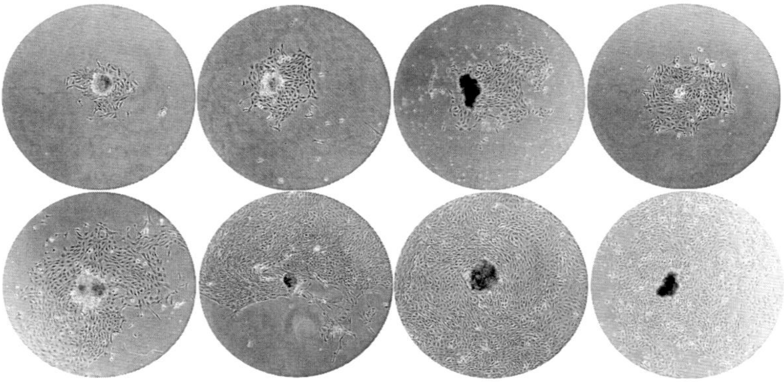

Fig. 1 Phase contrast images of eight separate prostate tumor stem cell (Pr*Tu*SC) colonies at 7 days after culturing in selective tumor stem cell conditions

cells, called a stem cell center (SCC). It was observed that every epithelial cell colony growth started in an SCC and expanded from there outwards (Fig. 2). No epithelial cell colony growth was initiated without roots in an SCC. Over several generations, SCCs were gradually lost from the cultures by differentiation into prostate epithelial cells and fewer SCCs were observed with each cell passage. Nevertheless, colonies maintained consistently epithelial cells and continued to divide by showing many mitotic figures in later passages. Epithelial cell growth rates were established by image analysis of specific colonies at successive time points. Pr*Tu*SC colonies divided with a doubling time of 20 h, as shown in Fig. 2. Adherent cultures were continually passaged until cells began to adopt the typical large, flattened morphology associated with senescence and the growth rate of the cells started to decline. Late-passage cells (after the eighth passage) stained positive

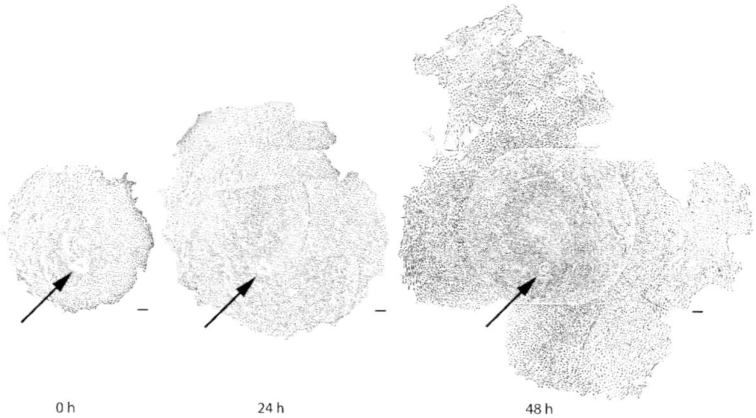

Fig. 2 Growth of a single *adult* prostate tumor stem cell (Pr*Tu*SC) colony at 0, 24, and 48 h after first observation of colony development. *Arrows* indicate stem cell centers (SCCs) in the same prostate tumor stem cell colony over time

for senescence-associated beta-X-galactosidase, indicating the onset of replicative senescence and cell death (data not shown) [33]. In summary, evidence is presented that supports the notion that in cell culture selectively isolated prostate stem cells are prostate *tumor* stem cells (Pr*Tu*SCs), and that each colony originates from a small stem cell niche, the so-called stem cell center (SCC).

Hypothesis of senescence: It is important to emphasize that SCC/Pr*Tu*SC colonies were associated exclusively with prostate early-stage carcinoma tissue and were absent from normal prostate tissue. Without exception, each Pr*Tu*SC colony originated from a SCC and cells in the SCCs differentiated into epithelial cells which grew consistently for eight passages, produced $1–2 \times 10^9$ cells, and then senesced. Hence, *adult* stem cells in early-stage human prostate tumors maintain the senescent phenotype of human somatic cells. *We propose that all human adult stem cells may possess a senescent phenotype.* In opposition to human embryonic stem cells, *adult* stem cells reside in the organs through the adulthood of humans and if physiological human *adult* stem cells from any organ were immortal and not kept under tight regulatory control by their niche, they would kill the host eventually, just like metastatic cancer stem cells [21]. There is no evidence in the literature for the symmetric self-renewal of human *adult* stem cells from any organ. The notion that human *adult* stem cells are of unlimited proliferation capacity, in other words immortal, is deeply ingrained in the scientific community. Although pervasive, it is probably erroneous. This characteristic is only appointed to metastatic *cancer* stem cells. As data presented suggest, even *adult* stem cells isolated from human early-stage prostate *tumors*, which have tumor-initiating potential and are not considered physiological, do not possess unlimited proliferation capability but possess a senescent phenotype [33]. More research is needed in the future, but the hypothesis of a declining number of *adult* stem cells through life might explain the declining capability of self-renewal and tissue repair in the human aging process.

Characterization of adult stem cells: Cultured cells from colonies grown from early-stage prostate tumors were characterized as adult stem cells by stem cell marker analysis and expression of stem cell-associated transcription factors. Since there were many cells in a growing Pr*Tu*SC colony, evidence was needed to distinguish between bona fide adult tumor stem cells and epithelial cells in the process of differentiation. One line of evidence relies on the expression of cell markers typically ascribed to and expressed by *adult* stem cells. Single colonies grown on laminin-coated glass cover slips were assayed for cell marker expression and associated transcription factors by RT-PCR, flow cytometry, and immunocytochemistry imaged with confocal microscopy [33]. Results are summarized in Fig. 3.

RT-PCR results indicated that cells from cultured Pr*Tu*SC colonies expressed the transcription factors Oct-4 and Sox-2, which are ascribed to *stem cell pluripotency* (Fig. 3g). Furthermore, a gene associated with the maintenance of *adult* stem cell renewal, Bmi-1, was expressed at levels similar to the positive control (Fig. 3g). As positive control, cells from a human teratoma cell line NTERA were used. SCC/Pr*Tu*SC colony cells did not express the *embryonic* stem cell marker hTERT (human telomerase reverse transcriptase) or the transcription factors Rex-1 and Nanog, which are associated with undifferentiated *embryonic* stem

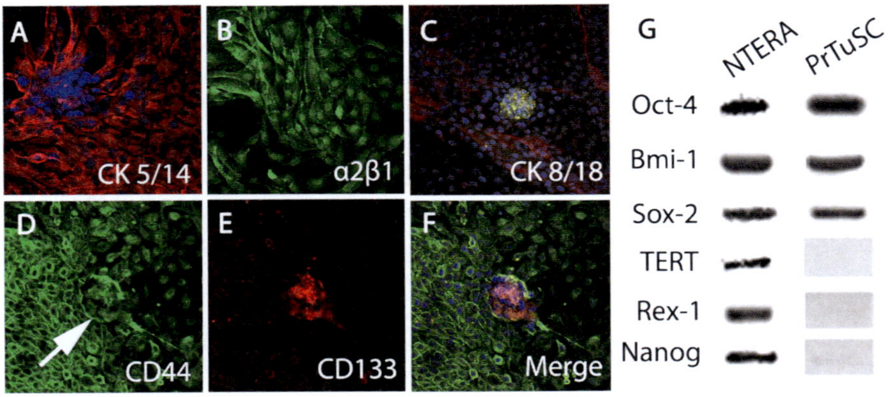

Fig. 3 Immunofluorescent stainings of cultured prostate tumor stem cell (Pr*Tu*SC) colonies with (**a**) CK5/14 cytokeratin marker, (**b**) Integrin $\alpha_{2\beta 1}$ marker, (**c**) CK8/18 cytokeratin marker, (**d**) CD44 cell maker, (**e**) CD133 cell marker, (**f**) CD44 and CD133 merged with nuclear DAPI (*blue*). Confocal microscopy images show that *all* cultured colony cells express CK5/14, Integrin $\alpha_{2\beta 1}$, and CD44 markers, whereas only cells in the SCCs (*arrow*) express CD133. CK8/18 is expressed preferably by cells in the close surroundings of SCCs, indicating that the actual Pr*Tu*SCs are located in the SCCs and from there originated cells differentiate into proliferating prostate epithelial cells in the surroundings of SCCs. (**g**) RT-PCR results from cultured Pr*Tu*SC colonies: cell markers associated with *adult* prostate tumor stem cells (Pr*Tu*SCs) on the *right* are compared to a positive control of NTERA cells on the *left*

cells (Fig. 3g). This detail is compatible with the senescent phenotype of *adult* Pr*Tu*SCs.

Flow cytometry results (data not shown) indicated that SCC/Pr*Tu*SC colony cells from early passages expressed the prostate *stem* cell markers CD44, CD133, CK5/14, and Integrin $\alpha_{2\beta 1}$ in virtually 100% of the cells [33]. Cells from colonies of continuing passages continued to express the *stem* cell markers CD44 and Integrin $\alpha_{2\beta 1}$, but the expression of high molecular weight cytokeratin CK5/14 and of CD133 was reduced with increasing passages in culture. Finally, expression of CD133 was lost completely in cells from later cultures [33]. CD133 (and also cytokeratin 5 in a lesser degree) is known to be expressed *only* by stem cells with a high renewal capacity and low proliferation index [42]. The loss of expression of CD133 over the course of multiple passages complied with the loss of bona fide adult stem cells and SCCs from later cultures and the differentiation of Pr*Tu*SCs into highly proliferating epithelial cells that was described earlier.

Immunocytochemistry results observed by confocal microscopy clarified further the decrease of CD133 expression in cells of later passages and the location of remaining cells with positive expression within the colony [33]. SCC/Pr*Tu*SC colonies were stained for the prostate *stem cell* markers CD44, CD133, cytokeratin CK5/14, and Integrin $\alpha_{2\beta 1}$, as well as the *differentiation* marker cytokeratin CK8/18. Markers from the cytokeratin 18 group are known to be expressed *only* by highly proliferating epithelial cells with little or no renewal capacity [42]. Data were recorded by confocal immunofluorescent microscopy techniques and are presented in Fig. 3a–f. Colonies from early passage cultures containing both SCCs and proliferating cells, which had grown out of a SCC into its periphery, expressed CK5/14 and Integrin $\alpha_{2\beta 1}$ in 100% of the cells (Fig. 3a, b). This confirmed that the selective cultured cells from early-stage prostate tumors were indeed adult prostate tumor *stem* cells with high renewal capacity. Cells of the stem cell center (SCC) region expressed the CK8/18 cytokeratin marker, associated with differentiated prostate luminal cells, only at a very *low* level (Fig. 3c), whereas the proliferating epithelial cells surrounding the SCCs expressed this marker at a significantly higher level. Nuclei were counterstained with DAPI (blue). These findings indicated that cells that originated from SCCs are capable of differentiation into prostate luminal cells. Remarkably, the prostate *stem* cell marker CD44 was expressed in all cells located in the SCCs as well as in the periphery of the cultured colonies (Fig. 3d, arrow indicates an SCC), whereas CD133 expression was negative in the epithelial cells that composed the majority of the colony, but was always localized in the tightly packed SCCs (Fig. 3e). Double-staining of CD133 with CD44 revealed the preferential expression of CD133 in SCCs (Fig. 3f). The localization of CD133-expressing cells in SCCs indicated that bona fide adult prostate stem cells were preferably located in the SCCs and the differentiation of Pr*Tu*SCs into proliferating prostate epithelial cells was taking place in the periphery of the SCCs. Other cell markers like VEGFR1 and c-kit, which are receptors for vascular endothelial growth factors and hematopoietic cytokines that promote angiogenesis (new vessel growth), were also co-expressed (data not shown) by colony cells directly around the SCCs [33]. VEGFR1 was also expressed in cells directly in the SCCs, whereas c-kit expression was limited to cells outside the SCCs in the close surroundings [33]. To be able to promote angiogenesis is a necessary characteristic for all tissue growth and in particular new growth originating from stem cells. The importance of angiogenesis will be discussed further in the window chamber method, a tool that can be used to evaluate and observe angiogenesis in vivo.

In summary, cell marker analysis and immunofluorescent confocal microscopy images showed that cells directly in the SCCs expressed the following *stem* cell markers: CD133, CD44, CK5/14, Integrin $\alpha_{2\beta 1}$, and VEGFR1. Marker expression by cultured colony cells in the SCCs is therefore compatible with the hypothesis that the bona fide adult prostate tumor stem cells are localized in the cultured prostate tumor stem cell niche, the SCC. These niche-contained stem cells then differentiate into rapid proliferating prostate epithelial cells, growing out of the SCCs into their surroundings, accompanied by the rapid loss of CD133 and VEGFR1 markers in cells outside the SCCs and in the periphery of the colonies in later cell passages. Besides the down-regulation of prostate *stem* cell markers, cell differentiation into prostate epithelial cells is accompanied by acquisition of the expression of c-kit markers outside the SCCs, which documents the natural up- and down-regulation of specific cell markers during differentiation of *adult* stem cells into rapidly proliferating functional epithelial cells that have been described in prostate tissue [43, 44].

It has been shown that *adult* stem cells can be isolated by carefully selected culture conditions of the stem cell niche. This method can be applied to adult stem cells of any origin. Furthermore, results indicate that *adult* stem cells possess a senescent phenotype, which makes them different from embryonic stem cells. An array of thoughtfully chosen stem cell markers and associated transcription factors are useful tools to identify the bona fide adult stem cells in cultured stem cell colonies. Finally, a distinction has to be made between metastatic *cancer* stem cells and adult *tumor* stem cells, found in early-stage primary tumors, which do not necessarily have to differentiate into a fully developed cancer, but can also have the potential to grow healthy tissues depending on the interaction with the environment of the stem cell niche. Re-implantation of cultured *adult* stem cells into tissues in vivo will explore some of the potential of *adult* stem cells in varying environments of different implantation sites. These methodologies will be introduced using the example of *adult* prostate tumor stem cells (Pr*Tu*SCs) in the following paragraphs.

4 Effects of *Adult* Stem Cell Re-implantation into Tissues

The cultured SCC/Pr*Tu*SC colonies displayed a significant group of distinctive prostate *stem* cell markers and stem cell-associated transcription factors without pre-selection beyond the culture conditions used, as well as differentiative behavior in culture. Another line of evidence to demonstrate that they are genuine stem cells, whether of tumor or normal phenotype, is to show their activity as stem cells in vivo. Therefore, cells from cultured SCC/Pr*Tu*SC colonies have been re-implanted into healthy tissues of severe combined immunodeficient (SCID) mice with two different xenografting methodologies. In the tissue recombination method, developed by Cunha and colleagues [45–47], prostate tissue cells have been grafted into the subcapsular area of the kidneys in SCID mice. Since the kidney is one of the most vascular organs in the body, the renal capsule implantation site has been known to grow organ rudiments in vivo and to maintain adult tissues for extended periods of time [48]. The interaction with embryonic rat urogenital mesenchyme (UGM) tissue via tissue recombination in this implantation site seems to play an important environmental role to bring out phenotypes of implanted cells and the development of organ tissues [47, 48]. The tissue recombination grafting method into the renal capsule presents a well-suited implantation site to explore the phenotype and the potential of organ development from prostate *tumor* stem cells (Pr*Tu*SCs), which were isolated, characterized, and cultured from early-stage prostate tumors. Another xenografting method lies in the re-implantation of Pr*Tu*SCs into the stem cell's origin organ: the prostate gland with its natural environment. In the orthotopic xenografting method the prostate *tumor* stem cells are grafted directly into the anterior prostate of recipient SCID mice [49] and observed for their development. Results of both re-implantation methods are presented in Fig. 4.

The tissue recombination method and implantation into the renal capsule. The biological activity of the SCC/Pr*Tu*SC cultures was tested in xenograft assays in vivo by tissue recombination with embryonic rat urogenital mesenchyme (UGM) and implantation under the kidney capsule of SCID mice. Tissue recombination

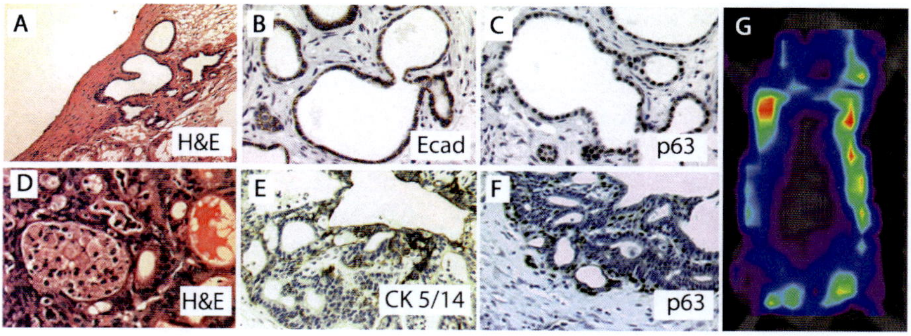

Fig. 4 Histology from in vivo assays. With the tissue recombination method and implantation into the renal capsule of SCID mice, cultured Pr*Tu*SCs gave rise to physiological benign, simple epithelial glands: (**a**) H&E staining, (**b**) benign glands stained positive for E-cadherin, confirming their epithelial nature, and (**c**) expressed the *prostate basal cell* marker p63. The orthotopic xenografting method and implantation into the anterior prostate gland generated features associated with typical human prostate cancer: (**d**) H&E staining, (**e**) cancer glands showed partly positive expression of the *stem cell* marker cytokeratin CK5/14, as well as (**f**) of the *prostate basal cell* marker p63, confirming that the cancer structures must have initiated from the implanted Pr*Tu*SCs. (**g**) Representative in vivo fluorescent image of an orthotopically engrafted SCID mouse after 7 weeks (head of the mouse in the *center top* and tail in the *center bottom*): eGFP-labeled Pr*Tu*SCs are localized in the prostatic lobes (*bottom left* and *right* of tail) and numerous auxillary lymph nodes along the sides of the body

with UGM mimics the embryonic microenvironment, in which the urogenital sinus develops into urinary and reproductive organs, and has proven its utility in enabling human prostate cells to differentiate in vivo. The resulting renal grafts were analyzed by standard histological methods 12 weeks after grafting. A total of 20 SCC/Pr*Tu*SC colonies were combined and hybridized with 10^5 UGM cells in a collagen matrix and grafted under the kidney capsule of recipient male SCID mice. Twelve weeks after grafting, the mice were killed and their tissues prepared for histological and immunocytological analyses (Fig. 4a–c). Classical histological H&E staining revealed the formation of physiological benign simple glandular structures in the recipients' kidney capsules with a single layer of epithelium (Fig. 4a). The glands were of human origin, not mouse or rat tissue, as shown by the use of a human mitochondria-specific monoclonal antibody staining (data not shown) [33]. Contrary to the expression profile of SCC/Pr*Tu*SC colony cultures in vitro, the glands were negative for the cell markers cytokeratin CK5/14, CD44, and CD133 (data not shown), which are typically expressed by *stem cells* with high renewal capacity, suggesting that in the process of differentiation and gland formation the *stem cell* markers were down-regulated [33]. Considering the inductive microenvironment of the kidney capsule, the results indicate that prostate stem cell cultures have progressed toward differentiation into prostate epithelial luminal cells. Positive expression of E-cadherin bridges confirmed that cells were of epithelial character (Fig. 4b). Glands were negative for prostate-specific antigen (PSA), indicating that these glandular cells did not yet secrete the luminal cell product of mature prostate epithelium (data not shown) [33]. This was expected, since the culture medium before implantation did not contain any detectable levels of androgen, which is needed to mediate the secretory function of prostate basal cells [50, 51]. Nevertheless, a high percentage of cells that lined the simple glands were positive for the *prostate basal cell* marker p63, which is associated with the proliferative potential of prostate basal cells and is required for prostate development (Fig. 4c) [52, 53]. Although the engrafted SCC/Pr*Tu*SCs were grown from pathologically diagnosed, early-stage prostate adenocarcinomas, the grafted simple glands did *not* express the anti-human coenzyme AMACR (data not shown) [33], which is strongly expressed by malignant human prostate tissue and considered a biomarker for prostate cancer [54].

In summary, histological results from the tissue recombination method confirmed the capacity of cultured SCC/Pr*Tu*SCs to generate human glands, as expected of bona fide adult prostate stem cells. Immunocytological results confirmed that human epithelial prostate glands were grown from cultured Pr*Tu*SCs and were of simple, physiological, benign appearance, although the cells originated from early-stage prostate carcinomas. If physiological benign glands can be generated in vivo from *tumor* stem cell material by simply placing the *adult* tumor stem cells in the correct environment, this methodology has a future for promising regenerative tissue and organ reconstruction approaches. Nevertheless, these experiments are only first results and more research is needed to determine the exact components of the environmental niche to possibly grow healthy tissues. However, when renal subcapsular grafts of Pr*Tu*SCs were recombined with embryonic UGM cells, the microenvironment imposed on the SCC/Pr*Tu*SCs led to apparent physiological benign differentiation of the implanted tumor stem cells [33, 55, 56]. In comparison, the differentiation of cultured SCC/Pr*Tu*SCs under conditions that did not impose embryonic mesenchyme-derived growth signals on the implanted Pr*Tu*SCs needed to be explored by implantation into the organ of cellular origin: the prostate gland.

The orthotopic xenografting method and implantation into the anterior prostate. Human early-stage prostate tumor-derived SCC/Pr*Tu*SC cultures were implanted into the anterior prostate glands of SCID mice. In orthotopic xenograft assays, *only* cultured SCC/Pr*Tu*SCs were engrafted; no other cells were injected in opposition to the tissue recombination method, where hybridized rat UGM-mesenchymal cells were co-injected. A total of 20 SCC/Pr*Tu*SC colonies, each containing about 100 epithelial cells surrounding an SCC of tightly packed additional 30–50 Pr*Tu*SCs, were grafted into anterior prostates of recipient male SCID mice. Four weeks after cell implantation, engrafted mice showed evidence of abdominal growths. At 7 weeks all engrafted mice with SCC/Pr*Tu*SCs were moribund and showed evidence of the presence of large abdominal tumors, up to a total mass of 12–15% of their body weight. The mice were killed and their tissues prepared for histological and immunocytological analyses (Fig. 4d–g). Classical histological H&E staining revealed features that clearly resembled typical human prostate cancers (Fig. 4d). Cancer glands stained partly positive for the high molecular weight *stem* *cell* marker cytokeratin CK5/14 (Fig. 4e) as well as the human *prostate basal cell* marker p63 (Fig. 4f), confirming that the cancer structures must have initiated from human *prostate tumor stem cells* (Pr*Tu*SCs), the cells that had been cultured and re-implanted. Prostate cancer structures also expressed strongly the anti-human coenzyme AMACR (data not shown) [33], a biomarker for malignant prostate cancer [54]. Figure 4g shows an in vivo fluorescent image of an orthotopically engrafted SCID mouse after 7 weeks. SCC/Pr*Tu*SCs were labeled in cell culture with eGFP (enhanced green fluorescent protein) for imaging purposes prior to in vivo implantation as described earlier. The GFP-positive SCC/Pr*Tu*SCs are localized at the sites of orthotopic engraftment, the prostatic lobes, and have spread to numerous auxillary lymph nodes along the sides of the body (Fig. 4 g).

In summary, histological and immunocytological results from the orthotopic xenografting method showed that in the anterior prostate, implanted SCC/Pr*Tu*SCs differentiated rapidly into a fully developed metastatic prostate cancer. This confirms the tumor-initiating phenotype of the implanted *adult* tumor stem cells and yet is another indication of the importance of the communication with the environment of the stem cell niche.

SCC/Pr*Tu*SC colonies were selectively grown only from human early-stage prostate tumors. Naturally tumors of this kind may result in malignant carcinomas *only after years* of observation or not at all, two possible outcomes, which modern medicine is not able to predict in advance. In this case of cultured re-implanted SCC/Pr*Tu*SC colonies, the outcome of *adult* stem cell differentiation was, on the one hand, a promptly cancerous and invasive prostate carcinoma, when very small numbers of the Pr*Tu*SCs were engrafted into the natural, orthotopic environment. On the other hand, when the same Pr*Tu*SCs were engrafted sub-renal combined with UGM cells in the tissue recombination method, the differentiation of *adult* stem cells yielded physiological benign simple human prostate glands. These results confirmed how highly consequential the importance of the role of the microenvironment is in the fate of *adult* stem cells and the development of *tumor* stem cells in vivo. These experiments also demonstrate that *adult* tumor stem cells are a new category of pluripotent human stem cells, of which the potential might be far-reaching in medicine for the prediction, prognosis, intervention, and treatment in cancer therapies.

Possible future applications in cancer diagnosis, for instance a prediction of the probability of tumor progression with SCC counts or how to create an immortal cell line to study the behavior and communication of *adult* stem cells with their environment in future research approaches, will be discussed in potential applications below. If it is important to study the behavior of *adult* stem cells, which specifically might be a target for cancer-inhibiting therapies such as anti-angiogenic treatments, a method to observe their ability to induce angiogenesis (new blood vessel growth) must be incorporated. An established technology for observation of angiogenesis and the behavior and differentiation of implanted cells in vivo regains importance in the new light of stem cell research: the dorsal mouse-skinfold window chamber.

5 The Dorsal Mouse-Skinfold Window Chamber Method in Stem Cell Research

The dorsal skinfold window chamber is not a new tool, having been developed over 65 years ago for the in vivo study of early cancer growth in mice [57, 58]. Subsequently, it has been used for the study of biocompatibility [59–63], ischemia/reperfusion injury [64, 65], lymphatics [66], biosensors [67, 68], cancer therapies [69–72], microvasculature [73, 74], and tumor angiogenesis [75–85]. The dorsal skinfold window chamber technique has been adapted for use in various animal models such as nude mice, laboratory rats [82], and hamsters [86]. The window chamber model is one of the few tools that allow detailed observations of the behavior and differentiation of implanted cells in vivo, as well as communication of implanted cells with natural or injected factors of the microenvironment. For example, a quantitative and qualitative evaluation of angiogenesis, which is induced by the implanted cells in the window chamber, can be directly observed, measured, and evaluated over time. Angiogenesis is one of the most important measurement tools to quantitatively evaluate the potential of cells to generate a complex tissue, as well as to qualitatively determine the type of generated tissue, as cancer tissue creates an unorganized, fast-growing blood vessel supply. In opposition to cancer-associated angiogenesis, physiological tissue development induces the growth of an orderly, fully developed mature blood vessel supply. The implantation of Pr*Tu*SCs into the dorsal skinfold window chamber would reveal more insights into the kind of vessel supply these stem cells will initiate in vivo and how well they can survive. In general, a modified dorsal mouse-skinfold window chamber technique can experience new life when utilized in the exciting new field of stem cell research [87, 88], and in particular to study cultured *adult* stem cells, their microenvironment dependencies, and their ability to initiate new blood vessel growth (angiogenesis). Without angiogenesis, new tissues, either tumor tissues or physiological organ tissues, *cannot* develop, repair, or survive [85, 89].

The importance of evaluation of angiogenesis initiated by stem cells: Most cells must be within 100–200 μm of a blood vessel to receive the necessary nutrients and oxygen for survival [89, 90]. Therefore, without a blood supply, stem cells and tumor stem cells would be unable to proliferate and differentiate adequately to form organs and tumors, respectively. Angiogenesis is the sprouting and subsequent development of new vessels from existing blood vessels. It occurs as a normal physiological process in wound-healing or female reproduction, natural events, wherein stem cells play a major role as well. Besides vascular and rheumatoid diseases, it is also seen in diseases such as diabetes [91], atherosclerosis [92, 93], macular degeneration [94], and cancer [95]. Other mechanical mechanisms, such as hypoxia [96–98], high blood pressure [99, 100], and also low blood pressure or blood stasis as seen in shock or stroke [101, 102], set into action molecular signaling pathways inducing increased angiogenesis. Most angiogenesis-inducing physiological and pathological circumstances go hand in hand with *adult* stem cell activation as the organs respond with repair or compensation strategies. Furthermore any kind of transplant, whether cellular, tissue, or whole organ implantation, must be supplied with adequate vascularization. Therefore, a graft of *adult* stem cells in varying implantation sites or into the dorsal skinfold window chamber has to induce angiogenesis as a sign of survival of the implanted cells. How fast, organized, and complete the vessel supply is sprouting toward the implanted adult stem cells reveals information about the type of tissue that might differentiate from the implanted stem cells, since these are the cells which provide the chemotaxic signals for angiogenesis to happen. Hence, the quantitative and qualitative evaluation of angiogenesis induced by *adult* stem cells brings knowledge about the potential of tissue that can be grown from these stem cells and may bring new insights into how stem cells interact with their environment.

To measure angiogenesis is even more important in cancer stem cell studies, since cancer tumors have a high demand of nutrients and metabolites, which need to be supplied by a blood vessel system. Cancer cells proliferate rapidly and often outgrow the local blood vessel supply. In order to keep the tumor supplied with oxygen and nutrients, the cancer cells need to be able to actively promote angiogenesis in parallel with tumor growth. Cancer tumors do not grow well at more than 100 μm distance from the blood vessel supply [89, 90]. Angiogenesis in prostate cancer has been correlated with tumor progression [103–105]. In fact, many new cancer therapies are being developed that focus on attacking the vasculature rather than the cancer cells directly, in order to "starve" the tumors [106, 107]. Angiogenesis is also critical in metastatic cancers as this is the path by which metastases often occur. Therefore, it is important to have an in vivo tool like the window chamber model in place that allows direct observation in order to study angiogenesis in response to the presence of *cancer* and other stem cells.

Since *adult* stem cells possess the ability to grow regular as well as cancerous glandular structures, as shown in the example of Pr*Tu*SCs described earlier [33], more details about the environmental factors that lead to the development of a specific kind of tissue can be investigated with the modified dorsal skinfold window chamber method. The ability to induce regular or tumor-associated angiogenesis is an important behavioral cell characteristic worth evaluating with this technique. Once the growth of implanted stem cells is established in the window chamber, many factors can be researched by injecting stem cell niche-altering components into the window chamber area. In summary, the dorsal skinfold window chamber method remains a viable tool for in vivo real-time visualization of implanted cells and their initiation of angiogenesis, because it allows observation with minimal damage to existing tissue and without exposing the tissue to the outside environment. The window chamber can be maintained for a long period of time with up to 6 weeks of in vivo observations, if installed under sterile conditions.

Implantation of human adult stem cells into the dorsal mouse-skinfold window chamber. Since nude mice are a bred, immune-suppressed species with furless transparent skin, this animal model was used for implantation of *adult* stem cells into the window chamber (Fig. 5). Makale et al. has provided a detailed description of the window chamber surgery [108, 109], so that the surgery technique will only be briefly summarized in this chapter. Nude mice were anesthetized using 50 mg ketamine and 1 mg medetomidine per 100 g animal body weight, injected

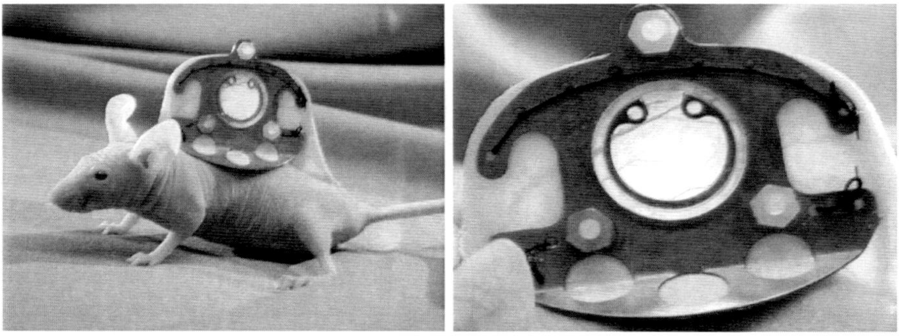

Fig. 5 Dorsal skinfold window chamber implanted onto a nude mouse. In the close-up image of the chamber, underlying blood vessels of the exposed inner layer of skin muscle (M. cutaneous max.) are visible through the cover glass in the center of the window chamber

intraperitoneally. The window chamber consists of two matching titanium halves (Fig. 6a), between which a fold of the elastic back skin of the mouse is sandwiched (Fig. 5). The right side of the window chamber (Fig. 6a right) shows a small technical modification with three nylon bolts held in place with threaded stainless steel spacers (about 0.5 mm thick) to prevent the sandwiched skinfold or any implanted constructs from being squeezed in the window chamber. Before surgery of the skin layers was started, triple antibiotic ointment was applied to the skinfold. The surgery was performed under sterile conditions to minimize the risk of infection. A circle of 15 mm diameter was carefully removed from the top skin layer down to an inner facia layer with small surgical and iridectomy scissors. This procedure leaves the

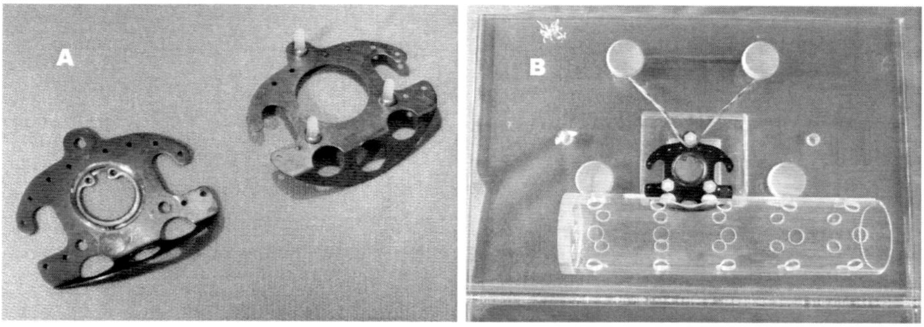

Fig. 6 Dorsal skinfold window chamber with (**a**) two matching titanium halves. The left half side shows where the cover glass is held in place by a retaining ring. The right half side shows three nylon bolts (about 0.5 mm long) held in place by threaded stainless steel spacers to prevent the sandwiched skinfold from being squeezed. (**b**) Imaging stage apparatus for microscopic observation of the window chamber. The acrylic tube with the window chamber protruding through a slit on top is used to immobilize the mouse during imaging. The window chamber is fixed over an acrylic block with a square hole in the middle to be stabilized under the microscope. Microscope imaging stage was a courtesy of Rui Juan Xiu, Institute of Microcirculation, Chinese Academy of Medical Sciences, Peking Union Medical College

inside layer of skin muscle (M. cutaneous max.) exposed and gives a clear view of the underlying blood vessels (Fig. 5). During surgery, the exposed tissue was irrigated with sterile saline containing 20% of the antibiotic tobramycin. Finally, the second half of the window chamber, holding a circular cover glass in its center (Fig. 6a left), was used to seal up the surgery area airtight. The exposed muscle layer and its vasculature can be viewed through the cover glass (Fig. 5). The window chamber was fixed permanently to the skin by sutures at the top and sides of the chamber. Animals were carefully observed during the recovery process and daily after the surgery to ensure minimal animal treatment stress. Three days after surgery the mice were anesthetized again as described before, and the cover glass, being held in place by a retaining ring, was temporarily removed. Cultured prostate *adult* tumor stem cells (Pr*Tu*SCs) on a collagen disk were placed into the window chamber under sterile conditions directly on the exposed muscular layer of the inner skin, before sealing the window with a new sterile cover glass. Implanted cells in the window chamber could be observed immediately and under in vivo conditions with a microscope. Figure 6b shows the complete imaging stage apparatus for microscopic observation of the window chamber. Once the chamber is installed onto the back skin of the mouse, the mouse is immobilized in the acrylic tube with the window chamber protruding through a slit on top of the tube (Fig. 6b). To keep the window chamber stable during microscopy, it is fixed over an acrylic block with a square hole in the middle (Fig. 6b) and can be viewed with a long-distance objective on a specially designed microscope stage. Often, the animals go to sleep in the tube while being imaged under the microscope.

Even though there are technical modifications required in the window chamber method, there are several reasons to implant human *adult* stem cells into the window chamber in a construct rather than directly as cell suspension. In general, immune protection of the implanted human cells in an animal model is a concern and can be maximized if the cells are located in a construct with a matrix, but in the case of immune-suppressed nude or SCID mice as a host this can be disregarded. Furthermore, encapsulation of the implanted stem cells in a construct is useful to study paracrine effects on the host tissue, since the matrix inhibits direct cell–cell contacts with host cells or unwanted cell incorporation effects as seen in macrophages. Finally and most important for quantitative observation of angiogenesis is to keep the implanted cells localized within the chamber, in order to measure the vessel sprouting toward the construct or disk over time. Unconstrained implanted cells tend to migrate in the window chamber, which makes it difficult to measure angiogenesis in regard to the implanted cells. In the past, implanted cells or angiogenic growth factors in the window chamber have been contained in alginate or collagen matrixes [110–112], which can be modeled to relatively small construct sizes. Since the nude mice model needs a smaller modified variation of the window chamber due to the small animal size, the construct *cannot* exceed 450 μm in thickness. Constructs thicker than these space limitations tend to either be crushed or put pressure upon the blood vessels underneath creating a hypoxic region. Furthermore, the construct should be small enough in spread to leave a region of unaffected vasculature in the window chamber some distance away from the implant as negative control region. Last but not least, any construct material naturally has to be biocompatible and should not have properties that might unduly affect cell growth. The implantation construct, used for human *adult* Pr*Tu*SC grafts into the dorsal skinfold window chamber in nude mice, was a collagen disk of about 4 mm in diameter and 0.45 mm in thickness containing 1.5×10^5 Pr*Tu*SCs in addition to rat urogenital sinus mesenchyme (UGM). This recombination construct was identical to the ones implanted under the kidney capsule in the tissue recombination experiments described earlier. A collagen disk of the same size, but without cells, was used as negative control. Implanted human *adult* Pr*Tu*SCs and their potential to initiate a vascular response with subsequent angiogenesis were imaged with bright-field, DIC (differential interference contrast), and fluorescent microscopy techniques (Olympus BX51) and analyzed using still photographic and video captures.

Bright-field images of implanted human *adult* Pr*Tu*SCs on collagen recombination disks in nude mouse-skinfold window chambers showed increased vascular sprouting 10 days after implantation (Fig. 7). The cloudy area in the images represents the recombination disk holding the implanted *adult* Pr*Tu*SCs. Single cells were not visible without fluorescent labeling in the bright-field images. A view of the entire visible window chamber area showed an increased angiogenic activity around the implanted disk (cloudy area), in comparison to an area of unaffected vascularization on the opposite side of the window (Fig. 7a). A magnification of the red square area with part of the collagen disk construct clearly indicated angiogenic sprouting in response to the implanted Pr*Tu*SCs pointing in direction and also growing into the collagen construct (Fig. 7b). Further magnification of the vessel sprouting (red rectangular area) showed the density of the sprouting vessels in great detail (Fig. 7c). Within another week, a fully developed healthy vessel system as seen in regular physiological angiogenesis was generated. These results support the findings of the tissue recombination graft experiments into the kidney capsule as described earlier [33]. As found in histological and immunostaining analyses from the tissue recombination grafts, the implanted *adult* Pr*Tu*SCs, although isolated from early-stage adenocarcinomas, differentiated into physiological benign prostate glandular structures [33]. A parallel development was expected to be seen of physiological angiogenesis with fully developed functional healthy blood vessels around the implant to supply nutrition and oxygen for the new tissue growth, as opposed to leaky, non-hierarchal, tortuous blood vessel system typical of tumor-associated angiogenesis. The modified dorsal mouse-skinfold window chamber method verified the growth of physiological healthy tissues including a healthy blood vessel network around implanted human *adult* Pr*Tu*SCs in a recombination construct model (Fig. 7).

On the other hand, there is evidence that *adult* Pr*Tu*SCs alone do not grow well in implantation sites other than the prostate [113]. Therefore, it is difficult to grow them in the dorsal mouse-skinfold window chamber without adding rat urogenital sinus mesenchyme (UGM) or other stromal cells to research possible cancer-associated angiogenesis as seen in the orthotopic xenografts. The perfect in vivo model for angiogenesis studies as induced by *adult* Pr*Tu*SCs in the dorsal mouse-skinfold

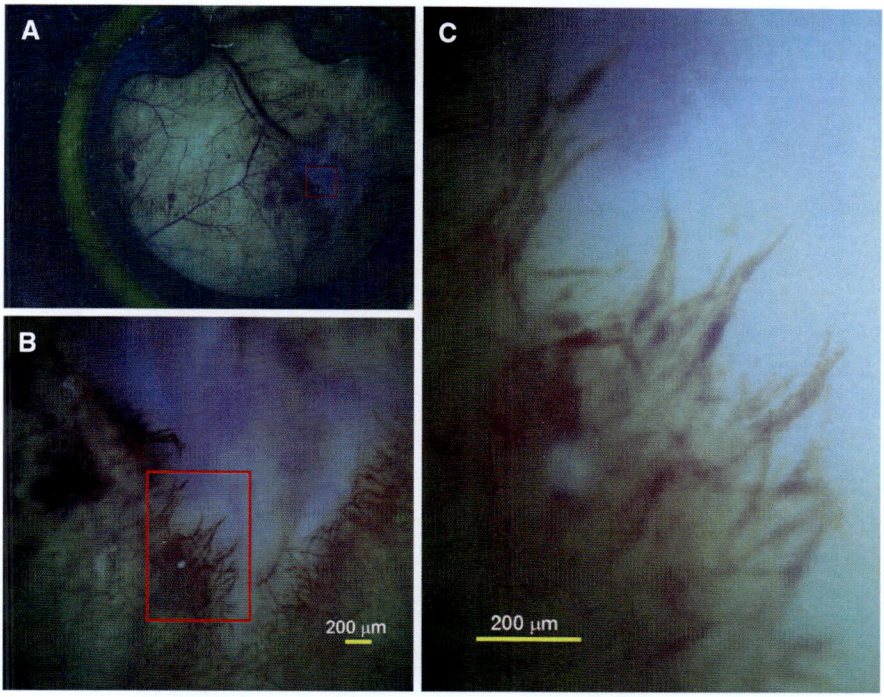

Fig. 7 Bright-field images of vascular sprouting due to the presence of implanted human *adult* Pr*Tu*SCs on a collagen recombination disk in a nude mouse-skinfold window chamber 10 days after implantation. (**a**) A view of the entire visible window chamber showed an increased angiogenic activity around the implanted construct (the *cloudy area* is representing the construct disk). (**b**) Magnification of the *red square area* with part of the collagen disk clearly showing angiogenic sprouting into the collagen construct. (**c**) Further magnification of the *red rectangular area* showed the density of the sprouting vessels in great detail, which later developed into a fully anastomosed healthy vessel system as seen in physiological angiogenesis, supporting the findings of benign prostate glandular differentiation including physiological angiogenesis in response to the implant of an *adult* Pr*Tu*SC construct

window chamber would be a replication of the microenvironment in which the stem cells naturally occur. The problem is that it is currently not known exactly what comprises the microenvironment of Pr*Tu*SCs in vivo. However, it is well known that there are paracrine interactions with stromal cells [44, 114, 115], which enable the Pr*Tu*SCs to survive. In case of the recombination method, urogenital sinus mesenchyme (UGM) provides the stromal component for the implanted Pr*Tu*SCs to proliferate and differentiate [45, 114]. Co-encapsulation with fibroblasts may also work [116]. One group successfully implanted prostate cancer cells in the window chamber on top of a slice of mouse prostate [113]. In any case, the dorsal skinfold window chamber provides an excellent model to study the environmental niche of stem cells. Different stem cell constructs composed of various features of the niche, such as companion cells, growth factors, cytokines, and extracellular matrix components, could be implanted into the window chamber. With this method, direct observation of resulting cell activity and adjacent tissue responses, including neovascularization, can bring new insights into the potential of stem cell differentiation and the interaction within the stem cell niche.

Quantification of angiogenesis in the dorsal mouse-skinfold window chamber. The dorsal skinfold model has been used extensively to study neovascularization, although most of the work has focused on cancer vasculature. With this tool, parameters like blood vessel diameter, density, tortuosity, and permeability can be quantified. With in vivo fluorescent videography, blood cells can be seen moving through the blood vessels in the window chamber, and therefore can be examined for functionality. Red blood cell velocity and functional microvascular density can be quantified. Microvascular density is defined as the total length of blood vessels in the region of interest, divided by the area of the region of interest, measured in units of cm^{-1}. This parameter has been correlated with prostate cancer metastasis [103–105]. Another parameter used to separate physiological intact vessels and cancer-associated vessel growth is permeability. Vessel permeability is tested to determine integrity of blood vessels, especially when they are developing in an angiogenic area with rapid vessel growth. A fluorescent vascular tracer of a known size is injected into the blood vessel system of the host. The decrease of tracer intensity is measured over time between tracer molecules visible inside the blood vessels and outside the blood vessels after the fluorescent tracer has leaked into the surrounding tissue. Cancer-associated blood vessels are known to be unfinished vessels with more tendencies to leak into the surrounding tissue, while physiological normal blood vessels are intact and show a very little degree of permeability [117]. Another blood vessel characteristic is vessel tortuosity, a parameter simply defined as the length of a blood vessel segment divided by its end-to-end distance, which can be measured directly in still photographs of the window chamber. Tumor-associated blood vessels are more tortuous than healthy blood vessels with an organized clear vessel hierarchy [80]. In summary, many parameters are established to quantify neovascularization in the skinfold window chamber in response to implanted cells to help decide between healthy and tumor-associated angiogenesis. This in turn may lead to early information about the type of tissue that might be developing from implanted stem cells.

The effect of wound healing after surgery and inflammation in the window chamber might be considered as influencing factors on angiogenesis and cell growth. To normalize quantitative measurement of angiogenesis, a mouse carrying a chamber without a cell construct can be used as baseline for wound-healing effects. A mouse carrying a chamber with a construct, but without cells in the construct, can be used as a negative control to account for material effects. The level of inflammation in the window chamber can be quantified by leukocyte adhesion tests [118]. A 5% Rhodamine 6G (Sigma) solution in sterile saline was injected weekly in the host mice via the tail veins. Leukocytes in the blood vessels are fluorescently labeled by Rhodamine 6G and their movement can be observed and quantified in videographic recordings. The number of leukocytes rolling and adhering to the blood vessel inner wall or flowing freely through a given blood vessel segment (per surface area unit of blood vessel wall) is quantified per time unit (e.g., 30 s). The more leukocytes that roll and adhere to the blood vessel wall, the higher the level of inflammation [118–120]. Data from leukocyte assays indicated very low levels of inflammation in the window chamber after recovery from surgery (Gough, unpublished data). On average, until up to 6 weeks after window chamber implantation, inflammation in the window chamber was minimal before unwanted tissue reactions indicated an increase of inflammation. In summary, inflammatory effects in the window chamber can be kept to a minimum with sterile surgery practices and can be measured with leukocyte adhesion tests. Wound-healing effects can be normalized with carefully designed negative controls for a baseline. The modified window chamber remains a useful tool to examine in vivo tissue responses to implanted stem cells in real-time visualization for up to 6 weeks of implantation time.

Fluorescent imaging of implanted human adult stem cells in the dorsal mouse-skinfold window chamber. Since single cells were not visible in the bright-field images of the window chamber, fluorescent labeling of implanted stem cells was needed to reveal further details. A combination of fluorescent labeling and DIC (differential interference contrast) microscopy were used to examine implanted Pr*Tu*SCs in the dorsal skinfold nude mouse window chamber. DIC microscopy was useful to enhance blood vessel contrast. Cells were fluorescently labeled with eGFP (enhanced green fluorescent protein) or the cell dye PKH-26 (Sigma) before implantation into the window chamber for comparison of the cell-labeling strategies. Use of two different fluorescent labels was helpful to image more than one structure in the window chamber at the same time; for example cells can be stained to appear in a red emission range and blood vessels can be injected with a vascular tracer to appear in a green emission range. Red-emitting structures were imaged with an Omega XF108-2 filter cube, which has an excitation wavelength of 525 nm and an emission wavelength of 595 nm and will excite cell stains like Rhodamine 6G or PKH-26. Green-emitting structures were imaged with an Olympus U-MNB2 filter cube, which has an excitation wavelength range of 470–490 nm and an emission wavelength of 515 nm and will excite cell labels like eGFP or vascular tracers like FITC (fluorescein isothiocyanate). Once labeled, Pr*Tu*SCs were implanted into the window chamber, the host mice were immobilized in the acrylic tube of the imaging stage apparatus (Fig. 6b), and the window chamber was examined under the microscope. Window chambers were photographed three times a week for the duration of their implantation to track blood vessel changes around the fluorescently labeled implanted stem cells. Additionally, once a week, 50–100 μl FITC-dextran solution (500 kDa, Sigma), a vascular tracer, was injected via the tail vein of the mice, and videos were taken with a high-performance monochrome CCD camera and Pinnacle Studio Plus V10.8 software. Videographic recordings were used to show *movement* of labeled cells in FITC-stained blood vessels for vessel functionality tests.

Fluorescent-labeled stem cells implanted in a collagen construct were easily located in the window chamber using different fluorescent filter cubes. Figure 8 shows a comparison of fluorescent labels in a series of images from the same implant construct in a window chamber. Underlying blood vessels of the window chamber can be seen in the background of the images as red lines filled with red blood cells. In bright-field images, the implanted cells were not visible (Fig. 8a). The same view,

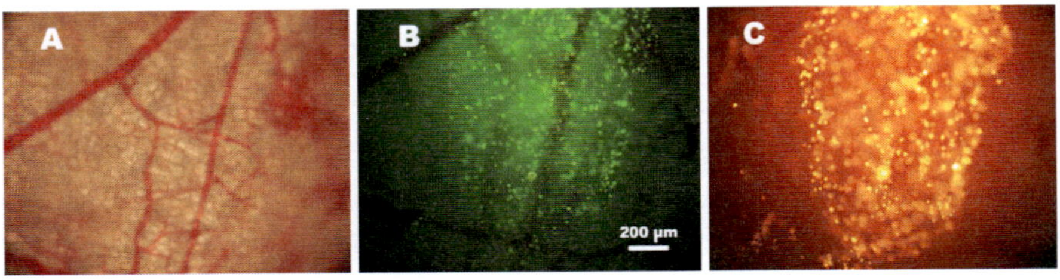

Fig. 8 Comparison of fluorescent stainings in the same collagen construct with double-fluorescent-labeled *adult* stem cells implanted in the dorsal mouse-skinfold window chamber: (**a**) bright-field image in which the implanted cells are not visible, (**b**) green filter cube image with eGFP-labeled implanted cells, (**c**) red filter cube image with PKH-26-labeled implanted cells. Though being a finite stain, PHK-26 signals appeared to be much brighter than eGFP signals, making this cell dye useful for monitoring and tracking of implanted stem cells in vivo

imaged with a green filter cube, shows eGFP-labeled implanted stem cells (Fig. 8b) and, imaged with a red filter cube, shows PKH-26-labeled implanted stem cells (Fig. 8c). In comparison to the bright-field images, implanted stem cells within the construct are clearly visible with eGFP as well as with PKH-26 fluorescent labels. PKH-26 cell dye consists of a fluorophore attached to a lipophilic tail that incorporates into the cell membrane. Therefore this is a finite stain; as the cells divide, the cell stain is divided between the two new cells. When labeled stem cells are proliferating, the signal from the PKH-26 decreases. In contrast, eGFP labeling is a transfection process wherein the green fluorescent protein is inserted retrovirally in the genome of the stem cells (described earlier). The stem cells continuously produce the eGFP protein, so when the cells proliferate, the signal grows stronger. However, as seen in Fig. 8, PKH-26 stain signals start out much brighter than eGFP signals in implanted stem cells, when initially labeled and implanted. Therefore, PKH-26 stain is useful to initially localize the stem cells in the window chamber. On the other hand, eGFP labeling of the implanted cells allows monitoring, in vivo, the proliferation of the stem cells over time by an increase in the eGFP signal. Double-staining of the implanted cells with PKH-26 and eGFP is also useful to distinguish the cells from the vascular tracer FITC signal, since eGFP and FITC are both emitting in the same green wavelength range. PKH-26-labeled cells emit in the red-orange emission wavelength range and can be imaged in the same window chamber view with green fluorescent, FITC-labeled blood vessels (an example is shown in Fig. 9).

A snapshot of a videographic recording with PKH-26 fluorescent-labeled stem cells, encapsulated in a collagen construct and implanted into the dorsal mouse-skinfold window chamber, is shown in Fig. 9. Blood vessel sprouting toward the implanted stem cells was visible with the green filter cube, since the vessels contain the fluorescent vascular tracer FITC-dextran, which had been injected into the host mouse's tail vein prior to imaging (Fig. 9a). Switching the filter cube to red emission enabled visualization of the PKH-26-labeled implanted stem cells. By overlaying both images (Fig. 9b) the relationship between single implanted stem cells and

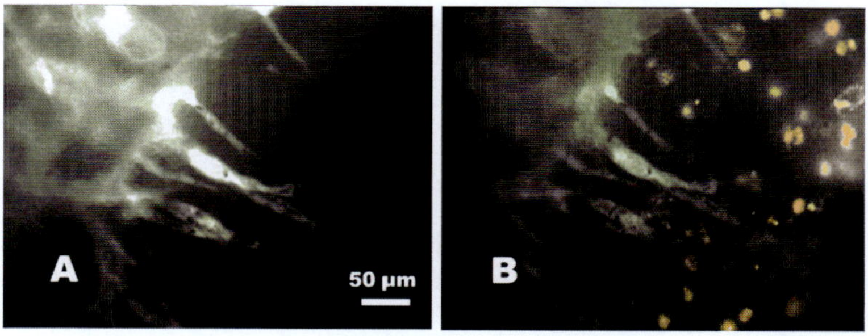

Fig. 9 Videographic images with PKH-26 fluorescent-labeled *adult* stem cells encapsulated in a collagen construct implanted into the dorsal mouse-skinfold window chamber. (**a**) Sprouting of angiogenic blood vessels perfused with the fluorescent, green-emitting, vascular tracer FITC-dextran. (**b**) Overlaying image A with a red-emitting image enabled visualization of the PKH-26-labeled single stem cells and showed the sprouting of angiogenic blood vessels directed toward the implanted Pr*Tu*SCs in great detail

angiogenic blood vessel sprouting toward the implanted cells became visible in great detail and could be quantified. In summary, fluorescent labeling of implanted stem cells in the dorsal mouse-skinfold window chamber provides detailed information about implanted *adult* stem cells and their ability to induce and direct angiogenesis and neovascularization. In the future, fluorescent proteins can be used to image and track stem cells of any kind implanted in the window chamber, and their effect on angiogenesis can be monitored over time.

The study of the formation of new blood vessels induced by implanted *adult* Pr*Tu*SC constructs in the mouse-skinfold window chamber showed a directional drive in the tips of the angiogenic blood vessels clearly advancing toward the implanted prostate stem cells (Fig. 9b). This observation supports the findings of Gerhardt et al., which showed that the guidance of tip growth in angiogenic blood vessel sprouts is dependent on a VEGF (vascular endothelial growth factor) gradient, while the proliferation of cells in the stalks of the sprouts depends on the VEGF concentration [121]. Like fibroblasts [122], Pr*Tu*SCs have been recently found to express VEGF [123], which could explain the directional growth of new vessels toward the prostate stem cells (Gough, unpublished data). Imaging implanted Pr*Tu*SCs every 3 days over the time period of the window chamber implantation up to 6 weeks showed a significant increase in vessel proliferation around the implanted and differentiating Pr*Tu*SCs over time (data not shown). After about 3 weeks, a well-developed hierarchy of a blood vessel network supported the implant of the prostate stem cell construct, verifying the observation from tissue recombination assays described earlier that implanted Pr*Tu*SC constructs have the ability to grow physiological regular prostate tissues including a physiological healthy vessel supply under certain microenvironmental conditions.

The modified dorsal mouse-skinfold window chamber method combined with fluorescent labeling of implanted *adult* stem cells remains a crucial tool to study stem cells in an in vivo environment. Adult stem cell differentiation, behavior, and interaction with the surrounding cells including blood vessels can be investigated in great detail and the environmental niche of all kinds of isolated *adult* stem cells can possibly be de-mystified with this method in the future.

Angiogenesis is an important tool to measure tissue growth, in either physiological or cancerous tissues. It can be used to directly observe the tissue response to implanted *adult* Pr*Tu*SCs and, vice versa, the reaction of the Pr*Tu*SCs to their new neighborhood. Whether the implanted stem cells differentiate into a tumorous or physiological healthy tissue could possibly be predicted with the dorsal skinfold window chamber method by direct observation of the architecture of the neovasculature built around the implanted cells. When Pr*Tu*SC recombination constructs were implanted into the window chamber, implanted stem cells stayed localized and produced a definite angiogenic response. In this model, the blood vessels sprouted, grew, and anastomosed into an orderly system of fully developed blood vessel network by the third week, which was not tortuous or permeable as typically seen in cancer-associated blood vessels. This result is consistent with the findings of the tissue recombination graft under the kidney capsule of SCID mice, which resulted in the growth of a physiological, non-cancerous prostate organ, although the *adult* Pr*Tu*SCs had been isolated from early-stage prostate adenocarcinoma specimens. Combining all introduced methods, this model presents a strong engineering approach to investigate the huge potential of isolated and cultured *adult* tumor stem cells in the future.

6 Potential Applications of *Adult* Stem Cells in the Future

Future applications of *adult* stem cells are twofold: near future medical advances and long-term approaches. In the near future, there is the potential to study the growth of Pr*Tu*SCs and other isolated tumor stem cells for advances in cancer diagnosis and therapy. But, a long hoped-for medical solution might lie in the possibility to use *adult* stem cells for long-term applications in tissue engineering approaches and regenerative medicine.

Near future applications in cancer research and treatment. In this chapter, stem cell engineering techniques were described that ventured into in vitro and in vivo investigations of the prostate tumor stem cell niche, but that can be modified for *adult* stem cells from other organs. The niche-forming stem cells were described in detail. New methodologies to isolate and grow adult stem cells in vitro and a variety of in vivo models have been introduced to investigate stem cells and their relationship with the microenvironment. Isolating stem cells via cell sorting has been recently discussed in the literature as problematic, since extensive technical treatment processes might change the biological characteristics of the isolated cells to question their identity as stem cells [124]. Using the stem cell niche to selectively isolate and culture adult stem cells in vitro serves to solve this issue in a novel way and opens up the possibility of isolating different kinds of stem cells, once stem cell niche conditions are established. Therefore, engineering tools like the modified skinfold window chamber model are essential to investigate conditions of the stem cell niche for varying kinds of stem cells in the near future. Furthermore, the fact that tumor stem cells can grow physiological as well as tumorous tissues, depending on the environmental conditions, brings even more attention to the importance of the exploration of mechanisms within the stem cell niche. Finally, the ability to harvest *adult* tumor stem cells from early-stage cancer specimen creates a whole new field of research with easy access to stem cells and verifies the uniqueness of *adult tumor* stem cells as cells with the potential to grow healthy tissues in certain conditions.

In the very near future, cancer diagnosis and therapy may benefit from the methodologies introduced in this chapter. Since Pr*Tu*SCs have been isolated and characterized, this could be a useful tool to study the metastatic potential of *tumor* stem cells in general. What are the mechanisms behind changing *tumor* stem cells into metastatic *cancer* stem cells? How can we avoid those conditions to prevent prostate and other cancers? Studies of influential factors from the microenvironment as well as specific genetic alterations in tumor and cancer stem cells can bring further clarification of the codes and mechanisms leading to metastatic cancer initiation.

Since each Pr*Tu*SC colony derived from an SCC niche, all undifferentiated prostate *tumor* stem cells were senescent in vitro like most other human adult stem cells studied [125]. Nonetheless, there was a difference in some early-stage prostate tumors, which were able to give rise to hundreds of SCC/Pr*Tu*SC colonies, while others only yielded a handful of such colonies. Therefore, there may be an in vivo mechanism whereby the bona fide prostate tumor stem cells with their SCC niche increase in number. One hypothesis is that *adult* prostate tumor stem cells can self-renew to a certain degree in vivo under specific circumstances via a microenvironment-mediated interaction with prostate stromal cells [44], which are known to play a central role in the generation of prostate cancer [43]. It may become possible to sustain the self-renewal capability of SCC/Pr*Tu*SC colonies in culture by implementation of environmental factors such as growth factors, 3D culture methods, extracellular matrix components, and/or suitable stromal cell feeder layers. Discovering the mechanisms behind tumor stem cell self-renewal and facilitating in vitro methods to maintain self-renewal of the SCC niche would be an important research goal. By blocking or inhibiting certain components of the stem cell niche, this would naturally lead to therapeutically applicable ways to inhibit self-renewal of prostate tumors in vivo. Clinical applications to treat prostate tumors by suppressing their self-renewal capacities in vivo are a most important aim in cancer research. The study of the microenvironmental niche is critical for understanding the complexity of tumor initiation.

Nevertheless the study of genetic alterations of the tumor and cancer stem cells themselves will bring further clarification in the tumor initiation process. The human adult prostate *tumor* stem cells (SCC/Pr*Tu*SCs), described in this chapter, differ significantly from *cancer* stem cells (CSCs) that have been isolated from human fully developed metastatic cancer sources like colorectal or brain carcinomas [126–130]. In opposition to *tumor* stem cells, CSCs grow indefinitely when successfully grown in culture, can form malignant tumors in animal models starting with very few cells, and are slow cycling, self-renewing, and capable of terminal differentiation. Primarily, CSCs have been isolated based on marker selection and are categorized by tumor initiation in xenographic implants [129, 130]. Some CSCs cannot be cultured in vitro, probably reflecting missing environmental requirements for self-renewal [40, 131]. One group isolated prostate CSCs from single clones of epithelial prostate cancer cultures, which were retrovirally transduced to express telomerase (TERT) to achieve cell immortality [132]. The properties of the isolated stem cells raised some doubts due to the immortalization procedures and extensive clonal selection steps used [124]. With the isolation method described in this chapter, prostate *tumor* stem cells were isolated from multiple tumor specimens, using no marker selection, but relying on serum-free medium that selectively supports the outgrowth of Pr*Tu*SCs and actively counter-selects other cell types. In addition, each colony of epithelial Pr*Tu*SCs developed from a stem cell niche, the SCC, without exception. Other epithelial cells that attached to the laminin-coated dish, and that did not develop from an SCC, underwent rapid apoptosis, while stem cells in the SCC differentiated into tight epithelial cell colonies around the SCCs. None of the SCC/Pr*Tu*SC colonies expressed telomerase (TERT gene) and they all senesced in vitro after several generations [33]. When comparing the properties of isolated adult prostate *tumor* stem cells from cultured SCCs to androgen-independent metastatic prostate *cancer* cells in late-stage prostate carcinomas, it is reasonable to hypothesize that *tumor* stem cells from cultured SCCs represent a kind of precursor cell to the class of prostate *cancer* stem cells, requiring additional conditioning to inherit malignancy. Therefore, besides studying the conditioning factors of the microenvironment, investigation of present genetic alterations in SCC/Pr*Tu*SCs can reveal some answers to questions of developing or preventing malignancy. Since methods to grow clonal colonies of *adult* prostate tumor stem cells are available now, cellular genetic mapping is approachable for Pr*Tu*SCs. It would also be revealing to investigate what genetic alterations of Pr*Tu*SCs would potentially endow them with a lethal, metastatic phenotype. Since cancer stem cells are immortal and pluripotent or at least plastic in their ability to differentiate (multi-potent), the induction of pluripotency and immortalization – in other words, indefinite growth potential – may distinguish prostate *tumor* stem cells from lethal prostate *cancer* stem cells. Indeed, in natural prostate cancer progression, immortalization and/or pluripotentiation of Pr*Tu*SCs may represent a crucial progression event, initiated through the microenvironmental niche. Studies to determine the effects of immortalization or pluripotentiation of isolated Pr*Tu*SCs via retroviral manipulation of several genes are under way (Haas, unpublished data). First results indicated that immortalization of prostate *tumor* stem cells via transduction to express Bmi-1 and TERT genes did not necessarily trigger a malignant cancer, when implanted in SCID mice. On the other hand, prostate *tumor* stem cells were retrovirally transfected using a four-gene altering model to induce pluripotency in a strategy that reprograms *adult* stem cells to a pluripotent *embryonic* stem cell state. The induction of pluripotency in these genetically altered prostate stem cells did trigger a highly invasive cancer, when implanted in SCID mice (Haas, unpublished data). These findings suggest that the acquisition of pluripotency triggers the transition from local *tumor* stem cells to metastatic cancer cells. With the combination of genetic engineering and the methodologies introduced in this chapter, research of the cancer-initiating triggers becomes available. Furthermore, in vitro selective stem cell culturing and in vivo transplantation methods of adult prostate *tumor* stem cells may enable the development of novel cancer therapeutics that suppress the transition of early-stage to late-stage prostatic disease. Further experimental evaluations of cellular genetic alterations and stem cell niche-influencing factors for understanding the development of malignancy are very promising in the near future with the use of cellular engineering tools and the techniques introduced here.

For cancer research in general, isolated *adult* prostate tumor stem cells can be genetically altered to create an immortal cell line [34]. As described earlier for fluorescent labeling, early Pr*Tu*SC colonies at a stage of only a few hundred cells can be retrovirally transduced to express Bmi-1 and telomerase (TERT) genes, a combination that allows epithelial cells to bypass senescence and become an immortal cell line that can be continuously cultured like commercially available cancer cell lines. Bmi-1 represses two tumor suppressor genes ($P19^{ARF}$ and $p16^{INK4a}$) and therefore represses senescence of the cell by maintaining self-renewal. The replicative lifespan of the cells can be indefinitely extended, when Bmi-1 is overexpressed by the cells [133]. Bmi-1 may also induce telomerase in epithelial cells [134, 135]. With the simultaneous induction of the TERT (telomerase reverse transcriptase) gene, the stem cells can be basically immortalized. In vitro, the mortality control, tightly regulated in vivo by the microenvironmental niche, is aborted. With the loss of the senescent phenotype, some in vivo potential of these unique *adult* tumor stem cells might be lost, but, by creating an immortal cell line, will facilitate cellular and molecular studies of prostate and other cancer stem cells and their behavior under different circumstances. Research products like "*adult* cancer stem cell lines" could be made commercially available for in vitro and in vivo cancer research studies, as it exists for many human cell lines today.

Further genetic manipulation of *adult* stem cells may provide the opportunity to create whole conditional stem cell systems, which can be studied in the window chamber. Once the conditional genetic system is in place, it will be possible to make these cells more or less adaptable at will by turning targeted genes on or off. In vivo, metastatic stem cells must be able to exit their niche, get access to bloodstream and lymph vessels, survive in the bloodstream, adhere and extravasate at alternative metastatic sites, and be able to proliferate and survive the resident cells at this particular location. So, the most important property of *cancer* stem cells is adaptability and plasticity. Immortal genetically manipulated *adult* stem cells may be able to be more or less adaptable at will in a conditional genetic system, which also incorporates the microenvironmental factors. The dorsal skinfold window chamber will be a useful tool to observe and explore conditional genetic systems under in vivo conditions.

Implantation of Pr*Tu*SCs in the window chamber can also help to determine the mechanisms leading to neovascularization. In companionship with in vitro protein inhibition studies, the window chamber model will help elucidate the pathways and proteins used by Pr*Tu*SCs to induce angiogenesis. It has already been shown that tumor-initiating cells express high levels of angiogenic proteins such as interleukin-8 [136, 137]. In vivo experiments in the dorsal skinfold window chamber can be used to demonstrate how these proteins contribute to angiogenesis and which signaling pathways they use. If the hypothesis that prostate carcinomas develop from prostate *cancer* stem cells [138, 139] is true, the effect of Pr*Tu*SCs on neovascularization seems to be an early necessary event in cancer progression. Therefore, this effect is very important and requires further exploration. Once the mechanism of angiogenesis is thoroughly understood, cancer therapies to inhibit cancer-associated angiogenesis can follow. Anti-angiogenic therapies in prostate cancer development have a lower probability of drug resistance development [140] and are a first approach to target the microenvironment instead of the cancer cell itself. Inhibitors of vascular growth factors such as vascular endothelial (VEGF) and basic fibroblast (bFGF) growth factors have been tested in clinical studies, but anti-angiogenic agents in prostate cancer have *not* been particularly successful at this point [141–143]. This is probably due to the fact that the multifunctionality of angiogenesis-supporting growth factors and their effect in the human body are not completely understood [107]. The interaction of this stem cell niche component with stem cells in vivo *cannot* be underestimated. The window chamber model is a very promising tool to further research these interconnections in vivo. Once the pathways of angiogenesis are understood, anti-angiogenic therapeutic interventions at the start of prostate cancer progression might become available in the near future.

Direct clinical applications in the near future might include SCC-colony counts as an in vitro surrogate method, which shows capability of predicting the probability of tumor re-occurrence and progression. Early-stage prostate carcinoma specimens gave rise to a characteristic number of SCC/Pr*Tu*SC colonies, as assayed 7 days after plating a standardized number of isolated prostate *tumor* stem cells as described earlier [33]. Although there was no obvious correlation between the differentiation and prognosis grade (Gleason Grade) of the prostate cancers and the specific number of cultured SCC/Pr*Tu*SC colonies, prostate carcinomas yielding low numbers (~2–4) of SCCs in the SCC/Pr*Tu*SC culture assays were distinct from carcinomas yielding high numbers (~400–500) of SCCs and succeeding colonies [33]. In addition, the complete absence of such SCCs and colonies in human *benign* prostate specimens emphasized the occurrences of SCCs grown from a specific early-stage prostate tumor as a reflection of the total number of cancer stem cell forces. In conclusion, the number of SCCs grown from a specific early-stage prostate carcinoma specimen could provide probability predictions of the cancer prognosis considering the tumorigenic phenotype of SCC/Pr*Tu*SC colonies in vivo. Combined with fluorescent labeling techniques that would allow for high-throughput image analysis, characteristic SCC counts in early Pr*Tu*SC colony culture assays could become a supplemental clinical method for prostate cancer prediction and may be shown to be prognostic for the course of the disease.

This prognostic tool may be expanded to early-stage carcinomas from other organs, since there is a high probability that stem cells similar to Pr*Tu*SCs can be isolated and cultured by the methods reviewed in this chapter. This would further facilitate studies of early events, mutations, and translocations associated with the onset of carcinomas. The importance of these findings is that the isolation of early-stage tumor stem cells via selective culturing would facilitate molecular studies of the events associated with progression of *adult tumor* stem cells into metastatic *cancer* stem cells, including – but not limited to – discovering cell markers, further mutations, and targets for detection, prognosis, and suppression of the lethal phase of cancer.

Potential long-term applications in tissue engineering and regenerative medicine. Every day, research is unveiling more information about *adult* stem cells and their dependence upon their niche. Based on these interactions within the stem cell niche, the window chamber can be used as a tool for exploring stem cell therapeutics. Great potential for therapeutic applications would exist in the creation of an artificial stem cell niche. This could be in the form of encapsulation using biopolymers that would contain supportive extra cellular matrix components as well as nurturing growth factors and cytokines required to induce stem cell differentiation along the desired pathway. The artificial stem cell niche could be used for cellular and tissue delivery for repair of damaged organs. In addition, creation of a whole new organ in vitro or in vivo may be possible one day in the future, when stem cell isolation and stem cell niche conditions have become standard knowledge. At this point, research about stem cells is still at its beginning, but surely has a promising outlook for tissue-regenerative approaches with the current results. If, as for one example demonstrated in this chapter, physiological benign glands can be generated in vivo from *tumor* stem cell material by placing the adult tumor stem cells in the correct environment, this methodology has a future for promising regenerative tissue and organ reconstruction approaches.

Furthermore, the isolation of *adult* tumor stem cells may be the start of a completely new research field investigating these unique kinds of pluripotent stem cells. The fact that *adult* tumor stem cells senesce in vitro and can generate physiologically healthy organs in vivo makes them attractive possibilities for tissue engineering and regenerative medicine. *Adult* tumor stem cells seem to have what is needed for tissue and organ engineering approaches: they are multi-potent and possess strong proliferative potential, but are *not* immortal and therefore not tumorigenic, even in a remote way. Transplanted stem cells from SCC populations furthermore were capable of plastic responses in different microenvironments. The field of *adult* tumor stem cell research is very promising for the hope in human medicine to be able to generate replacement tissues from human cells to prevent organ failure. A dream that has been worked on for almost a century now seems to be in reach. But, like in every other new research field, there will be many obstacles to overcome and it will take some time to reach this goal. At least, there is hope and a promising start.

In summary, there are many applications feasible in the near and long-term future for *adult* tumor stem cells. First, promising research results and methodologies, as introduced in this chapter, are already facilitating the creation of immortal tumor stem cell lines to study prostate cancer [34] and will facilitate more research to determine stem cell niche conditions for tumor stem cells of varying organs in the very near future. From there it will only take a few more steps to develop new cancer diagnostic tools and therapies. The application of SCC counts as a clinical prognostic tool for cancer diagnosis and progression is already one possibility with great potential. Long-term approaches to regenerative tissue and organ reconstructions are further away, but seem completely possible. The creation of an artificial stem cell niche for cancer therapeutics and/or cellular and organ reconstruction in tissue engineering approaches through *adult* stem cells are long-term goals that would bring modern medicine into a new era. Nevertheless, there are many unanswered questions and more work is needed. However, based on the current results presented in this chapter, the potential of cultured *adult* stem cells for re-implantation into tissues is enormous and, in human medicine, provides hope to cure lethal diseases.

References

1. Kim BH. Sanal theory. J Acad Med Sci. 1965; 108:39–62 (DPR Korea).
2. Peichev M, Naiyer AJ, Pereira D, Zhu Z, Lane WJ, Williams M, Oz MC, Hicklin DJ, Witte L, Moore MA, Rafii S. Expression of VEGFR-2 and AC133 by circulating human CD34(+) cells identifies a population of functional endothelial precursors. Blood. 2000; 95(3): 952–958.
3. Moore KA, Lemischka IR. Stem cells and their niches. Science. 2006; 311(5769): 1880–1885.
4. Cotsarelis G, Cheng SZ, Dong G, Sun TT, Lavker RM. Existence of slow-cycling limbal epithelial basal cells that can be preferentially stimulated to proliferate: implications on epithelial stem cells. Cell. 1989; 57(2):201–209.
5. Blanpain C, Horsley V, Fuchs E. Epithelial stem cells: turning over new leaves. Cell. 2007; 128(3):445–458.
6. Croagh D, Phillips WA, Redvers R, Thomas RJ, Kaur P. Identification of candidate murine esophageal stem cells using a combination of cell kinetic studies and cell surface markers. Stem Cells. 2007; 25(2):313–318.
7. Urbanek K, Cesselli D, Rota M, Nascimbene A, De Angelis A, Hosoda T, Bearzi C, Boni A, Bolli R, Kajstura J, Anversa P, Leri A. Stem cell niches in the adult mouse heart. Proc Natl Acad Sci USA. 2006; 103(24):9226–9231.
8. Maeshima A. Label-retaining cells in the kidney: origin of regenerating cells after renal ischemia. Clin Exp Nephrol. 2007; 11(4):269–274.
9. Bussolati B, Bruno S, Grange C, Buttiglieri S, Deregibus MC, Cantino D, Camussi G. Isolation of renal progenitor cells from adult human kidney. Am J Pathol. 2005; 166(2):545–555.
10. Hong KU, Reynolds SD, Giangreco A, Hurley CM, Stripp BR. Clara cell secretory protein-expressing cells of the airway neuroepithelial body microenvironment include a label-retaining subset and are critical for epithelial renewal after progenitor cell depletion. Am J Respir Cell Mol Biol. 2001; 24(6):671–681.
11. Borthwick DW, Shahbazian M, Krantz QT, Dorin JR, Randell SH. Evidence for stem-cell niches in the tracheal epithelium. Am J Respir Cell Mol Biol. 2001; 24(6): 662–670.
12. Duvillie B, Attali M, Aiello V, Quemeneur E, Scharfmann R. Label-retaining cells in the rat pancreas: location and differentiation potential in vitro. Diabetes. 2003; 52(8): 2035–2042.
13. Uchida N, Buck DW, He D, Reitsma MJ, Masek M, Phan TV, Tsukamoto AS, Gage FH, Weissman IL. Direct isolation of human central nervous system stem cells. Proc Natl Acad Sci USA. 2000; 97(26):14720–14725.

14. Lee A, Kessler JD, Read TA, Kaiser C, Corbeil D, Huttner WB, Johnson JE, Wechsler-Reya RJ. Isolation of neural stem cells from the postnatal cerebellum. Nat Neurosci. 2005; 8(6):723–729.
15. Kinbara H, Cunha GR, Boutin E, Hayashi N, Kawamura J. Evidence of stem cells in the adult prostatic epithelium based upon responsiveness to mesenchymal inductors. Prostate. 1996; 29(2):107–116.
16. Radtke F, Clevers H. Self-renewal and cancer of the gut: two sides of a coin. Science. 2005; 307(5717):1904–1909.
17. Cotsarelis G, Sun TT, Lavker RM. Label-retaining cells reside in the bulge area of pilosebaceous unit: implications for follicular stem cells, hair cycle, and skin carcinogenesis. Cell. 1990; 61(7):1329–1337.
18. Bendall SC, Stewart MH, Menendez P, George D, Vijayaragavan K, Werbowetski-Ogilvie T, Ramos-Mejia V, Rouleau A, Yang J, Bosse M, Lajoie G, Bhatia M. IGF and FGF cooperatively establish the regulatory stem cell niche of pluripotent human cells in vitro. Nature. 2007; 448(7157):1015–1021.
19. Reilly KM, Van Dyke T. It takes a (dysfunctional) village to raise a tumor. Cell. 2008; 135(3):408–410.
20. Sneddon JB, Werb Z. Location, location, location: the cancer stem cell niche. Cell Stem Cell. 2007; 1(6):607–611.
21. World-Health-Organization, WHO. Cancer (Fact sheet No 297). http://www.who.int/mediacentre/factsheets/fs297/en/index.html; Feb 2009.
22. Nowell PC. Mechanisms of tumor progression. Cancer Res. 1986; 46(5):2203–2207.
23. Pardal R, Clarke MF, Morrison SJ. Applying the principles of stem-cell biology to cancer. Nat Rev Cancer. 2003; 3(12):895–902.
24. Cho KR, Vogelstein B. Genetic alterations in the adenoma–carcinoma sequence. Cancer. 1992; 70(6 Suppl):1727–1731.
25. Fearon ER, Hamilton SR, Vogelstein B. Clonal analysis of human colorectal tumors. Science. 1987; 238(4824):193–197.
26. Yang J, Weinberg RA. Epithelial-mesenchymal transition: at the crossroads of development and tumor metastasis. Dev Cell. 2008; 14(6):818–829.
27. Thiery JP. Epithelial-mesenchymal transitions in development and pathologies. Curr Opin Cell Biol. 2003; 15(6):740–746.
28. Mani SA, Guo W, Liao MJ, Eaton EN, Ayyanan A, Zhou AY, Brooks M, Reinhard F, Zhang CC, Shipitsin M, Campbell LL, Polyak K, Brisken C, Yang J, Weinberg RA. The epithelial-mesenchymal transition generates cells with properties of stem cells. Cell. 2008; 133(4):704–715.
29. Brabletz T, Jung A, Reu S, Porzner M, Hlubek F, Kunz-Schughart LA, Knuechel R, Kirchner T. Variable beta-catenin expression in colorectal cancers indicates tumor progression driven by the tumor environment. Proc Natl Acad Sci USA. 2001; 98(18):10356–10361.
30. Barker N, Clevers H. Tumor environment: a potent driving force in colorectal cancer? Trends Mol Med. 2001; 7(12):535–537.
31. Brabletz T, Jung A, Spaderna S, Hlubek F, Kirchner T. Opinion: migrating cancer stem cells – an integrated concept of malignant tumour progression. Nat Rev Cancer. 2005; 5(9): 744–749.
32. Bissell MJ, Radisky D. Putting tumours in context. Nat Rev Cancer. 2001; 1(1):46–54.
33. Fiñones RR, Wu C, Yeargin JE, Le T, Baird SM, Haas M. Androgen-independent cancer cells cultured from stage 1 human prostate adenocarcinomas generate local prostate cancers in vivo. Prostate. 2010; 70(2):99–110.
34. Fiñones RR, Yeargin JE, Baird SM, Haas M. Inducing pluripotency and immortality in prostate tumor cells: A stem cell model of cancer progression. Material Science and Engineering. San Diego, University of California San Diego. Ph.D. (Finones 2009): 176.
35. Neuzil J, Stantic M, Zobalova R, Chladova J, Wang X, Prochazka L, Dong L, Andera L, Ralph SJ. Tumour-initiating cells vs. cancer 'stem' cells and CD133: what's in the name? Biochem Biophys Res Commun. 2007; 355(4):855–859.

36. Quintana E, Shackleton M, Sabel MS, Fullen DR, Johnson TM, Morrison SJ. Efficient tumour formation by single human melanoma cells. Nature. 2008; 456(7222):593–598.
37. Baker M. Cancer stem cells, becoming common. Nature Reports Stem Cells; Published online: 3 December 2008 | doi:10.1038/stemcells.2008.153.
38. Baker M. Melanoma in mice casts doubt on scarcity of cancer stem cells. Nature. 2008; 456(7222): 553.
39. Eaves CJ. Cancer stem cells: here, there, everywhere? Nature. 2008; 456(7222):581–582.
40. Roberts PE. Isolation and establishment of human tumor stem cells. Methods Cell Biol. 2008; 86:325–342.
41. Kiernan JA. Histological and histochemical methods: theory and practice. Ed. 3. Oxford: Butterworth Heinemann; 1999.
42. Litvinov IV, Vander Griend DJ, Xu Y, Antony L, Dalrymple SL, Isaacs JT. Low-calcium serum-free defined medium selects for growth of normal prostatic epithelial stem cells. Cancer Res. 2006; 66(17):8598–8607.
43. Tuxhorn JA, Ayala GE, Rowley DR. Reactive stroma in prostate cancer progression. J Urol. 2001; 166(6):2472–2483.
44. Gao J, Arnold JT, Isaacs JT. Conversion from a paracrine to an autocrine mechanism of androgen-stimulated growth during malignant transformation of prostatic epithelial cells. Cancer Res. 2001; 61(13):5038–5044.
45. Hayward SW, Haughney PC, Rosen MA, Greulich KM, Weier HU, Dahiya R, Cunha GR. Interactions between adult human prostatic epithelium and rat urogenital sinus mesenchyme in a tissue recombination model. Differentiation. 1998; 63(3):131–140.
46. Wang Y, Revelo MP, Sudilovsky D, Cao M, Chen WG, Goetz L, Xue H, Sadar M, Shappell SB, Cunha GR, Hayward SW. Development and characterization of efficient xenograft models for benign and malignant human prostate tissue. Prostate. 2005; 64(2):149–159.
47. Ishii K, Shappell SB, Matusik RJ, Hayward SW. Use of tissue recombination to predict phenotypes of transgenic mouse models of prostate carcinoma. Lab Invest. 2005; 85(9):1086–1103.
48. Brody J, Young P, Cunha GR. Renal capsule grafting. http://mammary.nih.gov/tools/mousework/Cunha001/index.html; 1998.
49. Wang Y, Xue H, Cutz J, Bayani J, Mawji NR, Chen WG, Goetz LJ, Hayward SW, Sadar MD, Gilks CB, Gout PW, Squire JA, Cunha GR, Wang YZ. An orthotopic metastatic prostate cancer model in SCID mice via grafting of a transplantable human prostate tumor line. Lab Invest. 2005; 85(11):1392–1404.
50. Cunha GR, Donjacour AA, Cooke PS, Mee S, Bigsby RM, Higgins SJ, Sugimura Y. The endocrinology and developmental biology of the prostate. Endocr Rev. 1987; 8(3):338–362.
51. Donjacour AA, Rosales A, Higgins SJ, Cunha GR. Characterization of antibodies to androgen-dependent secretory proteins of the mouse dorsolateral prostate. Endocrinology. 1990; 126(3):1343–1354.
52. Signoretti S, Waltregny D, Dilks J, Isaac B, Lin D, Garraway L, Yang A, Montironi R, McKeon F, Loda M. p63 is a prostate basal cell marker and is required for prostate development. Am J Pathol. 2000; 157(6):1769–1775.
53. Senoo M, Pinto F, Crum CP, McKeon F. p63 Is essential for the proliferative potential of stem cells in stratified epithelia. Cell. 2007; 129(3):523–536.
54. Rubin MA, Zhou M, Dhanasekaran SM, Varambally S, Barrette TR, Sanda MG, Pienta KJ, Ghosh D, Chinnaiyan AM. alpha-Methylacyl coenzyme A racemase as a tissue biomarker for prostate cancer. JAMA. 2002; 287(13):1662–1670.
55. Cunha GR, Hayashi N, Wong YC. Regulation of differentiation and growth of normal adult and neoplastic epithelia by inductive mesenchyme. Cancer Surv. 1991; 11:73–90.
56. Wong YC, Cunha GR, Hayashi N. Effects of mesenchyme of the embryonic urogenital sinus and neonatal seminal vesicle on the cytodifferentiation of the Dunning tumor: ultrastructural study. Acta Anat (Basel) . 1992; 143(2):139–150.

57. Algire GH. An adaptation of the transparent-chamber technique to the mouse. J Natl Cancer Inst. 1943; 4:1–11.
58. Algire GH, Legallais FY. The transparent chamber technique in the mouse in the study of tumor histo-physiology. Fed Proc. 1948; 7(1 Pt 1):268.
59. Druecke D, Langer S, Lamme E, Pieper J, Ugarkovic M, Steinau HU, Homann HH. Neovascularization of poly(ether ester) block-copolymer scaffolds in vivo: long-term investigations using intravital fluorescent microscopy. J Biomed Mater Res A. 2004; 68(1):10–18.
60. Laschke MW, Haufel JM, Thorlacius H, Menger MD. New experimental approach to study host tissue response to surgical mesh materials in vivo. J Biomed Mater Res A. 2005; 74(4):696–704.
61. Menger MD, Walter P, Hammersen F, Messmer K. Quantitative analysis of neovascularization of different PTFE-implants. Eur J Cardiothorac Surg. 1990; 4(4):191–196.
62. Menger MD, Hammersen F, Walter P, Messmer K. Neovascularization of prosthetic vascular grafts. Quantitative analysis of angiogenesis and microhemodynamics by means of intravital microscopy. Thorac Cardiovasc Surg. 1990; 38(3):139–145.
63. Rucker M, Laschke MW, Junker D, Carvalho C, Schramm A, Mulhaupt R, Gellrich NC, Menger MD. Angiogenic and inflammatory response to biodegradable scaffolds in dorsal skinfold chambers of mice. Biomaterials. 2006; 27(29):5027–5038.
64. Contaldo C, Meier C, Elsherbiny A, Harder Y, Trentz O, Menger MD, Wanner GA. Human recombinant erythropoietin protects the striated muscle microcirculation of the dorsal skinfold from postischemic injury in mice. Am J Physiol Heart Circ Physiol. 2007; 293(1):H274–H283.
65. Steinbauer M, Harris AG, Abels C, Messmer K. Characterization and prevention of phototoxic effects in intravital fluorescence microscopy in the hamster dorsal skinfold model. Langenbecks Arch Surg. 2000; 385(4):290–298.
66. Schacht V, Berens D, von Rautenfeld C. Abels, The lymphatic system in the dorsal skinfold chamber of the Syrian golden hamster in vivo. Arch Dermatol Res. 2004; 295(12):542–548.
67. Makale MT, Lin JT, Calou RE, Tsai AG, Chen PC, Gough DA. Tissue window chamber system for validation of implanted oxygen sensors. Am J Physiol Heart Circ Physiol. 2003; 284(6):H2288–H2294.
68. Makale MT, Chen PC, Gough DA. Variants of the tissue-sensor array window chamber. Am J Physiol Heart Circ Physiol. 2005; 289(1):H57–H65.
69. Borgstrom P, Hillan KJ, Sriramarao P, Ferrara N. Complete inhibition of angiogenesis and growth of microtumors by anti-vascular endothelial growth factor neutralizing antibody: novel concepts of angiostatic therapy from intravital videomicroscopy. Cancer Res. 1996; 56(17):4032–4039.
70. Huang Q, Shan S, Braun RD, Lanzen J, Anyrhambatla G, Kong G, Borelli M, Corry P, Dewhirst MW, Li CY. Noninvasive visualization of tumors in rodent dorsal skin window chambers. Nat Biotechnol. 1999; 17(10):1033–1035.
71. Lindner LH, Eichhorn ME, Eibl H, Teichert N, Schmitt-Sody M, Issels RD, Dellian M. Novel temperature-sensitive liposomes with prolonged circulation time. Clin Cancer Res. 2004; 10(6):2168–2178.
72. Tang Y, Borgstrom P, Maynard J, Koziol J, Hu Z, Garen A, Deisseroth A. Mapping of angiogenic markers for targeting of vectors to tumor vascular endothelial cells. Cancer Gene Ther. 2007; 14(4):346–353.
73. Cabrales P, Intaglietta M. Time-dependant oxygen partial pressure in capillaries and tissue in the hamster window chamber model. Antioxid Redox Signal. 2007; 9(7):845–853.
74. Menger MD, Laschke MW, Vollmar B. Viewing the microcirculation through the window: some twenty years experience with the hamster dorsal skinfold chamber. Eur Surg Res. 2002; 34(1–2):83–91.
75. Brown EB, Campbell RB, Tsuzuki Y, Xu L, Carmeliet P, Fukumura D, Jain RK. In vivo measurement of gene expression, angiogenesis and physiological function in tumors using multiphoton laser scanning microscopy. Nat Med. 2001; 7(7):864–868.

76. Chary SR, Jain RK. Direct measurement of interstitial convection and diffusion of albumin in normal and neoplastic tissues by fluorescence photobleaching. Proc Natl Acad Sci USA. 1989; 86(14):5385–5389.
77. Duda DG, Fukumura D, Munn LL, Booth MF, Brown EB, Huang P, Seed B, Jain RK. Differential transplantability of tumor-associated stromal cells. Cancer Res. 2004; 64(17):5920–5924.
78. Endrich B, Hammersen F, Gotz A, Messmer K. Microcirculatory blood flow, capillary morphology and local oxygen pressure of the hamster amelanotic melanoma A-Mel-3. J Natl Cancer Inst. 1982; 68(3):475–485.
79. Leunig M, Yuan F, Menger MD, Boucher Y, Goetz AE, Messmer K, Jain RK. Angiogenesis, microvascular architecture, microhemodynamics, and interstitial fluid pressure during early growth of human adenocarcinoma LS174T in SCID mice. Cancer Res. 1992; 52(23): 6553–6560.
80. Li CY, Shan S, Huang Q, Braun RD, Lanzen J, Hu K, Lin P, Dewhirst MW. Initial stages of tumor cell-induced angiogenesis: evaluation via skin window chambers in rodent models. J Natl Cancer Inst. 2000; 92(2):143–147.
81. Lichtenbeld HC, Barendsz-Janson AF, van Essen H, Struijker Boudier H, Griffioen AW, Hillen HF. Angiogenic potential of malignant and non-malignant human breast tissues in an in vivo angiogenesis model. Int J Cancer. 1998; 77(3):455–459.
82. Papenfuss HD, Gross JF, Intaglietta M, Treese FA. A transparent access chamber for the rat dorsal skin fold. Microvasc Res. 1979; 18(3):311–318.
83. Schneider G, Seidel R, Uder M, Wagner D, Weinmann HJ, Kramann B. In vivo microscopic evaluation of the microvascular behavior of FITC-labeled macromolecular MR contrast agents in the hamster skinfold chamber. Invest Radiol. 2000; 35(9):564–570.
84. Wu NZ, Klitzman B, Dodge R, Dewhirst MW. Diminished leukocyte-endothelium interaction in tumor microvessels. Cancer Res. 1992; 52(15):4265–4268.
85. Torres Filho IP, Hartley-Asp B, Borgstrom P. Quantitative angiogenesis in a syngeneic tumor spheroid model. Microvasc Res. 1995; 49(2):212–226.
86. Endrich B, Asaishi K, Gotz A, Messmer K. Technical report–a new chamber technique for microvascular studies in unanesthetized hamsters. Res Exp Med (Berl) . 1980; 177(2): 125–134.
87. Chen DC, Agopian VG, Avansino JR, Lee JK, Farley SM, Stelzner M. Optical tissue window: a novel model for optimizing engraftment of intestinal stem cell organoids. J Surg Res. 2006; 134(1):52–60.
88. Moosmann S, Hutter J, Moser C, Krombach F, Huss R. Milieu-adopted in vitro and in vivo differentiation of mesenchymal tissues derived from different adult human CD34-negative progenitor cell clones. Cells Tissues Organs. 2005; 179(3):91–101.
89. Torres Filho IP, Leunig M, Yuan F, Intaglietta M, Jain RK. Noninvasive measurement of microvascular and interstitial oxygen profiles in a human tumor in SCID mice. Proc Natl Acad Sci USA. 1994; 91(6):2081–2085.
90. Folkman J. Anti-angiogenesis: new concept for therapy of solid tumors. Ann Surg. 1972; 175(3):409–416.
91. Stitt AW, McGoldrick C, Rice-McCaldin A, McCance DR, Glenn JV, Hsu DK, Liu FT, Thorpe SR, Gardiner TA. Impaired retinal angiogenesis in diabetes: role of advanced glycation end products and galectin-3. Diabetes. 2005; 54(3):785–794.
92. Moulton KS. Plaque angiogenesis and atherosclerosis. Curr Atheroscler Rep. 2001; 3(3):225–233.
93. Kahlon R, Shapero J, Gotlieb AI. Angiogenesis in atherosclerosis. Can J Cardiol. 1992; 8(1):60–64.
94. Ohno-Matsui K, Morita I, Tombran-Tink J, Mrazek D, Onodera M, Uetama T, Hayano M, Murota SI, Mochizuki M. Novel mechanism for age-related macular degeneration: an equilibrium shift between the angiogenesis factors VEGF and PEDF. J Cell Physiol. 2001; 189(3):323–333.

95. Folkman J. Angiogenesis in cancer, vascular, rheumatoid and other disease. Nat Med. 1995; 1(1):27–31.
96. Schmid-Schonbein GW. Capillary plugging by granulocytes and the no-reflow phenomenon in the microcirculation. Fed Proc. 1987; 46(7):2397–2401.
97. Zweifach BW, Kovalcheck S, De Lano F, Chen P. Micropressure-flow relationships in a skeletal muscle of spontaneously hypertensive rats. Hypertension. 1981; 3(5): 601–614.
98. Zweifach BW, Lipowsky HH, eds. Pressure-flow relations in blood and lymph microcirculation. In: Renkin EM, Michel CC, eds. Handbook of physiology, section 2: the cardiovascular system. Bethesda: American Physiological Society; 1984, pp. 251–307.
99. Zhou A, Egginton S, Hudlicka O, Brown MD. Internal division of capillaries in rat skeletal muscle in response to chronic vasodilator treatment with alpha1-antagonist prazosin. Cell Tissue Res. 1998; 293(2):293–303.
100. Egginton S, Zhou AL, Brown MD, Hudlicka O. Unorthodox angiogenesis in skeletal muscle. Cardiovasc Res. 2001; 49(3):634–646.
101. Hueck IS, Rossiter K, Artmann GM, Schmid-Schonbein GW. Fluid shear attenuates endothelial pseudopodia formation into the capillary lumen. Microcirculation. 2008; 15(6):531–542.
102. Hueck IS, Hollweg HG, Schmid-Schonbein GW, Artmann GM. Chlorpromazine modulates the morphological macro- and micro-structure of endothelial cells. Am J Physiol Cell Physiol. 2000; 278(5):C873–C878.
103. Weidner N, Carroll PR, Flax J, Blumenfeld W, Folkman J. Tumor angiogenesis correlates with metastasis in invasive prostate carcinoma. Am J Pathol. 1993; 143(2):401–409.
104. Brawer MK, Deering RE, Brown M, Preston SD, Bigler SA. Predictors of pathologic stage in prostatic carcinoma. The role of neovascularity. Cancer. 1994; 73(3):678–687.
105. Borre M, Offersen BV, Nerstrom B, Overgaard J. Microvessel density predicts survival in prostate cancer patients subjected to watchful waiting. Br J Cancer. 1998; 78(7): 940–944.
106. Nicholson B, Schaefer G, Theodorescu D. Angiogenesis in prostate cancer: biology and therapeutic opportunities. Cancer Metastasis Rev. 2001; 20(3–4):297–319.
107. Kerbel RS. Tumor angiogenesis. N Engl J Med. 2008; 358(19):2039–2049.
108. Makale M. Intravital imaging and cell invasion. Methods Enzymol. 2007; 426:375–401.
109. Makale M. Noninvasive imaging of blood vessels, chapter 8. Methods Enzymol. 2008; 444:175–199.
110. Bingle L, Lewis CE, Corke KP, Reed MW, Brown NJ. Macrophages promote angiogenesis in human breast tumour spheroids in vivo. Br J Cancer. 2006; 94(1):101–107.
111. Read TA, Farhadi M, Bjerkvig R, Olsen BR, Rokstad AM, Huszthy PC, Vajkoczy P. Intravital microscopy reveals novel antivascular and antitumor effects of endostatin delivered locally by alginate-encapsulated cells. Cancer Res. 2001; 61(18):6830–6837.
112. Dellian M, Witwer BP, Salehi HA, Yuan F, Jain RK. Quantitation and physiological characterization of angiogenic vessels in mice: effect of basic fibroblast growth factor, vascular endothelial growth factor/vascular permeability factor, and host microenvironment. Am J Pathol. 1996; 149(1):59–71.
113. Frost GI, Lustgarten J, Dudouet B, Nyberg L, Hartley-Asp B, Borgstrom P. Novel syngeneic pseudo-orthotopic prostate cancer model: vascular, mitotic and apoptotic responses to castration. Microvasc Res. 2005; 69(1–2):1–9.
114. Cunha GR, Hayward SW, Dahiya R, Foster BA. Smooth muscle-epithelial interactions in normal and neoplastic prostatic development. Acta Anat (Basel) . 1996; 155(1):63–72.
115. Condon MS, Bosland MC. The role of stromal cells in prostate cancer development and progression. In Vivo. 1999; 13(1): 61–65.
116. Olumi AF, Grossfeld GD, Hayward SW, Carroll PR, Tlsty TD, Cunha GR. Carcinoma-associated fibroblasts direct tumor progression of initiated human prostatic epithelium. Cancer Res. 1999; 59(19):5002–5011.

117. Dvorak HF. Tumors: wounds that do not heal. Similarities between tumor stroma generation and wound healing. N Engl J Med. 1986; 315(26):1650–1659.
118. Schmid-Schonbein GW, Skalak R, Simon SI, Engler RL. The interaction between leukocytes and endothelium in vivo. Ann N Y Acad Sci. 1987; 516:348–361.
119. Suematsu M, Suzuki H, Delano FA, Schmid-Schonbein GW. The inflammatory aspect of the microcirculation in hypertension: oxidative stress, leukocytes/endothelial interaction, apoptosis. Microcirculation. 2002; 9(4):259–276.
120. Ohashi KL, Tung DK, Wilson J, Zweifach BW, Schmid-Schonbein GW. Transvascular and interstitial migration of neutrophils in rat mesentery. Microcirculation. 1996; 3(2): 199–210.
121. Gerhardt H, Golding M, Fruttiger M, Ruhrberg C, Lundkvist A, Abramsson A, Jeltsch M, Mitchell C, Alitalo K, Shima D, Betsholtz C. VEGF guides angiogenic sprouting utilizing endothelial tip cell filopodia. J Cell Biol. 2003; 161(6):1163–1177.
122. Houck KA, Leung DW, Rowland AM, Winer J, Ferrara N. Dual regulation of vascular endothelial growth factor bioavailability by genetic and proteolytic mechanisms. J Biol Chem. 1992; 267(36):26031–26037.
123. Eyler CE, Rich JN. Survival of the fittest: cancer stem cells in therapeutic resistance and angiogenesis. J Clin Oncol. 2008; 26(17): 2839–2845.
124. Hill RP. Identifying cancer stem cells in solid tumors: case not proven. Cancer Res. 2006; 66(4):1891–1895, discussion 1890.
125. Conti L, Pollard SM, Gorba T, Reitano E, Toselli M, Biella G, Sun Y, Sanzone S, Ying QL, Cattaneo E, Smith A. Niche-independent symmetrical self-renewal of a mammalian tissue stem cell. PLoS Biol. 2005; 3(9):e 283.
126. Dalerba P, Dylla SJ, Park IK, Liu R, Wang X, Cho RW, Hoey T, Gurney A, Huang EH, Simeone DM, Shelton AA, Parmiani G, Castelli C, Clarke MF. Phenotypic characterization of human colorectal cancer stem cells. Proc Natl Acad Sci USA. 2007; 104(24): 10158–10163.
127. Singh SK, Clarke ID, Terasaki M, Bonn VE, Hawkins C, Squire J, Dirks PB. Identification of a cancer stem cell in human brain tumors. Cancer Res. 2003; 63(18):5821–5828.
128. Singh SK, Hawkins C, Clarke ID, Squire JA, Bayani J, Hide T, Henkelman RM, Cusimano MD, Dirks PB. Identification of human brain tumour initiating cells. Nature. 2004; 432(7015):396–401.
129. Galli R, Binda E, Orfanelli U, Cipelletti B, Gritti A, De Vitis S, Fiocco R, Foroni C, Dimeco F, Vescovi A. Isolation and characterization of tumorigenic, stem-like neural precursors from human glioblastoma. Cancer Res. 2004; 64(19):7011–7021.
130. Prince ME, Sivanandan R, Kaczorowski A, Wolf GT, Kaplan MJ, Dalerba P, Weissman IL, Clarke MF, Ailles LE. Identification of a subpopulation of cells with cancer stem cell properties in head and neck squamous cell carcinoma. Proc Natl Acad Sci USA. 2007; 104(3):973–978.
131. Al-Hajj M. Cancer stem cells and oncology therapeutics. Curr Opin Oncol. 2007; 19(1): 61–64.
132. Gu Y, Li H, Miki J, Kim KH, Furusato B, Sesterhenn IA, Chu WS, McLeod DG, Srivastava S, Ewing CM, Isaacs WB, Rhim JS. Phenotypic characterization of telomerase-immortalized primary non-malignant and malignant tumor-derived human prostate epithelial cell lines. Exp Cell Res. 2006; 312(6):831–843.
133. Ben-Porath I, Weinberg RA. The signals and pathways activating cellular senescence. Int J Biochem Cell Biol. 2005; 37(5):961–976.
134. Dimri GP, Martinez JL, Jacobs JJ, Keblusek P, Itahana K, Van Lohuizen M, Campisi J, Wazer DE, Band V. The Bmi-1 oncogene induces telomerase activity and immortalizes human mammary epithelial cells. Cancer Res. 2002; 62(16):4736–4745.
135. Park IK, Morrison SJ, Clarke MF. Bmi1, stem cells, and senescence regulation. J Clin Invest. 2004; 113(2):175–179.

136. Kim SJ, Uehara H, Karashima T, McCarty M, Shih N, Fidler IJ. Expression of interleukin-8 correlates with angiogenesis, tumorigenicity, and metastasis of human prostate cancer cells implanted orthotopically in nude mice. Neoplasia. 2001; 3(1):33–42.
137. Murphy C, McGurk M, Pettigrew J, Santinelli A, Mazzucchelli R, Johnston PG, Montironi R, Waugh DJ. Nonapical and cytoplasmic expression of interleukin-8, CXCR1, and CXCR2 correlates with cell proliferation and microvessel density in prostate cancer. Clin Cancer Res. 2005; 11(11):4117–4127.
138. Collins AT, Berry PA, Hyde C, Stower MJ, Maitland NJ. Prospective identification of tumorigenic prostate cancer stem cells. Cancer Res. 2005; 65(23):10946–10951.
139. Gu G, Yuan J, Wills M, Kasper S. Prostate cancer cells with stem cell characteristics reconstitute the original human tumor in vivo. Cancer Res. 2007; 67(10):4807–4815.
140. Kenny PA, Lee GY, Bissell MJ. Targeting the tumor microenvironment. Front Biosci. 2007; 12:3468–3474.
141. Ross JS. The androgen receptor in prostate cancer: therapy target in search of an integrated diagnostic test. Adv Anat Pathol. 2007; 14(5):353–357.
142. Ross JS, Foster CS. Molecular oncology of prostate cancer. Sudbury: Jones and Bartlett Publishers; 2007.
143. Silvestre JS, Levy BI, Tedgui A. Mechanisms of angiogenesis and remodelling of the microvasculature. Cardiovasc Res. 2008; 78(2):201–202.

Enhanced Cardiac Differentiation of Mouse Embryonic Stem Cells by Electrical Stimulation

Paul R. Bidez III, J. Yasha Kresh, Yen Wei, and Peter I. Lelkes

Abstract Cardiovascular diseases account for more deaths than any other illness. Cardiac tissue engineering has turned to embryonic stem cells as a renewable source of myocytes for use in tissue replacement. Existing methods for stem cell differentiation toward the cardiac lineage are relatively non-specific, yielding low numbers of myocytes with varying contraction frequencies and strengths. Here we describe novel experimental approaches, utilizing an electrical stimulation regimen, aimed at increasing the efficiency of cardiac differentiation from mouse embryonic stem (mES) cells. These methods generate cardiac myocytes with functional characteristics that more closely resemble native tissues. The amplitude, duration, and frequency of the electrical stimulus as well as the timing of its onset are some of the critical experimental parameters that determine the enhancement of cardiac differentiation.

In order to form embryoid bodies, an optimum differentiation regime was followed incorporating the hanging drop method followed by suspension culture and subsequent post-plating on conductive slides with electrical stimulation. Approximately three times more stimulated mES cells exhibited evidence of cardiac differentiation than their non-stimulated counterparts, as determined by the expression of ventricular marker myosin light chain-2v. Spontaneous contractions of the stimulated cell populations began up to 1 day earlier and had an average beat frequency close to that of the stimulus applied during differentiation. The spontaneously contracting regions had larger areas of contraction, which beat more rhythmically, as determined by real-time digital imaging analysis.

Our results suggest that appropriate electrical stimulation generates greater numbers of more robust cardiac myocytes, which in turn may be better suited for repairing or regenerating an ailing heart and for use as 3D model systems for drug discovery.

P.I. Lelkes (✉)
Laboratory of Cellular Tissue Engineering, School of Biomedical Engineering, Science, and Health Systems, Drexel University, Philadelphia, PA, USA
e-mail: pilelkes@drexel.edu

G.M. Artmann et al. (eds.), *Stem Cell Engineering*,
DOI 10.1007/978-3-642-11865-4_5,

Keywords Mouse embryonic stem cells · Cardiac differentiation · Myosin light chain-2v · Electrical stimulation · Conductive surfaces · Indium-Tin-Oxide · Optical Recording · Multi-Electrode Assembly

1 Introduction

Cardiovascular diseases (CVDs) are reported by the American Heart Association (AHA) to have been responsible for ~34.2% of all deaths in 2006 or approximately 830,000 patients, in the USA alone [1]. Of these, 52% were directly caused by coronary heart disease, a primary result of infarcted ventricular tissue. Unlike most other cell types, adult cardiac myocytes, the functional components of the heart muscle, are incapable of replication or do so on a very small scale (<1% per year) [2]; however, the demand for regeneration far outstrips the supply of proliferative cardiomyocytes. As a result, when the heart is damaged through disease or injury, as in the case of a heart attack induced by coronary artery disease, there is limited internal mechanism for functional repair and the patient is left with decreased cardiac output, which rarely improves over time. Conservative treatments or interventional procedures such as angioplasty to open clogged arteries may temporarily alleviate some of the immediate causes of hypoxic damage to the myocardium but may also induce additional, free radical-based oxidative damage. However, current interventional treatments are unable to actually repair or regenerate an ailing heart. In some cases, the only option is for a rare total heart replacement or a mechanical ventricular assist implant. Unfortunately, the demand for exceeds the supply. To date, artificial hearts remain an emergency solution only, mostly in use as bridge to transplantation.

In recent years, the focus on cardiac repair has shifted to tissue engineering as a tool for generating cell-based "spare parts" for in vivo implantation, such as cardiac patches or heart valves, or for creating functional 3D myocardial tissue constructs as diagnostic platforms for enhancing in vitro drug discovery testing. A wide variety of approaches have been pursued, with most following the "classic" tissue engineering stratagem of culturing cells on a biomimetic scaffold that is then placed inside of a bioreactor for a period of time, generating a tissue construct.

Some of the main components for successful tissue engineering include the physiochemical properties and the ultrastructure of the scaffolds and their ability to mimic in vivo conditions. Among the major challenges of tissue engineering is the design of new bioactive materials with unique characteristics tailored to tissue-specific requirements, such as viscoelastic properties, pore size and interconnectivity of porous scaffolds, fiber diameter for fibrous scaffolds, and controlled biodegradability. Electrical conductivity of the scaffolds is potentially a critical component for cardiac and neuronal tissue engineering, considering that response to electrical stimulation is a key feature of neuron and myocyte differentiation and function [3, 4]. Three-dimensional scaffolds comprised of nanofibers, and incorporating specific biological components, including electrical signaling, will be uniquely suited to mimic some of the matrix-resident cues, which fine-tune cellular differentiation and function in vivo. Of late, porous and fibrous scaffolds or 2D substrates containing electrically conductive and electroactive polymers, such as polypyrrole or polyaniline, were shown to enhance neuronal and cardiac differentiation [5–7]. For example, polyaniline, after initial passivation treatment, supports cell growth and proliferation while remaining conductive for at least 100 h [7] while polyaniline films induced neuronal differentiation of PC12 cells in the absence of electrical stimulation [8]. Similarly, indium tin oxide (ITO) coated surfaces are electrically conductive and conducive to cellular growth and differentiation [9]. As an added benefit, ITO substrates are optically transparent allowing the cultures to be observed in real time during the course of the experiment. It is believed that the inclusion of tailored biomimetic elements in scaffolds, such as soluble growth factors, electrical conductivity, and biodegradable structural elements, will result in enhanced functional properties in the resultant tissues.

The ideal cell source for populating tissue-engineered constructs is widely debated. For cardiac tissue engineering, most preclinical studies in animal models have utilized primary isolates of adult, neonatal, and fetal cells of cardiac or skeletal origin. The use of fetal or neonatal heart cells clinically is not an option as the rat model requires killing up to 40 rat pups to isolate just $2–3 \times 10^8$ cardiomyocytes – still a far cry from the actual cell density needed to reconstruct a heart. The recent attempts to re-create a beating rat heart by infusing $50–75 \times 10^6$ isolated neonatal myocytes proved that this approach is feasible, in principle. However, probably because the cell density in the neo-construct was too low, the contractile strength of the reconstituted heart was significantly lower than that of a native heart [10].

In addition to designing smart materials for cardiac repair, adult and embryonic stem cells present the most promising option for cardiac regeneration [11, 12]. The latter are thought to be more plastic. However, both have been shown to successfully differentiate into contractile cardiac myocytes [13, 14]. Pluripotent mouse embryonic stem (mES) cells have been available for 25 years [15]. Mouse ES cells are particularly useful, as they are easier to maintain than human embryonic stem (hES) cells, raise fewer ethical and bureaucratic red flags, and share many properties with their human counterparts [16]. The availability of mES cells, their pluripotent differentiative properties, ease of culture, and the wealth of murine genetic information currently available make these cells an excellent model for cardiac differentiation studies. While all details of cardiac differentiation in vitro from embryonic stem cells are not yet known, the process is well characterized in terms of numerous stage/differentiation-specific markers. For recent reviews, see Bettiol et al. [17] or Mummery et al. [18].

The earliest marker of cardiac cellular commitment and differentiation is the homeobox gene Nkx-2.5. In the developing mouse embryo Nkx-2.5 is first expressed between days 3 and 6 post coitum (pc) [19] and it can be detected in vitro after 2 days of differentiation [20]. Nkx-2.5 appears to be critical for the initiation of cardiogenic differentiation, since in Nkx-2.5$^{-/-}$ mutants many of the downstream cardiac markers such as myosin light chain-2v (MLC-2v), cardiac ankyrin repeat protein (CARP), and HAND1 are greatly downregulated or completely undetectable [21–24]. Myosin, a key component of muscle contraction, is comprised of one heavy chain and two pairs of light chains, the first being essential and the other regulatory [25]. In the mouse, α- and β-isoforms of the myosin heavy chain (MHC) begin to be expressed at embryonic day (ED) 5. The β-isoform of MHC is the first to be expressed in both embryos and developing embryoid bodies (EBs), validating the use of EBs as a model of in vivo development [26]. In EBs, β-MHC has a bimodal expression pattern first appearing around days 3 and 4, missing or decreased in days 5–7, and strong expression thereafter [27, 28]. In contrast, α-MHC begins to show up on day 8 in beating EBs, but only in those that originally expressed β-MHC. Similarly, α-MHC in the developing embryos began to be expressed on day 7.5 pc. This chronologic expression allows these markers to be useful in determining both the rate and specificity of differentiation, with the β-isoform being particularly useful as an early predictor of ventricle-specific differentiation.

In cardiac muscle there are four myosin light chain forms, MLC-1a, -2a, -1v, and -2v, which are differentially expressed in the heart, with the "a" forms localized to the atria and the "v" forms in the ventricles of the adult [29]. MLC-2v is the earliest ventricle-specific marker [30]. Lack of MLC-2v leads to decreased ventricular function, sarcomere abnormalities, and total embryonic lethality by day 13.5 [31]. In the developing embryo, MLC-2v is expressed prior to looping morphogenesis from day 8.5 onwards. Its expression is restricted to areas of the developing ventricle [32, 28, 33]. In vitro, MLC-2v is expressed from day 7 onwards [20]. The restriction of expression to the ventricular regions makes MLC-2v an excellent marker for ventricular specialization.

2 Common Routes Toward Cardiac Differentiation of mES Cells

Methods of directed stem cell differentiation, cell selection, and tissue assembly, as well as delivery and host integration, remain prime areas of interest in tissue engineering and regenerative medicine. Given the focus of this chapter, we will concentrate on the issue of directed differentiation of embryonic stem cells targeted toward the cardiac lineage. The goal for efficient differentiation methods is to obtain sufficient numbers of differentiated tissue-specific parenchymal and/or lineage-specific progenitor cells for use in regenerative therapies. The safety issues surrounding stem cell transplantation highlight the urgency for completely eliminating any and all undifferentiated embryonic stem cells. Specifically, ES cells pose risks of teratoma formation if undifferentiated stem cells are not carefully removed or pre-differentiated before transplantation into the body [34].

Stem cell differentiation in vitro is typically initiated by culturing stem cells in embryoid bodies (EBs). This is accomplished through a variety of methods including hanging drops, suspension culture, methylcellulose culture, or cultures of single cells in rotating wall vessel (RWV) bioreactors [35, 36]. The suspension culture method creates a large number of embryoid bodies with no control over the size of the resultant aggregates. In contrast, the hanging drop method affords the greatest degree of control over cell numbers per aggregate by physically limiting the numbers of cells in each drop, and thus each aggregate, as seen in Fig. 1. The hanging drop method results in a very consistent, albeit tedious, method for creating

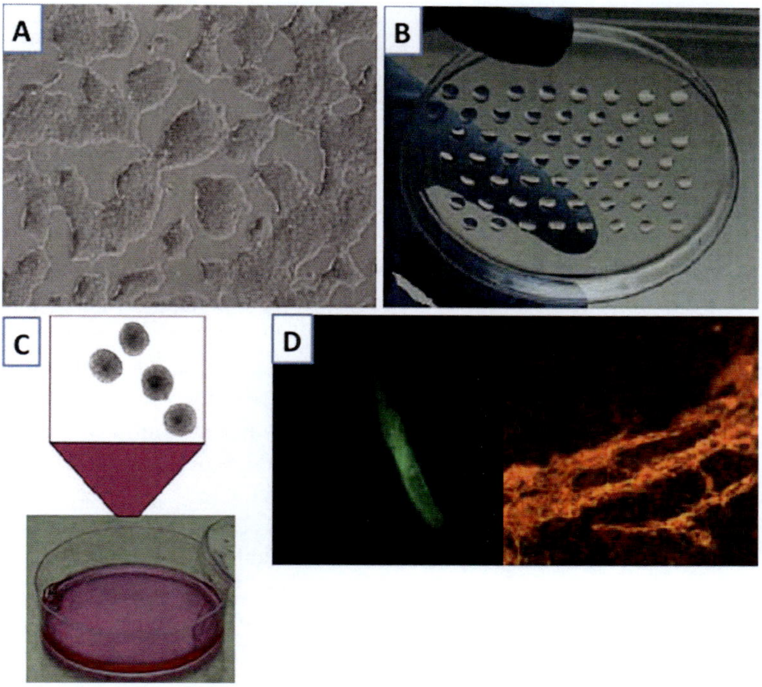

Fig. 1 Murine embryonic stem (mES) cell culture and cardiac differentiation. (**a**) Undifferentiated mES cells (original magnification ×40). (**b**) Hanging drop technology to generate embryoid bodies (EBs). (**c**) Inoculation of EBs into suspension culture, prior to plating. (**d**) *Left* – fluorescence micrograph of single GS-ES-CMV27 cell, expressing surrogate cardiac ventricular differentiation marker (GFP driven by MLC-2v promoter). *Right* – fluorescence micrograph of group of cardiac differentiated GS-ES cells, stained for MLC-2v

homogeneous populations of embryoid bodies. All of these methods have proven useful in the generation of spontaneously beating cardiac myocytes. Typically 8% of the EBs demonstrate spontaneous contractions within 6 days post-plating, which exceeds 40% by day 12 [37, 38]. Cardiac myocytes generated from embryonic stem cells display electrophysiological properties of nodal, ventricular, and atrial-like cells, resembling those of the embryonic heart tube [39, 40]. In vitro generated cardiac myocytes also express typical makers of functional cardiac tissue including, but not limited to, Nkx-2.5, myosin heavy chain, cardiac troponin, myosin light chains-1a, -1v, -2a, -2v, and connexins 40, 43, and 45 [20].

Efficiency and specificity of differentiation toward cardiac lineages have been enhanced through the addition of growth factors. For example, platelet-derived growth factor-BB (PDGF-BB) in concert with reduced serum or serum replacement medium has been shown to accelerate and enhance differentiation of mES cells toward cardiac myocytes [41]. Similarly, retinoic acid increases cardiac differentiation efficiency, specifically toward ventricle-like cardiomyocytes [42–44]. However, it is likely that retinoic acid acts non-specifically as it also exhibits potent effects in neuronal development [45, 46]. Interestingly, both retinoic acid and dimethyl sulfoxide (DMSO), while effective in mouse cells, are ineffective as enhancers of cardiac differentiation in human cells [47].

Human embryonic stem (hES) cells, cultured in hepatocellularcarinoma (HepG2) cell-conditioned medium, differentiate into beating cardiac myocytes in less time and in greater numbers than "normal" controls cultured in the absence of HepG2-conditioned medium [37]. The soluble factors secreted into the medium by the HepG2 cells resulted in the expression of Nkx-2.5, the earliest cardiac marker, 2 days earlier than the cells in control medium.

Enhancing the efficacy of cardiac derivation of ES cells involves both increasing the numbers of cells differentiated toward the cardiac lineage and selecting against those ES derivatives that have developed other phenotypes. Some separation methods rely on physical properties utilizing gradient centrifugation while others have taken a more biological approach. When murine ES cells were transfected with an α-cardiac myosin heavy chain (α-cMHC) promoter driving a neomycin resistance gene selection, enrichment was possible by the introduction of G418 which effectively killed off the non-α-cMHC-expressing cells, leaving the cardiac-specific lineage [48]. This method allows for purification of cell populations resulting from the non-homogeneous differentiation methods currently available.

Physical or non-biologic stimuli have also been shown to be potent activators or accelerators of cellular differentiation and maturation. For example, continuous stimulation of mES cells with a low frequency (50 Hz, 0.8 mT rms) magnetic field for 3–10 days resulted in an increase in the expression of the cardiac markers Nkx-2.5, MHC, and MLC-2v [49]. Chronic stretch has been shown to induce cardiac myocyte hypertrophy and increase contractile function in engineered heart tissue reconstituted from neonatal rat and chick myocytes [50].

In the past numerous attempts were undertaken to study the effects of electric stimulation on cells and tissue, frequently for therapeutic purposes, e.g., for accelerating bone healing [51, 52] or for promoting recovery after spinal cord injury [53, 54]. In the context of this chapter, electrical stimulation has predominantly been used to improve the function of tissue-engineered cardiac constructs made from neonatal cardiomyocytes [55]. Electrical stimulation increases interconnectivity and contractile function in engineered constructs of neonatal myocytes grown on collagen sponges and skeletal myoblasts grown on fibrous polyglycolic acid [55–57]. Evidence suggests that this increased intercellular communication is due to the increased activity of L-type calcium channels [56]. These stimulation protocols relied on the conductivity of the medium for transmission of the electrical stimulus; however, direct stimulation via a conductive substrate has also been explored.

To the best of our knowledge, Christine Schmidt and her colleagues were the first to demonstrate that electrical stimulation through a conductive surface enhanced neuritogenesis in PC12 pheochromocytoma cells cultured on conductive polypyrolle films [58]. Subsequently a few groups, including us, have started to reduce into practice the idea of modulating the differentiation of stem and progenitor cells using electrical currents and, hence, generating more efficiently "better" tissue-engineered constructs [59, 60, 6]. Given the importance of stem cells as general cell sources for tissue engineering and regenerative medicine, and specifically of electrical currents and stimulation for neural, cardiac, and musculoskeletal tissue engineering, it is astonishing that only very few groups have thus far published on optimizing stem cell differentiation by using electrical stimulation. For example, Abilez et al. [60] reported better coordination and larger numbers of spontaneously beating myocyte colonies following optimized electrical stimulation of P19 myogenic progenitor cells. Similarly, Chen et al. [61] described a narrow window of optimal electric stimulation regimen that was able to induce cardiac differentiation of mouse ES-D3 embryonic stem cells. While both these groups reported augmentation of spontaneous (cardiac) differentiation by electrical stimulation, no quantization was provided. Using a slightly different stimulation regiment, Yamada et al. [62] demonstrated that "mild" electrical stimulation modulates the fate determination of differentiating mES cells toward a neuronal phenotype. Interestingly, in contrast to chemically induced differentiation, electrically induced mES cells can differentiate into multipotent neurons that are plastic in terms of their capacity to differentiate into a wide variety of specific neuronal cell types.

Conductive polymers, in particular polypyrrole, have been studied as potential electroactive growth substrates for diverse cell- and tissue culture applications, with the implicit assumption that electrical stimulation, i.e., electrical currents, through these surfaces can control the shape and function of anchorage-dependent cells [63]. Biocompatibility of these conductive substrates was demonstrated by the fact that in the absence of electrical stimulation acetylcholine-stimulated catecholamine secretion of neural tube-derived adrenal medullary chromaffin cells grown on polypyrrole was equal to that of cells grown on collagen [64]. Electrical stimulation enhanced nerve growth factor (NGF)-induced neuronal differentiation of chromaffin cell-derived PC12 pheochromocytoma cells grown on polypyrrole as well. Specifically, a brief period of electrical stimulation (100 mV for 2 h) potentiated NGF-induced neuritogenesis, as assessed by the length of neurons in cells subjected to electrical stimulation [58]. More recently, this same group [65] reported that electrical stimulation of polypyrrole resulted in altered fibronectin adsorption to the conductive surface, which in turn enhanced neuritogenesis in PC12 cells.

The fundamental premise for our work is that stem cell fate and differentiation are influenced by numerous exogenous and endogenous factors in vivo, including electrical currents set up by ion gradients. We hypothesized that exogenous electrical stimulation will enhance/accelerate the tissue-specific differentiation of embryonic stem cells toward cardiac myocytes and that this effect is more pronounced on a substrate with innate electrical conductivity, as opposed to relying on the conductivity of the medium.

3 Electric Stimulation of mES Cell

Conductive polymers, such as polypyrrole and polyaniline, and composites, such as functionalized polypyrrole or polyaniline-reinforced polycaprolactone, are useful as substrates for cellular growth and effective stimulation [58, 7, 66, 67]. While these substrates are suitable for cell growth, they are difficult to work with; even thin films of these materials are opaque, rendering real-time visualization in a conventional inverted light microscope during cell culture impossible. However, cells grow and proliferate unimpeded on glass slides coated with a conductive layer of indium tin oxide (ITO, from Sigma, St. Louis, MO). The 150–300 Å coating of ITO on these slides has a resistance of 70–100 Ω and an optical transparency >87%, making them very useful to continually observe cells during the multi-week course of culture.

The electrical stimulation system housing the ITO-coated slides and used in the studies described below was designed and assembled in our lab (Fig. 2). The cells were grown on the ITO slides (Fig. 2a), assembled in two manifolds that can hold up to eight slides (Fig. 2b), and stimulated via an eight-channel system driven by a Grass stimulator S44. This system is capable of generating pulsed or continuous stimuli. The stimulator output has one channel with controls for stimulus rate, duration, and intensity, while the stimulus attenuator (Grass Telefactor SS2-SA1) splits and attenuates the signal into four channels with individual intensity controls. The Stimusplitter (Grass Telefactor SS2) further splits these four channels into a total of eight channels. The entire electronics is placed on top of a cell culture incubator (Fig. 2c). The wires are fed to the stimulating manifold through the gasket around the front door of the incubator.

In contrast to most prior studies that apply electrical stimulation via a conductive medium, our choice of a conductive surface leads to cell stimulation via the substratum, thus mimicking more realistically cell stimulation in situ through the extracellular matrix (ECM). We experimentally determined the resistance across the slide (on average 207.65 ± 27.72 Ω), while resistance of the fluid was 140 ± 2 kΩ. An approximate circuit diagram for the system is depicted in Fig. 3. Thus, we can safely assume that >99.5% of any applied current will travel primarily on the surface of the slide and not through the medium. The cells are resting on this

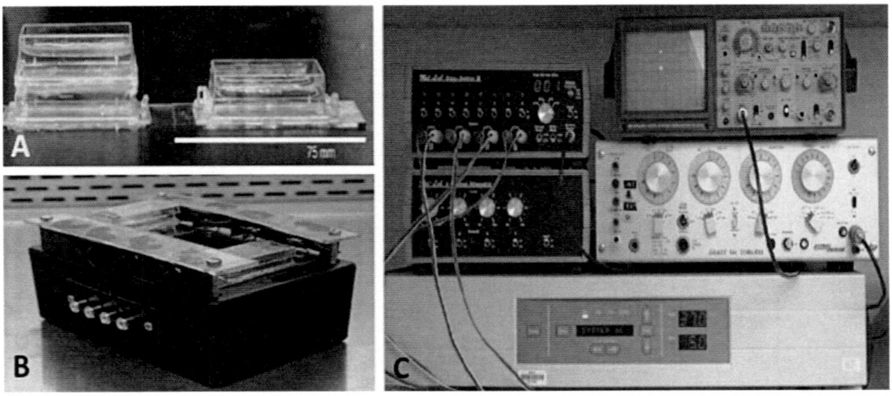

Fig. 2 Stimulation setup on top of cell culture incubator: Pictured are – starting at the *right* and moving clockwise – Grass stimulator S4, stimulus attenuator, Stimusplitter and oscilloscope for monitoring pulses (Fig. 2c), one of the manifolds that can hold 4 slides (Fig. 2b), and ITO-coated culture chambers (Fig. 2a) that are placed into the manifolds. For details, see text

The parallel resistance for this system is given by the following equation:

$$R_s = 1/(1/R_{slide} + 1/R_{medium})$$

$$R_s = 1/(1/207.65 + 1/140000) \quad = \quad \underline{207.34\ \Omega}$$

The current is given by:

$$I = \frac{V}{R}$$

$$I_{slide} = \frac{V}{R_{slide}} = \frac{0.1\ V}{207.65\ \Omega} = 4.82E\text{-}4\ A$$

$$I_{medium} = \frac{V}{R_{medium}} = \frac{0.1\ V}{140{,}000\ \Omega} = 7.14E\text{-}7\ A$$

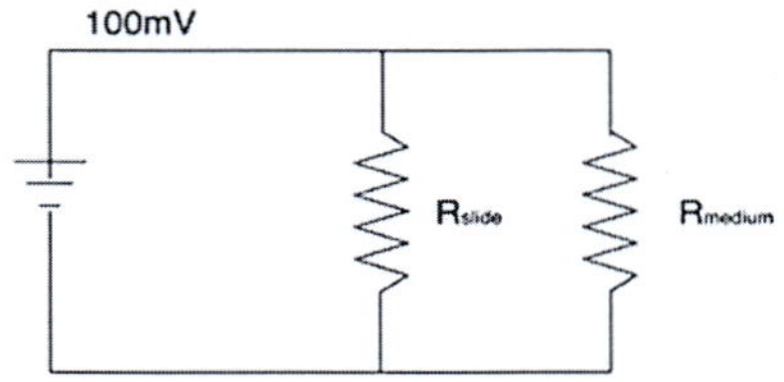

$$I_{combined} = I_{slide} + I_{medium}$$

$$I_{combined} = 4.82E\text{-}4\ A + 7.14E\text{-}7\ A = \underline{4.82E\text{-}4\ A}$$

Fig. 3 Circuit diagram for stimulated ITO slide: In addition to the simplified diagram the figure also denotes the calculation of the currents through the conductive surface and the culture medium, respectively. For details see text

surface and thus experience the stimulation primarily through direct contact with the stimulated surface.

For our studies we used two lines of mouse embryonic stem cells ES-D3 and GS-ES. The commercially available ES-D3 cell line (ATCC-1934) is capable of differentiating into functional cardiac myocytes [15, 27]. GS-ES-CMV27 cells were isolated from the same mouse strain as the ES-D3 [68]. However, this particular clone differentiates into cardiac myocytes more quickly than the ES-D3 cells, as assessed by the expression of MLC-2v [69]. In our experiments, spontaneous contractions began 3–5 days earlier, and the expression of MLC-2v was clearly visible ~5 days sooner than in ES-D3 cells. GS-ES-CMV27 cells are particularly useful for assessing cardiac differentiation, since they are transfected with enhanced green fluorescent protein (EGFP), the expression of which is driven by the MLC-2v promoter [69]. Upon differentiation into ventricular myocytes, cells expressing MLC-2v are detectable by fluorescent microscopy. Purified populations of myocytes can be obtained by sorting in a fluorescence-activated cell sorter (FACS). Since MLC-2v is exclusively expressed by ventricular myocytes, these cells have an onboard diagnostic system to aid in determining the efficiency of the differentiation method with respect to ventricular specification. For our study, all mES cells were maintained under standard feeder-free cell culture conditions in 10% ES-grade fetal bovine serum (FBS) containing maintenance medium (DMEM–high glucose) supplemented with LIF/ESGRO, see Fig. 1. The stemness, i.e., the undifferentiated state of the ES cells, was validated by immunostaining for OCT4 and SSAE-1, two established markers for undifferentiated pluripotent mES cells [70, 71]. For cell culture via the hanging drop method, the serum concentration was raised to 20%. Differentiation was initiated by lowering the serum levels to <1%. Cardiac differentiation was assessed by quantitative immunohistochemistry for specific markers and FACS analysis. Functional parameters of the ensuing cardiac myocytes/cardiac syncytium (contractility and electrical activity) were acquired through a combination of optical video recording (AVI files) and monitoring electrical activity in a MicroElectrode Array setup (MEA, Multi Channel Systems, Reutlingen, Germany). The AVI files were analyzed using the Video Spot Tracker v05.02 (NIH) analysis software and the results (contraction peak amplitude, frequency, etc.) were assessed by the Heart Rate Variability Analysis software provided by the Department of Applied Physics, University of Kuopio, Kuopio, Finland [72].

Based on the hypothesis that one of the key differentiation cues in the environment around the developing heart is the presence of electrical activity including electrical currents set up by ion gradients, we set out to re-create an environment with electrical cues similar to those found in a functional heart. The experimental hypothesis was that under optimal conditions electrical stimulation will enhance mES differentiation toward cardiac myocytes. In exposing mES to electrical stimulation there are several basic questions that have to be addressed first:

- What is the correct stimulus profile in terms of potential, frequency, intervals, etc.?
- When does one start to stimulate the cells (as EBs, in suspension, after plating)?
- How long does one continue to stimulate?

Preliminary studies quickly taught us that there is an optimal window for electrical stimulation of mES cells. Too little, infrequent stimuli do not do anything to alter the endogenous cell fate; too large, too frequent stimuli effectively kill the cells. After an arduous period of trial-and-error studies, we identified a stimulation optimum of 100 mV, 20 ms, 1.5 Hz as the most effective pattern for the post-plating differentiation of embryoid bodies. Premature differentiation/stimulation without the hanging drop/embryoid body (EB) phase and/or in low serum was also deleterious. The question then remained when to start electrical stimulation. To determine the optimal time point of differentiation in terms of marker expression and functionality/contraction, hanging drops (500 cells/20 μl) were formed for 2 days in 20% FBS medium (see Fig. 1b). EBs were then transferred for 2–5 days into suspension culture in 10% FBS medium (see Fig. 1c), after which they were plated onto gelatin-coated ITO slides. Electrical stimulation was initiated immediately, in the same medium. The medium was exchanged for DMEM with 0.2% FBS, 24 h after the onset of stimulation, with daily medium exchanges thereafter under continued electrical stimulation. As seen in Fig. 4 the optimal stimulation (in terms of endogenous MLC-2v–GFP expression) was a regimen of 2 days hanging drop and 3 days in suspension culture. Interestingly, both too short (<2 days) or too long (>4 days) a sojourn in suspension culture resulted in suboptimal priming of the cells for the subsequent stimulation. This (2–3 days) regimen was used in all subsequent experiments. In addition to the initial assessment of MLC-2v by GFP expression, we also labeled GS-ES cells with a commercially available monoclonal antibody against MLC-2v and quantified the MLC-2v positive areas in untreated controls and optimally treated EBs by quantitative image analysis. As seen in Figs. 5 and 6 there is an approximately 50% increase in the area of MLC-2v reactivity after 14 days.

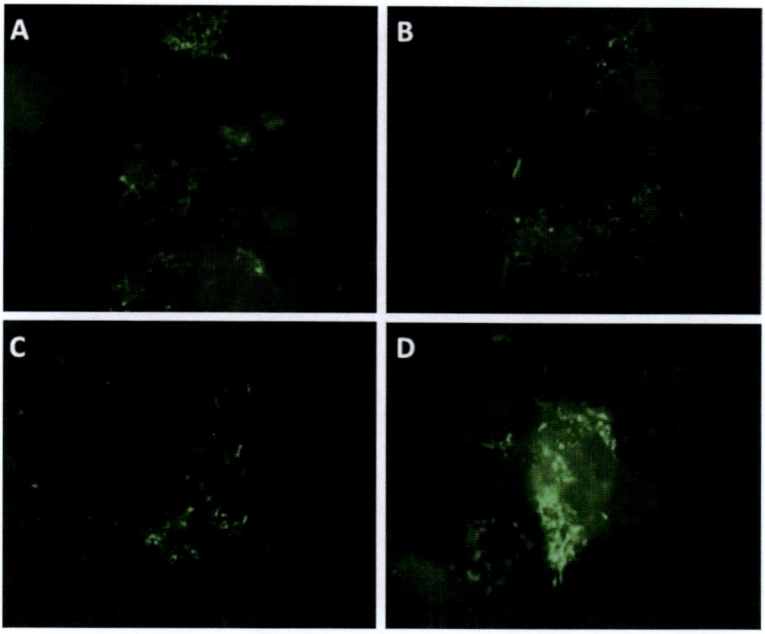

Fig. 4 Optimization of stimulation protocol: Endogenous, MLC-2v promoter-driven GFP expression was used to assess on day 18 of the experiments the effects of various stimulation protocols on cardiac differentiation of GS-ES-CMV27 cells. The different stimulation regimen entailed differences in the length of time of maintaining the cells as EBs, as well as the beginning and duration of electrical stimulation following plating. (**a**, **c**) Unstimulated, (**b**, **d**) stimulated. (**a**, **b**) 2 days as EB, beginning of stimulation at day 3 after plating. (**c**, **d**) 2 days as EB, beginning of stimulation at day 4 after plating

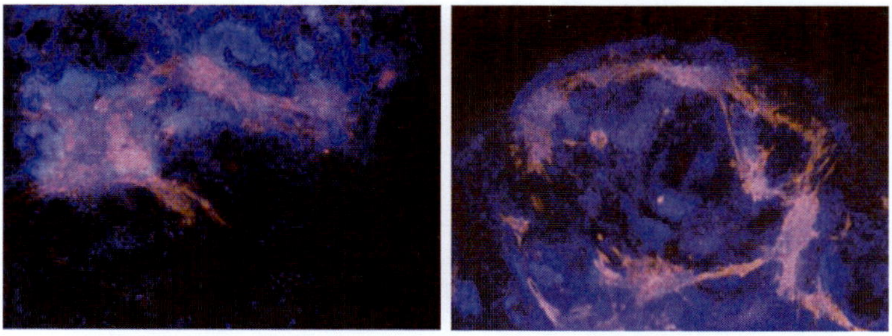

Fig. 5 Cardiac differentiation in day 14 (5 + 9) embryoid bodies from GS-ES cells: representative images of unstimulated (*left*) and stimulated EBs (*right*); cells/areas of cardiac/ventricular differentiation are labeled *red* with a fluorescent monoclonal antibody against MLC-2v. Cells are also stained with bisbenzimide, which lightly stains nuclear DNA *blue*. *Red areas* coincide with spontaneously contractile regions

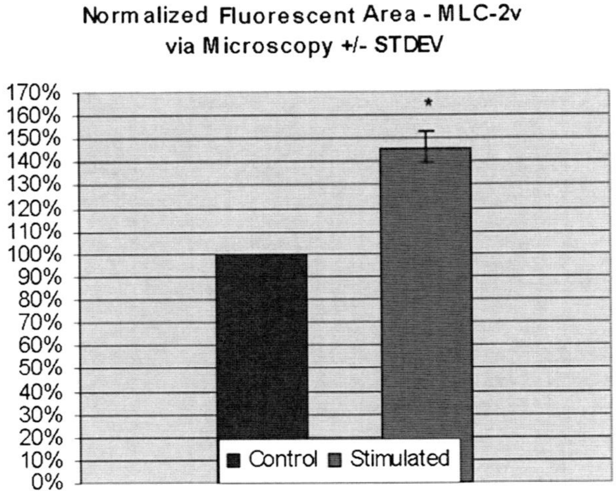

Fig. 6 Electrical stimulation enhances cardiac differentiation I: normalized area of MLC-2v positive GS-ES cells (left bar) as assessed by quantitative image analysis performed on 300 images similar to those depicted above, *red channel* only. Images from three separate experiments were stacked, thresholded, and the brightly fluorescent measured via ImageJ software. Taken together, the stimulated set (right bar) was found to have a statistically significant average increase of 46.1% in the MLC-2v expressing cells, vs. non-stimulated controls. $N = 3$, $P < 0.05$

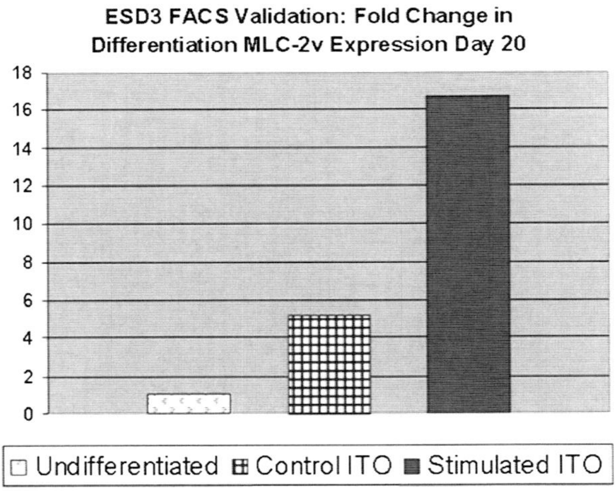

Fig. 7 Electrical stimulation enhances cardiac differentiation II: validation of cardiac differentiation of ES-DS cells by flow cytometry (FACS). The data are presented as fold increase in MLC-2v expression normalized to the level of expression of unstimulated controls growing on TC plastic. Note the ~5-fold increase in MLC-2v expression in the unstimulated cells cultured on ITO and the ~16-fold increase in MLC-2v expression following stimulation. For details see text

A more quantitative evaluation by flow cytometry was carried out with ES-D3 cells, which differentiate much more slowly than the GS-ES cells. These experiments were run for 20 days instead of 14. Under these conditions, the difference in MLC-2v expression between stimulated and non-stimulated cells was even more profound (Fig. 7). Surprisingly, cells growing on the conductive ITO surface had significantly elevated levels of MLC-2v expression compared to cells growing on control plastic, even without electrical stimulation. MLC-2v expression was further increased by approximately threefold when exposed to the optimized electrical stimulation regimen.

4 Functionality of mES Cell-Derived Cardiomyocytes

To assess the functionality of the mES cell-derived cardiomyocytes, either spontaneously or upon electrical stimulation, we used a combination of electrical tracing of the action potentials using the MicroElectrode Array setup (MEA, Multi Channel Systems, Reutlingen, Germany, see Fig. 8) and optical monitoring of the contractions. For video analysis of the contractions, 5 min videos were recorded as AVI files via a Hitachi video camera (KPD-50) at 30 frames per second, through a PixeLINK firewire converter (PLA-544) (PixeLINK, Ottawa, ON) attached to a Windows-based desktop computer running VirtualDub capture software (VirtualDub.org). Movement of the cells captured in the AVI files was then analyzed via Video Spot Tracker v05.02 (NIH) analysis software, to yield information on vital characteristics of the contraction, such as peak height and frequency. These data were then fed into the Heart Rate Variability (HRV) Analysis software which, for our cultured cells, generated the beat rate and beat rate variability (BRV). The video analysis system was validated against the 60 electrode MicroElectrode Array (MEA, Fig. 8). The MEA system records the actual electrophysiological activity of individual EBs or (groups of) cells plated in close proximity to an electrode. These data are presented as a typical EKG strip chart, as seen in Fig. 9, allowing for the

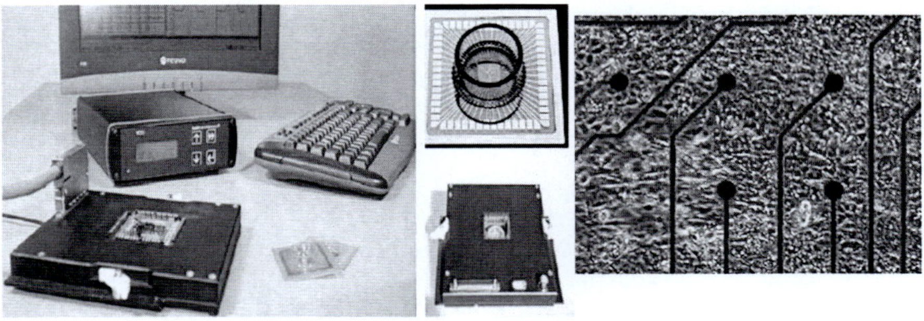

Fig. 8 *M*ulti*E*lectrode *A*ssembly (MEA) setup: *Left* – entire system; *Middle* – chip with 64 microelectrodes and below that the chip in holder with heating element. *Right* – Image of MEA during contraction (blurred cells in the lower left part of the photomicrograph are of contracting myocytes)

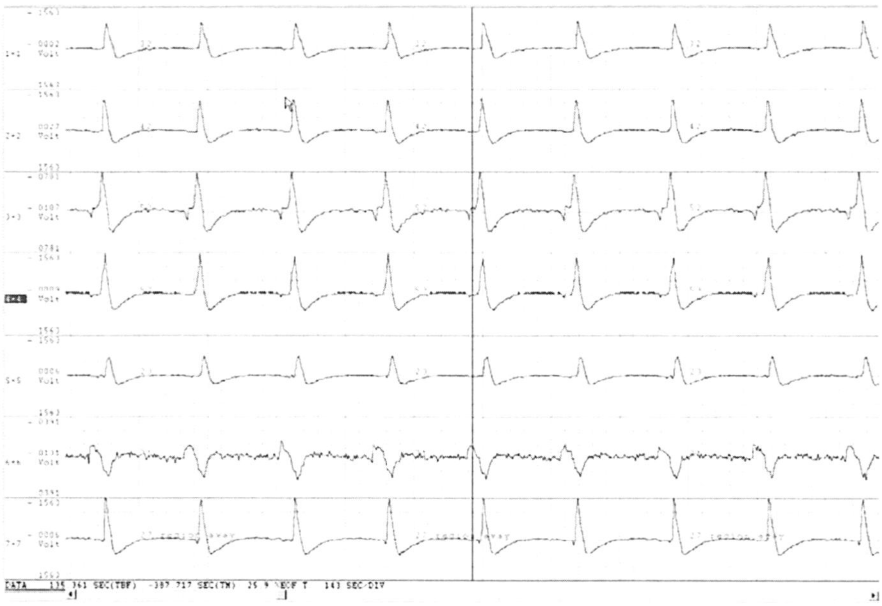

Fig. 9 Multi-channel strip chart recording of electrophysiological activity of beating cardiac myocytes from GS-ES-CMV27 embryonic stem cells: The individual traces represent independent readings from individual electrodes. For details see text

Table 1 Comparison of results obtained by optical and electrical recording

	MEA	Video
Beat rate (1/min)	81.8 ± 32.2	75.4 ± 15.7
Mean RR (s)	0.80 ± 0.19	0.85 ± 0.18
BRV	181.5	179.2

concomitant evaluation of the electrical activity and contractility in many individually identifiable cells. Data for the average (spontaneous) beat rate, mean RR values, and BRV obtained with the two analysis systems were in good agreement, as seen in Table 1.

Detailed analysis of the optical data revealed significant differences in the functionality and quality of the beating cardiomyocytes grown on control (plastic) or ITO-coated surfaces, with or without electric stimulation (1.5 Hz). The average rate of spontaneous contraction for the EBs on the tissue culture plastic substrate was 48.9 ± 12.3 BPM. The control ITO (non-stimulated) sample was slightly faster with a rate of 55.5 ± 11.1 BPM, while the rate for the EBs on the stimulated ITO was 75.4 ±10.4 BPM, which was significantly (~50%) faster than the rate of spontaneous contractions on the TC plastic surfaces (P <0.05) (Fig. 10). The R–R interval, viz. the time between contractions, was nearly identical between the plastic and ITO controls with 1.55 ± 0.9 and 1.49 ± 0.23 s, respectively. In contrast, the R–R

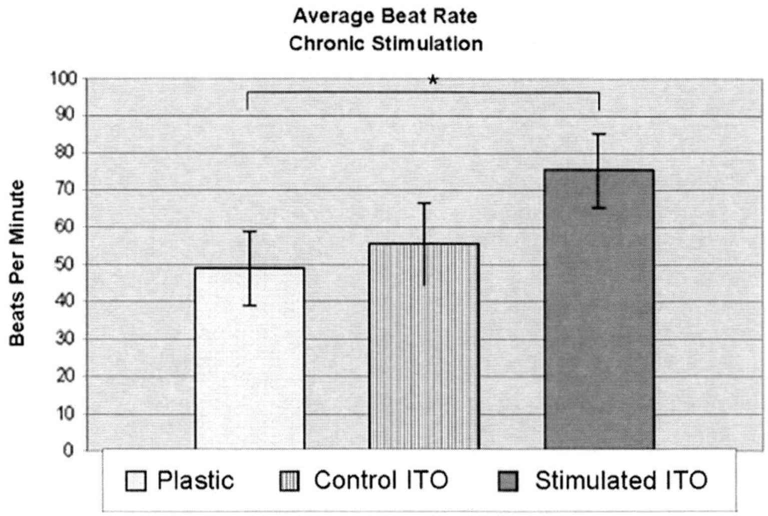

Fig. 10 Average spontaneous beat rate following chronic stimulation. Values are the average ±S.D. *Asterisk* denotes statistically significant differences. $P < 0.05$, $N = 4$. For details see text

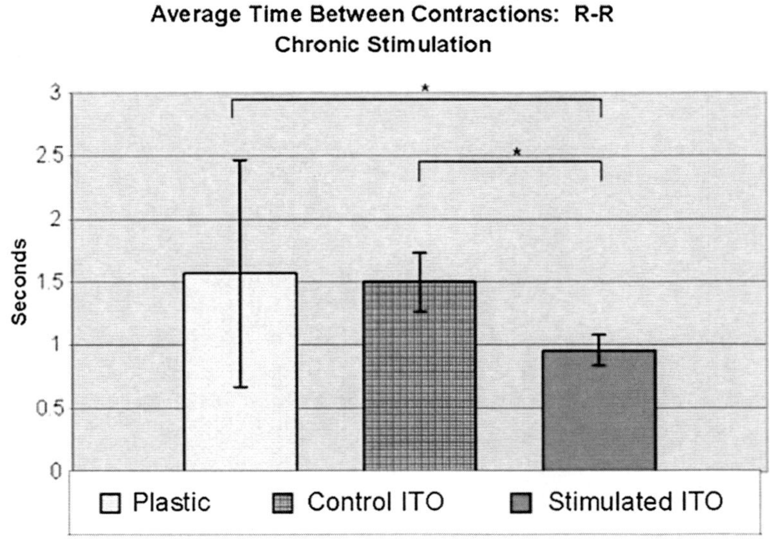

Fig. 11 Average R–R interval – chronic stimulation. Values are the average ±S.D. *Asterisk* denotes statistically significant differences. $P < 0.05$, $N = 4$. For details see text

interval for the stimulated ITO group was 0.97 ± 0.12 s, a statistically significant ($P < 0.05$) difference of approximately 0.5 s from both controls (Fig. 11). Similarly, the beat rate variability, or the analysis of the variations between beats, was very different in these samples. The stimulated sample's BRV was 116.7 ± 58 ms, which was half the variability of the control and nearly one-third that of the plastic sample which were 230.5 ± 178 ms and 294.7 ± 186.1 ms, respectively. Interestingly, the standard deviation of the variability in the stimulated samples was one-third that of the other groups.

5 Discussion and Outlook

Our studies provide solid evidence for the possibility to significantly enhance cardiac differentiation of mES cells by an optimized regimen of electrical stimulation. The rationale for this approach is based on developmental biological principles, which assign a functional role for electrical currents generated by ion gradients in the developing embryo [73–75]. More recently Rangappa et al. [76] reported that human mesenchymal stem cells, when juxtaposed and in direct (electrical) contact with human cardiac myocytes, can be reprogrammed into the cardiomyogenic lineage. Similarly, prior studies have suggested that specific differentiative cues could be tailored into artificial scaffolds to help efficiently direct stem cell fate decisions [77, 78, 59, 7, 6, 79].

As a prelude to testing the feasibility of generating 3D conductive scaffolds to enhance cardiac differentiation from stem cells, we cultured mouse ES-3D-derived embryoid bodies on ITO-coated glass slides. In contrast to our previous work, which was carried out on polyaniline-coated surfaces [7], these conductive surfaces are optically clear and allow for prolonged electrical stimulation. Innovative engineering aspects of this work are the multiplex stimulation capabilities, the combination of optical and electrical recordings, and the analysis of the results using a variety of software originally designed to evaluate heart rate variability. The contractions in the cells from embryoid bodies are influenced by pacemakers internal to the cells themselves or in developing Purkinje-like cells intermixed in the developing outgrowth, as determined by electrophysiological signals recorded via direct and invasive means such as patch clamp or the MEA-type systems [80, 81]. The electrophysiological characteristics of the signals have been characterized as they pertain to the class of myocyte–atrial, ventricular, or Purkinje-like cells [82, 43, 40, 83]. Contractile properties with respect to beat rate are beginning to be reported; however, the beat rate variability has not previously been reported for spontaneously contractile cardiac myocytes from embryoid bodies.

In order to measure beat rate and its variability in cultured ES-derived cardiac myocytes on a large scale and non-invasively, we devised a video system which allows for optical recordings of the contractions of the same cells over time (days and weeks) and without direct contact. This system allows for 5 min recordings to be captured and analyzed offline. The analysis of these 5 GB videos yields R–R intervals, or the time interval between contractions, which can be analyzed to determine variability of the contractions. The average beat rate of the cells in the stimulated ITO samples was 20–25 BPM faster than the control ITO or plastic and at 75 BPM it approached the applied stimulus of 90 BPM. The R–R interval for the stimulated sample was ~33% less (~1 s) than either non-stimulated controls (~1.5 s). The beat rate variability of the stimulated control was half that of the non-stimulated ITO and one-third that of the plastic control. Taken together, these parameters demonstrate a greatly enhanced contractility profile in the stimulated samples. The contractions of the stimulated embryoid bodies were more regular, indicating a more complete syncytium and more organized contractile mechanism, including inward and outward currents controlled by ion channels.

Careful analysis of the data, i.e., optical and electronic recordings of contractility and electrical activity, respectively, combined with immunocytochemical assessment, indicates a profound effect of electrical stimulation on several parameters involved in the differentiation toward the cardiac lineage. Using the appropriate stimulation parameters, the cells differentiate into cardiac myocytes earlier and in larger numbers and also express more robust functional features including higher spontaneous beating frequency and reduced beat rate variability.

At first glance the discrepancy in the results of cells cultured on plastic vs. electrically conductive surfaces is puzzling. Cardiac differentiation and functionality (Figs. 9, 10, and 11) are significantly enhanced on the conductive surface, even in the absence of an electrical stimulus. Given the fact that cardiac tissue is an interconnected set of cells relying on internal and external ion currents for signal propagation, it makes sense that during the early stages of ES cell differentiation and development of the interconnected cardiac-like syncytium, the electroactive/conductive properties of the surface act as additional venues for carrying the ionic current which enhance the differentiation toward ventricular cells, even in the absence of any exogenously applied current. The working hypothesis is that the local charge gradients generated by the cellular ionic currents were enhanced and/or amplified by the presence of a conductive substrate thus carrying and dispersing the potential further and affecting a greater area of developing tissue. The conductive coating on the slide thus acts as an additional carrier for the electric charge enabling a wider distribution of ion potential than a non-conductive plastic slide.

Embryoid bodies plated on the conductive, ITO-coated glass slides developed significantly more cardiac, and specifically ventricular, cells than those plated on plastic controls. Upon electrical stimulation, differentiation was further enhanced. For example, at day 12, we found by FACS analysis a 2.4-fold increase in the number of MLC-2v positive cells over the non-stimulated control ITO and a 4.6-fold increase over those on the tissue culture treated plastic. These 240 and 460% increases in cell numbers are not the only effect of the stimulus, as significant differences in the contractile properties were also observed. Thus, cells cultured on ITO without stimulation demonstrated nearly 30% less variability of the contractions than those on the plastic. When an external stimulus was applied the results were even more dramatic; the BRV of the stimulated cells exhibits approximately one-third the variability of the cells on either control substrate (Figs. 10 and 11).

Cardiac myocytes have outward currents which affect the surrounding environment and subsequently neighboring cells. Gryshchenko et al. [84] reported outward rectifier currents evoked from 7 + 2 – 4 day cardiac myocytes generated from EBs of ES-D3 cells, finding a quasi-linear relationship between increasing outward currents with an increase in the external stimulus with a maximal stimulus applied of 80 mV eliciting a 700 pA outward rectifying current. If we extrapolate these data to the 100 mV stimulus applied in our study, we may expect ~ 850 pA outward rectifying current in our system. According to Gryshchenko et al., these outward currents supply the majority of the aggregate whole-cell current and are instrumental in governing resting membrane potential and the frequency of action potentials [84]. These outward currents are believed to be essential in early cardiomyogenesis, although the specific mechanism has yet to be determined. It is known that in early (day 7 + 2 – 4) cardiomyocytes derived from embryonic stem cells the primitive pacemaker action potentials are generated by the voltage-dependent L-type Calcium channels and transient potassium channels [82].

Based on our results, we speculate that conductive properties of the ITO-coated surfaces facilitate the effective relay of these significant currents, thus affecting greater numbers of cells in the differentiating embryoid body outgrowths. Taken together with the enhancing effect of the electrical stimulus on ion channel activity, we can attempt an initial explanation for some of our results: The conductive ITO surface aids in the transmission of (endogenous) outward currents from the differentiating ES-derived cardiomyocytes to neighboring cells, contributing to a more efficient differentiation, as inferred from the larger numbers/areas of MLC-2v expressing contractile cells. While the increase in overall (area of) contractility could be, in part, due to the enhanced transmission of action potentials along the ITO surface (thus stimulating contractions in a larger array of differentiated cells), it is not likely that this is the only avenue for the increased areas of contraction. We conclude that a conductive/electroactive pericellular microenvironment, such as the ITO-coated surfaces used in this study, lends a major potential-dispersing, increased differentiation contribution early in the cardiomyogenesis process, facilitating the creation of larger numbers of contractile cells. Additionally, *after* the development of myocytes the ITO-coated surface may also contribute to the enhanced transmission of the action potential to larger areas of cells, as inferred from the mature, spontaneously contractile regions.

Taken together these results demonstrate the benefit of incorporating the functional cue of electrical conductivity into the growth substrate for the differentiation of stem cells toward ventricle lineages. Furthermore, when this surface is activated by the presence of an electrical stimulus with parameters near physiologic, the results are ever more pronounced. The increase in the number of myocytes and their enhanced contractile profile afforded by the stimulation program outlined in this study demonstrate a model for developing myocytes which may be better suited than traditional differentiation programs to applications in tissue-engineered constructs, direct implantation into an ailing heart, and for use in drug development models.

In terms of a clinical outlook, it is becoming increasingly apparent that even if we could overcome the social barriers inherent in implanting embryonic stem cells or cardiomyocytes derived from stem cells into the adult human heart, it is going to be especially difficult to make them behave "physiologically" when introduced into an injured or diseased cardiac microenvironment. Ultimately, the ability to integrate the donor cells both electrically and mechanically within the host myocardium is fundamental in restoring physiologically meaningful cardiac structure and function [85]. To date, human clinical trials using bone marrow and its cell derivatives have failed to show transdifferentiation to the cardiogenic phenotype [86]. It is generally understood that the observed short-term benefit in functional recovery from acute myocardial infarction is currently only attributable to paracrine factors (and/or neoangiogenesis). To date, long-term results have been largely ambiguous and marginal at best [87].

The approach that may have greater promise is to reprogram stem cells into maturing cardiomyocytes in vitro, prior to implantation, as demonstrated in this study. There is a growing consensus that merely injecting cells will not facilitate proper integration. A more viable approach may require building tissues in vitro that can be grafted onto a heart. Starting with stem cells, such tissues can be engineered to contain only properly organized (e.g., elongated, end-to-end coupled) cardiomyocytes that have a mature phenotype [85, 18]. These advances will ultimately facilitate bioengineering 3D tissue constructs [88] that functionally integrate into the host tissue and thereby promote normal cardiac conduction and perfusion and exhibit responsiveness to physiological demands (rate, contractility, relaxation).

Acknowledgments This work was supported, in part, by a Synergy Grant from Drexel University (JYK, YW, and PIL), grants from the Nanotechnology Institute of Southeastern Pennsylvania (YW and PIL), NIH (No. DE09848 to YW), and a NASA Graduate Student Research Fellowship (NAG9-138 to PRB). We are indebted to Dr. J. Hescheler for the generous gift of the MLC-2v expressing mES cell line. We thank Gregory Botta and Diane Keene for careful reading of the manuscript and for their constructive criticisms. We gratefully acknowledge the contributions of the late Professor Alan G. MacDiarmid of University of Pennsylvania (2000 Nobel laureate in Chemistry), his enthusiastic support, and numerous discussions of this project, before he passed away on February 7, 2007.

References

1. Lloyd-Jones D, et al. Heart disease and stroke statistics 2009 update: a report from the American Heart Association statistics committee and stroke statistics subcommittee. Circulation. 2009; 119:e21–e181.
2. Beltrami AP, Urbanek K, Kajstura J, Yan SM, Finato N, Bussani R, Nadal-Ginard B, Silvestri F, Leri A, Beltrami CA, Anversa P. Evidence that human cardiac myocytes divide after myocardial infarction. N Engl J Med. 2001; 344:1750–1757.
3. Gerecht-Nir S, Radisic M, Park H, Cannizzaro C, Boublik J, Langer R, Vunjak-Novakovic G. Biophysical regulation during cardiac development and application to tissue engineering. Int J Dev Biol. 2006; 50:233–243.
4. Zhang LI, Poo MM. Electrical activity and development of neural circuits. Nat Neurosci. 2001; 4 (Suppl):1207–1214.
5. Gomez N, Schmidt CE. Nerve growth factor-immobilized polypyrrole: bioactive electrically conducting polymer for enhanced neurite extension. J Biomed Mater Res A. 2007; 81:135–149.
6. Guo Y, Li M, Mylonakis A, Han J, MacDiarmid AG, Chen X, Lelkes PI, Wei Y. Electroactive oligoaniline-containing self-assembled monolayers for tissue engineering applications. Biomacromolecules. 2007; 8:3025–3034.

7. Bidez PR, Li S, MacDiarmid AG, Venancio EC, Wei Y, Lelkes PI. Polyaniline, an electroactive polymer, supports adhesion and proliferation of cardiac myoblasts. J Biomat Sci. 2005; 17:199–212.
8. Li M, Bidez P, Guterman-Tretter E, Gou Y, MacDiarmid AG, Lelkes PI, Yuan XB, Yuan XY, Sheng J, Li H, Song CX, Wei Y. Research progress on electroactive and electrically conductive polymers for tissue engineering scaffolds. Acta Acad Med Sin. 2006; 28:845–848.
9. Bieberich E, Anthony GE. Neuronal differentiation and synapse formation of PC12 and embryonic stem cells on interdigitated microelectrode arrays: contact structures for neuron-to-electrode signal transmission (NEST). Biosens Bioelectron. 2004; 19:923–931.
10. Ott HC, Matthiesen TS, Goh SK, Black LD, Kren SM, Netoff TI, Taylor DA. Perfusion-decellularized matrix: using nature's platform to engineer a bioartificial heart. Nat Med. 2008; 14:213–221.
11. Fukuda K, Yuasa S. Stem cells as a source of regenerative cardiomyocytes. Circ Res. 2006; 98:1002–1013.
12. Jawad H, Lyon AR, Harding SE, Ali NN, Boccaccini AR. Myocardial tissue engineering. Br Med Bull. 2008; 87:31–47.
13. Genovese J, Cortes-Morichetti M, Chachques E, Frati G, Patel A, Chachques JC. Cell based approaches for myocardial regeneration and artificial myocardium. Curr Stem Cell Res Ther. 2007; 2:21–27.
14. Roccio M, Goumans MJ, Sluijter JP, Doevendans PA. Stem cell sources for cardiac regeneration. Panminerva Med. 2008; 50:19–30.
15. Doetschman TC, Eistetter H, Katz M, Schmidt W, Kemler R. The in vitro development of blastocyst-derived embryonic stem cell lines: formation of visceral yolk sac, blood islands and myocardium. J Embryol Exp Morphol. 1985; 87:27–45.
16. Smith AG. Embryo-derived stem cells: of mice and men. Annu Rev Cell Dev Biol. 2001; 17:435–462.
17. Bettiol E, Clement S, Krause KH, Jaconi ME. Embryonic and adult stem cell-derived cardiomyocytes: lessons from in vitro models. Rev Physiol Biochem Pharmacol. 2006; 157:1–30.
18. Mummery C, van der Heyden MA, de Boer TP, Passier R, Ward D, van den Brink S, van Rooijen M, van de Stolpe A. Cardiomyocytes from human and mouse embryonic stem cells. Methods Mol Med. 2007; 140:249–272.
19. Lints TJ, Parsons LM, Hartley L, Lyons I, Harvey RP. Nkx-2.5: a novel murine homeobox gene expressed in early heart progenitor cells and their myogenic descendants. Development. 1993; 119:419–431.
20. Boheler KR, Czyz J, Tweedie D, Yang HT, Anisimov SV, Wobus AM. Differentiation of pluripotent embryonic stem cells into cardiomyocytes. Circ Res. 2002; 91:189–201.
21. Lyons I, Parsons LM, Hartley L, Li R, Andrews JE, Robb L, Harvey RP. Myogenic and morphogenetic defects in the heart tubes of murine embryos lacking the homeobox Gene Nkx2-5. Genes Dev. 1995; 9:1654–1666.
22. Tanaka M, Wechsler SB, Lee IW, Yamasaki N, Lawitts JA, Izumo S. Complex modular cis-acting elements regulate expression of the cardiac specifying homeobox gene Csx/Nkx2.5. Development. 1999; 126:1439–1450.
23. Zou Y, Evans S, Chen J, Kuo HC, Harvey RP, Chien KR. CARP, a cardiac ankyrin repeat protein, is downstream in the Nkx2-5 homeobox gene pathway. Development. 1997; 124:793–804.
24. Biben C, Harvey RP. Homeodomain factor Nkx2-5 controls left/right asymmetric expression of bHLH gene eHAND during murine heart development. Genes Dev. 1997; 11:1357–1369.
25. Katoh T, Lowey S. Mapping myosin light chains by immunoelectron microscopy. Use of anti-fluorescyl antibodies as structural probes. J Cell Biol. 1989; 109:1549–1560.
26. Robbins J, Gulick J, Sanchez A, Howles P, Doetschman T. Mouse embryonic stem cells express the cardiac myosin heavy chain genes during development in vitro. J Biol Chem. 1990; 265:11905–11909.

27. Sanchez A, Jones WK, Gulick J, Doetschman T, Robbins J. Myosin heavy chain gene expression in mouse embryoid bodies. An in vitro developmental study. J Biol Chem. 1991; 266:22419–22426.
28. Miller-Hance WC, LaCorbiere M, Fuller SJ, Evans SM, Lyons G, Schmidt C. In vitro chamber specification during embryonic stem cell cardiogenesis. Expression of the ventricular myosin light chain-2 gene is independent of heart tube formation. J Biol Chem. 1993; 268:25244–25252.
29. Zhao XS, Gallardo TD, Lin L, Schageman JJ, Shohet RV. Transcriptional mapping and genomic analysis of the cardiac atria and ventricles. Physiol Genomics. 2002; 12:53–60.
30. O'Brien TX, Lee KJ, Chien KR. Positional specification of ventricular myosin light chain 2 expression in the primitive murine heart tube. Proc Natl Acad Sci. 1993; 90:5157–5161.
31. Chen J, Kubalak SW, Minamisawa S, Price RL, Becker KD, Hickey R. Selective requirement of myosin light chain 2v in embryonic heart function. J Biol Chem. 1998; 273:1252–1256.
32. Faerman A, Shani M. The expression of the regulatory myosin light chain 2 gene during mouse embryogenesis. Development. 1993; 118:919–929.
33. Franco D, Markman MM, Wagenaar GT, Ya J, Lamers WH, Moorman AF. Myosin light chain 2a and 2v identifies the embryonic outflow tract myocardium in the developing rodent heart. Anat Rec. 1999; 254(1):135–146.
34. Fujikawa T, Oh SH, Pi L, Hatch HM, Shupe T, Petersen BE. Teratoma formation leads to failure of treatment for type I diabetes using embryonic stem cell-derived insulin-producing cells. Am J Pathol. 2005; 166:1781–1791.
35. Dang SM, Kyba M, Perlingeiro R, Daley GQ, Zandstra PW. Efficiency of embryoid body formation and hematopoietic development from embryonic stem cells in different culture systems. Biotechnol Bioeng. 2001; 78:442–453.
36. Gerecht-Nir S, Smadar Cohen S, Itskovitz-Eldor J. Bioreactor cultivation enhances the efficiency of Human Embryoid Body (hEB) formation and differentiation. Biotech Bioeng. 2004; 86(5):493–502.
37. Lake J, Rathjen J, Remiszewski J, Rathjen PD. Reversible programming of pluripotent cell differentiation. J Cell Sci. 2000; 113:555–566.
38. Kehat I, Kenyagin-Karsenti D, Snir M, Segev H, Amit M, Gepstein A. Human embryonic stem cells can differentiate into myocytes with structural and functional properties of cardiomyocytes. J Clin Invest. 2001; 108:407–414.
39. Fijnvandraat AC, van Ginneken ACG, de Boer PAJ, Ruijter JM, Christoffels VM, Moorman AFM, et al. Cardiomyocytes derived from embryonic stem cells resemble cardiomyocytes of the embryonic heart tube. Cardiovasc Res. 2003; 58:399–409.
40. He JQ, Ma Y, Lee Y, Thomson JA, Kamp TJ. Stem cells develop into multiple types of cardiac myocytes: action potential characterization. Circ Res. 2003; 93:32–39.
41. Sachinidis A, Gissel C, Nierhoff D, Hippler-Altenburg R, Sauer H, Wartenberg M. Identification of platelet-derived growth factor-BB as cardiogenesis-inducing factor in mouse embryonic stem cells under serum-free conditions. Cell Physiol Biochem. 2003; 13:423–429.
42. Wobus AM, Rohwedel J, Maltsev V, Hescheler J. Development of cardiomyocytes expressing cardiac-specific genes, action potentials, and ionic channels during embryonic stem cell-derived cardiogenesis. Acad Sci. 1995; 752:460–469.
43. Wobus AM, Kaomei G, Shan J, Wellner MC, Rohwedel J, Guanju J. Retinoic acid accelerates embryonic stem cell-derived cardiac differentiation and enhances development of ventricular cardiomyocytes. J Mol Cell Cardiol. 1997; 29:1525–1539.
44. Hidaka K, Lee JK, Kim HS, Ihm CH, Iio A, Ogawa M, et al. Chamber-specific differentiation of Nkx2.5-positive cardiac precursor cells from murine embryonic stem cells. FASEB J. 2003; 17:740–742.
45. Fujiwara S. Retinoids and nonvertebrate chordate development. J Neurobiol. 2006; 66:645–652.
46. Clagett-Dame M, McNeill EM, Muley PD. Role of all-trans retinoic acid in neurite outgrowth and axonal elongation. J Neurobiol. 2006; 66:739–756.

47. Xu C, Police S, Rao N, Carpenter MK. Characterization and enrichment of cardiomyocytes derived from human embryonic stem cells. Circ Res. 2002; 91:501–508.
48. Zandstra PW, Bauwens C, Yin T, Liu Q, Schiller H, Zweigerdt R, et al. Scalable production of embryonic stem cell-derived cardiomyocytes. Tissue Eng. 2003; 9:767–778.
49. Ventura C, Maioli M, Asara Y, Santoni D, Mesirca P, Remondini D. Turning on stem cell cardiogenesis with extremely low frequency magnetic fields. FASEB J. 2005; 19:155–157.
50. Fink C, Ergun S, Kralisch D, Remmers U, Weil J, Eschenhagen T. Chronic stretch of engineered heart tissue induces hypertrophy and functional improvement. FASEB J. 2000; 14:669–679.
51. Haupt HA. Electrical stimulation of osteogenesis. South Med J. 1984; 77:56–64.
52. Ciombor DM, Aaron RK. The role of electrical stimulation in bone repair. Foot Ankle Clin. 2005; 10:579–593.
53. Jaeger RJ. Principles underlying functional electrical stimulation techniques. J Spinal Cord Med. 1996; 19:93–96.
54. Hamid S, Hayek R. Role of electrical stimulation for rehabilitation and regeneration after spinal cord injury: an overview. Eur Spine J. 2008; 7:1256–1269.
55. Radisic M, Park H, Shing H, Consi T, Schoen FJ, Langer R. Functional assembly of engineered myocardium by electrical stimulation of cardiac myocytes cultured on scaffolds. Proc Natl Acad Sci. 2004; 101:18129–18134.
56. Pedrotty DM, Koh J, Davis BH, Taylor DA, Wolf P, Niklason LE. Engineering skeletal myoblasts: roles of three-dimensional culture and electrical stimulation. AJP Heart Circ Physiol. 2005; 288:H1620–H1626.
57. Tandon N, Cannizzaro C, Chao PH, Maidhof R, Marsano A, Au HT, Radisic M, Vunjak-Novakovic G. Electrical stimulation systems for cardiac tissue engineering. Nat Protoc. 2009; 4:155–173.
58. Schmidt CE, Shastri VR, Vacanti JP, Langer R. Stimulation of neurite outgrowth using an electrically conducting polymer. Proc Natl Acad Sci. 1997; 94:8948–8953.
59. Guterman E, Cheng S, Palouian K, Bidez PR, Lelkes PI, Wei Y. Peptide-modified electroactive polymers for tissue engineering applications. Polym Prepr Am Chem Soc Div Polym Chem. 2003; 43:766–767.
60. Abilez O, Benharash P, Miyamoto E, Gale A, Xu C, Zarins CK. P19 progenitor cells progress to organized contracting myocytes after chemical and electrical stimulation: implications for vascular tissue engineering. J Endovasc Ther. 2006; 13:377–388.
61. Chen MQ, Xie X, Hollis Whittington R, Kovacs GT, Wu JC, Giovangrandi L. Cardiac differentiation of embryonic stem cells with point-source electrical stimulation. Conf Proc IEEE Eng Med Biol Soc. 2008; 2008:1729–1732.
62. Yamada M, Tanemura K, Okada S, Iwanami A, Nakamura M, Mizuno H, Ozawa M, Ohyama-Goto R, Kitamura N, Kawano M, Tan-Takeuchi K, Ohtsuka C, Miyawaki A, Takashima A, Ogawa M, Toyama Y, Okano H, Kondo T. Electrical stimulation modulates fate determination of differentiating embryonic stem cells. Stem Cells. 2007; 25:562–570.
63. Wong JY, Langer R, Ingber DE. Electrically conducting polymers can noninvasively control the shape and growth of mammalian cells. Proc Natl Acad Sci USA. 1994; 91:3201–3204.
64. Aoki T, Tanino M, Sanui K, Ogata N, Kumakura K. Secretory function of adrenal chromaffin cells cultured on polypyrrole films. Biomaterials. 1996; 17:1971–1974.
65. Kotwal A, Schmidt CE. Electrical stimulation alters protein adsorption and nerve cell interactions with electrically conducting biomaterials. Biomaterials. 2001; 22:1055–1064.
66. Lee JW, Serna F, Nickels J, Schmidt CE. Carboxylic acid-functionalized conductive polypyrrole as a bioactive platform for cell adhesion. Biomacromolecules. 2006; 7:1692–1695.
67. Jun I, Jeong S, Shin H. The stimulation of myoblast differentiation by electrically conductive sub-micron fibers. Biomaterials. 2009; 30:2038–2047.
68. Heuchel R, Radtke F, Georgiev O, Stark G, Aguet M, Schaffner W. The transcription factor MTF-1 is essential for basal and heavy metal induced metallothionein gene expression. EMBO J. 1994; 13:2870–2875.

69. Muller M, Fleischmann BK, Selbert S, Ji GJ, Endle E, Middeler G. Selection of ventricular-like cardiomyocytes from ES cells in vitro. FASEB J. 2000; 14:2540–2548.
70. Loebel DA, Watson CM, De Young RA, Tam PP. Lineage choice and differentiation in mouse embryos and embryonic stem cells. Dev Biol. 2003; 264:1–14.
71. Saito S, Liu B, Yokoyama K. Animal embryonic stem (ES) cells: self-renewal, pluripotency, transgenesis and nuclear transfer. Hum Cell. 2004; 17:107–115.
72. Niskanen JP, Tarvainen MP, Ranta-Aho PO, Karjalainen PA. Software for advanced HRV analysis. Comput Methods Programs Biomed. 2004; 76:73–81.
73. Spitzer NC. Ion channels in development. Annu Rev Neurosci. 1979; 2:363–397.
74. Viatchenko-Karpinski S, Fleischmann BK, Liu Q, Sauer H, Gryshchenko O, Ji GJ, et al. Intracellular Ca^{2+} oscillations drive spontaneous contractions in cardiomyocytes during early development. Proc Natl Acad Sci USA. 1999; 96:8259–8264.
75. Rutenberg J, Cheng SM, Levin M. Early embryonic expression of ion channels and pumps in chick and Xenopus development. Dev Dyn. 2002; 225:469–484.
76. Rangappa S, Entwistle JWC, Wechsler AS, Kresh JY. Cardiomyocyte-mediated contact programs human mesenchymal stem cells to express cardiogenic phenotype. J Thoracic Cardiov Surg. 2003; 126(1):124–132.
77. Freed LE, Vunjak-Novakovic G, Biron RJ, Eagles DB, Lesnoy DC, Barlow SK, Langer R. Biodegradable polymer scaffolds for tissue engineering. Biotechnology. 1994; 12(7):689–693.
78. Vacanti CA, Mikos AG. Letter from the editors. Tissue Eng. 1995; 1:1–2.
79. Chai C, Leong KW. Biomaterials approach to expand and direct differentiation of stem cells. Mol Ther. 2007; 15:467–480.
80. Reppel M, Pillekamp F, Lu ZJ, Halbach M, Brockmeier K, Fleischmann BK, et al. Microelectrode arrays: a new tool to measure embryonic heart activity. J Electrocardiol. 2004; 37(Suppl):104–109.
81. Reppel M, Pillekamp F, Brockmeier K, Matzkies M, Bekcioglu A, Lipke T, et al. The Electrocardiogram of human embryonic stem cell-derived cardiomyocytes. J Electrocardiol. 2005; 38:166–170.
82. Maltsev VA, Wobus AM, Rohwedel J, Bader M, Hescheler J. Cardiomyocytes differentiated in vitro from embryonic stem cells developmentally express cardiac-specific genes and ionic currents. Circ Res. 1994; 75:233–244.
83. Kolossov E, Lu Z, Drobinskaya I, Gassanov N, Duan Y, Sauer H, et al. Identification and characterization of embryonic stem cell-derived pacemaker and atrial cardiomyocytes. FASEB J. 2005; 19:577–579.
84. Gryshchenko O, Lu ZJ, Fleischmann BK, Hescheler J. Outwards currents in embryonic stem cell-derived cardiomyocytes. Pflugers Arch Eur J Physiol. 2000; 439:798–807.
85. Kresh JY. Cell replacement therapy: the functional importance of myocardial architecture and intercellular gap-junction distribution. J Thorac Cardiovasc Surg. 2006; 131:1310–1313.
86. Lee MS, Makkar RR. Stem-cell transplantation in myocardial infarction: a status report. Ann Intern Med. 2004; 140:729–737.
87. Chien KR, Domian IJ, Parker KK. Cardiogenesis and the complex biology of regenerative cardiovascular medicine. Science. 2008; 322(5907):1494–1497.
88. Eschenhagen T, Zimmermann WH. Engineering myocardial tissue. Circ Res. 2005; 97:1220–1231.

The Therapeutic Potential of ES-Derived Haematopoietic Cells

Sabrina Gordon-Keylock and Lesley Forrester

Abstract Blood transfusions and bone marrow transplantations are cell-based therapies that have been used to treat diseases of the haematopoietic system for decades. However, such therapies are completely reliant on the availability of appropriate donors and fraught with the risk of transmissible infection. Can recent advances in pluripotent stem cell technology alleviate these problems? We review the development of protocols used in haematopoietic differentiation of mouse and human embryonic stem (ES) cells and the progress made in the production of haematopoietic stem cells capable of long-term reconstitution. Relatively pure populations of mature haematopoietic cells including erythrocytes, macrophage and dendritic cells have also been generated from ES cells and their potential use in the future treatment of disease is considered. However, even assuming the successful production of a desired cell type in the research laboratory, we are faced with significant challenges in the translation of these protocols into clinical grade processes and in the scale-up of these strategies that will allow the economic production of vast quantities of cells. We discuss briefly some of the technology that has been applied recently to the ES cell system as a first step in overcoming some of these challenges.

Keywords Embryonic stem cells · Haematopoietic differentiation · Cell therapy

1 Introduction

Haematopoiesis is the process whereby all the functional cells of the blood, including erythrocytes, lymphocytes, monocytes, neutrophils, platelets and megakaryoctes, are produced from the common haematopoietic stem cell (HSC) that is capable of extensive self-renewal and complete reconstitution of the haematopoietic

S. Gordon-Keylock (✉)
Centre for Regenerative Medicine, University of Edinburgh, Rodger Land Building, Kings Buildings, West main road, Edinburgh EH9 3JQ

G.M. Artmann et al. (eds.), *Stem Cell Engineering*,
DOI 10.1007/978-3-642-11865-4_6,

system [1, 2, 3, 4, 5]. Mature blood cells are short lived and must be produced continually to maintain the steady state and to respond to physiological or environmental changes. For example, red blood cells are normally produced at a rate of two million per second, but that rate must increase after severe blood loss or in response to changes in oxygen levels at high altitude. Haematological disorders, such as anaemia, immune deficiency and leukaemia, can occur when the tightly controlled haematopoietic regulation is disrupted. Cell therapy has been used for decades to treat these disorders; short-term therapies involve the transfusion of mature blood cells, such as erythrocytes, whereas long-term treatment is achieved by using HSCs derived either from the bone marrow, mobilized peripheral blood or umbilical cord blood [6]. Although widely used, there are problems associated with these treatments, including the shortage of appropriate donor supplies, the requirement for immune compatibility and the persistent threat of transmissible infectious diseases [7]. Could these problems be alleviated by the rapidly developing technologies surrounding the use of human embryonic stem (ES) and induced pluripotent (iPS) cells? Can these systems provide a potentially limitless, infection-free, immune-neutral source of cells for haematopoietic transfusion and transplantation?

2 Pluripotent Cells

ES cells are derived from the inner cell mass of the developing embryo and are considered to be pluripotent because they have the potential to differentiate into any cell type of the body [8–10]. The ability of ES cells to differentiate into any cell type in vitro provides an invaluable tool in regenerative medicine [11]. In theory, ES-derived cells could be used to treat diseases that are caused by a deficiency in a specific cell type. However, the application of human ES cell (hESC) technology in stem cell therapies poses ethical questions because the production of hES cells involves the use of human embryos. The ground-breaking discovery that pluripotency can be induced in somatic cells by the addition of four transcription factors, OCT4, SOX2, KLF4 and c-MYC [12, 13], has provided a strategy to create pluripotent cells that are ethically acceptable. Results from many independent groups have confirmed that induced pluripotent stem (iPS) cells can be produced from somatic cells of many species. They retain a normal karyotype and display many characteristics associated with ES cells including morphology, cell surface phenotype, transcriptional and epigenetic profile and the functional potential to differentiate into derivatives of all germ layers [14–17]. iPS cells have also been produced from the somatic cells of patients with a range of genetic disorders, providing an autologous source of cells for gene therapy as well as a tool to study disease mechanisms in mature cell types [18]. Production of iPS cells from an individual carrying the universal donor blood group, O Rh–, could provide a particularly valuable resource for haematopoietic transplantation. This review focuses on the differentiation of haematopoietic cells from ES cells, but many of the strategies have been replicated using iPS cells [19].

3 Production of Haematopoietic Cells from ES Cells

A seminal paper in 1985 first reported the production of haemoglobinized blood islands in three-dimensional aggregates (known as embryoid bodies [EBs]) from mouse ES cells [20]. Subsequent studies reported the morphological identification of mature haematopoietic cell types, including erythrocytes, macrophages, neutrophils and mast cells [21]. The first indication that multilineage haematopoietic progenitors could be produced from ES cells came from studies where EBs were dissociated and plated in a semi-solid matrix containing a cocktail of haematopoietic cytokines [22]. The timing of production of multilineage myeloid precursors capable of giving rise to macrophage, neutrophil and erythroid cells from ES cells in vitro appeared to recapitulate the timing of their appearance during embryonic haematopoiesis in vivo [23, 22].

Since these initial reports, one of the major challenges has been to establish well-defined and robust differentiation protocols to generate large numbers of cell types of interest. The frequency of haematopoietic progenitor production has been increased by co-culture of ES cells with extracellular matrices (ECM) or stromal cell feeder cell lines derived from adult bone marrow or from the embryonic haematopoietic tissues [24–29]. However, the precise mechanism of action of ECM and stromal cells has not been clearly defined and considerable variability in the levels of induction have been reported. Many of these studies are further complicated by the use of foetal calf serum that is present in the culture media that adds yet another source of complexity and variability. Chemically defined conditions using recombinant growth factors have been used in an effort to combat that variability and to define the cellular and molecular mechanisms involved in production of haematopoietic cells from ES cells [30–37].

3.1 Defining the Stages of Haematopoiesis In Vitro

Haematopoietic cells are descendents of the mesoderm during embryogenesis, so in vitro differentiation strategies have aimed to recreate that developmental process (Fig. 1). The mesodermal origin of ES-derived haematopoietic cells was confirmed using a murine ES cell line engineered to express the GFP reporter gene under the control of brachyury promoter [43, 44]. This was subsequently used to show that mesodermal specification of ES cells could be enhanced by treatment with BMP4 [34]. Downstream haematopoietic differentiation was stimulated by subsequent addition of recombinant bFGF, activin A and VEGF [31, 32, 33, 36]. BMP4 also supported haematopoietic mesoderm formation from human ES cells, with the addition of VEGF, SCF and FGF2 further enhancing the production of haematopoietic lineages [30, 35, 45].

Haematopoietic differentiation of ES cells in vitro has provided an excellent model to elucidate the early developmental events that give rise to haematopoietic cells that are inaccessible in vivo [46–51]. For example, the system has been

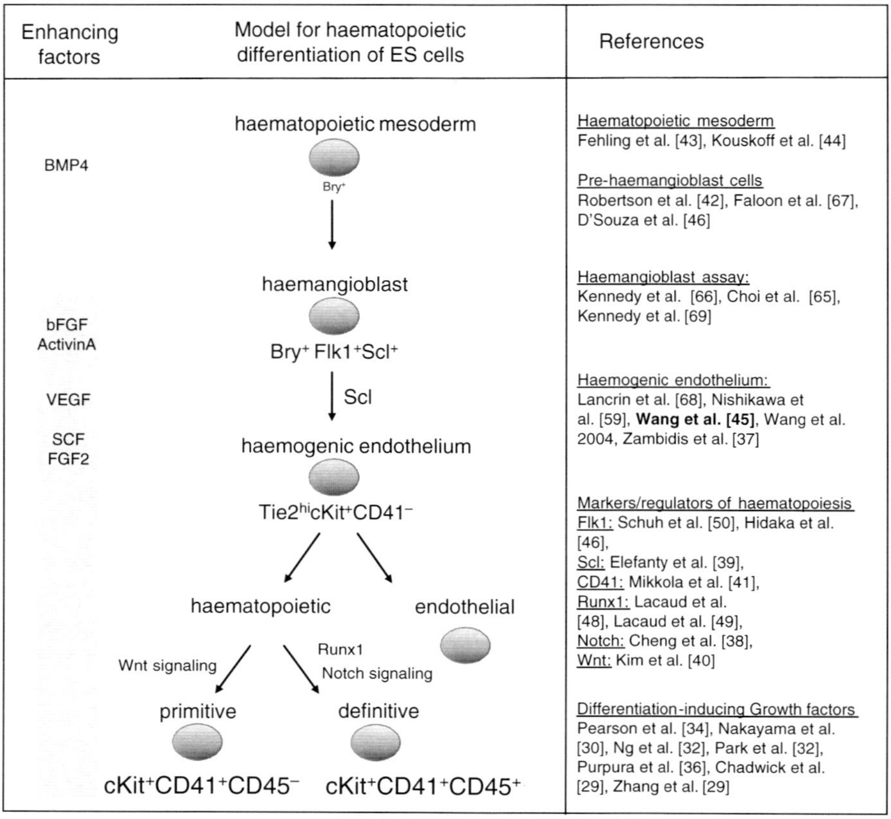

Fig. 1 Modelling the early stages of haematopoietic development in differentiating ES cells. References in *bold* refer to publications involving human ES cells [38–42]

used to address contradictory models concerning the developmental link between haematopoietic and endothelial lineages. During embryonic development in vivo, it had been proposed that haematopoietic lineages arose from a bipotent haemangioblast progenitor with both endothelial and haematopoietic potentials [51–53]. A contrasting hypothesis maintained that definitive haematopoiesis in vivo arose from specialized endothelium with haemogenic potential and/or from a precursor population residing in the sub-aortic mesenchyme [54–64]. Haemangioblast progenitors were identified in differentiating ES cells, using an in vitro "blast" colony assay (BL-CFC), and were detected as transient cells at a precise time point of EB differentiation [65, 66]. BL-CFC gave rise to definitive haematopoietic cells, primitive erythroid cells, endothelial and vascular smooth muscle cell lineages [65, 66] and were shown to be of mesodermal origin using the Bry-GFP reporter gene [43]. As *Bry*$^+$ pre-haemangioblast cells mature, Flk1 (a marker of endothelial cells) is expressed in a subset of cells and BL-CFC activity correlates with the Bry$^+$Flk1$^+$ fraction [67, 43, 44, 42]. An equivalent Brachyury$^+$Flk1$^+$ co-expressing cell type was isolated from the posterior primitive streak of murine embryos in vivo and shown to have both endothelial and haematopoietic potential validating the in vitro results [51]. ES-derived BL-CFCs can also give rise to a distinct cell type expressing markers indicative of haemogenic endothelium ($Tie2^{hi}cKit^{+}CD41^{-}$) and this population could give rise to haematopoietic cells expressing CD41 and the pan-haematopoietic marker, CD45 [68]. These complex in vitro experiments have, therefore, provided evidence for a direct developmental link between haemangioblast and haemogenic endothelium that had been considered in some theories as being independent cell lineages.

Cells with both endothelial and haematopoietic potential have also been identified in differentiating human ES cells cultured in the presence of BMP4 and bFGF, followed by addition of VEGF [69]. Transient FLK1expressing cells were sorted and seeded into methylcellulose supplemented with a combination of cytokines to stimulate blast colony formation. These cells are thought to arise from precursors co-expressing CD31, FLK1 and VE-cadherin, indicating endothelial origin of human haematopoietic lineages [45, 70, 37].

3.2 *Derivation of Transplantable ES–HSCs*

If ES-derived therapies are to be applied to the treatment of a wide range of haematopoietic disorders, the ultimate goal needs to be the generation of definitive ES–HSCs that are capable of complete, efficient and long-term reconstitution of an ablated haematopoietic system. ES cells do indeed have the potential to generate transplantable HSCs when allowed to differentiate within the embryonic microenvironment, as demonstrated by the fact that HSCs produced in completely ES-derived mouse embryos generated by tetraploid aggregation are capable of reconstituting the entire haematopoietic system [71, 72]. However, attempts to reproduce this embryonic microenvironment in the laboratory have proven rather challenging. This is perhaps not surprising since a very precise balance must be achieved between *driving* differentiation from the pluripotent (ES) state to a haematopoietic fate with effectively *inhibiting* the multipotent HSC from differentiating into lineage-restricted progenitors and mature cells.

Varying levels of multilineage repopulation have been achieved with human and mouse ES–HSCs using a range of differentiation and transplantation strategies. In initial murine studies, relatively low frequencies of multipotent HSCs capable of in vivo repopulation were detected [73–76] but the production of repopulating HSCs was improved by over-expression of haematopoietic genes [77–80]. ES cells engineered to over-express HOXB4 and/or CDX4 can enhance the reconstitution ability of mouse ES-derived HSCs [81, 82, 77, 78, 83, 80]. HOXB4 is thought to enhance the proliferation of ES-derived HSCs, similar to its effects on bone marrow-derived HSCs [84], but it is not clear whether it can also induce HSC production from non-haematopoietic mesoderm. Addition of HOXB4 either by over-expression or by addition of a recombinant tPTD-HOXB4 fusion protein can also increase haematopoietic differentiation of human ES cells in vitro [85–88] but does not appear to have a significant effect on the generation of transplantable cells [89, 87].

The generation of repopulating HSCs has also been improved by co-culturing both mouse and human ESC with bone marrow stromal cell lines [84, 90] and with cell lines derived from the aorta-gonad-mesonephros region (AGM) of the embryo [27], a tissue where definitive HSCs first emerge in vivo [91, 93]. This highlights the importance of providing the appropriate microenvironment in the production of definitive HSCs. Protocols that do not provide the appropriate signals generate ES–HSCs with characteristics of primitive yolk sac progenitors that do not have efficient long-term adult repopulating ability [93, 94]. Levels of in vivo reconstitution were improved by injecting ES-derived subpopulations, purification based on the expression of a number of different markers, suggesting that the presence of inhibitory cells might be responsible for the poor repopulating activity of unsorted cells [95, 83, 96, 70]. A detailed comparison of the cell surface phenotypes of haematopoietic repopulating cells from ES–HSC with foetal, embryonic and adult HSCs revealed that ES–HSCs have a unique surface phenotype indicating that they are a developmentally immature population of cells with features of both primitive and mature HSC [83]. Like HSCs generated in vivo, ES–HSCs express c-Kit but they also express CD41, which marks early embryonic HSCs, and CD150 which marks more mature HSCs. Direct intra-femoral injection of ES-derived cells significantly improved repopulation levels suggesting that ES–HSCs may differ from HSC generated in vivo in their ability to home to the adult bone marrow niche [97].

In summary, success in the production of repopulating ES–HSCs cannot be attributed to a single strategy. In recent studies the combination of HOXB4 overexpression, stromal cell culture and purification of a candidate pre-HSC population has resulted in relatively high levels of repopulation but the relative importance of each strategy has not been defined [95, 83]. Different aspects of the differentiation protocols likely contribute to the induction, the proliferation as well as the survival of ES–HSC to varying degrees. Now that the cell surface markers to identify repopulating ES–HSC have been defined, it should now be possible to determine the mechanisms involved in their generation and maintenance and to design better protocols for their production [83].

3.3 Derivation of Mature Haematopoietic Cells from ES Cells

In contrast to the complex strategies that have been employed to generate HSCs from ES cells in vitro, relatively simple and efficient strategies have been developed for the production of mature cell populations including macrophages, dendritic cells, neutrophils and red blood cells.

3.3.1 Macrophages

Macrophages represent a population of monocytic phagocytes that are involved in both innate and cell-mediated immunity [98, 99]. They are a critical cellular component in both the initiation and the resolution of inflammation via their secretion of pro- and anti-inflammatory cytokines and are intimately involved in the etiology of many diseases. Macrophage cell therapy has been shown to alleviate injury in inflammatory models [100, 101] and has been proposed as a treatment for inflammatory diseases, but progression in this field has been hampered by the availability of cells. Primary macrophages can be isolated from peripheral blood or bone marrow cultured in the presence of CSF1 and IL3, but this strategy generates heterogeneous and variable cell populations and is dependent on a limited supply of donors. By contrast, large numbers of relatively pure macrophages have been produced from both mouse and human ES cells by the simple adaptation of the protocols used in their isolation from bone marrow [102, 103]. Protocols involve the generation of EBs in the presence of IL3 and M-CSF, harvesting of a non-adherent cell population that contains myeloid progenitors and the subsequent plating of these non-adherent cells to select and promote terminal macrophage differentiation. ES-derived macrophages express the appropriate cell surface markers including CD11b and F4/80; they secrete specific cytokines in response to lipopolysaccharide, can be activated differentially with IFN-γ and IL-4 and can effectively phagocytose apoptotic cells [102, 103]. These ES-derived macrophages are currently being tested in models of inflammatory diseases with a view to translating this technology directly into the clinic.

3.3.2 Dendritic Cells

Dendritic cells (DC) have also been successfully generated from mouse and human ES cells using the relatively simple strategy of co-culturing differentiating ES cells on the OP9 stromal cell line followed by treatment with GM-CSF and IL4 [104–106]. Dendritic cells are considered as targets for immune intervention in the treatment of autoimmunity and allograft rejection as they are powerful antigen-presenting cells that can elicit immunity or self-tolerance. This limitless, infection-free and reproducible source of dendritic cells not only provides a valuable model for the study of antigen processing and presentation but also provides a scalable source of cells for DC vaccines or DC-mediated induction of immune tolerance [106].

3.3.3 Neutrophils

Protocols have been established that allows the production of large numbers of mature neutrophils from both mouse and human ES cells [107, 108]. Neutrophils make up the majority of circulating white blood cells and serve as the primary defence against infection. Over 10^{10} neutrophils are produced from progenitor cells on a daily basis and there is rapid induction of this production rate in response to infection and inflammatory stimulation. Congenital defects such as chronic granulomatous disease (CGD) or myeloablative anti-cancer chemotherapy can cause severe neutropenia resulting in a significant risk of serious infection. Current life-saving treatment can involve regular transfusion of G-CSF-mobilized granulocyte from donor blood but collection of these cells can involve some risk to the donor so an in vitro source of neutrophils could be beneficial [108]. An efficient feeder-free differentiation system has been developed for the production of large numbers of neutrophils from human ES cells. Haematopoietic differentiation was initiated by addition of VEGF, SCF, Flt3 ligand, thrombopoietin and G-CSF and the function of these ES-derived cells are comparable to their bone marrow counterparts, including their ability to migrate to sites of inflammation [107, 108].

3.3.4 Red Blood Cells

A number of protocols have been developed to produce red blood cell concentrates from hES that might in the future provide an essentially limitless supply of therapeutic cells guaranteed to be free from disease [109–113]. Successful protocols involve the initial induction of mesoderm and haemangioblast precursors using a combination of growth factors (BMP4, VEGF and bFGF) and the subsequent addition of SCF, thrombopoietin and Flt3 ligand [109]. However, the function of these ES-derived erythrocytes has not been tested in vivo and there are concerns that they are more akin to primitive, embryonic erythrocytes rather than the definitive, enucleated erythrocytes needed for clinical transfusion. Primitive and definitive erythrocytes express distinct haemoglobin profiles with globin switching occurring at precise time points during development. Some studies have reported very low levels of expression of adult beta globin in ES-derived red blood cells [109, 113, 114], whereas another reported that the appropriate globin switching can take place [110]. However, the oxygen-carrying capacity of hES-derived erythrocytes appear to be closer to that of erythrocytes derived from embryonic cord blood rather than adult peripheral blood in all studies, irrespective of their globin profile, raising concerns about their functional properties [109, 110]. The enucleation rate of ES-derived erythrocytes has also proven problematic. Primitive erythrocytes generally retain their nucleus, whereas fully functional definitive adult erythrocytes are enucleated. The level of enucleation in ES-derived erythrocytes seems to vary depending on the differentiation strategy with co-culturing on bone marrow or foetal liver stromal cell lines resulting in a significantly higher level [111, 112]. The precise role of the co-culturing process has not been defined but it is known that cell–cell interactions play a central role in the terminal maturation of erythroblasts [115]. The functional unit for definitive erythropoiesis is thought to be a multicellular structure composed of a central macrophage surrounded by developing erythroblasts and indeed co-culture of cord blood-derived erythrocytes with primary macrophages resulted in the production of fully mature enucleated erythrocytes [116]. Co-culture of hES-EPs with macrophages derived from the same ES cell line (as described in Sect. 3.3.1 of this review) might similarly increase the enucleation rate of these cells and resolve one of the outstanding challenges in the field.

4 Bioengineering and Processing

Protocols described for the differentiation of ES cells into haematopoietic lineages have all been developed in the research laboratory, but significant challenges must be overcome if these processes are to be translated into safe, robust and cost-effective therapeutic products.

Culture systems must be designed to produce expanded stem cells with uniform properties and to promote precisely controlled differentiation processes into the desired cell type with high purity. Scaling-up of culture conditions usually involves translating a static culture system into a dynamic system such as spinner flasks, stirred bioreactors and rotary culture where crucial culture conditions can be monitored and controlled. In these systems, cells must be provided with an adequate oxygen supply, their sensitivity to the build-up of metabolic byproducts and pH changes and the effects of shear stress that develops within the stirred cultures must all be considered [117]. A variety of bioreactor systems have been used to assess the expansion and differentiation of mouse and human ES cells [118–120], and it has been shown that cell growth of differentiating hES cells was significantly improved in four different dynamic culture systems [121]. Reproducible bioreactor cultures involving the generation of three-dimensional differentiating aggregates have now been established that allow the scale-up of embryoid body production, a critical step in haematopoietic differentiation [121, 122]. A micro-printing strategy integrated into a controlled bioreactor system is now able to generate homogeneous, size-specified ES cell aggregates [121]. Matrigel™ is bound to circular stamps (4–800 μm in diameter) of polymethylsoloxane that had been pre-spotted onto microscope slides using lithographic techniques. Single-cell suspensions of undifferentiated hES cells can be seeded onto Matrigel™ patterns and aggregates of uniform size developed [121]. This elegant system could be used to optimize the reproducibility of many differentiated cell types that rely on the initial generation of three-dimensional EBs from both ES and iPS cells.

5 Concluding Remarks

Significant progress has been made over the last 25 years in the production of mature haematopoietic cells and ES–HSC capable of long-term reconstitution as well as in defining cell surface markers for ES–HSC and the intermediate stages of their differentiation process. It is conceivable that these ES-derived cells could be used in the future to treat a number of haematological deficiencies. Applications range from short-term treatments of acute neutropenia or anaemia with ES-derived neutrophils or red blood cells to more long-term therapies involving transplantation of

ES-derived HSC. However, significant hurdles must be overcome before any ES-derived strategies could be translated into clinical practice. Cells to be transplanted into patients will have to be produced and maintained under strict GMP (Good Manufacturing Practice) conditions, free of animal products and their safety tested in large animal models. Although it has been possible to generate and maintain hES cells under such strictly defined conditions, hES–HSC and mature haematopoietic cell types have not been produced in this way to date with the majority of protocols still requiring the use of serum and co-culture. However, now that the surface phenotype of murine ES–HSC has been elucidated [83], it should be possible to perform high-throughput assays to define their production and maintenance conditions that will undoubtedly include the precisely timed addition of multiples cytokines at specific concentrations. As many aspects of mouse and human ES–HSC production in vitro are comparable, results of these studies will hopefully be translatable into the hES cell system.

ES–HSC, by definition, will have extensive proliferative capacity, so their tumour-forming capacity as well as the presence of any tumour-forming undifferentiated ES cells will have to be thoroughly tested in large animal models prior to their clinical use. If sufficient numbers of functional cells can be produced, we cautiously predict that the first therapy, if any, to reach clinical trials would be the transfusion of ES-derived erythrocytes. Enucleated red blood cells are short lived and unable to divide so the risk of tumour formation from transfusion of this cell population, if purified to a sufficient level, could be negligible. The vast amount of knowledge and experience in blood transfusion services worldwide on the storage and regulatory use of red blood cells can be directly applicable to ES-derived cells and should speed up the translational process.

Finally, it is important to note that in addition to their potential therapeutic value, ES-derived haematopoietic cells provide an invaluable model to study the molecular processes involved in mammalian haematopoietic development and provides a limitless resource of human blood cells that could be used in drug testing and discovery.

Acknowledgements The authors acknowledge funding from the Leukaemia Research and the European Commission (Framework 6) and Sandrine Prost and Melany Jackson for critical comments.

References

1. Keller G, Paige C, Gilboa E, Wagner EF. Expression of a foreign gene in myeloid and lymphoid cells derived from multipotent haematopoietic precursors. Nature. 1985; 318:149–154.
2. Lemischka IR, Raulet DH, Mulligan RC. Developmental potential and dynamic behavior of hematopoietic stem cells. Cell. 1986; 45:917–927.
3. Micklem, H.S., Lennon, J.E., Ansell, J.D., and Gray, R.A. (1987). Numbers and dispersion of repopulating hematopoietic cell clones in radiation chimeras as functions of injected cell does. Exp Hematol *15*, 251–257.

4. Morrison SJ, Weissman IL. The long-term repopulating subset of hematopoietic stem cells is deterministic and isolatable by phenotype. Immunity. 1994; 1:661–673.
5. Osawa M, Hanada K, Hamada H, Nakauchi H. Long-term lymphohematopoietic reconstitution by a single CD34-low/negative hematopoietic stem cell. Science. 1996; 273:242–245.
6. Burt RK, Loh Y, Pearce W, Beohar N, Barr WG, Craig R, Wen Y, Rapp JA, Kessler J. Clinical applications of blood-derived and marrow-derived stem cells for nonmalignant diseases. JAMA. 2008; 299:925–936.
7. Regan F, Taylor C. Blood transfusion medicine. Bmj. 2002; 325:143–147.
8. Evans MJ, Kaufman MH. Establishment in culture of pluripotential cells from mouse embryos. Nature. 1981; 292:154–156.
9. Martin GR. Isolation of a pluripotent cell line from early mouse embryos cultured in medium conditioned by teratocarcinoma stem cells. Proc Natl Acad Sci USA. 1981; 78:7634–7638.
10. Thomson JA, Itskovitz-Eldor J, Shapiro SS, Waknitz MA, Swiergiel JJ, Marshall VS, Jones JM. Embryonic stem cell lines derived from human blastocysts. Science. 1998; 282: 1145–1147.
11. Daley GQ, Scadden DT. Prospects for stem cell-based therapy. Cell. 2008; 132:544–548.
12. Takahashi K, Tanabe K, Ohnuki M, Narita M, Ichisaka T, Tomoda K, Yamanaka S. Induction of pluripotent stem cells from adult human fibroblasts by defined factors. Cell. 2007; 131:861–872.
13. Takahashi K, Yamanaka S. Induction of pluripotent stem cells from mouse embryonic and adult fibroblast cultures by defined factors. Cell. 2006; 126:663–676.
14. Kaji K, Norrby K, Paca A, Mileikovsky M, Mohseni P, Woltjen K. Virus-free induction of pluripotency and subsequent excision of reprogramming factors. Nature. 2009; 458: 771–775.
15. Nishikawa S, Goldstein RA, Nierras CR. The promise of human induced pluripotent stem cells for research and therapy. Nat Rev Mol Cell Biol. 2008; 9:725–729.
16. Wernig M, Meissner A, Foreman R, Brambrink T, Ku M, Hochedlinger K, Bernstein BE, Jaenisch R. In vitro reprogramming of fibroblasts into a pluripotent ES-cell-like state. Nature. 2007; 448:318–324.
17. Woltjen K, Michael IP, Mohseni P, Desai R, Mileikovsky M, Hamalainen R, Cowling R, Wang W, Liu P, Gertsenstein M, et al. piggyBac transposition reprograms fibroblasts to induced pluripotent stem cells. Nature. 2009; 458:766–770.
18. Park IH, Arora N, Huo H, Maherali N, Ahfeldt T, Shimamura A, Lensch MW, Cowan C, Hochedlinger K, Daley GQ. Disease-specific induced pluripotent stem cells. Cell. 2008; 134:877–886.
19. Choi KD, Yu J, Smuga-Otto K, Salvagiotto G, Rehrauer W, Vodyanik M, Thomson J, Slukvin I. Hematopoietic and endothelial differentiation of human induced pluripotent stem cells. Stem Cells. 2009; 27:559–567.
20. Doetschman TC, Eistetter H, Katz M, Schmidt W, Kemler R. The in vitro development of blastocyst-derived embryonic stem cell lines: formation of visceral yolk sac, blood islands and myocardium. J Embryol Exp Morphol. 1985; 87:27–45.
21. Wiles MV, Keller G. Multiple hematopoietic lineages develop from embryonic stem (ES) cells in culture. Development. 1991; 111:259–267.
22. Keller G, Kennedy M, Papayannopoulou T, Wiles MV. Hematopoietic commitment during embryonic stem cell differentiation in culture. Mol Cell Biol. 1993; 13:473–486.
23. Keller G. Embryonic stem cell differentiation: emergence of a new era in biology and medicine. Genes Dev. 2005; 19:1129–1155.
24. Kitajima K, Tanaka M, Zheng J, Sakai-Ogawa E, Nakano T. In vitro differentiation of mouse embryonic stem cells to hematopoietic cells on an OP9 stromal cell monolayer. Methods Enzymol. 2003; 365:72–83.
25. Krassowska A, Gordon-Keylock S, Samuel K, Gilchrist D, Dzierzak E, Oostendorp R, Forrester LM, Ansell JD. Promotion of haematopoietic activity in embryonic stem cells by the aorta-gonad-mesonephros microenvironment. Exp Cell Res. 2006; 312:3595–3603.

26. Ledran MH, Krassowska A, Armstrong L, Dimmick I, Renstrom J, Lang R, Yung S, Santibanez-Coref M, Dzierzak E, Stojkovic M, et al. Efficient hematopoietic differentiation of human embryonic stem cells on stromal cells derived from hematopoietic niches. Cell Stem Cell. 2008; 3:85–98.
27. Nakano T, Kodama H, Honjo T. Generation of lymphohematopoietic cells from embryonic stem cells in culture. Science. 1994; 265:1098–1101.
28. Weisel KC, Gao Y, Shieh JH, Moore MA. Stromal cell lines from the aorta-gonado-mesonephros region are potent supporters of murine and human hematopoiesis. Exp Hematol. 2006; 34:1505–1516.
29. Zhang WJ, Park C, Arentson E, Choi K. Modulation of hematopoietic and endothelial cell differentiation from mouse embryonic stem cells by different culture conditions. Blood. 2005; 105:111–114.
30. Chadwick K, Wang L, Li L, Menendez P, Murdoch B, Rouleau A, Bhatia M. Cytokines and BMP-4 promote hematopoietic differentiation of human embryonic stem cells. Blood. 2003; 102:906–915.
31. Nakayama N, Lee J, Chiu L. Vascular endothelial growth factor synergistically enhances bone morphogenetic protein-4-dependent lymphohematopoietic cell generation from embryonic stem cells in vitro. Blood. 2000; 95:2275–2283.
32. Ng YS, Ramsauer M, Loureiro RM, D'Amore PA. Identification of genes involved in VEGF-mediated vascular morphogenesis using embryonic stem cell-derived cystic embryoid bodies. Lab Invest. 2004; 84:1209–1218.
33. Park C, Afrikanova I, Chung YS, Zhang WJ, Arentson E, Fong Gh G, Rosendahl A, Choi K. A hierarchical order of factors in the generation of FLK1- and SCL-expressing hematopoietic and endothelial progenitors from embryonic stem cells. Development. 2004; 131:2749–2762.
34. Pearson S, Sroczynska P, Lacaud G, Kouskoff V. The stepwise specification of embryonic stem cells to hematopoietic fate is driven by sequential exposure to Bmp4, activin A, bFGF and VEGF. Development. 2008; 135:1525–1535.
35. Pick M, Azzola L, Mossman A, Stanley EG, Elefanty AG. Differentiation of human embryonic stem cells in serum-free medium reveals distinct roles for bone morphogenetic protein 4, vascular endothelial growth factor, stem cell factor, and fibroblast growth factor 2 in hematopoiesis. Stem Cells. 2007; 25:2206–2214.
36. Purpura KA, Morin J, Zandstra PW. Analysis of the temporal and concentration-dependent effects of BMP-4, VEGF, and TPO on development of embryonic stem cell-derived mesoderm and blood progenitors in a defined, serum-free media. Exp Hematol. 2008; 36:1186–1198.
37. Zambidis ET, Peault B, Park TS, Bunz F, Civin CI. Hematopoietic differentiation of human embryonic stem cells progresses through sequential hematoendothelial, primitive, and definitive stages resembling human yolk sac development. Blood. 2005; 106:860–870.
38. Cheng X, Huber TL, Chen VC, Gadue P, Keller GM. Numb mediates the interaction between Wnt and Notch to modulate primitive erythropoietic specification from the hemangioblast. Development. 2008; 135:3447–3458.
39. Elefanty AG, Robb L, Begley CG. Factors involved in leukaemogenesis and haemopoiesis. Baillieres Clin Haematol. 1997; 10:589–614.
40. Kim DJ, Park CS, Yoon JK, Song WK. Differential expression of the Wnt and Frizzled genes in $Flk1^+$ cells derived from mouse ES cells. Cell Biochem Funct. 2008; 26:24–32.
41. Mikkola HK, Fujiwara Y, Schlaeger TM, Traver D, Orkin SH. Expression of CD41 marks the initiation of definitive hematopoiesis in the mouse embryo. Blood. 2003; 101:508–516.
42. Robertson SM, Kennedy M, Shannon JM, Keller G. A transitional stage in the commitment of mesoderm to hematopoiesis requiring the transcription factor SCL/tal-1. Development. 2000; 127:2447–2459.
43. Fehling HJ, Lacaud G, Kubo A, Kennedy M, Robertson S, Keller G, Kouskoff V. Tracking mesoderm induction and its specification to the hemangioblast during embryonic stem cell differentiation. Development. 2003; 130:4217–4227.

44. Kouskoff V, Lacaud G, Schwantz S, Fehling HJ, Keller G. Sequential development of hematopoietic and cardiac mesoderm during embryonic stem cell differentiation. Proc Natl Acad Sci USA. 2005; 102:13170–13175.
45. Wang L, Li L, Shojaei F, Levac K, Cerdan C, Menendez P, Martin T, Rouleau A, Bhatia M. Endothelial and hematopoietic cell fate of human embryonic stem cells originates from primitive endothelium with hemangioblastic properties. Immunity. 2004; 21:31–41.
46. D'Souza SL, Elefanty AG, Keller G. SCL/Tal-1 is essential for hematopoietic commitment of the hemangioblast but not for its development. Blood. 2005; 105:3862–3870.
47. Hidaka M, Stanford WL, Bernstein A. Conditional requirement for the Flk-1 receptor in the in vitro generation of early hematopoietic cells. Proc Natl Acad Sci USA. 1999; 96: 7370–7375.
48. Lacaud G, Gore L, Kennedy M, Kouskoff V, Kingsley P, Hogan C, Carlsson L, Speck N, Palis J, Keller G. Runx1 is essential for hematopoietic commitment at the hemangioblast stage of development in vitro. Blood. 2002; 100:458–466.
49. Lacaud G, Kouskoff V, Trumble A, Schwantz S, Keller G. Haploinsufficiency of Runx1 results in the acceleration of mesodermal development and hemangioblast specification upon in vitro differentiation of ES cells. Blood. 2004; 103:886–889.
50. Schuh AC, Faloon P, Hu QL, Bhimani M, Choi K. In vitro hematopoietic and endothelial potential of flk-1(–/–) embryonic stem cells and embryos. Proc Natl Acad Sci USA. 1999; 96:2159–2164.
51. Huber TL, Kouskoff V, Fehling HJ, Palis J, Keller G. Haemangioblast commitment is initiated in the primitive streak of the mouse embryo. Nature. 2004; 432:625–630.
52. Vogeli KM, Jin SW, Martin GR, Stainier DY. A common progenitor for haematopoietic and endothelial lineages in the zebrafish gastrula. Nature. 2006; 443:337–339.
53. Yao H, Liu B, Wang X, Lan Y, Hou N, Yang X, Mao N. Identification of high proliferative potential precursors with hemangioblastic activity in the mouse aorta-gonad- mesonephros region. Stem Cells. 2007; 25:1423–1430.
54. Bertrand JY, Giroux S, Golub R, Klaine M, Jalil A, Boucontet L, Godin I, Cumano A. Characterization of purified intraembryonic hematopoietic stem cells as a tool to define their site of origin. Proc Natl Acad Sci USA. 2005; 102:134–139.
55. de Bruijn MF, Ma X, Robin C, Ottersbach K, Sanchez MJ, Dzierzak E. Hematopoietic stem cells localize to the endothelial cell layer in the midgestation mouse aorta. Immunity. 2002; 16:673–683.
56. de Bruijn MF, Peeters MC, Luteijn T, Visser P, Speck NA, Dzierzak E. CFU-S(11) activity does not localize solely with the aorta in the aorta-gonad-mesonephros region. Blood. 2000a; 96:2902–2904.
57. de Bruijn MF, Speck NA, Peeters MC, Dzierzak E. Definitive hematopoietic stem cells first develop within the major arterial regions of the mouse embryo. Embo J. 2000b; 19: 2465–2474.
58. Jaffredo T, Gautier R, Eichmann A, Dieterlen-Lievre F. Intraaortic hemopoietic cells are derived from endothelial cells during ontogeny. Development. 1998; 125:4575–4583.
59. Nishikawa SI, Nishikawa S, Hirashima M, Matsuyoshi N, Kodama H. Progressive lineage analysis by cell sorting and culture identifies $FLK1^{+}VE\text{-}cadherin^{+}$ cells at a diverging point of endothelial and hemopoietic lineages. Development. 1998; 125:1747–1757.
60. North TE, de Bruijn MF, Stacy T, Talebian L, Lind E, Robin C, Binder M, Dzierzak E, Speck NA. Runx1 expression marks long-term repopulating hematopoietic stem cells in the midgestation mouse embryo. Immunity. 2002; 16:661–672.
61. Taoudi S, Gonneau C, Moore K, Sheridan JM, Blackburn CC, Taylor E, Medvinsky A. Extensive hematopoietic stem cell generation in the AGM region via maturation of VE-$cadherin^{+}CD45^{+}$ pre-definitive HSCs. Cell Stem Cell. 2008; 3:99–108.
62. Taoudi S, Medvinsky A. Functional identification of the hematopoietic stem cell niche in the ventral domain of the embryonic dorsal aorta. Proc Natl Acad Sci USA. 2007; 104: 9399–9403.

63. Taoudi S, Morrison AM, Inoue H, Gribi R, Ure J, Medvinsky A. Progressive divergence of definitive haematopoietic stem cells from the endothelial compartment does not depend on contact with the foetal liver. Development. 2005; 132:4179–4191.
64. Ueno H, Weissman IL. Clonal analysis of mouse development reveals a polyclonal origin for yolk sac blood islands. Dev Cell. 2006; 11:519–533.
65. Choi K, Kennedy M, Kazarov A, Papadimitriou JC, Keller G. A common precursor for hematopoietic and endothelial cells. Development. 1998; 125:725–732.
66. Kennedy M, Firpo M, Choi K, Wall C, Robertson S, Kabrun N, Keller G. A common precursor for primitive erythropoiesis and definitive haematopoiesis. Nature. 1997; 386:488–493.
67. Faloon P, Arentson E, Kazarov A, Deng CX, Porcher C, Orkin S, Choi K. Basic fibroblast growth factor positively regulates hematopoietic development. Development. 2000; 127:1931–1941.
68. Lancrin C, Sroczynska P, Stephenson C, Allen T, Kouskoff V, Lacaud G. The haemangioblast generates haematopoietic cells through a haemogenic endothelium stage. Nature. 2009; 457:892–895.
69. Kennedy M, D'Souza SL, Lynch-Kattman M, Schwantz S, Keller G. Development of the hemangioblast defines the onset of hematopoiesis in human ES cell differentiation cultures. Blood. 2007; 109:2679–2687.
70. Wang L, Menendez P, Shojaei F, Li L, Mazurier F, Dick JE, Cerdan C, Levac K, Bhatia M. Generation of hematopoietic repopulating cells from human embryonic stem cells independent of ectopic HOXB4 expression. J Exp Med. 2005a; 201:1603–1614.
71. Forrester LM, Bernstein A, Rossant J, Nagy A. Long-term reconstitution of the mouse hematopoietic system by embryonic stem cell-derived foetal liver. Proc Natl Acad Sci USA. 1991; 88:7514–7517.
72. Nagy A, Rossant J, Nagy R, Abramow-Newerly W, Roder JC. Derivation of completely cell culture-derived mice from early-passage embryonic stem cells. Proc Natl Acad Sci USA. 1993; 90:8424–8428.
73. Gutierrez-Ramos JC, Palacios R. In vitro differentiation of embryonic stem cells into lymphocyte precursors able to generate T and B lymphocytes in vivo. Proc Natl Acad Sci USA. 1992; 89:9171–9175.
74. Hole N, Graham GJ, Menzel U, Ansell JD. A limited temporal window for the derivation of multilineage repopulating hematopoietic progenitors during embryonal stem cell differentiation in vitro. Blood. 1996; 88:1266–1276.
75. Muller AM, Dzierzak EA. ES cells have only a limited lymphopoietic potential after adoptive transfer into mouse recipients. Development. 1993; 118:1343–1351.
76. Palacios R, Golunski E, Samaridis J. In vitro generation of hematopoietic stem cells from an embryonic stem cell line. Proc Natl Acad Sci USA. 1995; 92:7530–7534.
77. Kyba M, Perlingeiro RC, Daley GQ. HoxB4 confers definitive lymphoid-myeloid engraftment potential on embryonic stem cell and yolk sac hematopoietic progenitors. Cell. 2002; 109:29–37.
78. Kyba M, Perlingeiro RC, Daley GQ. Development of hematopoietic repopulating cells from embryonic stem cells. Methods Enzymol. 2003; 365:114–129.
79. Schuringa JJ, Wu K, Morrone G, Moore MA. Enforced activation of STAT5A facilitates the generation of embryonic stem-derived hematopoietic stem cells that contribute to hematopoiesis in vivo. Stem Cells. 2004; 22:1191–1204.
80. Wang Y, Yates F, Naveiras O, Ernst P, Daley GQ. Embryonic stem cell-derived hematopoietic stem cells. Proc Natl Acad Sci USA. 2005b; 102:19081–19086.
81. Bonde S, Dowden AM, Chan KM, Tabayoyong WB, Zavazava N. HOXB4 but not BMP4 confers self-renewal properties to ES-derived hematopoietic progenitor cells. Transplantation. 2008; 86:1803–1809.
82. Chan KM, Bonde S, Klump H, Zavazava N. Hematopoiesis and immunity of HOXB4-transduced embryonic stem cell-derived hematopoietic progenitor cells. Blood. 2008; 111:2953–2961.

83. McKinney-Freeman SL, Naveiras O, Yates F, Loewer S, Philitas M, Curran M, Park PJ, Daley GQ. Surface antigen phenotypes of hematopoietic stem cells from embryos and murine embryonic stem cells. Blood. 2009; 114:268–278.
84. Antonchuk J, Sauvageau G, Humphries RK. HOXB4 overexpression mediates very rapid stem cell regeneration and competitive hematopoietic repopulation. Exp Hematol. 2001; 29:1125–1134.
85. Bowles KM, Vallier L, Smith JR, Alexander MR, Pedersen RA. HOXB4 overexpression promotes hematopoietic development by human embryonic stem cells. Stem Cells. 2006; 24:1359–1369.
86. Lee GS, Kim BS, Sheih JH, Moore M. Forced expression of HoxB4 enhances hematopoietic differentiation by human embryonic stem cells. Mol Cells. 2008; 25:487–493.
87. Unger C, Karner E, Treschow A, Stellan B, Felldin U, Concha H, Wendel M, Hovatta O, Aints A, Ahrlund-Richter L, Dilber MS. Lentiviral-mediated HoxB4 expression in human embryonic stem cells initiates early hematopoiesis in a dose-dependent manner but does not promote myeloid differentiation. Stem Cells. 2008; 26:2455–2466.
88. Lu, S.J., Feng, Q., Ivanova, Y., Luo, C., Li, T., Li, F., Honig, G.R., and Lanza, R. (2007). Recombinant HoxB4 fusion proteins enhance hematopoietic differentiation of human embryonic stem cells. Stem Cells Dev *16*, 547–559.
89. Ji J, Vijayaragavan K, Bosse M, Menendez P, Weisel K, Bhatia M. OP9 stroma augments survival of hematopoietic precursors and progenitors during hematopoietic differentiation from human embryonic stem cells. Stem Cells. 2008; 26:2485–2495.
90. Narayan AD, Chase JL, Lewis RL, Tian X, Kaufman DS, Thomson JA, Zanjani ED. Human embryonic stem cell-derived hematopoietic cells are capable of engrafting primary as well as secondary fetal sheep recipients. Blood. 2006; 107:2180–2183.
91. Medvinsky A, Dzierzak E. Definitive hematopoiesis is autonomously initiated by the AGM region. Cell. 1996; 86:897–906.
92. Medvinsky AL, Samoylina NL, Muller AM, Dzierzak EA. An early pre-liver intraembryonic source of CFU-S in the developing mouse. Nature. 1993; 364:64–67.
93. Cumano A, Ferraz JC, Klaine M, Di Santo JP, Godin I. Intraembryonic, but not yolk sac hematopoietic precursors, isolated before circulation, provide long-term multilineage reconstitution. Immunity. 2001; 15:477–485.
94. Yoder MC, Hiatt K. Engraftment of embryonic hematopoietic cells in conditioned newborn recipients. Blood. 1997; 89:2176–2183.
95. Matsumoto K, Isagawa T, Nishimura T, Ogaeri T, Eto K, Miyazaki S, Miyazaki J, Aburatani H, Nakauchi H, Ema H. Stepwise development of hematopoietic stem cells from embryonic stem cells. PLoS ONE. 2009; 4:e4820.
96. Miyagi T, Takeno M, Nagafuchi H, Takahashi M, Suzuki N. $Flk1^+$ cells derived from mouse embryonic stem cells reconstitute hematopoiesis in vivo in SCID mice. Exp Hematol. 2002; 30:1444–1453.
97. Burt RK, Verda L, Kim DA, Oyama Y, Luo K, Link C. Embryonic stem cells as an alternate marrow donor source: engraftment without graft-versus-host disease. J Exp Med. 2004; 199:895–904.
98. Gordon S. The macrophage: past, present and future. Eur J Immunol. 2007; 37(Suppl 1): S9–S17.
99. Gordon S, Taylor PR. Monocyte and macrophage heterogeneity. Nat Rev Immunol. 2005; 5:953–964.
100. Kluth DC. Pro-resolution properties of macrophages in renal injury. Kidney Int. 2007; 72:234–236.
101. Kluth DC, Ainslie CV, Pearce WP, Finlay S, Clarke D, Anegon I, Rees AJ. Macrophages transfected with adenovirus to express IL-4 reduce inflammation in experimental glomerulonephritis. J Immunol. 2001; 166:4728–4736.
102. Karlsson KR, Cowley S, Martinez FO, Shaw M, Minger SL, James W. Homogeneous monocytes and macrophages from human embryonic stem cells following coculture-free differentiation in M-CSF and IL-3. Exp Hematol. 2008; 36:1167–1175.

103. Odegaard JI, Vats D, Zhang L, Ricardo-Gonzalez R, Smith KL, Sykes DB, Kamps MP, Chawla A. Quantitative expansion of ES cell-derived myeloid progenitors capable of differentiating into macrophages. J Leukoc Biol. 2007; 81:711–719.
104. Fairchild PJ, Brook FA, Gardner RL, Graca L, Strong V, Tone Y, Tone M, Nolan KF, Waldmann H. Directed differentiation of dendritic cells from mouse embryonic stem cells. Curr Biol. 2000; 10:1515–1518.
105. Fairchild PJ, Nolan KF, Cartland S, Waldmann H. Embryonic stem cells: a novel source of dendritic cells for clinical applications. Int Immunopharmacol. 2005; 5:13–21.
106. Slukvin II, Vodyanik MA, Thomson JA, Gumenyuk ME, Choi KD. Directed differentiation of human embryonic stem cells into functional dendritic cells through the myeloid pathway. J Immunol. 2006; 176:2924–2932.
107. Lieber JG, Webb S, Suratt BT, Young SK, Johnson GL, Keller GM, Worthen GS. The in vitro production and characterization of neutrophils from embryonic stem cells. Blood. 2004; 103:852–859.
108. Saeki K, Nakahara M, Matsuyama S, Nakamura N, Yogiashi Y, Yoneda A, Koyanagi M, Kondo Y, Yuo A. A feeder-free and efficient production of functional neutrophils from human embryonic stem cells. Stem Cells. 2008; 10(3): 341–354.
109. Lu SJ, Feng Q, Park JS, Vida L, Lee BS, Strausbauch M, Wettstein PJ, Honig GR, Lanza R. Biologic properties and enucleation of red blood cells from human embryonic stem cells. Blood. 2008; 112:4475–4484.
110. Ma F, Ebihara Y, Umeda K, Sakai H, Hanada S, Zhang H, Zaike Y, Tsuchida E, Nakahata T, Nakauchi H, Tsuji K. Generation of functional erythrocytes from human embryonic stem cell-derived definitive hematopoiesis. Proc Natl Acad Sci USA. 2008; 105:13087–13092.
111. Qiu, C., Olivier, E.N., Velho, M., and Bouhassira, E.E. (2008). Globin switches in yolk sac-like primitive and fetal-like definitive red blood cells produced from human embryonic stem cells. Blood *111*, 2400–2408.
112. Lu, S.J., Feng, Q., Park, J.S., Vida, L., Lee, B.S., Strausbauch, M., Wettstein, P.J., Honig, G.R., and Lanza, R. (2008). Biologic properties and enucleation of red blood cells from human embryonic stem cells. Blood *112*, 4475–4484.
113. Olivier EN, Qiu C, Velho M, Hirsch RE, Bouhassira EE. Large-scale production of embryonic red blood cells from human embryonic stem cells. Exp Hematol. 2006; 34:1635–1642.
114. Qiu C, Olivier EN, Velho M, Bouhassira EE. Globin switches in yolk sac-like primitive and fetal-like definitive red blood cells produced from human embryonic stem cells. Blood. 2008; 111:2400–2408.
115. Soni S, Bala S, Gwynn B, Sahr KE, Peters LL, Hanspal M. Absence of erythroblast macrophage protein (Emp) leads to failure of erythroblast nuclear extrusion. J Biol Chem. 2006; 281:20181–20189.
116. Fujimi A, Matsunaga T, Kobune M, Kawano Y, Nagaya T, Tanaka I, Iyama S, Hayashi T, Sato T, Miyanishi K, et al. Ex vivo large-scale generation of human red blood cells from cord blood $CD34^+$ cells by co-culturing with macrophages. Int J Hematol. 2008; 87:339–350.
117. King JA, Miller WM. Bioreactor development for stem cell expansion and controlled differentiation. Curr Opin Chem Biol. 2007; 11:394–398.
118. Dang SM, Gerecht-Nir S, Chen J, Itskovitz-Eldor J, Zandstra PW. Controlled, scalable embryonic stem cell differentiation culture. Stem Cells. 2004; 22:275–282.
119. Dang SM, Zandstra PW. Scalable production of embryonic stem cell-derived cells. Methods Mol Biol. 2005; 290:353–364.
120. Fok EY, Zandstra PW. Shear-controlled single-step mouse embryonic stem cell expansion and embryoid body-based differentiation. Stem Cells. 2005; 23:1333–1342.
121. Niebruegge S, Bauwens CL, Peerani R, Thavandiran N, Masse S, Sevaptisidis E, Nanthakumar K, Woodhouse K, Husain M, Kumacheva E, Zandstra PW. Generation of human embryonic stem cell-derived mesoderm and cardiac cells using size-specified aggregates in an oxygen-controlled bioreactor. Biotechnol Bioeng. 2009; 102:493–507.
122. Yirme G, Amit M, Laevsky I, Osenberg S, Itskovitz-Eldor J. Establishing a dynamic process for the formation, propagation, and differentiation of human embryoid bodies. Stem Cells Dev. 2008; 17:1227–1241.

Genetic Modification of Human Embryonic and Induced Pluripotent Stem Cells: Viral and Non-viral Approaches

Nicole M. Kane, Chris Denning, and Andrew H. Baker

Abstract Human embryonic stem cells (hESCs) and induced pluripotent cells (iPSCs) are well suited for translational cell therapy. For hESC, the pluripotent phenotype, naturally occurring within the inner cell mass of the early embryo, bestows the capability to differentiate into any cell type of interest and this, coupled with their ability to remain in an undifferentiated state with indefinite proliferative capacity, means that essentially unlimited numbers of identical, well-defined and genetically characterised stem cells can be produced in culture for therapeutic applications. An understanding of the regulatory mechanisms responsible for pluripotency and differentiation potential of hESCs is critical for translating their potential in vitro to therapeutic use in vivo. Harnessing of this therapeutic potential in conjunction with modern genetic modification tools promises great advancement in the study of developmental and adult physiology and pathophysiology with a view towards implementation of genetically modified hESCs to advance regenerative medicine.

Keywords Human embryonic stem cells · Induced pluripotent stem cells · Reprogramming · Genetic modification · Non-viral · Viral · Lentivirus · Adenovirus

1 Introduction

Human embryonic stem cells (hESCs) provide a broad potential for cell-based therapy relating to conditions with unmet clinical need. However, owing to the rudimentary knowledge of the complex biological factors governing pluripotency and lineage determination, this therapeutic goal is not as yet attainable as evidenced by the relatively slow progression to clinical studies. Therefore, significant hurdles exist, including the production of high yields of the desired, fully differentiated

A.H. Baker (✉)
Faculty of Medicine, British Heart Foundation Glasgow Cardiovascular Research Centre, University of Glasgow, Glasgow, UK
e-mail: ab11f@clinmed.gla.ac.uk

G.M. Artmann et al. (eds.), *Stem Cell Engineering*,
DOI 10.1007/978-3-642-11865-4_7, © Springer-Verlag Berlin Heidelberg 2011

cell type that is effective and safe through which to treat a target human population. The availability of molecular tools to dissect the plethora of biological factors that govern the status of a stem cell is critical for basic biology and to potentially further improve cellular differentiation protocols and safety aspect of hESC translation. Major concerns relating to safety in humans include the homogeneity of the cells to be transplanted, their immunogenicity in patients and the potential of implanted cells to form tumours in humans. Notwithstanding this, the FDA has recently approved the first hESC clinical trial in the USA for spinal cord injury (http://www.isscr.org/press_releases/fda.html). Although more recent, one can envisage the translation of iPSC technology to the clinical setting since the immunogenicity issue is overcome by patient-matched cells. However, the other issues mentioned still remain.

Genetic modification of hESCs provides an approach to achieve the desirable result of reproducible, scalable, directed differentiation towards a specific developmental lineage in a controlled fashion [1]. This technology is not limited to directing hESC differentiation but has broader applications which will be pivotal to developing labelling and selection of desired lineages, cell tracking in vivo [2], silencing of endogenous genes or non-coding RNA, monitoring intracellular signalling activity during cellular transformation [3], promotion of post-transplantation survival and function [3], suicide gene ablation of transplanted cells [4], reducing or eliminating immunogenicity [5] and notably, the most recent application, generation of patient-specific induced pluripotent cells (iPS) from somatic tissue [6–11].

Despite huge expectation and intensive efforts the genetic modification of hESCs has been slow but necessitates the need for safe and efficient methodologies to exert genetic manipulation of stem cell behaviour and fate. In this context, this chapter provides a review of genetic modification approaches that may be used to study stem cell biology and iPS cell reprogramming and discusses their strengths and limitations. Gene delivery systems are varied and many have been developed and optimised over the last years by many gene therapy groups interested in both preclinical gene delivery and translation gene therapy in the clinical setting.

The efficiency of gene delivery to stem cells is crucial. Experiences from preceding years of clinical gene therapy has proven the importance for different vector systems from all gene delivery applications in basic and translational science to be subjected to rigorous analysis and, ideally, stringent testing in a myriad of recipient cell types in several independent laboratories. Depending on the desired outcome, e.g. use of stably expressed reporter genes for lineage tracking or ectopic expression and/or silencing of a gene to promote differentiation towards a specific lineage, different vector types and delivery methods of gene delivery will be more useful. Here, we discuss the most efficient vector techniques that have been employed for gene delivery to hESCs.

1.1 Non-viral Delivery Systems

Non-viral, low-cost delivery strategies that have been assessed for gene delivery to hESCs include chemical transfection with lipid or cationic polymer transfection reagents and a physical transfection using electroporation or nucleofection. These are described in more detail below.

Lipofection employs DNA encapsulation in cationic lipids comprising hydrophilic and hydrophobic domains, facilitating endocytosis-mediated DNA uptake [12]. Electroporation evokes a transient, reversible permeabilisation of the cell membrane. During the pulse duration, negatively charged DNA moves towards the anode and enters permeabilised cells in the process [12]. Nucleofection is a strategy based on the classic electroporation technology whereby a proprietary nucleofactor solution (Amaxa) is used in conjunction with specific electrical parameters to deliver plasmid DNA targeted directly to the nucleus with a greater efficiency and resulting in enhanced gene expression [13].

Despite intensive efforts and considerable advancement in non-viral gene delivery technologies since the introduction of hESCs into the scientific mainstream [14], many stem cell types have proven highly refractory to genetic modification via this route. This is exemplified with hESCs co-cultured on "feeder" cells, where extremely low transient transfection efficiencies have been reported [13, 15]. Until recently, the most efficient transfection methods have been optimised using two hESC cell lines H1 and H9 (WiCell Research Institute). Previous studies comparing transfection efficiencies in the H9 hESC line using the lipid-based transfection reagents lipofectamine 2000 (Invitrogen) and FuGENE (Roche), the cationic polymer-based reagent ExGen 500 (Fermentas) and electroporation using a reporter construct, the human elongation factor 1 α (EF1 α) promoter driving eGFP, reported that none of the methods yielded transfection efficiencies higher than 10%, with ExGen 500 reagent demonstrating a tenfold greater efficiency than the other methods [16]. These results were, however, in contrast to other studies whereby transfection efficiencies of 30–80% were reported in H9 hESCs using FuGENE or lipofectamine 2000 (28% efficiency) and an expression cassette including a cytomegalovirus (CMV) promoter and eGFP reporter gene [17, 18]. Additionally, another report demonstrated transfection efficiencies of H9 hESCs to be four times greater using lipofectamine compared with ExGen 500 [19]. These discrepancies in transfection efficiencies probably indicate that between labs and personnel handling cells the same hESC line may be cultured and transfected in a slightly different manner, whereas the low transfection efficiencies likely reflect that many transgenic promoters are methylated and silenced in hESC cultures typically maintained at high density as colonies in co-culture with "feeder" cells [14].

Each of these tangible issues has recently been addressed. The CAG promoter, comprising a cassette with CMV enhancer elements coupled to the chicken β-actin and the rabbit β-globin gene, was identified as the most effectual promoter in a systematic study comparing the effectiveness of five different promoters in four independent hESC lines [15]. Physical methods of transient transfection are thought to yield low transfection efficiencies owing to the resulting low cell viability [16, 20]. However, transfection efficiencies of ~40 and 65% in H9 hESCs (20% in H1 hESCs) have been reported using electroporation and nucleofection to transfer the eGFP gene driven by a CMV promoter, respectively [18]. An independent study also reported an increase in transfection efficiencies by 3.6-fold to 22% [13]. Interestingly, to date, electroporation remains the mainstay for production of gene targeted cell lines [21]. Perhaps, most successfully, standardised feeder-free culture methods, substituting mouse embryonic fibroblasts (MEFs) for Matrigel, a basement membrane matrix derived from Engelbreth-Holm-Swarm (EHS) mouse sarcoma, have been described that employ trypsin-based passaging [22]. This culture system tolerates the seeding of hESCs at low density enabling greater accessibility to chemical reagents. Using this strategy, highly variable transfection efficiencies with six lipid formulations were observed, with the non-toxic polyamine-based reagent, GeneJammer (Stratagene), reporter to be superior to all methods tested. A recent investigation of multiple technologies, including electroporation and lipofectamine, for inducing ectopic gene expression in 12 independently derived cell lines (HES-2; Envy; HUES1, 5, 7, 15; HESC-NL1, 2, 3, 4; and NOTT1, 2) successfully transferred to the aforementioned feeder-free culture conditions allows for direct comparison of the transfection efficiency depending on the presence or absence of feeder cells [23]. Using GeneJammer reagent to introduce a plasmid expressing GFP driven by the CAG promoter, low transfection efficiencies were reported in all lines cultured under the original conditions; however, once transferred to the feeder-free system, reproducible transfection efficiencies averaging 75% were achieved [23]. Additionally, transfection of an alexafluor 488-conjugated negative control siRNA using lipofectamine 2000 was demonstrated to achieve ~95% transfection efficiencies [23]. The enhanced transfection efficiencies for plasmid and siRNA may partly be attributable to the absence of feeder cells (which, although providing both nutritional support and a suitable extracellular matrix for adhesion, are also reported to sequester transfection reagents, reducing transfection efficiency [24]) and to the low density in which the cells were seeded in a monolayer, permitting optimal interaction with the transfection reagent, as evidenced by the inverse correlation observed between cell density and transfection efficiency [23].

Applications requiring continual, stable expression of a transgene, including assessment of potential genetic modification effects on hESCs phenotype and maintenance of gene expression during pluripotent self-renewal as well as during differentiation, necessitate the production of stable genetically modified hESCs. One characteristic of non-viral transfection strategies is that almost all the transgenes are not integrated into the genome, and thus the expression is transient, because the genome becomes diluted with each replication. The efficiency of generating stable genetically modified hESC lines, cultured on feeder cells, using transfection and selection for antibiotic resistance was reportedly very low (estimated to be 1×10^{-5}) [16, 19]; however, stable lines generated from cells cultured on a feeder-free system was denoted as approximately 2×10^{-4} [23].

Developments in gene transfer of DNA via non-viral technologies have now enabled applications to be realised, including transgenic enrichment of cardiomyocytes generated in an embryoid body (EB) differentiation system using the human myosin light chain-2v promoter fragment (560 bp), driving eGFP expression, or alveolar cells derived from hES cells [22, 25, 26]. Moreover, gene-targeted disruption of *HPRT1*, achieved using ExGen 500 and a non-isogenic DNA construct, has provided an in vitro model to study genetic diseases such as Lesch–Nyhan syndrome [27], while targeted knockin of fluorescent reporter genes to the *OCT4* [20] or *MIXL1* [28] loci has been performed using a promoter trap approach, with two constructs differing in the size of the 3′ homology arm, to generate viable readouts of pluripotency and mesoderm differentiation, respectively. Furthermore, the report of a non-viral transgenic integration methodology, permitting the introduction of large transgenes based on the bacteriophage C31 integrase, which recognises specific attachment (attP) sites meaning that integration events occur selectively at the pseudo-attP sites in the intronic sequences of the human genome, may advance the progress of non-viral genetic modification of hESC cells, with the potential of utilising this new methodology to map epigenetic changes as cells differentiate towards desired lineages [29].

1.2 Viral Delivery Systems

From the broad range of viruses available in the scientific community, many have been developed as vectors for transfer of genetic material to target cell populations. The efficiency of this procedure is critical to the success of the approach being undertaken. Cleary, viruses are natural pathogens and have evolved discrete and (usually) efficient cell infectivity pathways through which to manipulate in a gene delivery approach. For many cell types, many viruses have the ability to infect the target cell, yet some cells can be refractory to gene delivery. It is critical to realise that the gene delivery is the culmination of a series of steps that the virus undertakes – namely, cell binding, internalisation, trafficking, nuclear entry and integration or episomal maintenance. The efficiency of cell transduction per se is dependent on all of these processes. Here, we note the characteristics of some commonly used viral vectors and assess their potential uses in the context of stem cell gene transfer.

2 Adenovirus

Adenoviruses have been used for many years in the context of in vitro and in vivo gene delivery. The human adenovirus family consists of over 50 distinct serotypes, divided into a number of subgroups based phylogenetically. Adenovirus vectors are generally ineffective for gene transfer to hES cells, likely due to the availability of primary receptors and/or co-receptors required for cell binding and internalisation in vitro [30–32]. This is an important issue and reflects the need to understand (and potentially manipulate) basic cell infectivity mechanisms relating to the adenovirus. Due to the episomal maintenance of adenovirus in the recipient nucleus, adenovirus gene transfer is a very useful alternative to lentiviruses for the production of iPS cells (see later). As stated above, the commonly used adenovirus serotype 5 vectors (Ad5) bind with relatively high affinity to the coxsackie/adenovirus receptor (CAR) [30, 31], a protein usually found in tight junctions [33], with subsequent internalisation mediated through integrin engagement with the penton base of the virus

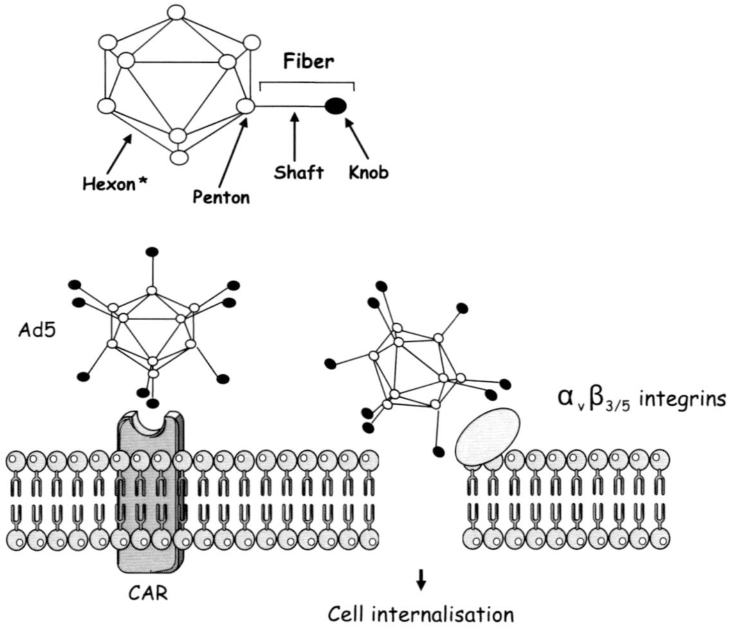

Fig. 1 Basic infectivity pathways for adenovirus serotype 5 vectors. Adenoviruses use capsid proteins fibre, penton and hexon for cell infectivity (see text). In vitro, and in certain in vivo settings, the fibre engages the CAR receptor to tether to the cell surface with subsequent internalisation through integrins. *In vivo, following exposure to blood, the hexon mediates hepatocyte infection via a bridging mechanism involving blood coagulation factor X

(Fig. 1). This system holds true in vitro for CAR-expressing cells since CAR binding is accessible. The interaction is mediated through the knob domain of the fibre protein (Fig. 1). In vivo, however, following intravascular injection (a commonly used route for many clinical applications of adenovirus-mediated gene therapy) the adenovirus uses a different cellular infection procedure altogether [34, 35]. This highlights the complexity of virus infectivity and the influence of environment that the virus is placed as to how cells are transduced [34].

With regard to stem cell gene delivery by adenoviruses, this is relatively weak compared to the level of gene delivery mediated by lentiviruses (see later). However, it is important to utilise the insights and vast prior research on adenoviruses that have shown the ability to re-engineer adenoviruses for highly efficient and selective delivery of cells, at least in vitro. It may therefore be possible to “create” more efficient adenoviruses for stem cell gene transfer. We discuss below a small number of strategies that may be useful. First, as stated, the adenovirus family is large and varied. Many other serotypes target other receptors. An example is the use of adenovirus serotype 35. The fibre of adenovirus 35, a subspecies B adenovirus, binds to CD46 as a high-affinity primary receptor [36, 37]. Whilst there are some technical difficulties creating vectors from the full serotype viruses, a relatively simple technique termed “pseudotyping” that is used to swap the fibre from adenovirus 35 can be

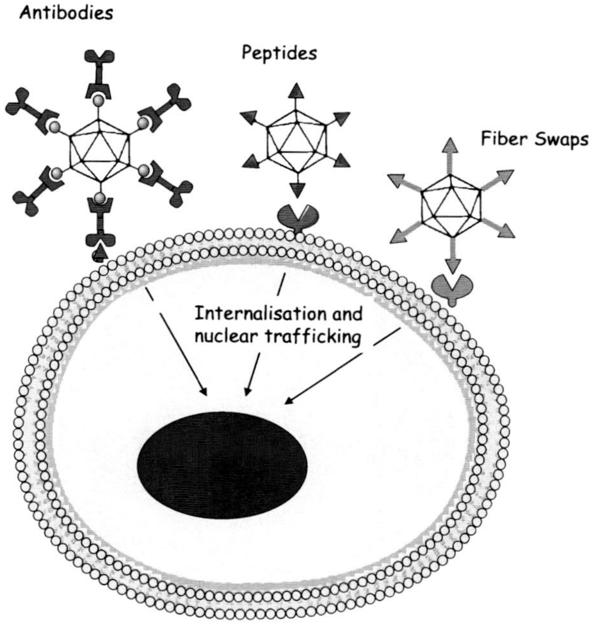

Fig. 2 Methodologies to alter cell infectivity of adenovirus serotype 5 vectors. There are three effective ways to alter Ad tropism: exchanging the fibre gene of one Ad serotype with another (fibre swapping), genetic mutation to the virion capsid proteins followed by the incorporation of peptide targeting motifs and the use of bispecific antibodies which recognise both the virus and target cells. Importantly, this directs the virus through novel pathways not normally used by the virus and so may affect the internalisation and nuclear trafficking routes

engineered to replace the adenovirus 5 on the adenovirus 5 virion, enabling the virus to infect cells through CD46. This strategy improves gene delivery to stem cells [38]. Although manipulation of such viruses is not simple, this technology affords the ability to transiently transduce stem cells with greater efficiency. Strategies used for producing genetically targeted Ads with altered tropism are now well established (Fig. 2). Insertion of targeting peptides into the HI loop of the virus fibre has proven to be an efficient way of expanding viral tropism by enabling the virus to bind a novel receptor [39–41]. Targeting peptides can efficiently be identified by phage display technology [42, 43]. By combining these types of modifications to the adenovirus capsid proteins it should be possible to produce a highly selective and efficient vector for any desired cell type. Little research in this regard has been attempted for stem cell gene transfer using adenovirus to date; however, with the advantages that adenoviruses afford over existing vector systems this is a potential future opportunity to improve stem cell gene transfer. The episomal nature of adenoviruses in the host nucleus can be both a disadvantageous (as it will be lost upon cell division) and advantageous (in the example of iPS generation).

3 Integrating Vectors for Stem Cell Research – The Use of Retroviruses and Lentiviruses in Stem Cell Genetic Modification

Retroviruses and lentiviruses have been used extensively in research and offer the ability to induce long-term expression through DNA integration (Fig. 3). The ability for retroviruses to achieve cell transduction is limited to actively dividing cells. This has, in the gene therapy community, meant that these systems have been used in limited but very effective clinical settings, including X-linked severe combined immunodeficiency disease [44]. Due to the random nature of integrations this has been associated with the development of leukaemia in a subset of patients [45, 46]. This has somewhat limited their appeal in the clinical setting, and novel safety features are being incorporated into the retroviral genome to limit insertational mutagenesis. This aspect of retroviruses has profound implications for iPS

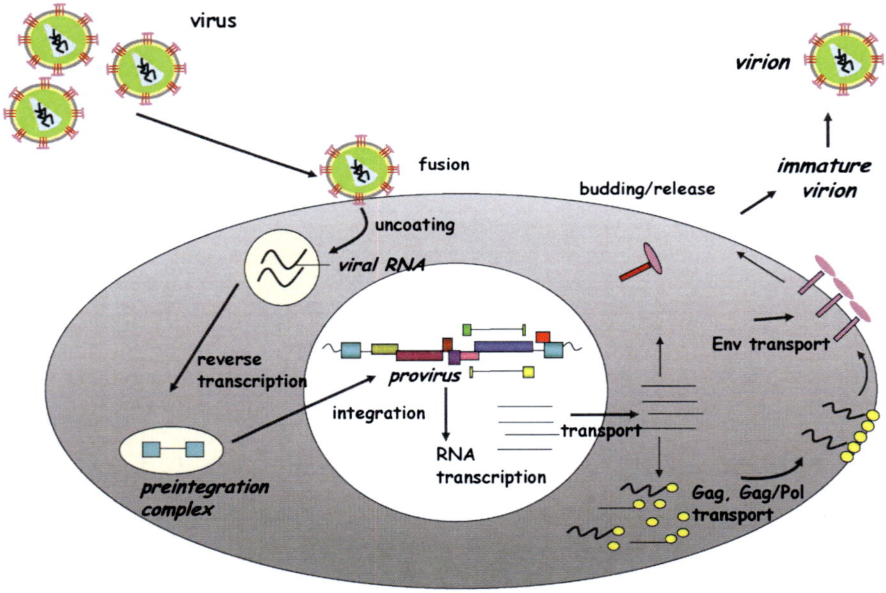

Fig. 3 Life cycle of a retrovirus. Retroviruses are RNA viruses which replicate through an RNA intermediate. Mature retrovirus virions, which contain two identical copies of a plus single-stranded RNA genome within its nucleocapsid core, interact with specific receptor/co-receptor complexes or group of proteins facilitating direct fusion with the target cell membrane, without first undergoing endocytosis. Following fusion the virion enters the cytoplasm where it undergoes uncoating by shedding the viral envelope; thereafter, deoxynucleoside triphosphates from the cytosol enter the nucleocapsid, where viral reverse transcriptase and other proteins rise to a single double-stranded (ds) DNA. This ds DNA then becomes a part of the pre-integration complex which traverses the nuclear membrane and integrates within the host genome using virally encoded integrases. The integrated viral DNA, referred to as the provirus, is transcribed by the host cell RNA polymerase, thereby generating mRNA and genomic RNA. Thereafter the viral mRNAs are translated into glycoproteins and nucleocapsid proteins which assemble with the genomic RNA to form the next generation of virions. After assembly of the viral structural and functional proteins at the cell surface, progeny virions are formed by budding off from the host cell membrane. These then undergo maturation

cell generation using retroviruses and lentiviruses. These viruses have been used to exceptional effect in the reprogramming of mouse and human dermal fibroblasts to iPS cells [9, 47] (see later). Due to the relative inefficiency of gene delivery on cell division mediated by retroviruses, self-inactivating lentivirus-based vectors have become very popular for hESC gene delivery (Fig. 4 for example). The attributes of lentiviruses are due to a number of factors:

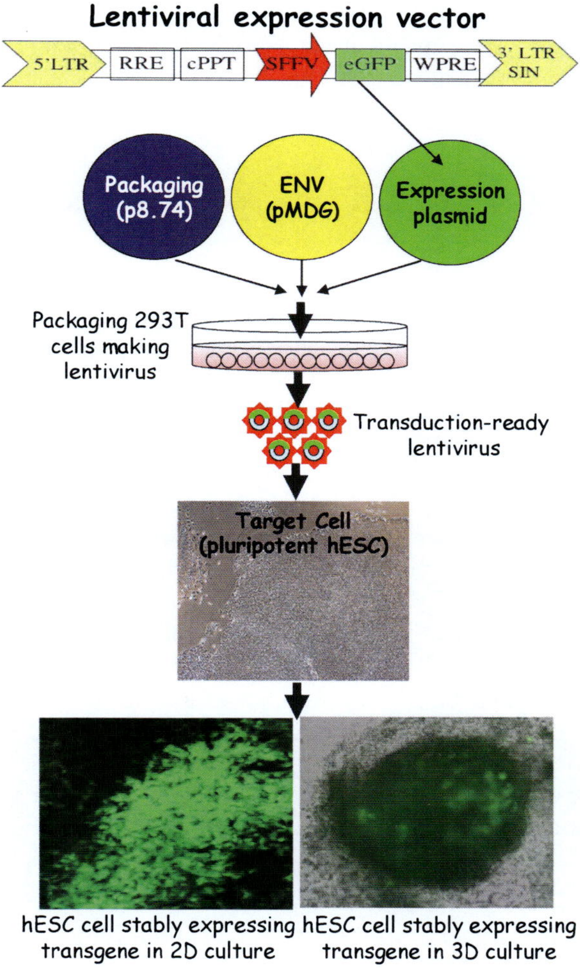

Fig. 4 hESC transduction with lentiviral vectors. Lentiviral vectors are produced by triple transient transfection of a packaging plasmid encoding viral structural and replicative proteins, an envelope plasmid, usually containing the glycoprotein of the vesicular stomatitis virus (VSVg) conferring a wide tropism, and an expression plasmid, containing all *cis*-acting regulatory sequences and the transgene of interest (in this case eGFP) driven by a promoter which may be constitutively active, as is in this example (spleen focus forming virus (SFFV) promoter) or a cell-specific promoter (facilitating lineage tracking). Once transduction-ready lentiviral particles have been collected and titred, hESC cells can be transduced in pluripotent culture conditions, with fluorescent transgene expression stably expressed throughout 2D pluripotent culture or 3D differentiation culture conditions

- Recombinant lentiviruses are generally pseudotyped with the highly effective, vesicular stomatitis virus glycoprotein (VSVg) [48]. Alternatives include rabies, ebola, baculovirus gp64 and makola virus envelope proteins but have not been extensively tested in stem cell gene delivery experiments.
- Lentiviruses are efficient at gene delivery to both dividing and non-dividing cells. Hence, vastly improved rates of cell transduction can be observed.
- Integration into the genome occurs, thus allowing long-term gene expression, even following cell division.
- Lentiviruses can be easily engineered to express transgenes from different promoters and have an acceptable insert capacity compared with other stable viral methods, permitting multiple transgene delivery within one lentiviral vector [49], and depending on the promoter used, gene expression can persist both in cells maintained in a pluripotent state and/or in cells directed towards differentiation [50–53].
- The production of lentiviruses is relatively simple (although a number of companies do sell fully purified infection particles).

Hence, very efficient, stable and reproducible gene delivery to hESCs can be achieved using the lentivirus system. These viruses have been used in a number of ways, for example, in the modulation of stem cell miRNAs [54].

The development of integrase-defective lentiviruses (IDLVs) (non-integrating vectors) has had a significant impact on the use of these vectors in the translational setting [55]. Non-integrating lentiviruses retain the high transduction efficiency of their integrating counterparts in the absence of potential problems relating to insertational mutagenesis or prolonged gene expression. They may also obviate variegation and silencing of expression that can be observed with integrating forms. Indeed, the introduction of IDLVs has permitted the use of zinc finger nucleases (ZFNs) to increase efficiency of gene targeting by several orders of magnitude. These ZFNs are engineered to recognise specific DNA sequences within or adjacent to a chromosomal target site, increasing the efficiency of homologous recombination. This has recently been demonstrated to increase GFP expression in hESC cells infected with a PGK-GFP cassette within IDLVs modified to express a pair of ZFNs specific for a site in the CCR5 gene when compared with hESC cells infected with the GFP donor-IDLV alone [56].

4 Induced Pluripotent Stem Cells: Gene Delivery Methods for Reprogramming Somatic Cells

The ultimate aim of generating reprogrammed somatic cells to patient-specific pluripotent stem cells, otherwise known as induced pluripotent stem (iPS) cells, is reliant upon manipulation of a select group of transcription factors (combinations of SOX2, LIN28, C-MYC, OCT3/4, KLF4 and NANOG) [6, 9, 47, 57]. The advent of iPS cell generation promises to circumvent some therapeutic limitations of hESC and adult stem cells, namely immunogenicity/allograft rejection and scalable production of desired cell lineages, respectively. One caveat to the therapeutic use of iPS cells at present is the need for genetic manipulation to reprogram the cells via ectopic expression of the aforementioned transcription factors, with potential oncogenicity issues for recipients [6, 9, 47, 57]. It should be noted, however, that many researchers are striving to develop efficient non-viral methods of reprogramming or reprogramming cell populations that express endogenous pluripotency genes already [58]. Nevertheless, iPS technology as it stands permits the generation of cell lines to research various genetic and degenerative diseases [10] and heralds the progression towards patient- and disease-specific pharmaceutical screening applications. Here, we review the current methodologies employed to reprogram adult somatic cells to iPS cells and adaptations potentially useful to drive this new cell type closer to the clinic (Current to February 2009).

As with embryonic stem (ES) cell research, iPS cell research originated in the mouse by reprogramming adult mouse dermal fibroblasts [47]. Several ensuing reports confirmed and extended this finding in murine cells demonstrating that iPS cells were extremely similar to ES cells [57, 59–61]. This technique was successfully applied to adult human fibroblast cells, initially in two independent laboratories [6, 9], with a superfluity of research laboratories now replicating the findings in both murine and human cells. Additionally, it has recently been demonstrated that gene expression differences existing between hESC and iPS cell lines are no larger than the variance present between different hESC lines in global expression profiles [62], with these reported iPS cell lines satisfying all original criteria for characterising hESC cells with the obvious exception of starting material [14].

The considerations involved in reproducibly establishing iPS cell lines, with regard to genetic modification, are

- The decision of which transcription factors to ectopically express
- The methods used to deliver the said transcription factors (summarised in Table 1)
- The methods used to identify and isolate reprogrammed cells

Other considerations include choice of target cell line; and culture techniques and conditions employed for deriving iPS cells. These are not discussed further but are reviewed elsewhere [63].

The mechanisms for reprogramming mouse and human somatic cells to iPS cells are subtly different, for example, KLF4 has been demonstrated to be a prerequisite for inducing pluripotency in mouse cells [64, 65], sharing many DNA targets with NANOG [66] whereas, although reported to enhance the efficiency, KLF4 has been demonstrated to be dispensable in the generation of human iPS cells [6–9, 62]. Indeed the inherent differences unique to each species is compounded by the fact that, to date, reprogramming of mouse somatic cells has not been reported by the original "human cocktail" of OCT4, SOX2, NANOG and LIN28; however, this may in part be owing to ectopic expression of NANOG reportedly not affecting the efficiency and/or frequency of murine somatic cell reprogramming [47] and SOX2

Table 1 Vector systems for generation of iPS cells

Methods of factor delivery for inducing reprogramming					
Method	Genetic material	Packaging capacity (kb)	Tropism	Advantages	Disadvantages
Retrovirus	RNA	8	Dividing cells only	• Reportedly silenced in pluripotent cells eliminating requirement for staged/timed factor withdrawal	• The pluripotency-associated silencing has since been demonstrated to not be complete, with expression reportedly maintained in iPSCs and transgene reactivation observed in germline transmission chimeric mice • Limited to dividing cells • Genomic integration resulting in increased risk of insertational mutagenesis • Low titres using standard production methods
Lentivirus	RNA	8	Broad (dependent on envelope glycoprotein used)	• Efficiency of gene transfer high • Capable of transducing both dividing and non-dividing cells with constitutive expression of transgene • Inducible, permitting temporal control over factor expression • Non-integrating vectors available, facilitating efficient expression without genomic integration	• Genomic integration resulting in increased risk of insertational mutagenesis • No silencing in pluripotent culture conditions • Inducible systems have a possibility of leaky expression • Non-integrated systems require multiple rounds of infection owing to lower expression levels and delayed kinetics of reprogramming than integrating counterparts

Table 1 (continued)

Adenovirus	dsDNA	≤ 36	Broad	• Low frequency of genomic integration • Easy to produce high titre	• Multiple rounds of infection required owing to delayed kinetics of reprogramming • Reports of tetraploid cells generated after infection
Non-viral transfection	dsDNA	> 20	Broad	• Low frequency of genomic integration • Easy to produce and technically simple transfection procedure	• Low transfection efficiency requiring multiple rounds of transfection • Low levels of expression than integrating systems resulting in delayed reprogramming kinetics

ds – double stranded

reported as dispensable when endogenous expression is high in the primary murine cell population [64, 65, 67]. It should be noted that OCT4 and SOX2 have been documented as essential for reprogramming human somatic cells, with NANOG, robustly, and LIN28, moderately, influencing the efficiency and frequency of reprogramming [6]. Indeed, several researchers have now published protocols for the generation of human iPS cells from adult tissues, with the only common transcription factors ectopically expressed being OCT4 and SOX2 [6–8, 62, 68]. OCT4, a member of the POU family of transcription factors, and SOX2, a member of the high mobility group (HMG) box transcription factors, are well characterised as having a critical role in the establishment and maintenance of pluripotency [69–71].

The initial generation of iPS cells from reprogrammed mouse and human dermal fibroblasts employed retroviruses to deliver the transcription factors. Utilised owing to their capability to integrate into the genome, providing sustained transgene expression in daughter cells upon division and the reported silencing of retroviral transgenes in the pluripotent state [8, 11, 72], retroviruses provided a potential method for maintaining expression of the transcription factors until true pluripotency was achieved. However, iPS cells reprogrammed using retrovirus-mediated transcription factor delivery have been reported to maintain viral transgene expression [8, 73]. Incidentally, due to their restriction of gene delivery to dividing cells (as discussed above), the range of cell types that can be reprogrammed is restricted. Additionally, it has also been demonstrated that silencing of the retroviral transgenes occurs gradually in concordance with progression of somatic cell reprogramming, thereby inducing reprogramming at a lower efficiency as compared with non-silencing viral methods, e.g. lentiviruses [74]. VSVg-pseudotyped integrating lentiviruses, which also integrate in a relatively random pattern (but with preference to regions of open chromatin), have been very effectively used for iPS cell reprogramming [6] owing to their ability to transduce both dividing and non-dividing cell types with high expression levels (see above). One limitation of integrating lentiviruses, with regard to utilisation as a vector for iPS generation, is sustained transgene expression in the pluripotent state [75], with consequences for progression of differentiation to desired cell lineages [76] and obvious impediment to therapeutic use. One modification of the technology, permitting temporal control over transcription factor/transgene expression, the so-called "drug-inducible lentiviruses" have led to a >100-fold increase in secondary iPS cell production from drug-induced reactivation of the viral transgenes in iPS cell-derived differentiated cells [77–79]. This modification may facilitate screening methods to identify small molecules capable of enhancing or indeed inducing somatic cell reprogramming. The major problems associated with integrating vectors (retroviral or lentiviral) include: alteration of gene function as a result of genomic insertion [80]; clonal evolution of cells with a growth advantage due to a vector integration event activating proto-oncogenic activity; and viral transgene reactivation [60]. This is reiterated by the fact that 15–20% iPS cell-derived chimeric mice and their offspring are prone to malignancy (or tumourigenicity) [81], most probably as a consequence of such insertational oncogenic activity or transgene reactivation [60]. In support of this observation others have reported incomplete silencing of KLF4 and C-MYC up to 2 months after induction of differentiation [82].

These limitations have prompted concerted, worldwide endeavours to generate more therapeutically viable iPS cells, with intense investigation focusing on the pursuit of safer vectors and non-integrating modalities. Recent advances include iPS cell selection based on cell morphology rather than virally inserted selection mechanisms, namely fluorescent reporter genes or antibiotic selection cassettes [83]; the generation of iPS cells in the absence of C-MYC, an improvement which markedly reduced tumourigenesis in iPS cell-derived mice [81]; the enhancement of iPS formation using small molecules [62, 65] or cell-permeant protein reagents [84]; and the development of non-integrating lentiviral vectors, generated via mutations in the integrase gene [85], thus alleviating some tumourigenic concerns and improving suitability for therapeutic application. In preclinical gene therapy applications non-integrating lentiviral vectors have been proven to be highly effective owing to the dilution of vector genomes with cell division [85]. Production of iPS cells using non-integration lentiviruses thus offers the efficiency of VSVg-mediated gene transfer but lacking the integration capacity. Certainly, recent reports analysing integration sites in iPS cells revealed no common targets or signalling pathways suggesting that genomic integration is not a prerequisite for successful reprogramming [57]. Moreover, the generation of mouse iPS cells using non-integrating approaches has confirmed that genomic integration is surplus to requirements [86]. Ergo, non-integrating gene delivery systems lend promise to the eventual therapeutic application of iPS cells.

The potential therapeutic applications of iPS cells are extensive and without limitation; however, understanding of the interactions that govern iPS cell generation at the molecular level is inadequate and the efficiency of the process is still low. Hence, the optimisation of genetic modification and consequential liaising with gene therapists will permit improvements in the derivation of iPS cells. Furthermore, to enable progression of stem cell therapy (embryonic or induced) towards the clinic, advances must be made towards understanding and manipulating differentiation processes for therapeutic interventions. To date, hESC-based therapy has been reported in animal models. One example demonstrates contractile improvement and alleviation of ventricular dilation after the engraftment of hESC-derived cardiomyocytes in rodent hearts following infarct [87–90]. Cardiomyocyte generation from iPS cells has been demonstrated to be highly similar to that from ES cells [82, 91], with recent reports demonstrating functional iPS cell-derived nodal, atrial and ventricular-like cardiomyocytes [92] highlighting the potential of iPS cell-based cellular cardiomyoplasty as a therapeutic reality, circumventing the potential immunogenicity obstacles associated with hESC cell therapy. Without considering the existing hurdles of integration and persistent transgene expression (discussed above) iPS cell therapy still faces significant impediments, not least in generating, sourcing or differentiating a suitable source of cells to compensate for the pathologic deficit and in their targeted delivery and integration towards injury to gain therapeutic benefit, notably the same impediments hindering hESC cell research.

iPS cell research and subsequent cell therapy additionally promises a gateway towards autologous cell therapy for genetic defects. Whilst hESCs can be derived from embryos determined by preimplantation genetic diagnosis (PGD) to carry genetic abnormalities [93], iPS cells can be generated without regard to ethical and practical obstacles, all the while retaining the advantage of encompassing the complex gene defects associated with the majority of inherited diseases, i.e. diseases not related to a single gene defect. The generation of iPS cell lines for research, if not therapy, is not dependant on existing aetiological knowledge. Induced pluripotent cell lines have been generated for a wide range of human genetic diseases including adenosine deaminase deficiency-related severe combined immunodeficiency, Shwachman-Bodian-Diamond syndrome, Gaucher disease type III, muscular dystrophy, Parkinson disease, Huntington disease, juvenile-onset, type 1 diabetes mellitus, Down syndrome/trisomy 21, the carrier state of Lesch–Nyhan syndrome [10] and, in the case of amyotrophic lateral sclerosis, without adverse effect on the ability of these iPS cells to terminally differentiate into motor neurons [73]. These disease model cells provide an opportunity to model both normal and pathologic physiology during human embryogenesis in vivo and tissue formation in vitro, aiding disease pathogenesis investigation and pharmacological interventions. Ultimately iPS technology may provide a source cell population to enable autologous transplantation in conjunction with gene modification strategies to revolutionise cell therapy, with promising results already observed in a humanised murine sickle cell anaemia model in which homologous recombination was employed to repair the β-haemoglobin allele in fibroblast-derived retrovirally generated iPS cells differentiated to the haematopoietic lineage and subsequently reimplanted [94].

Worldwide investigation of iPS technologies is likely to be intense with researchers pursuing iPS generation in a clinically viable manner. Vectors efficiently and transiently delivering transcription factors (without alteration to the genome via integration) will be the most exciting from a translational aspect with respect to regulatory approvals in the future. However, the ultimate goal will be the generation of iPS cells using gene transfer-independent techniques.

5 Conclusions

Genetic modification methodologies to delineate hESC basic biology or reprogramme somatic cells to iPS cells have fundamental importance for the advancement of clinical cell-based therapeutic strategies. In the laboratory setting several vector systems are in place to effectively answer pertinent biological questions. However, prior to translation from the bench to the bedside, safer gene delivery systems will be required, especially with regard to meeting regulatory body guidelines. Expertise must be drawn from the lessons learned during the last 15 years of clinical gene therapy, which has documented the paramount importance of safety and optimisation of gene transfer. In the years ahead, improvements in vector design and gene insertion technology must be achieved for therapeutic benefit. In the changing regulatory landscape, the use of genetic manipulation via gene transfer will continue to play a major role in stem cell biology and in time will provide an invaluable contribution to the field of regenerative medicine.

Acknowledgments The authors thank S. Mukherjee (Institute for Child Health, London) for help with Fig. 3.

References

1. Rovira M, Jane-Valbuena J, Marchand M, Savatier P, Real FX, Skoudy A. Viral-mediated coexpression of Pdx1 and p48 regulates exocrine pancreatic differentiation in mouse ES cells. Cloning Stem Cells. Fall 2007; 9(3):327–338.
2. Li Z, Suzuki Y, Huang M, et al. Comparison of reporter gene and iron particle labeling for tracking fate of human embryonic stem cells and differentiated endothelial cells in living subjects. Stem Cells. 2009 Mar: 18(2):205–14.
3. Xie X, Chan KS, Cao F, et al. Imaging of STAT3 signaling pathway during mouse embryonic stem cell differentiation. Stem Cells Dev. 2008 Jun 24.
4. Cao F, Drukker M, Lin S, et al. Molecular imaging of embryonic stem cell misbehavior and suicide gene ablation. Cloning Stem Cells. Spring 2007; 9(1):107–117.
5. Strulovici Y, Leopold PL, O'Connor TP, Pergolizzi RG, Crystal RG. Human embryonic stem cells and gene therapy. Mol Ther. 2007 May; 15(5):850–866.
6. Yu J, Vodyanik M, Smuga-Otto K, et al. Induced pluripotent stem cell lines derived from human somatic cells. Science. 2007 Dec 21; 318(5858):1917–20.
7. Lowry WE, Richter L, Yachechko R, et al. Generation of human induced pluripotent stem cells from dermal fibroblasts. Proc Natl Acad Sci U S A. 2008 Feb 26; 105(8):2883–2888.
8. Park IH, Zhao R, West JA, et al. Reprogramming of human somatic cells to pluripotency with defined factors. Nature. 2008 Jan 10; 451(7175):141–146.
9. Takahashi K, Tanabe K, Ohnuki M, et al. Induction of pluripotent stem cells from adult human fibroblasts by defined factors. Cell. 2007 Nov 30; 131(5):861–872.
10. Park IH, Arora N, Huo H, et al. Disease-specific induced pluripotent stem cells. Cell. 2008 Sep 5; 134(5):877–86.
11. Park IH, Lerou PH, Zhao R, Huo H, Daley GQ. Generation of human-induced pluripotent stem cells. Nat Protoc. 2008; 3(7):1180–1186.
12. Gao X, Kim KS, Liu D. Nonviral gene delivery: what we know and what is next. Aaps J. 2007; 9(1):E92–E104.
13. Lakshmipathy U, Pelacho B, Sudo K, et al. Efficient transfection of embryonic and adult stem cells. Stem Cells. 2004; 22(4):531–543.
14. Thomson JA, Itskovitz-Eldor J, Shapiro SS, et al. Embryonic stem cell lines derived from human blastocysts. Science. 1998 Nov 6; 282(5391):1145–1147.
15. Liew CG, Draper JS, Walsh J, Moore H, Andrews PW. Transient and stable transgene expression in human embryonic stem cells. Stem Cells. 2007 Jun; 25(6):1521–1528.
16. Eiges R, Schuldiner M, Drukker M, Yanuka O, Itskovitz-Eldor J, Benvenisty N. Establishment of human embryonic stem cell-transfected clones carrying a marker for undifferentiated cells. Curr Biol. 2001 Apr 3; 11(7):514–518.
17. Lebkowski JS, Gold J, Xu C, Funk W, Chiu CP, Carpenter MK. Human embryonic stem cells: culture, differentiation, and genetic modification for regenerative medicine applications. Cancer J. 2001 Nov-Dec ; 7(Suppl 2):S83–S93.
18. Siemen H, Nix M, Endl E, Koch P, Itskovitz-Eldor J, Brustle O. Nucleofection of human embryonic stem cells. Stem Cells Dev. 2005 Aug; 14(4):378–383.
19. Vallier L, Rugg-Gunn PJ, Bouhon IA, Andersson FK, Sadler AJ, Pedersen RA. Enhancing and diminishing gene function in human embryonic stem cells. Stem Cells. 2004; 22(1):2–11.
20. Zwaka TP, Thomson JA. Homologous recombination in human embryonic stem cells. Nat Biotechnol. 2003 Mar; 21(3):319–321.
21. Giudice A, Trounson A. Genetic modification of human embryonic stem cells for derivation of target cells. Cell Stem Cell. 2008 May 8; 2(5):422–433.

22. Anderson D, Self T, Mellor IR, Goh G, Hill SJ, Denning C. Transgenic enrichment of cardiomyocytes from human embryonic stem cells. Mol Ther. 2007 Nov: 15(11):2027–36.
23. Braam SR, Denning C, van den Brink S, et al. Improved genetic manipulation of human embryonic stem cells. Nat Methods. 2008 May; 5(5):389–392.
24. Denning C, Allegrucci C, Priddle H, et al. Common culture conditions for maintenance and cardiomyocyte differentiation of the human embryonic stem cell lines, BG01 and HUES-7. Int J Dev Biol. 2006; 50(1):27–37.
25. Huber I, Itzhaki I, Caspi O, et al. Identification and selection of cardiomyocytes during human embryonic stem cell differentiation. FASEB J. 2007 Aug; 21(10):2551–2563.
26. Wang D, Haviland DL, Burns AR, Zsigmond E, Wetsel RA. A pure population of lung alveolar epithelial type II cells derived from human embryonic stem cells. Proc Natl Acad Sci U S A. 2007 Mar 13; 104(11):4449–4454.
27. Urbach A, Schuldiner M, Benvenisty N. Modeling for Lesch-Nyhan disease by gene targeting in human embryonic stem cells. Stem Cells. 2004; 22(4):635–641.
28. Davis RP, Ng ES, Costa M, et al. Targeting a GFP reporter gene to the MIXL1 locus of human embryonic stem cells identifies human primitive streak-like cells and enables isolation of primitive hematopoietic precursors. Blood. 2008 Feb 15; 111(4):1876–1884.
29. Thyagarajan B, Liu Y, Shin S, et al. Creation of engineered human embryonic stem cell lines using phiC31 integrase. Stem Cells. 2008 Jan ; 26(1):119–126.
30. Bergelson JM, Cunningham JA, Droguett G, et al. Isolation of a common receptor for coxsackie B viruses and adenoviruses 2 and 5. Science. 1997; 275:1320–1323.
31. Tomko RP, Xu R, Philipson L. HCAR and MCAR: the human and mouse cellular receptors for subgroup C adenoviruses and group B coxsackieviruses. Proc Nat Acad Sci USA. 1997; 94:3352–3356.
32. Wickham TJ, Mathias P, Cheresh DA, Nemerow GR. Integrins alpha v beta 3 and alpha v beta 5 promote adenovirus internalization but not virus attachment. Cell. 1993 Apr 23; 73(2): 309–319.
33. Cohen CJ, Shieh JTC, Pickles RJ, Okegawa T, Hsieh J-T, Bergelson JM. The coxsackievirus and adenovirus receptor is a transmembrane component of the tight junction. PNAS. 2001 Dec 18; 98(26):15191–15196.
34. Waddington S, McVey J, Bhella D, et al. Adenovirus serotype 5 Hexon mediates liver gene transfer. Cell. 2008; 132(3):397–409.
35. Kalyuzhniy O, Di Paolo NC, Silvestry M, et al. Adenovirus serotype 5 hexon is critical for virus infection of hepatocytes in vivo. PNAS. 2008; 105(14):5483–5488.
36. Gaggar A, Shayakhmetov DM, Lieber A. CD46 is a cellular receptor for group B adenoviruses. Nat Med. 2003 Nov; 9(11):1408–1412.
37. Gaggar A, Shayakhmetov DM, Liszewski MK, Atkinson JP, Lieber A. Localization of regions in CD46 that interact with adenovirus. Virology. 2005; 79:7503–7513.
38. Tuve S, Wang H, Ware C, et al. A new Group B adenovirus receptor is expressed at high levels on human stem and tumor cells. Virology. 2006 Oct 04; 80(24):12109–12120.
39. Krasnykh V, Dmitriev I, Mikheeva G, Miller R, Belousova N, Curiel DT. Characterisation of an adenovirus vector containing a heterologous peptide epitope in the HI loop of the fiber knob. J Virol. 1998; 72(3):1844–1852.
40. Dmitriev I, Krasnykh V, Miller CR, et al. An adenovirus vector with genetically modified fibers demonstrates expanded tropism via utilization of a coxsackievirus and adenovirus receptor-independent cell entry mechanism. J Virol. 1998; 72(12):9706–9713.
41. Nicklin S, Von Seggern D, Work L, et al. Ablating adenovirus type 5 fiber-CAR binding and HI loop insertion of the SIGYPLP peptide generate an endothelial cell-selective adenovirus. Mol Ther. 2001; 4:534–542.
42. Nicklin S, White S, Watkins S, Hawkins R, Baker A. Selective targeting of gene transfer to vascular endothelial cells by use of peptides isolated by phage display. Circulation. 2000; 102:231–237.

43. Pasqualini R, Ruoslahti E. Organ targeting in vivo using phage display peptide libraries. Nature. 1996; 380:364–366.
44. Cavazzana-C M, Hacein-Bey S, Saint-Basile CD, et al. Gene therapy of human severe combined immunodeficiency (scid)-1x disease. Science. 2000; 288:669–672.
45. Hacein-Bey-Abina S, Kalle CV, Schmidt M, et al. LM02-associated clonal T cell proliferation in two patients after gene therapy for SCID-X1. Science. 2003; 302:415–419.
46. Hacein-Bey-Abina S, Garrigue A, Wang GP, et al. Insertional oncogenesis in 4 patients after retrovirus-mediated gene therapy of SCID-X1. J Clin Invest. 2008 Sept; 118(9): 3132–3142.
47. Takahashi K, Yamanaka S. Induction of pluripotent stem cells from mouse embryonic and adult fibroblast cultures by defined factors. Cell. 2006 Aug 25; 126(4):663–676.
48. Carneiro FA, Bianconi ML, Weissmuller G, Stauffer F, Da Poian AT. Membrane recognition by vesicular stomatitis virus involves enthalpy-driven protein-lipid interactions. J Virol. 2002 Mar 19; 76(8):3756–3764.
49. Clements MO, Godfrey A, Crossley J, Wilson SJ, Takeuchi Y, Boshoff C. Lentiviral manipulation of gene expression in human adult and embryonic stem cells. Tissue Eng. 2006 Jul; 12(7):1741–1751.
50. Ma Y, Ramezani A, Lewis R, Hawley RG, Thomson JA. High-level sustained transgene expression in human embryonic stem cells using lentiviral vectors. Stem Cells. 2003; 21(1):111–117.
51. Kobayashi N, Rivas-Carrillo JD, Soto-Gutierrez A, et al. Gene delivery to embryonic stem cells. Birth Defects Res C Embryo Today. 2005 Mar; 75(1):10–18.
52. Menendez P, Wang L, Bhatia M. Genetic manipulation of human embryonic stem cells: a system to study early human development and potential therapeutic applications. Curr Gene Ther. 2005 Aug; 5(4):375–385.
53. Suter DM, Cartier L, Bettiol E, et al. Rapid generation of stable transgenic embryonic stem cell lines using modular lentivectors. Stem Cells. 2006 Mar; 24(3):615–623.
54. Ivey K, Muth A, Arnold J, et al. MircoRNA regulation of cell lineages in mouse and human embryonic stem cells. Cell Stem Cell. 2008; 2:219–229.
55. Philpott N, Thrasher A. Use of nonintegrating lentiviral vectors for gene therapy. Hum Gene Ther. 2007 Jun; 18:483–489.
56. Lombardo A, Genovese P, Beausejour CM, et al. Gene editing in human stem cells using zinc finger nucleases and integrase-defective lentiviral vector delivery. Nat Biotechnol. 2007 Nov; 25(11):1298–1306.
57. Aoi T, Yae K, Nakagawa M, et al. Generation of pluripotent stem cells from adult mouse liver and stomach cells. Science. 2008 Aug 1; 321(5889):699–702.
58. Kim JB, Sebastiano V, Wu G, et al. Oct4-induced pluripotency in adult neural stem cells. Cell. 2009 Feb 6; 136(3):411–419.
59. Maherali N, Sridharan R, Xie W, et al. Directly reprogrammed fibroblasts show global epigenetic remodeling and widespread tissue contribution. Cell Stem Cell. 2007 Jun 7; 1(1):55–70.
60. Okita K, Ichisaka T, Yamanaka S. Generation of germline-competent induced pluripotent stem cells. Nature. 2007 Jul 19; 448(7151):313–317.
61. Wernig M, Meissner A, Foreman R, et al. In vitro reprogramming of fibroblasts into a pluripotent ES-cell-like state. Nature. 2007 Jul 19; 448(7151):318–324.
62. Huangfu D, Maehr R, Guo W, et al. Induction of pluripotent stem cells by defined factors is greatly improved by small-molecule compounds. Nat Biotechnol. 2008 Jul; 26(7):795–797.
63. Maherali N, Hochedlinger K. Guidelines and techniques for the generation of induced pluripotent stem cells. Cell Stem Cell. 2008 Dec 4; 3(6):595–605.
64. Eminli S, Jaenisch R, Hochedlinger K. Strategies to induce nuclear reprogramming. Ernst Schering Found Symp Proc. 2006; 5:83–98.

65. Shi Y, Do JT, Desponts C, Hahm HS, Scholer HR, Ding S. A combined chemical and genetic approach for the generation of induced pluripotent stem cells. Cell Stem Cell. 2008 Jun 5; 2(6):525–528.
66. Jiang J, Chan YS, Loh YH, et al. A core Klf circuitry regulates self-renewal of embryonic stem cells. Nat Cell Biol. 2008 Mar; 10(3):353–60.
67. Kim JB, Zaehres H, Wu G, et al. Pluripotent stem cells induced from adult neural stem cells by reprogramming with two factors. Nature. 2008 Jul 31; 454(7204):646–650.
68. Takahashi K, Tanabe K, Ohnuki M, et al. Induction of pluripotent stem cells from adult human fibroblasts by defined factors. Cell. 2007; 131:1–12.
69. Pesce M, Gross MK, Scholer HR. In line with our ancestors: Oct-4 and the mammalian germ. Bioessays. 1998 Sept; 20(9):722–732.
70. Fong H, Hohenstein KA, Donovan PJ. Regulation of self-renewal and pluripotency by Sox2 in human embryonic stem cells. Stem Cells. 2008 Aug; 26(8):1931–1938.
71. Boyer LA, Lee TI, Cole MF, et al. Core transcriptional regulatory circuitry in human embryonic stem cells. Cell. 2005 Sept 23; 122(6):947–956.
72. Wolf D, Goff SP. TRIM28 mediates primer binding site-targeted silencing of murine leukemia virus in embryonic cells. Cell. 2007 Oct 5; 131(1):46–57.
73. Dimos JT, Rodolfa KT, Niakan KK, et al. Induced pluripotent stem cells generated from patients with ALS can be differentiated into motor neutrons. Science. 2008 Aug 29; 321(5893):1218–21. Epub 2008 Jul 31.
74. Stadtfeld M, Maherali N, Breault DT, Hochedlinger K. Defining molecular cornerstones during fibroblast to iPS cell reprogramming in mouse. Cell Stem Cell. 2008 Mar 6; 2(3):230–240.
75. Lois C, Hong EJ, Pease S, Brown EJ, Baltimore D. Germline transmission and tissue-specific expression of transgenes delivered by lentiviral vectors. Science. 2002 Feb 1; 295(5556):868–872.
76. Brambrink T, Foreman R, Welstead GG, et al. Sequential expression of pluripotency markers during direct reprogramming of mouse somatic cells. Cell Stem Cell. 2008 Feb 7; 2(2):151–159.
77. Hockemeyer D, Soldner F, Cook EG, Gao Q, Mitalipova M, Jaenisch R. A drug-inducible system for direct reprogramming of human somatic cells to pluripotency. Cell Stem Cell. 2008 Sept 11; 3(3):346–353.
78. Maherali N, Ahfeldt T, Rigamonti A, Utikal J, Cowan C, Hochedlinger K. A high-efficiency system for the generation and study of human induced pluripotent stem cells. Cell Stem Cell. 2008 Sept 11; 3(3):340–345.
79. Wernig M, Lengner CJ, Hanna J, et al. A drug-inducible transgenic system for direct reprogramming of multiple somatic cell types. Nat Biotechnol. 2008 Aug; 26(8):916–924.
80. Kustikova O, Fehse B, Modlich U, et al. Clonal dominance of hematopoietic stem cells triggered by retroviral gene marking. Science. 2005 May 20; 308(5725):1171–1174.
81. Nakagawa M, Koyanagi M, Tanabe K, et al. Generation of induced pluripotent stem cells without Myc from mouse and human fibroblasts. Nat Biotechnol. 2008 Jan; 26(1):101–106.
82. Narazaki G, Uosaki H, Teranishi M, et al. Directed and systematic differentiation of cardiovascular cells from mouse induced pluripotent stem cells. Circulation. 2008; 118:498–506.
83. Meissner A, Wernig M, Jaenisch R. Direct reprogramming of genetically unmodified fibroblasts into pluripotent stem cells. Nat Biotechnol. 2007 Oct; 25(10):1177–1181.
84. Bosnali M, Edenhofer F. Generation of transducible versions of transcription factors Oct4 and Sox2. Biol Chem. 2008 July; 389(7):851–861.
85. Yanez-Munoz RJ, Balaggan KS, MacNeil A, et al. Effective gene therapy with nonintegrating lentiviral vectors. Nat Med. 2006 Mar; 12(3):348–353.
86. Okita K, Nakagawa M, Hyenjong H, Ichisaka T, Yamanaka S. Generation of mouse induced pluripotent stem cells without viral vectors. Science. 2008 Nov 7; 322(5903):949–953.
87. Caspi O, Huber I, Kehat I, et al. Transplantation of human embryonic stem cell-derived cardiomyocytes improves myocardial performance in infarcted rat hearts. J Am Coll Cardiol. 2007 Nov 6; 50(19):1884–1893.

88. Laflamme M, Chen K, Naumova A, et al. Cardiomyocytes derived from human embryonic stem cells in pro-survival factors enhance function of infarcted rat hearts. Nat Biotech. 2007; 25:1015–1024.
89. van Laake LW, Passier R, Doevendans PA, Mummery CL. Human embryonic stem cell-derived cardiomyocytes and cardiac repair in rodents. Circ Res. 2008 May 9; 102(9):1008–1010.
90. Yang L, Soonpaa MH, Adler ED, et al. Human cardiovascular progenitor cells develop from a KDR+ embryonic-stem-cell-derived population. Nature. 2008 May 22; 453(7194):524–528.
91. Mauritz C, Schwanke K, Reppel M, et al. Generation of functional murine cardiac myocytes from induced pluripotent stem cells. Circulation. 2008; 118:507–517.
92. Zhang J, Wilson GF, Soerens AG, et al. Functional cardiomyocytes derived from human induced pluripotent stem cells. Circ Res. 2009 Feb 27; 104(4):e30–41.
93. Verlinsky Y, Strelchenko N, Kukharenko V, et al. Human embryonic stem cell lines with genetic disorders. Reprod Biomed Online. 2005 Jan; 10(1):105–110.
94. Hanna J, Wernig M, Markoulaki S, et al. Treatment of sickle cell anemia mouse model with iPS cells generated from autologous skin. Science. 2007 Dec 21; 318(5858):1920–1923.

The Immune Barriers of Cell Therapy with Allogenic Stem Cells of Embryonic Origin

Olivier Preynat-Seauve, Karl-Heinz Krause, and Jean Villard

Abstract Human embryonic stem cells (ESC) provide a great hope for regenerative medicine in different diseases like neodegenerative disease, diabetes, heart, or liver failure. Immune rejection was not thought to be a major issue for cell therapy because of a low immunogenicity of fetal or embryonic cells in preliminary animal studies. However, increasing evidence suggests that this is not true, and controlling the immune response will be crucial for the success of ESC transplantation. The source of ESC is of crucial importance with regard to genetic difference between donors and recipients. Immune reaction against genetically identical origin (autologous or from identical twin) would be fully absent or negligible in contrast to immune response against allogenic transplanted cells. However, in the situation of ESC the origin of the cell would be necessarily allogenic and the immune reactivity against allogenic cells is expected to be similar to what has been learned from decades of research in the field of cell, tissue, or solid organ transplantation. In this chapter, we will first describe the risk of potential immune reactivity against ESC according to the origin of the cells. Second, we will review the immune mechanism of rejection and the current literature on the topic not only in animal models but also in humans. Finally, we will come with therapeutic approaches that can allow crossing genetic barriers of donor cells by preventing immune reaction that could lead to irreversible loss of the graft function.

Keywords Human embryonic stem cells (ESC) · Rejection · Immune response · T cells · Natural killer cells · Genetic barriers · HLA · Immune tolerance

---

J. Villard (✉)
Immunology and Transplant Unit, Division of Immunology and Allergology and Division of Laboratory Medicine, Geneva University Hospital and Medical School, Geneva, Switzerland
e-mail: Jean.Villard@hcuge.ch

G.M. Artmann et al. (eds.), *Stem Cell Engineering*,
DOI 10.1007/978-3-642-11865-4_8,

1 Introduction

Human embryonic stem cells (ESC) provide a great hope for regenerative medicine and are viewed as a potential source of transplantable cells in different diseases like neurodegenerative disease, diabetes, heart, or liver failure [1, 2]. The two major risks associated with the use of pluripotent stem cells for cell therapy are tumor/teratoma formation and immune rejection [3, 4]. Immune rejection was not thought to be a major issue for cell therapy of Parkinson's disease for two reasons: (i) immunoprivilege of the CNS and (ii) low immunogenicity of fetal or embryonic cells, even if genetically unrelated to the recipient [5, 6]. However, increasing evidence suggests that this is not true and that controlling the immune response will be crucial for the success of cell therapy. The control of the immune response will depend on several critical points like the origin of the ESC, the state of differentiation, and therapeutic regimen that will be able to control the immune reaction or induce tolerance when infused in the recipients [5, 6]. The source of ESC is of crucial importance with regard to genetic difference between donors and recipients [2, 7, 8]. Immune reaction against genetically identical origin (autologous or from identical twin) would be fully absent or negligible in contrast to immune response against allogenic transplanted cells. However, in the situation of ESC the origin of the cell would be necessarily allogenic and the immune reactivity against allogenic cells is expected to be similar to what has been learned from decades of research and clinical application in the field of cell, tissue, or solid organ transplantation [9, 10].

In this chapter, we will first describe the risk of potential immune reactivity against ESC according to the origin of the cells. Second, we will review the immune mechanism of rejection and the current literature on the topic not only in animal models but also in humans. Finally, we will come with therapeutic approaches that can allow crossing genetic barriers between donor cells and recipient by preventing immune reaction that could lead to irreversible loss of the graft function.

2 Stem Cells – Overview

Stem cells are found in most multicellular organisms. They are characterized by the ability to renew themselves through mitotic cell division and differentiating into a diverse range of specialized cell types. The two broad types of mammalian stem cells are as follows:

- Pluripotent stem cells, with the best known example being embryonic stem cells (ESC), that are found in blastocysts.
- Adult stem cells (ASC) that are found in adult tissues.

In a developing embryo, stem cells can differentiate into all of the specialized embryonic tissues. In adult organisms, stem cells and progenitor cells are supposed to act as a regeneration system for the body, not only replenishing specialized cells but also maintaining the normal turnover of regenerative organs, such as blood, skin, or intestinal tissues.

Highly plastic adult stem cells from a variety of sources, including umbilical cord blood, and bone marrow, are routinely used in medical therapies. Embryonic stem cells and adult stem cells have also been proposed as promising candidates for future therapies, including heart failure or neurodegenerative disorders.

The classical definition of a stem cell requires that it possess two properties:

- Self-renewal – the ability to go through numerous cycles of cell division while maintaining the undifferentiated state.
- Potency – the capacity to differentiate into specialized cell types. Potency can cover various properties: (i) Totipotency indicates the capacity to differentiate into embryonic and extraembryonic cell types. Such cells can construct a complete and viable organism. (ii) Pluripotency indicates the ability to differentiate into nearly all cells derived from any of the three germ layers. Multipotency is restricted to a closely related family of cells (e.g., neural stem cells differentiate into cells of the central nervous system: astrocytes, oligodendrocytes, neurons).

Figure 1 summarizes the various origins of stem cells available for transplantation

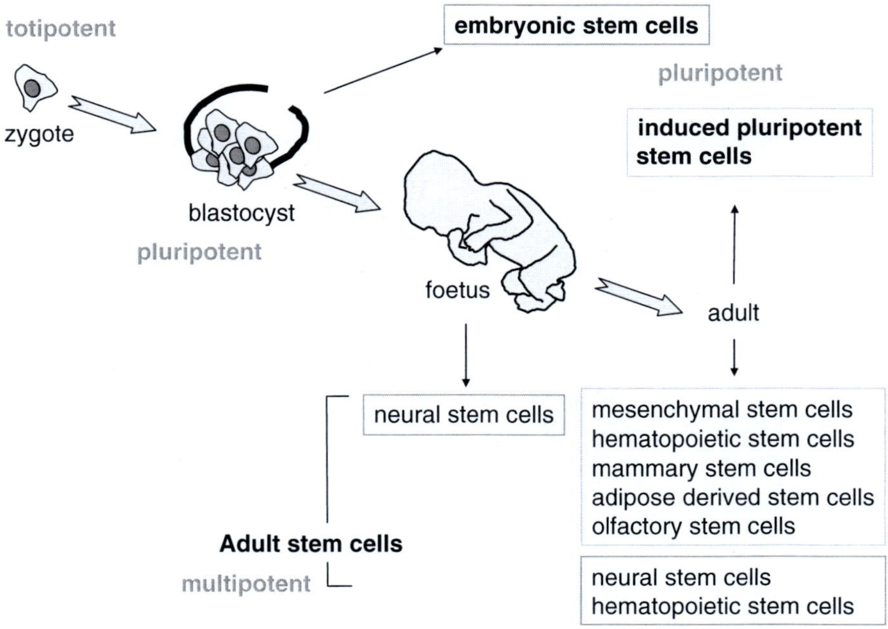

Fig. 1 Various sources of stem cells are available for transplantation

2.1 Embryonic Stem Cells

ESC lines are cultures of cells derived from the epiblast tissue of the inner cell mass of a blastocyst or earlier morula stage embryos. A blastocyst is an early stage embryo – approximately 4–5 days old in humans and consisting of 50–150 cells. ESC are pluripotent and during development give rise to all derivatives of the three primary germ layers: ectoderm, endoderm, and mesoderm. In other words, they can develop into each of the more than 200 cell types of the adult body when given sufficient and necessary stimulation for a specific cell type.

Human ESC in culture requires a specific environment in order to maintain an undifferentiated state. They are grown on a feeder layer of mouse or human embryonic fibroblasts and require the presence of basic fibroblast growth factor (bFGF or FGF-2).

Human ESC can be differentiated in vitro into various cell types of the body, opening the possibility of transplantation in the context of regenerative medicine. The currently most studied stem cell transplantation applications are heart failure, diabetes, and neurodegenerative disorders (e.g., Parkinson's disease, spinal cord injury).

Due to the unrelated genetic origin of ESC, differentiated therapeutic cells derived from ESC will be theoretically exposed to transplant rejection.

2.2 Adult Stem Cells

The term adult stem cell refers to any cell which is found in a developed organism that has two properties: the ability to divide and create another cell like itself and also divide and create a cell more differentiated than itself. They are also known as somatic stem cells and germline stem cells. Pluripotent adult stem cells are rare and generally small in number but can be found in a number of tissues including umbilical cord blood. Most adult stem cells are lineage restricted (multipotent) and are generally referred to by their tissue origin.

Adult stem cell transplantation has been successfully used for many years to treat bone marrow disorders.

In the case of allogenic adult stem cell transplantation, there are immunological questions related to graft/recipient rejection that must be considered. However, in some instances, adult stem cells can be obtained from the intended recipient. In these autograft situations, the risk of rejection is essentially non-existent.

- *Adipose-derived adult stem cells*: Adipose-derived stem cells (ASC) have been isolated from human fat, usually by liposuction. This cell population seems to be similar in many ways to mesenchymal stem cells derived from bone marrow. Human adipose-derived stem cells have been shown to differentiate into bone, cartilage, fat, and muscle, which makes ASCs a possible source for future applications in the clinic and they have been used to successfully repair a large cranial defect in a human patient. There are no immunological barriers for these cell types since they can be isolated from the recipient.

– *Dental pulp-derived stem cells*: Multipotent stem cells have been successfully recovered from dental pulp in the perivascular niche. They have been shown to have the same cellular markers and differential abilities of mesenchymal stem cells. They are not associated with an immunological barrier since they can be isolated from the recipient.
– *Hematopoietic stem cells*: Hematopoietic stem cells are found in the bone marrow and give rise to all the blood cell types. Allograft or autograft are routinely performed with these adult stem cells for bone marrow malignances or some severe immunodeficient syndromes.
– *Mammary stem cells*: Mammary stem cells provide the source of cells for growth of the mammary gland. Mammary stem cells have been isolated from human and mouse tissue as well as from cell lines derived from the mammary gland. Such single cells can give rise to both the luminal and myoepithelial cell types of the gland and have been shown to have the ability to regenerate the entire organ in mice.
– *Mesenchymal stem cells*: Mesenchymal stem cells are of stromal origin and may differentiate into a variety of tissues. They have been isolated from placenta, adipose tissue, lung, bone marrow and blood, and umbilical cord. MSCs are particularly attractive for clinical therapy not only due to their ability to differentiate into various cell types but also for their immunosuppressive properties. They are not systematically associated with an immunological barrier since they can be isolated from the recipient.
– *Neural stem cells*: The existence of stem cells in the fetal or adult brain has been postulated following the discovery that the process of neurogenesis, the birth of new neurons, continues into adulthood. Neural stem cells are commonly isolated and cultured in vitro as the so-called "neurospheres" – a kind of cell aggregate containing a large proportion of neural stem cells. They can be propagated for extended periods of time and differentiated into both astrocytes and neurons. Dopaminergic progenitors are neural stem cells from a mesencephalic origin that specifically differentiate into dopaminergic neurons. As Parkinson's disease is associated with a loss of dopaminergic neurons, transplantation of dopaminergic progenitors from a fetal origin has been achieved in humans.
– *Olfactory adult stem cells*: Olfactory adult stem cells have been successfully harvested from the human olfactory mucosa cells, which are found in the lining of the nose and are involved in the sense of smell. If they are given the right chemical environment, these cells have the same ability as embryonic stem cells to develop into many different cell types. Olfactory stem cells hold the potential for therapeutic applications and, in contrast to neural stem cells, can be harvested with ease without harm to the patient context. This means they can be easily obtained from all individuals, including older patients who might be most in need of stem cell therapies.

3 Induced Pluripotent Stem Cells Derived from Adult Somatic Cells

These are not really adult stem cells, but rather in vitro reprogrammed adult somatic cells with pluripotent capabilities. Using genetic reprogramming with protein transcription factors, pluripotent stem cells equivalent to ESC have been recently derived from human somatic tissues. Induced pluripotent stem cells (iPSC) are typically derived by transfection or transduction of certain stem cell-associated genes into non-pluripotent cells, such as adult fibroblasts. Transfected genes include the master transcriptional regulators Oct-3/4 and Sox2, although it is suggested that other genes enhance the efficiency of induction. Their main interest is to bypass the ethical and immunological barriers associated with ESC transplantation.

In this chapter we will focus on allogenic ESC and their precursors.

4 Immune Barriers and Mechanism of Rejection

Transplantation of organ cells or tissues from genetically unrelated individuals will lead to rejection in most of the cases. The mechanisms of rejection are well known and have been studied since many years in the context of solid organ transplantation and allogenic transplantation of different cell types [9, 10].

The different types of rejection can be classified according to several items like the mechanism of rejection, the timing of rejection episodes after the transplantation, or the genetic barriers (Fig. 2).

Antibody-mediated rejection: The first of rejection that could lead to irreversible graft damage is mediated by preformed antibody directed against the graft. The presence of anti-blood group ABO antibodies in the recipient against the graft was a major problem after solid organ transplantation leading to the so-called hyperacute rejection [11]. ABO antigens are not only expressed at the surface of red blood cells but also at the cell surface of other cell types like the endothelial cells which cover capillaries. In renal transplantation, for example, preformed anti-ABO antibody directed against the graft binds the ABO antigen and activates the coagulation cascade, complement, and induce inflammation. The final result is multiple thromboses that lead to the graft lost in a few hours. Interestingly, in hematopoietic stem cell transplantation the ABO incompatibility is not a big issue even if some studies suggest that ABO-incompatible transplantation is associated with more complications [12].

Other alloantibodies could be responsible for anti-mediated antibody like anti-HLA antibody. HLA molecules are strong antigens and the exposition to HLA antigens from unrelated individuals puts the recipient at high risk of developing anti-HLA antibody. Blood transfusions, pregnancy, and previous transplantation are the three situations to be exposed to allo-HLA antigen [10]. Most of the cells express HLA class I constitutively and if HLA class II is expressed only in specialized cells like dendritic cells and B cells, it can be expressed in any cell's surface upon

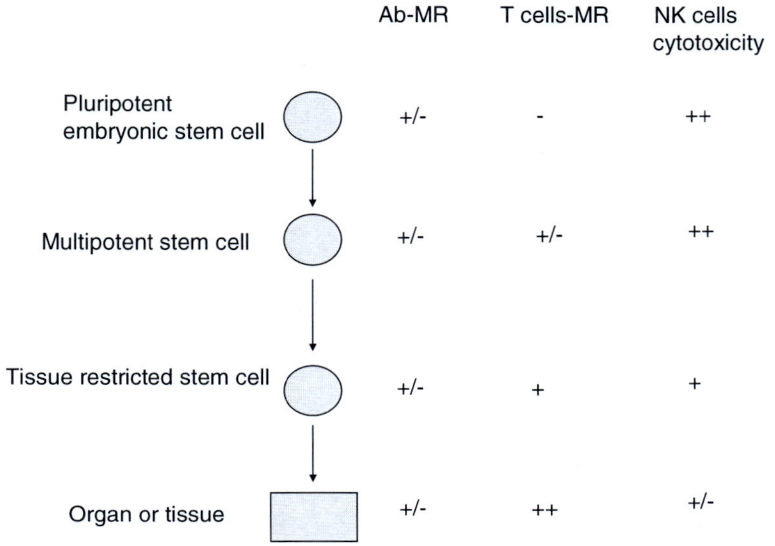

Fig. 2 Potential immune reactivity mediated by ABO or anti-HLA antibodies (Ab-MR), T cells (T cells-MR), or natural killer cells (NK) according to the state of differentiation

stimulation with pro-inflammatory cytokines like gamma interferon or TNF alpha. The presence of preformed anti-HLA antibody can also lead to hyperacute rejection or antibody-mediated acute rejection especially in kidney transplantation and the development of de novo anti-HLA antibody is associated with reduction of the graft function [11, 13].

In the case of allogenic ESC transplantation the graft rejection can also be theoretically mediated through a mismatch in ABO antigens between the donor and the recipient [14] due to the high immunogenicity of these antigens. To date, the ABO antigen expression status by ESC and their differentiated progeny is still under investigation but recent report suggests that blood group ABO antigen is expressed in human embryonic stem cells and in differentiated hepatocytes, but not in cardiomyocyte-like cells [15]. Nevertheless, the use of ESC for the purpose of transplantation can be ABO compatible with the development of ES cell line from the different blood groups. Alternatively, ESC from a universal blood group donor could circumvent the ABO mismatch potential problems. For HLA, the problem is similar. We will see in the next section that ESC express low levels of HLA class I and almost no HLA class II [6, 16]. But activation of any cells by inflammation would lead to the upregulation of MHC class I and most probably to expression of MHC class II.

To prevent antibody-mediated rejection in kidney transplantation the cross-match test is performed just before the transplantation. Peripheral blood cells from the donor and recent serum of the recipient are co-incubated in the presence of complement. The killing of donor cells signifies the presence of anti-HLA antibodies directed against the graft and is a contra-indication to perform the transplantation [17]. In the context of ESC-derived cell transplantation, the combination of ABO compatibility and a cross-match test would strongly decrease the risk of antibody-mediated rejection due to preformed antibody. Other types of alloantibodies have been described in organ transplantation like anti-endothelial antibodies, but previous exposition of such alloantigens in the context of ESC transplantation would be very unexpected as well as their expression at the surface of ESC.

After ESC transplantation, the development of de novo alloantigens against HLA and other antigens could theoretically be problematic [13]. However, the combination of the level of immunogenicity of ESC and immunosuppressive therapy could be sufficient to prevent antibody-mediated injury.

- *The major histocompatibility complex (MHC)*: The genetic barrier of the MHC class I and class II locus is impossible to cross without medical intervention. The MHC class I and class II are highly polymorphic and any MHC mismatch between donor and recipient elicits a strong immune reaction against the graft. The typing of the MHC class I and class II system at the allelic level has demonstrated thousands of different alleles that make a concept of MHC identity or compatibility of several MHC loci between ESC and recipient almost impossible in the context of ESC transplantation. In a very similar way to the worldwide registry of unrelated stem cell transplantation donor or to cord blood registries in the context of hematopoietic stem cell transplantation, we should imagine millions of different ESC cells typed at high/intermediate resolution for MHC and conserved in secured environment. Even if the expression of MHC antigen at the cell surface of ESC is low or absent, as mentioned before, inflammation could induce strong expression of MHC class I and MHC class II [6, 16]. A last critical point is still a matter of debate. Even if the expression of MHC class I is low or almost absent at the cell surface of ESC, MHC class II will be expressed at the cell surface of lineage derived from ES cells and it remains to be determined if the transplant of ESC or progenitors will be sufficient to induce the tolerance of MHC alloantigen when expressed at the cell surface of mature cells. Rodent models are encouraging but we have learnt from research in transplantation since many years that the transfer of mice model to human is often disappointing [18].
- *Minor antigens*: Minor antigens are polymorphic peptides that differ between individuals without modifying the function of the protein but that could be lead to antigenic determinant prone to activate the immune system and to induce rejection [19]. In contrast to the MHC, minor antigens do not elicit a strong immune response due to the low percentage of specific T cells directed against such antigen. But the immune response can be largely sufficient to injury the graft leading finally to a loss of function. Avoidance of minor antigens is impossible in the context of ESC genetically unrelated with the recipient but could be minimized by sex matching of ESC and recipient. ESC of male origin express minor antigens from the chromosome Y that could be recognized and induce a strong immune response by the immune system of females.

5 Effector Cells of the Immune Response

Several cell types are involved in the immune mechanisms of rejection. First, cells belonging to innate immunity include granulocytes, monocytes/macrophages, or dendritic cells (DC). Inflammation induced by the surgical procedure of cells or organ transplantation will activate and attract more cells of the innate immune system. Antigen-presenting cells (APC) or macrophages are great producers of pro-inflammatory cytokines like TNF-α, IL-1, IL-2, or INF-γ. The presence of such cytokines will contribute to the activation and maturation of DC which will migrate to the peripheral lymph nodes to activate naïve CD4 and CD8 T cells. DC are crucial to adaptive immune response which is the main factor mechanism of rejection. In the situation of ESC transplantation the inflammation induced by the transplantation will depend on the type of procedure. Local implantation of ESC or their progenitors by surgery even with minimal incision could be sufficient to induce inflammation per se, to activate local APC or macrophages to produce cytokines and chemotactic factors that will recruit granulocytes [4, 5]. If ESC or progenitor cells will be infused intravenously similar to hematopoietic stem transplantation, immunosuppressive condition regimen would certainly be recommended to make space and to inhibit immediate rejection by activated T cells. Inflammation produced by condition regimen which includes myeloablative drugs and radiotherapy is also pro-inflammatory in such situation and could be deleterious for the implanted ES cells. CD4 and CD8 T cells are key players of the immune response against allogenic tissue. After being activated and moved to peripheral lymph nodes, DC present allogenic peptides via MHC class II molecules to CD4 T cells. Specific CD4 T cells such as helper T cells activated CD8 T cells that will also proliferate. DC can also directly activate CD8 T cells by a mechanism called cross-presentation. Activation is mediated by the TCR which recognizes the MHC and allogenic peptide in the presence of co-stimulatory molecules such as B7 (APC)-CD28 (T cell) and CD 40 (APC)-CD40 ligand (T cell). The presence of IL-2 is required. After their activation and proliferation, CD8 and CD4 T cells move to the transplanted tissue which could be injured be the recognition of cytotoxic CD8 T cells via the MHC and peptide expressed by the graft. CD4+T cells have also the capacity to activate B cells leading to the production of antibody already described above [18].

Since ESC do not express MHC or at low level, the T-cell rejection mechanism could be strongly inhibited. However, natural killer (NK) cells are special cells that bridge the innate and adaptive immunity and are able to recognize and kill MHC-negative cells.

Natural killer (NK) cells play a critical role in host defense against certain viruses, by mediating cellular cytotoxicity and by producing chemokines and inflammatory cytokines [20, 21]. NK cells use "hard-wired" receptor systems, encoded by genes that do not undergo variable diversity joining recombination or sequence diversification in somatic cells. Also unlike B and T cells, but similar to other innate immune cells, NK cells use a multiple receptor recognition strategy, whereby an individual NK cell can be triggered through various receptors independently or in combination, depending on the ligands presented by the target cell in a given encounter. NK cell effector functions are regulated by a complex repertoire of activating and inhibitory receptors. The balance between activating and inhibitory signals is crucial in regulating NK cell effector functions. Interestingly, the ligands for most of the well-known inhibitory receptors are the MHC class I HLA-B and HLA-C loci [20, 21]. Therefore, the absence or downregulation of MHC class I expression (the concept of missing legend) can induce susceptibility of a potential target cell to NK cytotoxicity. The absence or weak expression of the MHC class I in ESC made such cells good targets for NK killing. In addition to its killing activity, NK cells are strong producers of INF-γ which is an important factor that amplifies the inflammation and the immune response [20, 21].

6 How to Monitor the Immune Response Against ESC

The analysis of the potential response of the immune system against ESC is a key element to anticipate potential failure of the graft function due to rejection. Monitoring immune responses to assess donor-specific immunity or tolerance uses different readouts such as proliferation, cytokine production, or cytotoxicity. The in vitro culture of ESC and their differentiation into progenitors give tolls to analyze the immune reactivity of T or NK cells against ESC. The presence of antibodies in the serum of recipients directed against ESCcould be monitored by the cross-match assay. In this assay, the serum of potential recipient is incubated with ESC or their progenitors and the binding of antibody directed against ABO antigen, HLA molecules, or other antigen could be analyzed by flow cytometer (FACS). Positive results indicate a potential risk of hyperacute rejection which would lead to an irremediable loss of the graft function. The alloreactivity of T cells could be analyzed by proliferation and cytotoxic assays. The mixed lymphocyte reaction (MLR), the cytotoxic T-lymphocyte assay (CTLp), or the helper lymphocyte assay (HTLp, which measure the secretion of IL-2 by CD4 T lymphocytes) has been used during many years in the context of bone marrow transplantation for the prediction of graft versus host disease (GVHD) and also for donor selection. Estimation of host-reactive MLR, CTLp, or HTLp in the peripheral blood of donors has been shown to be predictive of acute GVHD and survival in a series of studies [22, 23]. In the context of ESC, MLR has also been used already mainly in animal studies to test in vitro proliferation of T cells [24]. For CTLp or HTLP, because it is a dilution assay, the major problem is to have cells in suspension to plate enough wells. This is more difficult to obtain for ESC and probably more feasible with progenitors. In contrast, MLR can be performed on adherent irradiated cells as stimulator cells. For NK cells reactivity, a standard ^{51}Cr-release cytotoxicity assay can be performed. Target cells like ESC, their derived cells, or K562 cells (control that do not express MHC class I and class II molecules) are incubated with the NK cells in the presence of ^{51}Cr. The percentage of specific lysis is calculated from the formula: percentage of specific lysis = (experimental counts – spontaneous lysis)/(maximal lysis – spontaneous lysis)$\times$100. In parallel to the ^{51}Cr-release assay, the cytotoxicity can also be assessed by flow cytometry, using the biotin-conjugated anti-CD107a, a marker of NK cells degranulation [25]. Albeit intensive and complex labor, in vitro assays are still a valuable tool in monitoring donor-specific responses. Their specificity and relationship to clinical outcome have not been surpassed by any other assay to date. Their ability to unmask regulatory cell effects and the range of measurable readouts will ensure their continued usefulness.

Other approaches are available based on cytokine secretion by T and NK cells upon stimulation like the IFN-γ secretion assay. Whole blood, PBMC, or isolated T or NK cells are stimulated for a short period of time with specific peptides or allogenic cells like ESC. Secreted IFN-γ binds to the IFN-γ-receiving reagent on the positive secreting cells. These approaches to specific secretion of cytokines by T or NK cells can be combined with the technology available to measure cytokine secretion (by ELISA or ELISPOT) [26]. The phenotype of immune cells can also be analyzed at the mRNA level. Several technologies are available which detect a pattern of hundreds or thousands of genes by microarray [27] and/or to quantify gene expression by real-time polymerase chain reaction [10].

Such in vitro assay cannot replace animal studies which remain as an invaluable tool to analyze the immune response against autologous, allogenic, or xenogenic transplanted cells or tissue. However, years of research in the field of transplantation immunology have clearly demonstrated that rodent models, if certainly of great value, could not mimic accurately the human situation. Because the immune reconstitution of nude mice by the human immune system is not complete, this model will give important but incomplete information's about the human immune reactivity. Non-human primate models and finally phase I clinical studies are the way to go in most of the situation [18]. Therefore, in vitro assay with human sample is of great interest and is irreplaceable to give preliminary information's on the immune reactivity of the human cells against ESC.

On the long run, in vivo assays for immune rejection of transplanted cells will be needed. This would best allow adapting the immunosuppression to the individual needs of the patient. Presently, no such assays are available. However, on the long run, there will most likely be a development of approaches that will allow monitoring immune rejection of a cellular graft in an individual patient. PET-based technology is most likely to yield such an assay.

7 Immunogenicity of Undifferentiated ESC

ESC were described to express low levels of HLA class I, the latter being upregulated by IFN-γ stimulation or after differentiation into embryoid bodies or after teratoma formation [16, 24, 28]. HLA class II molecules, however, are not expressed on undifferentiated ESC. The expression of co-stimulatory molecules (CD40, CD80, CD86) has also been demonstrated to be low or absent on undifferentiated ESC [24, 28]. This suggests that ESC lack two prerequisites to function as antigen-presenting cells efficient to prime T lymphocytes.

In addition, molecules known to downregulate the immune response such as TGF-β, Fas-L, and IL-10 were not significantly found in undifferentiated ESC [28], suggesting that ESC do not harbor immunosuppressive properties mediated by these two well-defined pathways.

In xenogenic conditions, transplanted ESC under the kidney capsule of various immunodeficient mouse strains induce teratomas only seen in T-cell-deficient mice, suggesting that xenograft rejection of ESC was T-cell-mediated immune process [29].

In order to study the outcome of ESC in an allogenic context, a humanized mouse model was created in which the bone marrow was reconstituted with human peripheral blood mononuclear cells [29]. Transplantation of human cells or pieces of tissues was rejected in contrast to ESC, which developed into teratoma. It is noteworthy that although these mice can partially reject human cells, such humanized mice do not reflect the human in vivo situation since the immune system is not fully reconstituted and not functional. Thus, residual xenogenic conditions could explain the capacity of these mice to reject mature human cells/tissues. At least, in this comparison, ESC are suggested to be less immunogenic than mature cells.

In an allogenic in vitro test, ESC were loaded with a model antigen (peptide of influenza virus type A). Practically, ESC expressing the HLA-A2 allele of MHC class I were pulsed with a peptide of influenza virus and exposed to T lymphocytes that are specific for such antigen. In the absence of IFN-γ, T lymphocytes did not kill loaded ESC whereas IFN-γ-treated ESC were partially killed [29]. "Natural" infection of ESC with influenza virus confirmed this observation. Thus, direct recognition of ESC by T cells is not sufficient to kill them efficiently. This observation is consistent with the fact that MHC and co-stimulatory molecules are not expressed or at low level on the surface of ESC.

These data indicate that ESC have little potential to activate a direct allo-specific immune response. Other in vitro studies confirmed these observations [24]: In an MLR, ESC failed to induce the proliferation of allogenic human peripheral blood mononuclear cells and this lack of allogenic immune response was also seen when T-cell-enriched peripheral blood lymphocytes were used.

8 Immunogenicity of Differentiated Cells Derived In Vitro from ESC

Differentiation protocols of ESC to progenitors of many cell types are currently available [30]. For example, neural progenitor cells (NPC) derived in vitro from ESC [30] have the potential to "replace" the damaged nervous tissues in some neurological diseases after transplantation into the brain, being locally differentiated into mature neurons or other subtypes of neural cells. Because the origin of cells derived in vitro from ESC is genetically unrelated to the recipient, a major hurdle to cell therapy could be the host immune response toward the transplanted cells. In addition to the immune reaction that is expected in allogenic transplantation derived from ESC, products of animal origin used in the differentiation protocols have also been proposed to amplify a risk of xenogenic antigens inclusion (immunogenic non-human sialoproteins) increasing rejection in the recipient [31]. Multiple steps have been implemented to minimize the amount of animal components during the differentiation process including the replacement of bovine serum in the medium [32].

Nevertheless, the potential importance of immune rejection process against cells derived in vitro from ESC remains a subject of intense debate.

9 Therapeutical Approach to Overcome the Immune Barriers

The immunogenicity of human ESC remains a subject of intense debate [24]. But transplantation of allogenic cells in human would lead with a good probability to a risk of rejection. Preliminary experiment on human ESC for realizing cardiac regeneration has shown that direct differentiation of ESC into genuine cardiomyocytes was more efficient than ESC non-cardiomyocyte-derived cell types [33]. This suggests that the best source of transplanted tissue will be more progenitors or mature cells than ESC and imply a higher risk of rejection. Similar evidence has been demonstrated for neuron derived from ESC. Neural progenitor would be the best source of cells for potential clinical application of stem cells therapy to cure neurodegenerative disease [4, 5]. In both cases, such progenitors or mature cells would be at higher risk to be rejected by the immune system of the recipient. By targeting the immune system rejection could be prevented. The situation of the brain is of special interest because the brain is supposed to be an anatomical site with a certain degree of immune privilege [34]. Transplantation of ESC or neural progenitors points the question of the relevance of immune rejection. Although the alloreactivity in the brain is poorly described, a recipient immune reaction against transplanted cells has been demonstrated since at least 20 years [35, 36]. For these reasons, the first clinical trial protocols of fetal tissue transplantation included immunosuppressive drugs for the transplanted patients [37, 38]. The easiest way to target the immune system would be the use of immunosuppressive regimen. Classical drugs like steroid or calcineurin inhibitors (i.e., cyclosporine, tacrolimus) are the logical immunosuppressive regimen to propose and have been shown to be very efficient to prevent rejection in solid organ or tissue transplantation since decades [39]. In addition, to the side effect of the drugs, the state of immunodeficiency induced by the inhibition of the immune system puts the patient at higher risk of infection. Recent data suggest that in vitro NK cells could not be significantly affected by immunosuppressive drugs in humans [40–42], which could be an advantage if considering the potential of NK cells to kill teratoma [43] but also a disadvantage if considering the potential risk of NK cell-mediated rejection of MHC class I negative cells. Finally, a critical point is the potential deleterious effect of not only immunosuppressive drugs but also any other drugs on the final step of differentiation of ESC or their progenitor after the transplantation. The risk that such drugs would inhibit or modify the differentiation of progenitor into mature cells in vivo should be carefully evaluated. Different immunosuppressive regimens have been tested in vitro and in animal model before going to clinical studies.

A more sophisticated approach to ensure that a patient accepts an ESC transplant from a genetically unrelated donor over the long term would be the use of the patient's own immunological mechanisms to induce tolerance to donor antigens [44].

Tolerance could be induced at the central level or at the periphery [9]. Central tolerance is established during the maturation of T lymphocytes in the thymus. T lymphocytes that are too reactive against auto-antigens are eliminated during a process called negative selection. The success of inducing tolerance to a foreign transplant will rely mainly on allowing uptake of donor cells (for example, ESC or progenitors (HSCs)) and therefore antigen presentation of the donor's antigens in the thymus, where the patient's T cells that recognize these antigens, can be eliminated. Because the thymus is supposed to be non-functional at adult age, this involves a step of restoring the thymic function. However, recent data suggest that central tolerance could be achieved even in adults after combined hematopoietic stem cells and kidney transplantation or after liver transplantation [45]. Such an approach remains complex because the patient receives a partial myeloablative conditioning regimen that includes cyclophosphamide, anti-thymoglobulin polyclonal antibody, and thymic irradiation. Peripheral tolerance could be achieved by different mechanisms that include anergy, ignorance, or regulation. Anergy signified that the T cells are able to recognize the allogenic cells but are unable to be activated. Blocking T cells' co-stimulatory molecules like CD28 or CD40 has been shown to be able to induce peripheral tolerance in animal models, but it is usually coupled to classical immunosuppressive drugs already described [46]. In human, the utilization of such approach in organ transplantation is still under evaluation. Regulatory $CD4^+CD25^+FOXP3^+$T cells can help to induce peripheral tolerance and several protocols have been tried to use this elegant way of inhibiting specifically the immune response against alloantigens [47, 48]. Induction of central or peripheral tolerance has been of great hope in solid organ or tissue transplantation but the transplantation from promising animal model to human is far more complicated as expected [18]. Therefore, in the next decade it is predictable that classical immunosuppressive drugs will remain the major apache to inhibit the immune system after transplantation which indicates that it should also be validated in the context of ESC or its progenitor's transplantation.

As a long-term perspective, the inclusion of a local slow-release immunosuppressant system in the graft is also an attractive option. For example, compounds that antagonize interferon-induced MHC class I upregulation might be included in such a system. As there are good reasons to believe that immune rejection will be most virulent during the initial period after cell transplantation, the time limitations of such a local system are not necessarily a major problem.

10 Conclusion

ESC and their progenitors offer a potential for many different diseases that need to regenerate damaged cells or tissue. Such regenerative medicine approaches are ethically controversial surrounding the use of ESC but also fully differentiated somatic cells that are reprogrammed. In any case transplantation with such cells coming from genetically unrelated donors would lead to the risk of rejection. Therefore, a better understanding of the immunogenicity of the transplanted cells is mandatory before hoping to start any clinical trial. Through in vitro assay and animal models, we believe that the immunogenicity of ESC and their progenitor should be analyzed in deep. The optimization of immunosuppressive approaches via classical drugs or tolerance induction with a special focus onto ESC or their progenitor should be the subject of intense research. The generation of iPSC has the potential to produce an autologous source of pluripotent stem cells which could bypass the problem of allogenic rejection. However, many hurdles need to be overcome before the appropriate regulatory approval will be granted for their use in the clinic like non-viral delivery systems, the effect of the differentiation state, or the predisposition to malignant transformation. Therefore, the use of allogenic ESC still remains an available option.

Finally, most transplantation protocols would use single-cell suspensions. This could be accompanied by extensive cell death, inadequate integration, and inflammation. In a hostile immunoreactive, ischemic, or necrotic environment, survival of the grafted tissue even derived from genetically identical donor like iPSC is a major challenge. Reduction of the injury and inflammation induced by the transplantation procedure should also be carefully evaluated.

References

1. Jones JM, Thomson JA. Human embryonic stem cell technology. Semin Reprod Med. 2000; 18:219–223
2. Thomson JA, et al. Embryonic stem cell lines derived from human blastocysts. Science. 1998; 282:1145–1147
3. Guillaume DJ, Zhang SC. Human embryonic stem cells: a potential source of transplantable neural progenitor cells. Neurosurg Focus. 2008; 24:E3
4. Winkler C, Kirik D, Bjorklund A. Cell transplantation in Parkinson's disease: how can we make it work? Trends Neurosci. 2005; 28:86–92
5. Morizane A, Li JY, Brundin P. From bench to bed: the potential of stem cells for the treatment of Parkinson's disease. Cell Tissue Res. 2008; 331:323–336
6. Drukker M, Benvenisty N. The immunogenicity of human embryonic stem-derived cells. Trends Biotechnol. 2004; 22:136–141
7. Hoffman LM, Carpenter MK. Characterization and culture of human embryonic stem cells. Nat Biotechnol. 2005; 23:699–708
8. Wilmut I, Schnieke AE, McWhir J, Kind AJ, Campbell KH. Viable offspring derived from fetal and adult mammalian cells. Nature. 1997; 385:810–813
9. Rogers NJ, Lechler RI. Allorecognition. Am J Transplant. 2001; 1:97–102
10. Villard J. Immunity after organ transplantation. Swiss Med Wkly. 2006; 136:71–77
11. Kissmeyer-Nielsen F, Olsen S, Petersen VP, Fjeldborg O. Hyperacute rejection of kidney allografts, associated with pre-existing humoral antibodies against donor cells. Lancet. 1966; 2:662–665
12. Remberger M, et al. Major ABO blood group mismatch increases the risk for graft failure after unrelated donor hematopoietic stem cell transplantation. Biol Blood Marrow Transplant. 2007; 13:675–682
13. McKenna RM, Takemoto SK, Terasaki PI. Anti-HLA antibodies after solid organ transplantation. Transplantation 2000 Feb 15; 69(3):319–326.

14. Cooper DK. Clinical survey of heart transplantation between ABO blood group-incompatible recipients and donors. J Heart Transplant. 1990; 9:376–381
15. Molne J, et al. Blood group ABO antigen expression in human embryonic stem cells and in differentiated hepatocyte- and cardiomyocyte-like cells. Transplantation. 2008; 86:1407–1413
16. Drukker M, et al. Characterization of the expression of MHC proteins in human embryonic stem cells. Proc Natl Acad Sci USA. 2002; 99:9864–9869
17. Terasaki PI, Cai J. Humoral theory of transplantation: further evidence. Curr Opin Immunol. 2005; 17:541–545
18. Lechler RI, Sykes M, Thomson AW, Turka LA. Organ transplantation–how much of the promise has been realized? Nat Med. 2005; 11:605–613
19. Simpson E, et al. Minor H antigens: genes and peptides. Eur J Immunogenet. 2001; 28: 505–513
20. Lanier LLNK. cell recognition. Annu Rev Immunol. 2005; 23:225–274
21. Parham P. MHC class I molecules and KIRs in human history, health and survival. Nat Rev Immunol. 2005; 5:201–214
22. Spencer A, et al. Cytotoxic T lymphocyte precursor frequency analyses in bone marrow transplantation with volunteer unrelated donors. Value in donor selection. Transplantation. 1995; 59:1302–1308
23. Speiser DE, Loliger CC, Siren MK, Jeannet M. Pretransplant cytotoxic donor T-cell activity specific to patient HLA class I antigens correlating with mortality after unrelated BMT. Br J Haematol. 1996; 93:935–939
24. Li L, et al. Human embryonic stem cells possess immune-privileged properties. Stem Cells. 2004; 22:448–456
25. Uhrberg M. The CD107 mobilization assay: viable isolation and immunotherapeutic potential of tumor-cytolytic NK cells. Leukemia. 2005; 19:707–709
26. Miyahira Y, et al. Quantification of antigen specific $CD8^{+}T$ cells using an ELISPOT assay. J Immunol Methods. 1995; 181:45–54
27. Chua MS, Sarwal MM. Microarrays: new tools for transplantation research. Pediatr Nephrol. 2003; 18:319–327
28. Grinnemo KH, et al. Human embryonic stem cells are immunogenic in allogeneic and xenogeneic settings. Reprod Biomed Online. 2006; 13:712–724
29. Drukker M. Immunogenicity of embryonic stem cells and their progeny. Methods Enzymol. 2006; 420:391–409
30. Petersdorf EW, et al. Optimizing outcome after unrelated marrow transplantation by comprehensive matching of HLA class I and II alleles in the donor and recipient. Blood. 1998; 92:3515–3520
31. Takemoto SK, Terasaki PI, Gjertson DW, Cecka JM. Twelve years' experience with national sharing of HLA-matched cadaveric kidneys for transplantation. N Engl J Med. 2000; 343:1078–1084
32. Opelz G. New immunosuppressants and HLA matching. Transplant Proc. 2001; 33:467–468
33. Passier R, van Laake LW, Mummery CL. Stem-cell-based therapy and lessons from the heart. Nature. 2008; 453:322–329
34. Simpson E. A historical perspective on immunological privilege. Immunol Rev. 2006; 213:12–22
35. Mason DW, et al. The fate of allogeneic and xenogeneic neuronal tissue transplanted into the third ventricle of rodents. Neuroscience. 1986; 19:685–694
36. Widner H, Brundin P. Immunological aspects of grafting in the mammalian central nervous system. A review and speculative synthesis. Brain Res. 1988; 472:287–324
37. Olanow CW, et al. A double-blind controlled trial of bilateral fetal nigral transplantation in Parkinson's disease. Ann Neurol. 2003; 54:403–414
38. Freed CR, et al. Transplantation of embryonic dopamine neurons for severe Parkinson's disease. N Engl J Med. 2001; 344:710–719
39. Kaufman DB, et al. Immunosuppression: practice and trends. Am J Transplant 2004; 4(Suppl 9):38–53

40. Chiossone L, et al. Molecular analysis of the methylprednisolone-mediated inhibition of NK-cell function: evidence for different susceptibility of IL-2-versus IL-15-activated NK cells. Blood. 2007; 109:3767–3775
41. Wang H, et al. The unexpected effect of cyclosporin A on CD56$^+$. Blood. 2007; 110: 1530–1539
42. Wai LE, Fujiki M, Takeda S, Martinez OM, Krams SM. Rapamycin, but not cyclosporine or FK506, alters natural killer cell function. Transplantation. 2008; 85:145–149
43. Dressel R, et al. The tumorigenicity of mouse embryonic stem cells and in vitro differentiated neuronal cells is controlled by the recipients' immune response. PLoS ONE. 2008; 3:e2622
44. Billingham RE, Brent L, Medawar PB. Actively acquired tolerance of foreign cells. Nature. 1953; 172:603–606
45. Kawai T, et al. HLA-mismatched renal transplantation without maintenance immunosuppression. N Engl J Med. 2008; 358:353–361
46. Vincenti F, et al. Costimulation blockade with belatacept in renal transplantation. N Engl J Med. 2005; 353:770–781
47. Miyara M, Sakaguchi S. Natural regulatory T cells: mechanisms of suppression. Trends Mol Med. 2007; 13:108–116
48. Sakaguchi S. Naturally arising Foxp3-expressing CD25$^+$CD4$^+$ regulatory T cells in immunological tolerance to self and non-self. Nat Immunol. 2005; 6:345–352

Reponses of Mesenchymal Stem Cells to Varying Oxygen Availability In Vitro and In Vivo

Frank R. Kloss, Sarvpreet Singh, and Günter Lepperdinger

Abstract Oxygen is vital for cellular metabolism in higher organisms. Explanted somatic cells are regularly cultured at ambient oxygen conditions, which often appear stressful for primary cells, thereby leading to accelerated aging or premature senescence in vitro. Therefore, sophisticated instrumentation for ex vivo cell manipulation has been introduced and is now increasingly being used for the propagation and subsequent differentiation of various types of stem cells. In this particular context also primary mesenchymal stromal cells (MSC) have been investigated. Their developmental fate greatly depends on oxygen availability.

In this contribution, we describe the current knowledge about MSC properties according to varying oxygen tension in vitro: while reduced oxygen tension supports their proliferation, differentiation potential is reversibly attenuated. This finding is relevant for various in vivo situations such as wound healing or distinct pathologic alterations. Inappropriate oxygenation is actually not impacting on MSC viability, but greatly diminishes their differentiation capabilities, in due course leading to irreversible changes in tissue structure including functional alterations involved.

Keywords Mesenchymal stem cells · Multipotential differentiation capacity · Tissue oxygenation · Radio necrosis · Bone healing

1 Introduction

Atmospheric oxygen (O_2) is fundamental for life, and the respective concentration available in various tissues is thought to be an important signal for virtually all cellular processes. However, physiological O_2 levels are rarely taken into account when explanting and culturing primary cells. Hence, this contribution highlights biological responses of mesenchymal stromal cells (MSC) with respect to varying levels of disposable oxygen both in vitro and in vivo.

G. Lepperdinger (✉)
Institute for Biomedical Aging Research, Austrian Academy of Sciences, Innsbruck, Austria
e-mail: guenter.lepperdinger@oeaw.ac.at

G.M. Artmann et al. (eds.), *Stem Cell Engineering*,
DOI 10.1007/978-3-642-11865-4_9, © Springer-Verlag Berlin Heidelberg 2011

MSC contain a rare population of uncommitted progenitor cells that exhibit multipotential differentiation capacity. These specialized cells are often termed mesenchymal stem cells. The designation "mesenchymal stem cell" is used by many investigators. However, up till recently, the definition of what a "mesenchymal stem cell" is has been sketchy. Most blood cell types emerge from mesodermal derivatives. Therefore, hematopoietic stem cells (HSC) should be considered of mesenchymal origin. In contrast, the term mesenchymal stem cell is by all rules assigned to non-hematopoietic cells. Moreover, doubts about the appropriateness of the term "mesenchymal *stem* cell" have been raised, in particular, because it appears to be scientifically inaccurate and potentially misleading. Instead, the term multipotent mesenchymal stromal cell (which could be similarly abbreviated "MSC") has been proposed recently [1].

By all means, an important feature of a stem cell is its capability to self-renew, that is, to divide and create additional stem cells. In addition to that, stem cells need to give rise to committed progeny, which differentiates further on into at least one somatic lineage [2]. Apart from bone marrow [3], MSC can be derived from various tissues [4]. Interestingly, most tissue-specific MSC exhibit a broad range of differentiation potential in vitro. It is, however, assumed that besides intrinsic stem cell factors, the respective stem cell niche provides cell–cell and cell–matrix interactions, which are controlling tissue-specific lineage commitment. There are many reasons to believe that the delicate balance between self-renewal and generation of progeny through asymmetric cell division is tightly controlled within the niche; having said that, it is also assumed that a proper niche comforts and protects stem cells from adverse influences throughout life.

As described by others [5], also in our hands, MSC exhibit a stable undifferentiated phenotype: they proliferate extensively while retaining the potential to differentiate along osteogenic, adipogenic, and chondrogenic lineages in vitro. MSC are part of a highly specialized microenvironment and thus generally believed to participate in the regulation of hematopoietic precursor cell differentiation into mature progeny [6, 7]. They reside in a complex three-dimensional network, which comprises of a plethora of other cell types, such as bone marrow, HSC, adipocytes, and endothelial cells, supported by an extracellular matrix consisting mainly of fibronectin, collagen I and IV, heparin sulfate, chondroitin sulfate, and hyaluronan [8]. Hence, there are lots of opportunities for MSC to interact at both the cellular and matrix level, and all of these elements may modify MSC behavior, in particular in respect to pathological alterations, traumatic insults, as well as organismic or cellular aging [9, 10].

Up to now, common characteristics of MSC could only be insufficiently specified. The cell isolates are heterogeneous and comprise of more than one mesenchymal cell type. Due to the fact that there has not been a distinct and universal antigenic definition of an MSC, analogous to $CD34^+$ cells for HSC, and as there is no universal assay around, analogous to hematopoietic re-population assays, novel methods for the isolation of MSC with both self-renewal and multipotential differentiation capacity are currently being developed. Arnold Caplan was the first to propose the term "mesenchymal stem cell" for an adherent fibroblastic cell isolated by Percoll® density centrifugation that expresses antigens reactive with the monoclonal antibodies SH2 (CD105) and SH3 (CD73) [3]. Pittenger and colleagues have reported that CD29, CD44, and CD90 are important determinants [11]. Prockop and coworkers discovered a subpopulation of MSC, termed RS (rapidly self-renewing) cells, which exhibit enhanced proliferation rates, yet cannot be clearly distinguished from other adherent MSC solely by expression of specific surface markers [12]. At the same time, Simmons and colleagues described the so-called STRO-1 antibody, which identifies an immature population of mesenchymal cells [13]. Many other determinants have been discovered [14]. Given these observations, the following minimum specifications in order to generally define MSC have been proposed only recently by a group of peers in the field: (i) plastic adherence when maintained in standard culture conditions; (ii) expression of surface molecules CD105, CD73, and CD90 and lack of expression of CD45, CD34, CD14 or CD11b, CD79 alpha or CD19, and HLA-DR, as determined by the Mesenchymal and Tissue Stem Cell Committee of the International Society for Cellular Therapy; and last but not least (iii) differentiation potential toward osteogenic, adipogenic, and chondrogenic cell types in vitro [15].

2 Normoxic Conditions

Most often, cells are cultured at ambient atmospheric conditions of about 20% O_2. In vivo, however, the O_2 tension (pO_2) depends on the level of vascularization and the type of microenvironment within the respective organ. Cancellous bone is well vascularized. To date it is technically unfeasible to assess spatial differences in pO_2 within cancellous bone and marrow. It is known that blood in human bone marrow contains about 7% O_2 [16], from which O_2 gradients in surrounding tissues have been further extrapolated applying mathematical modeling. Thereby, O_2 gradients have been calculated in bone marrow to be highest (~5%) near the sinuses and lowest (~1%) at the inner surface of cortical bone [17, 18].

The importance and influence of low O_2 concentration for various cell types has been independently described. Results of these studies have shown that animal cells proliferate with elevated rates under reduced O_2 conditions [19–21]. This incomplete listing of examples greatly argues for O_2 availability being involved in controlling the fate and commitment of undifferentiated cells and tissue primordia.

2.1 MSC at Varying Oxygen Levels In Vivo

It is assumed that the developmental fate of MSC within a particular tissue and their progeny depends on environmental influences occurring in the respective biological niche [22, 23]. Data regarding the microenvironment and/or niche for the adult hematopoietic [24, 25], neural [26], and epithelial stem cell [27] have previously been communicated. Taken together, it is now generally accepted that maintenance of the stem cell niche depends on endogenous regulators and external signals.

Only recently, data became available showing the existence of mesenchymal stem and progenitor cells in perivascular locations, and therefore these cells have often been called pericytes [28]. In view of these data, MSC, which are generally believed to contribute to tissue homeostasis by, e.g., supporting bone regeneration or fulfilling permissive action on hematopoiesis, appear to exert yet another function: maintaining blood vessel integrity [29]. It can further be envisaged that upon tissue damage and injury, MSC/pericytes are either being activated or released from the perivascular niche and, thus, support wound healing and tissue regeneration. To fully understand the effects of physiological niches, it is important to consider injury-derived changes to the microenvironments such as hypoxia. Anoxic conditions may occur after injury and during wound healing [30]. For instance, after bone fracture, O_2 levels are known to transiently drop to 0–2% at the center of the wound [31, 32].

2.2 Regulation of O_2 in Cell Culture

As mentioned above, it has well been documented over the past few years that exposure to variable O_2 conditions affects cellular physiology in dramatic ways when compared to cells cultured at traditional conditions (20% O_2/5% CO_2) throughout an experiment or during long-term expansion. These days, however, controlled-environment glove boxes can be employed, which offer maximum flexibility for working at reduced ambient O_2 atmospheres (Fig. 1a). Depending on the device, O_2, CO_2, and temperature as well as humidity levels can be adjusted (Fig. 1b).

Hypoxic chambers or incubators were introduced in the early 1990s. They are hermetically sealed and use automatic oxygen sensors and controllers, which monitor O_2 levels, to initiate a purge procedure, which, when required, injects a background gas (usually nitrogen) or O_2 gas through appropriate valves. Alternatively, there is also a low-cost option, in which premixed gas containing O_2 at a desired concentration together with the standard 5% is used in small, closed culture systems. However, the gas mixtures are usually more expensive than the bulk-compressed, single-mixture gas tanks, and therefore routines involving atmospheric changes result in increased costs over time. In combination with a leak-free airlock system as a transfer or shuttle method for exchanging samples, supplies, pipettes, and tools, the chambers allow uninterrupted culture at chosen O_2 conditions. Further sophistications in addition to the basic system may include the monitoring of pO_2 within cell culture media or controlled humidified incubation, the latter preventing desiccation of cell culture media.

Employing such chambers, long-term culture paradigms have been reliably conducted. Furthermore, phenotypic changes of in vitro models with regard to varying pO_2 can be approached and phenomena associated with ischemia/reperfusion can be investigated. Occasionally, these devices are also used for studies with live animals, which are residing in the chamber for prolonged periods.

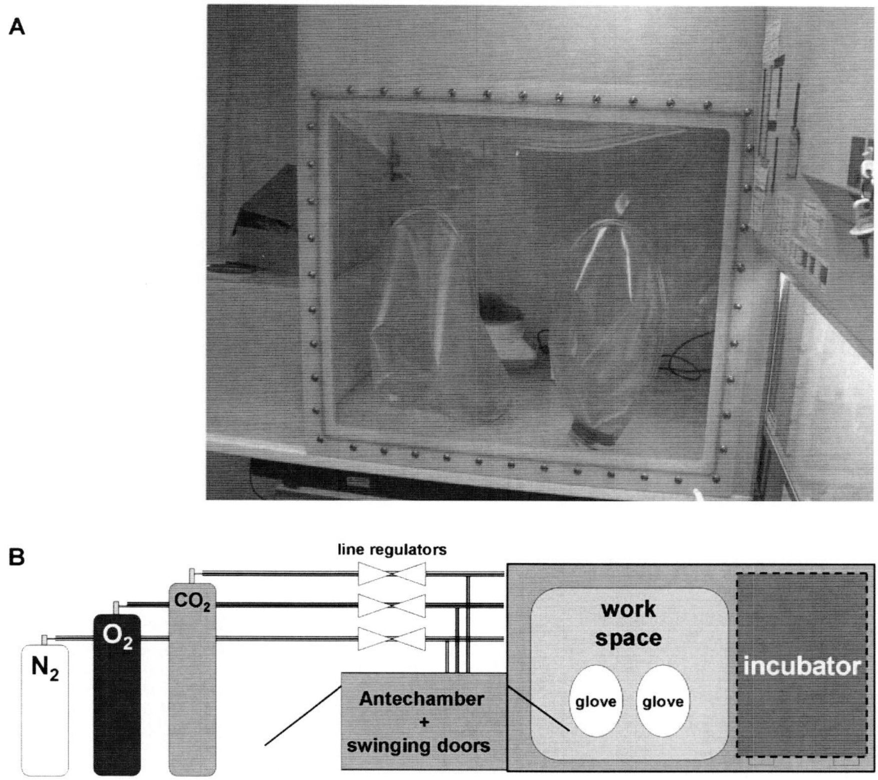

Fig. 1 Controlled environment for long-term cell cultivation. (**a**) Glove box and workstation for uninterrupted cultivation at set conditions; (**b**) schematic drawing of gas-controlled system consisting of a pass-through to deliver material into the workspace for manipulation. The gadget also houses a temperature-controlled and humidified incubator, in which cell cultures reside after manipulation

2.3 MSC at Varying O_2 Levels In Vitro

Explanted mesenchymal progenitors display the following characteristic variations when being incubated at physiological O_2 levels during ex vivo cultures [33]: (i) investigation of the transcriptional profile of mesenchymal progenitors revealed particular changes in the expression level of a subset of functionally interesting genes; however, when culturing MSC at 3% O_2, hypoxia-induced transcriptional pathways are not activated; (ii) with regard to replication potential and in vitro aging, 3% O_2 is beneficial for cultured MSC (Fig. 2); (iii) in comparison to traditional culture conditions, in vitro differentiation into adipogenic progeny was found to be markedly reduced, when O_2 content is lowered to 3% during induction. No osteogenic differentiation was evident at 3% O_2 (Fig. 3). This suppression of differentiation can, however, be overcome by increasing the level of O_2 to ambient air condition. Still in most cases MSC are expanded under ambient atmospheric conditions of ~20%

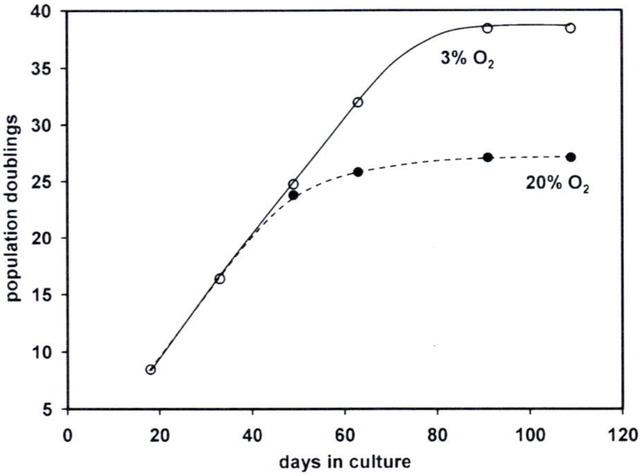

Fig. 2 Proliferation potential of primary human MSC. Growth rate of cells propagated in long-term culture either at 3 or at 20% O_2

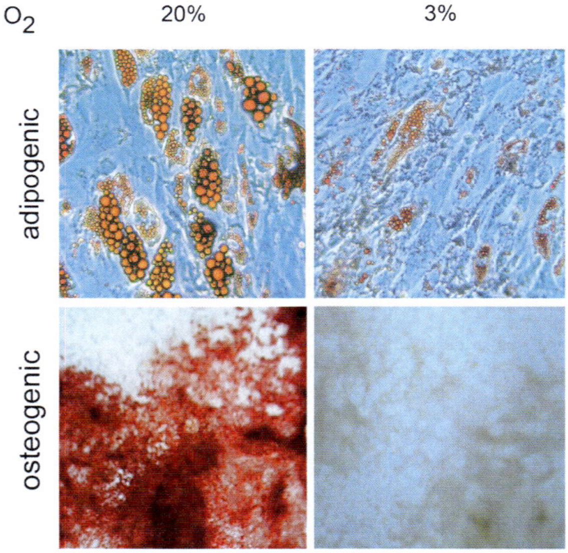

Fig. 3 In vitro differentiation of primary human MSC. Adipogenic and osteogenic differentiation was conducted either at 3 or 20% O_2. Adipogenesis was envisaged by *Oil Red* staining and is indicated by lipid-filled vesicles within the cells; osteogenesis was unveiled by *Alizarin Red* staining of the mineralized matrix deposited in the extracellular space

O_2 for later use in biomedicine and tissue engineering. These findings, however, greatly suggest that it is pivotal to propagate MSC undisturbed at normal physiologic pO_2. Working along these lines this approach may also permit modeling of common disease processes, which may be further used as potential drug discovery tools.

3 Iatrogen-Induced Dysoxygenation

Several circumstances lead to hypoxic conditions of the organism during its lifetime. For instance, decreased oxygen saturation of distinct organs and tissues follows from atherosclerotic changes of vessels. Also particular clinical treatments result in iatrogen-induced hypoxia of certain organs.

3.1 Irradiation-Induced Changes of Bone Anatomy

Ionizing radiation is but one choice in treating malignancies, which otherwise can only be medicated by means of surgery or chemotherapy. Several treatment protocols have been introduced and standardized. Irradiation of tumors is either applied before or after surgery with the intent to raze malignant cells. Inevitably, radiotherapy (RT) also impacts healthy locoregional cells and often leads to irreversible damage and concomitant deterioration of blood vessels. In due course blood circulation is greatly reduced in the respective areas, and due to deprivation of O_2 and other essential factors, the tissue gradually degenerates. Furthermore, the regenerative capacity of surrounding tissues is often greatly affected.

In the case of bone, in particular, when the mandible is being irradiated, osteoradionecrosis (ORN) is a most dreaded consequence of RT (Fig. 4). Its appearance for the head and neck varies from small asymptomatic bone exposures that may remain stable for months to years, for which healing with conservative management is possible. Severe necroses need more drastic interventions, which most often involve surgical bone reconstruction. RT with hyperfractionated irradiation at a dose of 72–80 Gy, as well as moderately accelerated fractionated radiation with doses of 64–72 Gy, was reported to result in ORN in up to 5% of the cases [34]. Besides this severe pathological consequence, irradiation most often leads to changes of bone metabolism that negatively influence healing processes (Fig. 5) [35].

The pathophysiologic concept of radiation-induced bone changes was first published by *Marx* [36], and it is by now generally accepted that the etiology involves hypovascularity, tissue hypoxia, and hypocellularity. Due to RT, irradiated bone, periosteum, and overlying soft tissue undergo hyperemia, inflammation, and endarteritis. These conditions ultimately lead to thrombosis, cell death, progressive hypovascularity, and fibrosis. In consequence, the irradiated area is in a hypoxic state, and as a consequence thereof, it becomes hypocellular eventually being devoid of fibroblasts, osteoblasts, and undifferentiated osteocompetent cells [37]. In a histological study of irradiated osteoradionecrotic mandibles, several characteristic changes became apparent. The inferior alveolar artery (the predominant arterial blood supply to the body of the mandible) as well as the periosteal arteries showed significant intimal fibrosis and thrombosis, and normal marrow was replaced by dense fibrous tissue. This was further paralleled by the loss of osteocytes [38].

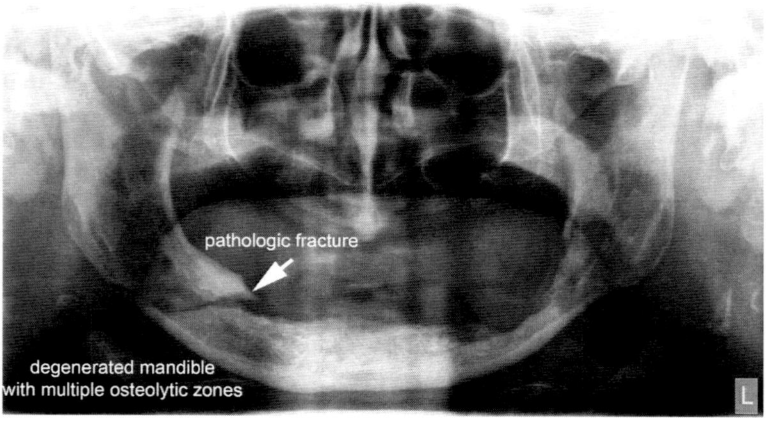

Fig. 4 Panoramic X-ray of an irradiated mandible, suffering from osteoradionecrosis. Multiple osteolytic zones with interrupted margins are encountered at the rim of the jaw's compact bone leading to spontaneous fracture (*arrow*)

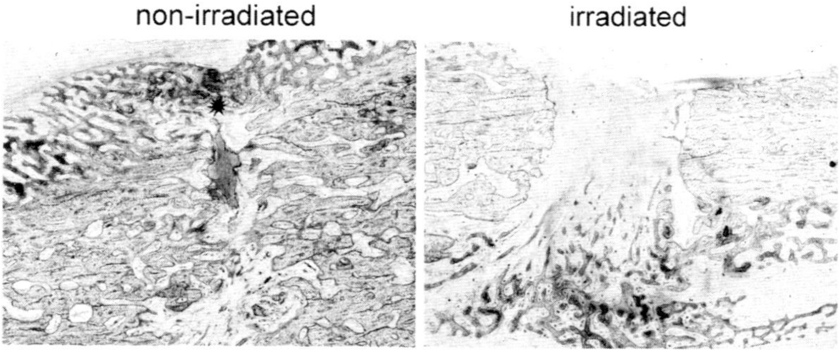

Fig. 5 Histology of irradiated pig mandible. A fracture in a pig mandible, which had preoperatively undergone fractionated irradiation of 2 × 9 Gy, was simulated by sawing the *corpus mandibulae*. The mandible was then refixated with a reconstruction plate. The specimens depicted here were derived from experimental animals, which had been sacrificed 2 weeks post-operation. At that time point, irradiated bone showed only little to no regeneration compared to control specimen, which displayed well-commenced osseous healing

3.2 Irradiation-Induced Alteration of MSC Properties

Little is known about the fate of MSC after irradiation in vivo. Somatic cells within irradiated tissue are dramatically affected due to DNA damage or through indirect cues such as the consequences of a hypoxic environment in situ. Tissue hypoxia is paralleled by an increase of reactive oxygen intermediates resulting in increased cellular oxidative stress, which leads to DNA damage, consequently yielding chromosomal instability and eventually driving cells into apoptosis [39]. Until recently only little attention has been paid whether ionizing radiation compromises MSC quality and viability. It was, however, noted that after patient irradiation and subsequent successful allogeneic bone marrow transplantation, MSC remained host-derived [40]. Further interesting in this context is the finding that bone healing could be greatly improved when applying growth factors such as bone morphogenetic protein-2 to irradiated bone in vivo [41, 42]. These observations imply that MSC remain unaffected in irradiated bone.

Irradiation in vitro has a great impact on MSC: (i) the apoptosis rate for bone marrow mononuclear cells increases from 8 to 20% after irradiation with 4 Gy [43], while MSC irradiated in vitro with 9 Gy showed an apoptosis rate of only 5.2% [44]; (ii) MSC proliferation rate is decreased and the cells show a greatly diminished response to differentiation stimuli [45]. The effect of irradiation on MSC differentiation capacity, however, remains unchanged. While a sustained inhibitory effect on the proliferation of MSC was apparent in culture for up to 2 weeks after the treatment, some of the surviving cells regained a normal proliferation potential thereafter [46]. Similarly, a few MSC were able to survive a single dose of 12 Gy, which is far more than normally applied clinically [47]. These findings indicate that a small number of cells within the stromal population hold an enhanced repair capacity or other unprecedented means to protect themselves from irradiation-induced damages. Indeed, MSC exhibit pronounced antioxidant activities; they provide an ample reactive oxygen species-scavenging capacity and possess an active double-strand break repair [44]. These features may bring by all means radioresistance to perfection and thus yield an increased survival rate for MSC after irradiation.

At least to our knowledge, there is no evidence available to date regarding the quality and fitness of MSC after irradiation in vivo. We, therefore, set out to evaluate various parameters such as number, proliferation, and differentiation capacity of MSC derived from bone that was irradiated with a total dose of 18 Gy in comparison to MSC which were irradiated in vitro only. Interestingly, there was no significant change in MSC number in bone after irradiation, and moreover, MSC explanted from irradiated bone exhibited equivalent proliferation rates (Fig. 6), as well as potent osteogenic differentiation capacities in vitro. These results suggest that MSC can cope with relatively high doses of irradiation, and repair or overcome cellular damage efficiently. Irradiation of in vitro proliferating MSC stalls their cell cycle, while the cells are still capable of differentiating into adipocytes or osteoblasts. This implies that either MSC are perfectly protected in their in vivo niche or, alternatively, dormant stem cells stalled in quiescence exhibit extraordinary repair capabilities. Yet another possibility is controlled migration of MSC toward lesioned areas. In mice irradiated with 3.5 Gy, intravenously administered human MSC showed enhanced engraftment into bone marrow, muscle, brain, heart, lungs, and liver [48]. Concordant results were obtained after low-dose irradiation of tumors in mice, which lead to MSC attraction to the tumor microenvironment [49].

In light of the latest results, however, which demonstrated that pericytes share a high degree of resemblance with MSC, it is further conceivable that MSC, which are attached to blood vessels, which are about to degenerate, appear to cope rather well with direct irradiation-induced damage as well as with the enhanced production of reactive oxygen species within the local environment. The latter greatly impacts on endothelial cells, which in turn initiate acute immune response through releasing

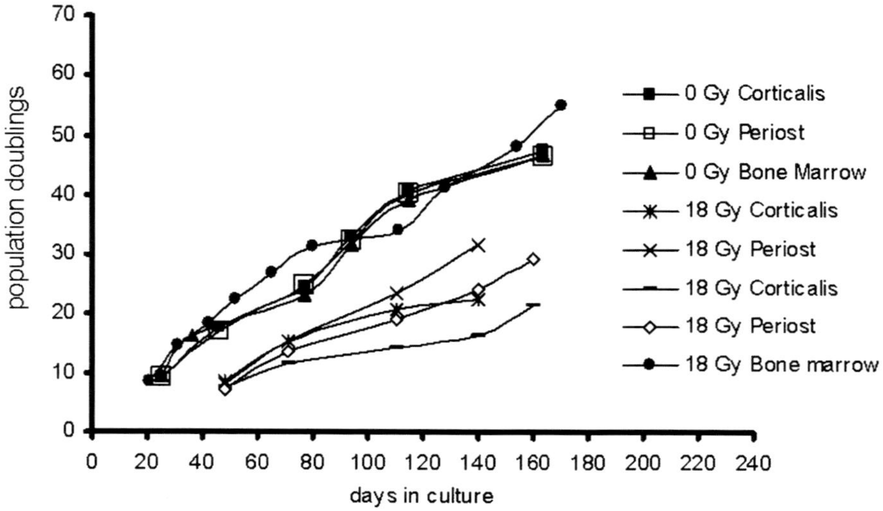

Fig. 6 Proliferation potential of primary porcine MSC. Cells were explanted from various sites of bone biopsies, which had undergone fractionated irradiation of 2 × 9 Gy. MSC which firmly attached to plastic were propagated in long-term culture and their proliferation was accounted

cytokines and further production of ROS via the recruitment of phagocytes. Vascular thrombosis and endothelial cells destruction result in microvessel necrosis, local ischemia, and consequently tissue loss. Because the endothelial cell barrier is lacking, cytokines can disseminate easier and induce differentiation of fibroblasts into myofibroblasts. Myofibroblasts proliferate excessively and produce abnormal ECM components, and the ability to destroy those products is reduced. These vastly changing environmental cues, in particular resulting in hypoxia, are furthermore predisposing the tissue to necrosis, which however only commences after a secondary impact.

4 Conclusions

The developmental fate of MSC is controlled by the availability of oxygen. Reduced pO_2, which is still high enough not to be sensed as a hypoxic insult, greatly supports proliferation of this fibroblastoid cell type. At the same time, differentiation into specialized progeny is suppressed. This concept appears to be also recapitulated during wound healing: after formation of a clot, MSC/pericytes face reduced O_2 availability. Only after revascularization of the wounded area, the propagated cells diverge from their uncommitted character, and due to the local environment, i.e., through tissue-specific cues, they are shaping up to more specialized cells, which are thus being functionally integrated into the newly formed tissue. Notably, dysoxygenated conditions do not lead to an incapacitation of tissue-inherent MSC. However, without appropriate oxygenation, MSC are restrained and thus refrain from commencing to differentiate, which in due course may cause fibrotic lesion formation.

Acknowledgments The authors are grateful for sustained collaboration with, as well as guidance and supervision during their clinical research by, Dr. Robert Gaßner. GL's work is supported by the Austrian Science Fund (FWF), NRN project S9305, and by the Austrian Research Agency (FFG). The authors greatly acknowledge the support by the Jubilee Fund of the Austrian National Bank (OeNB).

References

1. Horwitz EM, Le Blanc K, Dominici M, Mueller I, Slaper-Cortenbach I, Marini FC, Deans RJ, Krause DS, Keating A. Clarification of the nomenclature for MSC: the international society for cellular therapy position statement. Cytotherapy. 2005; 7:393–395.
2. Watt FM, Hogan BL. Out of Eden: stem cells and their niches. Science. 2000; 287:1427–1430.
3. Caplan AI. Mesenchymal stem cells. J Orthop Res. 1991; 9:641–650.
4. Barry FP, Murphy JM. Mesenchymal stem cells: clinical applications and biological characterization. Int J Biochem Cell Biol. 2004; 36:568–584.
5. Sakaguchi Y, Sekiya I, Yagishita K, Muneta T. Comparison of human stem cells derived from various mesenchymal tissues: superiority of synovium as a cell source. Arthritis Rheum. 2005; 52:2521–2529.
6. Muguruma Y, Yahata T, Miyatake H, Sato T, Uno T, Itoh J, Kato S, Ito M, Hotta T, Ando K. Reconstitution of the functional human hematopoietic microenvironment derived from human mesenchymal stem cells in the murine bone marrow compartment. Blood. 2005; 107: 1878–1887.
7. Prockop DJ. Marrow stromal cells as stem cells for nonhematopoietic tissues. Science. 1997; 276:71–74.
8. Dorshkind K. Regulation of hemopoiesis by bone marrow stromal cells and their products. Annu Rev Immunol. 1990; 8:111–137.
9. Fehrer C, Lepperdinger G. Mesenchymal stem cell aging. Exp Gerontol. 2005; 40:926–930.
10. Sethe S, Scutt A, Stolzing A. Aging of mesenchymal stem cells. Ageing Res Rev. 2006; 5:91–116.
11. Pittenger MF, Mackay AM, Beck SC, Jaiswal RK, Douglas R, Mosca JD, Moorman MA, Simonetti DW, Craig S, Marshak DR. Multilineage potential of adult human mesenchymal stem cells. Science. 1999; 284:143–147.
12. Prockop DJ, Sekiya I, Colter DC. Isolation and characterization of rapidly self-renewing stem cells from cultures of human marrow stromal cells. Cytotherapy. 2001; 3:393–396.
13. Simmons PJ, Torok-Storb B. Identification of stromal cell precursors in human bone marrow by a novel monoclonal antibody, STRO-1. Blood. 1991; 78:55–62.
14. Horwitz EM, Keating A. Nonhematopoietic mesenchymal stem cells: what are they? Cytotherapy. 2000; 2:387–388.
15. Dominici M, Le Blanc K, Mueller I, Slaper-Cortenbach I, Marini F, Krause D, Deans R, Keating A, Prockop D, Horwitz E. Minimal criteria for defining multipotent mesenchymal stromal cells. The International Society for Cellular Therapy position statement. Cytotherapy. 2006; 8:315–317.
16. Ishikawa Y, Ito T. Kinetics of hemopoietic stem cells in a hypoxic culture. Eur J Haematol. 1988; 40:126–129.
17. Antoniou ES, Sund S, Homsi EN, Challenger LF, Rameshwar P. A theoretical simulation of hematopoietic stem cells during oxygen fluctuations: prediction of bone marrow responses during hemorrhagic shock. Shock. 2004; 22:415–422.
18. Chow DC, Wenning LA, Miller WM, Papoutsakis ET. Modeling pO(2) distributions in the bone marrow hematopoietic compartment. I. Krogh's model. Biophys J. 2001; 81:675–684.

19. Cooper PD, Burt AM, Wilson JN. Critical effect of oxygen tension on rate of growth of animal cells in continuous suspended culture. Nature. 1958; 182:1508–1509.
20. Packer L, Fuehr K. Low oxygen concentration extends the lifespan of cultured human diploid cells. Nature. 1977; 267:423–425.
21. Zwartouw HT, Westwood JC. Factors affecting growth and glycolysis in tissue culture. Br J Exp Pathol. 1958; 39:529–539.
22. Ankeny DP, McTigue DM, Jakeman LB. Bone marrow transplants provide tissue protection and directional guidance for axons after contusive spinal cord injury in rats. Exp Neurol. 2004; 190:17–31.
23. Cui JH, Park K, Park SR, Min BH. Effects of low-intensity ultrasound on chondrogenic differentiation of mesenchymal stem cells embedded in polyglycolic acid-an in vivo study. Tissue Eng. 2006; 12:75–82.
24. Calvi LM, Adams GB, Weibrecht KW, Weber JM, Olson DP, Knight MC, Martin RP, Schipani E, Divieti P, Bringhurst FR, et al. Osteoblastic cells regulate the haematopoietic stem cell niche. Nature. 2003; 425:841–846.
25. Zhang J, Niu C, Ye L, Huang H, He X, Tong WG, Ross J, Haug J, Johnson T, Feng JQ, et al. Identification of the haematopoietic stem cell niche and control of the niche size. Nature. 2003; 425:836–841.
26. Shen Q, Goderie SK, Jin L, Karanth N, Sun Y, Abramova N, Vincent P, Pumiglia K, Temple S. Endothelial cells stimulate self-renewal and expand neurogenesis of neural stem cells. Science. 2004; 304:1338–1340.
27. Cotsarelis G, Sun TT, Lavker RM. Label-retaining cells reside in the bulge area of pilosebaceous unit: implications for follicular stem cells, hair cycle, and skin carcinogenesis. Cell. 1990; 61:1329–1337.
28. Crisan M, Yap S, Casteilla L, Chen CW, Corselli M, Park TS, Andriolo G, Sun B, Zheng B, Zhang L, et al. A perivascular origin for mesenchymal stem cells in multiple human organs. Cell Stem Cell. 2008; 3:301–313.
29. da Silva Meirelles L, Caplan AI, Nardi NB. In search of the in vivo identity of mesenchymal stem cells. Stem Cells. 2008; 26:2287–2299.
30. Jauniaux E, Watson A, Ozturk O, Quick D, Burton G. In-vivo measurement of intrauterine gases and acid-base values early in human pregnancy. Hum Reprod. 1999; 14:2901–2904.
31. Heppenstall RB, Grislis G, Hunt TK. Tissue gas tensions and oxygen consumption in healing bone defects. Clin Orthop Relat Res. 1975; 106:357–365.
32. Knighton DR, Silver IA, Hunt TK. Regulation of wound-healing angiogenesis-effect of oxygen gradients and inspired oxygen concentration. Surgery. 1981; 90:262–270.
33. Fehrer C, Brunauer R, Laschober G, Unterluggauer H, Reitinger S, Kloss F, Gully C, Gassner R, Lepperdinger G. Reduced oxygen tension attenuates differentiation capacity of human mesenchymal stem cells and prolongs their lifespan. Aging Cell. 2007; 6:745–757.
34. Studer G, Gratz KW, Glanzmann C. Osteoradionecrosis of the mandibula in patients treated with different fractionations. Strahlenther Onkol. 2004; 180:233–240.
35. Grünert J, Kloss D, Kloss F. Strahlenschäden. In: Krupp S, Rennekampff HO, ed. Plastische Chirurgie. Berlin: Ecomed Verlag; 2004, pp. 1–16.
36. Marx RE. Osteoradionecrosis: a new concept of its pathophysiology. J Oral Maxillofac Surg. 1983b; 41:283–288.
37. Marx RE. A new concept in the treatment of osteoradionecrosis. J Oral Maxillofac Surg. 1983a; 41:351–357.
38. Bras J, de Jonge HK, van Merkesteyn JP. Osteoradionecrosis of the mandible: pathogenesis. Am J Otolaryngol. 1990; 11:244–250.
39. Redpath JL, Gutierrez M. Kinetics of induction of reactive oxygen species during the post-irradiation expression of neoplastic transformation in vitro. Int J Radiat Biol. 2001; 77: 1081–1085.
40. Koc ON, Peters C, Aubourg P, Raghavan S, Dyhouse S, DeGasperi R, Kolodny EH, Yoseph YB, Gerson SL, Lazarus HM, et al. Bone marrow-derived mesenchymal stem cells

remain host-derived despite successful hematopoietic engraftment after allogeneic transplantation in patients with lysosomal and peroxisomal storage diseases. Exp Hematol. 1999; 27:1675–1681.

41. Hasse A, Porksen M, Schultze S, Engel A, Feyerabend T. Effect of bFGF on regeneration of distracted mandibles after radiation. Mund Kiefer Gesichtschir. 2000; 4(Suppl 2):S423–S427.
42. Wurzler KK, DeWeese TL, Sebald W, Reddi AH. Radiation-induced impairment of bone healing can be overcome by recombinant human bone morphogenetic protein-2. J Craniofac Surg. 1998; 9:131–137.
43. Bertho JM, Mathieu E, Lauby A, Frick J, Demarquay C, Gourmelon P, Gorin NC, Thierry D. Feasibility and limits of bone marrow mononuclear cell expansion following irradiation. Int J Radiat Biol. 2004; 80:73–81.
44. Chen MF, Lin CT, Chen WC, Yang CT, Chen CC, Liao SK, Liu JM, Lu CH, Lee KD. The sensitivity of human mesenchymal stem cells to ionizing radiation. Int J Radiat Oncol Biol Phys. 2006; 66:244–253.
45. Schonmeyr BH, Wong AK, Soares M, Fernandez J, Clavin N, Mehrara BJ. Ionizing radiation of mesenchymal stem cells results in diminution of the precursor pool and limits potential for multilineage differentiation. Plast Reconstr Surg. 2008; 122:64–76.
46. Li J, Kwong DL, Chan GC. The effects of various irradiation doses on the growth and differentiation of marrow-derived human mesenchymal stromal cells. Pediatr Transplant. 2007; 11:379–387.
47. Dickhut A, Schwerdtfeger R, Kuklick L, Ritter M, Thiede C, Neubauer A, Brendel C. Mesenchymal stem cells obtained after bone marrow transplantation or peripheral blood stem cell transplantation originate from host tissue. Ann Hematol. 2005; 84:722–727.
48. Francois S, Bensidhoum M, Mouiseddine M, Mazurier C, Allenet B, Semont A, Frick J, Sache A, Bouchet S, Thierry D, et al. Local irradiation not only induces homing of human mesenchymal stem cells at exposed sites but promotes their widespread engraftment to multiple organs: a study of their quantitative distribution after irradiation damage. Stem Cells. 2006; 24:1020–1029.
49. Klopp AH, Spaeth EL, Dembinski JL, Woodward WA, Munshi A, Meyn RE, Cox JD, Andreeff M, Marini FC. Tumor irradiation increases the recruitment of circulating mesenchymal stem cells into the tumor microenvironment. Cancer Res. 2007; 67:11687–11695.

Endothelial Progenitor Cells and Nitric Oxide: Matching Partners in Biomedicine

Stefanie Keymel, Burcin Özüyaman, Marijke Grau, Malte Kelm, and Petra Kleinbongard

Abstract Regeneration of damaged tissue by embryonic or adult stem cells has been and still is a topic of highest interest in experimental and clinical medicine. Adult stem cells have the ability to differentiate into a number of different cell types in dependence on their environment. Endothelial progenitor cells (EPCs) represent a population of adult stem cells. They contribute to the renewal of the largest organ of the body: the endothelium which is the inner layer of the vessel wall.

After an introduction into the topic of EPCs presenting the origin, fate, and biology of EPCs, we describe techniques of cell culture and bioassays for in vitro and in vivo testing of EPCs. Their applications in animal models and in humans are described to demonstrate the relevance of EPCs in cardiovascular medicine. On the other hand, nitric oxide (NO), a key player in cardiovascular biology, is introduced and the first findings describing the apparently powerful interactions between EPCs and NO are discussed. EPCs and NO seem to be the "Yin and Yang" of endothelial function and regeneration. We would like to illustrate recent findings of EPC biology with a special interest in their interaction with NO. Finally, we open the wide and interesting spectrum of further research activities in this field of biomedicine. If the differentiation of EPCs could be controlled in the laboratory, these cells may become the basis of cell-based therapies for cardiovascular disease. An interdisciplinary approach and the cooperation between scientists, engineers, and physicians will be able to transfer the findings of basic science into daily clinical application. Initial ideas and first steps for the realization of new concepts in EPC–NO research and therapeutic strategies with a special reference to bioengineering concepts are discussed.

Keywords Endothelial progenitor cells · Vasculogenesis · Nitric oxide · Cardiovascular disease

---

P. Kleinbongard (✉)
Institute of Pathophysiology, University of Essen Medical School, Essen, Germany
e-mail: petra.kleinbongard@uk-essen.de

G.M. Artmann et al. (eds.), *Stem Cell Engineering*,
DOI 10.1007/978-3-642-11865-4_10,

Abbreviations

BMS	bare metal stents
BM	bone marrow
BOOST	Bone Marrow Transfer to Enhance ST-Elevation Infarct Regeneration Trial
CVRFs	cardiovascular risk factors
CFUs	colony-forming units
CAD	coronary artery disease
CRP	C-reactive protein
cGMP	cyclic guanosine monophosphate
DDAH	dimethylarginine dimethylaminohydrolase
DES	drug-eluting stents
ECs	endothelial cells
eNOS; NOS III	endothelial NOS
EPCs	endothelial progenitor cells
$Enos^{-/-}$	eNOS knockout mice
FAD	flavin–adenine dinucleotide
FMN	flavin mononucleotide
FACS	fluorescence-activated cell sorting
GTP	guanosine triphosphate
Hb	hemoglobin
iNOS; NOS II	inducible NOS
LDL	low-density lipoprotein
MRI	magnetic resonance imaging
MMP-9	matrix metalloproteinase-9
MNCs	mononuclear cells
nNOS; NOS I	neuronal NOS
ADMA	N^G,N^G-dimethylarginine (asymmetric dimethylarginine)
NHA	N^G-hydroxyl-L-arginine
L-NMA	N^G-monomethyl-L-arginine
L-NAME	N^G-nitro-L-arginine-methylester
NADPH	nicotinamide adenine dinucleotide phosphate
NO	nitric oxide
NOS	nitric oxide synthase
PCI	percutaneous coronary intervention
PAD	peripheral artery disease
RBCs	red blood cells
SMCs	smooth muscle cells
SNO-Hb	*S*-nitrosated derivative of oxygenated hemoglobin
SDF-1	stromal cell-derived factor 1
O_2^-	superoxide anions
SOD	superoxide dismutase
BH_4	Tetrahydrobiopterin
TACT	therapeutic angiogenesis using cell transplantation

TOPCARE-AMI	Transplantation of Progenitor Cells and Regeneration Enhancement in Acute Myocardial Infarction
VEGF	vascular endothelial growth factor
WT	wild-type mice

1 The Role of Endothelial Progenitor Cells In Vivo

Endothelial progenitor cells (EPCs): the name of these cells tells us exactly about their fate. EPCs differentiate into endothelial cells (ECs) which in whole constitute the endothelium. The endothelium covers the luminal side of the complete vasculature, thereby covering a total surface area of about 1–7 m^2. The location of ECs on the luminal side of blood vessels allows intense interaction with blood components or other cell types of the vessel wall, e.g., smooth muscle cells (SMCs). In the current view, the endothelium is understood as a dynamic, heterogeneous, disseminated organ that possesses vital secretory, synthetic, metabolic, and immunologic functions. The endothelium also plays a pivotal role in regulating vascular tone, in modulating anti-thrombotic and anti-adhesive properties of the vessel wall, and in controlling vascular permeability.

1.1 Neovascularization

The process of blood vessel formation occurs by two different mechanisms: angiogenesis and vasculogenesis. Angiogenesis describes blood vessel growth from existing blood vessels by sprouting of differentiated ECs or intussusception of existing capillaries [1, 2]. In contrast, vasculogenesis is defined as blood vessel growth from in situ differentiating angioblasts [3]. In all living mammalians, vasculogenesis starts during embryonic development by the creation of blood islands from the mesoderm of the yolk sac. The cells in the center of these blood islands differentiate into hematopoietic stem cells, while the peripheral cells generate endothelial stem cells, the angioblasts [3]. Angioblasts are able to proliferate, migrate, and differentiate into mature ECs (Fig. 1).

Until recently, it has been assumed that vasculogenesis is limited to embryogenesis. But in 1997, Asahara et al. first described the existence of EPCs in the peripheral blood of adults. Further investigations confirmed the role of vasculogenesis during the process of postnatal neovascularization. Vasculogenesis by EPCs contributes to the physiological and pathological neovascularization of the embryonic and adult organism.

1.2 Endothelial Progenitor Cells

In accordance with the definition of the angioblast during embryonic development, EPCs are considered as bone marrow-derived committed progenitor cells which

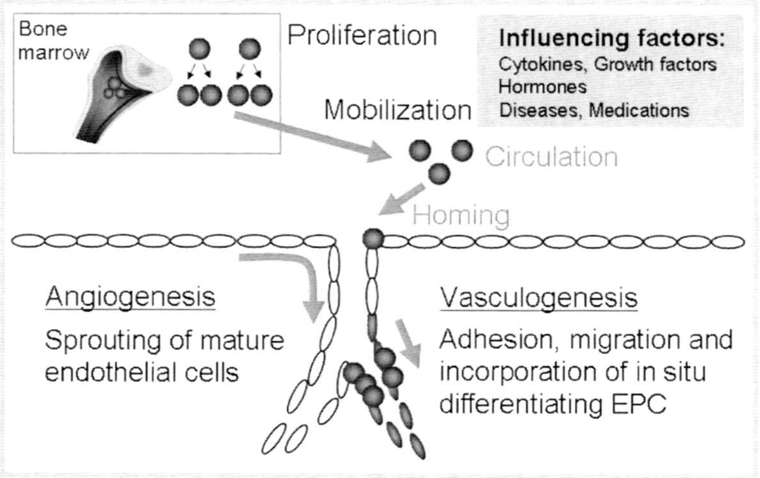

Fig. 1 Neovascularization by vasculogenesis. Vasculogenesis is defined as the formation of new blood vessels from in situ differentiating EPCs. After mobilization of EPCs from the bone marrow by endogenous or exogenous factors, EPCs circulate in the peripheral blood. EPCs adhere at sites of endothelial injury, migrate into the tissue, and differentiate in situ into mature ECs

have the ability to differentiate into mature ECs during postnatal vasculogenesis. Endothelial cell injury, e.g., after tissue ischemia or vascular injury, initiates physiological processes of reparation and regeneration. Shortly, the initial step is the mobilization of EPCs from the bone marrow (BM) into peripheral blood which is followed by the recruitment of EPCs to the site of tissue ischemia or vascular injury. Locally, vasculogenesis occurs after adhesion and migration of EPCs into the newly formed vascular network and differentiation into mature ECs (Fig. 1).

EPCs can be isolated from the peripheral blood of adults, from BM, adipose tissue, or cord blood, and represent a fraction of the mononuclear cells (MNCs). Usually, the identification and isolation of stem or progenitor cells is based on the characteristic expression pattern of surface proteins. Typical markers for the identification of EPCs include the cell surface proteins CD34, CD133, and the vascular endothelial growth factor (VEGF) receptor 2 (KDR or flk-1). Thereby, CD34 and CD133 identify the stem cell character and KDR indicates the endothelial lineage of EPCs. In most of the studies, the co-expression of CD34/KDR or CD133/KDR has been used for the quantification and isolation of EPCs. In healthy adults, circulating EPCs represent less than 0.1% of total MNCs [4–6]. Expressing this in absolute numbers means a cell count of about 10 cells/μl of blood (Fig. 2).

Specific cell culture conditions enable the enrichment of EPCs to about 70% [7]. For this purpose, isolated MNCs are cultured on fibronectin-coated culture dishes in EC-specific medium. After 7–10 days of cell culture the cells exhibit endothelial-like characteristics such as the typical spindle-shaped morphology, binding of Ulex europaeus I agglutinin, and phagocytosis of acetylated low-density lipoprotein (LDL) [7, 8]. In addition, an increased expression of characteristic endothelial cell

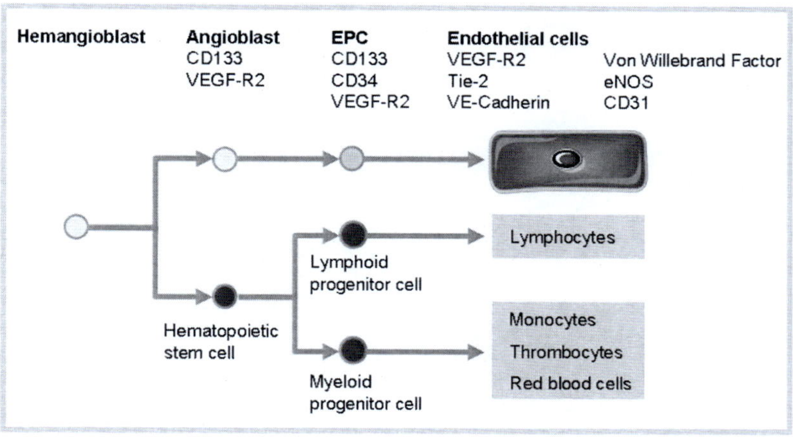

Fig. 2 From the hemangioblast to the mature endothelial cell. It is thought that the vascular and the blood system derive from a common stem cell, the hemangioblast. It forms hematopoietic stem cells for the development of blood cells and the angioblast for the endothelial cells. During the maturation process the angioblast becomes an endothelial progenitor cell which further differentiates into ECs

proteins such as CD31 [7, 8], KDR [7, 8], VE-cadherin [7], endothelial NO synthase [8], and von Willebrand factor [4] has been shown.

1.3 Mobilization of Endothelial Progenitor Cells

A successful mobilization or release of EPCs from the BM into the peripheral blood is a critical step during the process of vasculogenesis. Although the exact mechanisms of EPC mobilization into peripheral blood are poorly understood, different endogenous and exogenous factors have been identified to mobilize EPCs into the peripheral blood and therefore increase the number of circulating EPCs [9]. In accordance with the role of EPCs during the process of neovascularization in ischemic disease, tissue ischemia is considered as one of the strongest factors for EPC mobilization into peripheral blood. Correspondingly, an increased number of circulating EPCs has been found after induction of hindlimb ischemia or myocardial infarction [10, 11]. These results suggest that endogenous EPC mobilization as a response to tissue ischemia represents the initial step of physiological repair mechanisms, exaggerating the vasculogenic potential of the organism.

It has been assumed that ischemia-induced mobilization of EPCs is regulated by cytokines, e.g., VEGF. In patients suffering from acute myocardial infarction, elevated plasma concentrations of VEGF have been found [11]. In addition, in patients with burn injury or after coronary artery bypass operation, the mobilization of EPCs was associated with an increased plasma concentration of VEGF [12]. Besides the endogenous release, exogenous supply of cytokines has been used. Using an animal model, the number of circulating and cultured EPCs was significantly higher after intraperitoneal injection of recombinant VEGF than under control conditions [13]. In a clinical study treating patients with severe peripheral artery disease [14] or chronic myocardial ischemia [15], a mobilization of EPC was achieved by VEGF gene transfer. Remarkably, VEGF gene transfer also enhanced EPC function in addition to an increase in circulating cells.

Furthermore, other cytokines such as G-CSF [4] and GM-CSF [10], stromal cell-derived factor-1 (SDF-1) [16], angiopoietin [17], erythropoietin [18], estrogens [19], and HMG-CoA-reductase inhibitors (statins) [20] and physical activity [21] induce mobilization or recruitment of EPCs.

1.4 Homing of Endothelial Progenitor Cells

After mobilization of EPCs from the BM into the circulating peripheral blood, EPCs have to be recruited to sites of endothelial injury or tissue ischemia to contribute to endothelial regeneration or neovascularization. Then, EPCs adhere to the area of damaged endothelium and transmigrate into the injured or hypoxic tissue to form a new endothelial monolayer and new blood vessels. The exact mechanisms of homing of EPCs are poorly understood. It is known that EPCs clearly find their way. In a murine model of hindlimb ischemia, ex vivo culture-expanded EPCs have been stained with a fluorescent marker before transfusing them intravenously. Immunohistological analyses revealed that EPCs have been mainly recruited to the ischemic hindlimb and less in the vasculature of other organs [7]. It is thought that endothelial injury or tissue hypoxia induces the local secretion of cytokines and growth factors. The relevance of cytokines and their receptors in EPC homing was underlined by a murine model of carotid injury. For this purpose, EPCs treated with neutralizing antibodies against the cytokine receptor CXCR2 have been systemically transfused into mice after induction of an endothelial injury of the carotid artery. The receptor CXCR2 and its ligand – the proangiogenic keratinocyte-derived chemokine – are typically expressed at sites of endothelial dysfunction and contribute to endothelial recovery after endothelial injury. In these experiments, endothelial regeneration after carotid injury in animals receiving EPCs pretreated with the CXCR2-neutralizing antibody was less effective than endothelial regeneration by native EPCs [22]. Another chemokine receptor, CXCR4, is essential for migration and homing of hematopoietic stem cells. Inhibition of the SDF-1/CXCR4 interplay partially blocked the homing of progenitor and stem cells to the ischemic myocardium. Likewise, inhibition of CXCR4 in cell culture experiments significantly reduced SDF-1-induced adhesion of EPCs to mature endothelial cell monolayers and the migration of EPCs. Moreover, overexpression of SDF-1 in EPCs enhanced cell homing and incorporation into ischemic tissues. Homing of EPCs to sites of neovascularization or endothelial regeneration represents a complex process depending on a timely and spatially orchestrated interplay between chemokines, e.g., SDF-1, chemokine receptors, intracellular signaling, adhesion molecules (selectins and integrins), and proteases (reviewed by Chavakis et al. [23]).

2 In Vitro Testing of EPC and Its Use in Clinical Studies

In vitro testing of EPC properties such as EPC number and EPC function offers essential tools for the characterization of EPCs and understanding of the mechanisms in the process of vasculogenesis in experimental and clinical situations. The number of EPCs is mainly determined by flow cytometry. In accordance with the definition of EPCs, functional properties that have been described for in vitro testing include the proliferative capacity, the migration, the adhesion, or the formation of tube-like structures (Fig. 3).

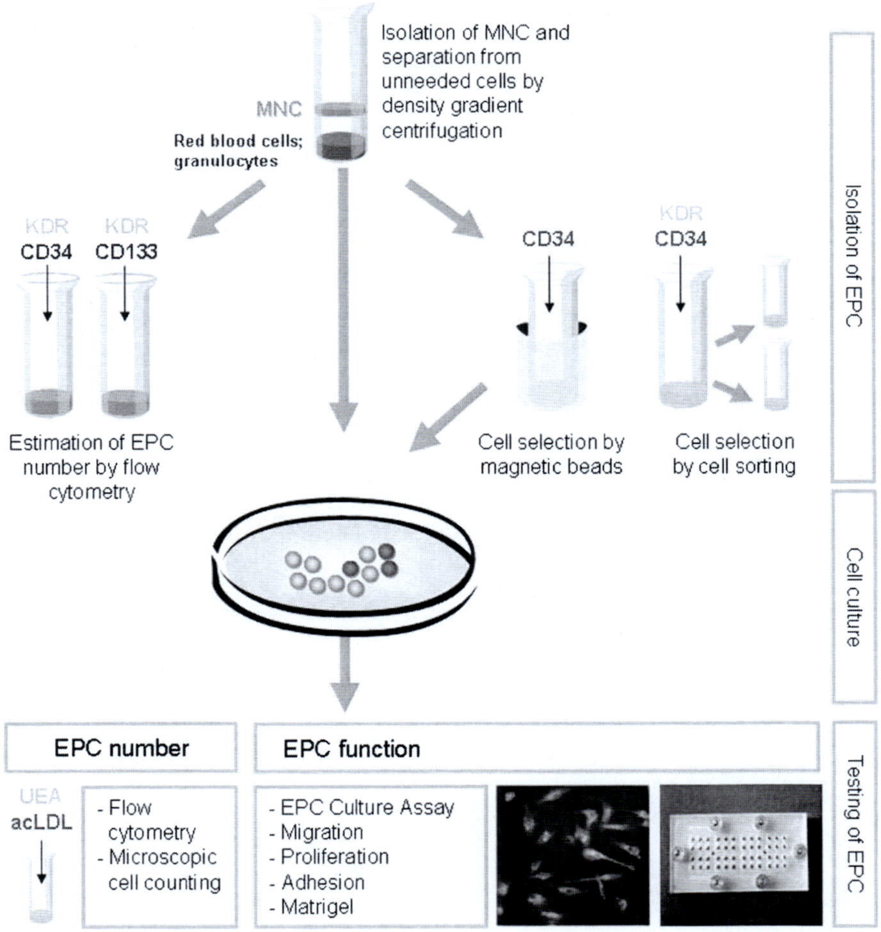

Fig. 3 Testing of endothelial progenitor cells. The figure gives an overview of the general procedures of in vitro EPC testing. First, MNCs are isolated from peripheral blood by density gradient centrifugation. Then, EPCs are further separated from MNCs by magnetic beads or by automatic cell sorting. The number of circulating EPCs is usually determined at this point. After isolation of MNCs, the cells are cultured under EC-specific conditions and can be harvested after 1 week of culture to assess the number or the function of EPCs

2.1 Methods of In Vitro Testing

Cell Number

The quantification of EPCs is mainly performed by flow cytometry. For the cellular characterization and counting, flow cytometry analyzes scattered and fluorescent light at the single-cell level. The forward and side-scattered light are dependent on cell size and cell granularity, thereby identifying different cell populations in the analyzed cell suspension. The measurement of fluorescent light properties after incubation of the cell suspension with monoclonal fluorescence-bound antibodies is used for the analysis of the expression pattern of proteins at single-cell level. In accordance with the most widely accepted definition of EPCs as CD34/KDR or CD133/KDR double-expressing cells, the quantification of EPCs by flow cytometry is carried out after incubation with the corresponding fluorescence-bound antibodies and counting of those cells expressing both markers.

Cell Culture

As EPCs represent a rather small fraction of MNCs in peripheral blood, cord blood, BM, or peripheral tissue, e.g., spleen, an enrichment of EPCs by cell culture is reasonable before further EPC testing. For this purpose, MNCs have to be isolated from the sample which is usually performed by density gradient centrifugation. After isolation of MNCs, a further selection of EPC is possible by magnetic bead selection or cell sorting. Both selection methods are based on characteristic surface markers of EPCs also applied for quantification of EPC cell number. Magnetic beads are coated with antibodies, and after binding of the beads on the target cells, they are separated by a magnet. A common protocol for EPC culture has not been established yet. Although the conditions and techniques of the cell culture may differ between study groups, they all aim to enrich endothelial-like cells (reviewed by Hirschi et al. [24]). In general, isolated MNCs or selected cells are plated on fibronectin-coated culture dishes in EC-specific medium. This cell culture medium contains a cocktail of different cytokines and growth factors that are critical for the differentiation of ECs, for example, VEGF, fibroblast growth factor, insulin-like growth factor, and human epidermal growth factor. After 7–10 days in culture, EPCs exhibit typical EC morphology and expression patterns as described above and culture-enriched EPCs can be harvested for EPC testing (Fig. 4).

Proliferation

The proliferative potential and self-renewal of progenitor or stem cells represents a specific criterion for this group of cells. In comparison to pluripotent stem cells, EPCs possess a rather low proliferative potential. A number of methods have been developed to study cell viability and proliferation in different cell populations. In general, measurement of proliferation can be performed either by analysis of metabolic activity [25] or by quantification of DNA synthesis, e.g., detection of incorporated labeled DNA precursors after DNA synthesis.

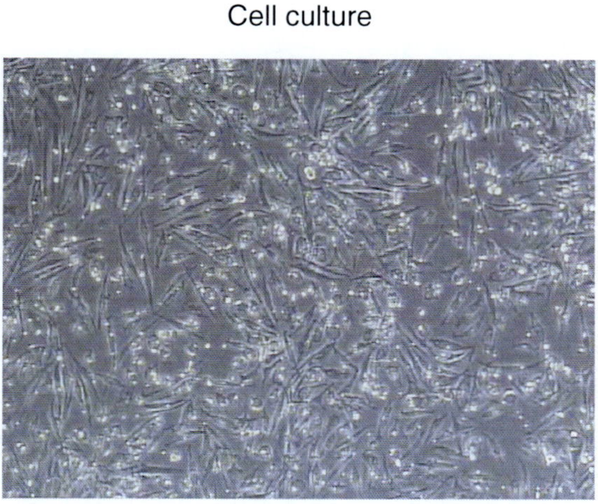

Fig. 4 Cell culture of EPCs. After isolation, the cells are *round shaped* and mostly non-attached. After 7 days of EC-specific culture conditions, EPCs exhibit typical spindle-shaped morphology and are firmly attached to the fibronectin-coated culture plate

Colony-Forming Units

The number of colony-forming units (CFUs) as a functional assay for the characterization of EPCs has been first described by Jonathan Hill in 2003 [26]. The morphology of CFUs resembles the blood islands in the yolk sac during embryonic development. Accordingly, a CFU consists of multiple thin, flat cells radiating from a central cluster of rounded cells. To test the formation of CFUs, isolated peripheral blood MNCs are cultured in growth medium on fibronectin-coated dishes and non-adherent cells are replated after 2 days. After additional 5 days of cell culture, the number of CFUs can be counted (Fig. 5).

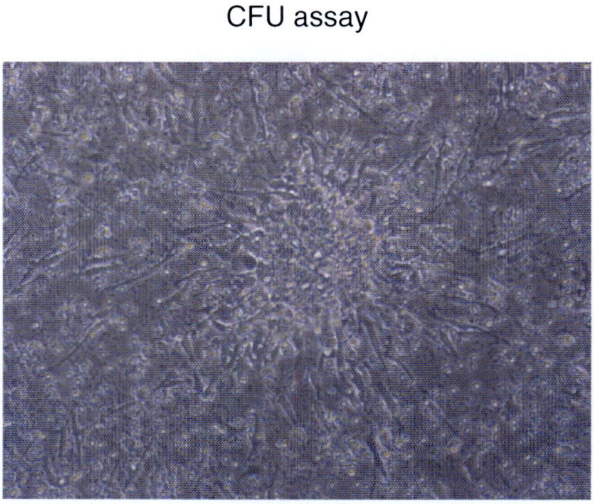

Fig. 5 Colony-forming units. A CFU consists of a center of round cells which are surrounded by elongated cells spreading from the center to the periphery

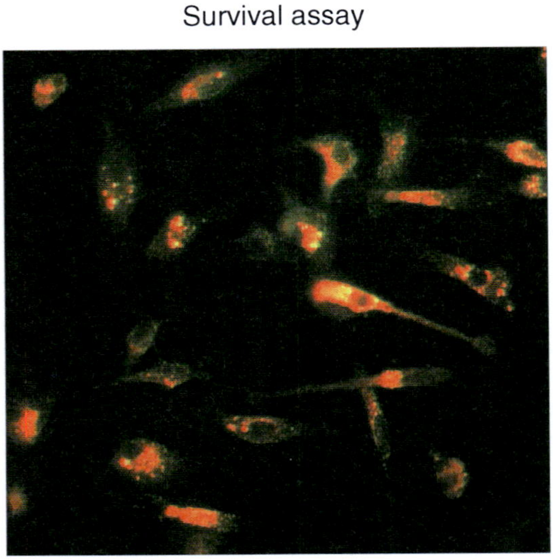

Fig. 6 Double staining of EPCs for the survival assay. The number of cells present after a defined time of cell culture can be microscopically counted by the co-expression of Ulex europaeus I agglutinin (*green*) and acetylated LDL (*red*)

Survival

The EPC survival assay detects the number of EPCs after a specific period of cell culture time, thereby reflecting a composite of EPC properties such as proliferation, apoptosis, and adhesion of EPCs on the culture dish and differentiation [7]. The principle of the survival assay includes the seeding of a defined number of EPCs and the culturing for 4–7 days under EC-specific conditions. Then, attached cells are identified as EPCs by their co-expression of Ulex europaeus I agglutinin and acetylated LDL. The number of double-positive cells can be counted by fluorescence microscopy (Fig. 6) or flow cytometry. Of note, some study groups apply the described CFU assay and survival assay for the quantification of EPCs instead of functional characterization.

Migration

Migration of EPCs is assumed as one of the key functions of EPCs. Testing EPC migration under in vitro conditions has been performed in diverse experimental and clinical studies. In general, the migratory capacity is defined as the migration of EPCs through a separating layer toward a chemoattractive substance. A modified Boyden chamber or microchemotaxis chambers can be used for this purpose [27]. The lower chambers are filled with the chemoattractant substance and the upper chambers are filled with the cell suspension containing a defined number of cells. The upper chambers are constructed without a bottom so that upper and lower chambers are separated only by the membrane with a defined pore size. After incubation, migrated cells in the lower chamber or migrated cells adherent to the lower side of the membrane can be counted.

Adhesion

In the process of vasculogenesis, homing and adhesion of EPCs represents the initial step of blood vessel formation. Although EPC adhesion plays a key role in vasculogenesis, only few studies investigated the relevance and regulation of EPC adhesion. The functional testing of EPC adhesion and the analysis of the expression pattern of adhesion molecules on the surface of EPCs can be distinguished. Functional testing of EPC adhesion can be performed under static or under flow conditions. Adhesion behavior can be tested as attachment of EPCs on extracellular matrix or on cells, e.g., ECs. To test EPC adhesion under static conditions, a culture plate is covered with an extracellular matrix or a cell monolayer. Then, a defined number of EPCs are transferred to the culture plate. After incubation, the non-adherent cells are carefully washed and the adherent cells are microscopically counted [28]. For testing EPC adhesion under flow conditions, so-called flow chambers are applied [22]. These flow chambers are connected to a pump that enables a regular, one-way flow of the cell suspension over the adhesion material. Applying video-microscopic analyses, the complex behavior of cell adhesion such as rolling, attachment, and firm adhesion can be recorded.

Analysis of the expression pattern of adhesion molecules occurs at mRNA or protein level. As adhesion molecules are expressed on the surface of the EPCs for communication with other cells, flow cytometry seems to be the most convenient method for the characterization of the expression of adhesion molecules. Typical markers for EPCs are, e.g., the fibronectin receptor $\alpha_5\beta_1$ and the integrin $\alpha_v\beta_5$.

Matrigel Assay

The Matrigel assay is used for in vitro assessment of the formation of tube-like structures. Matrigel consists of a mixture of gelatinous protein secreted by mouse tumor cells. The structure of Matrigel is similar to that of the extracellular matrix and therefore provides an adequate environment for ECs to form three-dimensional tube-like structures that are highly suggestive of the microvascular capillary network. At low temperatures, Matrigel appears as a gel and can be easily dispensed into a culture plate. When incubated at 37°C, Matrigel becomes a compact structure forming scaffolding for the co-incubated EPCs. The formation of tube-like structures by EPCs can be enhanced by additional culture of mature ECs but needs a permanent staining of EPCs for later analysis. The extent of network formation is semi-quantitatively assessed by microscopic analysis [28].

2.2 Testing the Endothelial Progenitor Cell Number and Function for Risk Stratification in Patients

Cardiovascular risk factors (CVRFs) such as hypercholesterolemia, arterial hypertension, diabetes mellitus, and smoking or the presence of atherosclerotic disease are believed to affect the properties of EPCs. Early reports indicated an inverse correlation between the number of EPCs and the number of CVRFs with decreasing EPC numbers in individuals with increasing CVRFs [6, 26]. Several clinical studies investigated the prognostic value of the EPC number in peripheral blood in patients with coronary artery disease (CAD). It was demonstrated that the level of CD34/KDR-positive EPCs predicted the occurrence of cardiovascular events and death from cardiovascular causes underlining the potential use of EPC as a biomarker for cardiovascular disease [29].

A variety of studies associated the *lipid metabolism* with the properties of human EPCs. Hypercholesterolemia leads to early changes of the vessel wall and the ECs. CFU numbers of ex vivo cultured EPCs are significantly reduced in study subjects with elevated serum cholesterol levels [26]. In patients suffering from CAD, LDL cholesterol inversely correlated with the number of circulating EPCs [6]. In addition, the functional characteristics of isolated EPCs, such as proliferation, migration, adhesion, and in vitro vasculogenic capacity, were also impaired in patients with hypercholesterolemia [5, 30]. *Arterial hypertension* forces the blood pushing against the arterial walls and therefore induces vascular damages. Among other risk factors, hypertension has been shown to be the strongest predictor of EPC migration [6]. *Diabetes mellitus* is known to present an impaired capacity of ischemia-induced neovascularization, for example, after myocardial infarction [31]. It has been demonstrated that EPC numbers are reduced in both type 1 and type 2 diabetes [28]. Interestingly, such conditions impaired EPC proliferation, adhesion, and angiogenic properties [28]. *Smoking* is a predictor of reduced EPC numbers [6]. The number of circulating EPCs inversely correlated with the number of cigarettes consumed [32]. The EPCs of massive smokers died during early phases of culture [32]. After smoking cessation, the number of EPCs increased again [32]. However, if smoking was resumed, EPC number rapidly decreased to levels seen before smoking cessation [32]. Interestingly, nicotine dose-dependently affected EPCs with cytotoxic effects at higher nicotine concentrations [33]. *Aging* represents a CVRF that cannot be modified but should always be considered in the interpretation of EPC properties or in planning of EPC-based therapy. The number of circulating EPCs showed an inverse correlation with age [6]. In addition, a functional impairment of EPCs such as reduced proliferative or migratory activity occurred with aging [6, 34, 35].

Interestingly, CVRFs also influenced vascular nitric oxide (NO) levels. Increased numbers of CVRFs were associated with a decrease in NO bioavailability [36–38]. The correlation between CVRFs, the NO metabolism, and EPCs is discussed in Sect. 5.2.

3 In Vivo Testing of Endothelial Progenitor Cells

While in vitro testing of EPCs allows detailed information about distinct EPC properties, in vivo testing displays the functional role of EPCs in a whole living organism. Using in vivo models allows the investigation of EPC mobilization and cell tracking of endogenously mobilized or exogenously applied EPCs to the site of incorporation. Cell tracking is possible via transplantation of genetically modified BM or by ex vivo staining of EPCs. EPCs can be identified at the site of incorporation by immunohistological staining, thereby evaluating the contribution of EPCs to the endothelial regeneration or neovascularization. In addition, morbidity and mortality of study objects display the relevance of EPCs for the organism in different disease states. Most of the experimental studies investigated the regenerative potential of EPCs in ischemic diseases as a state of acute or chronic tissue hypoxia by obstruction of the feeding artery. Animal and human studies showed that EPCs contribute to the neovascularization and rescue of ischemic organs, which is described in detail in the following sections.

3.1 Animal Studies and the Relevance of Endothelial Progenitor Cells in Neovascularization

The number of different animal models that investigate the role of EPCs in cardiovascular diseases is high and goes beyond the scope of this chapter. Typical in vivo models for testing EPCs include animal models investigating ischemic disease and endothelial injury. In the murine model of *hindlimb ischemia*, human EPCs were transplanted after induction of unilateral hindlimb ischemia. In comparison to the transplantation of mature human microvascular EC or culture medium (control mice), the transplantation of EPCs reduced the extent of cell death and the number of autoamputations. The transplantation of EPCs also resulted in increased capillary density as a result of successful neovascularization and increased blood flow. Interestingly, the transplanted EPCs were nearly completely found in the ischemic hindlimb [7]. Accordingly, in a murine model of *focal cerebral ischemia*, EPCs are found in the boundary of the cerebral ischemic lesions [39]. After these promising results, EPC therapy was performed in nude rats after *acute myocardial infarction*, induced by the ligation of coronary arteries. Intravenous administration of ex vivo expanded human EPCs inhibited fibrosis and preserved cardiac function. Comparable to the hindlimb model, the infused EPCs were recruited to the ischemic area and contributed to new vessel formation [40]. Besides vasculogenesis, which is usually understood as the formation of blood vessels from circulating progenitor cells, EPCs are also responsible for the endothelial regeneration or re-endothelialization after vascular injury [41–43]. As an example, in a murine model of *endothelial injury* of the carotid artery, the transfusion of EPCs resulted in an accelerated re-endothelialization of the injured site and subsequently to a reduced extent of neointima formation [43]. In addition, EPCs colonized the surface of vascular implants or left ventricular assistant devices [4], thereby inhibiting the adhesion of leukocytes or thrombus formation. The first positive results support the importance of endothelial progenitors in ischemic disease such as peripheral artery ischemia, stroke, and myocardial infarction and lead to the initiation of several preclinical and clinical studies.

3.2 Human Studies and the Potential Role of EPCs as a Therapeutic Target

The number of clinical studies testing EPCs under in vivo conditions is rather low until now. EPCs have been applied as a therapeutic target for acute or chronic ischemic disease. Since the initial preclinical studies were promising, small-scale clinical trials of BM, MNC, and EPC transplantation for the treatment of myocardial infarction, peripheral limb ischemia, and heart failure as a result of myocardial infarction were conducted (Fig. 7). These studies provided preliminary evidence of feasibility and safety of cell transplantation. First approaches were conducted in patients suffering from myocardial infarction as a result of coronary artery occlusion. The easiest way of applying such cells is the direct injection of the cells into the infarcted myocardial areas. This was performed by applying autologous BM cells into the viable myocardium during bypass vessel surgery, while the chest of the patients was open (Fig. 8). Three of five patients experienced a long-term improvement of myocardial perfusion [45]. Autologous $CD133^{+}$ BM cells injected into the infarct border led to an improvement of global left ventricular heart function, and blood perfusion

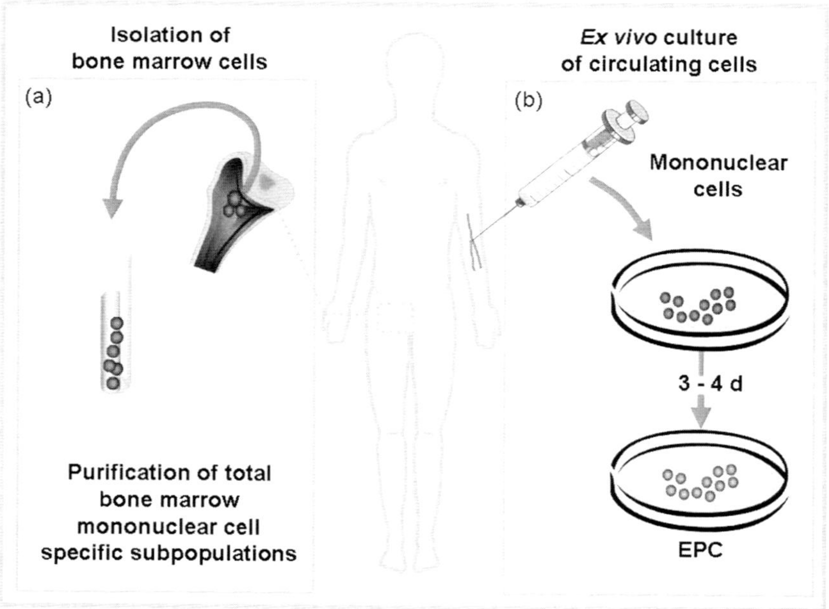

Fig. 7 Enhancement of stem cells and EPCs for therapeutic application. (**a**) BM-derived MNCs include an enriched number of stem cells. MNCs are withdrawn from the BM by puncturing the pelvic bone. Then, cells are isolated using standard centrifugation procedures and rapidly re-infused into the patient. (**b**) EPCs are isolated from peripheral blood. The MNC fraction is gained by density gradient centrifugation of the blood. Afterward, these cells are cultured for 4–7 days and then applied for review see Dimmeler et al. [44])

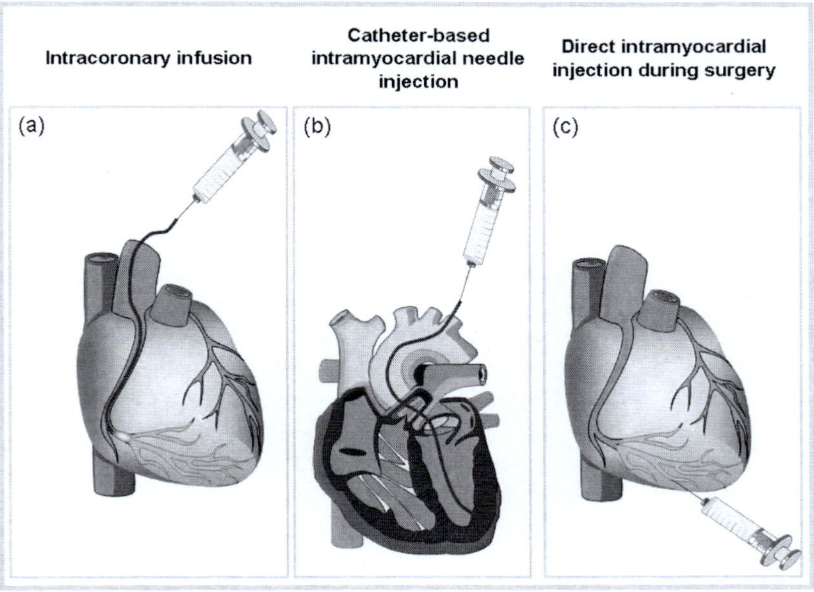

Fig. 8 Stem cell application procedures to the heart. To date, three different application procedures exist: (a) the intracoronary infusion, (b) the intramyocardial application, and (c) the direct injection during surgery (for review see Dimmeler et al. [44])

of the infarcted area was evident. Adverse effects of cell application, e.g., cardiac arrhythmia, could be disclosed even after 9 months of follow-up [46].

Intracoronary application of the cells during cardiac catheterization improved global left ventricular heart contractility in comparison to patients treated with standard therapy for myocardial infarction [47]. In the more elaborate TOPCARE-AMI (Transplantation of Progenitor Cells and Regeneration Enhancement in Acute Myocardial Infarction) trial [48, 49], ex vivo culture-expanded BM-derived MNCs or culture-enriched EPCs derived from peripheral blood MNCs were infused into the coronary arteries of a randomized group of 20 patients with acute myocardial infarction. After 4 months, left ventricular contractility and wall motion in the infarct zone were significantly enhanced and the coronary blood flow was improved. The final 1-year results confirmed a sustained improvement in left ventricular contractility as well as a reduction in areas of dysfunctional myocardium in treated groups [48]. The beneficial effects of MNC transplantation on postinfarct heart contractility restoration seemed to be highly correlated with the migratory capacity of these cells [50]. The randomized, controlled BOOST (Bone Marrow Transfer to Enhance ST-Elevation Infarct Regeneration Trial) [51] evaluated the therapeutic effect of BM cell transfer in 60 patients with myocardial infarction undergoing percutaneous coronary intervention (PCI). In this trial, global left ventricular contractility increased after 6 months of autologous BM-derived MNC transfer [51]. However, the difference in left ventricular contractility between the groups was not significant after 18 months of follow-up, despite an acceleration of left ventricular function recovery by cell therapy. Another group published similar results of intracoronary infusion of BM-derived MNCs with estimated numbers of $CD34^+$, $CD117^+$, and $CD133^+$ subpopulations in patients with myocardial infarction [52]. Significant improvement of the heart function (left ventricular contractility) was found as measured by magnetic resonance imaging at 6 months [53]. Although stem cell transplantation was associated with a reduction in infarct size, no significant improvement of left ventricular function was observed during the 4-month follow-up period. However, the cells were delivered within 24 h after myocardial infarction which is significantly earlier when compared to above-mentioned studies. This points toward the importance of timing of stem cell transplantation. In another prospective study which included patients with *end-stage ischemic heart disease*, the transendocardial intramyocardial injections of autologous BM-derived MNCs confirmed the safety and effectiveness of the treatment. Interestingly, at the 2- and 4-month follow-up periods, 21 patients in the treatment group experienced less heart failure and fewer anginal symptoms as well as an enhancement in myocardial perfusion and ventricular function [54]. Another important manifestation of atherosclerosis besides CAD is *peripheral arterial disease* (PAD). Leg ischemia caused by severe PAD was the target of the TACT (Therapeutic Angiogenesis Using Cell Transplantation) study [55]. Four weeks after autologous bone marrow-derived MNC injection into lower leg muscles, the blood perfusion in the legs significantly improved in patients treated with cells but not in patients treated with placebo. In line with this result, an increase in the number of collateral vessels in the treated legs during the 24-week duration of the study was demonstrated. Pain-free walking as a result of the improved blood supply significantly improved during the follow-up period.

3.3 How Do EPCs Act?

The mechanisms by which adult progenitor cells acquire their cell fate are discussed controversially. Differentiation, transdifferentiation, and cell fusion have been proposed as possible modes of cell type changes. The *differentiation* of EPCs results in the generation of endothelial cells, which has been definitively proved by the typical morphology and expression of EC-specific markers. The *transdifferentiation* into cardiomyocytes has been originally proposed as another mechanism underlying their above-mentioned therapeutic action [40, 56, 57], whereas other investigators have failed to detect permanent engraftment and transdifferentiation of transplanted BM-derived MNCs [58, 59]. To date, it was not possible to reproducibly induce a functional cardiomyocyte phenotype in BM-derived MNCs. *Cellular* or *nuclear fusion* is considered as an alternative for cell reprogramming. In cell culture, a cell fusion of BM-derived donor cells with recipient cardiomyocytes has been reported and suggested to be a contributory mechanism [60, 61]. Co-culture experiments of EPCs with cardiomyocytes detected ultrafine intercellular connections between the cell types. These nanotubular structures may be a sign for cell fusion and might lead to an exchange of macromolecules. However, the frequency of cell fusion is also debated [62]. The latest findings propose that the functional benefits might be

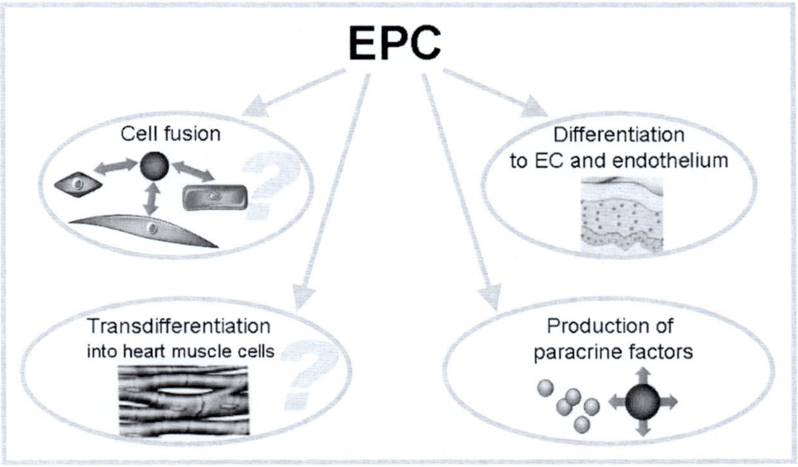

Fig. 9 Proposed mechanism of EPC action. EPCs seem to differentiate into ECs and lead to the production of paracrine factors. However, differentiation into cardiomyocytes or the cellular or nuclear fusion is still a matter of debate

related to the secretion of soluble factors that, acting in a paracrine fashion, protect the heart, induce neovascularization, and promote tissue regeneration [63, 64] (Fig. 9).

4 A Short Survey of Nitric Oxide and Its General Effects

NO or nitrogen monoxide is a short-living diatomic free radical. This gas represents an important signaling molecule in the body of mammals including humans. NO is highly reactive and diffuses across cell membranes [65]. Therefore, NO is able to act in a paracrine (between adjacent cells) and an autocrine (within a single cell) manner after its liberation [66]. As an important messenger molecule, NO is involved in many physiological and pathological processes within the mammalian body, both beneficial and detrimental [67, 68]. An appropriate level of NO production is essential for homeostasis of organ structure and function, and NO plays a key role in the regulation of vascular reactivity and blood flow. NO is also important for neurotransmission, immunomodulation, axon outgrowth, cellular growth, apoptosis, proliferation, and differentiation. However, sustained levels of NO production result in direct tissue toxicity and contribute to the vascular collapse associated with septic shock. A chronic expression of NO is associated with various carcinomas [69].

NO as a "Second Messenger Molecule"

NO affects the biology of living cells via several mechanisms. On the cellular basis it exhibits multiple activities such as the direct interaction with proteins in terms of oxidation of iron-containing proteins such as ribonucleotide reductase and aconitase, activation of the soluble guanylate cyclase, ADP ribosylation of proteins, nitrosylation of protein sulfhydryl group, and activation of iron regulatory factor [70, 71]. Another biologically important reaction of NO is the *S*-nitrosylation, a conversion of thiol groups – including cysteine residues in proteins – to form *S*-nitrosothiols. *S*-Nitrosylation is a mechanism for dynamic, post-translational regulation of proteins. Additionally, NO has been demonstrated to be an important transcription factor [72, 73]. After the identification of NO as a fundamental player in the fields of cardiovascular science, neuroscience, physiology, and immunology, NO was proclaimed "Molecule of the Year" in 1992 [74].

NO Determines Vascular Function

In 1998, Furchgott, Ignarro, and Murad won the Nobel Prize for their contribution in determining NO-dependent functions. NO functions as an "endothelium-derived relaxing factor" by relaxing the vascular SMCs, which results in vasodilation and increased blood flow. NO diffuses into the vascular SMCs adjacent to the endothelium where it binds to the heme moiety of the enzyme guanylyl cyclase and leading to its activation. This enzyme catalyzes the dephosphorylation of guanosine triphosphate (GTP) to cyclic guanosine monophosphate (cGMP), which serves as a second messenger for diverse cellular functions, particularly for signaling smooth muscle relaxation. In the vasculature, further important actions of NO are well known. Reduced NO availability results in (I) vasoconstriction (e.g., coronary vasospasm, elevated systemic vascular resistance, arterial hypertension) and (II) thrombosis due to platelet aggregation and adhesion to vascular endothelium or (III) inflammation due to upregulation of leukocyte and endothelial adhesion molecules.

4.1 The Enzymatic Source of Nitric Oxide in Mammals

NO synthases (NOS) are the primary sites of endogenous NO synthesis in mammals. NOS catalyzes monooxidation of the first substrate, L-arginine, and forms N^G-hydroxyl-L-arginine (NHA). The enzyme then catalyzes monooxidation of the second substrate NHA and generates NO and L-citrulline as final products (Fig. 10).

NOS Isoforms and Cofactors

The NOS family consists of three isoforms: neuronal NOS (nNOS; NOS I), endothelial NOS (eNOS; NOS III), and inducible NOS (iNOS; NOS II). All isoforms are composed of two domains: the oxygenase domain and the reductase domain. The oxygenase domain contains a cytochrome P450-like heme–thiol complex, substrate, and cofactor tetrahydrobiopterin (BH_4) binding sites and the reductase domain contains nicotinamide adenine dinucleotide phosphate (NADPH), flavin–adenine dinucleotide (FAD), and flavin mononucleotide (FMN) binding sites. A

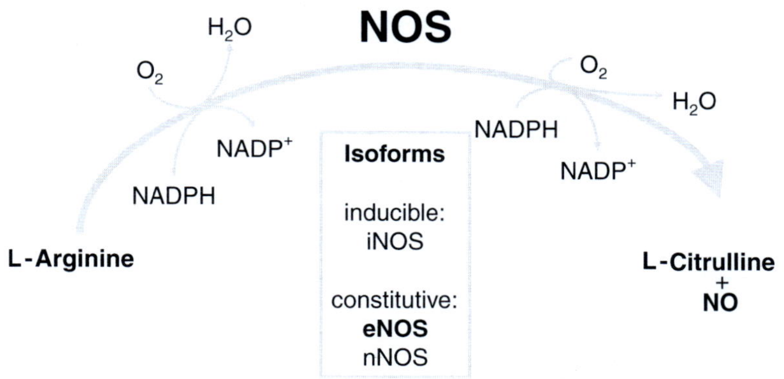

Fig. 10 The L-arginine–NO pathway. Nitric oxide is produced from L-arginine in a process of five electron oxidation of the terminal guanidino nitrogen of L-arginine. This process involves molecular oxygen and NADPH as co-substrates and requires other redox cofactors

calmodulin (CaM) binding site is located between these two domains. NOS catalyzes the oxidation of L-arginine, resulting in the formation of NO and L-citrulline (Fig. 10). Active NOS enzymes are composed of a dimeric structure and their activity requires additional factors such as NADPH, BH_4, FAD, FMN, heme, and CaM. The expression of eNOS was first observed in ECs but it is also present in other cell types. Additionally, a constitutive NOS activity under resting conditions is also described in red blood cells (RBCs) [75]. Concerning functional aspects, RBC NOS is similar to eNOS but the exact structure of RBC NOS is still under investigation. Interestingly, studies have shown that eNOS is also expressed in EPCs (see Sect. 5.1).

"Dysregulation" of NO Synthase in Disease

Under certain pathophysiological conditions, e.g., in cardiovascular disease, a NOS cofactor-dependent upregulation or downregulation occurs. For example, a reduction of the essential cofactor BH_4 is associated with a so-called eNOS uncoupling which means that the eNOS dimer is disrupted. Under this condition, eNOS produces superoxide anions (O_2^-) instead of NO. Then, O_2^- forms highly reactive peroxynitrite which induces nitration of proteins [76, 77]. This process becomes particularly significant when the superoxide dismutase (SOD) is unable to scavenge superoxide, which is often described as oxidative stress. Uncoupling of eNOS has been shown in a variety of vascular disease states, for example, in diabetes mellitus. Diabetes mellitus is associated with decreased NO bioavailability, increased superoxide production, and disrupted eNOS dimer formation within the vascular wall, while eNOS mRNA and/or protein levels are maintained or even increased [78].

Another crucial regulation factor of eNOS is the circulating endogenous inhibitor N^G,N^G-dimethyl-L-arginine (or asymmetric dimethyl-L-arginine; ADMA). ADMA is an arginine analogue and therefore competes with the substrate L-arginine for the binding sites of the NOS. ADMA is metabolized to L-citrulline by the intracellular enzyme dimethylarginine dimethylaminohydrolase (DDAH). As the concentration of DDAH in the blood changes in disease states, it affects ADMA concentration and by this NOS activity. Increased concentration of ADMA in cardiovascular patients reduces the NO bioavailability during disease processes [79].

4.2 NO Synthase-Independent Pathways of NO

NO reacts in several ways, including the oxidation to nitrite and further to nitrate, nitrosation, or the formation of iron–nitrosyl compounds [80]. RBCs are believed to be a major sink for NO caused by the rapid co-oxidation of NO with oxygenated Hb to form methylated Hb and nitrate [81]. Another possibility is the reaction of NO or a higher oxidation product such as NO_2 or N_2O_3 with cysteine-93 of the β-globin chain of Hb, leading to the formation of SNO-Hb, the *S*-nitrosated derivative of oxygenated Hb [82]. SNO-Hb has been suggested to participate in the regulation of blood flow [83] which converted NO from the inactive scavenger into a "transporter" of bioactive NO. NO is also generated during the reaction of nitrite with Hb [84].

4.3 The Role of Nitric Oxide in Cell Viability

NO affects cell differentiation and cell vitality as it possesses both anti- and pro-apoptotic properties, depending on the concentration and flux of NO and depending on the source of NO delivery. At low concentrations, NO enhances cell differentiation and at high concentrations, it inhibits differentiation and development (Fig. 11).

High NO concentrations promote apoptosis in most cases; low NO concentrations can result in resistance to cell apoptosis. NO induces apoptosis in macrophages, thymocytes, pancreatic islets, certain neurons, and tumor cells. The factors affecting cell-specific sensitivity to NO-mediated apoptosis can be

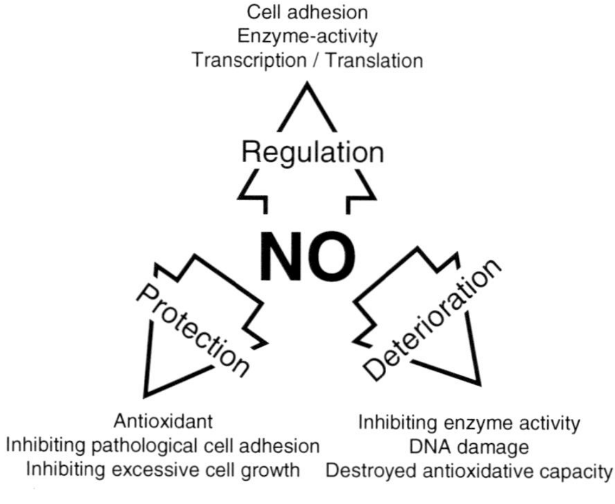

Fig. 11 Effects of NO on cell viability

associated with the redox state within the cells, the activation of the apoptotic signaling cascade such as caspases, or the release of mitochondrial cytochrome *c*. But the threshold of the NO level triggering apoptosis is different between cell types. NO produces peroxynitrite in response to a rapid reaction with superoxide, a powerful oxidant with sufficient stability to diffuse through cells and react with targets. Enhanced formation of peroxynitrite induces DNA damage. In the case of apoptosis promotion, the physiological relevant concentration of peroxynitrite induces apoptosis in human leukemia cells but fails to affect normal human ECs and MNCs [85–87].

NOS containing ECs, macrophages, and the cell components of the BM play an active role in the modulation of stem cell growth. The endogenous concentration of NO stimulates cell differentiation. Therefore, the different components of human BM such as fibroblasts, fat cells, macrophages, and ECs may provide the preferable microenvironment for a rapid expansion of the lineage-restricted progenitor cells. Accordingly eNOS-deficient mice show a defect in progenitor cell mobilization [88]. These findings indicate that eNOS, expressed by the endothelium, determines the recruitment of stem and progenitor cells.

5 Nitric Oxide and Endothelial Progenitor Cells

A potential relevance of NO for EPCs is conceivable since NO is one of the most important endothelial mediators. Further, EPCs are a cell population within the NOS expressing mononuclear cell fraction of the blood. Therefore, the first investigations in the field of EPC research included the detection of a potential NOS and its involvement in basic EPC biology. Currently, the interplay between NO availability and EPC function in disease states is under investigation.

5.1 A Nitric Oxide Synthase in Endothelial Progenitor Cells

Asahara et al. first described EPCs in humans and demonstrated that EPCs express NOS mRNA which is translated into a functionally active protein [8]. mRNA as well as the detected protein complied the endothelial-type isoform [8]. The finding of eNOS mRNA expression during EPC cultivation was confirmed by other groups [89]. Shear stress was applied to EPCs to receive more information about this NOS type. In the vascular system, shear stress is understood as the force which is applied parallel or tangential to the vascular wall and largely depends on blood flow. Shear stress is one of the major stimuli of endothelial NOS. Interestingly, application of shear stress on EPCs in vitro induced an upregulation of eNOS expression [6]. This finding underlines that the eNOS protein in EPCs represents a functionally active enzyme. Indeed, NO production in response to the classical receptor-dependent eNOS activators acetylcholine and VEGF was detected within differentiated EPCs [8]. In late endothelial outgrowth cells which were cultured for 4 weeks, a direct

NO release was detected by DAF-2, a fluorescent dye that detects liberated NO. Enzymatic NO liberation was shown by the inhibition of the NO signal after application of the pharmaceutical NOS inhibitor N^{G}-nitro-L-arginine-methylester (L-NAME) [90].

5.2 EPCs and NO In Vitro and In Vivo

To understand the relevance and mechanisms of NO action on EPC biology, different in vitro tests have been performed. A well-structured research of NO-dependent regulation of EPC number and function is still pending. Conducted in vitro assays include the analysis of expression patterns and biochemistry as well as the determination of EPC number and function by bioassays in indirect or direct dependence on NO. Decreased NO availability was exclusively achieved by inhibiting enzymatic NO production, and increased NO availability was observed indirectly by adding NOS activity-stimulating agents. In cell culture experiments the NOS inhibitor L-NAME was added to EPCs without showing any influence on CFU formation as a marker of EPC number or EPC proliferation [91]. But, an impairment of EPC function was observed when the C-reactive protein (CRP) was added. CRP is usually synthesized by the liver during inflammation. When it was added under in vitro conditions, EPC microtubule formation was impaired accompanied by elevated eNOS mRNA expression. The treatment of EPCs with the NO donor DETO-NO attenuated the detrimental effects of CRP on in vitro angiogenesis, an effect that was prevented by the addition of the NOS inhibitor L-NAME [92]. In contrast, statins, a drug class that is used by almost all patients suffering from cardiovascular disease and that primarily lowers blood cholesterol levels, dose-dependently increased EPC migration. This phenomenon was reversed by PI3K inhibitors, which are an important second messenger of statins, or the NOS inhibitor N^{G}-monomethyl-L-arginine (L-NMA) [93]. Taken together, there is a plethora of reports that clearly demonstrate NOS actions within EPCs. It might be possible that in reports failing to demonstrate such an action, the EPC culture conditions may vary resulting in functional ineffective progenitors.

While in vitro experiments helped to understand the role of NO for distinct properties of EPCs, in vivo experiments are needed to comprehensively assess the relevance in the complex organism. To demonstrate the importance of the NO metabolism in in vivo situations, eNOS knockout mice ($eNOS^{-/-}$) represent an excellent animal model. A knockout mouse is a genetically engineered mouse in which one or more genes have been turned off through a gene knockout. Knockout mice are important animal models for studying the role of genes which have been sequenced but have unknown functions. The animal model $eNOS^{-/-}$ mouse gave first impressions of the relevance of a basal constitutive eNOS expression to EPC number and vessel function. NO levels are significantly lower and vascular function is reduced in comparison to wild-type (WT) animals although an NO production by iNOS or nNOS is still possible. The basal numbers of EPCs in $eNOS^{-/-}$ mice are not different from WT mice. But it could be demonstrated that EPCs from eNOS$^{-/-}$ animals showed defective neovascularization. In these animals, matrix metalloproteinase-9 (MMP-9) was reduced in the BM after myeloablation [94]. A stimulation of EPC mobilization in response to VEGF did not occur in eNOS$^{-/-}$ mice. Implementing acute NOS inhibition in this animal model demonstrated that NO itself is responsible for key processes of EPC mobilization. An increased number of EPCs in BM of eNOS$^{-/-}$ or L-NAME-treated mice could be detected but the cell count within the peripheral compartments did not differ.

In a murine model of hindlimb ischemia, it has been shown that eNOS$^{-/-}$ mice show impaired neovascularization and therefore an increased rate of autoamputation of the hindlimb. Interestingly, a rescue of the impaired hindlimb perfusion in eNOS$^{-/-}$ mice was achieved by transfusion of EPCs from wild-type animals, but not by transplantation of wild-type BM [94]. In a murine model of retinal ischemia, it has been shown that eNOS$^{-/-}$ animals, which have had BM transplantation with WT BM, show impaired neovascularization. At the site of retinal ischemia, the vessels were relatively large, unbranched, and poorly perfused [95].

Taken together, these data provide evidence that NO is a determinant of EPC homeostasis. Lack of NO, either acute or chronic, led to a diminished mobilization of EPCs into the peripheral circulation and distinctively modulated EPC function and consecutively impairing differentiation of this particular cell population [88].

Human Nitric Oxide-Mediated Cardiovascular Diseases and EPCs

In general, the number of EPCs is associated with the presence of cardiovascular risk factors. A list of conditions, including those commonly associated as risk factors for atherosclerosis such as arterial hypertension and hypercholesterolemia, are associated with diminished release of NO because of either impaired synthesis of NO or excessive oxidative degradation. Diminished NO bioactivity may cause constriction of arteries and contribute to provocation of ischemia. Additionally, diminished NO bioactivity may facilitate vascular inflammation that could lead to lipoprotein oxidation and foam cell formation, the precursor of the atherosclerotic plaque. Numerous therapies have been investigated to assess the possibility of reversing endothelial dysfunction by enhancing the release of NO from the endothelium. This may occur either by stimulation of NO synthesis or by protection of NO from oxidative inactivation and conversion into toxic molecules such as peroxynitrite [96, 97]. Reduced NO availability and increased reactive oxygen species are typical in patients with cardiovascular disease and important determinants of impaired EPC availability, function, and differentiation [98]. Clinical studies have shown that EPC properties are associated with vascular function and structure [26, 34, 35, 99]. Improving the NO-reactive oxygen species balance in patients may have a beneficial effect on EPC availability and function and may be crucial for progenitor cell-based therapies. Another very interesting approach should be the improvement of endothelial regeneration by EPCs, thereby enhancing the diminished NO availability and consequently endothelial function. Accordingly, causal relationships between improved

NO availability and endothelial function in combination with EPCs have been investigated in a few human studies. But NO-based therapies improving EPC function and therefore organ function in vivo are rare until now.

Endogenously determined NOS inhibition – measured as high ADMA levels – in patients with cardiovascular disease leads to endothelial dysfunction and impaired angiogenesis [100]. Interestingly, circulating ADMA levels inversely correlated with the number of progenitor cells and impaired EPC function. In line, the numbers of CFUs that were cultivated from peripheral blood MNCs were reduced when endogenous ADMA levels were elevated. Under in vitro conditions, the delivery of ADMA resulted in the reduction of tube formation and eNOS activity of EPCs [101].

Patients suffering from diabetes mellitus with a persistent hyperglycemic metabolism exhibit a high risk of developing cardiovascular diseases. In vitro, the cultivation of endothelial progenitors in hyperglycemic medium resembling the in vivo situation led to a decrease in eNOS phosphorylation associated with a decrease in NO production [102]. The importance of NO on EPC action is further supported by patients suffering from CAD. For instance, a strong correlation of the vascular NO concentration and EPC numbers could be observed after exercise with high numbers of EPCs within individuals with high NO levels and vice versa [103]. In diabetes, the number and function of circulating EPCs was impaired and the adhesion and migration of EPCs into damaged tissue was decreased due to poor cutaneous microcirculation. Recent evidence indicated that hyperoxia achieved by therapeutic hyperbaric oxygen can increase the mobilization of EPCs from the BM into the peripheral blood via NOS activity [104].

However, it would be a favorable aim to improve endogenous EPC levels without performing potentially harmful interventions. Such interventions might include lifestyle modifications and the use of protective cardiovascular drugs. It has been demonstrated that physical activity led to higher endogenous EPC levels, a mechanism that depends on endogenous NO levels [21]. Moreover, the use of statins resulted in an increase of EPC level [105]. Once again this effect was largely dependent on NOS activity. Further experimental and clinical studies have suggested that angiotensin-converting enzyme inhibitors, angiotensin II type 1 receptor blockers, PPAR-γ agonists, and erythropoietin (all therapeutic agents, applied in cardiovascular medicine) increase the number and functional activity of EPCs [106]. The underlying mechanisms still remain to be clarified [107].

6 Bioengineering in EPC Research

EPC application is gaining more importance in the clinical routine, e.g., as supportive therapy for tissue ischemia. But there are two essential limitations: (1) the number of circulating EPCs in peripheral blood is rather low and (2) the function of EPCs in patients is impaired. Therefore, scientists and biomedical engineers are evaluating methods to improve EPC isolation, to enlarge their function and number under culture conditions, to improve their application technique, and to attract them to places where they are needed. Altogether, there is still much work to do.

6.1 Optimized Cell Sorting

Concerning the isolation procedures, there is still a debate on which EPCs have the highest potential to improve tissue regeneration. This debate is primarily related to cell surface antigens such as CD34, CD117, and CD133. Further research is needed to find the ideal combination of surface marker for the identification and isolation of EPCs with the highest vasculogenic potential. The currently used methods to isolate the EPCs, magnetic bead isolation and FACS, are different in cost and quality. While FACS is more expensive, the quality of cell separation has been considerably improved by this method. However, further improvements are necessary to obtain high-throughput systems with an excellent gain of the cells.

6.2 Optimal Culture Conditions

The isolation of the cells is usually followed by cell culture for a variable period of time. This leads to further selection and enlargement of EPC numbers depending on culture media and growth factors within the media. Cell number and cell function should be improved by additional supplemental factors. For this purpose, statins have already been used. The addition of statins into the cell culture medium has been shown to inhibit the onset of EPC senescence and to increase EPC proliferation [108], thereby augmenting the number of EPCs in cell culture. In addition, statins improve functional properties such as the migratory capacity [109] that may enhance EPC incorporation after cell transplantation. Concerning clinical studies, this approach for improving the cell culture conditions has been applied in the TOPCARE-AMI study where the culture medium was supplemented with atorvastatin [49]. Additional improvement of cell culture conditions is very important for the in vivo application and needs further investigation.

6.3 Enhancement of EPC Properties

Several investigators tried to augment EPC function via gene transfer. In general, gene transfer is used to increase the expression of a selected protein. This can be obtained by introducing a vector which carries the genetic information of the protein into cells. The target within the host cell will be transcribed and translated into a protein. One gene of great interest in the biology of EPCs is the gene encoding the VEGF. Therefore, Iwaguro et al. tried to enhance EPC number and function ex vivo before cell transplantation by gene transfer of VEGF. In vitro, genetically modified or "transfected" EPCs showed an enhanced proliferation, adhesion, and incorporation into endothelial monolayers [110]. In a murine model of hindlimb ischemia, EPC transplantation of VEGF-transfected EPCs resulted in an increased salvage ratio of the ischemic hindlimbs in comparison to unmodified EPCs. Histological investigations revealed an increased capillary density and incorporation of EPCs into the ischemic hindlimbs of mice receiving genetically modified EPCs. The number of transplanted cells was reduced 30-fold by applying genetically modified EPCs [110], which represents an important fact concerning the applicability in humans. Kong et al. tested the effect of transfusion of eNOS-overexpressing EPCs on the re-endothelialization after vascular injury by balloon angioplasty in the carotid artery in the rabbit [111]. Transfusion of eNOS-overexpressing EPCs resulted in an enhanced covering of the injured, denuded area of the coronary vessels and consequently in a reduced pathological formation of neointimal hyperplasia in comparison to the transfusion of control cells [111]. As presented by animal models of tissue ischemia and vascular injury, it was demonstrated that gene-modified EPCs exhibited enhanced vasculoprotective effects. However, gene transfer into cells using, for example, virus-based vectors is still in early stages of development and needs further improvement. But as this approach modifies genetic information, it raises severe ethical concerns.

6.4 Pinpoint Application of Endothelial Progenitor Cells In Vivo

The application method of isolated and cultured EPCs is very important (Fig. 8). PCI is a therapeutic procedure to treat the stenotic coronary arteries. PCI includes the dilation of the narrowed coronary segment by a balloon or the implantation of a coronary stent, a small mesh tube that stabilizes the dilated vessels. The application techniques and materials such as catheters are constantly improved with much work of bioengineers. As one example, it is important to develop catheters that are minimally invasive and that are applicable within magnetic resonance imaging (MRI). MRI is of growing interest in cardiovascular imaging as it does not use X-ray but uses the nuclear magnetization of protons by a magnetic field with high contrast. MRI is an imaging technique used to visualize the structure and function of organs in living mammals. But, as it works with powerful magnetic field, both materials and patients must be free of any metal. Currently, the development of an MRI-compatible stent is under investigation and the application in combination with MRI has many advantages. A direct comparison of functional changes during intervention is one of the outstanding characteristics.

Furthermore, the local delivery of transplanted cells to the cardiac system by a PCI catheter is considered as favorable as it increases the cell concentration at the site of endothelial injury or tissue ischemia. For further improvement of EPC action, it may be of interest to keep the transplanted cells at the site of endothelial injury. How could this be done? EPCs may be concentrated in a certain area by biological scaffolding that serves as a physical support. In the field of tissue engineering, biodegradable scaffoldings, e.g., extracellular matrix or fibrin gel, have been developed to carry cells for the construction of bioprosthetic valves. EPCs may be attached to those biodegradable scaffoldings and delivered at the site of injury.

Locally, EPCs may be released from the scaffolding and contribute to endothelial repair and neovascularization.

6.5 Local Enrichment of Endothelial Progenitor Cells In Vivo

A fascinating field of bioengineering is currently the development of new stents which are metallic vascular implants. To date, two major different stent types exist: (I) bare metal stents (BMS) and (II) drug-eluting stents (DES). BMS enable a fast re-endothelialization of the inner stent lining which decreases the period of anti-clotting therapy (platelet-inhibiting therapy) down to 1 month. On the other hand, DES are coated with anti-cancer drugs such as sirolimus or paclitaxel which inhibit the proliferation and immigration of SMCs into the stent, thereby preventing the reocclusion of the stent. However, the disadvantage of DES is the prolonged re-endothelialization of the inner stent lining necessitating a prolonged intake of platelet-inhibiting drugs for at least 1 year. Taken together, the "ideal stent" would be one that inhibits neointimal formation and enhances re-endothelialization. Therefore, a stent that recruits EPCs into areas of stent-induced vessel denudation might represent an ideal combination of a device that maintains an open vessel and seals the injured area. EPCs, after homing to such areas, might favor the production of endothelial factors such as prostacyclin and NO. This may lead to stabilization of SMCs and the formation of neointima is inhibited. Further, EPCs might enhance re-endothelialization and lower the risk for platelet-induced vessel occlusion via rapid differentiation into endothelial cells. Using a porcine model, the transfusion of EPCs led to an enhanced endothelial coverage of coronary stents coated with an integrin that binds to the cyclic Arg-Gly-Asp peptide on the inner surface. Coating of stents with this a EPC-attracting substance reduced neointimal hyperplasia and may represent a useful strategy for reducing in-stent restenosis by accelerating endothelialization [112]. In line with these assumptions and the first results obtained in the animal model, some research groups started implanting stents that were coated with antibodies directed against CD34. Using such stents in patients reduced the time of anti-platelet therapy and lowered the rates of stent-induced restenosis of targeted vessels after 6 months. However, the investigators could not exclude an enhanced neointima formation after this period because CD34 is an unspecific antigen potentially attracting SMC progenitors. Therefore, finding antigens that solely attract EPCs is a favorable aim but still under investigation.

References

1. Carmeliet P. Mechanisms of angiogenesis and arteriogenesis. Nat Med. 2000 Apr; 6(4): 389–395.
2. Risau W. Mechanisms of angiogenesis. Nature. 1997 Apr 17; 386(6626):671–674.
3. Risau W, Flamme I. Vasculogenesis. Annu Rev Cell Dev Biol. 1995; 11:73–91.
4. Peichev M, Naiyer AJ, Pereira D, et al. Expression of VEGFR-2 and AC133 by circulating human CD34(+) cells identifies a population of functional endothelial precursors. Blood. 2000; 95:952–958.

5. Vasa M, Fichtlscherer S, Adler K, et al. Increase in circulating endothelial progenitor cells by statin therapy in patients with stable coronary artery disease. Circulation. 2001; 103: 2885–2890.
6. Vasa M, Fichtlscherer S, Aicher A, et al. Number and migratory activity of circulating endothelial progenitor cells inversely correlate with risk factors for coronary artery disease. Circ Res. 2001; 89:e1–e7.
7. Kalka C, Masuda H, Takahashi T, et al. Transplantation of ex vivo expanded endothelial progenitor cells for therapeutic neovascularization. Proc Natl Acad Sci USA. 2000; 97:3422–3427.
8. Asahara T, Murohara T, Sullivan A, et al. Isolation of putative progenitor endothelial cells for angiogenesis. Science. 1997; 275:964–967.
9. Aicher A, Zeiher AM, Dimmeler S. Mobilizing endothelial progenitor cells. Hypertension. 2005; 45:321–325.
10. Takahashi T, Kalka C, Masuda H, et al. Ischemia- and cytokine-induced mobilization of bone marrow-derived endothelial progenitor cells for neovascularization. Nat Med. 1999; 5:434–438.
11. Shintani S, Murohara T, Ikeda H, et al. Mobilization of endothelial progenitor cells in patients with acute myocardial infarction. Circulation. 2001; 103:2776–2779.
12. Gill M, Dias S, Hattori K, et al. Vascular trauma induces rapid but transient mobilization of VEGFR2(+)AC133(+) endothelial precursor cells. Circ Res. 2001; 88:167–174.
13. Asahara T, Takahashi T, Masuda H, et al. VEGF contributes to postnatal neovascularization by mobilizing bone marrow-derived endothelial progenitor cells. EMBO J. 1999; 18: 3964–3972.
14. Kalka C, Masuda H, Takahashi T, et al. Vascular endothelial growth factor(165) gene transfer augments circulating endothelial progenitor cells in human subjects. Circ Res. 2000; 86:1198–1202.
15. Kalka C, Tehrani H, Laudenberg B, et al. VEGF gene transfer mobilizes endothelial progenitor cells in patients with inoperable coronary disease. Ann Thorac Surg. 2000 Sept; 70(3):829–834.
16. Yamaguchi J, Kusano KF, Masuo O, et al. Stromal cell-derived factor-1 effects on ex vivo expanded endothelial progenitor cell recruitment for ischemic neovascularization. Circulation. 2003; 107:1322–1328.
17. Hattori K, Dias S, Heissig B, et al. Vascular endothelial growth factor and angiopoietin-1 stimulate postnatal hematopoiesis by recruitment of vasculogenic and hematopoietic stem cells. J Exp Med. 2001; 193:1005–1014.
18. Heeschen C, Aicher A, Lehmann R, et al. Erythropoietin is a potent physiologic stimulus for endothelial progenitor cell mobilization. Blood. 2003; 102:1340–1346.
19. Strehlow K, Werner N, Berweiler J, et al. Estrogen increases bone marrow-derived endothelial progenitor cell production and diminishes neointima formation. Circulation. 2003; 107:3059–3065.
20. Dimmeler S, Aicher A, Vasa M, et al. HMG-CoA reductase inhibitors (statins) increase endothelial progenitor cells via the PI 3-kinase/Akt pathway. J Clin Invest. 2001; 108: 391–397.
21. Laufs U, Werner N, Link A, et al. Physical training increases endothelial progenitor cells, inhibits neointima formation, and enhances angiogenesis. Circulation. 2004; 109: 220–226.
22. Hristov M, Zernecke A, Bidzhekov K, et al. Importance of CXC chemokine receptor 2 in the homing of human peripheral blood endothelial progenitor cells to sites of arterial injury. Circ Res. 2007 Mar 2; 100(4):590–597.
23. Chavakis E, Urbich C, Dimmeler S. Homing and engraftment of progenitor cells: a prerequisite for cell therapy. J Mol Cell Cardiol. 2008 Oct; 45(4):514–522.
24. Hirschi KK, Ingram DA, Yoder MC. Assessing identity, phenotype, and fate of endothelial progenitor cells. Arterioscler Thromb Vasc Biol. 2008; 28:1548–1595.

25. Mosmann T. Rapid colorimetric assay for cellular growth and survival: application to proliferation and cytotoxicity assays. J Immunol Methods. 1983 Dec 16; 65(1–2):55–63.
26. Hill JM, Zalos G, Halcox JPJ, et al. Circulating endothelial progenitor cells, vascular function, and cardiovascular risk. N Engl J Med. 2003; 348:593–600.
27. Falk W, Goodwin RH Jr., Leonard EJA. 48-Well micro chemotaxis assembly for rapid and accurate measurement of leukocyte migration. J Immunol Methods. 1980; 33(3):239–247.
28. Tepper OM, Galiano RD, Capla JM, et al. Human endothelial progenitor cells from type II diabetics exhibit impaired proliferation, adhesion, and incorporation into vascular structures. Circulation. 2002; 106:2781–2786.
29. Werner N, Kosiol S, Schiegl T, et al. Circulating endothelial progenitor cells and cardiovascular outcomes. N Engl J Med. 2005 Sept 8; 353(10):999–1007.
30. Imanishi T, Hano T, Matsuo Y, Nishio I. Oxidized low-density lipoprotein inhibits vascular endothelial growth factor-induced endothelial progenitor cell differentiation. Clin Exp Pharmacol Physiol. 2003; 30:665–670.
31. Abaci A, Oguzhan A, Kahraman S, et al. Effect of diabetes mellitus on formation of coronary collateral vessels. Circulation. 1999; 99:2239–2242.
32. Kondo T, Hayashi M. Smoking cessation rapidly increases circulating progenitor cells in peripheral blood in chronic smokers. Arterioscler Thromb Vasc Biol. 2004; 24:1442–1447.
33. Wang X, Zhu J, Chen J, Shang Y. Effects of nicotine on the number and activity of circulating endothelial progenitor cells. J Clin Pharmacol. 2004; 44:881–889.
34. Heiss C, Keymel S, Niesler U, Ziemann J, Kelm M, Kalka C. Impaired progenitor cell activity in age-related endothelial dysfunction. J Am Coll Cardiol. 2005; 45:1441–1448.
35. Keymel S, Kalka C, Rassaf T, Yeghiazarians Y, Kelm M, Heiss C. Impaired endothelial progenitor cell function predicts age-dependent carotid intimal thickening. Basic Res Cardiol. 2008; 103:582–586.
36. Kleinbongard P, Dejam A, Lauer T, et al. Plasma nitrite concentrations reflect the degree of endothelial dysfunction in humans. Free Radic Biol Med. 2006; 40:295–302.
37. Rassaf T, Kleinbongard P, Kelm M. Circulating NO pool in humans. Kidney Blood Press Res. 2005; 28:341–348.
38. Heiss C, Lauer T, Dejam A, et al. Plasma nitroso compounds are decreased in patients with endothelial dysfunction. J Am Coll Cardiol. 2006; 47:573–579.
39. Zhang ZG, Zhang L, Jiang Q, Chopp M. Bone marrow-derived endothelial progenitor cells participate in cerebral neovascularization after focal cerebral ischemia in the adult mouse. Circ Res. 2002 Feb 22; 90(3):284–288.
40. Kawamoto A, Gwon H-C, Iwaguro H, et al. Therapeutic potential of ex vivo expanded endothelial progenitor cells for myocardial ischemia. Circulation. 2001; 103:634–637.
41. Walter DH, Rittig K, Bahlmann FH, et al. Statin therapy accelerates reendothelialization: a novel effect involving mobilization and incorporation of bone marrow-derived endothelial progenitor cells. Circulation. 2002; 105:3017–3024.
42. Werner N, Priller J, Laufs U, et al. Bone marrow-derived progenitor cells modulate vascular reendothelialization and neointimal formation. Effect of 3-hydroxy-3-methylglutaryl coenzyme a reductase Inhibition. Arterioscler Thromb Vasc Biol. 2002; 22:1567–1572.
43. Werner N, Junk S, Laufs U, et al. Intravenous transfusion of endothelial progenitor cells reduces neointima formation after vascular injury. Circ Res. 2003; 93:e17–e24.
44. Dimmeler S, Zeiher AM, Schneider MD. Unchain my heart: the scientific foundations of cardiac repair. J Clin Invest. 2005 Mar; 115(3):572–583.
45. Hamano K, Nishida M, Hirata K, et al. Local implantation of autologous bone marrow cells for therapeutic angiogenesis in patients with ischemic heart disease – Clinical trial and preliminary results. Jpn Circ J. 2001 Sept; 65(9):845–847.
46. Stamm C, Westphal B, Kleine HD, et al. Autologous bone-marrow stem-cell transplantation for myocardial regeneration. Lancet. 2003 Jan 4; 361(9351):45–46.
47. Strauer BE, Brehm M, Zeus T, et al. Repair of infarcted myocardium by autologous intracoronary mononuclear bone marrow cell transplantation in humans. Circulation. 2002; 106:1913–1918.

48. Schächinger V, Assmus B, Britten MB, et al. Transplantation of progenitor cells and regeneration enhancement in acute myocardial infarction: final one-year results of the TOPCARE-AMI Trial. J Am Coll Cardiol. 2004 Oct; 19(44):1690–1699.
49. Assmus B, Schaechinger V, Teupe C, et al. Transplantation of progenitor cells and regeneration enhancement in acute myocardial infarction (TOPCARE-AMI). Circulation. 2002; 106:3009–3017.
50. Britten MB, Abolmaali ND, Assmus B, et al. Infarct remodeling after intracoronary progenitor cell treatment in patients with acute myocardial infarction (TOPCARE-AMI): mechanistic insights from serial contrast-enhanced magnetic resonance imaging. Circulation. 2003; 108:2212–2218.
51. Wollert KC, Meyer GP, Lotz J, et al. Intracoronary autologous bone-marrow cell transfer after myocardial infarction: the BOOST randomised controlled clinical trial. Lancet. 2004; 364:141–148.
52. Fernandez-Aviles F, San Roman JA, Garcia-Frade J, et al. Experimental and clinical regenerative capability of human bone marrow cells after myocardial infarction. Circ Res. 2004 Oct 1; 95(7):742–748.
53. Janssens S, Dubois C, Bogaert J, et al. Autologous bone marrow-derived stem-cell transfer in patients with ST-segment elevation myocardial infarction: double-blind, randomised controlled trial. Lancet. 2006; 367:113–121.
54. Perin EC, Dohmann HFR, Borojevic R, et al. Transendocardial, autologous bone marrow cell transplantation for severe, chronic ischemic heart failure. Circulation. 2003; 107:2294–2302.
55. Tateishi-Yuyama E, Matsubara H, Murohara T, et al. Therapeutic angiogenesis for patients with limb ischaemia by autologous transplantation of bone-marrow cells: a pilot study and a randomised controlled trial. Lancet. 2002; 360:427–435.
56. Orlic D, Kajstura J, Chimenti S, et al. Mobilized bone marrow cells repair the infarcted heart, improving function and survival. Proc Natl Acad Sci USA. 2001; 98:10344–10349.
57. Pittenger M, Martin BJ. Mesenchymal stem cells and their potential as cardiac therapeutics. Circ Res. 2004; 95:9–20.
58. Balsam LB, Wagers AJ, Christensen JL, Kofidis T, Weissman IL, Robbins RC. Haematopoietic stem cells adopt mature haematopoietic fates in ischaemic myocardium. Nat Med. 2004; 428:668–673.
59. Murry CE, Soonpaa MH, Reinecke H, et al. Haematopoietic stem cells do not transdifferentiate into cardiac myocytes in myocardial infarcts. Nature. 2004 Apr; 8(428):664–668.
60. Alvarez-Dolado M, Pardal R, Garcia-Verdugo JM, et al. Fusion of bone-marrow-derived cells with Purkinje neurons, cardiomyocytes and hepatocytes. Nature. 2003; 425:968–973.
61. Nygren JM, Jovinge S, Breitbach M, et al. Bone marrow-derived hematopoietic cells generate cardiomyocytes at a low frequency through cell fusion, but not transdifferentiation. Nat Med. 2004 May; 10(5):494–501.
62. Kajstura J, Rota M, Whang B, et al. Bone marrow cells differentiate in cardiac cell lineages after infarction independently of cell fusion. Circ Res. 2005; 96:127–137.
63. Gnecchi M, He H, Liang OD, et al. Paracrine action accounts for marked protection of ischemic heart by Akt-modified mesenchymal stem cells. Nat Med. 2005; 11:367–368.
64. Gnecchi M, He H, Noiseux N, et al. Evidence supporting paracrine hypothesis for Akt-modified mesenchymal stem cell-mediated cardiac protection and functional improvement. FASEB J. 2006; 20:661–669.
65. Bruckdorfer R. The basics about nitric oxide. Mol Aspects Med. 2005; 26:3–31.
66. Gutteridge JM, Halliwell B. Free radicals and antioxidants in the year 2000. A historical look to the future. Ann N Y Acad Sci. 2000; 899:136–147.
67. Valko M, Leibfritz D, Moncol J, Cronin MT, Mazur M, Telser J. Free radicals and antioxidants in normal physiological functions and human disease. Int J Biochem Cell Biol. 2007; 39(1):44–84.
68. Darley-Usmar V, Wiseman H, Halliwell B. Nitric oxide and oxygen radicals: a question of balance. FEBS Lett. 1995 Aug 7; 369(2–3):131–135.

69. Brennan PA, Moncada S. From pollutant gas to biological messenger: the diverse actions of nitric oxide in cancer. Ann R Coll Surg Engl. 2002 Mar; 84(2):75–78.
70. Bouton C, Drapier JC. Iron regulatory proteins as NO signal transducers. Sci STKE. 2003 May 13; 2003(182):e17.
71. Domachowske JB. The role of nitric oxide in the regulation of cellular iron metabolism. Biochem Mol Med. 1997 Feb; 60(1):1–7.
72. Hughes MN. Chemistry of nitric oxide and related species. Methods Enzymol. 2008; 436: 3–19.
73. Searles CD. Transcriptional and posttranscriptional regulation of endothelial nitric oxide synthase expression. Am J Physiol Cell Physiol. 2006 Nov; 291(5):C803–C816.
74. Gibaldi M. What is nitric oxide and why are so many people studying it? J Clin Pharmacol. 1993 Jun; 33(6):488–496.
75. Kleinbongard P, Schulz R, Rassaf T, et al. Red blood cells express a functional endothelial nitric oxide synthase. Blood. 2006; 107:2943–2951.
76. Landmesser U, Dikalov S, Price SR, et al. Oxidation of tetrahydrobiopterin leads to uncoupling of endothelial cell nitric oxide synthase in hypertension. J Clin Invest. 2003; 111:1201–1209.
77. Münzel T, Daiber A, Ullrich V, Mülsch A. Vascular consequences of endothelial nitric oxide synthase uncoupling for the activity and expression of the soluble guanylyl cyclase and the cGMP-dependent protein kinase. Arterioscler Thromb Vasc Biol. 2005; 25:1551–1557.
78. Endemann DH, Schiffrin EL. Nitric oxide, oxidative excess, and vascular complications of diabetes mellitus. Curr Hypertens Rep. 2004; 6:85–89.
79. Böger R. When the endothelium cannot say 'NO' anymore. ADMA, an endogenous inhibitor of NO synthase, promotes cardiovascular disease. Eur Heart J. 2003; 24:1901–1902.
80. Rassaf T, Kleinbongard P, Preik M, et al. Plasma nitrosothiols contribute to the systemic vasodilator effects of intravenously applied NO: experimental and clinical study on the fate of NO in human blood. Circ Res. 2002; 91:470–477.
81. Doyle MP, Hoekstra JW. Oxidation of nitrogen oxides by bound dioxygen in hemoproteins. J Inorg Biochem. 1981; 14:351–358.
82. Stamler JS, Jia L, Eu JP, et al. Blood flow regulation by S-nitrosohemoglobin in the physiological oxygen gradient. Science. 1997; 276:2034–2037.
83. McMahon TJ, Stone AE, Bonaventura J, Singel DJ, Stamler JS. Functional coupling of oxygen binding and vasoactivity in *S*-nitrosohemoglobin. J Biol Chem. 2000; 275:16738–16745.
84. Gladwin MT, Crawford JH, Patel RP. The biochemistry of nitric oxide, nitrite, and hemoglobin: role in blood flow regulation. Free Radic Biol Med. 2004; 36:707–717.
85. Taylor EL, Megson IL, Haslett C, Rossi AG. Nitric oxide: a key regulator of myeloid inflammatory cell apoptosis. Cell Death Differ. 2003; 10:418–430.
86. Yao K, Shida S, Selvakumaran M, et al. Macrophage migration inhibitory factor is a determinant of hypoxia-induced apoptosis in colon cancer cell lines. Clin Cancer Res. 2005; 11:7264–7272.
87. Pietraforte D, Matarrese P, Straface E, et al. Two different pathways are involved in peroxynitrite-induced senescence and apoptosis of human erythrocytes. Free Radic Biol Med. 2007; 42:202–214.
88. Özüyaman B, Ebner P, Niesler U, et al. Nitric oxide differentially regulates proliferation and mobilization of endothelial progenitor cells but not of hematopoietic stem cells. Thromb Haemost. 2005; 94:770–772.
89. Hur J, Yoon CH, Kim HS, et al. Characterization of two types of endothelial progenitor cells and their different contributions to neovasculogenesis. Arterioscler Thromb Vasc Biol. 2004; 24:288–293.
90. Gulati R, Jevremovic D, Peterson TE, et al. Diverse origin and function of cells with endothelial phenotype obtained from adult human blood. Circ Res. 2003 Nov; 28(93):1023–1025.
91. Hoetzer GL, Irmiger HM, Keith RS, Westbrook KM, DeSouza CA. Endothelial nitric oxide synthase inhibition does not alter endothelial progenitor cell colony forming capacity or migratory activity. J Cardiovasc Pharmacol. 2005 Sept; 46:387–389.

92. Verma S, Kuliszewski MA, Li S-H, et al. C-reactive protein attenuates endothelial progenitor cell survival, differentiation, and function. Further evidence of a mechanistic link between C-reactive protein and cardiovascular disease. Circulation. 2004; 109:2058–2067.
93. Urbich C, Dernbach E, Zeiher AM, Dimmeler S. Double-edged role of statins in angiogenesis signaling. Circ Res. 2002 Apr; 5(90):737–744.
94. Aicher A, Heeschen C, Mildner-Rihm C, et al. Essential role of endothelial nitric oxide synthase for mobilization of stem and progenitor cells. Nat Med. 2003; 9:1370–1376.
95. Guthrie SM, Curtis LM, Mames RN, Simon GG, Grant MB, Scott EW. The nitric oxide pathway modulates hemangioblast activity of adult hematopoietic stem cells. Blood. 2005 Mar 1; 105(5):1916–1922.
96. Heiss C, Kleinbongard P, Dejam A, et al. Acute consumption of flavanol-rich cocoa and the reversal of endothelial dysfunction in smokers. J Am Coll Cardiol. 2005; 46:1276–1283.
97. Balzer J, Rassaf T, Heiss C, et al. Sustained benefits in vascular function through flavanol-containing cocoa in mediated diabetic patients: a double-masked, randomized, controlled trial. J Am Coll Cardiol. 2008; 51:2141–2149.
98. Thum T, Fraccarollo D, Schultheiss M, et al. Endothelial nitric oxide synthase uncoupling impairs endothelial progenitor cell mobilization and function in diabetes. Diabetes. 2007; 56:666–674.
99. Fadini GP, Coracina A, Baesso I, Agostini C, Tiengo A, Avogaro A. Peripheral blood CD34+KDR+ endothelial progenitor cells are determinants of subclinical atherosclerosis in a middle-aged general population. Stroke. 2006; 37:2277–2282.
100. Schnabel R, Blankenberg S, Lubos E, et al. Asymmetric dimethylarginine and the risk of cardiovascular events and death in patients with coronary artery disease: results from the AtheroGene Study. Circ Res. 2005; 97:e53–e59.
101. Thum T, Tsikas D, Stein S, et al. Suppression of endothelial progenitor cells in human coronary artery disease by the endogenous nitric oxide synthase inhibitor asymmetric dimethylarginine. J Am Coll Cardiol. 2005 Nov; 1(46):1693–1701.
102. Kränkel N, Adams V, Linke A, et al. Hyperglycemia reduces survival and impairs function of circulating blood-derived progenitor cells. Arterioscler Thromb Vasc Biol. 2005; 25: 698–703.
103. Steiner S, Niessner A, Ziegler S, et al. Endurance training increases the number of endothelial progenitor cells in patients with cardiovascular risk and coronary artery disease. Atherosclerosis. 2005 Aug; 181(2):305–310.
104. Gallagher KA, Goldstein LJ, Thom SR, Velazquez OC. Hyperbaric oxygen and bone marrow-derived endothelial progenitor cells in diabetic wound healing. Vascular. 2006 Nov; 14(6):328–337.
105. Landmesser U, Engberding N, Bahlmann FH, et al. Statin-induced improvement of endothelial progenitor cell mobilization, myocardial neovascularization, left ventricular function, and survival after experimental myocardial infarction requires endothelial nitric oxide synthase. Circulation. 2004 Oct 5; 110(14):1933–1939.
106. Thum T, Fraccarollo D, Galuppo P, et al. Bone marrow molecular alterations after myocardial infarction: impact on endothelial progenitor cells. Cardiovasc Res. 2006; 70:50–60.
107. Besler C, Doerries C, Giannotti G, Luscher TF, Landmesser U. Pharmacological approaches to improve endothelial repair mechanisms. Expert Rev Cardiovasc Ther. 2008 Sept; 6(8):1071–1082.
108. Assmus B, Urbich C, Aicher A, et al. HMG-CoA reductase inhibitors reduce senescence and increase proliferation of endothelial progenitor cells via regulation of cell cycle regulatory genes. Circ Res. 2003; 92:1049–1055.
109. Spyridopoulos I, Haendeler J, Urbrich C, et al. Statins enhance migratory capacity by upregulation of the telomere repeat-binding factor TRF2 in endothelial progenitor cells. Circulation. 2004; 110:3136–3142.
110. Iwaguro H, Yamagushi J, Kalka C, et al. Endothelial progenitor cell vascular endothelial growth factor gene transfer for vascular regeneration. Circulation. 2002; 105:732–738.

111. Kong D, Melo LG, Mangi AA, et al. Enhanced inhibition of neointimal hyperplasia by genetically engineered endothelial progenitor cells. Circulation. 2004; 109:1769–1775.
112. Blindt R, Vogt F, Astafieva I, et al. A novel drug-eluting stent coated with an integrin-binding cyclic Arg-Gly-Asp peptide inhibits neointimal hyperplasia by recruiting endothelial progenitor cells. J Am Coll Cardiol. 2006; 47:1786–1795.

Skeletal Stem Cells and Controlled Nanotopography

Matthew J. Dalby and Richard O.C. Oreffo

Abstract Cells respond strongly to the shape of their environment and this is called contact guidence. For many years this has been studies at the microscale and with terminally differentiated cell types. With developments in materials technology studies at the nanoscale are now possible. Hand-in hand with these advances are developments in stem cell culture. This review looks at the adult stem cell/nanoscale interface considering cell adhesion, motility, gene expression (mechanotransduction) and ultimately differentiation. Particularly of interest is the reorganisation of the interface nucleus and possible links to differentiation.

Keywords Nanobioscience · Mesenchymal stem cell · Filopodia · Mechanotransduction

1 Introduction

It has been known for almost a century that cells will respond to shape or topography. However, our knowledge of the breadth of cell response rose exponentially in the 1980s with the refinement of miniaturisation techniques, such as photolithography, essentially for the microelectronics industry. These developments allowed the reproducible fabrication of construct sizes relevant to cells – i.e. microtopography. All cell types tested, including platelets, responded to changes in adhesion, migration and alterations in cytoskeletal organisation. Over the last 20 years, techniques such as electron beam lithography (EBL, providing patterns as small as 10 nm in *X*- and *Y*-axes [1]) became available to biologists and thus giving access to defined nanoscale topographies. Linkage of cell behaviour on nanoscale topographies together with post-genomic techniques such as microarray demonstrated

---

M.J. Dalby (✉)
Centre for Cell Engineering, Joseph Black Building, Institute of Biomedical and Life Sciences, University of Glasgow, Glasgow, G12 8QQ, Scotland, UK

G.M. Artmann et al. (eds.), *Stem Cell Engineering*,
DOI 10.1007/978-3-642-11865-4_11,

the breadth and importance of nanotopography modulating widespread genomic changes [2, 3].

Photolithography and EBL [2, 3], as well as other techniques such as colloidal lithography [4–8], are 'top-down' techniques where large constructs, like a wafer of silicon, are taken and a small pattern etched on. An alternative and potentially promising route to even smaller structures is 'bottom-up' fabrication or the manipulation of individual molecules (or possibly atoms) to produce ultra-small structures. Examples of this have been used in cell experiments, such as phase separation (or polymer demixing) [9–12] whereby incompatible polymers spontaneously form nanostructures when cast in a thin film. However, whilst single-molecule manipulations are now possible, such strategies are far from routine and remain aspirations for the future of cell biology. Nevertheless, it is important to realise the potential of such an approach – The natural environment of the cell will contain topographical information from microscale as presented by large protein chains, other cells and from other matrix constituents, e.g. mineral in bone. Furthermore, the topography may also present at the nanoscale to cells from protein folding and banding as well as nanocrystalline elements.

2 Skeletal or Mesenchymal Stem Cells

Postnatal bone retains an inherent capacity for controlled growth, remodelling in response to mechanical stimuli and regeneration upon damage providing, alongside skeletal development, a natural paradigm that can inform and approaches to skeletal regeneration. This is largely due to the directed differentiation of mesenchymal cells into osteogenic cells, a process subject to exquisite regulation and complex interplay by a variety of hormones, differentiation factors and environmental cues present within the bone matrix [13–17]. The osteoblast, the cell responsible for bone formation, is derived from a multipotent bone marrow stem cell. This population of marrow stromal fibroblastic stem cells is also referred to as osteogenic stem cells, marrow stromal fibroblastic cells, bone marrow stromal stem cells, mesenchymal stem cells, stromal precursor cells and, more recently, skeletal stem cells. While the non-specific phrase mesenchymal stem cells (MSCs) (undifferentiated multipotent cells of the mesenchyme) appears to have gained wider acceptance in the last 10–15 years and for ease of the reader will be used in this review, it should be borne in mind in a skeletal context, the more defined term, that skeletal stem cell is preferable [16]. These MSCs can give rise to cells of the adipogenic, reticular, osteoblastic, myoblastic and fibroblastic lineages and generate progenitors committed to one or more cell lines with an apparent degree of plasticity or interconversion [18–21]. Thus, the mesenchymal stem cell gives rise to a hierarchy of bone cell populations, artificially, divided into a number of developmental stages including stem cell, determined osteoprogenitor cell (DOPC), preosteoblast, osteoblast and ultimately, the terminally differentiated osteocyte [16, 17, 22–25].

3 Cell Filopodia

It is likely that filopodia are one of the cells' main sensory tools. Gustafson and Wolpert [26] first described filopodia in living cells in 1961 observing mesenchymal cells migrating up the interior wall of the blastocoelic cavity in sea urchins and noted that the filopodia produced appeared to explore the substrate. From these observations, the authors speculated that the filopodia were being used by the cells to gather spatial information.

When considering filopodial sensing of topography, fibroblasts have been described as using filopodia to sense and align the cells to microgrooves [27]. Macrophages have been reported to sense grooves down to a depth of 71 nm by actively producing many filopodia and elongating in response to the shallow topography [28]. Whilst distinctly different from the filopodia of the aforementioned cell types, neuronal growth cone filopodia have also been described as firstly sensing microgrooves and then aligning neurons to the grooves [29, 30].

Cytoskeletal actin bundles are key in driving the filopodia, and as the filopodia encounter a favourable guidance cue, they are stabilised following the recruitment of microtubules and accumulation of actin in a direction predictive of the future turn if a cell is to experience contact guidance.

Once cells locate a suitable feature using the filopodia present on the cells' leading edge, lamellipodia are formed which move the cell to the desired site [31]. These actions require G-protein signalling and actin cytoskeleton. Specifically of interest are Rho, Rac and Cdc42. Rho induces actin contractile stress fibre assembly to allow the cell to pull against the substrate, Rac induces lamellipodium formation and Cdc42 activation is required for filopodial assembly [32]. Both Rho and Rac are required for cell locomotion; however, cells can translocate when Cdc42 is knocked out. Interestingly, cells lacking Cdc42 cannot sense chemotactic gradients and will simply migrate in a random manner [33], suggesting a pivotal role for filopodial involvement in cell sensing.

We have recently produced evidence that MSCs use filopodia to probe their nanoenvironment (Fig. 1). Filopodia on these progenitor cells are more highly sensitive to nanotopography than those on other mature cell types [34]. To date, the smallest feature that fibroblasts have been observed to respond to are 10-nm-high polymer-demixed islands [31]. Skeletal progenitor cells have now been shown to produce filopodia in response to a variety of nanotopographies, the smallest being 40-nm-high polymer-demixed islands [35], although it is safe to assume that this is not the limit of skeletal progenitor sensory capacity, rather the limit of the emerging cell testing so far.

Filopodia appear to be likely candidates for nanotopographical sensing from a second perspective, that of size. Filopodia are approximately 100 nm in diameter, thus putting them on a similar scale to the topographies now being fabricated. MSCs are large cells and when spread and adhered can have lengths and widths of tens to hundreds of micrometres. Many of the topographies tested remain several orders of magnitude smaller than the cell, and for a long time it appeared and was thought unlikely that skeletal progenitor cells would notice them or indeed that

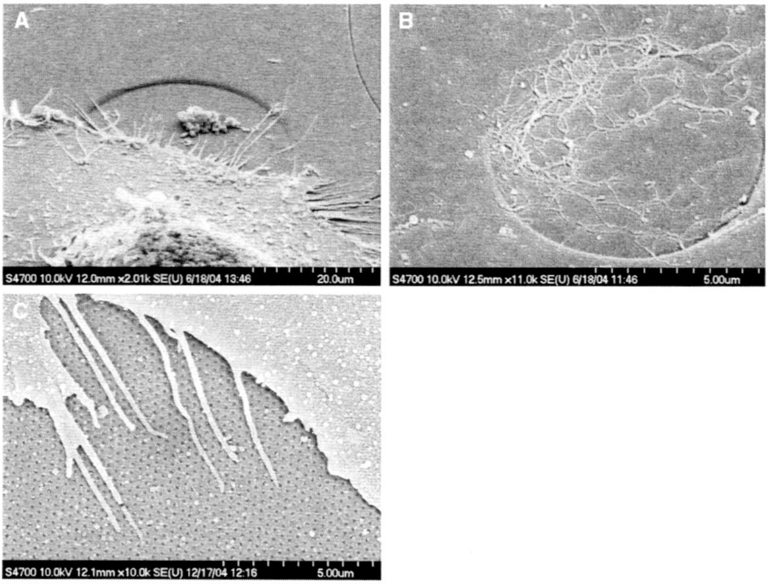

Fig. 1 Filopodial interaction with topographical surfaces. (**a**) Photolithographically patterned pit (30 μm wide, 300 nm deep) probed by osteoprogenitor filopodia. (**b**) Photolithographically produced pit (10 μm wide, 300 nm deep) filled with collagen after an osteoprogenitor has contacted the pit. (**c**) Filopodial interaction with electron beam lithographically fabricated pits (120 nm diameter, 100 nm deep, 300 nm centre–centre spacing, hexagonal pattern); note the similar dimensions to the tips of the filopodia

such topography would have any functional relevance. Thus far, topographies that induce filopodial interactions and cell spreading have been discussed. A second mode of nanotopographical action on cells is reduction of adhesion and spreading [36]. This is typically achieved using highly ordered pits fabricated by EBL. Such pits (e.g. 120 nm diameter, 100 nm deep and 300 nm centre–centre spacing in a square arrangement) have been shown to reduce adhesion of epitenon cells [37], fibroblasts [38] and now MSCs [39] with filopodia playing a key role – this time in avoiding the features. Immuno-TEM of vinculin has been used to show that on such EBL pits, fibroblasts will use their filopodia to locate adhesions as far from pits as possible [39]. Immuno-SEM of vinculin of osteoblasts on these structures shows that the adhesion forms only along interpit spaces [40].

4 Cell Cytoskeleton and Cellular Adhesions

The ability of a cell to form adhesions will effect organisation of the cytoskeletons and alter a cells' mechanotransductive pathways. Nanotopography can change the ability of skeletal stem cells and progenitors to form focal adhesions (this has regularly been observed with differentiated cell types and it has recently been shown with osteoblasts that nanotopography can change the type of adhesion formed, from very small and immature focal complexes to large super-mature adhesions that align to endogenous matrix proteins [41]). These super-mature adhesions are sometimes described as in vivo fibrillar adhesions running along long 'cables' of filamentous protein.

The cytoskeleton is anchored to the adhesions, with actin microfilaments linked directly. The cytoskeleton is a network of protein filaments extending through the cell cytoplasm within eukaryotic cells. The cytoskeleton is of fundamental importance in the control of many aspects of cell behaviour, movement and metabolism including proliferation, intracellular signalling, movement and cell attachment. The three main cytoskeletal components are microfilaments (actin), microtubules (tubulin) and intermediate filaments [e.g. vimentin (MSCs), cytokeratin, desmin depending on the cell type].

There are two types of microfilament bundles, involving different bundling proteins. One type of bundle, containing closely spaced actin filaments aligned in parallel, supports projections in the cell membrane (filopodia or mikrospikes involved in cell sensing and lamellipodia involved in cell crawling). The second type is composed of more loosely spaced filaments and is capable of contraction; these are the stress fibres. The ability of cells to move across substrate surfaces is a function of the actin cytoskeleton. The microtubules provide a system by which vesicles and other membrane-bound organelles may travel and regulate cell shape, movement and the plane of cell division. Microtubules, composed of tubulin, exist as single filaments that radiate outward through the cytoplasm from the centrosome, near the nucleus. Microtubules emanating from the centrosome (microtubule-organising centre) act as a surveying device that is able to find the centre of the cell.

Intermediate filaments are tough protein fibres in the cell cytoplasm. They extend from the nucleus in gently curving arrays to the cell periphery and are particularly dominant when the cells are subject to mechanical stress. The intermediate filaments are classified into four broad bands: type I are keratin-based proteins, type II includes vimentin and desmin, type III includes neurofilament proteins and type IV are nuclear lamins. In cells of mesenchymal origin, vimentin is the main intermediate filament protein providing mechanical support for the cell and nucleus [42–45]. Thus cells on nanotopography templates will produce large-scale alterations in cytoskeletal organisation [46–48]. Such alterations in mesenchymal stem cell cytoskeletons can take the form of almost total collapse of the cytoskeleton on highly organised EBL structures, an increase in cytoskeletal organisation on polymer-demixed nanoislands and alignment of the cytoskeleton along nanodepth grooves fabricated by photolithography [35, 49]. The key question that follows is, what are the implications of such cytoskeletal changes for stem cells?

5 Mechanotransduction

Changes in cytoskeletal organisation result in alterations in proliferation and differentiation as a number of signalling cascades are regulated by the cytoskeleton. These mechanotransductive signalling events may be chemical, e.g. kinase based linked to focal adhesions influenced by cytoskeletal contraction. For example, integrin gathering as an adhesion is formed will activate myosin light chain kinase (MCLK) which will generate actin–myosin sliding and, in turn, will change focal adhesion kinase (FAK) activity. It is known that contact guidance leads to FAK upregulation in mesenchymal stem cells and that in $FAK^{-/-}$, fibroblasts do not align well to grooves (Dalby and Yarwood, unpublished observation). In addition, cytoskeletal involvement in contraction against adhesions will alter calcium influx and G-protein events (e.g. Rho, Rac and CDC42), a process of indirect mechanotransduction [44].

Recent genomic reports using nanoscale topography and mesenchymal or skeletal stem cells point to several key pathways being critical to bone development. These are integrin-related signalling (which includes FAK), actin-medicated signalling, mitogen-activated protein kinase p38 (MAPK P38) and the extracellular receptor kinase (ERK). The role of integrin signalling is thus clear due to the involvement in the formation of integrin-containing adhesions at the cell/material interface. Activation of this pathway impacts on actin-related signalling controlling filopodial and lamellipodial movement, adhesion formation (in feedback) and cell contraction and further activates MAPK/ERK pathways which impact on a wide range of transcription factors including the osteogenic master transcription factor Cbfa1/Runx2 with obvious implications for cell differentiation into osteogenic or non-osteogenic lineages.

Another form of mechanotransduction, direct mechanotransduction, is considered to be transduced by the cytoskeletons as an integrated unit (i.e. microfilaments, microtubules and intermediate filaments working together). The cytoskeleton is firstly thought to provide the cytoplasm with the inhomogeneity required for force transduction to the nucleus [50]. This is what can be described as a percolation network [51] and is associated with cellular tensegrity whereby the cells' mechanical structure is explained via tensional integrity [52–58]. Through this tensegrity structure, tensional forces from the extracellular environment (e.g. from tissue loading or changes in cell spreading) are potentially conferred to the nucleus and alter genome regulation [59–63]. In addition, intermediate filaments of the cytoskeleton are linked to the lamin intermediate filaments of the nucleoskeleton. It is known that the telomeric ends of the interphase chromosomes are intimately linked to the peripheral lamins [64]. Thus, tension directed through the cytoskeleton may be passed directly to the chromosomes during gene transcription. Not long ago, the interphase nucleus was considered to contain randomly arranged chromosomal DNA which condensed during mitosis. However, Heslop-Harrison introduced ideas of relative consistency of positioning of the chromosomes and more recently chromosomal territories have been identified [62, 65]. Thus, changes in chromosomal three-dimensional arrangement may effect transcriptional events such as access to the genes by transcription factors and polymerases (thought to reside in specific transcription factories [66]). Furthermore, changes in DNA tension can also cause polymerase enzymes to slow down, speed up or even stall completely [67]. These may be mechanisms by which changes in cell spreading, as seen by MSCs on nanotopographies, can change differentiation events (Fig. 2).

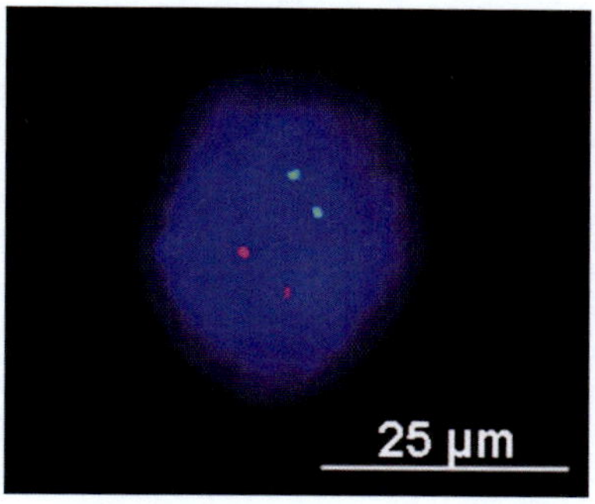

Fig. 2 Fibroblast nucleus (*blue*) with centromeres for chromosomes 3 (*dark*) and 16 (*light*) stained by FISH. It is considered that there is a relative consistency of position in the nucleus and it is possible that by changing this relative positioning through alterations in cell spreading, genomic output could be altered. We have shown that there can be changes in centromere distances resulting from changing surface topography [68].

Nanoimprinting into cells by nanofeatures is a recent discrete theory of how nanotopography may alter cytoskeletal mechanotransductive event by nanofeatures [69, 70]. This describes a phenomenon seen in platelets and for which some evidence has been provided in more complex cell types. For nanoimprinting to occur, the pattern of the topography must be transferred to the cytoskeletal filaments. As with filopodia, the dimensions of the cytoskeletal elements are small enough to warrant consideration of influence by nanoscale features. There is evidence that this imprinting leads to increased 'attempted' endocytosis, i.e. the cells recognise the features as being in the correct size range to attempt to endocytose the features and to form clathrin-coated pits [71]. Such endocytotic vesicles are moved by actin cables – suggesting that it is this mechanism that is causing the topography to mimic actin patterning described by Curtis and others [70].

Nanoimprinting has been shown to be integrin specific. Beta-1 integrin blocking resulted in no reduction in the effect; however, blocking the beta-3 subunit reduced the effect to background [72], although screening of alpha units is required to show which extracellular matrix proteins are required for nanoimprinting and subsequent cell response.

6 Differentiation

When MSCs or skeletal progenitor cells are cultured in vitro on planar materials without chemical treatment, they are unlikely to differentiate to mature osteoblastic cells capable of producing osteocalcin and bone mineral. This is an important issue for orthopaedic biomaterials for load-bearing bone replacement (reviewed in Ref. [73]). The inability of materials to induce osteogenic differentiation in orthopaedic implant materials has obvious implications with weak repair, disorganised tissue formation and implant encapsulation in soft tissue as a consequence of the absence of cues from the materials to specify cell differentiation. Thus, it follows, can nanotopography alter cellular differentiation? The answer, from emergent studies, is unequivocally yes with nanoscale topography providing discrete cues, resulting in, for example, subtle adhesion shifts, shape of adhesions and alignment of adhesions [74]. Thus, nanoscale grooves may not necessarily impact on the whole cell plasma membrane but rather align discrete foci, which in turn align the cell. For example, EBL pits have been shown to influence adhesion length – with highly organised pits resulting in many small focal complexes (<2 μm, transient and involved in cell motility), whereas adding a degree of disorder results in super-mature adhesion formation (> 5 μm in length) [75]. This topographical cue has resulted in perhaps the single most important observation in the nano/stem cell area – namely highly ordered pit arrangements resulting in small adhesions produce fibroblastic differentiation, whereas controlled disorder (±50 nm from central position) arrangements produce large adhesions, resulting in osteoblastic differentiation of skeletal progenitor and stem cells and random arrangements produce negligible changes from planar control (Fig. 2).

This is perhaps not to be unexpected as increasing adhesion size will increase integrin, actin, MAPK/ ERK signalling and increase cytoskeletal tension with implications therein for mechanotransduction and perhaps activation of Cbfa1/Runx2. Furthermore, from a biomimetic perspective, mature osteoblasts are large cells with larger adhesions in comparison to mature fibroblasts. Whilst on reflection, this would appear clear, the discovery was the culmination of taking a middle approach – EBL fabrication whilst producing high order will deliver low adhesion and the use of blasting, etching and other techniques to produce random roughness failed to create an appropriate level of disorder critical, it appears for skeletal stem cell differentiation along the osteogenic lineage [16, 17].

7 Summary

There is a wealth of evidence to suggest that whilst cells have the same genome, transcriptional machinery and transcription factor pool, niche environment is critical in cell function and activity and, within such an environment, the ability of a cell to sense and respond to the environment will be key. Nanoscale topography has emerged as an important player together with chemical cues (interleukins, hormones, surrounding matrix proteins) in modulating stem cell commitment to different phenotypical lineages through mechanisms of spatial sensing and cytoskeletal organisation [76]. Work from the Chen and Discher/Engler groups suggest that environmental conditions have similar effects to nanotopography and that a wide variety of differentiation pathways can be elicited with MSCs using different material conditions [77, 78]. Thus, changes in cytoskeletal tension result in changes in MSC commitment and recent studies have shown that alteration of the modulus of the environment from soft to hard will modulate MSC differentiation between adipocyte and osteoblast lineages [77]. With the emergent role of nanotopography on stem cell function arise exciting health-care opportunities – for example, the creation of 'next-generation' materials capable of eliciting desired stem cell function through discrete nanotopographical cues/imprints possibly as part of the extracellular filamentous protein percolation network. Forgacs [51, 79] has likened this as being rather like a spider in a web – the spider's body being the cell nucleus, the legs, the cytoskeleton, and the web the extracellular matrix. The vistas are endless with opportunities to use nanotopography as powerful non-invasive tools to understand stem cells. The next decade will provide new challenges and paradigm shifts, not least an opportunity for biologists to use materials to test cells as well as cells to test materials and the exploitation of discrete topographical cues in the development of new health-care strategies for hard and soft tissue regeneration.

Acknowledgements Matthew Dalby is supported by awards from the BBSRC. Richard Oreffo is supported by awards from the BBSRC and EPSRC.

References

1. Vieu C, Carcenac F, Pepin A, Chen Y, Mejias M, Lebib A, et al. Electron beam lithography: resolution limits and applications. Appl Surf Sci. 2000; 164:111–117.
2. Wilkinson CD. Making structures for cell engineering. Eur Cell Mater. 2004 Oct; 22(8): 21–26.
3. Wilkinson CDW, Riehle M, Wood M, Gallagher J, Curtis ASG. The use of materials patterned on a nano- and micro-metric scale in cellular engineering. Mater Sci Eng. 2002; 19:263–269.
4. Andersson AS, Backhed F, von Euler A, Richter-Dahlfors A, Sutherland D, Kasemo B. Nanoscale features influence epithelial cell morphology and cytokine production. Biomaterials. 2003 Sep; 24(20):3427–3436.
5. Andersson AS, Brink J, Lidberg U, Sutherland DS. Influence of systematically varied nanoscale topography on the morphology of epithelial cells. IEEE Trans Nanobiosci. 2003; 2(2):49–57.
6. Andersson AS, Olsson P, Lidberg U, Sutherland D. The effects of continuous and discontinuous groove edges on cell shape and alignment. Exp Cell Res. 2003 Aug 1; 288(1):177–188.
7. Denis FA, Hanarp P, Sutherland DS, Dufrene YF. Fabrication of nanostructured polymer surfaces using colloidal lithography and spin coating. Nanoletters. 2002; 2:1419–1425.
8. Denis FA, Hanarp P, Sutherland DS, Gold J, Mustin C, Rouxhet PG, et al. Protein adsorption on model surfaces with controlled nanotopography and chemistry. Langmuir. 2002; 18(3):819–828.
9. Affrossman S, Henn G, O'Neill SA, Pethrick RA, Stamm M. Surface topography and composition of deuterated polystyrene – poly(bromostyrene) blends. Macromolecules. 1996; 29(14):5010–5016.
10. Affrossman S, Jerome R, O'Neill SA, Schmitt T, Stamm M. Surface structure of thin film blends of polystyrene and poly(*n*-butyl methacrylate). Colloid Polym Sci. 2000; 278:993–999.
11. Affrossman S, O'Neill SA, Stamm M. Topography and surface composition of thin films of blends of polystyrene with brominated polystyrenes: effects of varying the degree of bromination and annealing. Macromolecules. 1998; 31:6280–6288.
12. Affrossman S, Stamm M. The effect of molecular weight on the topography of thin films of blends of poly(4-bromostyrene) and polystyrene. Colloid Polym Sci. 2000; 278:888–893.
13. Bianco P, Riminucci M, Gronthos S, Robey PG. Bone marrow stromal stem cells: nature, biology, and potential applications. Stem Cells. 2001; 19(3):180–192.
14. Bianco P, Robey PG. Stem cells in tissue engineering. Nature. 2001; 414:118–121.

15. Oreffo ROC. Growth factors for skeletal reconstruction and fracture repair. Curr Opin Investig Drugs. 2004; 5(4):419–423.
16. Tare RS, Babister JC, Kanczler J, Oreffo RO. Skeletal stem cells: phenotype, biology and environmental niches informing tissue regeneration. Mol Cell Endocrinol. 2008 Jun 25; 288(1–2):11–21.
17. Dawson JI, Oreffo RO. Bridging the regeneration gap: stem cells, biomaterials and clinical translation in bone tissue engineering. Arch Biochem Biophys. 2008 May 15; 473(2): 124–131.
18. Yang X, Tare RS, Partridge KA, Roach HI, Clarke NM, Howdle SM, et al. Induction of human osteoprogenitor chemotaxis, proliferation, differentiation, and bone formation by osteoblast stimulating factor-1/pleiotrophin: osteoconductive biomimetic scaffolds for tissue engineering. J Bone Miner Res. 2003 Jan; 18(1):47–57.
19. Rose FR, Oreffo RO. Bone tissue engineering: hope vs hype. Biochem Biophys Res Commun. 2002 Apr 22; 292(1):1–7.
20. Triffitt JT, Oreffo RO. Osteoblast lineage. In: Zaidi M, ed. Advances in organ biology: molecular and cellular biology of bone. Stamford: JAI Press; 1998, pp. 475–498.
21. Friedenstein AJ. Precursor cells of mechanocytes. Int Rev Cytol. 1976; 47:327–359.
22. Mirmalek-Sani SH, Roach HI, Wilson DI, Hanley NA, Oreffo ROC. Characterisation of human fetal populations: a comparative model for skeletal and stem cell differentiation. J Bone Miner Res. 2005; 20:1292.
23. Locklin RM, Oreffo RO, Triffitt JT. Modulation of osteogenic differentiation in human skeletal cells in vitro by 5-azacytidine. Cell Biol Int. 1998; 22(3):207–215.
24. Oreffo RO, Bord S, Triffitt JT. Skeletal progenitor cells and ageing human populations. Clin Sci (Lond). 1998 Jun; 94(5):549–555.
25. Oreffo RO, Triffitt JT. Future potentials for using osteogenic stem cells and biomaterials in orthopedics. Bone. 1999 Aug; 25(2 Suppl):5S–9S.
26. Gustafson T, Wolpert L. Studies on the cellular basis of morphogenesis in the sea urchin embryo. Directed movements of primary mesenchyme cells in normal and vegetalized larvae. Exp Cell Res. 1961 Jun; 24:64–79.
27. Clark P, Connolly P, Curtis AS, Dow JA, Wilkinson CD. Cell guidance by ultrafine topography in vitro. J Cell Sci. 1991 May; 99(Pt 1):73–77.
28. Wojciak-Stothard B, Madeja Z, Korohoda W, Curtis A, Wilkinson C. Activation of macrophage-like cells by multiple grooved substrata – topographical control of cell behavior. Cell Biol Int. 1995; 19:485–490.
29. Rajnicek A, McCaig C. Guidance of CNS growth cones by substratum grooves and ridges: effects of inhibitors of the cytoskeleton, calcium channels and signal transduction pathways. J Cell Sci. 1997 Dec; 110 (Pt 23):2915–2924.
30. Rajnicek A, Britland S, McCaig C. Contact guidance of CNS neurites on grooved quartz: influence of groove dimensions, neuronal age and cell type. J Cell Sci. 1997 Dec; 110 (Pt 23):2905–2913.
31. Dalby MJ, Riehle MO, Johnstone H, Affrossman S, Curtis AS. Investigating the limits of filopodial sensing: a brief report using SEM to image the interaction between 10 nm high nano-topography and fibroblast filopodia. Cell Biol Int. 2004; 28(3):229–236.
32. Schmitz AA, Govek EE, Bottner B, Van Aelst L. Rho GTPases: signaling, migration, and invasion. Exp Cell Res. 2000 Nov 25; 261(1):1–12.
33. Jones GE, Allen WE, Ridley AJ. The Rho GTPases in macrophage motility and chemotaxis. Cell Adhes Commun. 1998; 6(2–3):237–245.
34. Hart A, Gadegaard N, Wilkinson CDW, Oreffo ROC, Dalby MJ. Osteoprogenitor response to low-adhesion nanotopographies originally fabricated by electron beam lithography. J Mater Sci Mater Med. 2007 Jun; 18(6):1211–1218. Epub 2007 Feb 3.
35. Dalby MJ, McCloy D, Robertson M, Agheli H, Sutherland D, Affrossman S, et al. Osteoprogenitor response to semi-ordered and random nanotopographies. Biomaterials. 2006 May; 27(15):2980–2987.

36. Dalby MJ. Cellular response to low adhesion nanotopographies. Int J Nanomed. 2007; 2(3):373–381.
37. Gallagher JO, McGhee KF, Wilkinson CDW, Riehle MO. Interaction of animal cells with ordered nanotopography. IEEE Trans Nanobiosci. 2002; 1(1):24–28.
38. Dalby MJ, Gadegaard N, Riehle MO, Wilkinson CD, Curtis AS. Investigating filopodia sensing using arrays of defined nano-pits down to 35 nm diameter in size. Int J Biochem Cell Biol. 2004 Oct; 36(10):2015–2025.
39. Dalby MJ, Biggs MJP, Gadegaard N, Kalna G, Wilkinson CDW, Curtis ASG. Nanotopographical stimulation of mechanotransduction and changes in interphase centromere positioning. J Cell Biochem. 2007; 100(2):326–338.
40. Biggs MJ, Richards RG, Gadegaard N, McMurray RJ, Affrossman S, Wilkinson CD, et al. Interactions with nanoscale topography: adhesion quantification and signal transduction in cells of osteogenic and multipotent lineage. J Biomed Mater Res A. 2008 Sep 23; 91(1): 195–208.
41. Biggs MJ, Richards RG, Gadegaard N, Wilkinson CD, Dalby MJ. Regulation of implant surface cell adhesion: Characterization and quantification of S-phase primary osteoblast adhesions on biomimetic nanoscale substrates. J Orthop Res. 2007 Feb; 25(2):273–282.
42. Alberts B, Bray D, Lewis J, Raff M, Watson J. Molecular biology of the cell. New York: Garland; 1994.
43. Cooper GM. Cell. Sunderland: Sinauer Associates; 2000.
44. Burridge K, Chrzanowska-Wodnicka M. Focal adhesions, contractility, and signaling. Annu Rev Cell Dev Biol. 1996; 12:463–518.
45. Amos LA, Amos WB. Molecules of the cytoskeleton. London: MacMillan; 1991.
46. Dalby MJ, Riehle MO, Johnstone H, Affrossman S, Curtis AS. In vitro reaction of endothelial cells to polymer demixed nanotopography. Biomaterials. 2002 Jul; 23(14):2945–2954.
47. Dalby MJ, Childs S, Riehle MO, Johnstone HJ, Affrossman S, Curtis AS. Fibroblast reaction to island topography: changes in cytoskeleton and morphology with time. Biomaterials. 2003 Mar; 24(6):927–935.
48. Dalby MJ, Riehle MO, Johnstone HJ, Affrossman S, Curtis AS. Polymer-demixed nanotopography: control of fibroblast spreading and proliferation. Tissue Eng. 2002 Dec; 8(6):1099–1108.
49. Dalby MJ, McCloy D, Robertson M, Wilkinson CDW, Oreffo ROC. Osteoprogenitor response to defined topographies with nanoscale depths. Biomaterials. 2006; 27:1306–1315.
50. Wang N, Suo Z. Long-distance propagation of forces in a cell. Biochem Biophys Res Commun. 2005 Mar 25; 328(4):1133–1138.
51. Forgacs G. On the possible role of cytoskeletal filamentous networks in intracellular signaling: an approach based on percolation. J Cell Sci. 1995 Jun; 108 (Pt 6):2131–2143.
52. Ingber DE, Tensegrity II. How structural networks influence cellular information processing networks. J Cell Sci. 2003 Apr 15; 116(Pt 8):1397–1408.
53. Ingber DE, Tensegrity I. Cell structure and hierarchical systems biology. J Cell Sci. 2003 Apr 1; 116(Pt 7):1157–1173.
54. Ingber DE. Mechanosensation through integrins: cells act locally but think globally. Proc Natl Acad Sci USA. 2003 Feb 18; 100(4):1472–1474.
55. Maniotis AJ, Chen CS, Ingber DE. Demonstration of mechanical connections between integrins, cytoskeletal filaments, and nucleoplasm that stabilize nuclear structure. Proc Natl Acad Sci USA. 1997 Feb 4; 94(3):849–854.
56. Maniotis AJ, Bojanowski K, Ingber DE. Mechanical continuity and reversible chromosome disassembly within intact genomes removed from living cells. J Cell Biochem. 1997 Apr; 65(1):114–130.
57. Ingber DE. Cellular tensegrity: defining new rules of biological design that govern the cytoskeleton. J Cell Sci. 1993 Mar; 104 (Pt 3):613–627.
58. Charras GT, Horton MA. Single cell mechanotransduction and its modulation analyzed by atomic force microscope indentation. Biophys J. 2002 Jun; 82(6):2970–2981.

59. Dalby MJ. Topographically induced direct cell mechanotransduction. Med Eng Phys. 2005; 27(9):730–742.
60. Dalby MJ, Riehle MO, Sutherland DS, Agheli H, Curtis AS. Use of nanotopography to study mechanotransduction in fibroblasts–methods and perspectives. Eur J Cell Biol. 2004 May; 83(4):159–169.
61. Heslop-Harrison JS. Comparative genome organization in plants: from sequence and markers to chromatin and chromosomes. Plant Cell. 2000 May; 12(5):617–636.
62. Heslop-Harrison JS, Leitch AR, Schwarzacher T. The physical organisation of interphase nuclei. In: Heslop-Harrison JS, Flavell RB, eds. The chromosome. Oxford: Bios; 1993, pp. 221–232.
63. Mosgoller W, Leitch AR, Brown JK, Heslop-Harrison JS. Chromosome arrangements in human fibroblasts at mitosis. Hum Genet. 1991 Nov; 88(1):27–33.
64. Foster HA, Bridger JM. The genome and the nucleus: a marriage made by evolution. Genome organisation and nuclear architecture. Chromosoma. 2005 Sep; 114(4):212–229.
65. Cremer T, Cremer C. Chromosome territories, nuclear architecture and gene regulation in mammalian cells. Nat Rev Genet. 2001 Apr; 2(4):292–301.
66. Osborne CS, Chakalova L, Brown KE, Carter D, Horton A, Debrand E, et al. Active genes dynamically colocalize to shared sites of ongoing transcription. Nat Genet. 2004 Oct; 36(10):1065–1071.
67. Bustamante C, Bryant Z, Smith SB. Ten years of tension: single-molecule DNA mechanics. Nature. 2003 Jan 23; 421(6921):423–427.
68. Dalby MJ, Gadegaard N, Herzyk P, Sutheraland DS, Agheli H, Wilkinson CDW, et al. Nanomechanotransduction and interphase nuclear organisation influence on genomic control. J Cell Biochem. 2007; 102:1234–1244.
69. Curtis ASG, Dalby MJ, Gadegaard N. Cell signaling arising from nanotopography: implications for nanomedical devices. Nanomedicine. 2006; 1(1):67–72.
70. Curtis ASG, Dalby MJ, Gadegaard N. Nanoimprinting onto cells. J R Soc Interface. 2006; 3:393–398.
71. Dalby MJ, Berry CC, Riehle MO, Sutherland DS, Agheli H, Curtis AS. Attempted endocytosis of nano-environment produced by colloidal lithography by human fibroblasts. Exp Cell Res. 2004 May 1; 295(2):387–394.
72. Wood MA, Bagnaninchi P, Dalby MJ. The beta integrins and cytoskeletal nanoimprinting. Exp Cell Res. 2008 Feb 15; 314(4):927–935.
73. Balasundaram G, Webster TJ. Nanotechnology and biomaterials for orthopaedic medical applications. Nanomedicine. 2006; 1(2):169–176.
74. Biggs MJ, Richards RG, McFarlane S, Wilkinson CD, Oreffo RO, Dalby MJ. Adhesion formation of primary human osteoblasts and the functional response of mesenchymal stem cells to 330 nm deep microgrooves. J R Soc Interface. 2008 Mar; 5(27):1231–1242.
75. Dalby MJ, Gadegaard N, Tare R, Andar A, Riehle MO, Herzyk P, et al. The control of human mesenchymal cell differentiation using nanoscale symmetry and disorder. Nat Mater. 2007 Dec; 6(12):997–1003.
76. Getzenberg RH. Nuclear matrix and the regulation of gene expression: tissue specificity. J Cell Biochem. 1994 May; 55(1):22–31.
77. Engler AJ, Sen S, Sweeney HL, Discher DE. Matrix elasticity directs stem cell lineage specification. Cell. 2006 Aug 25; 126(4):677–689.
78. McBeath R, Pirone DM, Nelson CM, Bhadriraju K, Chen CS. Cell shape, cytoskeletal tension, and RhoA regulate stem cell lineage commitment. Dev Cell. 2004 Apr; 6(4):483–495.
79. Shafrir Y, Forgacs G. Mechanotransduction through the cytoskeleton. Am J Physiol Cell Physiol. 2002 Mar; 282(3):C479–C486.

Part III
Clinical Applications

Cells and Vascular Tissue Engineering

John Paul Kirton, Tsung-Neng Tsai, and Qingbo Xu

Abstract Cardiovascular disease is the leading cause of death worldwide. As such, vascular reconstruction and graft bypass surgery are in high demand to slow the rate of morbidity following the onset of this pathology. However, the use of non-biocompatible synthetic grafts, especially those of small diameter (6 mm or less), frequently leads to thrombosis and occlusion of the vessel. The field of vascular tissue engineering provides an alternative method to generate small (and large)-diameter bypass grafts that can support cell growth and are expected to exhibit long-term patency. Following the generation of the first in vitro blood vessel over 30 years ago, there has been considerable progress in this area in terms of scaffold availability, construction of the vessels and application of stem cells. This chapter will specifically focus on the use of stem cells in the generation of vascular grafts. Here we will highlight the current scaffolds available for seeding cells, the alternate stem cell sources and their isolation, the methods used to differentiate stem cells into vascular lineages and their application in generating blood vessels in vitro. The future hurdles that must be overcome before tissue-engineered blood vessels can be applied to a clinical setting will also be discussed.

Keywords Stem Cell · iPS cell · Vascular · Tissue Engineering

1 Introduction

According to the World Health Organization, cardiovascular disease is the number one cause of death in the world, responsible for over 30% of fatalities (www.who.int/cardiovascular_diseases/en). In 2007, a British Heart Foundation's report supported these figures with their latest statistics, confirming that cardiovascular disease is accountable for more than one-third (36%) of mortalities

Q. Xu (✉)
Cardiovascular Division, King's College London BHF Centre, London, UK
e-mail: qingbo.xu@kcl.ac.uk

G.M. Artmann et al. (eds.), *Stem Cell Engineering*,
DOI 10.1007/978-3-642-11865-4_12, © Springer-Verlag Berlin Heidelberg 2011

in the United Kingdom each year (http://www.heartstats.org/uploads/documents%5C2007.chapter1.pdf). Coronary artery disease is the main form of cardiovascular disease and the most common cause of premature death in the United Kingdom. Coronary artery disease is the inability of the coronary arteries to deliver oxygen-rich blood to the heart muscles due to the development of an atherosclerotic plaque and is surgically treated by two methods. Either autologous vessels from the patient can be harvested and used as a bypass graft or coronary balloon angioplasty is conducted to widen the narrowed vessel whilst concomitantly depositing a metal stent to maintain the vessel integrity. However, autologous grafts may not always be used due to damage/pre-existing disease (reviewed in [1]) and the use of stents (or synthetic grafts in replacement of autologous counterparts) can often lead thrombosis. Although synthetic grafts have been produced, these materials are generally incompatible for cell growth and their application often results in failure due to the development of stenosis, thromboembolism, infection and calcification (reviewed in [1]). This is most evident when artificial materials are used to bypass arteries that are less than 6 mm in diameter, with thrombosis rates greater than 40% seen 6 months post-implantation [2], and hence the generation of small-diameter vessel substitutes is urgently required [3].

Tissue engineering has emerged with the aim of creating transplantable tissue in the laboratory. Langer and Vacanti [4] described this technology as ‘an interdisciplinary field that applies the principles of engineering and life sciences toward the development of biological substitutes that restore, maintain, or improve tissue function or a whole organ.’ In terms of vascular tissue engineering, the ultimate aim is to generate biocompatible vessel grafts in vitro (Fig. 1). Ideal biological grafts should possess a confluent endothelium, differentiated, quiescent, smooth muscle cells (SMCs) and the mechanical integrity to tolerate pulsatile arterial pressures so that it may function successfully when applied in vivo. If this can be achieved, tissue-engineered vessels may replace current techniques and treatments used to treat patients exhibiting vascular-related pathologies such as coronary artery disease.

Recent advances made in stem cell biology have led to increased interest in their potential clinical use in repairing tissue damage throughout the body and, with regard to the focus of this chapter, their application to vascular tissue engineering. When considering the application of stem cells in generating tissue-engineered vessels, a number of factors should be addressed. These include (i) clarification of potential scaffolds available to seed cells, (ii) identification of an abundant obtainable source of cells, (iii) the ability to direct cells into the required functional vascular phenotype, (iv) what the immunological implications of their use may be, (v) establishment of optimal cell-seeding conditions and (vi) assessment of the durability and patency of the tissue-engineered vessels in vivo. This chapter focuses on highlighting the current scaffolds available in tissue engineering, potential sources of stem cells and the techniques used to differentiate them into vascular phenotype. Where applicable, the success and failures of their use in tissue-engineered models will be discussed, along with the latest advances and future directions in this exciting field.

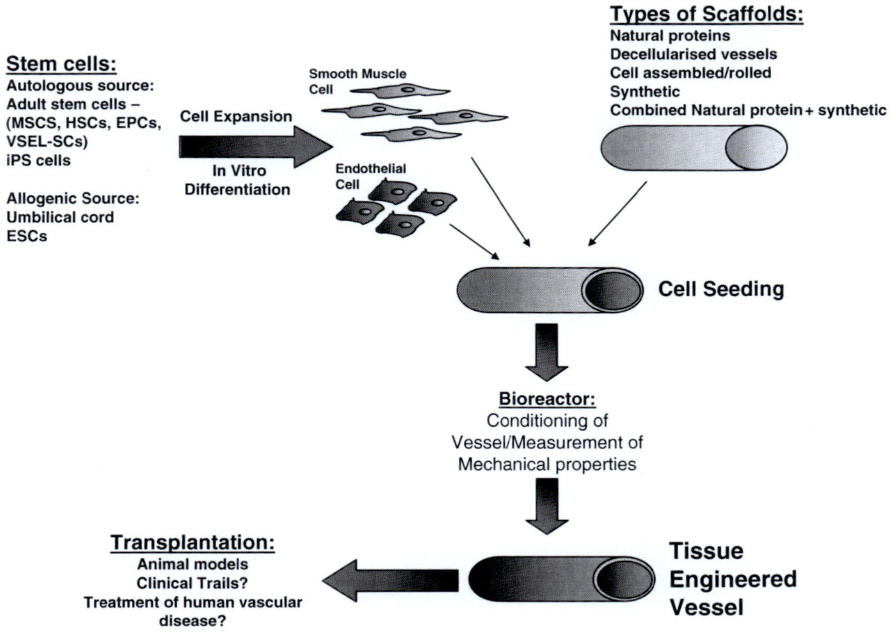

Fig. 1 Schematic image showing the different stages and options available when generating tissue-engineered blood vessels

1.1 Components of a Blood Vessel

Blood vessels are composed of three main layers. The innermost region (or the luminal side of the vessel) is called the intima. This is surrounded by the medial layer, which in turn is encircled by the adventia. Endothelial cells (ECs) line the intima, SMCs populate the media and fibroblasts the adventitia. The size of each layer varies greatly depending on the vessel type, such as large vs. small arteries or veins, in addition to microvessels (e.g. capillaries), which are composed only of endothelial tube sparsely covered by mural cell (pericytes/SMCs). Both ECs and SMCs are vital in maintaining vessel structure and repair following damage within the body [5], as such, the success of a functional tissue-engineered vessel cannot be achieved without these cell types [1]. Consequently, the differentiation of stem cells into these cell phenotypes will be discussed in later sections.

2 Methodology for Producing Vascular Grafts

In 1986, Weinberg and Bell first reported the construction of a blood vessel in vitro using collagen and cultured bovine vascular cells. This multi-layered structure resembled an artery and exhibited an EC-lined lumen and SMCs within the wall of the construct [6]. Following this report, a number of different materials have been used to generate the framework or scaffold of vascular grafts (reviewed in [7]). Scaffolds provide a three-dimensional structure to which the desired cells adhere in vitro; however, if tissue-engineered vessels are to be successfully used in humans, methods should aim to ensure that the engineered grafts are biocompatible, supporting cell growth, migration and secretion of extracellular matrix (ECM), can withstand in vivo physiological pressures and ideally be biodegradable. The following section evaluates the alternate materials available thus far to generate vascular tissue-engineered scaffolds.

2.1 Natural Protein-Based Grafts

In addition to the cellular components, native blood vessels are composed of a number of ECM proteins including proteoglycans, fibronectin and predominantly collagen and elastin. As such, Weinberg and Bell tried to replicate a natural blood vessel by using a collagen-based scaffold and generated a tissue-engineered vessel that displayed a vessel-like phenotype. However, the physical properties of the vessel were poor, bursting under low pressure even when multiple collagen layers were cast along with incorporation of a Dacron mesh [6]. The low mechanical strength was attributed to the inherent physical weakness of collagen. However, it was noted that the burst strength increased if the scaffold was used 3–6 weeks after casting, a property thought to equate to the cross-linking of the collagen. Subsequent studies have developed methods to increase the strength of collagen-based scaffolds. Researchers have used glycation, dynamic mechanical stimulation, incorporation of synthetic materials such as Dacron meshes and bioreactor systems to do this; however, the mechanical strengths of the scaffolds (rupture strengths of 500 mmHg) are still below that of a native artery (rupture strengths of more than 2,000 mmHg) (reviewed in [3, 1]). As mentioned above, elastin is also found in abundance within blood vessels. It is a vital structural and regulatory matrix protein that plays a critical role by conferring elasticity to the vessel wall in addition to regulating SMC activity and phenotype [8]. Consequently, researches have investigated the properties of grafts when this structural protein was incorporated. Using electrospinning technology, elastin/collagen vascular constructs which support cell growth and differentiation have been produced [9]; however, although the tissue-engineered vessels displayed increased tensile strength and viscoelastic properties when compared to collagen scaffolds, their burst strength was still below that of control arteries [10].

The recent use of fibrin-based gels as scaffolds has proved more successful in generating tissue-engineered vessels capable of withstanding pulsatile blood flow and haemodynamic pressures. Small-diameter blood vessels, generated using both primary and stem cell-derived vascular cells, were embedded in fibrin gel scaffold and implanted into jugular veins of lambs. Tissue-engineered vessels maintained a uniform endothelial monolayer, exhibited ECM remodelling with production of collagen and elastin fibres, displayed SMC orientation perpendicular to the direction of blood flow, gained considerable mechanical strength and were patent when analysed 5 (stem cell-derived grafts) and 15 (primary cell-derived grafts) weeks post-implantation [11, 12]. Recent data from the same group has also suggested that the mechanical strength can be further increased by using a two-layered tissue-engineered vessel which exhibited a burst pressure that was 10-fold higher than the single-layered tissue-engineered vessel [13]. Further research is required to ensure that mechanical strength can be maintained over extended periods before natural protein tissue-engineered vessels can be used as replacement vascular structures.

2.2 Decellularized Vessels

Due to the lack of strength exhibited by the majority of natural protein scaffolds, many researches turned to the method of decellularization of native vessels. The cellular component of transplanted allogeneic or xenogeneic tissues is the primary initiator of the immune response and as such removal of foreign cells from a blood vessel would leave a suitable ECM scaffold from which to recreate a new tissue. To this end, decellular vessels have now been applied in numerous vascular tissue-engineered models (reviewed in [14, 15, 1]). Decellularization can be achieved using physical methods (e.g. snap freezing, sonication and agitation), chemical agents (e.g. ionic or non-ionic detergent, acid/alkaline treatments) and enzymatic treatment (e.g. endo- and exo-nucleases). The most effective protocol includes a combination of all three approaches, although it must be noted that the biological and mechanical properties of the decellularized vessel can vary between alternate treatments [14]. Studies have shown that decellularized vessels treated with heparin alone (to prevent thrombus formation) can be successfully applied in animal models; however, it is accepted that this approach would not be applicable to the human setting and an EC layer is essential for the prevention of graft thrombosis or atherosclerosis (reviewed in [1]). Consequently, investigators have focused on seeding both human ECs and SMCs onto decellularized grafts [16–18], generating tissue-engineered vessels that possess morphological and functional characteristics of human small-calibre vessels. Alternate studies have concentrated on modifying the decellularization protocol to preferentially remove certain ECM proteins and producing a scaffold enriched with a desired protein (e.g. collagen or elastin) [10, 19]. This was done as cell migration into decellularized scaffolds was thought to be inadequate due to the compact ECM organization within the vessel. Data has shown that these scaffolds have enhanced potential for repopulation in addition to their ability to support de novo ECM synthesis to replace the removed proteins [20]. Successful generation of tissue-engineered vessels using decellularized human umbilical cord [21] and saphenous vein has also been documented [22]. Analysis of the decellularized saphenous vein showed unchanged collagen morphology, elastin expression and compared with fresh vein had similar in vitro burst potential. When this technique was applied in a canine model, 8-week post-implantation decellularized saphenous vein exhibited satisfactory strength to maintain arterial flow, reduced antigenicity compared to fresh allograft and the ability to support cellular repopulation [23]. This area of research is highly relevant to the clinical setting for patients requiring distal arterial bypass or coronary artery bypass but lacking adequate autologous vessels for their surgeries. Saphenous veins are relatively easy to procure from transplant tissue donors, are readily available, can act as small-diameter tissue-engineered vessels once decellularized and importantly would eradicate the immunogenic issues associated with xenografts (e.g. using decellularized porcine vessels).

2.3 Cell-Assembled Vascular Grafts

In 1998, L'Heureux and colleagues presented a novel approach to generate a tissue-engineered vessel, exclusively using cultured human cells without any synthetic or exogenous biomaterials [24, 15]. Here, SMCs and fibroblasts were cultured in the presence of ascorbic acid to produce a cohesive cellular sheet. The SMC sheet was rolled around a tubular support and surrounded by the fibroblast sheet, generating an in vitro medial and adventitial layer. After maturation, the inner region of the cellular tube was seeded with ECs, generating an intima-like region. Importantly, this construct was the first completely biological tissue-engineered vessel to display burst strengths of over 2,000 mmHg, comparable to those of human vessels. In subsequent studies, human tissue-engineered vessels with internal diameters of up to 4.5 mm were xenografted into immunosuppressed canine (short-term) and rat (long-term) models [25]. The tissue-engineered vessels were anti-thrombogenic and mechanically stable for 8 months in vivo. It must be noted that in latter studies, the SMC layer was excluded and replaced by a decellularized fibroblast sheet; however, a population of α-smooth muscle actin (α-SMA)-positive cells developed within the tissue-engineered vessel, suggesting the regeneration of the medial layer. Although this model has produced exciting results, the time required to generate the vessel (28 weeks) would clearly prohibit application in urgent clinical settings. However, many individuals who lack autologous vessels for revascularization procedures can be identified months before surgical intervention and as such sheet-based tissue-engineered vessel may serve an integral role in future vascular medicine.

2.4 Biodegradable Synthetic Vascular Grafts

The most commonly used biodegradable synthetic material to generate vascular scaffolds is polyglycolic acid (PGA); however, application of poly(L-lactic acid) (PLLA), poly(D,L-lactic-*co*-glycolic acid) (PLGA) and polycaprolactone (PCL) has also been documented [26–30]. Niklason and colleagues generated the first autologous tissue-engineered vessel using PGA scaffolds seeded with SMCs. These grafts were exposed to pulsatile stress and once conditioned were injected with porcine ECs to generate an endothelialized luminal region. The tissue-engineered vessels were then implanted into the saphenous artery of a swine and displayed patency when analysed 24 days later [26]. Alternatively, Iwasaki and colleagues adapted the cell-assembled graft technique to produce hybrid tissue-engineered vessels. Bovine

SMCs and fibroblasts, seeded in PGA and PCL sheets, were rolled around a support tube to generate an SMC layer surrounded by a fibroblast layer. The lumen was then seeded with ECs and a bioreactor used to generate pulsatile pressure over a 2-week period. These tissue-engineered vessels acquired similar appearance, elasticity and importantly tensile strength when compared to native arteries [31].

Genetically modified human SMCs from both young and elderly donors and more recently non-modified human bone marrow-derived SMCs (discussed further below) have been applied to successfully generate tissue-engineered vessels using this method; however, the inability of the humanized tissue-engineered vessel to withstand chronic physiological stress (burst strength still below that of standard vessel) still remains an issue [32–34].

The degradation products of these polymers are removed by natural metabolic pathways within the body. As these are acidic compounds, the effect on local environment can be detrimental to cell viability [27]. Physical and chemical modifications have improved the biocompatibility of these scaffolds (reviewed in [1]), but further research is required to confirm the long-term effects of these acidic by-products on tissue-engineered vessels and surrounding vasculature.

2.5 Combined Natural and Synthetic Vascular Grafts

The vascular engineering field has also recently tested the potential of generating a small-diameter (<6 mm) scaffolds by combining natural proteins and biodegradable synthetic polymers. This has been done using different methods, one of which is electrospinning [35–37]. These hybrid grafts are biocompatible, supporting cell growth, and when compared to natural protein-spun scaffolds, these display increased strength. This has been demonstrated by Lee et al., whose PCL/collagen scaffolds had greater tensile strength and exhibited higher burst pressure properties than native vessels [37], and in their earlier studies they showed that similar tissue-engineered vessels elicited no local or systemic toxic effects when implanted in vivo [38]. However, further studies are required to clarify how these tissue-engineered vessels perform in long-term studies when implanted in vivo.

2.6 Bioreactors

Within the vascular system, a tissue-engineered vessel would encounter a dynamic environment such as cyclic strain and shear stress. Consequently, bioreactors are often used to condition the engineered vessel by generating biochemical and physiological stimuli, replicating those it would encounter in vivo [39]. The use of bioreactors prior to implantation is vital for the success of the tissue-engineered vessel as it can aid in cell seeding, differentiation, maturation and ECM secretion [40, 41], cell alignment and adhesion, in addition to improving the mechanical properties of the construct ([35]; reviewed in [42, 7]).

3 Stem Cell Sources for Tissue-Engineered Vessels

An ideal cell source would be easy to isolate and expand in culture, non-immunogenic and able to differentiate into functional vascular cells. To date, there are three main sources of stem/progenitor cells that have been hypothesized to be potential candidates for use in generating tissue-engineered vessels: (1) adult stem cells, (2) embryonic stem (ES) cells, (3) induced pluripotent stem (iPS) cells. It is important to remember that although the name of each cell type is associated with the word 'stem', a word in cell biology often linked with pluripotency, adult stem cells have a limited potential of cell phenotypes they can acquire as opposed to embryonic and iPS cells that are the capable of unlimited proliferation and differentiation into any cell type. The research using each thus far has shown promising results, identifying important factors required for successful differentiation into ECs/SMCs and in the application to generate tissue-engineered vessels. The following section will now discuss each cell source in more detail and conclude the potential for their use in future vascular tissue-engineering processes.

3.1 Adult Stem Cells

Several reports have described multi- or pluripotent cells derived from numerous adult tissues such as the muscle, spleen, liver, adipose and blood vessel ([43]; reviewed in [44]). Additionally the bone marrow harbours heterogeneous populations of haematopoietic stem cells and non-haematopoietic stem cells, such as mesenchymal stem cells, multi-potent adult progenitor cells, very small embryonic-like stem cells and endothelial progenitor cells. The following section will describe these potential sources along with the cell isolation and characterization techniques used to clarify their stem cell-like qualities.

3.2 Mesenchymal Stem Cells

Mesenchymal stem cells are non-haematopoietic multi-potent cells [45]. Certain literatures have also referred to mesenchymal stem cells as 'marrow stromal cells', as it was thought that these cells 'arose from the complex array of structures found within the marrow' [46]. In 1966, Friedenstein and colleagues [47] were the first to show their stem cell-like qualities as mesenchymal stem cells acquired an osteogenic phenotype. Resulting studies further clarified their capabilities to differentiate into many mesenchymal cell types. When whole bone marrow aspirates are placed in culture dishes, mesenchymal stem cells are isolated due to their capability to adhere to plastic and exhibit a fibroblastic-like morphology, whereas non-adherent haematopoietic stem cells are removed. Using different culture conditions, in addition to bone formation, differentiation into chondrocytes, adipocytes and myoblasts has now been shown ([48–51]; reviewed in [46]). Friedenstein et al.

also demonstrated that after 20 or 30 cell doublings in culture, mesenchymal stem cells retain the multi-potent potential, in vitro and in vivo [52]. Although these cells have been studied for a number of decades, there is still no specific mesenchymal stem cell marker to aid in isolation procedures. As a minimal prerequisite for mesenchymal stem cell classification, cells must be plastic adherent when maintained in standard culture conditions and be able to differentiate into the cell types mentioned above. Also they must express CD73, CD90 and CD105 [53] (although CD29, CD44, CD124, CD166 are also classed as markers), and lack expression of markers indicative of other lineages such as CD11, CD14, CD34 and CD45 (haematopoietic) or adhesion molecules such as CD31 (platelet/endothelial cells), CD18 (leukocyte function associated antigen-1) and CD56 (neuronal cell adhesion molecule-1) (reviewed in [54, 45]).

In contrast to original suggestions, mesenchymal stem cells can be categorized into two groups, bone marrow derived and non-bone marrow derived. Human bone marrow-derived mesenchymal stem cells can be isolated from samples that undergo separation by density gradient centrifugation and are harvested from the iliac crest (containing mononuclear cells) [55, 56]; however, within the bone marrow, these cells are thought to be found at a frequency of 0.01–0.0001% of all nucleated cells (reviewed in [54]). Additionally, within the last decade, isolation and characterization of human and murine multi-potent adult progenitor cells from the bone marrow have been documented. Multi-potent adult progenitor cells co-purify with bone marrow mesenchymal stem cells but can differentiate into cells with mesoderm, neuroectoderm and endoderm characteristics [57, 58]. These are characterized as vascular endothelial cadherin (VE-cadherin or CD144), CD19, CD34, CD44, CD45, Sca-1, MHC I, MHC II negative, CD13, CD133, VEGFR2, VEGFR1 and SSEA1 positive [59, 58]. Like mesenchymal stem cells, these cells can also be expanded in culture without obvious senescence over high passages (for more than 80 population doublings), which would be an important factor if expansion of these cells was required for tissue-engineering purposes.

Non-bone marrow mesenchymal stem cells can alternatively be found within the vascular system, adipose tissue in addition to the umbilical cord blood and connective tissue [60–67] and mesenchymal stem-like cells within the umbilical cord tissues. Within the cardiovascular system, isolation and characterization of cells with stem cell qualities have been well documented. Indeed early work from our group provided the first evidence that the adventia contains a large number of cells expressing stem cell markers, e.g. $Sca1^+$ (21%), c-kit^+ (9%) and $CD34^+$ (15%) [61]. In addition, a subpopulation of arterial medial cells express mesenchymal stem cell-like markers [68] and pericytes, found primarily within the microvasculature surrounding the ECs, exhibit the same differentiation potential as bone marrow-derived mesenchymal stem cells both in vitro and in vivo [69, 70]. Interestingly in a recent review, Caplan [71], who is a mesenchymal stem cell investigator for over two decades, suggested that all mesenchymal stem cells are a sub-type of pericytes. Indeed the presence of the vascular system within all adult tissues in which mesenchymal stem cells are thought to reside indicates that a common progenitor may be source of these multi-potent cells. However, the use of these progenitors for vascular tissue-engineering purposes is very unlikely as isolation, culture and expansion of these cells would require sacrifice of a healthy vessel from patients who already present with vascular illness. Obtaining mesenchymal stem cells from the bone marrow, adipose or umbilical cord blood, however, provides feasible cell sources for generating tissue-engineered vessels due to the ease of procurement and expansion potential of these cells. Importantly, mesenchymal stem cells also display storage potential; therefore cells could be harvested, cryogenically stored, thawed and expanded at a later date if required for therapeutic uses. However the use of bone marrow-derived mesenchymal stem cells has practical limitations that may limit their widespread use as autogenous cells for tissue-engineered vessel generation. These include invasive and painful harvesting techniques of cells, general requirement for ex vivo cell expansion and associated significant decrease in cell number and proliferative/differentiation capacity with age of patient (reviewed in [72]). In comparison, adipose tissue-derived mesenchymal stem cells can be easily isolated in large quantities by minimally invasive liposuction with a significantly higher yield of autogenous progenitor cells per volume compared with bone marrow [73]. Alternatively, as the umbilical cord is a post-natal organ discarded after birth, the collection of cells does not require an invasive procedure. On this note, a recent publication documented the isolation of a sub-type of mesenchymal stem cells from the umbilical cord matrix (Wharton's Jelly), termed human extra-embryonic mesoderm stem cells, and cord blood, termed unrestricted somatic stem cells [74, 67]. Both cell types exhibited mesenchymal stem cell marker expression, characteristic differentiation and clonogenic potential, in addition to maintenance of long telomeres in high passaged cells. Intriguingly, human extra-embryonic mesoderm stem cells also expressed markers of embryonic stem cells (molecules responsible for the maintenance of an undifferentiated state and self-renewal), namely Oct1, Oct4 and Nanog, along with the ability to express transcription factors involved in the development of mesoderm- and endoderm-derived lineages [67] and unrestricted somatic stem cells displayed potential to develop into mesodermal, endodermal and ectodermal cell fates both in vitro and in vivo [74]. In similarity, cells termed human multi-potent adult stem cells, obtained from liver, heart and bone marrow, also share mesenchymal stem cell properties but express embryonic stem cell pluripotent state-specific transcription factors Oct4, Nanog and Rex-1, and display telomerase activity, clonogenic ability and differentiation potential along all three germ layers [43]. Further research will undoubtedly determine the full differentiation potential of this novel cell type and determine if it may play a role in therapeutic processes such as vascular tissue engineering.

3.2.1 Mesenchymal Stem Cell Differentiation into Vascular Cells

In order to generate a mesenchymal stem cell-derived tissue-engineered vessel which resembles its native counterpart, differentiation into ECs (to line the lumen) and SMCs (to generate a medial layer) would be essential. To this end, Oswald and colleagues provided conclusive evidence that human bone marrow mesenchymal stem cells could differentiate into ECs in defined culture conditions [75]. Expression of mesenchymal stem cell markers, co-commitment lack of haematopoietic stem cell markers (see above) and the ability to differentiate into osteo-, chondro- and adipogenic cell lineages confirmed a mesenchymal stem cell population. EC differentiation was stimulated following culture for 7 days with vascular endothelial growth factor (VEGF). Analysis revealed that these cells acquired an EC phenotype expressing endothelial-specific markers such as Flk-1 (VEGF receptor 2, VEGFR2) Flt-1 (VEGFR1) and von Willebrand factor (vWF), and the ability to form capillary-like structures in vitro. In an alternate studies, bone marrow mesenchymal stem cell differentiated into ECs and SMCs in vivo [76]; [77]. Thirty days following injection of murine bone marrow-derived mesenchymal stem cells into an infarcted myocardium, engrafted cells acquired both EC and SMC phenotypes, expressing CD31 and α-SMA, respectively [76]. In vitro, bone marrow mesenchymal stem cells exposed to cyclic strain also acquire an SMC phenotype expressing α-SMA and calponin [78]. Additionally, Liu et al. demonstrated a method to isolate highly purified population of bone marrow-derived smooth muscle progenitor cells. Here, ovine bone marrow-derived mesenchymal stem cells were transiently transfected with a plasmid in which an SMC-specific promoter (α-SMA) drove expression of GFP allowing isolation of α-SMA-positive cells via fluorescence-activated cell sorting (FACS). Bone marrow-derived smooth muscle progenitors showed high proliferative potential and displayed morphological and phenotypic properties of SMCs as shown by the expression of early and late smooth muscle markers such as α-SMA, calponin, myosin heavy chain, caldesmon, SM22 and smoothelin [12]. The same authors also isolated ECs from bone marrow-derived mesenchymal stem cells following culture on fibronectin-coated dishes and removal of non-adherent cells. Endothelial growth medium supplemented with EGF and βFGF gave rise to cells positive for Dil-Ac-LDL uptake, CD31, CD144 and vWF. Alternatively human and murine bone marrow-derived mesenchymal stem cells can differentiate into SMCs when cultured with media supplemented with TGF-β1 along with attachment to dishes coated with elastin, fibronectin or collagen I/IV [79, 80, 34]. These studies highlight the importance of specific cytokine addition to stimulate either differentiation or replication into the required phenotype. Indeed Gong and Niklason also found that PDGF-BB and βFGF repressed SMC differentiation but enhanced proliferation of these cells [34].

Studies have described that murine and human adipose-derived stromal cells were able to differentiate into endothelial cells both in vitro [81] and in vivo [82], with the latter study showing incorporation into vessels, promotion of post-ischemic neovascularization and formation of vessel-like structure in Matrigel plug experiments. In must be noted that although the isolated cells were termed mesenchymal stem-like cells due to the lack of expression of haematopoietic stem cell markers (CD45 and CD14) and differentiated endothelial cell markers (CD144 and/or CD31), a large proportion of the progenitor cells were CD34 and VEGFR2 positive and as such could have represented tissue-resident endothelial progenitors (see below); however further experiments are required to fully characterize these cells. Indeed Moon and colleagues also initially isolated CD34-positive cells from human adult adipose tissue but discovered that within three passages, marker expression had been downregulated. These passage 3 cells expressed mesenchymal stem cell markers CD29, CD44, CD90, CD105 but not haematopoietic stem cell markers CD34, CD14, CD45 or EC markers CD31 and VEGFR2. The adipose-derived mesenchymal stem cells induced a significant increase in the blood flow and capillary density in the ischemic hindlimb of nude mice when transplanted in vivo, results which mirrored those of the unpassaged CD34-positive fraction of cells. In addition, they could differentiate into ECs in vitro and produce pro-angiogenic cytokines and inhibited apoptotic cell death of human aortic ECs [83]. Murine adipose-derived mesenchymal stem cells also exhibited similar differentiation potential and secretion of pro-angiogenic factors such as VEGF and hepatocyte growth factor, which would be a viable asset if applied to tissue-engineered vessel [84]. In accord with other mesenchymal progenitors, multi-potent adult progenitor cells [59, 85], umbilical cord blood and tissue-derived mesenchymal stem cells [86–88] also display the ability to differentiate in vitro and in vivo into ECs and SMCs.

3.2.2 Tissue-Engineering Models Using Mesenchymal Stem Cells

Due to their autologous origin, adult stem cells avoid both immunorejection and ethical issues that surround the use of ES cells. They do not exhibit tumourigenic capacity that ES cell and some iPS cell lines (see following section) present but do have high proliferative capacity and can differentiate into vascular cells, making them ideal candidates for generating tissue-engineered vessels. To investigate the success of these adult stem cell-derived tissue-engineered vessels in vivo, animal models have been used, providing insight into the integrity of bone marrow-derived vascular cell-seeded tissue-engineered vessels when exposed to physiological conditions [89, 90].

Ovine bone marrow-derived SMCs, implanted into fibrin-based grafts, were wrapped around a mandrel to form a cylindrical tissue and cultured for 2 weeks. The lumen was then seeded with bone marrow-derived ECs prior to implantation into jugular vein of lambs. Histological analysis revealed a patent vessel 5 weeks post-implantation. The tissue-engineered vessels exhibited enhanced ECM remodelling when compared to neovessels generated with primary SMC including higher amounts of synthesized elastin, which was organized in a fibrillar network similar to that of native vessels [12]. Recent data has also shown successful tissue-engineered vessel generation using PGA scaffolds seeded with human bone marrow-derived SMCs and human umbilical cord-derived ECs, cultured in a two-phase (proliferation and differentiation) culture system within a bioreactor [34]; however the in vivo potential of this vessel was not tested. Co-implantation of human bone marrow-derived mesenchymal stem and endothelial cells into SCID mice did however demonstrate that mesenchymal stem cell-derived SMCs can generate stable and functional vessels for over 130 days [91]. This finding was supported by alternate studies in which umbilical cord-derived or bone marrow-derived mesenchymal stem cells suspended in Matrigel with endothelial progenitors (see below) and implanted into SCID mice were able to differentiate into mural cells and form functional vessel that anastomosed with host vasculature [88].

3.3 Haematopoietic Stem Cells

Developmental biology has determined that haematopoietic stem cells derived from the haemangioblast found within primitive streak/aorta gonado mesonephros of the developing embryo and pre-haematopoietic stem cells initially populate the foetal liver. In response to an SDF-1 gradient, haematopoietic stem cells, which express the corresponding receptor CXCR4, migrate to and reside within bone marrow, establish adult-type haematopoiesis (reviewed in [92]) but constitute only 0.01% of the total cells (reviewed in [44]). They possess the capability of self-renewal and the multi-potent potential of differentiating into eight major haematopoietic lineages and can repopulate the haematopoietic system in myeloablated recipients (reviewed in [93]). Haematopoietic stem cells can be isolated from the population of mononuclear cells from the bone marrow, mobilized peripheral blood or cord blood samples which undergo ficoll gradient centrifugation, hypotonic lysis and leucapheresis (reviewed in [92]). As there is no specific haematopoietic stem cell marker available, a number of surface antigens/receptors are used when extracting a haematopoietic stem cell fraction. For human haematopoietic stem cells these include expression of CD34 (although side populations of CD34-negative haematopoietic stem cells do exist), CD45, CD133, CXCR4, CD150 and lack of lineage differentiation marker (Lin^-), CD48, CD244 and low accumulation/staining of DNA dye (e.g. Hoechst 3342), whereas murine haematopoietic stem cells are also Flk-1, Sca-1, c-kit and $Thy1.1^{low}$ positive. However, it must be noted that antigen expression can vary with haematopoietic stem cells isolated from alternate sources (bone marrow vs. peripheral blood), exposure to specific cytokines and cell cycle status (reviewed in [54, 92]). Recent data has also suggested that the adipose tissue also appears to harbour haematopoietic stem cells as indicated by the marker expression such CD45 and CD11 [81]. Although the techniques for isolating these adult stem cells are readily available, haematopoietic stem cells, like mesenchymal stem cells, display a correlation of decreased ability to self-renew and differentiate with increasing age [94, 93, 72], which could act as a cell source limitation for vascular tissue-engineering applications.

3.3.1 Haematopoietic Stem Cell Differentiation into Vascular Lineages

It has been proposed that haematopoietic stem cells also have the ability to differentiate into both ECs and SMCs and as such are seen as an obtainable autogenous cell source for tissue-engineered vessel construction (reviewed in [54]). Evidence of this differentiation capacity has been shown by a number of groups. Sata et al. co-cultured haematopoietic stem cells (c-kit/Sca-1 positive/Lin negative) with mature SMCs and found that these cells acquired an SMC phenotype expressing α-SMA and calponin. In addition, when these cells were injected into mice with arterial injury, they were shown to contribute to vascular remodelling, differentiating into SMCs and ECs [95]. Jackson and colleagues showed that when side populations of murine haematopoietic stem cells (CD34 negative, Sca-1/c-kit positive) were transplanted into lethally irradiated mice with myocardial infarction, these cells became incorporated into vessel structures and expressed ECs markers [96]. Additionally lineage-negative/c-kit-positive haematopoietic stem cells differentiated into ECs and SMCs in an alternate myocardial infarction model [97]. Alternatively, many studies have isolated CD34-positive cells to investigate their differentiation potential. For example, human CD34-positive mononuclear peripheral blood cells displayed the ability to differentiate into SMCs and ECs when cultured on collagen I and exposed to PDGF-BB or VEGF, respectively [98]. The addition of VEGF was also essential for EC differentiation in an alternate study using human CD34-positive cells isolated from the bone marrow, mobilized peripheral blood and cord blood [99]. However, it must be noted that the majority of these studies did not use an extensive panel of haematopoietic stem cell markers and as such potential contamination from alternate cell types may have contributed to the finding. Indeed, as haematopoietic stem cells share many isolation markers with endothelial progenitor cells (CD34/CD133; see below), many studies may in fact be describing this cell type. With regard to vascular differentiation, it has been proposed that endothelial progenitor cells occupy the middle of the differentiation spectrum between haematopoietic stem cells and ECs [54] and as such many recent studies have concentrated on their differentiation and application to tissue-engineered vessels (below).

3.4 Endothelial Progenitor Cells

In 1997 Asahara et al. [100] reported the presence of CD34-positive haematopoietic progenitors from peripheral blood. These heterogeneous population cells, termed endothelial progenitor cells (EPCs), have the capacity to proliferate, migrate and differentiate into mature ECs. These cells can be expanded for over 20 passages without losing their differentiation potential. As such, they provide an ideal target for ex vivo expansion and application to tissue-engineered vessels. In addition, it is now known that these cells can also be isolated from cord blood [101–103]. At present, EPCs are generally isolated as the CD133- and/or VEGFR2-positive subpopulation of CD34-positive cells [104] but lose CD133/CD34 expression and gain high expression of CD31, vascular endothelial cadherin, VEGFR1 and vWF as they differentiate into a mature state (reviewed in [44]). These cells have multiple origins including the haemangioblast, in addition to non-haematopoietic mesenchymal precursors from either the bone marrow-derived or resident stem cells in adult tissues such as the adipose and spleen (reviewed in [44, 105]). Indeed data from our group has clearly shown the presence of non-bone marrow-derived EPCs and their ability to contribute to endothelium repair in vessel grafts. In these studies, two different models of vessel graft atherosclerosis revealed that the reconstituted endothelial monolayer is formed from recipient-circulating progenitor cells. Importantly, when Balb/c aorta was allografted into the carotid artery of chimeric mice with bone marrow derived from Tie2-LacZ mice, β-gal activity was seen on the surface of allografts 4 weeks after surgery. Quantification of the obtained data indicated that more than 70% of the regenerated endothelial cells were derived from non-bone marrow tissues [106, 107]. Furthermore, EPC number within the peripheral blood can be enhanced by cytokines/drugs, such as G-CSF [108], VEGF and statins (reviewed in [44]) and their presence within the blood system is thought to be inversely correlated to cardiovascular risk [109]. Details regarding EPC origin and mobilization will not be discussed further; for additional information, readers are directed to an excellent review article by Zampetaki et al. [105].

3.4.1 EPC Differentiation into Mature ECs

The method defined by Asahara stated that mature ECs could be obtained if CD34-positive peripheral blood cells were plated on fibronectin and cultured in the presence of VEGF. The attaching cells initially acquired spindle-shaped morphology before forming tube-like structures. Commitment to an EC lineage was confirmed by FACS, PCR and immunocytochemistry which showed expression of EC markers such as CD31, VEGFR2, eNOS and uptake of low-density lipoprotein [100].

3.4.2 Tissue-Engineering Models Using EPCs

EPCs isolated from both human and animals (e.g. rabbit, canine) have been used to seed and successfully endothelialize synthetic vessel grafts [110–112, 102]. In addition, EPC-derived tissue-engineered vessels have provided successful data when applied in vivo. Kaushal and colleagues seeded decellularized small-diameter porcine vessels with sheep EPCs to construct a tissue-engineered vessel. The EPC-seeded grafts remained patent for 130 days when implanted into sheep, whereas non-seeded grafts occluded within 15 days. Explanted vessels displayed contractile activity and nitric oxide-mediated vascular relaxation that were similar to native carotid arteries, suggesting that EPCs can function similarly to native arterial ECs even in the presence of physiological stimuli. These properties along with their availability and autologous nature indicate that EPCs represent an excellent source of ECs when generating tissue-engineered vessels [113]. A recent study also provided intriguing data for tissue-engineered vessel generation. Decellularized porcine vessels were seeded with SMCs, cultured with pulsatile stress within a bioreactor for 6 weeks to generate a medial layer and then seeded with EPC-differentiated ECs. Half of the EPCs used were transfected to convey *A20* gene expression, a gene that had previously been shown to be essential for prevention of inflammation and prevention of atherogenesis in blood vessel grafts. Both histological analysis and mechanical testing revealed that the tissue-engineered vessel was comparable to a native vessel [114]. To test their anti-atherosclerotic potential, EPC-derived grafts were transplanted into rats. One month post-implantation, 50% of control tissue-engineered vessels (non-*A20*-transfected cells) displayed evidence of stenosis and by 6 months no grafts were patent (20/20). Alternatively all *A20*-transfected grafts were patent following 6-month implantation [114], suggesting that expression of specific anti-atherosclerotic genes or drugs in tissue-engineered vessels may be advantageous to convey long-term patency.

3.5 Very Small Embryonic-Like Stem Cells

In 2006, Kucia et al. described the isolation of a rare homogenous, non-haematopoietic stem cell population from murine bone marrow (approximately 0.02% of bone marrow-derived mononuclear fraction) [115]. These cell types, now also isolated from human bone marrow, are characterized by lack of CD29, CD45, CD90 and CD105 (haematopoietic and mesenchymal stem cell markers) and concomitant positive expression of Sca-1 (mouse cells), CD133, CXCR4 (also found on haematopoietic stem cells) and, unlike haematopoietic/mesenchymal stem cells, express markers of pluripotent ES cells such as SSEA-1 (mouse), SSE1-4 (human), Oct4, Nanog, Rex-1 and telomerase activity (reviewed in [116, 117]). Further analysis revealed that these cells are small (2–4 μm; haematopoietic stem cells, 8–10 μm), possess an ES cell-like large nuclei surrounded by a narrow rim of cytoplasm and contain open-type chromatin, whereas haematopoietic stem cells, for example, display scattered chromatin and a prominent nuclei. In vitro, cultures of these cells are able to form structures that resemble embryoid bodies (5–10%), which can subsequently differentiate into all three germ-layer lineages ([115]; reviewed in [118]). Kucia and others have hypothesized that these cells are epiblast-derived pluripotent stem cells deposited during development in the bone marrow. As such, they may act as precursors of haematopoietic and mesenchymal stem cells, contributing to haematopoiesis and following mobilization, and contribute to tissue-specific stem cells throughout the body. They may also aid in organ regeneration and potentially become a less-controversial source of stem cells (as opposed to ES cells) for tissue regeneration ([116, 118]; reviewed in [92]). Furthermore, these small stem cells have now also been isolated from murine foetal liver, spleen and thymus in addition to human umbilical cord and peripheral blood (reviewed in [116, 117]) and can be mobilized and respond robustly to SDF-1 (which binds to CXCR4). Additionally it has been suggested that due to their origin of isolation along with adherent fibroblastic culture qualities, the ES cell-like features displayed by alternate novel cell lines (e.g. multi-potent adult progenitor cells/multi-potent adult stem cells) may be attributable to 'contamination' by these small rare cells (reviewed in [118]).

These cells display great potential as a novel autologous cell source for (vascular) tissue-engineering purposes; however. they do present potential obstacles that must be considered before therapeutic application. At present, culture conditions which allow differentiation of these cells without supportive feeder cells have not been established (reviewed by Kucia et al. [118]). Additionally, they represent only 1 in 10^4–10^5 of bone marrow mononuclear cells and this number decreases with age. Indeed, the percentage of small stem cells obtained from murine bone marrow dropped from 0.052 to 0.003% between 2-month- and 3-year-old specimens [117], which also coincided with the ability to form embryoid body-like structures. Further research will clarify whether these cells can be adapted for tissue regeneration and application to vascular engineering.

3.6 Embryonic Stem Cells

In 1981, the first embryonic stem cells (ES cells) were derived from the mouse inner cell mass of embryos [119, 120], and in 1998 human ES cells were also isolated [121]. The following studies confirmed that ES cells could provide an unlimited source of cells for transplantation therapies such as tissue-engineered vessels, exhibiting unlimited differentiation potential and high proliferation capabilities and providing the demand for high cell numbers in this process. Nevertheless, there are still several major problems that must be addressed before clinical use of ES cells, including the immunological rejection, potential tumourigenesis and ethical issues. To address immunogenic and tumourigenic issues, the establishment of personalized pluripotent ES cells (using methods such as somatic cell nuclear transfer) has been demonstrated. However, future application of this method would still have to bypass ethical issues formed by the general public.

3.6.1 Embryonic Stem Cell Differentiation into a Vascular Lineage

In vitro, embryoid body formation, an aggregation of ES cells, initiates spontaneous differentiation into endoderm, ectoderm and mesoderm lineages, including progenitors of a vascular lineage. Levenberg and colleagues were the first to demonstrate that human ES cells can be differentiated into ECs in vitro. They isolated PECAM1-positive cells from day 13 embryoid bodies by FACS and following characterization they found that these cells expressed endothelial marker including vWF, CD34, VEGFR2 and CD144 and the ability to uptake Dil-Ac-LDL. In addition, when the PECAM1-positive cells were cultured in Matrigel, they formed tube-like structure. To assess their functional properties, ES cell-derived ECs were seeded onto PLA–PLGA biodegradable polymer scaffolds and implanted subcutaneously into SCID mice. Two weeks following implant, histological examination found that microvessels were immunoreactive with human PECAM1 and CD34 [122], indicating that human ES cell may act as a cell source for future vascular tissue engineering. Also, CD34-positive cells isolated from human ES cell-derived embryoid bodies were also shown to differentiate into endothelial-like and smooth muscle-like cells following culture with endothelial growth medium containing VEGF (50 ng/ml) or PDGF-β (50 ng/ml), respectively. Interestingly, the ES cell-derived SMCs not only expressed the SMC-specific marker (α-SMA, SM myosin heavy chain, calponin, caldesmon and SM alpha-22) but also showed the ability to contract or relax when carbachol or atropine was administrated [123], indicating a functional phenotype. VEGFR2-positive cells isolated from mouse ES cells also differentiate into ECs and SMCs when treated with VEGF and PDGF-β, respectively [124]. Recently, the same group also showed that human VEGFR2-positive ES cells can differentiate into ECs and SMCs [125]. Alternatively mouse and human ES cells can be differentiated into SMC, expressing phenotypic markers and contraction capabilities, following treatment with retinoic acid and/or dibutyryl cyclic adenosine monophosphate (db-cAMP) [126, 127].

However, inmost of these reports, the cell purity of differentiated cells was not mentioned or low.

Alternatively, data from our group has clearly demonstrated that the Sca-1-positive progenitors sorted from ES cells can be differentiated into both EC and SMCs [128, 129]. When cultured in the presence of VEGF for approximately 21 days, these Sca-1 progenitors displayed EC-like morphology and marker expression (CD31, CD106, CD144, VEGFR1, VEGFR2 and vWF). These cells were shown to be functional in vivo; ES cell-derived ECs mixed with Matrigel and injected into mice were shown to form highly vascularized structures [128]. Additionally, these cells could decrease neointimal formation in a femoral artery mouse model (Fig. 2a–d) and also promote re-endothelialization (Fig. 3a–i). These studies also discovered that histone deacetylase 3 (HDAC3) expression is essential for VEGF-induced EC differentiation from Sca-1 progenitor cells [128]. This data was further supported by subsequent data which showed laminar flow-enhanced ES cell differentiation into ECs by activating HDAC3 via an Flk-1–PI3K–Akt pathway [130]. Furthermore, Xiao et al. also showed that Sca-1-positive progenitors sorted from mouse ES cells and cultured on collagen IV in the presence or the absence of PDGF can generate highly enriched populations of differentiated SMCs cultures (>95%, expression markers such as α-SMA, SM-MHC and calponin; see Fig. 4a–c). They subsequently demonstrated that SMC differentiation on collagen IV involves the integrin (α1, β1, α5)-FAK/paxilin–PI3K–MEK–ERK/JNK pathways and the PDGF-β signalling and that the collagen produced by the culturing cells has a

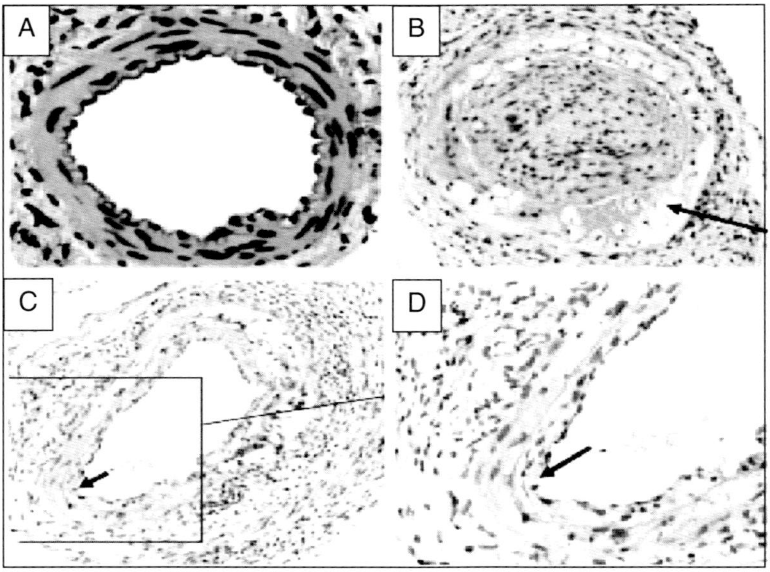

Fig. 2 ES cell-derived ECs can prevent neointimal formation in a murine-injured femoral artery model. Two weeks post-cell transfer, injured arteries were harvested, fixed, sectioned and stained with H&E. Images show uninjured artery (**a**), wire-injured artery (**b**) and ES cell derived EC-treated (**c** and **d**), wire-injured femoral arteries. Magnification 100× (**a–c**); 200× (**d**)

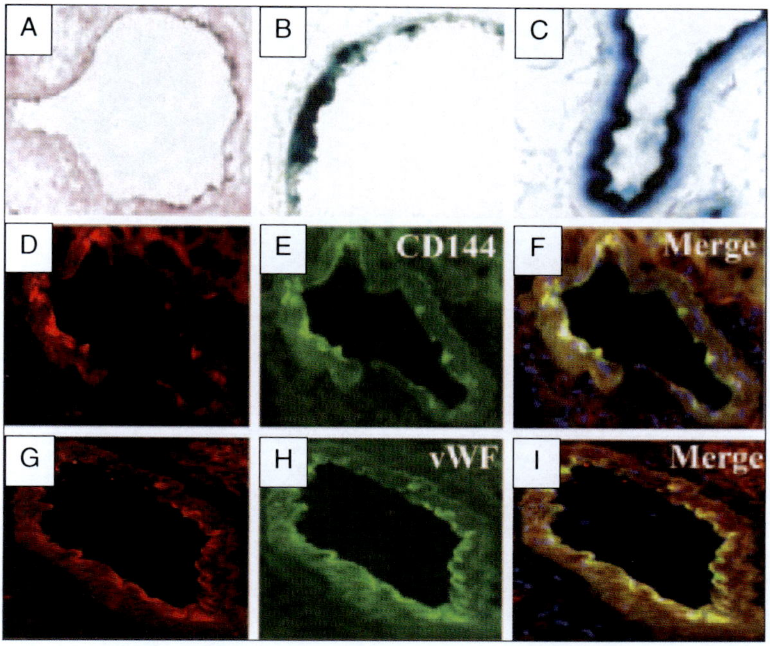

Fig. 3 Transplanted ES cell-derived ECs contribute to re-endothelialization in a murine-injured femoral artery model. ES cell-derived ECs infected with adenovirus encoding LacZ were transferred into murine-injured femoral artery model. Arterial samples were harvested 3 (**b**) or 7 (**c**) days after cell transfusion and stained with X-gal to detect LacZ. Images show negative control (**a**), partial re-endothelialization (**b**) and complete re-endothelialization (**c**) following X-gal staining. Double immunofluorescence staining also shows double-positive cells for β-gal (**d** and **g**) and EC-specific markers, CD144 (**e**) and vWF (**h**), and nuclear counterstaining with DAPI (*blue*). Merged images of the double-positive stain cells appear *yellow* (**f** and **i**). Magnification 100× (**a**, **d**–**i**); 200× (**b** and **c**)

functional role in the differentiation [129]. As highly purified populations of differentiated vascular cells would be required for tissue-engineered vessel generation, such methods will ultimately contribute to the success of this exciting field.

3.6.2 Tissue-Engineering Models Using Embryonic Stem Cells

Although ES cell differentiation into vascular lineages has been well studied, at present, their use as a cell source for in vitro-generated tissue-engineered vessels has been limited. However, Shen and colleagues have demonstrated ES cell-derived EC application in a small-diameter (3 mm) tissue-engineered vessel. Initially SMCs seeded onto non-woven biodegradable PGA fibres and wrapped around a silicon tube were subcutaneously implanted into nude mice and after 8 weeks formed a vessel-like structure. Murine ES cell-derived ECs were then injected onto the luminal region of the structure and re-implanted into the mouse for 5 days. Analysis

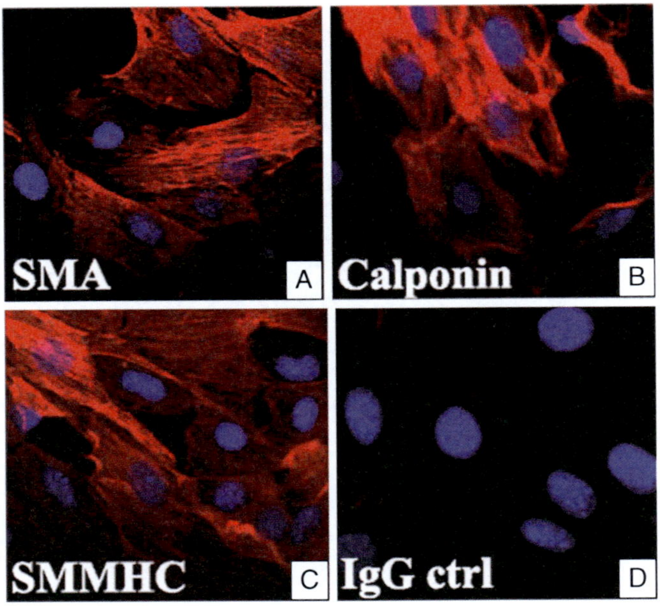

Fig. 4 ES cell-derived SMCs (passage 10) exhibiting expression of SMC-specific markers. Cells were labelled with antibodies to α-SMA (**a**), calponin (**b**), smooth muscle myosin heavy chain (**c**) and control mouse isotype IgG (**d**), and gene expression was visualized following incubation with a secondary antibody conjugated to phycoerythrin. Nuclei were counterstained with 4′,6-diamidino-2-phenylindole (*blue*); magnification 400×

revealed a typical blood vessel structure, including an EC lining of the luminal surface and the presence of SMC and collagen in the wall along with degradation of the PGA fibres [131]. However, as the tissue-engineered vessels were not implanted into the circulatory system, the mechanical and functional properties, in addition to whether the vessel can maintain patency in an in vivo setting, cannot be discussed.

ES cells provide an excellent source of cells for generating tissue-engineered vessels. However, problems such as ethical issues in addition to immunogenic and tumourigenic problems must be overcome before their application in tissue engineering can be fully exploited. Indeed, although ES cell-derived ECs/SMCs express EC- and SMC-specific marker, there still remains doubt that these cells are fully differentiated. As such any future application should aim to confirm terminal differentiation of the cells before application to tissue-engineered vessels, decreasing tumourigenic formation by these cells. Recent technology has also highlighted the potential use of parthenogenesis-derived stem cells for tissue-engineering applications. Parthenogenesis is the process in which an embryo develops from an egg in the absence of sperm, following a stimulus such as electroporation. This process has been successfully used to generate human parthenogenetic ES cell lines (reviewed in [132]). Recent work has suggested that cells derived from this method could provide an alternative source of ES cells avoiding some of the political and ethical concerns surrounding this cell type [133].

3.7 Induced Pluripotent Stem (iPS) Cells

Research conducted in the field of somatic cell fusion with ES cells indicated that specific transcription factors expressed by ES cells had the potential to de-differentiate lineage committed cells to a stem cell-like phenotype. These findings prompted Takahashi and Yamanaka to analyse the effect on 24 strategically chosen transcription factors when exogenously expressed in different combinations in fibroblasts [134]. They concluded that retroviral-mediated introduction of four specific transcription factors (Oct4, Sox2, Klf4, c-Myc) and consequent culture under ES cell conditions were sufficient to reprogramme mouse embryonic and adult tail-tip fibroblasts into an embryonic-like state. These cells, termed 'induced pluripotent stem (iPS) cells', ignited a huge interest in this novel technique and their potential use in cell therapy. iPS cells, selected via expression of Fbx (expressed by mouse ES cells), were shown to be similar to ES cells in morphology and could form tumours which consisted of tissues from all three germ layers when subcutaneously transplanted into mice but were different with regard to gene expression, DNA methylation patterns and failure to produce adult chimeric mice. This issue was quickly resolved by three independent groups using fibroblasts obtained from mice in which an antibiotic-resistant reporter gene was knocked into the endogenous loci of either Oct4 or Nanog [135–137]. These second-generation iPS cells exhibited improved reprogramming capacity (displaying increased ES cell-like gene expression and DNA methylation patterns) and the ability to generate chimeric mice, contributing to the germ line when cells were injected into a blastocyst and transplanted into the uteri of pseudo-pregnant mice, the most stringent criteria to evaluate pluripotency and stem cell capacity. In addition, all groups reported that during the process, exogenous retroviral genes had been silenced and endogenous expression of the introduced genes 'switched on'. The main issue which hindered all of the researchers was the efficiency of reprogramming fibroblasts; this was shown to be low, estimated to be a maximum of 0.03–0.1% of fibroblast used per culture to generate iPS cells [136, 137]. Okita and colleagues also reported that approximately 20% of chimeric mice generated developed tumours within their head and neck, which was contributed to the re-emergence of the exogenous expression of the oncogene *c-Myc*. It is now known that iPS cells can be generated without introduction of exogenous c-Myc retrovirus [138, 139]. Omission of c-Myc results in a less-efficient method than when all four transcription factors are used. However, chimeric mice generated from these iPS cells did not develop tumours following analysis at 4 month, whereas 15% of chimeric mice generated from iPS cells using the standard method developed tumours in the same time period [138] and 50% developed tumours after 8 months [139].

3.7.1 Advances in Murine iPS Cell Biology

This reproducible way of 'creating' stem cells within the laboratory provided many scientific questions, such as the following: Are there chemical factors/alternate factors that can aid in reprogramming? Can the efficiency of the process be increased?

Is it possible to generate iPS cells without retroviral use/without selection markers which leads to genetically modified cells? Is reprogramming restricted to fibroblasts or can this method be applied to any somatic cell? Is iPS cell differentiation limited to certain lineages? Can Human iPS cells be generated? Could these cells be used in a cell therapy-based models?

Within only a few months of generating germ line-competent mice from iPS cells, researchers addressed many of these questions. Activation of the canonical Wnt pathway can increase iPS cells generated without c-Myc transduction by almost 20-fold [140]. Furthermore, studies showed that reprogramming somatic cell to an ES-like phenotype requires 'turning off' of somatic genes with concomitant 'turning on' of ES cell-specific genes. Genomic analysis conducted by two groups clearly showed that successful reprogramming is dependent upon demethylation (activation) of promoter regions of ES cell-specific genes and methylation (inactivation) of promoters governing somatic-specific genes [135, 141]. As mentioned above, iPS generation is an inefficient, lengthy process (up to 30 days), with many cells failing to be re-programmed [135, 142]; [142–144]. To gain insights into why the efficiency of this method is so low, Mikkelsen and colleagues compared stable iPS clones and partially re-programmed somatic cells, which although display ES-like marker expression or morphology, never spontaneously give rise to fully re-programmed cells. They showed that fully re-programmed cells had similar gene expression and epigenetic states to ES cells; however, partially re-programmed cells display hypermethylation of pluripotent loci (e.g. Oct4 and Nanog promoters). Treatment of these cells with chemicals such as DNA methyltransferase inhibitor 5-azacytidine induced transition to a fully re-programmed state [141], indicating that the low efficiency of the process was due to incomplete epigenetic changes, and treatment with reagents such as histone deacetylases/DNA methyltransferase inhibitors can enable cells that are 'trapped' in a partially de-differentiated state to acquire an iPS cell genetic phenotype. Indeed, the use of alternate chemical compounds has been shown to aid in reprogramming [145, 141, 146], as addition of chromatin-modifying agents such as trichostatin A, which affects histone acetylation, BIX-01294, a histone methyltransferase, and valproic acid, a histone deacetylase inhibitor, all increases the efficiency of iPS cell formation.

Alternate research also showed that iPS cells could be obtained without using drug selection and based upon morphological criteria alone [147, 142, 138]. In addition to the use of retroviruses, third-generation iPS cells have also been generated. This system uses drug-inducible lentiviral vectors which, once integrated into the genome of the somatic cell, allow expression of the required transcription factors to be easily turned on/off [143, 144], providing a more efficient method than its predecessors and negating the need for either drug selection or the potential selection of false-positive colonies present in drug-negative systems. Also many independent laboratories have shown that reprogramming is not just restricted to fibroblasts or cells of a mesodermal lineage. Indeed, stomach, liver [148] and pancreatic cells [149], as well as terminally differentiated B lymphocytes [150], have all been successfully re-programmed. Additionally, ectodermal-derived neural stem cells (derived from embryonic neural tissue or from regions of the adult brain) express high levels of endogenous Sox2 and c-Myc and have been used to generate iPS cells. This was done using viral introduction of only three (Oct4/Klf4/c-Myc) or two factors (Oct4/Klf4), indicating that identification of somatic cells that endogenously express 'reprogramming genes' may aid in the iPS cell generation [151–153].

Generation of iPS-derived chimeric mice and the ability to form tissues from all three germ layers when subcutaneously implanted into mice indicate that iPS cells could form any cell type. In vitro studies thus far have shown iPS cells can differentiate into haematopoietic cells [154, 155] and neuronal cells [151, 156], in addition to cells of a cardiovascular lineage (cardiomyocytes, ECs and SMCs) [151, 157, 158, 155, 159, 160]. Indeed using mouse embryoid body-mediated differentiation and beating cardiomyocytes were obtained in culture which expressed cardiac markers connexin-43 and tropomyosin-I [151], indicating that these could act as functional cells in vivo. The differentiation process into endothelial and smooth muscle lineages will be discussed in more detail below.

Research into the therapeutic potential of iPS cells has also been initiated. Using fibroblasts obtained from a humanized mouse model of sickle cell anaemia, iPS cells were generated and the human sickle haemoglobin gene was corrected by gene-specific transfer. Haematopoietic progenitors were then obtained in vitro from the autologous iPS cells and shown to rescue donor mice following transplantation [154]. Additionally, in a rat model of Parkinson's disease, iPS-derived dopaminergic neurons were shown to functionally integrate and improve the parkinsonian behaviour [156].

3.7.2 Human iPS Cell Biology

Within months of creating murine iPS cells, three independent groups generated human iPS cells from human embryonic, foetal and adult dermal fibroblasts [161–163]. Interestingly, whilst two groups used retroviral transduction of Oct4, Sox2, Klf4 and c-Myc [161, 163], Yu and colleagues demonstrated that lentiviral transduction of somatic cells with an alternate combination of factors, namely Oct4, Sox2, Nanog (transcription factor that promotes pluripotency) and Lin28 (RNA-binding protein which blocks the processing of microRNAs which promote differentiation in cancer stem cells), could also generate iPS cells [162, 164, 165]. Using morphological criteria, based upon resemblance to human ES cell colonies, iPS cells were selected 2–3 weeks post-viral transfection. In addition to morphology, expanded human iPS were found to be similar to human ES cells with regard to their proliferation ability, surface antigens and gene expression, epigenetic status of pluripotent-specific ES cell genes, telomerase activity and importantly, the ability to differentiate into cells from all three germ layers in vitro, forming mature teratomas when injected into immunodeficient mice. Addition groups, including our own, have now successfully generated human iPS cells confirming this as a reproducible reprogramming technique [166–169]. Indeed data from our group clearly shows that fibroblasts acquire an ES cell-like morphology following reprogramming (Fig. 5a) and once isolated these cells express undifferentiated markers (alkaline phosphatase, Oct4 and Tra-1–60) associated with human ES cells in culture (Fig. 5b–d). Park

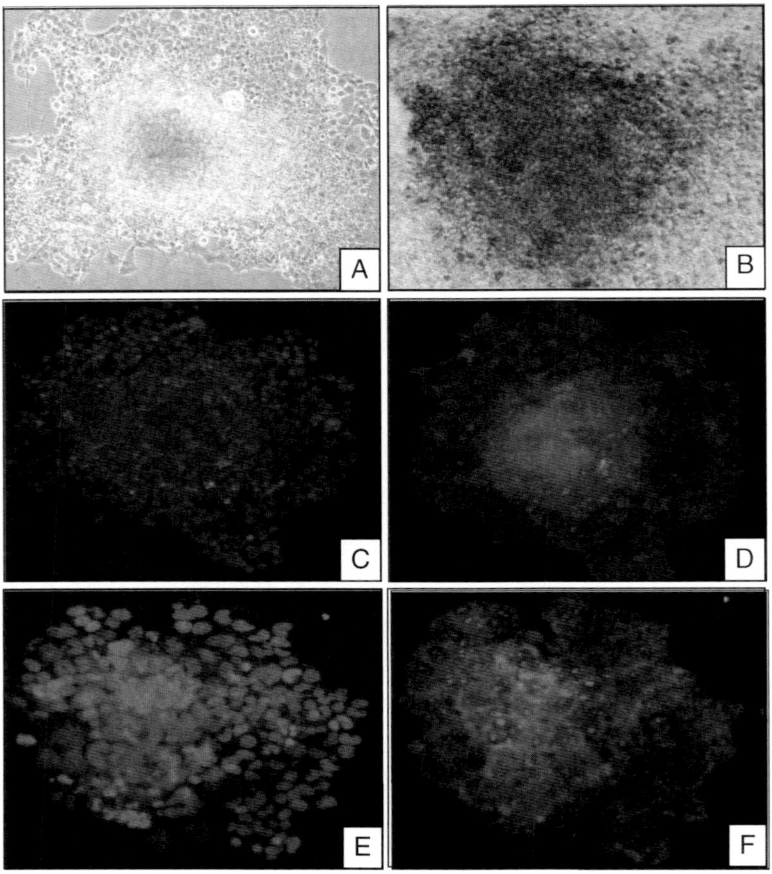

Fig. 5 Human iPS cell morphology and gene expression. Human fibroblasts transfected with lentiviral constructs containing Oct4, Sox2, Klf4 and c-Myc generated acquire an ES cell-like morphology (**a**) and gene expression, alkaline phosphatase (**b**), Oct4 (**d**) and Tra-1–60 (**f**). Nuclei were counterstained (**c** and **e**) with 4′,6-diamidino-2-phenylindole (images donated by Dr A. Margariti)

et al. also stated the introduction of the catalytic subunit of human telomerase and SV40 large T-antigen-aided reprogramming of adult fibroblasts. This was further supported in an alternate study in which the addition of SV40 large T antigen in combination of aforementioned reprogramming factors was seen to enhance efficiency of iPS cell production from both human adult and foetal fibroblasts by a higher rate (23–70-fold) [168]. However, unlike in the study by Park et al., this huge increase in reprogramming ability was also accompanied by integration of large T-antigen sequence into the iPS cell genome and acquisition of an abnormal karyotype.

Recent data also provided evidence that patient-specific iPS cells may be generated. Fibroblasts obtained from an 82-year-old woman diagnosed with a familial form of amyotrophic lateral sclerosis were successfully re-programmed into iPS cells by Moloney-based retroviral transduction of Oct4 Sox2, Klf4 and c-Myc without the need of hTERT or SV40 large T antigen [169] and consequently differentiated in motor neurons, the cell type destroyed in amyotrophic lateral sclerosis. Additionally at the time of writing, reprogramming of human fibroblasts with only two factors, Oct4 and Sox2, along with valproic acid treatment enabled successful iPS generation [145], giving support to the theory that future reprogramming may be possible through purely chemical means without the introduction of tumourigenic factors into the host cell genome and making iPS cells safer and applicable to therapeutic application.

3.7.3 iPS Cell Differentiation into a Vascular Lineage

The application of iPS cell biology to vascular tissue engineering would require established models of differentiation into SMCs and ECs. Currently, investigators have presented murine models to produce either iPS cell-derived vascular progenitor cells which can become SMCs or ECs or direct differentiation into SMCs [157, 155, 160]. Undifferentiated iPS cells cultured for 4 days on collagen IV gave rise to a population of VEGFR2-positive cells, which once sorted via FACS and cultured in differentiation media containing either PDGF-BB or VEGF gave rise to SMCs and ECs, respectively. Acquisition of the correct morphology and expression of α-SMA, calponin and caldesmon in addition to exhibiting similar contractile abilities when compared to mature SMCs indicated successful differentiation into SMCs [155]. Alternatively, VEGFR2-positive-derived ECs displayed typical cobblestone morphology and expressed EC-associated markers, including CD31, CD144, Flk-1, CD34, vWF and eNOS, and the ability to take up acetylated low-density lipoprotein. When compared to mature ECs, differences were seen during characterization; however, iPS cell-derived ECs displayed similar in vitro tube-forming behaviour when seeded on Matrigel, suggesting comparable angiogenic potential [155]. Using a similar method to isolate VEGFR2-positive cells from iPS cells, Narazaki and colleagues drove differentiation into venous, arterial or lymphatic EC phenotypes following culture with VEGF, VEGF plus activation of the cAMP pathway or co-culture with OP9 cells ([158]; reviewed in [159]). In addition, Xie et al. used retinoid acid to induce differentiation of monolayered iPS cell into SMCs and compared them to SMCs derived from ES cell cultures [160]. Interestingly the authors also showed differences between alternate iPS cell lines in the ability to form SMC and in SMC-specific-gene expression patterns between SMCs derived from iPS cells and ES cells. It was suggested that these differences may be due to the lack of understanding with regard to mechanisms that regulate ES cell and iPS cell differentiation as well as the random integration of viruses that occurred when these different iPS cell lines were generated.

To date, generation and study of human iPS cell-derived smooth muscle and endothelial cells have not been achieved. However, directed differentiation into cardiac and neural-like cells has been demonstrated, in addition to cells expressing SMC markers α-SMA and desmin, indicating that specific culture conditions can derive a required phenotype [161]. Additionally, no data has been published describing the application of iPS cell-derived vascular cells in an in vivo system or tissue-engineered vessels. However, insight into their ability to act as a functional cell in vivo was given by Schenke-Layland and colleagues. They analysed sections taken from iPS cell-derived chimeric mice and confirmed that the re-programmed cells contributed to the formation of the blood vessels, differentiating into SMCs and ECs, in addition to other cells of the cardiovascular system such as mature cardiomyocytes [155].

4 Summary and Future Directions

In recent years, huge advances have been made in vascular tissue engineering. Successful generation of adult and ES cell-derived tissue-engineered vessels, which are functional and remain patent in vivo, indicates that stem cell-derived small (and large)-diameter vessels may be applied in the clinical setting and act as an alternate treatment for diseases such as coronary artery disease. The recent emergence of small stem cells and iPS cells has provided great excitement as to their potential use in cell therapy and tissue engineering. As these cells could be derived on an individual basis, they would not generate immunogenic implications. Unlike mesenchymal stem cells, obtaining the cells would not require an invasive procedure and their abundance, proliferation and differentiation potential would not decrease with age. Additionally their application would not generate the ethical issue debate that surrounds ES cell use.

If iPS cells are to be applied in a clinical setting, there are still many issues that must be addressed. In addition to the oncogenic potential of c-Myc, Klf4 and Oct4 have been shown to either transform or stimulate dysplastic growth in epithelial cells, respectively, and Sox2/Oct4 levels are markedly expressed in a number of tumours (reviewed in [165]). As previously mentioned, alternate factors (Lin28/Nanog) can successfully be used to replace some of the original four reprogramming factors and the addition of specific chemicals during reprogramming of somatic cells can be used instead of c-Myc/Klf4. As such, future studies must aim to generate human iPS cells without viral integration into the cell genome. To this end, two recent articles have shown that murine iPS cells can be generated using transient modes of gene expression [171, 172] and using repeat transfection of either replication-deficient adenoviruses or plasmids containing the cDNAs of specific transcription factors required for reprogramming. These cells displayed no evidence of adenoviral/plasmid integration into the host genome and displayed all the pluripotent potential of lenti- or retroviral-generated iPS cells. This data addresses a critical safety concern that was held over the potential use of iPS cells in regenerative medicine/tissue engineering. It now remains to be seen whether human iPS cells can be generated in a similar manner.

As mentioned, generation of a personalized iPS cell lines would eliminate immunorejection issues; however, time, cost and safety validation of each cell line would undoubtedly delay their application, deeming them unusable for disease types such as heart disease which require intervention within months. Nevertheless, it has been suggested by Nakatsuji and colleagues that a human leukocyte antigen (HLA)-haplotyped iPS cell banks could be established [173]. Intestinally they estimated that generation of only 50 iPS cell lines, selected from approximately 24,000 individuals, would be required to genetically match nearly 91% of the Japanese population based on the frequencies of three HLA haplotype loci, although mismatch of the remaining three HLA loci would consequently lead to the use of immunosuppressive drugs.

Ultimately, for stem cell therapies to become clinically relevant, in vitro differentiated cells must share the same functional qualities as their in vivo counterparts. Although studies have shown that ES cell-derived ECs share characteristics of mature in vivo ECs, they have also highlighted their lack of terminal differentiation as cells retain markers of alternate phenotypes including immature endothelium [174]. In addition, when EPCs were compared to mature ECs, similar findings were also found [175]. Variation in cellular characteristics between iPS and ES cell-derived SMCs has also been documented, indicating the diverse population of vascular cells generated during in vitro differentiation models from different stem cell sources [160]. Consequently, future studies must ensure that terminal differentiation of stem cell-derived vascular cells occurs, ensuring generation of fully functional tissue-engineered vessels in addition to decreasing the possibility of de-differentiation of the applied vascular cell.

The use of iPS cells in tissue-engineered vessel models has yet to be investigated and ultimately any model generated would require long-term clinical trials before applied for general treatment of vascular disease. Additionally, advances in scaffold generation will be critical. The recent application of nanotechnology to generate tissue-engineered vessels has shown promising results. This methodology has provided novel approaches to produce thromboresistant nanosurfaces, obtain enhanced endothelialization of grafts both in vitro and in vivo and obtain the ability to direct differentiation of stem cells into a vascular cell phenotype. Indeed, it has been suggested that the application of the nanotechnological field will ultimately supersede the time and cost-consuming bioreactor-based methods of tissue-engineered vessel generation, using strategies such as magnetic force-driven engineering and simultaneous electrospinning of scaffolds and stem cell-derived (iPS cells) vascular cells [176].

Acknowledgements This work was supported by grants from the British Heart Foundation and Oak Foundation.

References

1. Zhang WJ, Liu W, et al. Tissue engineering of blood vessel. J Cell Mol Med. 2007; 11(5):945–957.
2. Sayers RD, Raptis S, et al. Long-term results of femorotibial bypass with vein or polytetrafluoroethylene. Br J Surg. 1998; 85(7):934–938.
3. Gong Z, Niklason LE. Blood vessels engineered from human cells. Trends Cardiovasc Med. 2006; 16(5):153–156.
4. Langer R, Vacanti JP. Tissue engineering. Science. 1993; 260:920–926.

5. Mandal K, Jahangiri M, et al., eds. Progenitor cells and vascular repair. Cardiovascular regeneration and stem cell therapy. Malden: Blackwell; 2007.
6. Weinberg CB, Bell E. A blood vessel model constructed from collagen and cultured vascular cells. Science. 1986; 231(4736):397–400.
7. Isenberg BC, Williams C, et al. Small-diameter artificial arteries engineered in vitro. Circ Res. 2006; 98(1):25–35.
8. Patel A, Fine B, et al. Elastin biosynthesis: the missing link in tissue-engineered blood vessels. Cardiovasc Res. 2006; 71(1):40–49.
9. Boland ED, Matthews JA, et al. Electrospinning collagen and elastin: preliminary vascular tissue engineering. Front Biosci. 2004; 9:1422–1432.
10. Berglund JD, Nerem RM, et al. Incorporation of intact elastin scaffolds in tissue-engineered collagen-based vascular grafts. Tissue Eng. 2004; 10(9–10):1526–1535.
11. Swartz DD, Russell JA, et al. Engineering of fibrin-based functional and implantable small-diameter blood vessels. Am J Physiol Heart Circ Physiol. 2005; 288(3):H1451–H1460.
12. Liu JY, Swartz DD, et al. Functional tissue-engineered blood vessels from bone marrow progenitor cells. Cardiovasc Res. 2007; 75(3):618–628.
13. Yao L, Liu J, et al. Composite fibrin scaffolds increase mechanical strength and preserve contractility of tissue engineered blood vessels. Pharm Res. 2008; 25(5):1212–1221.
14. Gilbert TW, Sellaro TL, et al. Decellularization of tissues and organs. Biomaterials. 2006; 27(19):3675–3683.
15. Stegemann JP, Kaszuba SN, et al. Review: advances in vascular tissue engineering using protein-based biomaterials. Tissue Eng. 2007; 13:2601–2613.
16. McFetridge PS, Daniel JW, et al. Preparation of porcine carotid arteries for vascular tissue engineering applications. J Biomed Mater Res A. 2004; 70(2):224–234.
17. Amiel GE, Komura M, et al. Engineering of blood vessels from acellular collagen matrices coated with human endothelial cells. Tissue Eng. 2006; 12:2355–2365.
18. Yu XX, Wan CX, et al. Preparation and endothelialization of decellularised vascular scaffold for tissue-engineered blood vessel. J Mater Sci Mater Med. 2008; 19(1):319–326.
19. Lu Q, Ganesan K, et al. Novel porous aortic elastin and collagen scaffolds for tissue engineering. Biomaterials. 2004; 25(22):5227–5237.
20. Simionescu DT, Lu Q, et al. Biocompatibility and remodeling potential of pure arterial elastin and collagen scaffolds. Biomaterials. 2006; 27(5):702–713.
21. Daniel J, Abe K, et al. Development of the human umbilical vein scaffold for cardiovascular tissue engineering applications. Asaio J. 2005; 51(3):252–261.
22. Schaner PJ, Martin ND, et al. Decellularized vein as a potential scaffold for vascular tissue engineering. J Vasc Surg. 2004; 40(1):146–153.
23. Martin ND, Schaner PJ, et al. In vivo behavior of decellularized vein allograft. J Surg Res. 2005; 129(1):17–23.
24. L'Heureux N, Paquet S, et al. A completely biological tissue-engineered human blood vessel. FASEB J. 1998; 12:47–56.
25. L'Heureux N, Dusserre N, et al. Human tissue-engineered blood vessels for adult arterial revascularization. Nat Med. 2006; 12(3):361–365.
26. Niklason LE, Gao J, et al. Functional arteries grown in vitro. Science. 1999; 284(5413): 489–493.
27. Sung H-J, Meredith C, et al. The effect of scaffold degradation rate on three-dimensional cell growth and angiogenesis. Biomaterials. 2004; 25(26):5735–5742.
28. Hoerstrup SP, Cummings Mrcs I, et al. Functional growth in tissue-engineered living, vascular grafts: follow-up at 100 weeks in a large animal model. Circulation . 2006; 114:I-159–I-166.
29. Nelson GN, Mirensky T, et al. Functional small-diameter human tissue-engineered arterial grafts in an immunodeficient mouse model: preliminary findings. Arch Surg. 2008; 143: 488–494.
30. Roh JD, Nelson GN, et al. Small-diameter biodegradable scaffolds for functional vascular tissue engineering in the mouse model. Biomaterials. 2008; 29(10):1454–1463.

31. Iwasaki K, Kojima K, et al. Bioengineered three-layered robust and elastic artery using hemodynamically-equivalent pulsatile bioreactor. Circulation. 2008; 118(14 Suppl): S52–S57.
32. McKee JA, Banik SS, et al. Human arteries engineered in vitro. EMBO Rep. 2003; 4(6): 633–638.
33. Poh M, Boyer M, et al. Blood vessels engineered from human cells. Lancet. 2005; 365(9477):2122–2124.
34. Gong Z, Niklason LE. Small-diameter human vessel wall engineered from bone marrow-derived mesenchymal stem cells (hMSCs). FASEB J. 2008; 22(6):1635–1648.
35. Jeong SI, Kim SY, et al. Tissue-engineered vascular grafts composed of marine collagen and PLGA fibers using pulsatile perfusion bioreactors. Biomaterials. 2007; 28(6):1115–1122.
36. Heydarkhan-Hagvall S, Schenke-Layland K, et al. Three-dimensional electrospun ECM-based hybrid scaffolds for cardiovascular tissue engineering. Biomaterials. 2008; 29(19):2907–2914.
37. Lee SJ, Liu J, et al. Development of a composite vascular scaffolding system that withstands physiological vascular conditions. Biomaterials. 2008; 29(19):2891–2898.
38. Stitzel J, Liu J, et al. Controlled fabrication of a biological vascular substitute. Biomaterials. 2006; 27(7):1088–1094.
39. Sodian R, Lemke T, et al. Tissue-engineering bioreactors: a new combined cell-seeding and perfusion system for vascular tissue engineering. Tissue Eng. 2002; 8(5):863–870.
40. Williams C, Wick TM. Perfusion bioreactor for small diameter tissue-engineered arteries. Tissue Eng. 2004; 10(5–6):930–941.
41. McFetridge PS, Abe K, et al. Vascular tissue engineering: bioreactor design considerations for extended culture of primary human vascular smooth muscle cells. Asaio J. 2007; 53(5):623–630.
42. Barron V, Lyons E, et al. Bioreactors for cardiovascular cell and tissue growth: a review. Ann Biomed Eng. 2003; 31(9):1017–1030.
43. Beltrami AP, Cesselli D, et al. Multipotent cells can be generated in vitro from several adult human organs (heart, liver, and bone marrow). Blood. 2007; 110(9):3438–3446.
44. Sales KM, Salacinski HJ, et al. Advancing vascular tissue engineering: the role of stem cell technology. Trends Biotechnol. 2005; 23(9):461–467.
45. Chamberlain G, Fox J, et al. Concise review: mesenchymal stem cells: their phenotype, differentiation capacity, immunological features, and potential for homing. Stem Cells. 2007; 25(11):2739–2749.
46. Prockop DJ. Marrow stromal cells as stem cells for nonhematopoietic tissues. J Sci. 1997; 276:71–74.
47. Friedenstein AJ, Piatetzky S II, et al. Osteogenesis in transplants of bone marrow cells. J Embryol Exp Morphol. 1966; 16(3):381–390.
48. Friedenstein AJ, Gorskaja JF, et al. Fibroblast precursors in normal and irradiated mouse hematopoietic organs. Exp Hematol. 1976; 4(5):267–274.
49. Ashton BA, Allen TD, et al. Formation of bone and cartilage by marrow stromal cells in diffusion chambers in vivo. Clin Orthop Relat Res. 1980; 151:294–307.
50. Beresford JN, Bennett JH, et al. Evidence for an inverse relationship between the differentiation of adipocytic and osteogenic cells in rat marrow stromal cell cultures. J Cell Sci. 1992; 102(Pt 2):341–351.
51. Wakitani S, Saito T, et al. Myogenic cells derived from rat bone marrow mesenchymal stem cells exposed to 5-azacytidine. Muscle Nerve. 1995; 18(12):1417–1426.
52. Friedenstein AJ, Chailakhyan RK, et al. Bone marrow osteogenic stem cells: in vitro cultivation and transplantation in diffusion chambers. Cell Tissue Kinet. 1987; 20(3):263–272.
53. Dominici M, Le Blanc K, et al. Minimal criteria for defining multipotent mesenchymal stromal cells. The International Society for Cellular Therapy position statement. Cytotherapy. 2006; 8(4):315–317.
54. Riha GM, Lin PH, et al. Review: application of stem cells for vascular tissue engineering. Tissue Eng. 2005; 11(9–10):1535–1552.

55. Pittenger MF, Mackay AM, et al. Multilineage potential of adult human mesenchymal stem cells. Science. 1999; 284(5411):143–147.
56. Colter DC, Class R, et al. Rapid expansion of recycling stem cells in cultures of plastic-adherent cells from human bone marrow. Proc Natl Acad Sci USA. 2000; 97(7):3213–3218.
57. Jiang Y, Jahagirdar BN, et al. Pluripotency of mesenchymal stem cells derived from adult marrow. Nature. 2002; 418(6893):41–49.
58. Jiang Y, Jahagirdar BN, et al. Pluripotency of mesenchymal stem cells derived from adult marrow. Nature. 2007; 447(7146):880–881.
59. Reyes M, Dudek A, et al. Origin of endothelial progenitors in human postnatal bone marrow. J Clin Invest. 2002; 109(3):337–346.
60. Zuk PA, Zhu M, et al. Human adipose tissue is a source of multipotent stem cells. Mol Biol Cell. 2002; 13(12):4279–4295.
61. Hu Y, Zhang Z, et al. Abundant progenitor cells in the adventitia contribute to atherosclerosis of vein grafts in ApoE-deficient mice. J Clin Invest. 2004; 113(9):1258–1265.
62. Wang HS, Hung SC, et al. Mesenchymal stem cells in the Wharton's jelly of the human umbilical cord. Stem Cells. 2004; 22(7):1330–1337.
63. Wagner W, Wein F, et al. Comparative characteristics of mesenchymal stem cells from human bone marrow, adipose tissue, and umbilical cord blood. Exp Hematol. 2005; 33(11):1402–1416.
64. Baksh D, Yao R, et al. Comparison of proliferative and multilineage differentiation potential of human mesenchymal stem cells derived from umbilical cord and bone marrow. Stem Cells. 2007; 25(6):1384–1392.
65. Schaffler A, Buchler C. Concise review: adipose tissue-derived stromal cells–basic and clinical implications for novel cell-based therapies. Stem Cells. 2007; 25(4):818–827.
66. Bieback K, Kern S, et al. Comparing mesenchymal stromal cells from different human tissues: bone marrow, adipose tissue and umbilical cord blood. Biomed Mater Eng. 2008; 18(1 Suppl):S71–S76.
67. La Rocca G, Anzalone R, et al. Isolation and characterization of Oct-4(+)/HLA-G (+) mesenchymal stem cells from human umbilical cord matrix: differentiation potential and detection of new markers. Histochem Cell Biol. 2008 Feb; 131(2):267–282.
68. Tintut Y, Alfonso Z, et al. Multilineage potential of cells from the artery wall. Circulation. 2003; 108(20):2505–2510.
69. Doherty MJ, Ashton BA, et al. Vascular pericytes express osteogenic potential in vitro and in vivo. J Bone Miner Res. 1998; 13(5):828–838.
70. Farrington-Rock C, Crofts NJ, et al. Chondrogenic and adipogenic potential of microvascular pericytes. Circulation. 2004; 110(15):2226–2232.
71. Caplan AI. All MSCs are pericytes? Cell Stem Cell. 2008; 3(3):229–230.
72. Roobrouck VD, Ulloa-Montoya F, et al. Self-renewal and differentiation capacity of young and aged stem cells. Exp Cell Res. 2008; 314(9):1937–1944.
73. Strem BM, Hicok KC, et al. Multipotential differentiation of adipose tissue-derived stem cells. Keio J Med. 2005; 54(3):132–141.
74. Kogler G, Sensken S, et al. A new human somatic stem cell from placental cord blood with intrinsic pluripotent differentiation potential. J Exp Med. 2004; 200(2):123–135.
75. Oswald J, Boxberger S, et al. Mesenchymal stem cells can be differentiated into endothelial cells in vitro. Stem Cells. 2004; 22(3):377–384.
76. Davani S, Marandin A, et al. Mesenchymal progenitor cells differentiate into an endothelial phenotype, enhance vascular density, and improve heart function in a rat cellular cardiomyoplasty model. Circulation. 2003; 108(Suppl 1):II253–II258.
77. Gojo S, Gojo N, et al. In vivo cardiovasculogenesis by direct injection of isolated adult mesenchymal stem cells. Exp Cell Res. 2003; 288(1):51–59.
78. Hamilton DW, Maul TM, et al. Characterization of the response of bone marrow-derived progenitor cells to cyclic strain: implications for vascular tissue-engineering applications. Tissue Eng. 2004; 10(3–4):361–369.

79. Kinner B, Zaleskas JM, et al. Regulation of smooth muscle actin expression and contraction in adult human mesenchymal stem cells. Exp Cell Res. 2002; 278(1):72–83.
80. Seruya M, Shah A, et al. Clonal population of adult stem cells: life span and differentiation potential. Cell Transplant. 2004; 13(2):93–101.
81. Prunet-Marcassus B, Cousin B, et al. From heterogeneity to plasticity in adipose tissues: site-specific differences. Exp Cell Res. 2006; 312(6):727–736.
82. Planat-Benard V, Silvestre JS, et al. Plasticity of human adipose lineage cells toward endothelial cells: physiological and therapeutic perspectives. Circulation. 2004; 109(5):656–663.
83. Moon MH, Kim SY, et al. Human adipose tissue-derived mesenchymal stem cells improve postnatal neovascularization in a mouse model of hindlimb ischemia. Cell Physiol Biochem. 2006; 17(5–6):279–290.
84. Nakagami H, Maeda K, et al. Novel autologous cell therapy in ischemic limb disease through growth factor secretion by cultured adipose tissue-derived stromal cells. Arterioscler Thromb Vasc Biol. 2005; 25(12):2542–2547.
85. Ross JJ, Hong Z, et al. Cytokine-induced differentiation of multipotent adult progenitor cells into functional smooth muscle cells. J Clin Invest. 2006; 116(12):3139–3149.
86. Gang EJ, Jeong JA, et al. In vitro endothelial potential of human UC blood-derived mesenchymal stem cells. Cytotherapy. 2006; 8(3):215–227.
87. Wu KH, Zhou B, et al. In vitro and in vivo differentiation of human umbilical cord derived stem cells into endothelial cells. J Cell Biochem. 2007; 100(3):608–616.
88. Melero-Martin JM, De Obaldia ME, et al. Engineering robust and functional vascular networks in vivo with human adult and cord blood-derived progenitor cells. Circ Res. 2008; 103(2):194–202.
89. Cho SW, Lim SH, et al. Small-diameter blood vessels engineered with bone marrow-derived cells. Ann Surg. 2005; 241(3):506–515.
90. Cho SW, Park HJ, et al. Vascular patches tissue-engineered with autologous bone marrow-derived cells and decellularized tissue matrices. Biomaterials. 2005; 26(14):1915–1924.
91. Au P, Tam J, et al. Bone marrow-derived mesenchymal stem cells facilitate engineering of long-lasting functional vasculature. Blood. 2008; 111(9):4551–4558.
92. Ratajczak MZ. Phenotypic and functional characterization of hematopoietic stem cells. Curr Opin Hematol. 2008; 15(4):293–300.
93. Metcalf D. Concise review: hematopoietic stem cells and tissue stem cells: current concepts and unanswered questions. Stem Cells. 2007; 25(10):2390–2395.
94. Geiger H, Van Zant G. The aging of lympho-hematopoietic stem cells. Nat Immunol. 2002; 3(4):329–333.
95. Sata M, Saiura A, et al. Hematopoietic stem cells differentiate into vascular cells that participate in the pathogenesis of atherosclerosis. Nat Med. 2002; 8(4):403–409.
96. Jackson KA, Majka SM, et al. Regeneration of ischemic cardiac muscle and vascular endothelium by adult stem cells. J Clin Invest. 2001; 107(11):1395–1402.
97. Orlic D, Kajstura J, et al. Bone marrow cells regenerate infarcted myocardium. Nature. 2001; 410(6829):701–705.
98. Simper D, Stalboerger PG, et al. Smooth muscle progenitor cells in human blood. Circulation. 2002; 106(10):1199–1204.
99. Shi Q, Rafii S, et al. Evidence for circulating bone marrow-derived endothelial cells. Blood. 1998; 92(2):362–367.
100. Asahara T, Murohara T, et al. Isolation of putative progenitor endothelial cells for angiogenesis. Science. 1997; 275(5302):964–967.
101. Murohara T, Ikeda H, et al. Transplanted cord blood-derived endothelial precursor cells augment postnatal neovascularization. J Clin Invest. 2000; 105(11):1527–1536.
102. Schmidt D, Asmis LM, et al. Engineered living blood vessels: functional endothelia generated from human umbilical cord-derived progenitors. Ann Thorac Surg. 2006; 82(4):1465–1471, discussion 1471.

103. Wu X, Lensch MW, et al. Hemogenic endothelial progenitor cells isolated from human umbilical cord blood. Stem Cells. 2007; 25(11):2770–2776.
104. Hristov M, Erl W, et al. Endothelial progenitor cells: isolation and characterization. Trends Cardiovasc Med. 2003; 13(5):201–206.
105. Zampetaki A, Kirton JP, et al. Vascular repair by endothelial progenitor cells. Cardiovasc Res. 2008; 78(3):413–421.
106. Hu Y, Davison F, et al. Endothelial replacement and angiogenesis in arteriosclerotic lesions of allografts are contributed by circulating progenitor cells. Circulation. 2003; 108(25):3122–3127.
107. Xu Q, Zhang Z, et al. Circulating progenitor cells regenerate endothelium of vein graft atherosclerosis, which is diminished in ApoE-deficient mice. Circ Res. 2003; 93(8):e76–e86.
108. Powell TM, Paul JD, et al. Granulocyte colony-stimulating factor mobilizes functional endothelial progenitor cells in patients with coronary artery disease. Arterioscler Thromb Vasc Biol. 2005; 25(2):296–301.
109. Werner N, Kosiol S, et al. Circulating endothelial progenitor cells and cardiovascular outcomes. N Engl J Med. 2005; 353(10):999–1007.
110. Griese DP, Ehsan A, et al. Isolation and transplantation of autologous circulating endothelial cells into denuded vessels and prosthetic grafts: implications for cell-based vascular therapy. Circulation. 2003; 108(21):2710–2715.
111. He H, Shirota T, et al. Canine endothelial progenitor cell-lined hybrid vascular graft with nonthrombogenic potential. J Thorac Cardiovasc Surg. 2003; 126(2):455–464.
112. Shirota T, He H, et al. Human endothelial progenitor cell-seeded hybrid graft: proliferative and antithrombogenic potentials in vitro and fabrication processing. Tissue Eng. 2003; 9(1):127–136.
113. Kaushal S, Amiel GE, et al. Functional small-diameter neovessels created using endothelial progenitor cells expanded ex vivo. Nat Med. 2001; 7(9):1035–1040.
114. Zhu C, Ying D, et al. Development of anti-atherosclerotic tissue-engineered blood vessel by A20-regulated endothelial progenitor cells seeding decellularized vascular matrix. Biomaterials. 2008; 29(17):2628–2636.
115. Kucia M, Reca R, et al. A population of very small embryonic-like (VSEL) CXCR4(+)SSEA-1(+)Oct-4^{+} stem cells identified in adult bone marrow. Leukemia. 2006; 20(5):857–869.
116. Kucia M, Zuba-Surma EK, et al. Adult marrow-derived very small embryonic-like stem cells and tissue engineering. Expert Opin Biol Ther. 2007; 7(10):1499–1514.
117. Ratajczak MZ, Zuba-Surma EK, et al. Very small embryonic-like (VSEL) stem cells in adult organs and their potential role in rejuvenation of tissues and longevity. Exp Gerontol. 2008; 43(11):1009–1017.
118. Kucia M, Wysoczynski M, et al. Identification of very small embryonic like (VSEL) stem cells in bone marrow. Cell Tissue Res. 2008; 331(1):125–134.
119. Evans MJ, Kaufman MH. Establishment in culture of pluripotential cells from mouse embryos. Nature. 1981; 292(5819):154–156.
120. Martin GR. Isolation of a pluripotent cell line from early mouse embryos cultured in medium conditioned by teratocarcinoma stem cells. Proc Natl Acad Sci USA. 1981; 78(12): 7634–7638.
121. Thomson JA, Itskovitz-Eldor J, et al. Embryonic stem cell lines derived from human blastocysts. Science. 1998; 282(5391):1145–1147.
122. Levenberg S, Golub JS, et al. Endothelial cells derived from human embryonic stem cells. Proc Natl Acad Sci USA. 2002; 99(7):4391–4396.
123. Ferreira LS, Gerecht S, et al. Vascular progenitor cells isolated from human embryonic stem cells give rise to endothelial and smooth muscle like cells and form vascular networks in vivo. Circ Res. 2007; 101(3):286–294.
124. Yamashita J, Itoh H, et al. Flk1-positive cells derived from embryonic stem cells serve as vascular progenitors. Nature. 2000; 408(6808):92–96.

125. Sone M, Itoh H, et al. Pathway for differentiation of human embryonic stem cells to vascular cell components and their potential for vascular regeneration. Arterioscler Thromb Vasc Biol. 2007; 27(10):2127–2134.
126. Drab M, Haller H, et al. From totipotent embryonic stem cells to spontaneously contracting smooth muscle cells: a retinoic acid and db-cAMP in vitro differentiation model. FASEB J. 1997; 11(11):905–915.
127. Huang H, Zhao X, et al. Differentiation of human embryonic stem cells into smooth muscle cells in adherent monolayer culture. Biochem Biophys Res Commun. 2006; 351(2):321–327.
128. Xiao Q, Zeng L, et al. Sca-1+ progenitors derived from embryonic stem cells differentiate into endothelial cells capable of vascular repair after arterial injury. Arterioscler Thromb Vasc Biol. 2006; 26(10):2244–2251.
129. Xiao Q, Zeng L, et al. Stem cell-derived Sca-1+ progenitors differentiate into smooth muscle cells, which is mediated by collagen IV–integrin alpha1/beta1/alphav and PDGF receptor pathways. Am J Physiol Cell Physiol. 2007; 292(1):C342–C352.
130. Zeng L, Xiao Q, et al. HDAC3 is crucial in shear- and VEGF-induced stem cell differentiation toward endothelial cells. J Cell Biol. 2006; 174(7):1059–1069.
131. Shen G, Tsung HC, et al. Tissue engineering of blood vessels with endothelial cells differentiated from mouse embryonic stem cells. Cell Res. 2003; 13(5):335–341.
132. Cheng L. More new lines of human parthenogenetic embryonic stem cells. Cell Res. 2008; 18:215–217.
133. Koh CJ, Delo DM, Lee JW, Siddiqui MM, Lanza RP, Soker S, Yoo JJ, Atala A. Parthenogenesis-derived multipotent stem cells adapted for tissue engineering applications. Methods. 2009 Feb; 47(2):90–97.
134. Takahashi K, Yamanaka S. Induction of pluripotent stem cells from mouse embryonic and adult fibroblast cultures by defined factors. Cell. 2006; 126:663–676.
135. Maherali N, Sridharan R, et al. Directly reprogrammed fibroblasts show global epigenetic remodeling and widespread tissue contribution. Cell Stem Cell. 2007; 1(1):55–70.
136. Okita K, Ichisaka T, et al. Generation of germline-competent induced pluripotent stem cells. Nature. 2007; 448:313–317.
137. Wernig M, Meissner A, et al. In vitro reprogramming of fibroblasts into a pluripotent ES-cell-like state. Nature. 2007; 448(7151):318–324.
138. Nakagawa M, Koyanagi M, et al. Generation of induced pluripotent stem cells without Myc from mouse and human fibroblasts. Nat Biotech. 2008; 26(1):101–106.
139. Wernig M, Meissner A, et al. c-Myc is dispensable for direct reprogramming of mouse fibroblasts. Cell Stem Cells. 2008; 2:11–12.
140. Marson A, Foreman R, et al. Wnt signaling promotes reprogramming of somatic cells to pluripotency. Cell Stem Cell. 2008; 3(2):132–135.
141. Mikkelsen TS, Hanna J, et al. Dissecting direct reprogramming through integrative genomic analysis. Nature. 2008; 454(7205):794
142. Meissner A, Wernig M, et al. Direct reprogramming of genetically unmodified fibroblasts into pluripotent stem cells. Nature Biotechnol. 2007; 25:1177–1181.
143. Brambrink T, Foreman R, et al. Sequential expression of pluripotency markers during direct reprogramming of mouse somatic cells. Cell Stem Cell. 2008; 2(2):151–159.
144. Stadtfeld M, Maherali N, et al. Defining molecular cornerstones during fibroblast to iPS cell reprogramming in mouse. Cell Stem Cell. 2008; 2(3):230–240.
145. Huangfu D, Osafune K, et al. Induction of pluripotent stem cells from primary human fibroblasts with only Oct4 and Sox2. Nat Biotechnol. 2008; 26(11):1269–1275.
146. Shi Y, Do JT, et al. A combined chemical and genetic approach for the generation of induced pluripotent stem cells. Cell Stem Cell. 2008; 2(6):525–528.
147. Blelloch R, Venere M, et al. Generation of induced pluripotent stem cells in the absence of drug selection. Cell Stem Cell. 2007; 1(3):245–247.
148. Aoi T, Yae K, et al. Generation of pluripotent stem cells from adult mouse liver and stomach cells. Science. 2008; 321(5889):699–702.

149. Stadtfeld M, Brennand K, et al. Reprogramming of pancreatic [beta] cells into induced pluripotent stem cells. Curr Biol. 2008; 18(12):890–894.
150. Hanna J, Markoulaki S, et al. Direct reprogramming of terminally differentiated mature B lymphocytes to pluripotency. Cell. 2008; 133(2):250–264.
151. Di Stefano B, Prigione A, et al. Efficient genetic reprogramming of unmodified somatic neural progenitors uncovers the essential requirement of Oct4 and Klf4. Stem Cells Dev. 2009 Jun; 18(5):707–716.
152. Eminli S, Utikal JS, et al. Reprogramming of neural progenitor cells into iPS cells in the absence of exogenous Sox2 expression. Stem Cells. 2008 Oct; 26(10):2467–2474.
153. Kim JB, Zaehres H, et al. Pluripotent stem cells induced from adult neural stem cells by reprogramming with two factors. Nature. 2008; 454(7204):646–650.
154. Hanna J, Wernig M, et al. Treatment of sickle cell anemia mouse model with iPS cells generated from autologous skin. Science. 2007; 318(5858):1920–1923.
155. Schenke-Layland K, Rhodes KE, et al. Reprogrammed mouse fibroblasts differentiate into cells of the cardiovascular and hematopoietic lineages. Stem Cells. 2008; 26(6):1537–1546.
156. Wernig M, Zhao JP, et al. Neurons derived from reprogrammed fibroblasts functionally integrate into the fetal brain and improve symptoms of rats with Parkinson's disease. Proc Natl Acad Sci USA. 2008; 105(15):5856–5861.
157. Mauritz C, Schwanke K, et al. Generation of functional murine cardiac myocytes from induced pluripotent stem cells. Circulation. 2008; 118(5):507–517.
158. Narazaki G, Uosaki H, et al. Directed and systematic differentiation of cardiovascular cells from mouse induced pluripotent stem cells. Circulation. 2008; 118(5):498–506.
159. Tulloch NL, Pabon L, et al. Get with the (re)program: cardiovascular potential of skin-derived induced pluripotent stem cells. Circulation. 2008; 118(5):472–475.
160. Xie C, Huang H, et al. A comparison of murine smooth muscle cells generated from embryonic versus induced pluripotent stem cells. Stem Cells Dev. 2009 Jun; 18(5):741–748.
161. Takahashi K, Tanabe K, et al. Induction of pluripotent stem cells from adult human fibroblasts by defined factors. Cell. 2007; 131(5):861–872.
162. Yu J, Vodyanik MA, et al. Induced pluripotent stem cell lines derived from human somatic cells. Science. 2007; 318(5858):1917–1920.
163. Park I-H, Zhao R, et al. Reprogramming of human somatic cells to pluripotency with defined factors. Nature. 2008; 451(7175):141–146.
164. Newman MA, Thomson JM, et al. Lin-28 interaction with the Let-7 precursor loop mediates regulated microRNA processing. RNA. 2008; 14(8):1539–1549.
165. Zhao R, Daley GQ. From fibroblasts to iPS cells: induced pluripotency by defined factors. J Cell Biochem. 2008 Nov 1; 105(4):949–955.
166. Liao J, Wu Z, et al. Enhanced efficiency of generating induced pluripotent stem (iPS) cells from human somatic cells by a combination of six transcription factors. Cell Res. 2008; 18(5):600–603.
167. Lowry WE, Richter L, et al. Generation of human induced pluripotent stem cells from dermal fibroblasts. Proc Natl Acad Sci USA. 2008; 105(8):2883–2888.
168. Mali P, Ye Z, et al. Improved efficiency and pace of generating induced pluripotent stem cells from human adult and fetal fibroblasts. Stem Cells. 2008; 26(8):1998–2005.
169. Masaki H, Ishikawa T, et al. Heterogeneity of pluripotent marker gene expression in colonies generated in human iPS cell induction culture. Stem Cell Res. 2008; 1(2):105–115.
170. Dimos JT, Rodolfa KT, et al. Induced pluripotent stem cells generated from patients with ALS can be differentiated into motor neurons. Science. 2008; 321(5893):1218–1221.
171. Okita K, Nakagawa M, et al. Generation of mouse induced pluripotent stem cells without viral vectors. Science. 2008; 322(5903):949–953.
172. Stadtfeld M, Nagaya M, et al. Induced pluripotent stem cells generated without viral integration. Science. 2008; 322(5903):945–949.
173. Nakatsuji N, Nakajima F, et al. HLA-haplotype banking and iPS cells. Nat Biotech. 2008; 26(7):739–740.

174. McCloskey KE, Smith DA, et al. Embryonic stem cell-derived endothelial cells may lack complete functional maturation in vitro. J Vasc Res. 2006; 43(5):411–421.
175. Bagley RG, Walter-Yohrling J, Cao X, Weber W, Simons B, Cook BP, Chartrand SD, Wang C, Madden SL, Teicher BA. Endothelial precursor cells as a model of tumor endothelium: characterization and comparison with mature endothelial cells. Cancer Res. 2003 Sep 15; 63(18):5866–5873.
176. Mironov V, Kasyanov V, et al. Nanotechnology in vascular tissue engineering: from nanoscaffolding towards rapid vessel biofabrication. Trends Biotechnol. 2008; 26(6):338–344.

Endothelial Progenitor Cells for Vascular Repair

Melissa A. Brown, Cindy S. Cheng, and George A. Truskey

Abstract Endothelial progenitor cells (EPCs), present in the blood and bone marrow, represent a potential source of endothelial cells for repair of injured blood vessels, neovascularization, and tissue engineering. EPCs are present at low levels in peripheral blood, although their numbers increase in response to cytokines, VEGF, and statins. There are at least two types of EPCs characterized following in vitro culture: colony-forming unit ECs (CFU-ECs) and endothelial colony-forming cells (ECFCs). CFU-ECs appear early in culture, have limited ability to proliferate, and share markers for endothelial cells and monocytes. In contrast, ECFCs appear later in culture, grow rapidly, and to large numbers express only endothelial cell markers. This chapter examines the properties of these EPCs, in vitro and in vivo studies using these two cell types, and the potential of these EPCs for therapeutic applications.

Keywords Endothelial progenitor cells · Rascular grafts · Cell adhesion

1 Introduction

1.1 Problem

Cardiovascular disease is the leading cause of death in the world. Ischemic heart disease and cerebrovascular disease accounted for over 35% of all deaths in the United States in 2005 [1]. Atherosclerosis, the underlying pathology behind most cardiovascular disease and vascular disease in large arteries, involves inflammation and intimal accumulation of lipids, macrophages, and smooth muscle cells. The disease specifically affects large-sized and medium-sized arteries. The underlying cause of atherosclerosis is endothelial dysfunction [2] induced by risk factors such as

G.A. Truskey (✉)
Department of Biomedical Engineering, Duke University, Durham, NC, USA
e-mail: gtruskey@duke.edu

G.M. Artmann et al. (eds.), *Stem Cell Engineering*,
DOI 10.1007/978-3-642-11865-4_13,

elevated plasma cholesterol, hypertension, diabetes, smoking, rheumatoid arthritis, genetics, age, and gender [3–5]. Endothelial dysfunction results in the adhesion and transmigration of monocytes, which become macrophages. Macrophages reside in the arterial intima, accumulating lipid and releasing reactive oxygen species that modify lipoproteins and extracellular matrix proteins. The combined effect of macrophage accumulation and smooth muscle cell migration into the intima leads to the formation of atherosclerotic lesions. Clinical symptoms such as shortness of breath, angina, and damage to cardiac muscle arise from occlusion of the vessel or thrombosis [3–5].

Current treatments for atherosclerosis involve addition of drugs that lower cholesterol and plaque formation, angioplasty with stent placement to reopen vessels, or bypass grafting for calcified lesions or multiple sites of occlusion. Suitable vessels are not always available due to the need for repeat procedures. Currently, autologous blood vessels are the primary option to replace small diameter arteries during bypass procedures. For peripheral atherosclerosis in the legs, synthetic grafts are used due to the larger diameter of the vessel, although their performance is not as good as native vessel grafts.

With saphenous vein grafts, there is a 10–15% loss of patency after 1 year of coronary bypass surgery, which is associated with increased loss of endothelium from the vein grafts [6]. Saphenous vein grafts undergo remodeling to adapt to the arterial hemodynamic environment and the vein endothelial cells lose the venous endothelium marker ephrin B4 receptor, but the endothelium does not express the arterial marker tyrosine kinase ephrin B2 [7]. Failure of saphenous vein grafts is associated with endothelial dysfunction which is correlated with levels of the marker of systemic inflammation, C reactive protein [8]. Further, ECs detach, exposing the subendothelium that leads to thrombosis [9].

Several solutions are being developed to replace or repair diseased or damaged arteries: a tissue-engineered blood vessel, synthetic grafts or extracellular matrix grafts seeded with endothelium, and the direct injection of endothelial cells to adhere to damaged regions of arteries in vivo both pre- and post-angioplasty and stent procedures.

An optimal therapy requires a confluent and fully functional endothelium. Current treatments such as angioplasty and bypass surgery often damage the endothelium. Hence, replacement of the endothelial layer could prevent further complications. Coverage of a synthetic or extracellular matrix grafts with endothelium is conceptually appealing, but progress in developing a functional graft with a confluent layer of endothelial cells has been slow owing to a lack of understanding of key issues and the complexity of the available approaches. In a number of animal species, endothelial cells can cover a significant portion of the graft due to growth from the adjacent vessel, but very little coverage occurs with grafts in humans [10]. Due to the high arterial shear stresses, seeding in vivo has been examined in a few cases. While growth of endothelial cells (ECs) that are placed in contact with the adventitia can be promoted after placement of stents in native vessels [11], efforts to promote endothelialization from surrounding tissue or growth of progenitor cells did not reduce intimal hyperplasia [12, 13]. As a result, the cells must be seeded onto the grafts prior to implantation. However, a suitable and reliable cell source must be identified for these treatment methods.

While a number of sources for ECs have been considered, endothelial progenitor cells (EPCs) derived from umbilical cord blood or peripheral blood (PB) are an especially promising source of cells for cardiovascular applications because of the suitability for transplantation and ease of availability. Cord blood-derived EPCs specifically have an improved capacity to proliferate and show an improved response to stress compared to progenitor cells derived from blood [14]. We have recently cultured cord blood-derived EPCs and found that they exhibit firm adhesion and respond to steady laminar flow in a very similar fashion to aortic endothelial cells, hence we believe they can and should be pursued for cardiovascular treatment applications.

In this chapter, we first review the structure of the arterial wall and the role of the vascular endothelium. This is followed by a summary of previous studies to seed vascular grafts with vascular endothelium, highlighting the need for a large number of cells that can be used for tissue-engineering applications. The term EPCs is used to describe several different types of cells that can be isolated from bone marrow or whole blood. The characterization of these different cell types is presented followed by a summary of in vitro and in vivo studies to use these cells in regenerative medicine applications.

2 Artery Structure and Function of Endothelial Cells

Arteries consist of three layers: the intima, media, and adventitia. The intima is the innermost layer and consists of endothelium overlying a basement membrane. Beneath the endothelium, the intimal basement membrane consists of a network of collagen, elastin, fibronectin, thrombospondin, laminins, and heparin sulfate proteoglycan [15]. ECs bind to extracellular matrix (ECM) proteins in the basement membrane through integrins. Integrins are heterodimeric transmembrane proteins composed of various alpha and beta subunits. ECs express the following integrins: $\alpha_1\beta_1$, $\alpha_2\beta_1$, $\alpha_3\beta_1$, $\alpha_4\beta_1$, $\alpha_5\beta_1$, $\alpha_6\beta_1$, $\alpha_V\beta_1$, $\alpha_V\beta_3$, $\alpha_6\beta_4$, $\alpha_V\beta_5$, $\alpha_V\beta_6$ [16] and the Lutheran antigen that binds to laminin [17]. EC binding to specific interins can affect function [18]. The media consists of alternating bands of smooth muscle cells (SMCs) and an ECM of fibrillar collagens and glycoproteins [19, 20]. The adventitia, the outermost layer, consists of loose connective tissue and capillaries [15]. In thicker blood vessels, the capillary network extends into the media.

ECs form a confluent monolayer with a low rate of proliferation and play an important role in maintaining the physiological and biochemical conditions in blood vessels [15, 21–24]. They provide a structural barrier between the circulation and the surrounding tissue, secrete vasodilators and vasoconstrictors which regulate vascular hemodynamics and vascular tone, modulate vascular wall remodeling through regulation of the behavior of SMCs [21, 23], regulate vascular permeability, present an active anti-thrombogenic surface that inhibits platelet adhesion and clotting and promotes fibrinolysis, and express receptors and mediators that regulate the interaction of circulating leukocytes with the vessel wall [10, 25].

The endothelium exhibits considerable variability in function [26]. For example, permeability is lowest in the brain capillary endothelium where tight junctions are most abundant and basal permeability is higher where tight junctions are less frequent and fenestrae are present. In most vascular beds, leukocyte recruitment occurs in the post-capillary venules and involves P-selectin, E-selectin, and ICAM-1, but in the liver and lung recruitment occurs in the capillaries and is mediated by ICAM [26]. These functional variations arise, in part, from the interaction of the endothelium with underlying cell types and the matrix that they secrete, as well as the hemodynamic environment. In arteries, variations in the local flow environment are important factors influencing the onset of atherosclerosis [27].

Further complicating the characterization of endothelium, few molecules are expressed solely by endothelium and present in all endothelium [26]. For example, von Willebrand factor (vWF) is exclusively expressed by endothelium, but is absent or expressed in low levels by capillary endothelium of lung, kidney, and spleen [28]. Platelet endothelial cell adhesion molecule (PECAM-1 or CD31), VE-cadherin (CD144), and CD 146 [29] are present in endothelial junctions and play a role in leukocyte transmigration, permeability, and mechanosensing [30], although CD146 is also present on T cells. CD31 is expressed in all vascular beds except the spleen [28]. CD34 is present on endothelium and a number of cells of the immune system, as well as medial smooth muscle cells [28]. The transcription factor Kruppel-like factor 2 (KLF-2) regulates antioxidative, anticoagulant, and antiinflammatory responses to shear stress [31]. This transcription factor is expressed by ECs and naïve T cells, but is not expressed by SMC in mature blood vessels [32] and its expression does not induce expression of lineage-specific genes [26]. VEGF receptor 1 or Flt-1 is expressed by ECs and monocytes and binds VEGFA [33]. VEGF receptor 2 or Flk1 is expressed by endothelial cells and a number of hematopoietic cells. Nitric oxidase synthase III (NOS III or eNOS) is one of the few endothelial-specific molecules. In spite of the limited number of endothelial-specific markers, the endothelial phenotype is often identified by specific responses to flow and the combined expression of PECAM, CD34, VE-cadherin, NOS III, and other markers.

ECs are activated by endotoxin from bacteria, interleukin-1, tumor necrosis factor α (TNF α), thrombin, low oxygen tension, and spatial gradients in shear stress or oscillatory shear stress [15]. Activation of endothelium causes production of chemoattractants and adhesion molecules that bind to receptors on leukocytes. Activated endothelial cells cease to produce antithrombotic and anticoagulant factors such as tissue plasminogen activator (tPA) and thrombomodulin and start to produce tissue factor, vWF, type 1 plasminogen activator inhibitor, thrombospondin, collagen, and platelet-activating factor (PAF) [25]. Dysfunctional ECs induce thrombus and plaque formation in the vessel wall. The formed thrombi release many bioactive substances, including thrombin, platelet-derived growth factor, and fibroblast growth factor [21]. These bioactive substances shift SMCs from a non-proliferating contractile state into a proliferative and migrating state [21, 24], contributing to the development of intimal hyperplasia [21].

The non-thrombogenic nature of ECs makes them an attractive option for cardiovascular applications. Many researchers focus on seeding synthetic grafts and tissue-engineering vessels with ECs in order to produce a confluent, non-thrombogenic lining. Under normal circumstances, ECs limit clot formation by deactivating thrombin via the antithrombin III and thrombomodulin systems. ECs also suppress platelet activation by producing prostacyclin and play a role in dissolution of clots. Endothelial cells affect SMC differentiation, in part by activating transforming growth factor beta (TGF-β) and inhibiting SMC growth.

3 Endothelial Cell Seeding of Vascular Grafts

Synthetic vascular grafts seeded with a confluent endothelium represent a practical way to provide a non-thrombogenic alternative to native vessels. Requirements for a suitable EC source to seed synthetic vascular grafts are that the cells are readily available, are in sufficient numbers or can be grown to high density suitable for graft applications (2–16 $\times$ 10^6 cells/cm^2), and there is minimal contamination with other cell types. Ideally, the graft should be seeded with endothelium shortly before implantation in a single-stage procedure rather than in a two-stage procedure that involves surgery to isolate the cells and an extensive period of culture of ECs on the graft surface. Many vascular EC sources have been investigated, such as jugular or saphenous veins and microvascular ECs [34]. The number of venous cells obtained is limited. Adipose-derived microvascular ECs are plentiful and can be expanded to sufficient numbers. However, all of these cells are isolated through methods that either require an invasive surgical procedure or require an extensive in vitro expansion. Although ECs from adipose tissue are more abundant than venous ECs, adipose-derived microvascular cell cultures are often contaminated with other cell types (e.g., macrophages and fibroblasts) which are responsible for intimal hyperplasia and inflammation [35]. When these cells are removed during cell isolation, there is a reduction in tissue factor expression and intimal hyperplasia [36]. Mesothelial cells are also present, but these cells can inhibit thrombosis and do not seem to present any problems during grafting [10]. Given the difficulty of isolating a pure population of cells and resulting intimal hyperplasia produced in the presence of an EC monolayer derived from adipose tissue, adipose-derived ECs are not suitable for vascular graft applications.

In 1978, Herring and colleagues used pre-clotted blood to promote EC attachment to Dacron, thereby establishing that ECs could attach to vascular grafts [37]. Initially, coverage was low due to the small number of endothelial cells that could be obtained from veins and weak cell adhesion [38, 39]. Subsequent studies focused upon improving the isolation procedure to reduce cell damage, culturing cells to increase cell number, and promoting EC attachment using various ECM proteins. These changes increased the time between tissue harvest from the patient and implantation of the graft. Consequently, no clinically effective single-stage procedure is available [10]. Two-stage seeding methods involve EC culture to produce seeding densities between 2 and 16 $\times$ 10^5 cells/cm^2 of graft surface area, depending on the type of graft [40]. A clinical study of venous EC seeding on ePTFE grafts treated with fibrin glue found a 5-year patency of 70% for above-knee procedures [41], which is comparable to results obtained with vein grafts. These results suggest that maintaining a viable endothelium on the graft is the primary factor influencing graft survival and that graft compliance may be a secondary issue. Below-knee patency of seeded grafts was 60%. Larger 7-mm inner diameter (I.D.) grafts had a higher patency than 6 mm I.D. grafts [41]. Given the sensitivity of the shear stress to the vessel diameter [42], this result suggests that the graft hemodynamics may be a key feature affecting the survival of the seeded graft.

4 Culture and Characterization of EPCs

4.1 Methods of EPC Isolation and Culture

The term endothelial progenitor cell is used to represent a very diverse group of cells that express a number of endothelial antigens. However, not all of these cells may differentiate into functional endothelium. There are three commonly used methods to isolate and culture EPCs, which result in two very different cell types (Fig. 1). The original method of culturing EPCs was developed by Asahara et al. [43] and later modified by Hill et al. [44]. First, adult peripheral blood or human umbilical cord blood mononuclear cells (MNCs) attach to fibronectin-coated surfaces. After 48 h, the cells that have not adhered are removed and subsequently re-plated on another fibronectin-coated surface. From this sample, colonies are formed in 5–9 days. The cells in these colonies are round with some spindle-shaped cells protruding around the periphery. Cells isolated by this method of culture are known as colony-forming unit ECs (CFU-ECs) or early outgrowth EPCs [45]. A kit (Endocult) is available from StemCell Technologies to isolate CFU-ECs based on the protocol developed by Hill et al. [44].

The second method involves culturing whole-blood MNCs for several days before removing non-adherent cells. While these cells are very similar in surface marker expression and function to the CFU-ECs found in Method A, they are also called circulating angiogenic cells (CACs). CACs are different from CFU-ECs because they have different colony morphology and are much more prominent in the whole mononuclear population (Table 1) [45].

In the third method, adult peripheral blood or umbilical cord blood mononuclear cells are seeded onto plates coated with collagen I. Similar to CAC culture, the non-adherent cell population is discarded; however, this is done within the first 24 h in order to remove all contaminating cells, and the targeted adherent population is cultured. After about 10–21 days, colonies form from these cells and are called endothelial colony-forming cells (ECFCs). In shape and size, ECFCs are very similar to cultured ECs [45] (Fig. 2). ECFCs can be isolated by capture of CD34-positive cells on beads followed by attachment and growth of the cells on gelatin [46].

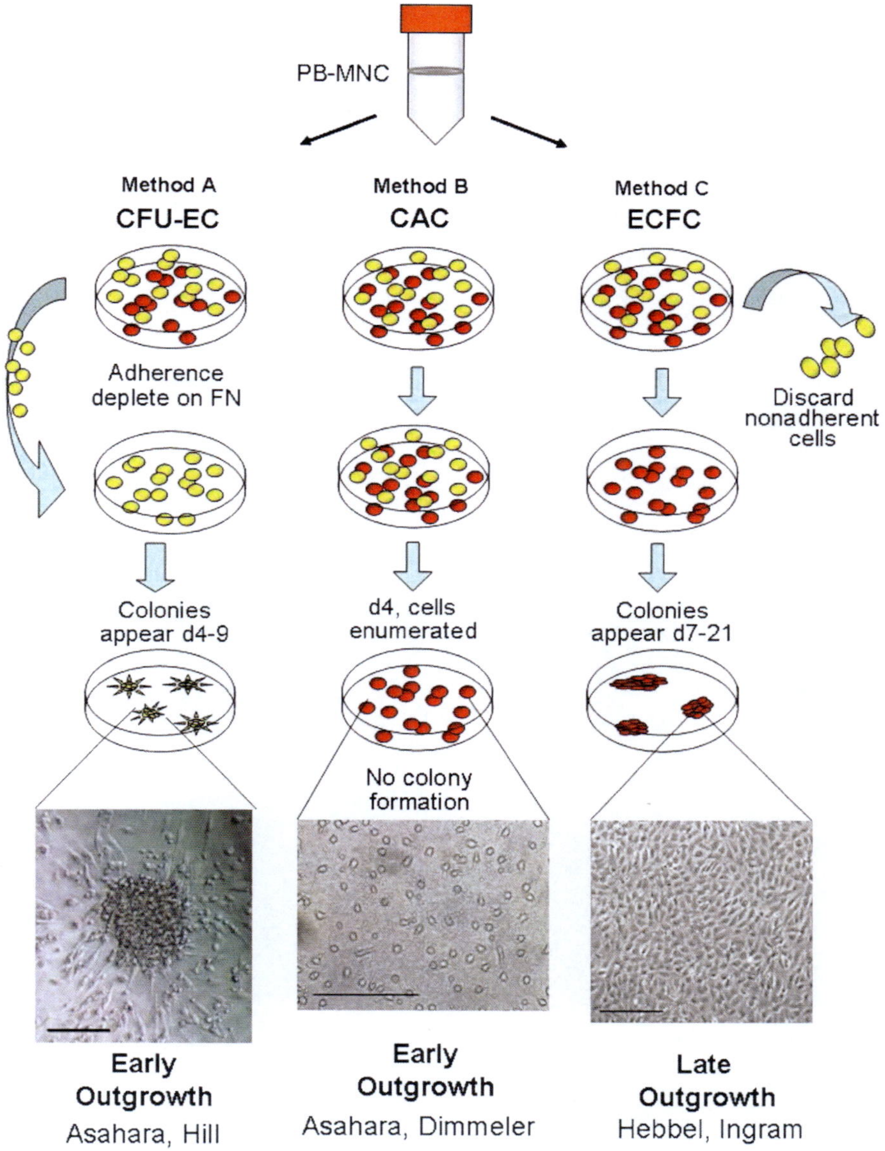

Fig. 1 Common methods of EPC culture. Culture of colony-forming unit-endothelial cells (CFU-EC, Method A, scale bar = 100 μm) includes a 5-day process wherein non-adherent mononuclear cells (MNCs) give rise to the EPC colony. Circulating angiogenic cells (CAC, Method B, scale bar = 200 μm) are the adherent MNCs of a 4- to 7-day culture procedure. CAC cultures typically do not display colony formation. Endothelial colony-forming cells (ECFCs, Method C, scale bar = 400 μm) are derived from adherent MNCs cultured for 7–21 days in endothelial conditions and colonies display a cobblestone morphology. From Prater DN et al. [45] with permission

Table 1 Comparison of EPCs [51, 80, 123]

	Angiogenic cells, ACs or CFU-ECs	ECFCs or endothelial outgrowth cells
Endothelial antigens		
VEGFR1	Yes	Yes
VEGFR2	No	Yes
vWF	No	Yes
CD146	No	Yes
Acetylated LDL uptake	Yes	Yes
Macrophage antigens (e.g., CD14 and CD45) and phagocytosis	Yes	No
Proliferative potential	Limited	High
Form vessels in vivo	No	Yes
Display EC morphology	No	Yes
Grow from clones	No	Yes
CD34 expression	Yes	Yes
CD133 expression	Yes	Yes

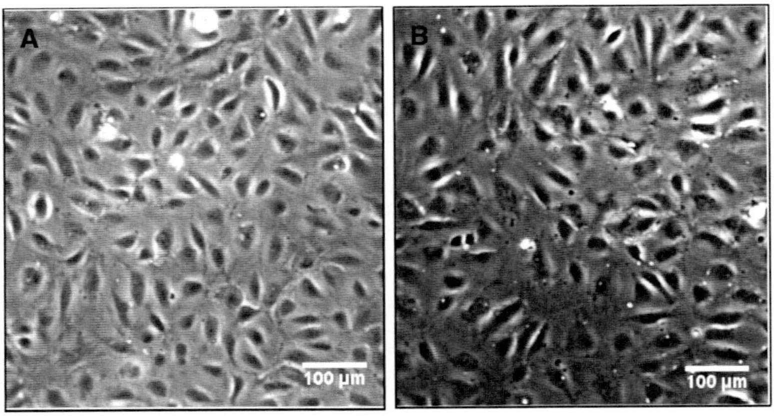

Fig. 2 Phase contrast images of (**a**) cord blood ECFCs and (**b**) human aortic ECs show similar cobblestone morphology

4.2 Characterization of CFU-ECs and ECFCs

CFU-ECs and hematopoietic stem cells (HSCs) are both derived from a common precursor believed to be a hemangioblast present in the bone marrow [47, 48] (Fig. 3). CFU-ECs and HSCs share a number of common cell markers, such as CD133 and CD34 [49, 50]. Asahara and colleagues [43] initially identified CFU-ECs by the expression of vascular endothelial growth factor receptor-2 (VEGFR-2, Flk-1, or KDR) and CD34. They found that CFU-ECs are derived from a precursor population in the bone marrow that expresses CD133, CD45, CD34, and the

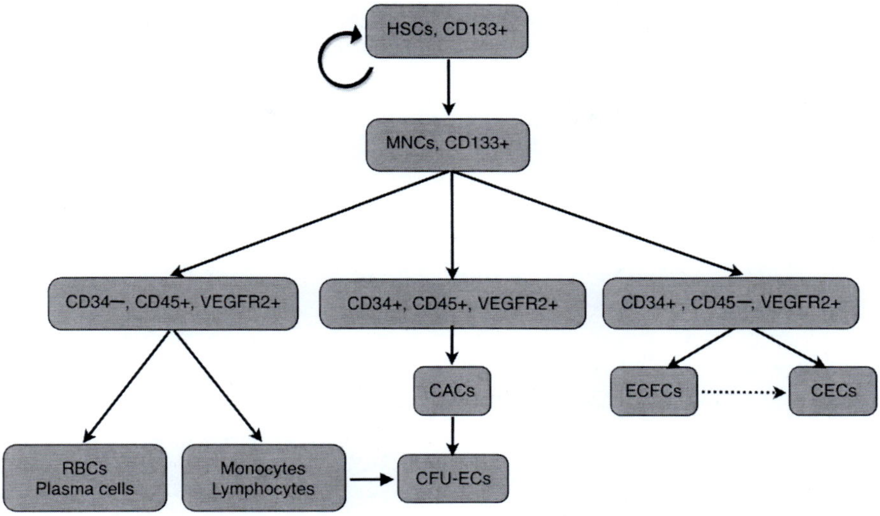

Fig. 3 Schematic of the origins of the various types of EPCs found in the blood. Hematopoietic stem cells (HSCs) in the bone marrow either undergo self-renewal or differentiate into mononuclear cells. Mononuclear cells (MNCs) can be separated into hematopoietic stem cells (HSCs) and circulating endothelial cells (CECs) by the presence of CD45. Circulating angiogenic cells are a subset of HSCs, and CACs can be separated further into colony-forming unit ECs (CFU-ECs) and hematopoietic progenitor cells (HPCs). Endothelial colony-forming cells (ECFCs) can be isolated from CECs

VEGF receptor Flk-1. After release into the bloodstream, these cells lose CD133 and express CD31 (PECAM) and von Willebrand factor (vWF) and bind acetylated LDL.

Within 5 days of isolation these cells form clusters comprised of round cells located centrally and sprouts of spindle-shaped cells located at the periphery. The purity of the starting population of these cells was measured by typical EC markers and found to be 15.7% CD34, 27.6% VEGFR-2, and 94.1% CD45 (CD45 is found on all HSCs, monocytes, and macrophages). Seven days after attachment 27.2% of the cells still expressed CD45 and there was increasing expression of CD31, CD34, VEGFR-2, Tie-2 (receptor tyrosine kinases found on ECs and HSCs), and E-selectin (an adhesion molecule involved in leukocyte adhesion and found on activated endothelium) [43]. These cells, initially termed EPCs, are now commonly called CFU-ECs [51].

Lin et al. identified ECFCs from peripheral blood and termed them late blood outgrowth ECs or endothelial outgrowth cells (EOCs) since they appear after CFU-ECs and because they have such a remarkable proliferative capacity [52]. These ECFCs can be serially passaged and have the potential to grow for 30 population doublings, whereas CFU-ECs may undergo more limited growth. In addition, cord blood ECFCs may be derived from a different population of progenitors than peripheral blood-derived ECFCs, since they are younger and have a greater potential to proliferate. This conjecture is supported by the 15-fold increase in ECFCs from cord blood, the earlier emergence of ECFCs from cord blood than adult blood, and their ability to expand for at least 100 population doublings (compared to peripheral ECFC's 30 doublings) without signs of senescence [53]. Single cells from cord blood can also divide and form 100-fold greater number of cells compared to peripheral blood and have a significantly higher amount of telomerase activity [14].

4.3 Challenges to the Use of EPCs

Two issues affecting the therapeutic use of EPCs are the low cell density of EPCs in blood and the identification of the populations that are capable of differentiating into ECs.

Mice and rats have higher numbers of CFU-ECs than humans. While the overall number of hematopoietic stem cells increases with average mass of a species, the number of active cells is small. Consequently, smaller animals have a larger reserve of stem cells and, by extension, EPCs [54]. Under normal conditions, only 0.01% of the circulating MNCs in human peripheral blood are CFU-ECs. The number of CFU-ECs can be elevated 2- to 10-fold by a variety of signals, such as chemokines and cytokines/growth factors: vascular endothelial growth factor (VEGF) [48, 55–59], granulocyte macrophage colony-stimulating factor [48, 55, 57–59], granulocyte colony-stimulating factor (G-CSF) [55], stromal cell-derived factor-1 (SDF-1) [60], fibroblast growth factor-2 [55], and angiopoietins [55]; hormones (erythropoietin, estrogen) [57, 58]; and signaling molecules: NO and Akt [57, 58]. Drugs (such as statins) [57, 58] and physical exercise have also been shown to increase CFU-EC levels in both humans and mice [58].

Conversely, CFU-EC production declines with age [61]. $CD34^{+}$/Flk-1^{+} CFU-EC levels are inversely related to the risk of cardiovascular disease severity [61] or survival after a heart attack [62]. Patients with a number of risk factors for atherosclerosis have a reduced level of circulating CFU-ECs [55, 58, 63, 64]. When cardiovascular risk factors are present, the circulating CFU-ECs have a reduced ability to proliferate and differentiate [50]. Patients with acute vascular injury and inflammation have an increased level of CFU-ECs, while patients that experienced in-stent restenosis have a decreased level [65]. CFU-ECs may respond to vessel damage in an attempt to promote healing, but patients with decreased levels of CFU-ECs due to risk factors for diabetes or cardiovascular disease may still have restenosis of the vessel. In contrast, circulating endothelial cells (CECs), identified by the presence of CD146, while normally present at low levels (10–40 cells/ml), increase in a number of disease states [66, 67].

Possible reasons why the level of CFU-ECs declines as a person ages include a reduced capacity to mobilize or differentiate, and/or exhaustion of stem/progenitor cells in the bone marrow with time [65]. CFU-ECs may have an increased susceptibility to apoptosis because of an imbalance between pro- and anti-apoptotic factors and/or a decline in antioxidant defense. With advancing age, HSCs lose their ability to react to stress-induced mobilization [58]. Further, production of VEGF and NO, which help both the mobilization and the proliferation of CFU-ECs, also decreases with age [65].

A recent clinical study found that patients that experienced stent restenosis had a decreased number of CFU-ECs that would form colonies in vitro and had an increase in senescent CFU-ECs, although the number of circulating CFU-ECs in patients with and without restenosis was similar [68]. Further, CFU-ECs in patients with restenosis have a reduced ability to bind to fibronectin which may lead to reduced adhesion [69]. Therefore, it appears that the number of CFU-ECs in circulation does not always correlate with reendothelialization, but functional assessments of CFU-ECs are needed to determine effectiveness in a given clinical situation.

Baboon EPCs, isolated using the CFU-EC protocol and grown on vascular grafts, exhibited the typical cobblestone appearance of normal endothelium [70]. The cells were positive for vWF, VE-cadherin, CD31, and VEGFR2. Protein staining for CD146, thrombomodulin, and NOSIII was less intense but comparable to levels on baboon and human ECs. Western blotting showed reduced levels of NOSIII relative to ECs.

The identification of EPCs as distinct from other hematopoietic cells is challenging. The rarity of EPCs in peripheral blood makes the process even more difficult. Current methods of EPC identification consist of flow cytometry and immunolabeling of the surface markers with antigens. Due to the lack of known surface markers specific to EPCs, a collection of markers have been used to help with identification. CD34 and CD133 have been used to identify EPCs in flow cytometry. Cells with CD34 expression have morphological similarities with ECs and have surface markers also present on ECs. CD34, however, is not specific to EPCs but is found on CECs as well as other hematopoietic cells (Fig. 3). CD133, a glycoprotein cell surface marker found on HSCs and EPCs, is also used in flow cytometry and stains cells that have the potential to differentiate into endothelial cells. CD133-positive cells can differentiate into cells that express the EC markers CD34, CD31, VE-cadherin, and vWF [71, 72].

Many groups isolate EPCs using flow cytometry to identify early outgrowth cells that express CD34 and CD133 [50]. However, these molecules are also expressed on hematopoietic stem/progenitor populations [45, 73]. The heterogeneity of the population is reduced if additional cell surface markers are included in the sorting process, such as CD31, CD146, VE-cadherin, or acetylated LDL.

In vitro, CFU-ECs respond to fluid shear stress and adopt a phenotype closer to that of ECs. After 24 h exposure to 2.5 dyne/cm^2, CFU-ECs orient in the direction of flow and increase levels of VEGFRR1, VEGFR2, and VE-cadherin relative to cells under static conditions [74]. Shear stress promoted tube formation [74]. Cells increase NO release and production of NOSIII mRNA [75]. Exposure of CFU-ECs to shear stress between 0.5 and 2.5 dyne/cm^2 for 6–24 h caused a decrease in expression of ephrin B4 mRNA which is present in venous ECs and an increase in expression of ephrin B2 mRNA which is found in arterial ECs [76]. Ephrin B2 expression increased proportionately with shear stress

[76]. Thus, after adhesion and spreading onto a surface, the local fluid dynamic environment can shift the phenotype of CFU-ECs more toward that of arterial endothelium.

CFU-ECs fail to form perfused microvessels in vivo, undergo a limited amount of expansion, and survive in culture for a short time, while ECFCs do form perfused microvessels in vivo and display a robust proliferative potential. CFU-ECs isolated using a commercially available kit express genes and cell surface markers similar to those expressed by T lymphocytes [77, 78]. Monocytes that are CD14 and VEGFR2 positive differentiate into a cell type that expresses NOSIII, and CD34 in vitro as well as CD45, and the monocytes marker CD11b [79]. These cells could form tubes in a Matrigel assay. Mixtures of purified T cells and monocytes exhibited similar behavior [78]. When CD14- and VEGFR2-positive monocytes injected into the femoral artery of mice after removal of the endothelium by balloon catheterization and allowed to attach under static conditions, the cells remained adherent and the femoral artery did not dilate [79]. CD14-positive monocytes that were negative for VEGFR2 did not exhibit any of these properties. These studies suggest that CFU-ECs can originate from monocytes and lymphocytes which may have a repair function in vivo.

The evidence suggests that the ECFCs that Yoder et al. isolated [45] should be considered 'true EPCs,' while cells from other isolation techniques may have EPC qualities but are most likely a heterogeneous population of monocytes and macrophages. CFU-ECs closely resemble cells of a myeloid lineage and fail to exhibit the properties of ECs [80]. CFU-ECs do have the capacity to function in arterial repair, although it is not known whether their monocytic properties affect the progression of atherosclerosis.

Since ECFCs are quite rare (~0.2 cells/ml [45]), mobilization of the resident population of ECFCs and their EPC precursors may not be sufficient to not raise the concentration of ECFCs to a level that will enable coverage blood vessels or graft surface to be functionally important. Thus, ex vivo culture of ECFCs can provide sufficient numbers of cells for therapeutic applications.

4.4 Other Sources of ECs or EPCs

There is interest in using stem cells to generate ECs, but this research is in an early stage. Flk-1-positive embryonic stem cells [81] and mesenchymal stem cells [82] can be induced to express EC markers. Mouse embryonic stem cells exposed to shear stress aligned with the direction of flow and exhibited increased proliferation due to the expression of Flk-1 [81]. (In contrast, steady or pulsatile shear stress without any significant flow reversal inhibits EC proliferation by causing cell arrest in G_0/G_1 [83].) The effect of shear stress upon proliferation was more pronounced at higher shear stresses. Shear-induced proliferation was independent of Flk-1 binding to VEGF but significantly reduced with SU1498 which inhibits Flk-1 kinase. Shear stress also increased expression of EC-specific proteins such as PECAM, Flt-1, VE-cadherin, and Tie-2. Blocking Flk-1 activity with SU1498 reduced the levels of all of these molecules. Further, shear stress stimulates the PI3K-Akt pathway [84], which may represent the mechanism by which proliferation is stimulated. Human embryonic stem cells that differentiated to form ECs aligned with the flow direction [85]. Exposure to 15 dyne/cm^2 for 24 h led to increased mRNA levels of MMP1 COX2 and TGFb reduced levels of ICAM-1, tPA, and MCP-1 relative to cells under static conditions [85].

Mesenchymal stem cells (MSC) derived from bone marrow can differentiate into SMCs [86] and MSCs have been grown in scaffolds of nanoporous poly-L-lactic acid to create engineered blood vessels [87]. These vessels were antithrombotic in vitro. The vessels were patent for 60 days in a rat common carotid graft model, even though the MSC had not proliferated and were replaced with vessel wall cells from the host [87]. The suitability of embryonic stem cells and MSCs for human-engineered vessels is not known since there is little regrowth of endothelium and MSCs have limited mechanical strength [88].

An increase in bone marrow-derived cells (BMCs) in the blood may regenerate lost myocardium in the ischemic heart of the mouse [89, 90]. Addition of cytokines, such as stem cell factor and granulocyte colony-stimulating factor [89], chemokines, VEGF-1, stromal-derived factor, or platelet-derived growth factor [91] increases the number of BMCs in blood, facilitating repair. However, addition of these molecules may have independent effects that may improve repair. Homing of BMCs to the damaged area has not been established and the BMCs may randomly distribute in tissues with only those cells that enter the damaged area undergoing rapid proliferation and transdifferentiation [89]. In a canine model of vascular graft healing, transplanted bone marrow-derived CD34$^+$ hematopoietic precursor cells participated in the endothelialization of a Dacron graft [57], suggesting another possible therapeutic approach. In humans, the number of cells that can be mobilized, even after addition of cytokines, is modest. Further, the mobilized or injected cells are a diverse population that may have multiple positive and negative effects on the repair process. Use of a pure cell population, while more complex, enables a more targeted therapeutic approach.

5 Applications

5.1 Adhesion of EPCs to the Vessel Wall or Graft Surface

Once EPCs are released into the circulation they must home to and differentiate at sites without endothelium or with dysfunctional endothelium [92]. Hemodynamics influences the attachment of white blood cells to the vessel wall and should influence EPC adhesion. Leukocytes adhere via a sequential process of rolling involving selectins and their ligands followed by arrest, which involves integrins. In some blood cells, such as monocytes, arrest can occur without rolling. The molecular properties in bond formation and dissociation appear to determine whether the adhesive event produces rolling or arrest [93]. By increasing the contact area, cell deformation modulates adhesion, particularly at higher shear stresses [94–96]. In vitro studies with leukocytes show that at shear stresses in the range of those present in arteries, the adhesion frequency in one-dimensional shear flow is inversely related to shear stress. The presence of secondary flows affects the transport of blood cells to the vessel wall [97]. This resembles the observed distribution of monocytes in the rabbit aorta after cholesterol feeding [98]. Simulations of cell adhesion show the importance of local fluid dynamics in transporting cells to sites of activated endothelium.

While the detailed mechanism of leukocyte adhesion to vascular endothelium is well studied, less is known about the manner by which EPCs adhere to the vessel wall. CFU-ECs adhere to activated endothelium or exposed subendothelium. TNF-α treatment of EC increases adhesion of CFU-ECs [69, 99]. CFU-ECs adhesion to activated microvascular ECs is mediated by c-kit expressed on circulating CFU-ECs. ECs overlying atherosclerotic plaques expressed the c-kit ligand, mbKitL, while it is absent from normal endothelium. CFU-ECs from diabetics showed reduced adhesion to TNF-α-activated endothelium [100]. CFU-ECs express VLA4, VLA5, and LFA1, which are normally expressed on leukocytes, and the levels of these receptors change through the cell cycle [101]. In addition, since CFU-ECs contain CD18, they adhere to ICAM-1 in vitro and in vivo; blocking ICAM-1 limits neovascularization [102].

CD34-positive cord blood-derived mononuclear cells adhere to adsorbed VCAM-1 and Fn [103]. Adhesion to each ligand varies with the stage of the cell cycle; binding to Fn increases and VCAM-1 decreases as cells progress through the cycle. CFU-EC isolation involves adhesion to fibronectin, while ECFC isolation involves adhesion to gelatin. We have found that ECFCs do not adhere to activated endothelium but bind to smooth muscle cells in vitro via the $\alpha_5\beta_1$ and $\alpha_v\beta_3$ receptors and respond to shear stress similar the ECs [104].

5.2 EPCs and Atherosclerosis

Atherosclerosis develops as a response to injury induced by risk factors such as elevated plasma cholesterol, hypertension, diabetes, smoking, rheumatoid arthritis, age, and gender [105]. In concert with hemodynamics, risk factors selectively activate ECs at lesion-prone sites, initiating a localized inflammatory response that attracts monocytes. Two key features of this model are that the response involves repair induced by vessel wall cells or infiltrating blood monocytes and that due to inefficiencies in the repair process, over time, the repair produces a positive feedback that results in the formation of atherosclerotic lesions.

Animal studies suggest that bone marrow-derived EPCs limit the development of atherosclerosis [106]. In this view, atherosclerosis arises when the demand for vessel wall repair exceeds the rate at which EPCs are produced [106]. ApoE$^{-/-}$ mice, which are genetically altered to develop atherosclerosis, on a high-cholesterol diet that had their bone marrow removed by irradiation, had reduced atherosclerosis when they received bone marrow from young mice [107]. In contrast, when the bone marrow was replaced with marrow from older mice, atherosclerosis was not produced [107]. These results, which are consistent with studies showing reduced CFU-EC function in older humans and animals, suggest that diseases, which damage endothelium, may deplete endothelial progenitors and alter their function.

The transfer of bone marrow cells and spleen-derived CFU-ECs in young $ApoE^{-/-}$ mice actually accelerated atherosclerosis and reduced the expression of markers associated with plaque stability [108]. Based on mRNA expression data, the transfer of cells may induce a proinflammatory response that influences atherosclerotic plaque size and stability [108]. A limitation of many of these studies is that the cell population is impure and the effects could be mediated by cytokines released by contaminating cells.

5.3 Potential Therapeutic Applications

The application of EPCs to tissue engineering and regenerative medicine has been studied in many animal models. EPCs have been used to treat disorders such as hindlimb ischemia and neointimal hyperplasia in animals and may have the potential to do so in humans as well.

Kalka et al. [109] injected into nude mice 5×10^5 human CFU-ECs that were grown in culture and tagged with fluorescent carbocyanine DiI dye. Limb perfusion was drastically reduced in the control groups that received human microvascular ECs or media from CFU-EC culture plates. Over the course of the 28-day study, mice that received CFU-ECs gradually restored perfusion levels to their ischemic hindlimbs to levels comparable to those in nonischemic hindlimbs. Peak CFU-EC integration into the tissue occurred at 3–7 days post-injection and labeled CFU-ECs were found in $56 \pm 4.7\%$ of vessels and capillary density had increased. Therefore, ex vivo cell therapy of CFU-ECs may promote neovascularization of ischemic tissues [110]. Subsequent studies indicate that mononuclear cells and CFU-ECs do not incorporate into new blood vessels but release chemokines and cytokines that promote blood vessel formation [111].

Reendothelialization after angioplasty is another possible application of EPCs. Gulati et al. [112], Griese et al. [113], Kong et al. [114], and Nowak et al. [115] all achieved increased reendothelialization and decreased intimal thickening after MNCs from the bone marrow [112] and CFU-ECs derived from peripheral blood [113–115] were seeded on injured vessels in vivo. These studies involved blocking blood flow for up to 30 min to allow attachment of exogenous MNCs and CFU-ECs. This lengthy interruption of blood flow is not feasible for coronary vessels, and the balloons used to deliver EPCs may cause additional damage to the vessel.

A simpler and less invasive method of using exogenous EPCs is through intravenous transfusion. This method effectively delivers EPCs to sites of vascular injury [4]. Werner et al. achieved a significant increase in reendothelialization and a reduction in neointima formation when MNCs derived from the spleen (possibly CFU-ECs) were intravenously transfused after induction of vessel damage. However, this reendothelialization was only observed when the spleen of the mice was removed, otherwise the MNCs would home to the spleen instead of the injury site [4]. In addition, intravenous transfusion of bone marrow cells of a young healthy donor mouse has been shown to prevent progression of atherosclerosis in ApoE$^{-/-}$ mice fed a high-fat atherosclerotic diet [63, 107] and is localized more in the aorta (58%) compared to wild-type mice [64].

One approach to seed stents or synthetic grafts in vivo is to attach a protein or peptide that is specific to EPCs. In this way, EPCs or ECs specifically adhere to the stent surface. A number of different coatings have been developed to promote EC or EPC attachment and minimize nonspecific cell adhesion, but few approaches have been tried clinically [116]. Anti-CD34-coated stents are safe with no stent thrombosis after 6 months or immunoreactivity to the murine antibodies [117]. A second study with 129 patients showed similar results after 1 year [118]. No data are available about EPC coverage or adhesion of other cells to the stent and a clinical trial has yet to be conducted.

Due to limitations with CD34 capture, Markway et al. [119] examined EC capture in vitro to surfaces containing an anti-VEGFR2 antibody at a shear rate of 100 s^{-1}. Adhesion was specific to VEGFR2 and was enhanced by oriented arrangement of the antibody on the surface. The number or cells attached was proportional to the number of ECs flowing in solution, even in the presence of cells that do not express KDR. Adhesion was firm since cells did not detach when the shear rate was raised 10-fold.

Using antibodies to capture EPCs is promising but the antibodies used to date will capture a wide range of monocytic cells. This approach could be used in a combination release of antithrombotic drugs from the stent. Further, the level of graft coverage and the function of the EPCs attached to the graft need to be established.

The use of EPCs for therapeutic applications is promising. Unfortunately the extremely low yield of ECFCs in peripheral blood and subsequent cell isolation have impeded the translation of EPC therapies into clinical studies [120]. However, ECFCs have been successfully expanded and reinjected into mice. While these cells promote revascularization [121], the process is enhanced with the addition of smooth muscle cells or mesenchymal progenitor cells isolated from cord blood or bone marrow [122].

5.4 Future Studies and Applications

EPCs show great therapeutic potential. ECFCs are functionally similar to native endothelial cells and are most suited for therapeutic applications. The major challenge is to develop efficient and inexpensive methods to expand these cells in vitro or promote their mobilization in vivo. While autologous cells are preferable for many applications, the high growth rate of ECFCs from umbilical cord blood and success with cord blood in various treatments, the development of EPC cell banks should be considered. Rapid antigen typing may permit the development of single-stage graft procedures for stored EPCs derived from cord blood. If successful, these cells would be ideal for tissue-engineering applications. The ability to culture EPCs and mesenchymal progenitor cells from cord blood raises the possibility of developing a tissue-engineered blood vessel of high mechanical strength and a non-thrombogenic intima.

While CFU-ECs are clearly modified form of monocytes and lymphocytes, they may still have therapeutic functions. The long-term fate of these cells needs to be established to ascertain whether they limit or promote atherosclerosis.

Acknowledgments This work was supported by NIH Grants RO1 HL57446, RO1 HL 88825, and NIH Biotechnology Training Grant (GM8555) fellowship to C.S.C. and an American Heart Association Fellowship to M.A.B.

References

1. Lloyd-Jones D, Adams R, Carnethon M, De Simone G, Ferguson TB, Flegal K, Ford E, Furie K, Go A, Greenlund K, Haase N, Hailpern S, Ho M, Howard V, Kissela B, Kittner S, Lackland D, Lisabeth L, Marelli A, McDermott M, Meigs J, Mozaffarian D, Nichol G, O'Donnell C, Roger V, Rosamond W, Sacco R, Sorlie P, Stafford R, Steinberger J, Thom T, Wasserthiel-Smoller S, Wong N, Wylie-Rosett J, Hong Y, For the American Heart Association Statistics, C, and Stroke Statistics, S.. Heart disease and stroke statistics – 2009 update: a report from the American Heart Association Statistics Committee and Stroke Statistics Subcommittee. Circulation 2009; 119:e21–e181.
2. Karra R, Vemullapalli S, Dong C, Herderick EE, Song X, Slosek K, Nevins JR, West M, Goldschmidt-Clermont PJ, Seo D. Molecular evidence for arterial repair in atherosclerosis. Proc Natl Acad Sci USA. 2005; 102:16789–16794.
3. Werner N, Priller J, Laufs U, Endres M, Bohm M, Dirnagl U, Nickenig G. Bone marrow-derived progenitor cells modulate vascular reendothelialization and neointimal formation: effect of 3-hydroxy-3-methylglutaryl coenzyme a reductase inhibition. Arterioscler Thromb Vasc Biol. 2002; 22:1567–1572.
4. Werner N, Junk S, Laufs U, Link A, Walenta K, Bohm M, Nickenig G. Intravenous transfusion of endothelial progenitor cells reduces neointima formation after vascular injury. Circ Res. 2003; 93:e17–e24.
5. Tan H, Jiang X, Yang F, Li Z, Liao D, Trial J, Magera MJ, Durante W, Yang X, Wang H. Hyperhomocysteinemia inhibits post-injury reendothelialization in mice. Cardiovasc Res. 2006; 69:253–262.
6. Manchio JV, Gu J, Romar L, Brown J, Gammie J, Pierson RN III, Griffith B, Poston RS. Disruption of graft endothelium correlates with early failure after off-pump coronary artery bypass surgery. Ann Thorac Surg. 2005; 79:1991–1998.
7. Kudo FA, Muto A, Maloney SP, Pimiento JM, Bergaya S, Fitzgerald TN, Westvik TS, Frattini JC, Breuer CK, Cha CH, Nishibe T, Tellides G, Sessa WC, Dardik A. Venous identity is lost but arterial identity is not gained during vein graft adaptation. Arterioscler Thromb Vasc Biol. 2007; 27:1562–1571.
8. Momin A, Melikian N, Wheatcroft SB, Grieve D, John LC, El Gamel A, Marrinan MT, Desai JB, Driver C, Sherwood R, Shah AM, Kearney MT. The association between saphenous vein endothelial function, systemic inflammation, and statin therapy in patients undergoing coronary artery bypass surgery. J Thorac Cardiovasc Surg. 2007; 134:335–341.
9. Torsney E, Mayr U, Zou Y, Thompson WD, Hu Y, Xu Q. Thrombosis and neointima formation in vein grafts are inhibited by locally applied aspirin through endothelial protection. Circ Res. 2004; 94:1466–1473.

10. Zilla P, Deutsch M, Meinhart J. Endothelial cell transplantation. Semin Vasc Surg. 1999; 12:52–63.
11. Nugent HM, Sjin RT, White D, Milton LG, Manson RJ, Lawson JH, Edelman ER. Adventitial endothelial implants reduce matrix metalloproteinase-2 expression and increase luminal diameter in porcine arteriovenous grafts. J Vasc Surg. 2007; 46:548–556.
12. Walpoth BH, Zammaretti P, Cikirikcioglu M, Khabiri E, Djebaili MK, Pache JC, Tille JC, Aggoun Y, Morel D, Kalangos A, Hubbell JA, Zisch AH. Enhanced intimal thickening of expanded polytetrafluoroethylene grafts coated with fibrin or fibrin-releasing vascular endothelial growth factor in the pig carotid artery interposition model. J Thorac Cardiovasc Surg. 2007; 133:1163–1170.
13. Rotmans JI, Heyligers JM, Verhagen HJ, Velema E, Nagtegaal MM, de Kleijn DP, de Groot FG, Stroes ES, Pasterkamp G. In vivo cell seeding with anti-CD34 antibodies successfully accelerates endothelialization but stimulates intimal hyperplasia in porcine arteriovenous expanded polytetrafluoroethylene grafts. Circulation 2005; 112:12–18.
14. Ingram DA, Mead LE, Tanaka H, Meade V, Fenoglio A, Mortell K, Pollok K, Ferkowicz MJ, Gilley D, Yoder MC. Identification of a novel hierarchy of endothelial progenitor cells using human peripheral and umbilical cord blood. Blood 2004; 104:2752–2760.
15. Brown M, Wallace CS, Truskey GA. Vascular and capillary endothelium. In: Akay M, ed. Wiley encyclopedia of biomedical engineering. New York: Wiley; 2006.
16. Iivanainen E, Kahari V-M, Heino J, Elenius K. Endothelial cell-matrix interactions. Microsc Res Tech. 2003; 60:13–22.
17. Vainionpaa N, Kikkawa Y, Lounatmaa K, Miner JH, Rousselle P, Virtanen I. Laminin-10 and Lutheran blood group glycoproteins in adhesion of human endothelial cells. Am J Physiol. 2006; 290:C764–C775.
18. Orr AW, Sanders JM, Bevard M, Coleman E, Sarembock IJ, Schwartz MA. The subendothelial extracellular matrix modulates NF-{kappa}B activation by flow: a potential role in atherosclerosis. J Cell Biol. 2005; 169:191–202.
19. Jaffe EA. Cell biology of endothelial cells. Hum Pathol. 1987; 18:234–239.
20. Iivanainen E, Kahari VM, Heino J, Elenius K. Endothelial cell-matrix interactions. Microsc Res Tech. 2003; 60:13–22.
21. Shirota T, Yasui H, Matsuda T. Intralumenal tissue-engineered therapeutic stent using endothelial progenitor cell-inoculated hybrid tissue and in vitro performance. Tissue Eng. 2003; 9:473–485.
22. Hutter R, Carrick FE, Valdiviezo C, Wolinsky C, Rudge JS, Wiegand SJ, Fuster V, Badimon JJ, Sauter BV. Vascular endothelial growth factor regulates reendothelialization and neointima formation in a mouse model of arterial injury. Circulation 2004; 110:2430–2435.
23. Consigny PM, Vitali NJ. Resistance of freshly adherent endothelial cells to detachment by shear stress is matrix and time dependent. J Vasc Interv Radiol. 1998; 9:479–485.
24. Thompson MM, Budd JS, Eady SL, Underwood MJ, James RF, Bell PR. The effect of transluminal endothelial seeding on myointimal hyperplasia following angioplasty. Eur J Vasc Surg. 1994; 8:423–434.
25. Cines DB, Pollak ES, Buck CA, Loscalzo J, Zimmerman GA, McEver RP, Pober JS, Wick TM, Konkle BA, Schwartz BS, Barnathan ES, McCrae KR, Hug BA, Schmidt AM, Stern DM. Endothelial cells in physiology and in the pathophysiology of vascular disorders. Blood 1998; 91:3527–3561.
26. Aird W. Phenotypic heterogeneity of the endothelium. Circ Res. 2007; 100:158–173.
27. Passerini AG, Polacek DC, Shi C, Francesco NM, Manduchi E, Grant GR, Pritchard WF, Powell S, Chang GY, Stoeckert CJ, Davies PF. Coexisting proinflammatory and antioxidative endothelial transcription profiles in a disturbed flow region of the adult porcine aorta. Proc Natl Acad Sci USA. 2004; 101:2482–2487.
28. Pusztaszeri MP, Seelentag W, Bosman FT. Immunohistochemical expression of endothelial markers CD31, CD34, von Willebrand factor, and Fli-1 in normal human tissues. J Histochem Cytochem. 2006; 54:385–395.

29. Solovey AN, Gui L, Chang L, Enenstein J, Browne PV, Hebbel RP. Identification and functional assessment of endothelial P1H12. J Lab Clin Med. 2001; 138:322–331.
30. Tzima E, Irani-Tehrani M, Kiosses W, Dejana E, Schultz D, Engelhardt B, Cao G, DeLisser H, Schwartz M. A mechanosensory complex that mediates the endothelial cell response to fluid shear stress. Nature 2005; 437:426–431.
31. Parmar K, Larman H, Dai G, Zhang Y, Wang E, Moorthy S, Kratz J, Lin Z, Jain M, Gimbrone M, Garcia-Cardena G. Prolonged fluid shear stress induces a distinct set of endothelial cell genes, most specifically lung Kruppel-like factor (KLF2). J Clin Invest. 2006; 116:49–58.
32. Dekker R, van Soest S, Fontijn R, Salamanca S, de Groot P, VanBavel E, Pannekoek H, Horrevoets A. Prolonged fluid shear stress induces a distinct set of endothelial cell genes, most specifically lung Kruppel-like factor (KLF2). Blood 2002; 100:1689–1698.
33. Tchaikovski V, Fellbrich G, Waltenberger J. The molecular basis of VEGFR-1 signal transduction pathways in primary human monocytes. Arterioscler Thromb Vasc Biol. 2008; 28:322–328.
34. Jarrell BE, Williams SK, Stokes G, Hubbard FA, Carabasi RA, Koolpe E, Greener D, Pratt K, Moritz MJ, Radomski J, et al. Use of freshly isolated capillary endothelial cells for the immediate establishment of a monolayer on a vascular graft at surgery. Surgery 1986; 100:392–399.
35. Arts CH, Hedeman Joosten PP, Blankensteijn JD, Staal FJ, Ng PY, Heijnen-Snyder GJ, Sixma JJ, Verhagen HJ, de Groot PG, Eikelboom BC. Contaminants from the transplant contribute to intimal hyperplasia associated with microvascular endothelial cell seeding. Eur J Vasc Endovasc Surg. 2002; 23:29–38.
36. Arts CH, Blankensteijn JD, Heijnen-Snyder GJ, Verhagen HJ, Hedeman Joosten PP, Sixma JJ, Eikelboom BC, de Groot PG. Reduction of non-endothelial cell contamination of microvascular endothelial cell seeded grafts decreases thrombogenicity and intimal hyperplasia. Eur J Vasc Endovasc Surg. 2002; 23:404–412.
37. Herring M, Gardner A, Glover J. A single-staged technique for seeding vascular grafts with autogenous endothelium. Surgery 1978; 84:498–504.
38. Herring M, Smith J, Dalsing M, Glover J, Compton R, Etchberger K, Zollinger T. Endothelial seeding of polytetrafluoroethylene femoral popliteal bypasses: the failure of low-density seeding to improve patency. J Vasc Surg. 1994; 20:650–655.
39. Bhat VD, Klitzman B, Koger K, Truskey GA, Reichert WM. Improving endothelial cell adhesion to vascular graft surfaces: clinical need and strategies. J Biomater Sci Polym Ed. 1998; 9:1117–1135.
40. Alobaid N, Salacinski HJ, Sales KM, Hamilton G, Seifalian AM. Single stage cell seeding of small diameter prosthetic cardiovascular grafts. Clin Hemorheol Microcirc. 2005; 33: 209–226.
41. Meinhart JG, Deutsch M, Fischlein T, Howanietz N, Froschl A, Zilla P. Clinical autologous in vitro endothelialization of 153 infrainguinal ePTFE grafts. Ann Thorac Surg. 2001; 71:S327–S331.
42. Truskey GA, Yuan F, Katz DF. Transport phenomenon in biological systems. Upper Saddle River: Pearson/Prentice Hall; 2004.
43. Asahara T, Murohara T, Sullivan A, Silver M, van der Zee R, Li T, Witzenbichler B, Schatteman G, Isner JM. Isolation of putative progenitor endothelial cells for angiogenesis. Science 1997; 275:964–967.
44. Hill JM, Zalos G, Halcox JPJ, Schenke WH, Waclawiw MA, Quyyumi AA, Finkel T. Circulating endothelial progenitor cells, vascular function, and cardiovascular risk. N Engl J Med. 2003; 348:593–600.
45. Prater DN, Case J, Ingram DA, Yoder MC. Working hypothesis to redefine endothelial progenitor cells. Leukemia 2007; 21:1141–1149.

46. Bompais H, Chagraoui J, Canron X, Crisan M, Liu XH, Anjo A, Tolla-Le Port C, Leboeuf M, Charbord P, Bikfalvi A, Uzan G. Human endothelial cells derived from circulating progenitors display specific functional properties compared with mature vessel wall endothelial cells. Blood 2004; 103:2577–2584.
47. Urbich C, Dimmeler S. Endothelial progenitor cells functional characterization. Trends Cardiovasc Med. 2004; 14:318–322.
48. Ishikawa M, Asahara T. Endothelial progenitor cell culture for vascular regeneration. Stem Cells Dev. 2004; 13:344–349.
49. Rossig L, Urbich C, Dimmeler S. Endothelial progenitor cells at work: not mature yet, but already stress-resistant. Arterioscler Thromb Vasc Biol. 2004; 24:1977–1979.
50. Zhang L, Yang R, Han ZC. Transplantation of umbilical cord blood-derived endothelial progenitor cells: a promising method of therapeutic revascularisation. Eur J Haematol. 2006; 76:1–8.
51. Yoder MC, Mead LE, Prater D, Krier TR, Mroueh KN, Li F, Krasich R, Temm CJ, Prchal JT, Ingram DA. Redefining endothelial progenitor cells via clonal analysis and hematopoietic stem/progenitor cell principals. Blood 2007; 109:1801–1809.
52. Lin Z, Kumar A, SenBanerjee S, Staniszewski K, Parmar K, Vaughan DE, Gimbrone MA Jr., Balasubramanian V, Garcia-Cardena G, Jain MK. Kruppel-like factor 2 (KLF2) regulates endothelial thrombotic function. Circ Res. 2005; 96:e48–e57.
53. Ingram DA, Caplice NM, Yoder MC. Unresolved questions, changing definitions, and novel paradigms for defining endothelial progenitor cells. Blood 2005; 106:1525–1531.
54. Dingli D, Traulsen A, Pacheco JM. Dynamics of haemopoiesis across mammals. Proc R Soc B Biol Sci. 2008; 275:2389–2392.
55. Luttun A, Carmeliet G, Carmeliet P. Vascular progenitors: from biology to treatment. Trends Cardiovasc Med. 2002; 12:88–96.
56. Vasa M, Fichtlscherer S, Adler K, Aicher A, Martin H, Zeiher AM, Dimmeler S. Increase in circulating endothelial progenitor cells by statin therapy in patients with stable coronary artery disease. Circulation 2001; 103:2885–2890.
57. Zammaretti P, Zisch AH. Adult 'endothelial progenitor cells'. Renewing vasculature. Int J Biochem Cell Biol. 2005; 37:493–503.
58. Dimmeler S, Zeiher AM. Vascular repair by circulating endothelial progenitor cells: the missing link in atherosclerosis? J Mol Med. 2004; 82:671–677.
59. Ikenaga S, Hamano K, Nishida M, Kobayashi T, Li TS, Kobayashi S, Matsuzaki M, Zempo N, Esato K. Autologous bone marrow implantation induced angiogenesis and improved deteriorated exercise capacity in a rat ischemic hindlimb model. J Surg Res. 2001; 96:277–283.
60. Aicher A, Heeschen C, Mildner-Rihm C, Urbich C, Ihling C, Technau-Ihling K, Zeiher AM, Dimmeler S. Essential role of endothelial nitric oxide synthase for mobilization of stem and progenitor cells. Nat Med. 2003; 9:1370–1376.
61. Schmidt-Lucke C, Rossig L, Fichtlscherer S, Vasa M, Britten M, Kamper U, Dimmeler S, Zeiher AM. Reduced number of circulating endothelial progenitor cells predicts future cardiovascular events: proof of concept for the clinical importance of endogenous vascular repair. Circulation 2005; 111:2981–2987.
62. Werner N, Kosiol S, Schiegl T, Ahlers P, Walenta K, Link A, Bohm M, Nickenig G. Circulating endothelial progenitor cells and cardiovascular outcomes. N Engl J Med. 2005; 353:999–1007.
63. Wassmann S, Werner N, Czech T, Nickenig G. Improvement of endothelial function by systemic transfusion of vascular progenitor cells. Circ Res. 2006; 99:e74–e83.
64. Vemulapalli S, Metzler SD, Akabani G, Petry NA, Niehaus NJ, Liu X, Patil NH, Greer KL, Jaszczak RJ, Coleman RE, Dong C, Goldschmidt-Clermont PJ, Chin BB. Cell therapy in murine atherosclerosis: in vivo imaging with high-resolution helical SPECT. Radiology 2007; 242:198–207.

65. Dzau VJ, Gnecchi M, Pachori AS, Morello F, Melo LG. Therapeutic potential of endothelial progenitor cells in cardiovascular diseases. Hypertension 2005; 46:7–18.
66. Woywodt A, Blann AD, Kirsch T, Erdbruegger U, Banzet N, Haubitz M, Dignat-George F. Isolation and enumeration of circulating endothelial cells by immunomagnetic isolation: proposal of a definition and a consensus protocol. J Thromb Haemost. 2006; 4:671–677.
67. Dignat-George F, Sampol J. Circulating endothelial cells in vascular disorders: new insights into an old concept. Eur J Haematol. 2000; 65:215–220.
68. Matsuo Y, Imanishi T, Hayashi Y, Tomobuchi Y, Kubo T, Hano T, Akasaka T. The effect of senescence of endothelial progenitor cells on in-stent restenosis in patients undergoing coronary stenting. Intern Med. 2006; 45:581–587.
69. George J, Herz I, Goldstein E, Abashidze S, Deutch V, Finkelstein A, Michowitz Y, Miller H, Keren G. Number and adhesive properties of circulating endothelial progenitor cells in patients with in-stent restenosis. Arterioscler Thromb Vasc Biol. 2003; 23:e57–e60.
70. Hinds MT, Ma M, Tran N, Ensley AE, Kladakis SM, Vartanian KB, Markway BD, Nerem RM, Hanson SR. Potential of baboon endothelial progenitor cells for tissue engineered vascular grafts. J Biomed Mater Res A 2008; 86:804–812.
71. Gehling UM, Ergun S, Schumacher U, Wagener C, Pantel K, Otte M, Schuch G, Schafhausen P, Mende T, Kilic N, Kluge K, Schafer B, Hossfeld DK, Fiedler W. In vitro differentiation of endothelial cells from AC133-positive progenitor cells. Blood 2000; 95:3106–3112.
72. Khan SS, Solomon MA, McCoy JP Jr.. Detection of circulating endothelial cells and endothelial progenitor cells by flow cytometry. Cytometry B Clin Cytom. 2005; 64:1–8.
73. de Boer HCet al. Fibrin and activated platelets cooperatively guide stem cells to a vascular injury and promote differentiation towards an endothelial cell phenotype. Arterioscler Thromb Vasc Biol. 2006; 26:1653–1659.
74. Yamamoto K, Takahashi T, Asahara T, Ohura N, Sokabe T, Kamiya A, Ando J. Proliferation, differentiation, and tube formation by endothelial progenitor cells in response to shear stress. J Appl Physiol. 2003; 95:2081–2088.
75. Tao J, Yang Z, Wang J-M, Tu C, Pan S-R. Effects of fluid shear stress on eNOS mRNA expression and NO production in human endothelial progenitor cells. Cardiology 2006; 106:82–88.
76. Obi S, Yamamoto K, Shimizu N, Kumagaya S, Masumura T, Sokabe T, Asahara T, Ando J. Fluid shear stress induces arterial differentiation of endothelial progenitor cells. J Appl Physiol. 2009; 106:203–211.
77. Desai A, Glaser A, Liu D, Raghavachari N, Blum A, Zalos G, Lippincott M, McCoy JP, Munson PJ, Solomon MA, Danner RL, Cannon RO III. Microarray-based characterization of a colony assay used to investigate endothelial progenitor cells and relevance to endothelial function in humans. Arterioscler Thromb Vasc Biol. 2009; 29:121–127.
78. Rohde E, Bartmann C, Schallmoser K, Reinisch A, Lanzer G, Linkesch W, Guelly C, Strunk D. Immune cells mimic the morphology of endothelial progenitor colonies in vitro. Stem Cells 2007; 25:1746–1752.
79. Elsheikh I, Uzunel M, He Z, Holgersson J, Nowak G, Sumitran-Holgersson S. Only a specific subset of human peripheral-blood monocytes has endothelial-like functional capacity. Blood 2005; 106:2347–2355.
80. Timmermans F, Van Hauwermeiren F, De Smedt M, Raedt R, Plasschaert F, De Buyzere ML, Gillebert TC, Plum J, Vandekerckhove B. Endothelial outgrowth cells are not derived from $CD133^+$ cells or $CD45^+$ hematopoietic precursors. Arterioscler Thromb Vasc Biol. 2007; 27:1572–1579.
81. Yamamoto K, Sokabe T, Watabe T, Miyazono K, Yamashita JK, Obi S, Ohura N, Matsushita A, Kamiya A, Ando J. Fluid shear stress induces differentiation of Flk-1-positive embryonic stem cells into vascular endothelial cells in vitro. Am J Physiol Heart Circ Physiol. 2005; 288:H1915–H1924.

82. Wang H, Riha GM, Yan S, Li M, Chai H, Yang H, Yao Q, Chen C. Shear stress induces endothelial differentiation from a murine embryonic mesenchymal progenitor cell line. Arterioscler Thromb Vasc Biol. 2005; 25:1817–1823.
83. Guo D, Chien S, Shyy JYJ. Regulation of endothelial cell cycle by laminar versus oscillatory flow: distinct modes of interactions of amp-activated protein kinase and akt pathways. Circ Res. 2007; 100:564–571.
84. Zeng L, Xiao Q, Margariti A, Zhang Z, Zampetaki A, Patel S, Capogrossi MC, Hu Y, Xu Q. HDAC3 is crucial in shear- and VEGF-induced stem cell differentiation toward endothelial cells. J Cell Biol. 2006; 174:1059–1069.
85. Metallo CM, Vodyanik MA, de Pablo JJ, Shukvin II, Placek SP. The response of human embryonic stem cell-derived endothelial cells to shear stress. Biotechnol Bioeng. 2008; 100:830–837.
86. Stephan S, Ball SG, Williamson M, Bax DV, Lomas A, Shuttleworth CA, Kielty CM. Cell-matrix biology in vascular tissue engineering. J Anat. 2006; 209:495–502.
87. Hashi CK, Zhu Y, Yang GY, Young WL, Hsiao BS, Wang K, Chu B, Li S. Antithrombogenic property of bone marrow mesenchymal stem cells in nanofibrous vascular grafts. Proc Natl Acad Sci USA. 2007; 104:11915–11920.
88. Gong Z, Niklason LE. Small-diameter human vessel wall engineered from bone marrow-derived mesenchymal stem cells (hMSCs). FASEB J. 2008; 22:1635–1648.
89. Orlic D, Kajstura J, Chimenti S, Limana F, Jakoniuk I, Quaini F, Nadal-Ginard B, Bodine DM, Leri A, Anversa P. Mobilized bone marrow cells repair the infracted heart, improving function and survival. Proc Natl Acad Sci USA. 2001; 98:10344–10349.
90. Kocher AA, Schuster MD, Szabolcs MJ, Takuma S, Burkhoff D, Wang J, Homma S, Edwards NM, Itescu S. Neovascularization of ischemic myocardium by human bone-marrow-derived angioblasts prevents cardiomyocyte apoptosis, reduces remodeling and improves cardiac function. Nat Med. 2001; 7:430–436.
91. Rafii DC, Psaila B, Butler J, Jin DK, Lyden D. Regulation of vasculogenesis by platelet-mediated recruitment of bone marrow-derived cells. Arterioscler Thromb Vasc Biol. 2008; 28:217–222.
92. Szmitko PE, Wang CH, Weisel RD, de Almeida JR, Anderson TJ, Verma S. New markers of inflammation and endothelial cell activation: Part I. Circulation 2003; 108:1917–1923.
93. Chang KC, Tees DF, Hammer DA. The state diagram for cell adhesion under flow: leukocyte rolling and firm adhesion. Proc Natl Acad Sci USA. 2000; 97:11262–11267.
94. Chen S, Springer TA. An automatic braking system that stabilizes leukocyte rolling by an increase in selectin bond number with shear. J Cell Biol. 1999; 144:185–200.
95. Rinker KD, Prabhakar V, Truskey GA. Effect of contact time and force on monocyte adhesion to vascular endothelium. Biophys J. 2001; 80:1722–1732.
96. Khismatullin DB, Truskey GA. A 3D numerical study of the effect of channel height on leukocyte deformation and adhesion in parallel-plate flow chambers. Microvasc Res. 2004; 68:188–202.
97. Buchanan JR, Kleinstreuer C, Hyun S, Truskey GA. Hemodynamics simulation and identification of susceptible sites of atherosclerotic lesion formation in a model abdominal aorta. J Biomech. 2003; 36:1185–1196.
98. Malinauskas RA, Herrmann RA, Truskey GA. The distribution of intimal white blood cells in the normal rabbit aorta. Atherosclerosis 1995; 115:147–163.
99. Dentelli P, Rosso A, Balsamo A, Colmenares Benedetto S, Zeoli A, Pegoraro M, Camussi G, Pegoraro L, Brizzi MF, C-KI T. by interacting with the membrane-bound ligand, recruits endothelial progenitor cells to inflamed endothelium. Blood 2007; 109:4264–4271.
100. Tepper OM, Galiano RD, Capla JM, Kalka C, Gagne PJ, Jacobowitz GR, Levine JP, Gurtner GC. Human endothelial progenitor cells from type II diabetics exhibit impaired proliferation, adhesion, and incorporation into vascular structures. Circulation 2002; 106:2781–2786.

101. Huygen S, Giet O, Artisien V, Di Stefano I, Beguin Y, Gothot A. Adhesion of synchronized human hematopoietic progenitor cells to fibronectin and vascular cell adhesion molecule-1 fluctuates reversibly during cell cycle transit in *ex vivo* culture. Blood 2002; 100:2744–2752.
102. Wu Y, Ip JE, Huang J, Zhang L, Matsushita K, Liew CC, Pratt RE, Dzau VJ. Essential role of ICAM-1/CD18 in mediating EPC recruitment, angiogenesis, and repair to the infarcted myocardium. Circ Res. 2006; 99:315–322.
103. Huygen S, Giet O, Artisien V, Di Stefano I, Beguin Y, Gothot A. Adhesion of synchronized human hematopoietic progenitor cells to fibronectin and vascular cell adhesion molecule-1 fluctuates reversibly during cell cycle transit in *ex vivo* culture. Blood 2002; 100:2744–2752.
104. Brown MA, Wallace CS, Angelos MA, Truskey GA. Characterization of umbilical cord blood derived late outgrowth endothelial progenitor cells exposed to laminar shear stress. Tissue Eng. 2009; 35:3575–3587.
105. Ross R. The pathogenesis of atherosclerosis: a perspective for the 1990s. Nature 1993; 362:801–809.
106. Seo D, Wang T, Dressman H, Herderick EE, Iversen ES, Dong C, Vata K, Milano CA, Rigat F, Pittman J, Nevins JR, West M, Goldschmidt-Clermont PJ. Gene expression phenotypes of atherosclerosis. Arterioscler Thromb Vasc Biol. 2004; 24:1922–1927.
107. Rauscher FM, Goldschmidt-Clermont PJ, Davis BH, Wang T, Gregg D, Ramaswami P, Pippen AM, Annex BH, Dong C, Taylor DA. Aging, progenitor cell exhaustion, and atherosclerosis. Circulation 2003; 108:457–463.
108. George J, Afek A, Abashidze A, Shmilovich H, Deutsch V, Kopolovich J, Miller H, Keren G. Transfer of endothelial progenitor and bone marrow cells influences atherosclerotic plaque size and composition in apolipoprotein E knockout mice. Arterioscler Thromb Vasc Biol. 2005; 25:2636–2641.
109. Kalka C, Masuda H, Takahashi T, Kalka-Moll WM, Silver M, Kearney M, Li T, Isner JM, Asahara T. Transplantation of *ex vivo* expanded endothelial progenitor cells for therapeutic neovascularization. Proc Natl Acad Sci USA. 2000; 97:3422–3427.
110. Kalka C, Masuda H, Takahashi T, Kalka-Moll WM, Silver M, Kearney M, Li T, Isner JM, Asahara T. Transplantation of *ex vivo* expanded endothelial progenitor cells for therapeutic neovascularization. Proc Natl Acad Sci USA. 2000; 97:3422–3427.
111. Gnecchi M, Zhang Z, Ni A, Dzau VJ. Paracrine mechanisms in adult stem cell signaling and therapy. Circ Res. 2008; 103:1204–1219.
112. Gulati R, Jevremovic D, Peterson TE, Witt TA, Kleppe LS, Mueske CS, Lerman A, Vile RG, Simari RD. Autologous culture-modified mononuclear cells confer vascular protection after arterial injury. Circulation 2003; 108:1520–1526.
113. Griese DP, Ehsan A, Melo LG, Kong D, Zhang L, Mann MJ, Pratt RE, Mulligan RC, Dzau VJ. Isolation and transplantation of autologous circulating endothelial cells into denuded vessels and prosthetic grafts: implications for cell-based vascular therapy. Circulation 2003; 108:2710–2715.
114. Kong D, Melo LG, Mangi AA, Zhang L, Lopez-Ilasaca M, Perrella MA, Liew CC, Pratt RE, Dzau VJ. Enhanced inhibition of neointimal hyperplasia by genetically engineered endothelial progenitor cells. Circulation 2004; 109:1769–1775.
115. Nowak G, Karrar A, Holmen C, Nava S, Uzunel M, Hultenby K, Sumitran-Holgersson S. Expression of vascular endothelial growth factor receptor-2 or Tie-2 on peripheral blood cells defines functionally competent cell populations capable of reendothelialization. Circulation 2004; 110:3699–3707.
116. de Mel A, Jell G, Stevens MM, Seifalian AM. Biofunctionalization of biomaterials for accelerated in situ endothelialization: a review. Biomacromolecules 2008; 9:2969–2979.
117. Aoki J, Serruys PW, van Beusekom H, Ong ATL, McFadden EP, Sianos G, van der Giessen WJ, Regar E, de Feyter PJ, Davis HR, Rowland S, Kutryk MJBK. Endothelial progenitor cell

capture by stents coated with antibody against CD34 the HEALING-FIM (Healthy endothelial accelerated lining inhibits neointimal growth-first in man) registry. J Am Coll Cardiol. 2005; 45:1574–1579.
118. Co M, Tay E, Lee CH, Poh KK, Low A, Lim J, Lim IH, Lim YT, Tan HC. Use of endothelial progenitor cell capture stent (Genous Bio-Engineered R Stent) during primary percutaneous coronary intervention in acute myocardial infarction: intermediate- to long-term clinical follow-up. Am Heart J. 2008; 155:128–132.
119. Markway BD, Mccarty OJT, Marzec UM, David W, Courtman DW, Stephen R, Hanson SR, Hinds MT. Capture of flowing endothelial cells using surface-immobilized anti-kinase insert domain receptor antibody. Tissue Eng Part C 2008; 14:97–105.
120. Gurtner GC, Chang E. "Priming" endothelial progenitor cells: a new strategy to improve cell based therapeutics. Arterioscler Thromb Vasc Biol. 2008; 28:1034–1035.
121. Melero-Martin JM, Khan ZA, Picard A, Wu X, Paruchuri S, Bischoff J. In vivo vasculogenic potential of human blood-derived endothelial progenitor cells. Blood 2007; 109:4761–4768.
122. Melero-Martin JM, De Obaldia ME, Kang S-Y, Khan ZA, Yuan L, Oettgen P, Bischoff J. Engineering robust and functional vascular networks in vivo with human adult and cord blood-derived progenitor cells. Circ Res. 2008; 103:194–202.
123. Hirschi KK, Ingram DA, Yoder MC. Assessing identity, phenotype, and fate of endothelial progenitor cells. Arterioscler Thromb Vasc Biol. 2008; 28:1584–1595.

Regenerating Tubules for Kidney Repair

W.W. Minuth, L. Denk, and A. Roessger

Abstract Stem/progenitor cells are in the focus of regenerative medicine for a future therapy of acute and chronic renal failure. However, broad knowledge about parenchymal regeneration in kidney is lacking. For that reason developmental pathways leading from stem/progenitor cells to newly formed tubules have to be investigated. A new technique promotes renal stem/progenitor cells to form numerous tubules between layers of polyester fleeces. This artificial interstitium replaces coating by extracellular matrix proteins, supports spatial extension of renal tubules, and can be used with chemically defined Iscove's modified Dulbecco's medium (IMDM) during a long-term culture period of 13 days. The development of tubules is stimulated by aldosterone and depends on the applied hormone concentration. The tubulogenic effect cannot be mimicked by precursors of the aldosterone synthesis pathway or by other steroid hormones. Antagonists such as spironolactone or canrenoate prevent the development of tubules, which indicates that the mineralocorticoid receptor (MR) is involved. Administration of geldanamycin, radicicol, quercetin, or KNK 437 in combination with aldosterone blocks development of tubules by disturbing the contact between MR and heat-shock proteins. Transmission electron microscopy (TEM) further demonstrates that generated tubules exhibit a junctional complex between the apical and the lateral plasma membrane. At the basal aspect a continuously developed basal lamina is present. Immuno-label for Troma I (cytokeratin Endo-A) shows isoprismatic cells, while label for laminin $\gamma 1$, occludin, and Na/K-ATPase $\alpha 5$ confirms typical features of a polarized epithelium. Finally, the introduced system makes it possible to pile and pave renal stem/progenitor cells, so that the spatial development of tubules can be systematically investigated.

Keywords Artificial interstitium · Kidney · Stem/progenitor cells · Aldosterone · Perfusion culture

---

W.W. Minuth (✉)
Department of Molecular and Cellular Anatomy, University of Regensburg, Regensburg, Germany
e-mail: will.minuth@vkl.uni-regensburg.de

G.M. Artmann et al. (eds.), *Stem Cell Engineering*,
DOI 10.1007/978-3-642-11865-4_14,

1 Promoting Regeneration in Kidney

The multitude of patients with chronic or acute renal failure shows that the kidney has in comparison to other organs a decreased capability for functional regeneration. In view of this clinical background the question arises why the diseased kidney is not able to regenerate new nephron segments. Of special experimental interest is the question which cell-biological process inhibits the regeneration of parenchyme.

An ideal form of therapy for the future could be to induce a process of regeneration and to steer it therapeutically [1, 2]. One could imagine that the renewal of parenchyme occurs via a stimulation of non-diseased parenchymal cells or via an activation of tissue-specific stem/progenitor cells. Independent of the chosen strategy it is apparent that an application or an activation of stem/progenitor cells alone is not sufficient to trigger the optimal course of therapy. Most important, one has to learn to promote the spatial development of tubules and their integration in a diseased environment. However, at present the knowledge about these processes is minimal and the complex micro-architecture of the kidney makes it difficult to investigate such processes (Fig. 1).

For that reason controlled in vitro experiments are performed to learn about optimizing stem/progenitor cell development and to steer their spatial growth so that finally three-dimensional structured tubules arise. Only with this basic knowledge it will be possible in future to steer regeneration within a diseased kidney.

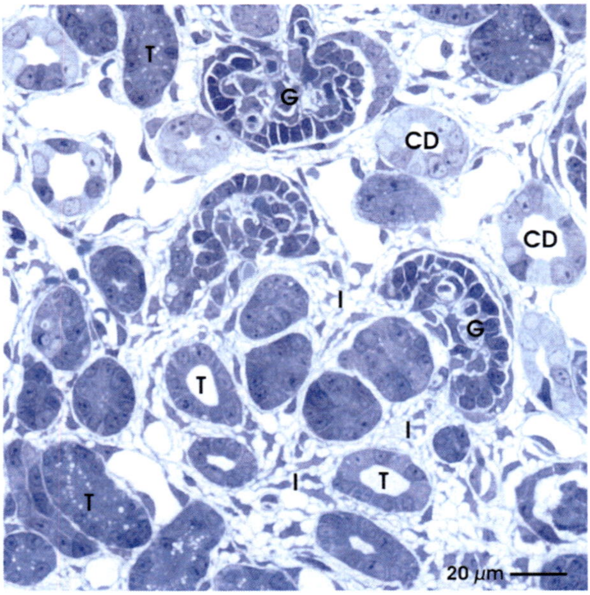

Fig. 1 Horizontal semi-thin section of the cortex from neonatal rabbit kidney. The illustration shows developing glomeruli (*G*), tubules (*T*), and collecting ducts (*CD*) embedded in the interstitium (*I*)

2 Viewing Renal Stem/Progenitor Cell Niches

In our model system the niche for stem/progenitor cells can be demonstrated in the outer cortex of neonatal rabbit kidney (Fig. 2). This region is developing for days still after birth. Beyond the capsule two kinds of very different stem/progenitor cell populations are recognized. The collecting duct tubule (CD) develops from the ampullae (A) located below the renal organ capsule (CF). Each collecting duct ampulla contains in its tip epithelial stem/progenitor cells. At the basal aspect of the ampullar epithelium mesenchymal stem/progenitor cells are found, which develop into the different nephron segments.

After a reciprocal interaction between cells in the tip of the collecting duct ampulla (A) and the nephrogenic mesenchymal stem/progenitor cells, first signs of nephron anlagen as comma-shaped and further matured S-shaped bodies are recognized. By an unknown mechanism the capsule (CF)-orientated wing of the S-shaped body forms all the tubule portions of the nephron, while the medulla-oriented wing is the origin of the glomerulus. The tubular portion develops further into the proximal, intermediate, and distal nephron segments.

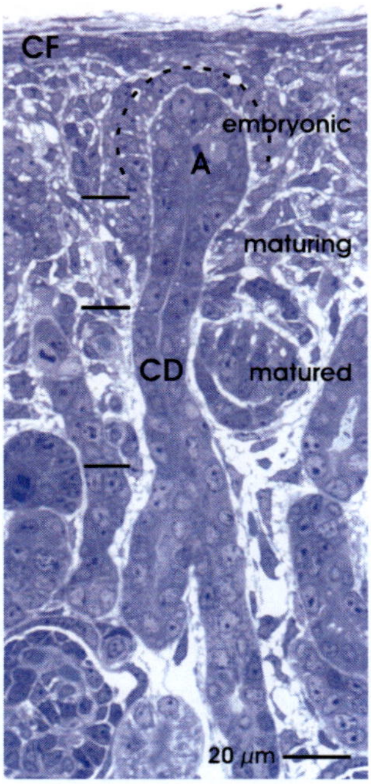

Fig. 2 Vertical section through the cortex of neonatal rabbit kidney. Beyond the capsula fibrosa (*CF*) epithelial stem/progenitor cells are found within the ampulla (*A*) tip of the collecting duct (*CD*). Nephrogenic mesenchymal cells surround the basal aspect of the ampulla (*dotted line*). The outer cortex can be divided into an embryonic, maturing, and matured zone

3 Telling Stem/Progenitor Cells to Form Structured Tubules

At the first view the development of a three-dimensional tubule derived from renal stem/progenitor cells appears simple, but in reality it is a rather complex cell-biological process, which is up to date not understood [3, 4]. As seen from a theoretical sight the intricate development comprises the sprouting of cells for reaching the necessary amount, the formation of a lumen, and the elongation of the tubule segment (Fig. 3) [5–8]. Thus, renal stem/progenitor cells first transdifferentiate, then they are multiplied to form a polarized epithelium. The epithelium is integrated into a structured tubule exhibiting exact geometrical dimensions such as length and inner and outer diameter. For renal tubules it is unknown if the process forming a lumen occurs by wrapping, budding, cavitation, or hollowing. The development continues at the basal side of the tubule by the development of a basal lamina. During the

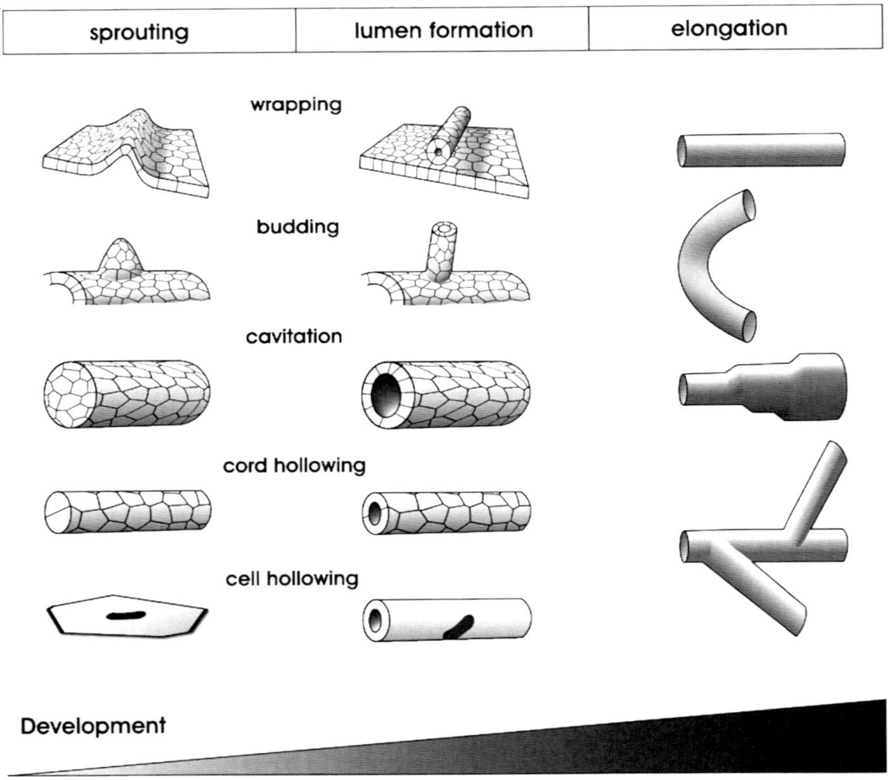

Fig. 3 Schematic development of a polarized tubule comprising sprouting, lumen formation, and elongation. During this process embryonic cells transdifferentiate and develop into a three-dimensional tubule with a defined inner and outer diameter. Further on a straightforward development, a convolution, or an arborization is triggered

following growth phase, geometrical properties such as straight course, length, convolution, or branching of the tubule are realized. Finally a structured tubule with a defined inner and outer diameter arises.

It is obvious that a tubule does not appear automatically after the application of a single growth factor. In contrast, the complex development depends on parameters such as different morphogenic factors, selected adhesion substrates, intercellular communication, and the fluid environment. All these factors have to complement one another to promote histotypical development and to prevent arise of atypical features.

The molecular interactions during the process of reciprocal induction between renal stem/progenitor cells are intensively investigated over a period of decades. However, only little knowledge is available concerning the development of the different tubule segments. For example, the morphogenetic factors are unknown, which trigger the segmentation into the proximal tubule, the loop of Henle, the distal tubule, the connecting tubule, and finally the heterogeneously composed collecting duct. It is further unclear why parts of the proximal and distal tubule develop convolutes, while all the other segments show a straightforward orientated course. In this coherence it is also unknown why some of the nephron segments are homogeneously composed, while the connecting and collecting duct tubules exhibit a heterogeneous cell population.

4 Registering Development of Epithelial Stem/Progenitor Cells

When renal stem/progenitor cells are taken into culture it is important to know about their current embryonic, maturing, and matured states of development. In running experiments the degree of differentiation can be compared between the in vivo and in vitro situation. To register the development potent histochemical markers are required, which make it possible to distinguish between an embryonic, a mature, and an adult state of differentiation. This difference can be visualized, for example, by soybean agglutinin (SBA) labeling. The lectin recognizes terminal *N*-acetylgalactosamine (GalNAcα1) residues on glycoproteins. In comparison to an antibody label the staining with a lectin saves 2 h of incubation time.

Cryosections of the neonatal rabbit kidney reveal that cells within the tip of the collecting duct ampulla (A) lack SBA label. This site of the collecting duct ampulla (A) is known to be the niche for epithelial stem/progenitor cells. It is found below the capsula fibrosa (CF) in the outer cortex of the neonatal rabbit kidney (Fig. 4). In contrast, the maturing cells on the ampulla neck gain SBA label, while the ampulla shaft and further matured collecting duct (CD) tubules reveal an intensive SBA label. Thus, comparison between the ampulla tip, the neck, the shaft, and the functional tubule illustrates a unique developmental gradient showing renal stem/progenitor cells and maturing, respectively, matured epithelial cells within the collecting duct tubule.

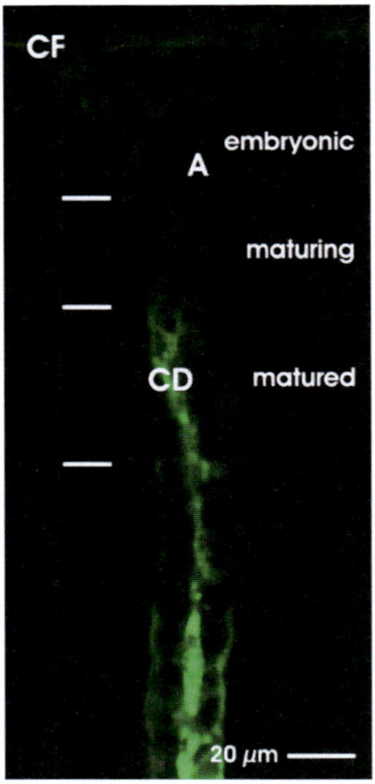

Fig. 4 SBA label demonstrates the developmental gradient between the ampulla (*A*), the neck, and the shaft of the collecting duct (*CD*) tubule. The epithelial stem/progenitor cells within the ampulla are not recognized by the lectin, while the neck shows first signs of reaction. Intensive reaction is found in the matured collecting duct

5 Isolating Renal Stem/Progenitor Cell Containing Tissue

To investigate regeneration of tubules under in vitro conditions a suitable source of renal stem/progenitor cells is needed. Due to the limited size of embryonic mouse or rat specimens, neonatal rabbit kidney is selected as a cell-biological model, since even after birth the embryonic cortex of the organ contains numerous stem cell niches in their original extracellular environment [9]. Further on, the embryonic tissue layer is easily accessible for isolation and can be harvested in sufficient amounts for culture and cell-biological analysis (Fig. 5a).

For the generation of renal tubules embryonic explants from the outer cortex of newborn rabbit kidney can be used. Stripping off the capsula fibrosa (CF) with fine forceps a thin layer of embryonic tissue of constant thickness adheres to the explant (Fig. 5b). Besides the isolated epithelial stem/progenitor cells located in the tip of the collecting duct ampulla, surrounding nephrogenic mesenchymal stem/progenitor cells and S-shaped bodies are found. By this simple micro-surgical method a thin embryonic tissue layer of up to 1 cm^2 can be harvested. Up to date no other species is known, which makes it possible to harvest embryonic tissue containing renal stem/progenitor cells (rS/Pc) in such an amount by this simple preparation technique.

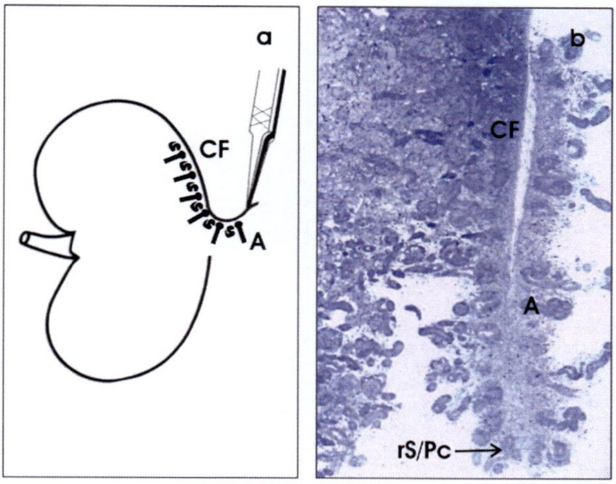

Fig. 5 Isolation of renal stem/progenitor cells from neonatal rabbit kidney. (**a**) Using fine forceps the renal capsula fibrosa (*CF*) is stripped off with an adherent layer of embryonic tissue. (**b**) Semithin section of an isolated explant consisting of renal organ capsule (*CF*), collecting duct ampullae (*A*), renal mesenchymal stem/progenitor cells (*rS/Pc*), and S-shaped bodies

6 Engineering a Micro-environment for Structural Development

When renal stem/progenitor cells are used in future for the repair of parenchyme in acute or chronic renal failure, it is important to investigate their developmental capacity and to learn about formation of tubules. Unsolved issues comprise a suitable implantation technique, the controlled application of growth factors, and the steering of tubule development in a diseased spatial environment of the kidney. However, up to date the growth of stem/progenitor cells is unpredictable. It is, for example, unknown if the complex histoarchitecture of the diseased kidney promotes or even inhibits regeneration of tubules. For that reason sophisticated in vitro experiments have to be performed to obtain exact information about mechanisms involved in the formation of new parenchyme. Such culture experiments have to meet the physiological needs of renal stem/progenitor cells under in vitro conditions. They have to support their development in a spatial environment, to avoid the formation of unstirred layers of medium, and finally to be able to harvest the necessary amount of tissue for cell-biological analysis. Since earlier applied culture methods do not fulfill these needs, an innovative technique for the generation of renal tubules was elaborated [10].

The technical solution is to culture renal stem/progenitor cells between layers of polyester fleece, which replaces the coating by extracellular matrix proteins. For culture the isolated embryonic renal tissue is placed between two punched out layers of polyester fleece (Walraf, Grevenbroich, Germany) measuring 5 mm in diameter

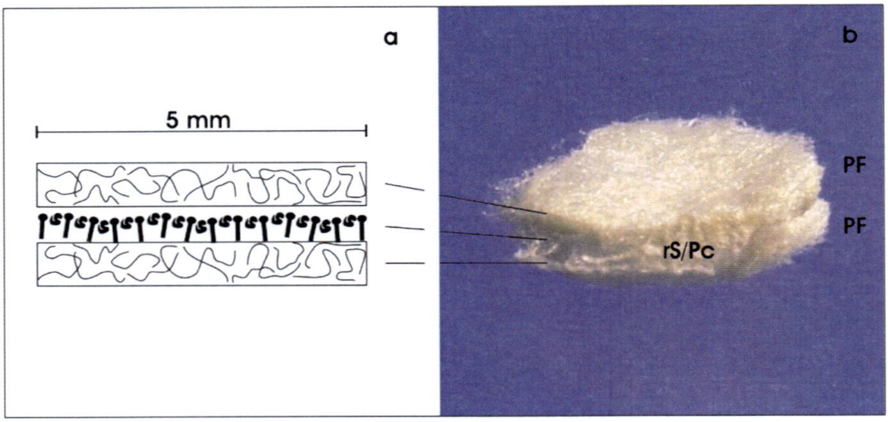

Fig. 6 Creation of an artificial interstitium for the culture of renal stem/progenitor cells. (**a**) Schematic illustration shows that the isolated tissue is placed between two layers of polyester fleece. (**b**) Illustration demonstrates basic sandwich set-up containing isolated embryonic tissue containing renal stem/progenitor cells (*rS/Pc*) between layers of polyester fleece (*PF*)

(Fig. 6a). This arrangement results in a basic sandwich set-up configuration with the freshly isolated embryonic tissue in the middle and layers of polyester fleece covering the outer sides (Fig. 6b).

The interface between the layers of polyester fleece is used as an artificial interstitium, which promotes the spatial development of tubules. Further on, the space between the fleece fibers is an advantage for continuous exchange of culture medium and respiratory gas [11, 12].

7 Offering a Suitable Bioreactor Housing

The basic sandwich set-up containing renal stem/progenitor cells has to be held in its position inside a perfusion culture container to prevent damage to the developing tissue during culture (Fig. 7a). A base ring of a Minusheet® tissue carrier with 13 mm inner diameter is transferred to a perfusion culture container with horizontal flow characteristics (Minucells and Minutissue, Bad Abbach, Germany; www.minucells.de) .

To mount the basic sandwich set-up first a polyester fleece measuring 13 mm in diameter is placed into this tissue carrier (Fig. 7a). Then the basic sandwich set-up containing renal stem/progenitor cells measuring 5 mm in diameter is inserted (Figs. 6, 7a). Finally, a polyester fleece 13 mm in diameter is placed on top of the sandwich as a cover (Fig. 7a, b). After closing the lid of the perfusion container the basic sandwich set-up is fixed in an exact position (Fig. 7c). The spatial area for tubule formation between the polyester fleece layers is 5 mm in diameter and up to 250 μm in height. The specific interface between the fleece layers produces an

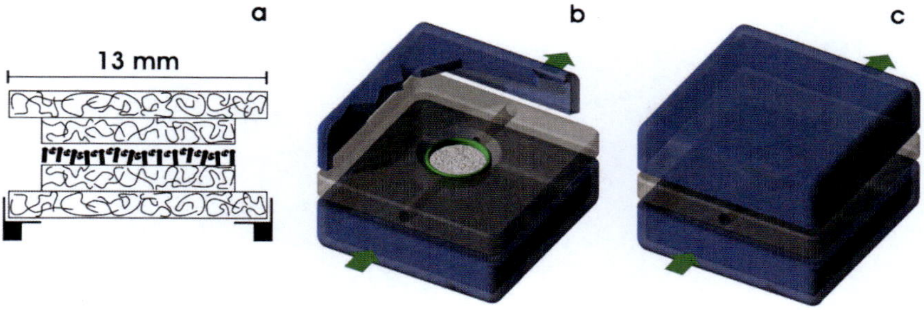

Fig. 7 Microreactor design for the culture of a basic sandwich set-up at the interface of an artificial interstitium. (**a**) The basic sandwich set-up containing renal stem/progenitor cells is mounted in a tissue carrier. (**b**) The perfusion container consists of a base plate to hold the tissue carrier with the developing tissue embedded between layers of polyester fleece. (**c**) The container is sealed for culture with a lid

artificial interstitium providing an optimal microenvironment for the development of tubules during the entire culture period.

The principal design of the perfusion culture container consists of a base plate and a lid made of polycarbonate. In the basic version a tissue carrier with developing tissue is placed onto a base plate (Fig. 7b). Then a lid of the container is secured on top. Lid and/or base plate features a medium inlet and outlet on opposing ends. The container is sealed by clamping the lid onto a silicon gasket on the base plate (Fig. 7c). The advantage of such a container is the creation of an artificial interstitium and a decreased overall height that entails a reduction in dead volume and optimizes medium exchange.

8 Providing Always Fresh Culture Medium

To generate renal tubules chemically defined IMDM (Iscove's modified Dulbecco's medium including Phenolred, GIBCO/Invitrogen, Karlsruhe, Germany) is used. In order to maintain a constant pH of 7.4 under atmospheric air containing 0.3% CO_2 HEPES (50 mmol/l) is added to the medium. To induce tubulogenic development of renal stem/progenitor cells aldosterone (1×10^{-7} M, Fluka, Taufkirchen, Germany) is administered and to prevent infections an antibiotic–antimycotic cocktail (1%, GIBCO) is applied in the culture medium.

Dynamic culture is performed throughout the experimental phase of 13 days. Always fresh medium is perfused at a rate of 1.25 ml/h with an IPC N8 peristaltic pump (Ismatec, Wertheim, Germany). To maintain a constant temperature of 37°C, the culture container is placed on a thermoplate (Medax-Nagel, Kiel, Germany) and covered with a transparent lid (Fig. 8).

To avoid unstirred layers of fluid developing tissue at the interphase of an artificial interstitium is supplied with a continuous flow of always fresh culture medium.

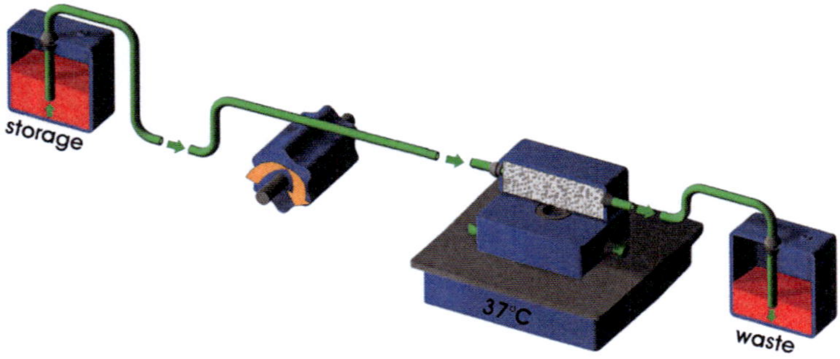

Fig. 8 Dynamic culture containing an artificial interstitium is performed with several modules that are assembled into a working line. A peristaltic pump maintains a flow (1.25 ml/h) of always fresh medium from the storage bottle to the culture container. Used medium that leaves the culture container is not recycled but is collected in a waste bottle

Medium is saturated to 190 mmHg oxygen during transportation. It guarantees an optimal supply for the growing tissue. The high content of oxygen in the medium is reached by a long thin-walled silicone tube. The tubing is highly gas permeable and guarantees optimal diffusion of gases between culture medium and surrounding atmosphere. In this way it is possible to adjust the gas partial pressures within the medium under absolutely sterile conditions because the medium does not get in direct contact with the gases. By maintaining a defined carbon dioxide concentration in the medium this method can be employed to control medium pH via the bicarbonate buffer and irrespective if the system runs in a CO_2 incubator or under atmospheric air. In consequence, the enrichment of oxygen is optimized and the formation of gas bubbles is minimized in this procedure.

9 Visualizing Pattern of Generated Tubules

After a culture period of 13 days at the interface of an artificial interstitium the generated tissue can be analyzed. In order to visualize tubules the artificial interstitium is opened by tearing off the layers of the polyester fleece. Then the specimens are fixed in 70% ethanol and labeled by fluorescent soybean agglutinin (SBA) to analyze the growth pattern (Fig. 9). While the isolated tissue does not exhibit any cellular SBA label, acquisition of SBA binding indicates maturation as it can be recognized during development of the collecting duct within the neonatal kidney (Fig. 4).

Fluorescence microscopy of whole-mount specimens demonstrates that numerous tubules are growing in a spatial arrangement (Fig. 9a, b). When the tubules are developing closely attached to each other the fluorescence is so bright that the details cannot be clearly recognized (Fig. 9a). For that reason areas on the fleece are selected, where the tubules are growing more or less separately from each other (Fig. 9b). Cases are registered, where tubules demonstrate a singular straightforward occurrence, while others reveal a dichotomous branching or curling. When

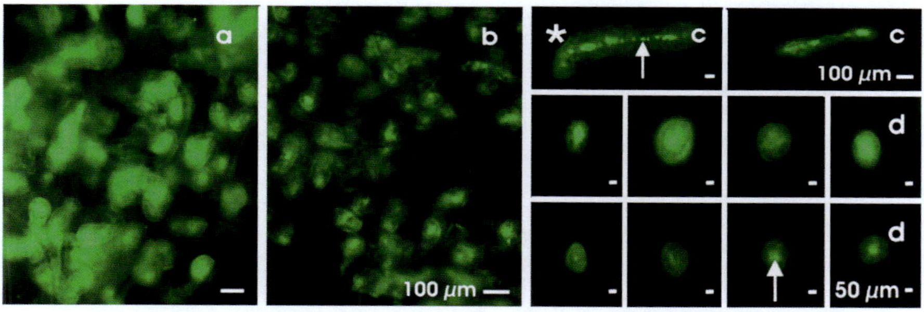

Fig. 9 Fluorescence microscopy of whole-mount specimens labeled by SBA. (**a**, **b**) Opening the artificial interstitium after 13 days of culture reveals a field of densely packed tubules. (**c**) Longitudinal and (**d**) vertical view of generated tubules demonstrates a lumen (*arrow*) and a basal lamina (*asterisk*)

the tubules are not leaving the optical plain, it is possible to follow the longitudinal growth over a distance between 300 and 400 μm. Longitudinal (Fig. 9c) and vertical (Fig. 9d) view depicts that the tubules contain in their center a clearly recognizable lumen, while at the outer surface a smooth basal lamina is present.

10 Scanning the Interface of an Artificial Interstitium

To analyze the covering surface of generated tubules scanning electron microscopy (SEM) was performed (Fig. 10). Special interest is focused on the basal aspect of generated tubules and especially on their contact with surrounding fleece fibers. In contrast to experiments earlier performed by other groups, tubules were generated without coating them by extracellular matrix proteins. This makes it for the first time possible to analyze the basal aspect of generated tubules without the interference of extracellular matrix proteins derived from a coating process [13].

SEM of the fleece without cells demonstrates numerous polyester fibers in a three-dimensional extension (Fig. 10a). The fibers are running in a longitudinal, transversal, and oblique course. They appear to be of homogeneous composition and show a smooth surface without recognizable protrusions or roughness. The average diameter of a polyester fiber is 10 μm. Chemical cross-linking between the polyester fibers cannot be observed.

The area of the polyester fleece used for tissue development exhibits numerous tubules (Fig. 10b). Part of the tubules develop in a parallel fashion in the vicinity of the fleece fibers, some show a curling growth, while others exhibit a dichotomous branching. All of the tubules are covered by a continuously developed basal lamina. The overall view further demonstrates that the tubules have only loose contact with the fibers of the polyester fleece. On the surface of the tubules single interstitial cells and bundles consisting of newly synthesized extracellular matrix proteins are recognized.

Fig. 10 SEM at the interphase of an artificial interstitium. (**a**) The fibers of the polyester fleece (*PF*) are detected in a longitudinal, transversal, and oblique course. They exhibit a homogeneous composition, a smooth surface without recognizable protrusions or roughness. (**b**) Generated tubules (*T*) grow in close vicinity of the polyester fibers. On the surface of tubules single interstitial cells and thin fibers consisting of extracellular matrix are observed

11 Linking Collagen Between Tubules and Polyester Fibers

In the kidney the tubules are not in contact with others but are separated by the interstitium (Fig. 1). It consists of a network of collagen type III fibers surrounding the tubules. To gain insights in the microenvironment of generated tubules SEM and immunohistochemistry for collagen type III were performed.

SEM shows that bundles of synthesized fibers are spanning between the basal lamina of generated tubules and toward the neighboring polyester fibers (Fig. 11a).

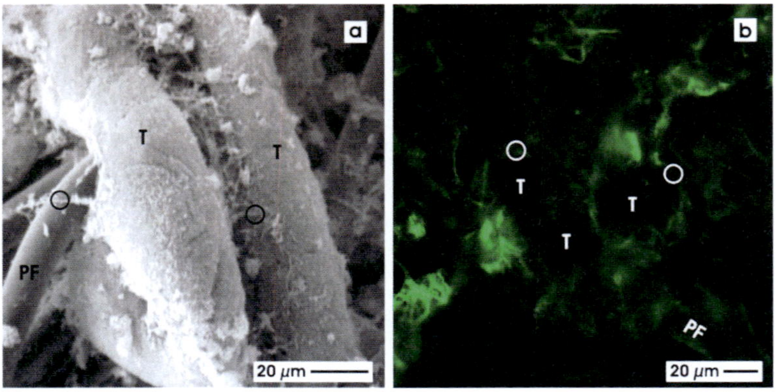

Fig. 11 SEM analysis and immuno-label for collagen type III. (**a**) Synthesized collagen (*O*) is spanning between the basal lamina of both tubules and toward neighboring polyester fibers (*PF*). (**b**) Collagen type III (*O*) is found on the basal lamina of generated tubules (*T*) and on fibers lining toward the polyester fibers (*PF*) of the artificial interstitium

Since the generation of tubules was performed without any coating by extracellular matrix proteins, the basal aspect of generated tubules can also be analyzed by immunohistochemical methods without the interference of proteins derived from a coating process. The label on cryosections demonstrates that collagen type III is contained in the basal lamina of generated tubules and in the surrounding interstitial space (Fig. 11b). This result points out that special attention has to be given to the development of extracellular matrix during generation of tubules, since the cellular differentiation of epithelial cells is strongly correlated with the synthesis of an intact basal lamina and interstitial proteins such as collagen type III.

12 Looking to the Ultrastructure of Generated Tubules

Whole-mount label by soybean agglutinin (SBA, Fig. 9) reveals that generated tubules exhibit a clearly visible lumen in the center and a basal lamina at the outer surface. To obtain further morphological insights transmission electron microscopy (TEM) of the generated tubule epithelium was performed (Fig. 12). Low magnification reveals that tubules generated at the interface of an artificial interstitium show a lining epithelium, a lumen, and a consistently developed basal lamina (Fig. 12a). The tubules occur in close neighborhood to polyester fibers, but it appears that they avoid close contact with them. In the surrounding of a tubule synthesized extra-cellular matrix, single cells and cellular debris are noticed.

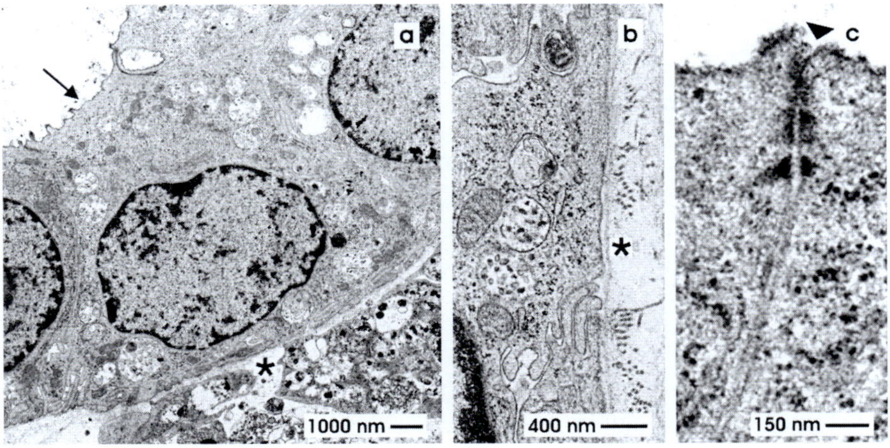

Fig. 12 Transmission electron microscopy of tubules generated at the interface of an artificial interstitium. (**a**) Low magnification reveals that generated tubules contain a polarized epithelium. The apical side borders a lumen (*arrow*), while the basal side contains a basal lamina (*asterisk*). (**b**) The basal lamina consists of a lamina rara interna, a lamina densa, and a lamina fibroreticularis (*asterisk*). (**c**) Between the apical and lateral plasma membrane a junctional complex is developed (*arrow head*). It consists of a zonula occludens, zonula adherens, and a desmosome indicating that a sealing epithelium is established

TEM further demonstrates that an isoprismatic epithelium is established (Fig. 12a). The cells exhibit a large nucleus, which is located in the center of the cytoplasm. In the apical and basal cytoplasm numerous lysosomal elements are found. Small, medium-sized, and large vacuoles are filled to a various degree with electron-dense material. The vacuoles suggest that the containing material has been phagocytosed. At the basal aspect of the epithelium a completely developed basal lamina consists of a lamina rara interna, a lamina densa, and an extended lamina fibroreticularis (Fig. 12b). Neighboring epithelial cells are in close contact with each other. Some microvilli or microplicae of the lateral plasma membrane project into the intercellular space. The basal slits of plasma membranes between neighboring cells are narrow. Higher magnification of TEM illuminates that the luminal and lateral plasma membranes are separated by a typical junctional complex (Fig. 12c). It consists of a zonula occludens, zonula adherens, and a desmosome indicating that a polarized epithelium is established. Thus, all these ultrastructural data show that during culture an obviously sealing epithelium within the generated tubules is developed.

13 Featuring Cell Differentiation

In the presented experiments renal stem/progenitor cells containing tissue was kept in culture for a period of 13 days. Most important is to determine the degree of acquired differentiation in generated tubules. One has to compare the initial state of development during isolation with the harvested tissue at the end of culture. This result further can be compared with the embryonic state, the process of maturation, and finally the adult state in the functional kidney.

To analyze the degree of cell differentiation cryosections were made. For orientation, the surface view of a toluidine blue-stained section demonstrates the distribution of developed tubules at the interface of the artificial interstitium covered by polyester fleeces on both sides (Fig. 13a). Histochemical label for SBA (Fig. 13b) and cyclooxygenase 2 (Cox2, Fig. 13c) reveals an intensive cellular label. Immuno-label for Troma I (Fig. 13d), cytokeratin 19 (Fig. 13e), and E-cadherin (Fig. 13f) exhibits intensive label on tubule cells. Label for occludin demonstrates the development of a junctional belt recognized as faint label in the luminal portion of generated tubules (Fig. 13g). It indicates a functional polarization in the form of a tight junction at the border between the luminal and the lateral plasma membrane of cells within the generated tubules. Immuno-label for Na/K-ATPase α5 reveals an intensive fluorescence at the basolateral aspect of generated tubules (Fig. 13h). All these data show that the generated tubules exhibit a polarized epithelium. Labeling the tissue for laminin γ1 finally exhibits an intensive reaction at the basal aspect of generated tubules indicating that a basal lamina is developed during 13 days of culture (Fig. 13i).

SBA, cyclooxygenase 2, Troma I, cytokeratin 19, E-cadherin, occludin, Na/K-ATPase α5, and laminin γ1 are naturally found on cells of the adult renal collecting

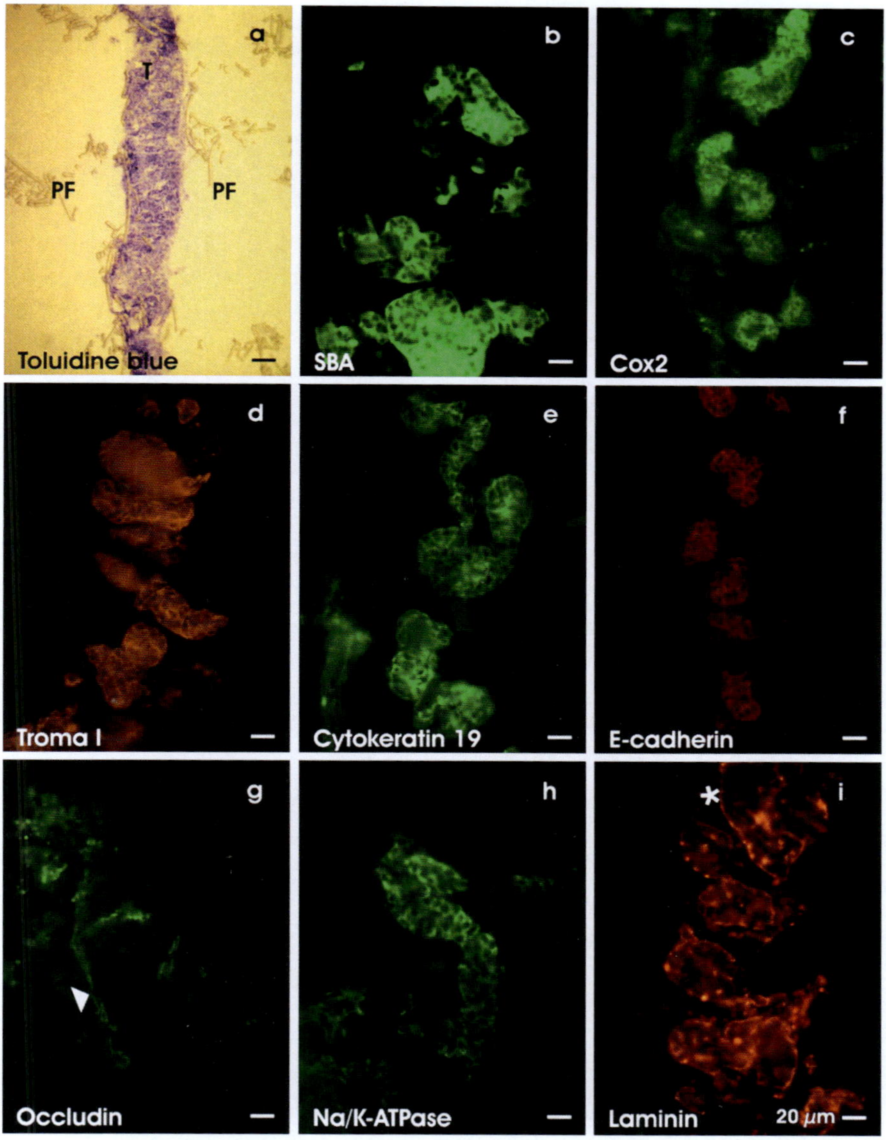

Fig. 13 Histochemical label of tubules generated for 13 days at the interface of an artificial interstitium. (**a**) Toluidine blue stain of a cryosection demonstrates numerous tubules (*T*) grown between fleeces of polyester. Histochemical label for (**b**) SBA and (**c**) cyclooxygenase 2 (Cox2) reveals an intensive cellular label. Label for (**d**) Troma I, (**e**) cytokeratin 19, and (**f**) E-cadherin shows intensive reaction on tubule cells. (**g**) Occludin (*arrow head*) is found in the luminal portion of generated tubules, while (**h**) Na/K-ATPase $\alpha5$ is detected at the basolateral aspect. (**i**) Laminin $\gamma1$ exhibits that a basal lamina (*asterisk*) is formed

duct within the kidney. Since the same markers are detected on tubules generated at the interface of an artificial interstitium (Fig. 13), it is most probable that renal collecting duct-derived tubules are developed.

14 Inducing Tubulogenic Development

Experiments show that SBA-labeled tubules are not developed when the culture of renal stem/progenitor cells is performed without application of aldosterone (Fig. 14a). Cryostat sections show in this case a disintegration of tissue. Only faint rows of cells are developed and in none of the samples structured tubules are observed.

In contrast, administration of aldosterone (1×10^{-7} M) to the standard medium (IMDM) completely changes the developmental pattern of renal stem/progenitor cells generated at the interface of an artificial interstitium (Fig. 14b) [14]. After a culture period of 13 days numerous tubules become visible.

To elaborate a tubulogenic profile of aldosterone the hormone was applied in concentrations ranging from 1×10^{-10} to 1×10^{-5} M. The low dose of 1×10^{-10} M does not stimulate the development of tubules, while concentrations of 1×10^{-9} and 1×10^{-8} M start to induce outgrowth of SBA-labeled cells that form long rows and clusters but not structured tubules. Intact formation of tubules is obtained by the use of 1×10^{-7} and 1×10^{-6} M aldosterone. In contrast, the application of 1×10^{-5} M does not further stimulate a better development of tubules.

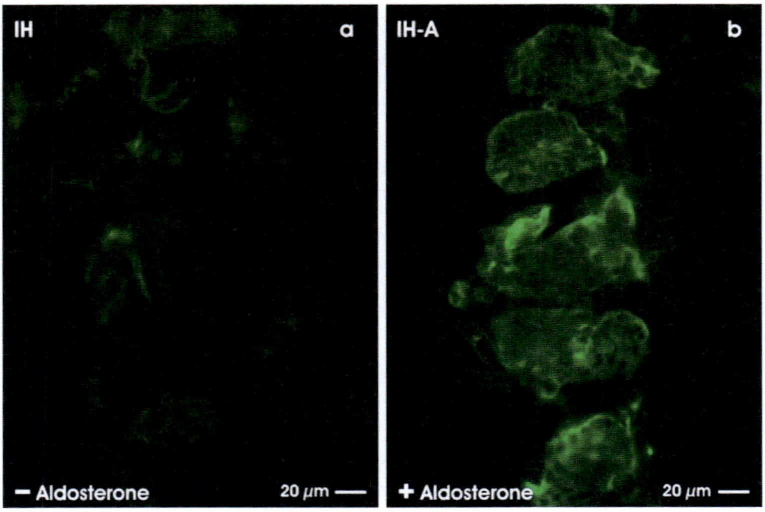

Fig. 14 Culture of embryonic renal tissue at the interphase of an artificial interstitium. (**a**) Generation of tubules cannot be recognized when aldosterone is omitted in IMDM. (**b**) Numerous tubules are observed after application of aldosterone (1×10^{-7} M) in IMDM after 13 days of culture

It was a fully unexpected finding that aldosterone exhibits tubulogenic activity in isolated explants containing renal stem/progenitor cells (Fig. 14b). In the adult kidney the hormone stimulates the Na^+ transport by genomic effects via the mineralocorticoid receptor (MR). A morphogenic effect of aldosterone on renal stem cells sounds on the one hand curiously, but on the other hand, a comparable morphogenic action of the steroid hormone was found on other cells. For example, development of hippocampal neurons is induced by aldosterone via stimulation of the mineralocorticoid receptor (MR), while an activation of the glucocorticoid receptor (GR) is suppressing the development. T37i cells show after application of aldosterone differentiation into cells of brown adipose tissue. This effect can be inhibited by antagonists such as spironolactone or RU-26752. Experiments with cells of the adenohypophysis producing growth hormone demonstrate that their development is triggered by MR and is inhibited by the application of spironolactone. During embryogenesis the blockade of MR with ZK 91587 leads to a reduction of somite number, to reduced growth, and to an alteration of blood vessels in the umbilical cord. In adult cardiac fibroblasts aldosterone increases levels of collagen type I and III synthesis promoting myocardial fibrosis. Finally, for the kidney it was described that aldosterone promotes the synthesis of fibronectin through a Smad2-dependent TGF-β1 pathway in mesangial cells during the development of glomerular sclerosis.

15 Scoring the Development of Tubules

To investigate the influence of hormones or growth factors on the generation of tubules it is most important to determine the degree of differentiation. Depending on the applied substances many or few tubules, cell islets, or even extended cell clusters may arise. For that reason a special score for the development of tubules was elaborated.

The space for tubule development in a basic sandwich set-up is 5 mm in diameter and up to 250 μm in height (Fig. 6). For example, labeling of specimens by SBA reveals numerous tubules exhibiting polarized cells, a visible lumen, and a basal lamina (Fig. 15). The number of generated tubules can be determined with a WCIF ImageJ program (Bethesda, Maryland, USA) by counting each SBA-labeled tubule. Applying this technique on generated specimens, between 41 and 77 tubules are detected within a microscopic opening of 620×930 μm.

Besides the overall appearance of tubules, the staining intensity, the formation of a lumen, the development of a basal lamina, their distribution, and their length are registered (Fig. 15). When each of these criteria is scored with 1 point it results in a maximum of 6 points reflecting an ideal development of generated tubules. However, this result will only be obtained when aldosterone is added as tubulogenic factor to the culture medium. When the cultures are properly developed, no cell clusters, no filopodia, and no overgrowth of epithelial cells on the polyester fibers of the artificial interstitium are observed.

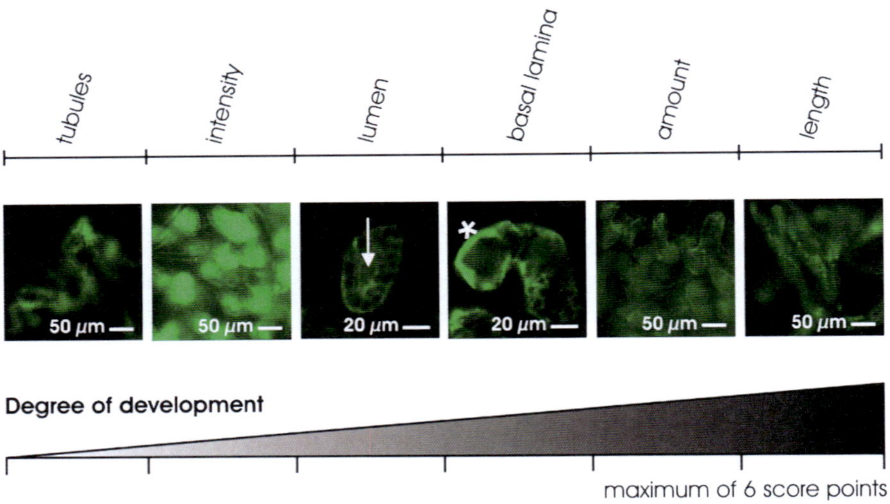

Fig. 15 Scoring the development of tubules. Beside the appearance of tubules, the staining intensity, the formation of a lumen (*arrow*), the development of a basal lamina (*asterisk*), the overall distribution, and the length can be registered

16 Specifying the Morphogenic Action of Aldosterone

From adult kidney it is known that not only aldosterone but also some of its molecular precursors have an affinity to the mineralocorticoid receptor (MR) and influence thereby physiological functions. Consequently it was investigated if also precursors of the aldosterone synthesis pathway show effects on the development of renal stem/progenitor cells [14].

The synthesis of aldosterone starts from cholesterol, which is metabolized over pregnenolone to progesterone, 11-deoxycorticosterone, corticosterone, and 18-hydroxycorticosterone. Physiological experiments with adult kidneys show, for example, that 11-deoxycorticosterone is as effective as aldosterone on the mineralocorticoid receptor, while corticosterone is 100 times less potent.

Culture experiments with precursors of the aldosterone synthesis pathway (Fig. 16, each 1×10^{-7} M) were performed and graded according to the scoring profile (Fig. 15). It is demonstrated that application of cholesterol or pregnenolone does not result in the formation of any SBA-positive tubules (0 point). Treatment with progesterone leads to the development of few tubules but without intensive SBA label (1 point). When 11-deoxycorticosterone is used, only few tubules with a faint SBA label can be detected (1 point). In contrast, administration of corticosterone does not reveal any development of tubules. Instead, numerous SBA-labeled cell clusters are observed in close contact to polyester fibers (0 point). Data for 18-hydroxycorticosterone are missing, since this substance is not commercially available. Administration of aldosterone results in numerous SBA-positive tubules

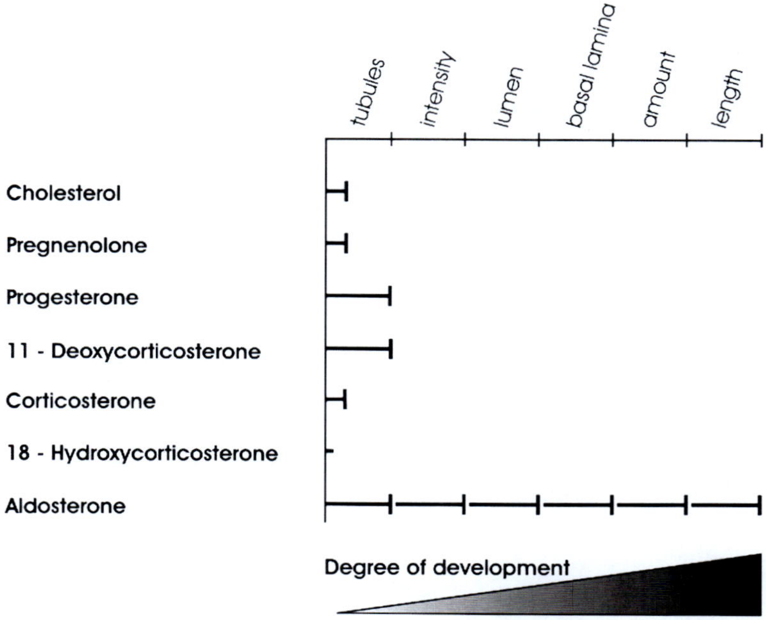

Fig. 16 Tubulogenic effect of aldosterone and its molecular precursors on renal tubule development. Only aldosterone reaches a maximum of 6 points in the score scale. It depicts that only aldosterone but not its precursors lead to the development of SBA-labeled tubules

exhibiting a distinct lumen and a clearly recognizable basal lamina (maximum = 6 points).

The tubulogenic action of aldosterone may be triggered in cooperation via the glucocorticoid receptor (GR). Consequently, the glucocorticoid dexamethasone instead of aldosterone was tested. While the administration of aldosterone (1×10^{-7} M) results in the development of numerous SBA-labeled tubules, the use of dexamethasone (1×10^{-7} M) produces huge clusters of non-polarized cells. Thus, the progress of structured tubule development appears specific for aldosterone by stimulating the mineralocorticoid receptor (MR).

17 Antagonizing the Tubulogenic Signal on MR

It was demonstrated that the generation of renal tubules depends specifically on the presence of aldosterone in the culture medium (Fig. 14). Further it was shown that the tubulogenic effect is dependent on the applied hormone concentration and cannot be mimicked by precursors of the aldosterone synthesis pathway (Fig. 16). In the following experiments it was further investigated if the tubulogenic effect of aldosterone is exclusively mediated via the mineralocorticoid receptor (MR) or if it is related to an unspecific side effect of the steroid hormone (Fig. 17) [15]. To interfere the binding between aldosterone and MR, antagonists such as spironolactone and canrenoate were used in the culture experiments.

Low dose of spironolactone (1×10^{-7} M) in the presence of aldosterone does not affect the development of SBA-labeled tubules. However, application of a higher concentration of spironolactone (1×10^{-5} M) demonstrates first inhibitory effects and leads to a switch of tubule development. The number of structured tubules is reduced and SBA-labeled cells start to form extended cell clusters. However, presence of 1×10^{-4} M spironolactone in aldosterone-containing medium completely prevents the development of SBA-labeled tubules (Fig. 17b).

Canrenoate has the same inhibitory profile on the tubulogenic action of aldosterone as it is observed with spironolactone. Application of 1×10^{-7} M canrenoate in the aldosterone-containing medium does not affect the development of tubules. However, administration of 1×10^{-6} and 1×10^{-5} M canrenoate drastically reduces SBA-labeled structures. The use of 1×10^{-4} M canrenoate results in a complete lack of SBA-labeled cells and tubules (Fig. 17c).

Thus, the simultaneous administration of aldosterone in combination with spironolactone (Fig. 17b) or canrenoate (Fig. 17c) demonstrates that the tubulogenic effect is inhibited in a dose-dependent manner. The result further shows that the tubulogenic effect of aldosterone is mediated via the mineralocorticoid receptor (MR).

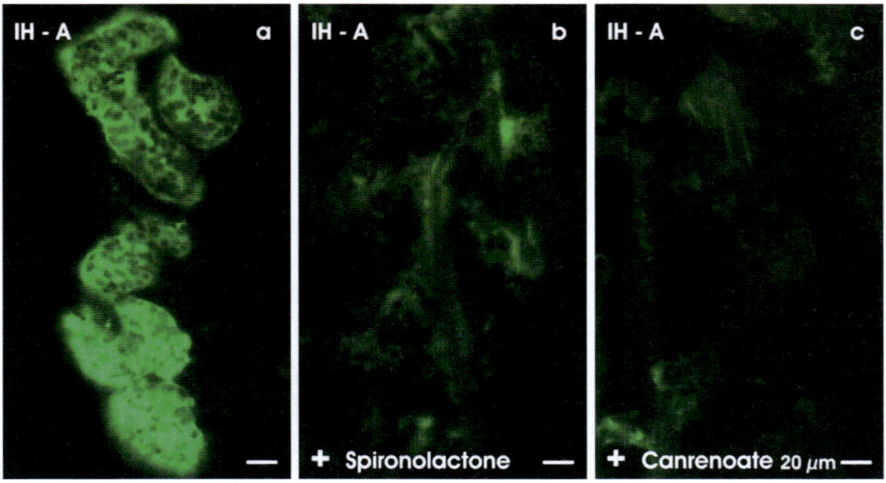

Fig. 17 Antagonizing the tubulogenic action of aldosterone on the receptor level. (**a**) For control, aldosterone generates numerous SBA-labeled tubules. Application of (**b**) 1×10^{-4} M spironolactone or (**c**) 1×10^{-4} M canrenoate in the presence of aldosterone completely inhibits the development of tubules

18 Interfering the Tubulogenic Signaling in the Cytoplasm

Earlier experiments demonstrated that MR is not randomly distributed within the cytoplasm of the target cell but stays in close molecular contact with heat shock proteins (hsp) 90 and 70 (Fig. 18a). Both proteins, in turn, are located in close proximity to immunophilins, especially FKBP 12. To obtain insights in the molecular signaling renal stem/progenitor cells were treated with aldosterone in combination

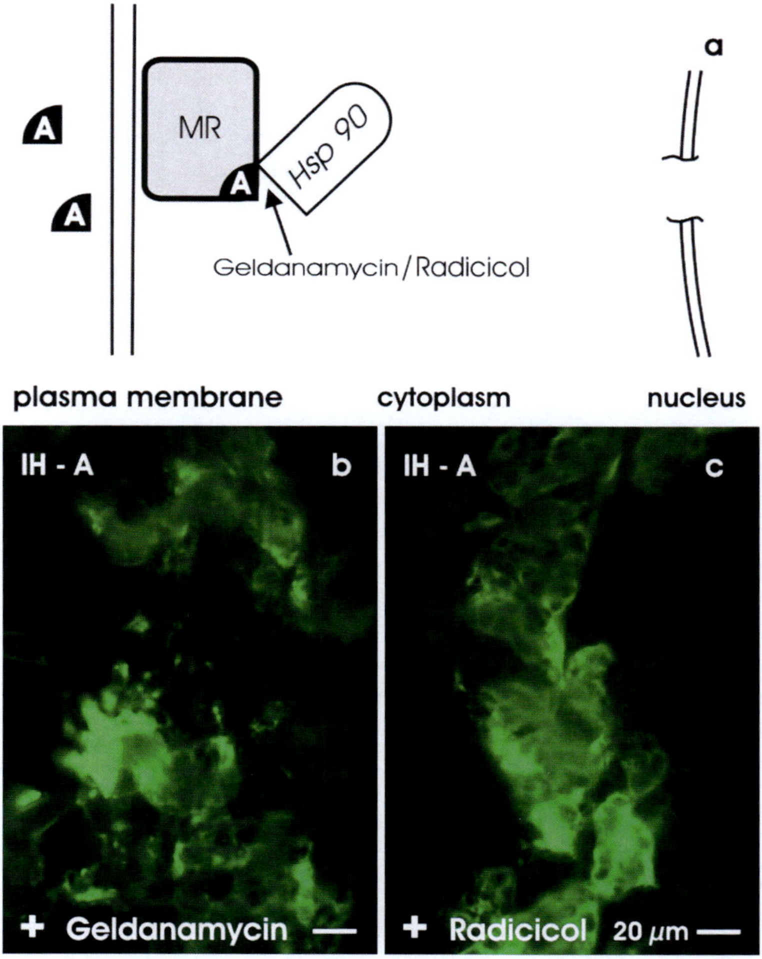

Fig. 18 Cytoplasmic interference of the tubulogenic signal (**a**). Schematic illustration of the aldosterone (*A*)-stimulated mineralocorticoid receptor (*MR*) in relation to *hsp 90*. Fluorescence microscopy on renal tissue generated for 13 days with aldosterone (1×10^{-7} M) in combination with (**b**) 3.6×10^{-6} M geldanamycin or (**c**) 1×10^{-6} M radicicol shows multiple SBA-labeled cells within clusters but solid formation of tubules is lacking, when the binding between *MR* and *hsp 90* is interfered

with substances that disrupt the cytoplasmic interaction between MR, hsp 90, hsp 70, and immunophilins [16].

To disrupt the contact between MR and hsp 90, renal stem/progenitor cells were cultured in IMDM containing geldanamycin (3.6×10^{-6} M) in combination with aldosterone (1×10^{-7} M) for 13 days (Fig. 18b). Geldanamycin specifically binds to hsp 90, thereby blocking the ATP-binding site due to its higher affinity compared to ATP. In this way, it disturbs the contact between hsp 90 and activated MR. The culture experiments show that structured tubules are not found in this series of experiments. Instead numerous SBA-labeled cells are localized in extended clusters.

Radicicol is a macrocyclic antifungal substance that binds in the same way as geldanamycin. It hinders ATP-dependent conformational changes that are required for cytoplasmic interactions with target proteins such as MR. Culture of renal stem/progenitor cells with radicicol (1×10^{-6} M) in combination with aldosterone (1×10^{-7} M) produces only few structured tubules, but numerous SBA-labeled cells in the form of extended clusters (Fig. 18c). In consequence, experiments with geldanamycin (Fig. 18b) and radicicol (Fig. 18c) show that the tubulogenic pathway of aldosterone is blocked by interfering the contact to hsp 90.

19 Producing Renal Superstructures

The presented experiments show the feasibility to generate tubules at the interface of an artificial interstitium (Fig. 9). Further it is demonstrated that aldosterone exhibits a tubulogenic effect on renal stem/progenitor cells (Figs. 14, 15, 16).

The presented new technique opens an experimental way to increase systematically the amount of generated tubules, since the arrangement of the basic sandwich set-up is most advantageous for extending the spatial environment, the supply of medium, and respiratory gas. Since the developing tissue is covered by layers of polyester fleece always fresh medium is transported through the space between the fibers of the polyester fleece separating the tissue layers (Fig. 19). Using this experimental design unstirred layers of fluid are minimized so that for the first time piling and paving of basic sandwich set-ups becomes possible [17].

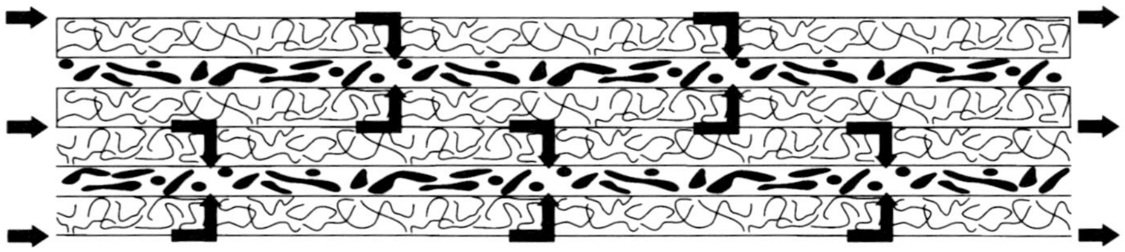

Fig. 19 Piling and paving of basic sandwich set-ups. The supply with culture medium between the tissue layers is optimal, since nutrition and respiratory gases are continuously exchanged through the space between the polyester fibers. *Arrows* indicate the possible flow of medium during dynamic culture

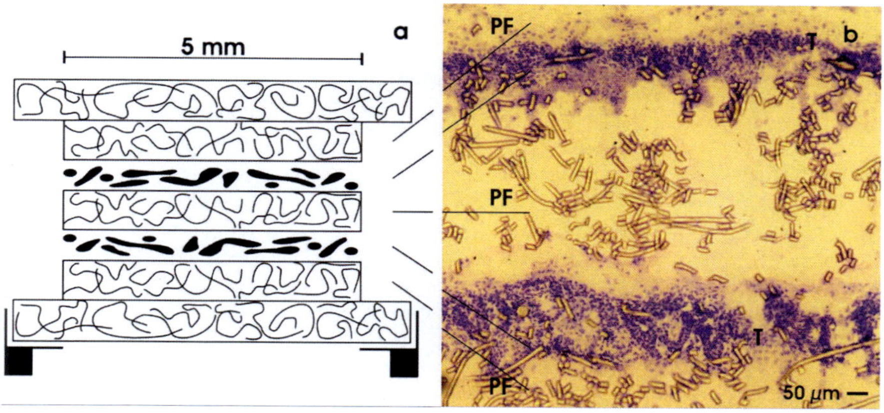

Fig. 20 Schematic illustration of (**a**) piling basic sandwich set-ups. (**b**) Toluidine blue stain of a cryosection shows two piled rows of generated tubules (*T*) cultured for 13 days at the interphase of an artificial interstitium made of polyester fibers (*PF*)

Thus, to raise renal superstructures basic sandwich set-ups containing renal stem/progenitor cells are piled like bricks (Fig. 20). Running experiments demonstrate that tubules in such piled specimens show the same degree of differentiation as it is found in experiments with single rows (Fig. 13).

The advantages of piling and paving basic sandwich set-ups containing renal stem/progenitor cells are numerous (Fig. 20). One could imagine that these superstructures can be used to generate organoid configurations. The use of an artificial interstitium in combination with piling of basic sandwich set-ups may further be advantageous for micro-vascularization. Endothelial cells could be added to the culture medium and infused along the fluid transportation path within the polyester fleeces. The presence of widely distributed endothelial cells will consequently support rapid vascularization of piled and paved basic sandwich set-ups. It is further imaginable that the procedure for piling and paving is not performed by forceps in hand but can be processed in a computerized robotic process. Such superstructures could be raised to investigate under controlled in vitro conditions better the development of organoid structures as it was done in the past. This technique may be also used to generate complex tissue constructs for the minimal invasive implantation of stem/progenitor cells into a diseased kidney.

20 Summing Up

Using renal stem/progenitor cells it was demonstrated that tubules can be generated under controlled perfusion culture at the interphase of an artificial interstitium. The development of tubules formation in a spatial environment is induced by aldosterone. The special experimental configuration makes it possible to pile and pave basic sandwich set-ups containing renal stem/progenitor cells. These new techniques appear to be a successful step for further investigation of the spatial development of tubules. It appears also as a realistic experimental strategy for a future regeneration of parenchyme within the diseased kidney.

References

1. Mansilla E, Drago H, Sturla F, Bossi S, Salas E, Marin GH, Ibar R, Soratti C. Matrix superhighways configurations: new concepts for complex organ regeneration. Transplant Proc. 2007; 39(7):2431–2433.
2. Humphreys BD, Bonventre JV. The contribution of adult stem cells to renal repair. Nephrol Ther. 2007; 3:3–10.
3. Hammerman MR. Cellular therapies for kidney failure. Expert Opin Biol Ther. 2006; 6:87–97.
4. Maeshima A, Sakurai H, Nigam SK. Adult kidney tubular cell population showing phenotypic plasticity, tubulogenic capacity, and integration capability into developing kidney. J Am Soc Nephrol. 2006; 17(1):188–198.
5. Hogan BL, Koledziej M. Molecular mechanisms of tubulogenesis. Nat Rev. 2002; 3:513–523.
6. Lubarsky B, Krasnow MA. Tube morphogenesis: making and shaping biological tubes. Cell 2003; 112:19–28.
7. Karihaloo A, Nickel C, Cantley LG. Signals which build a tubule. Nephron Exp Nephrol. 2005; 100:40–45.
8. Walid S, Eisen R, Ratcliffe DR, Dai K, Hussain M, Ojakian GK, The PI. 3-kinase and mTOR signaling pathways are important modulators of epithelial tubule formation. J Cell Physiol. 2008; 216:469–479.
9. Minuth WW. Neonatal rabbit kidney cortex in culture as tool for the study of collecting duct formation and nephron differentiation. Differentiation 1987; 36:12–22.
10. Minuth WW. Patent DE 199 52 847. Vorrichtung zum Kultivieren und/oder Differenzieren und/oder Halten von Zellen und/oder Geweben; 2006.
11. Heber S, Denk L, Minuth WW. Modulating the development of renal tubules growing in serum-free culture medium at an artificial interstitium. Tissue Eng. 2007; 13:281–291.
12. Hu K, Denk L, de Vries U, Minuth WW. Chemically defined medium environment for the development of renal stem cells into tubules. Biotechnol J. 2007; 2:992–995.
13. Blattmann A, Denk L, Strehl R, Castrop H, Minuth WW. The formation of pores in the basal lamina of regenerated tubules. Biomaterials 2008; 29(18):2749–2756.
14. Minuth WW, Denk L, Hu K. The role of polyester interstitium and aldosterone during structural development of renal tubules in serum-free medium. Biomaterials 2007; 28:8–28.
15. Minuth WW, Denk L, Hu K, Castrop H, Gomez-Sanchez C. Tubulogenic effect of aldosterone is attributed to intact binding and intracellular response of the mineralocorticoid receptor. CEJB 2007; 3:307–325.
16. Minuth WW, Blattmann A, Denk L, Castrop H. Mineralocorticoid receptor, heat shock proteins and immunophilins participate in the transmission of the tubulogenic signal of aldosterone. J Epithel Biol Pharmacol. 2008; 1:24–34.
17. Minuth WW, Denk L, Castrop H. Generation of tubular superstructures by piling of renal stem/progenitor cells. Tissue Eng C Methods 2008; 14(1):3–13.

Stem Cells in Tissue Engineering and Cell Therapies of Urological Defects

Christoph Becker, Katrin Montzka, and Gerhard Jakse

Abstract *Objectives*: This chapter focuses on advances in regenerative therapies using stem cells in urology. Different stem cell types will be introduced due to their classification and hierarchic order.

Methods: A detailed literature search has been performed using the PubMed database of the National Center of Biotechnology Information (NCBI). Publications of experimental investigations and in vivo trials using stem cells in reconstructive urology have been summarized and critically reviewed.

Results: Tissue engineering and autologous cell therapy techniques have been developed in order to generate prostheses for different urologic tissues and organ systems. During the last decade increasing numbers of studies have described stem cells in the context of therapeutic tools. The ability of adult and embryonic stem cells as well as progenitors to improve bladder wall architecture, to improve renal tubule formation or to promote restoration of spermatogenesis or recovery of continence has been investigated in a number of animal models. Another promising stem cell source, the so-called induced pluripotent stem cells, was recently generated, but not further investigated yet. Although results have been encouraging, to date, none of these stem cell-based therapies could reach clinical trials.

Conclusions: Review of the current literature has revealed several populations of adult stem cells and progenitor cells as useful cellular sources in the treatment and reconstruction of urologic organs. However, considerable basic research still needs to be performed to ensure the controlled differentiation and long-term fate of stem cells following transplantation.

Keywords Adult stem cells · Embryonic stem cells · iPS cells · Progenitors · Urinary tract · Kidney · Gonads · Rhabdosphincter · Regenerative medicine

C. Becker (✉)
Department of Urology, University Hospital and Medical Faculty, RWTH Aachen University, Aachen, Germany
e-mail: cbecker@dgu.de

G.M. Artmann et al. (eds.), *Stem Cell Engineering*,
DOI 10.1007/978-3-642-11865-4_15, © Springer-Verlag Berlin Heidelberg 2011

1 Introduction

Tissue and organ replacement surgery often raises complications because the human immune system detects and antagonizes artificial prostheses and donor organs. The suppression of the patient's immune system often causes further unwanted side effects. Tissue replacement using autologous (stem) cell-generated material may help to overcome these problems (Fig. 1).

The genito-urinary system is exposed to a variety of possible injuries from the time of foetal development. Since most urinary organs are mainly composed of smooth muscle and uroepithelial cells, these tissues might be repaired by autologous cell therapy techniques. In fact, in the last decade several attempts to engineer urologic tissue have been published. These include examples of biohybrid prostheses for the urethra [1], bladder [2, 3] and ureter [4]. Furthermore, trials have been undertaken to reconstruct even higher complex tissues, such as renal structures [5], corpora cavernosa [6] or vaginal tissue [7]. However, only some of these approaches have advanced beyond animal experiments to human clinical studies. Recently, Atala and co-workers reported a clinical trial of "de novo" tissue-engineered urinary bladders [8]. Bioartificial organs were created with autologous bladder cells seeded onto collagen–polyglycolic acid scaffolds and transplanted with an omental wrap in patients requiring cystoplasty. The engineered bladders displayed a physiologic tri-layered morphology and clinical parameters such as bladder capacity, compliance, intravesical pressure and renal function were stable at least over a 5-year period and conformed to normal bladder values [8]. Such data have emphasized the outstanding relevance of autologous cell therapy in modern regenerative medicine. However, adult organ-specific cells have several limitations such as difficulty in harvesting, low proliferative capacity and reduced functional quality resulting from in

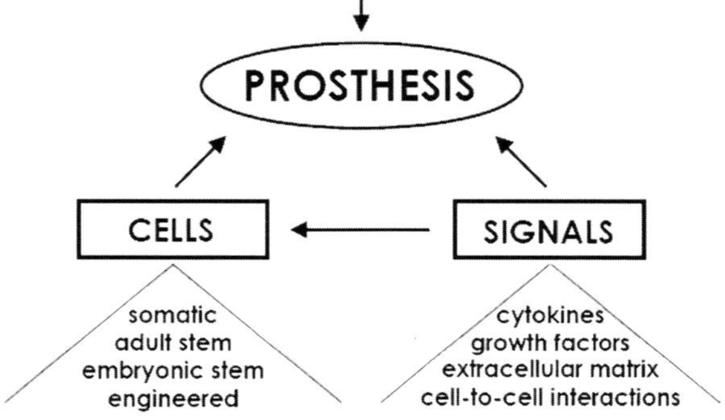

Fig. 1 The tissue-engineering triad. Tissue engineering is a multidisciplinary approach which requires material sciences to provide scaffolds and life sciences to provide living cells. To generate functional biohybrid prostheses from these substantial components, specific biologic signals provide desired phenotypes and behaviour of the cells. Increasing attention has been directed to different stem cell types

vitro cultivation. Furthermore, adult organ-specific cells cannot be used in malignant conditions. Therefore, increasing attention has been paid to pluripotent and multipotent stem cells, capable of self-renewal and tissue-specific differentiation.

2 Stem Cells

Stem cells are undifferentiated cells that are defined by their ability of both self-renewal and differentiation, producing mature progeny consisting of both non-renewing progenitors and terminally differentiated effector cells [9, 10].

2.1 Stem Cell Populations

The problem in classifying stem cells is due to their lack of defined morphologic and molecular characteristics [11]. Therefore, stem cells are classified according to their potency, giving rise to cells, tissues, organs or organisms. Accordingly, the

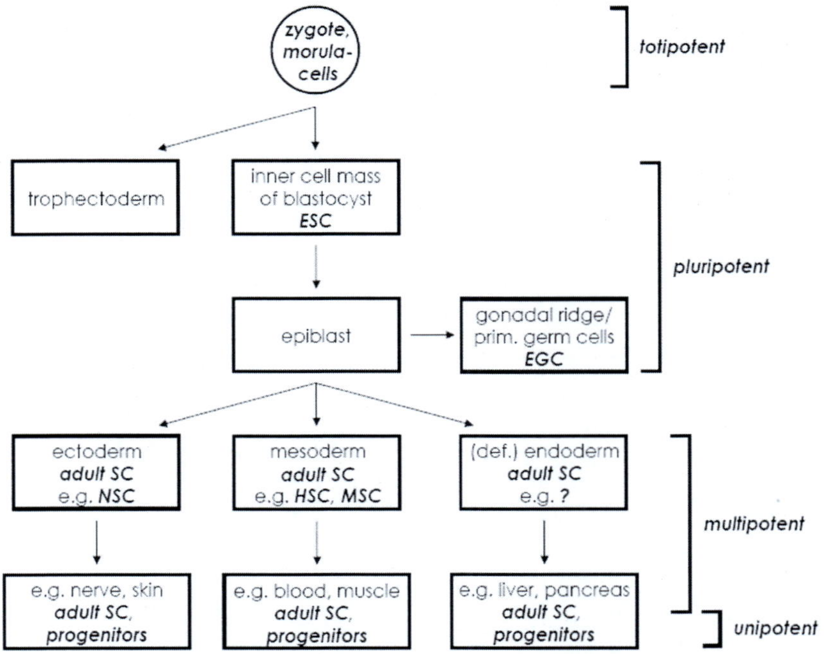

Fig. 2 Simplified scheme of stem cell populations. A hierarchical organization is derived according to the differentiation capacity of the particular stem cell type. The zygote and cells of the morula stage can give rise to both embryonic and extra-embryonic tissues and hence can generate a complete embryo. The three germ layers, as well as embryonic germ cells, originate from embryonic stem cells from the inner cell mass of the blastocyst. Adult stem cells produce progenitor cells and differentiated tissue (figure modified pursuant to Gordon Keller [75])

hierarchic order of stem cells ranges from totipotency to pluri- and multipotency, to unipotency (Fig. 2). The zygote and its offspring cells of the morula are not only capable of forming cells of the ectoderm, mesoderm and (definitive) endoderm layers, as well as the gonadal ridge, but also capable of supporting trophoblast required for the survival of the developing embryo [12]. These cells are therefore at the top of the hierarchy of stem cells and have been termed "totipotent". Embryonic stem cells (ESC) isolated from the inner cell mass of the blastocyst [13, 14] and embryonic germ cells (EGC) derived from primordial germ cells of an early embryo [15] can give rise to cells of the three germ layers, and to those of the gonadal ridge, but not to extra-embryonic tissues. Such cells have therefore been termed "pluripotent". Stem cells isolated from the developing germ layers [16] and/or its descended adult organs [17] are capable of self-renewal and differentiate into multiple organ-specific cell types. These cells have been termed "multipotent". Examples of such multipotent adult stem cells include haematopoietic stem cells (HSC) [18], mesenchymal stromal cells (MSC) [19] and neural stem cells (NSC) [20]. Progenitor cells or precursor cells that have been reported to exhibit limited or no capacity for self-renewal and differentiate into only one defined cell type, e.g. epithelial cells. These cells have been termed "unipotent" [21].

2.2 Embryonic Stem and Germ Cells

The derivation of murine embryonic stem cell lines was described for the first time in 1981 [13] and has since proved to be a very useful tool to study mammalian development (which is characterized by both pluripotency and differentiation). About 20 years later, human embryonic stem cell lines were successfully generated [14]. ESC can potentially be maintained in an undifferentiated state indefinitely in vitro and can theoretically be directed to differentiate into any cell type of the body. This opened visionary possibilities, such as laboratory-grown tissues or even artificial organs being created, in order to treat a variety of diseases. Similar potencies have been described for human embryonic germ cells [15]. Both ESC and EGC have therefore been regarded as very interesting populations of cells for preclinical studies aimed at developing their potential for clinical applications [22]. For example, mouse ESC have been induced to differentiate into midbrain cells (including dopaminergic neurons) by exposure to the signalling factors, sonic hedgehog and FGF-8 [23]. These cells were found to be functional when transplanted into a model of Parkinson's disease. In another approach, rodent ESC have been transplanted into the injured myocardium of infarcted hearts [24]. The phenotypes of transplanted cells ranged from striated cardiomyocytes to vascular smooth muscle cells and endothelial cells. Importantly, post-infarcted remodelling of the hearts was reduced and cardiac functions were generally improved. Although the potential of ESC in regenerative medicine is clear, several methodological problems and issues of control of differentiation following transplantation need to be solved before their benefit can be realized in human regenerative therapies.

2.3 Adult Stem Cells

Less controversial, but equally promising, are stem cells derived from adult human tissues, which probably have much wider differentiation potential than was previously thought [25]. According to the predestination theory, adult stem cells were thought to be developmentally committed and restricted to differentiate only into cell lineages from the tissue in which they reside, e.g. neural stem cells give rise to nerve cells and glia, haematopoietic stem cells produce blood. This dogma has been challenged over recent years by reports that demonstrated that adult stem cells, under certain microenvironmental conditions, may differentiate into cell types other than those of their tissue of origin. For example, HSC have been reported to generate liver cells [26, 27], NSC generate haematopoietic precursors [28] and MSC generate neuronal cell types [29]. However, this capacity of crossing lineage boundaries, which were formerly thought to be uncrossable, is still under discussion [30–32]. Additionally, the biological mechanisms responsible for the possible broad developmental potential of stem cells derived from adult tissues are only poorly understood. Although undermined by cell fusion theories [33], a widely accepted model for this switch of cell fate is that of transdifferentiation, in which certain tissues produce and suspend restrictive signals which in turn induce the stem cells to run certain gene-expression profiles and thus to undergo specific differentiation towards a certain phenotype [34]. Among the diversity of tissues reported, the bone marrow represents a major source for tissue-derived adult stem cells. This complex haematopoietic cellular system contains both haematopoietic stem cells and mesenchymal stromal cells [35]. HSC have become the most widely investigated and best understood population of adult stem cells. The transplantation of purified histocompatible HSC can completely reconstitute the entire haematopoietic cell line in a patient after whole body radiation or chemotherapy [36]. However, increasing attention has been paid to MSC, due to their versatility. MSC have the ability to differentiate, both in vivo and in vitro, into a variety of adult mesenchymal tissue cell types (such as bone, cartilage, adipose or muscle cells), and contradictory reports describe transdifferentiation into non-mesenchymal tissue cell types (such as endodermal or neural cells) [19, 31, 37]. The relatively wide range of differentiation, coupled to the high proliferation rate and relative ease of isolation of MSC (e.g. from the iliac crest), has made these cells suitable candidates for all kinds of tissue engineering and cell therapy applications. Accordingly, various case studies and clinical trials have been published that employed mesenchymal stromal cells. These reports demonstrated MSC in the therapeutic treatment of myocardial infarction [38], large bone defects [39], peripheral ischaemia [40] or in the treatment of bone marrow suppression after high-dose chemotherapy for breast cancer [41].

A new and very promising approach of getting embryonic stem cell-like cells is the generation of induced pluripotent stem cells (the so-called iPS cells). Therefore any somatic cell can be taken and reprogrammed using a defined and limited set of four (OCT4, SOX2, KLF4 and c-MYC) [42, 43] and later even less factors by retroviral transduction [44, 45]. This method of reprogramming has been reproduced from several independent groups. In contrast to embryonic stem cells, there are no ethical concerns regarding the use for research as well as for future clinical application. Consequently, this method is attractive especially for researchers in countries in which the use of embryonic stem cells is restricted. However, the detailed mechanism for this reprogramming and the resulting consequences are still unknown.

This technique has been successfully applied for the generation of human iPS cells, but the efficiency is still low [46, 47]. However, researchers regard iPS cells for the ideal source for cell therapy, since they can be easily generated from any tissue of the patient itself and consequently, any immune response will be avoided. Since this field of research is fairly new, iPS cells have not been differentiated into any specific cell type or organ, although gene-expression profiles demonstrated the existence of different lineage markers. However, this technique will revolutionize the possibilities for tissue engineering and regenerative medicine, although several hurdles still need to be taken.

3 Stem Cells in Urology

Research activity that explores the possible applications of stem cells in the field of urology has been increasing in the last decade. However, clinical transplantation studies and cell-based intervention strategies using stem cells are still lacking.

3.1 Urinary Tract Tissue

Since urinary tract organs, such as bladder, ureter and urethra, are mainly composed of two cell types, an obvious challenge would be to obtain differentiated smooth muscle and urothelial cells from stem cells or progenitors for regenerative therapies.

In vitro studies have revealed that human fat-derived MSC can increase smooth muscle gene expression in response to dexamethasone and hydrocortisone [48]. Other researchers have emphasized that transforming growth factor (TGF) beta-1 is capable of driving rat neural crest stem cells towards a smooth muscle fate [49]. Furthermore, we have investigated the potential of several stimuli to induce smooth muscle gene expression by bone marrow-derived MSC [50]. In these studies, we applied different cocktails of growth factors and corticosteroids. Our results, in part, confirmed data from other groups. Moreover, we applied epithelial–mesenchymal interactions which have been suggested to play a critical role in the development and maintenance of a smooth muscle phenotype during both embryogenesis and adulthood [51]. The combination of stimulation by humoral factors and co-culture with primary urothelial cells resulted in a further significant increase of smooth muscle-specific gene expression in the treated MSC (Fig. 3). However, we also demonstrated that the cells only underwent a partial differentiation (i.e. when gene-expression patterns of smooth muscle genes were compared with those of primary isolated smooth muscle cells from bladder by real-time PCR). On the basis of all

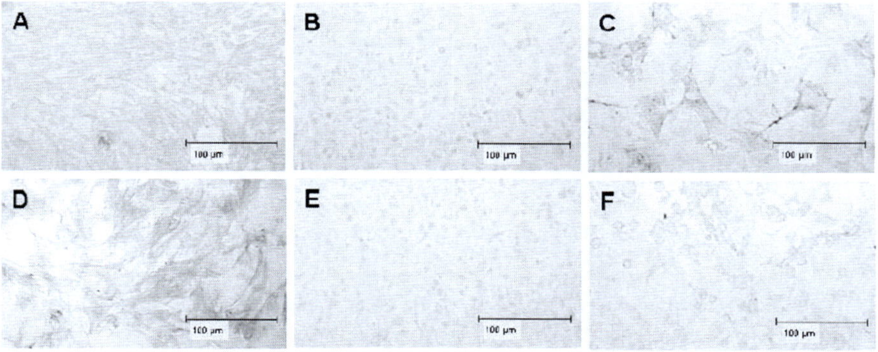

Fig. 3 In vitro-myogenic differentiation of bone marrow stromal cells (BMSC) [50]. (**a+d**) Primary smooth muscle cells. (**b+e**) Undifferentiated BMSC. (**c+f**) Myogenic differentiated BMSC. (**a–c**) Immunocytochemical staining of α-smooth muscle actin. (**d–f**) Immunocytochemical staining of desmin. (**c+f**) BMSC were stimulated for 3 days with TGFβ-1 + co-culture with urothelial cells

these data, one is drawn to the conclusion that complete smooth muscle differentiation in vitro, at present, is hardly achievable. This is more comprehensible when considering the complexity of events leading to a smooth muscle differentiation during embryogenesis in vivo [52].

The second important cell type of the urinary organs is the urothelium. However, little is known about the differentiation of stem cells towards an urothelial cell phenotype. Embryonic stem cells were reported to differentiate into urothelium by interaction with foetal bladder mesenchyme acting as an inductive tissue [53]. During the differentiation process the expression of specific endodermal transcription factors (Foxa1 and Foxa2) was found to be expressed. Additionally, terminal differentiation was reported by the expression of uroplakin. More recently, the same group reported the differentiation of bone marrow-derived mesenchymal stromal cells towards urothelium using their co-culture model described above (Fig. 4) [54]. However, since MSCs were claimed to differentiate in several other tissues

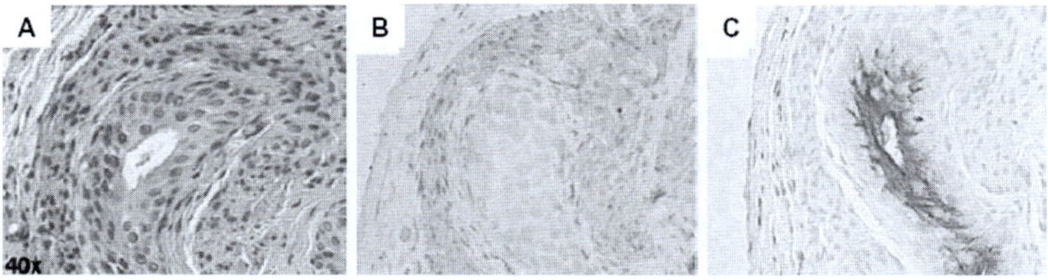

Fig. 4 Mouse cultured bone marrow mesenchymal stromal cells plus rat foetal bladder mesenchyme graft [54]. (**a**) HE staining. (**b**) Immunostaining of smooth muscle actin. (**c**) Immunostaining of uroplakin (figure modified pursuant to Anumanthan et al. [54])

as described above, one needs to be cautious regarding such data and a lot more research in this area is definitively needed.

The lack of suitable in vitro differentiation protocols for adult stem cells has led to strategies which apply "native" stem cells, "pre-differentiated" stem cells or committed precursors for transplantation. A tissue engineering attempt using undifferentiated MSC from rat bone marrow has been performed by Chung et al. [55]. Stem cells seeded on porcine small intestinal submucosa (SIS) were augmented in rats after partial cystectomy and long-term follow-up observations were undertaken. After 3 months, the biohybrids exhibited three-layered cellular constituents including urothelial cells and smooth muscle cells as demonstrated by immunostaining and real-time PCR of cytokeratins 8 and 19 and smooth muscle myosin heavy chain, respectively. Although a three-layered cellular architecture was also observed in control experiments using unseeded SIS, only the stem cell-seeded biohybrid exhibited gene-expression levels similar to those of sham-operated animals. These results suggest that the transplanted stem cells had enhanced the sprouting of host bladder cells from the edges of the residual native tissue and probably supported the phenotypes of those cells, rather than transdifferentiated into bladder-specific cell types themselves.

An autologous approach (and therefore more relevant for clinical application) was performed by Shukla et al. in a porcine model using MSC and SIS [56]. MSC were isolated from aspirat obtained from the iliac crest, expanded and analysed for their in vitro differentiation capacity towards a smooth muscle phenotype. Undifferentiated labelled MSC were seeded onto SIS and transplanted as a patch onto the excised bladder area. Interestingly, the implanted material had become even thicker than the surrounding bladder tissue. The internal surface of the patch was found to be covered by urothelium. However, the labelled MSC were only rarely (2–3%) found to differentiate into smooth muscle cells in vivo. Hence it has to be assumed that MSC contribute to tissue regeneration by other mechanisms than differentiation into smooth muscle cells.

3.2 Renal Tissue

Stem cells have also been taken into account for their contribution to the formation, cellular turnover and repair of more complex tissues, such as kidney. Interestingly, MSC were found to home into the kidney after damage to the renal tissue, with no indication that this occurs in the absence of damage [57]. Poulsom and colleagues provided evidence that adult stem cells could differentiate into renal tubular epithelial cells and associated stromal cells [58]. The authors described the transplantation of whole bone marrow from male donor mice to irradiated female recipients. The presence of Y-chromosome positive staining (FISH) in recipient's kidney structures indicated a contribution by circulating bone marrow stromal cells to renal epithelial cell turnover. Moreover, integration of extra-renal circulating cells into kidney structures has also been found in humans. A series of biopsies from female kidneys that had been transplanted into male recipients has revealed the presence of

Y-chromosome- and CAM5.2-positive cells, indicating the presence of cortical tubular epithelial cells derived from male recipient.

Chen and colleagues demonstrated the isolation of MSC from adult mouse kidney and reported the capacity of these local MSC to participate in renal tissue protection by paracrine effects [59]. The factor responsible for the accelerated functional recovery was identified as vascular endothelial growth factor (VEGF). Since tissue protection was found to be due to a factor produced by local cells, the small percentage of resident MSC seems not to be sufficient for self-regene ration.

These findings underline the plasticity of adult stem cells and illustrate a new axis for kidney regeneration which suggests possible implications for renal therapy. Accordingly, the use of renal progenitor cells to generate complex renal structures through tissue engineering approaches has been attempted. Humes and colleagues presented the concept of a bioartificial kidney composed of a standard dialysis device and a biohybrid containing renal tubule cells [60]. In an extracorporeal circuit, a synthetic haemofilter cartridge was connected to a renal tubule assist device

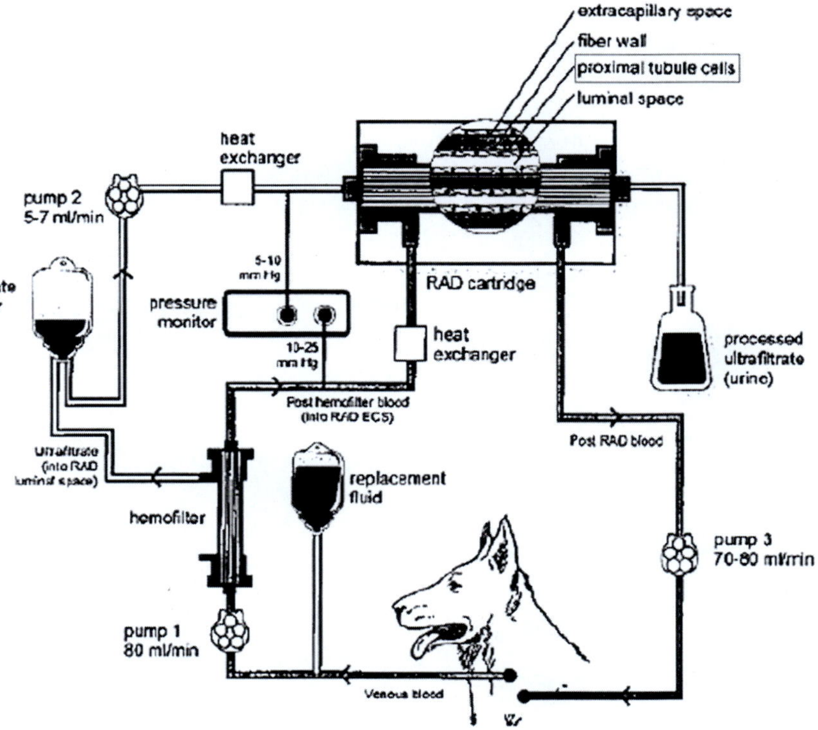

Fig. 5 Schematic diagram of an extracorporeal haemoperfusion circuit employing a stem cell-related bioartificial kidney [60]. A standard synthetic haemofilter cartridge is placed in series with a bioartificial renal tubule assist device cartridge, consisting of porcine renal tubule progenitor cells (figure modified pursuant to Humes et al. [60])

cartridge (Fig. 5). The cells incorporated into the device were derived from porcine renal tubule progenitor cells. This system successfully replaced filtration, transport, metabolic and endocrinologic functions of the kidney in acutely uraemic dogs, as assessed by plasma values and tubular fluid/ultra-filtrate ratios.

Lanza and co-workers applied the technique of somatic cell nuclear transfer (SCNT), also known as therapeutic cloning, to produce histocompatible progenitor cells in a bovine model [61]. Embryonic tissue was harvested 12 weeks after blastocyst transfer and cloned metanephric progenitors were seeded onto cylindrical polycarbonate membranes. Biohybrids with collecting devices for excreted fluid were transplanted subcutaneously for 6 weeks. Histological examination of retrieved implants revealed extensive vascularization and self-organization of the cells into glomeruli- and tubule-like structures (Fig. 6). A clear continuity between glomeruli, tubules and the polycarbonate membrane was noted that allowed the passage of fluid into the collecting reservoir. Importantly, secreted fluid exhibited urine-like levels for urea nitrogen/creatinine, electrolytes and glucose, indicating that the biohybrid possessed filtration, reabsorption and secretory capabilities.

An in vitro approach to generate a bioartificial kidney with renal progenitor cells has been presented by Minuth and colleagues [62]. Renal progenitor cells from the embryonic cortex of neonatal rabbits were harvested and placed into specific tissue holders and incubated in vitro in a perfusion culture container exhibiting an artificial interstitium made of a polyester fleece. After 14 days of incubation, the generation of a wide network of numerous renal tubules at the interface of the fleece and the embryonic tissue could be observed.

Although these reports sound very promising regarding a therapeutical application to patients with acute renal failure, the relatively simple configuration of the developing systems and the short incubation times are in some respects inconsistent with the complexity of nephrogenesis during embryonic development. Ongoing investigations are required to explore further the molecular mechanisms underlying the generation of bioartificial renal devices.

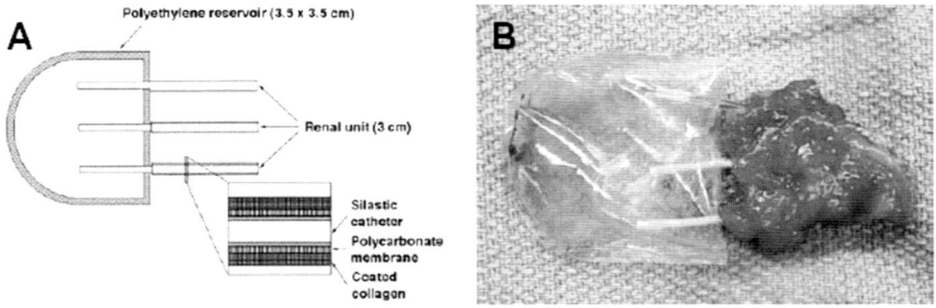

Fig. 6 Tissue-engineered renal unit [61]. (**a**) Illustration of biohybrid renal unit. (**b**) Retrieved biohybrid renal unit containing bovine therapeutic cloning-derived metanephric progenitor cells 3 months after implantation, showing the accumulation of urine-like fluid (figure modified pursuant to Lanza et al. [61])

3.3 Gonadal Tissue

Brinster and colleagues have described the successful transplantation of adult spermatogonial stem cells into a mouse model of male infertility [63]. Donor progenitor cells from digested testicular tubules were microinjected into the seminiferous tubules of sterile recipients. After 1 month, recovery of spermatogenesis was observed in the transplanted animals, with normal architecture of the testes. The origin of the cells responsible for functional spermatogenesis, spermatogonia, spermatocytes and spermatids could be assigned to the transplanted cells since these cells were derived from mice transgenic for beta-galactosidase (Fig. 7, blue staining cells). Lo and colleagues have embarked on a strategy in applying adult Leydig cell progenitors. Cells were isolated from adult testes of healthy mice and transplanted into the testes of luteinizing hormone receptor knockout recipients [64]. Such transplantations resulted in increased circulating testosterone levels and the restoration of spermatogenesis. It is possible that both of these landmark experiments could be developed for human infertility therapy, e.g. to conserve the fertility of (juvenile) patients undergoing radiotherapy during cancer treatment. Autologous spermatogonial stem cells could be harvested and cryopreserved prior to irradiation and subsequently retransplanted.

Other approaches to generate gametes were reported some years ago by several researchers using embryonic stem cells. For example, Toyooka and colleagues successfully isolated differentiating germ cells from embryoid body cultures which were derived from murine ESC [65]. Treatment with the transcription factor bone morphogenetic factor 4 (BMP-4) enhanced the development of the embryoid body cells towards the germ cell lines. When these cells were transplanted into the seminiferous tubules of sterile recipient mice, functional spermatogenesis could be observed. Others have demonstrated the stimulatory effect of retinoic acid on cultured murine ESC [66]. Stimulated ESC exhibited gene-expression profiles specific to male germ cells. Cells were able to undergo meiosis and the resulting haploid gametes were able to fertilize oocytes and develop into blastocysts in vitro. Although

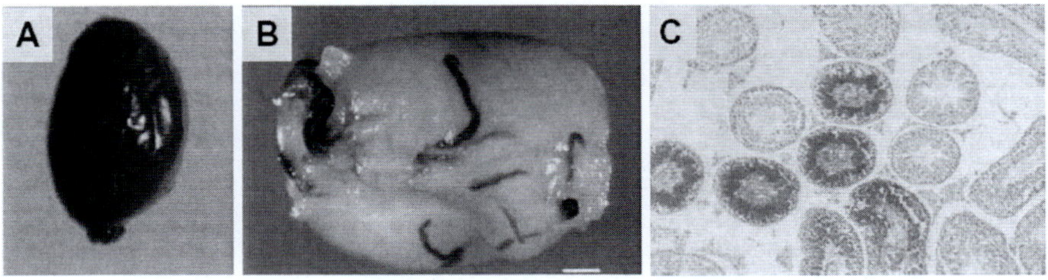

Fig. 7 Treatment of infertility by transplantation of adult spermatogonial stem cells [63]. (**a**) *darker* staining of donor testis with X-Gal indicates the presence of the transgene in all cells, including stem cell areas of spermatogenesis. (**b**) *darker* tubules indicate spermatogenesis from ZFlacZ donor spermatogonial stem cells in the sterile recipient testis. (**c**) *darker*-stained tubule cross-sections represent spermatogenesis from donor stem cells (figure modified pursuant to Brinster and Avarbock [63])

these investigations with ESC are highly impressive, several arguments contradict the use of ESC in the treatment of human infertility. These include fundamental challenges and risks related to embryonic stem cells such as discussed below (see "The Potential of Selectively Cultured Adult Stem Cells Re-implanted in Tissues", Challenges and risks). Furthermore, no studies have reported the successful generation of a viable embryo from ESC-derived gametes, probably due to the lack of the cells necessary for the generation of trophoblast, the cell type required for entire embryogenesis.

3.4 Sphincter Muscle Tissue

To develop suitable techniques for the treatment of urinary incontinence, Chancellor and colleagues have established a therapeutical strategy which depends on autologous muscle-derived progenitor cells (MDPC; Fig. 8) [67]. Isolated MDPC from biopsies of gastrocnemius muscle of female rats were obtained by a preplate technique [68]. Animals underwent sciatic nerve lesion resulting in almost complete loss of fast-twitch muscle contraction and partial impairment of smooth muscle contractility. Propagated progenitor cells were injected at two positions into the urethra

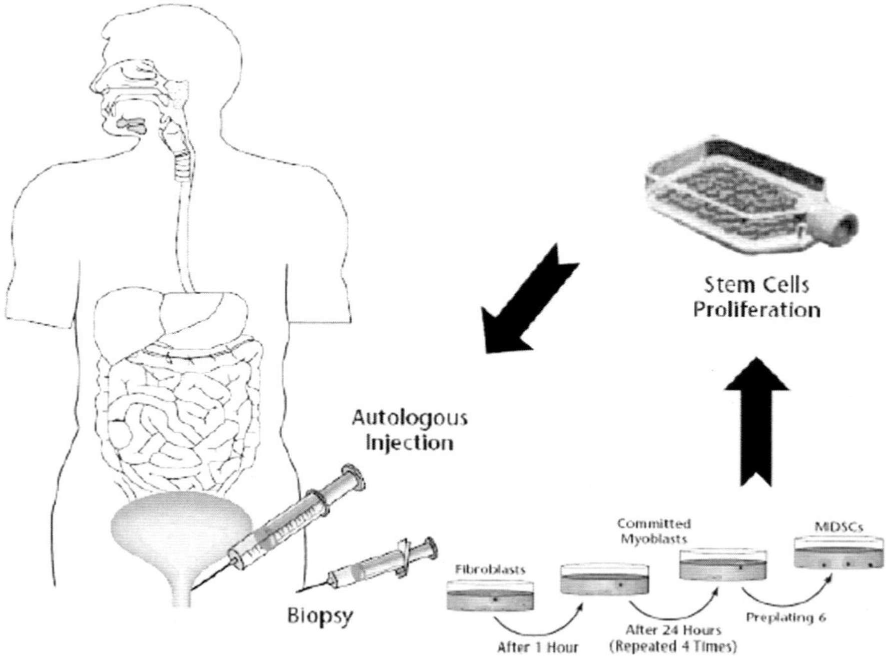

Fig. 8 Vision for stem cell therapy of urinary incontinence [67]. Cells can be obtained from skeletal muscle biopsy. Muscle-derived stem cells (MDSC) are enriched by preplating technique (figure modified pursuant to Furuta et al. [67])

of the incontinent animals and restoration of deficient urethral sphincter function could be observed 2 weeks after transplantation. The experimental data suggest that urinary incontinence can be treated by this method, supporting the notion that a therapeutic concept using autologous adult stem cells may represent a very promising modality in the future treatment of a range of disorders including urologic diseases.

4 Challenges and Risks

The use of stem cells to treat and reconstruct urologic organs appears to be a very powerful and promising approach for the future. Due to the multitude of unanticipated problems including the handling, development and long-term fate of stem cells, only limited applications in the field of urology have progressed beyond the experimental domain.

The issues of harvesting and use of pluripotent embryonic stem cells have been extremely controversial and intensely debated in both the scientific and the public arenas. Ethical and political debates have revolved around the issue since a number of human ES cell lines were made available from "excess" embryos from in vitro fertilization clinics. Under similar scrutiny and debate is the putative application of hESCs derived from somatic cell nuclear transfer ("therapeutic cloning"). Nonetheless, the enormous proliferative capability and cell differentiation capacity of hESC have resulted in formulation of a number of methods and protocols for their potential applications in basic developmental biology and regenerative medicine. Furthermore, strategies using specific growth factor combinations, cell–cell and cell–extracellular matrix induction systems are being explored for the controlled differentiation of stem cells along a desired lineage. The current lack of sufficient differentiation protocols is limiting the rate progress since residual undifferentiated ESC can pose a significant health risk of tumourigenesis following implantation into patients. In fact, there is strong evidence that hESC transplanted into immunodeficient mice form complex teratomas consisting of a range of differentiated somatic cells, some of which appear to be highly organized, closely resembling tissues in the developing embryo and adult [69]. This exemplifies the absolute need to identify and characterize the mechanisms that control self-renewal and differentiation to improve approaches for the purification and differentiation of hESC prior to their application in future clinically relevant intervention strategies. Another critical issue is the morphological and developmental difference that exists between murine and human ESC, suggesting that murine models may not be adequate for direct translation to the human situation.

On the other hand, certain adult multipotent stem cells also present a number of different disadvantages, principally due to the lack of unique cell surface markers, preventing the isolation of homogenous, well-defined populations of stem cells [70, 71]. Furthermore adult stem cells run the risk of senescence and epigenetic modifications, which in turn could alter the fate of the cells [70, 72].

The enormous proliferative capability and cell differentiation capacity of ESC and the absence of any ethical concerns are combined by the newly generated induced pluripotent stem cells [42, 43]. These cells offer a broad spectrum of applications in tissue engineering and regenerative medicine in future. Since this field is fairly new, a lot more research is needed concerning the side effects of reprogramming.

Although several methodological and bio-ethical obstacles need to be overcome, recent progress in experimental and clinical urology suggests that stem cells may deliver great benefits in regenerative strategies, specifically for urologic structures. Further experiments and trials are therefore warranted. However, there is still much unknown territory in the field of urology, particularly regarding the introduction of stem cell-based therapies. For example, up to now, no reports have been published using stem cells in andrology. In this context, stem cells might prove to be beneficial in the treatment of penile deformations or erectile dysfunction. Another, as yet, unmet challenge is the development of a stem cell-based strategy for the treatment of vesicoureteral reflux (VUR). Since the transurethral injection of autologous chondrocytes has already been demonstrated as a successful therapy of VUR [73], chondrogenic differentiated stem cells could be used to guide cartilage formation in the submucosal layer [74]. The list of further putative applications of stem cells in urology can readily be extended to almost any conceivable indication.

It seems clear from the current chapter that the development of stem cell-based strategies for the treatment of urological diseases or disorders is in its early stages. This provides an ample opportunity for both basic scientists and clinicians to enter this area of research.

Acknowledgements We would like to thank Dr. Gary Brook for linguistic revision of this manuscript.

References

1. Bach AD, Bannasch H, Galla TJ, Bittner KM, Stark GB. Fibrin glue as matrix for cultured autologous urothelial cells in urethral reconstruction. Tissue Eng. 2001; 7:45–53.
2. Oberpenning F, Meng J, Yoo JJ, Atala A. De novo reconstitution of a functional mammalian urinary bladder by tissue engineering. Nat Biotechnol. 1999; 17:149–155.
3. Fraser M, Thomas DF, Pitt E, Harnden P, Trejdosiewicz LK, Southgate J. A surgical model of composite cystoplasty with cultured urothelial cells: a controlled study of gross outcome and urothelial phenotype. BJU Int. 2004; 93:609–616.
4. Matsunuma H, Kagami H, Narita Y, Hata K, Ono Y, Ohshima S, Ueda M. Constructing a tissue-engineered ureter using a decellularized matrix with cultured uroepithelial cells and bone marrow-derived mononuclear cells. Tissue Eng. 2006; 12:509–518.
5. Humes HD, MacKay SM, Funke AJ, Buffington DA. Tissue engineering of a bioartificial renal tubule assist device: in vitro transport and metabolic characteristics. Kidney Int. 1999; 55:2502–2514.
6. Kershen RT, Yoo JJ, Moreland RB, Krane RJ, Atala A. Reconstitution of human corpus cavernosum smooth muscle in vitro and in vivo. Tissue Eng. 2002; 8:515–524.
7. De Filippo RE, Yoo JJ, Atala A. Engineering of vaginal tissue in vivo. Tissue Eng. 2003; 9:301–306.
8. Atala A, Bauer SB, Soker S, Yoo JJ, Retik AB. Tissue-engineered autologous bladders for patients needing cystoplasty. Lancet 2006; 367:1241–1246.

9. Watt FM, Hogan BL. Out of Eden: stem cells and their niches. Science 2000; 287:1427–1430.
10. Blau HM, Brazelton TR, Weimann JM. The evolving concept of a stem cell: entity or function? Cell 2001; 105:829–841.
11. Shostak S. (Re)defining stem cells. Bioessays 2006; 28:301–308.
12. Monk M. A stem-line model for cellular and chromosomal differentiation in early mouse-development. Differentiation 1981; 19:71–76.
13. Evans MJ, Kaufman MH. Establishment in culture of pluripotential cells from mouse embryos. Nature 1981; 292:154–156.
14. Thomson JA, Itskovitz-Eldor J, Shapiro SS, Waknitz MA, Swiergiel JJ, Marshall VS, Jones JM. Embryonic stem cell lines derived from human blastocysts. Science 1998; 282: 1145–1147.
15. Shamblott MJ, Axelman J, Wang S, Bugg EM, Littlefield JW, Donovan PJ, Blumenthal PD, Huggins GR, Gearhart JD. Derivation of pluripotent stem cells from cultured human primordial germ cells. Proc Natl Acad Sci USA. 1998; 95:13726–13731.
16. Minasi MG, Riminucci M, De Angelis L, Borello U, Berarducci B, Innocenzi A, Caprioli A, Sirabella D, Baiocchi M, De Maria R, et al. The meso-angioblast: a multipotent, self-renewing cell that originates from the dorsal aorta and differentiates into most mesodermal tissues. Development 2002; 129:2773–2783.
17. Conrad C, Huss R. Adult stem cell lines in regenerative medicine and reconstructive surgery. J Surg Res. 2005; 124:201–208.
18. Park IK, He Y, Lin F, Laerum OD, Tian Q, Bumgarner R, Klug CA, Li K, Kuhr C, Doyle MJ, et al. Differential gene expression profiling of adult murine hematopoietic stem cells. Blood 2002; 99:488–498.
19. Reyes M, Verfaillie CM. Characterization of multipotent adult progenitor cells, a subpopulation of mesenchymal stem cells. Ann N Y Acad Sci. 2001; 938:231–233, discussion 233–235.
20. Gritti A, Vescovi AL, Galli R. Adult neural stem cells: plasticity and developmental potential. J Physiol Paris 2002; 96:81–90.
21. Young HE, Black AC Jr.. Adult stem cells. Anat Rec A Discov Mol Cell Evol Biol. 2004; 276:75–102.
22. Semb H. Human embryonic stem cells: origin, properties and applications. Apmis 2005; 113:743–750.
23. Kim JH, Auerbach JM, Rodriguez-Gomez JA, Velasco I, Gavin D, Lumelsky N, Lee SH, Nguyen J, Sanchez-Pernaute R, Bankiewicz K, et al. Dopamine neurons derived from embryonic stem cells function in an animal model of Parkinson's disease. Nature 2002; 418:50–56.
24. Singla DK, Hacker TA, Ma L, Douglas PS, Sullivan R, Lyons GE, Kamp TJ. Transplantation of embryonic stem cells into the infarcted mouse heart: formation of multiple cell types. J Mol Cell Cardiol. 2006; 40:195–200.
25. Lakshmipathy U, Verfaillie C. Stem cell plasticity. Blood Rev. 2005; 19:29–38.
26. Petersen BE, Bowen WC, Patrene KD, Mars WM, Sullivan AK, Murase N, Boggs SS, Greenberger JS, Goff JP. Bone marrow as a potential source of hepatic oval cells. Science 1999; 284:1168–1170.
27. Khurana S, Mukhopadhyay A. In vitro transdifferentiation of adult hematopoietic stem cells: an alternative source of engraftable hepatocytes. J Hepatol. 2008; 49:998–1007.
28. Bjornson CR, Rietze RL, Reynolds BA, Magli MC, Vescovi AL. Turning brain into blood: a hematopoietic fate adopted by adult neural stem cells in vivo. Science 1999; 283:534–537.
29. Kopen GC, Prockop DJ, Phinney DG. Marrow stromal cells migrate throughout forebrain and cerebellum, and they differentiate into astrocytes after injection into neonatal mouse brains. Proc Natl Acad Sci USA. 1999; 96:10711–10716.
30. Lian G, Wang C, Teng C, Zhang C, Du L, Zhong Q, Miao C, Ding M, Deng H. Failure of hepatocyte marker-expressing hematopoietic progenitor cells to efficiently convert into hepatocytes in vitro. Exp Hematol. 2006; 34:348–358.

31. Liu Y, Rao MS. Transdifferentiation – fact or artifact. J Cell Biochem. 2003; 88:29–40.
32. Wells WA. Is transdifferentiation in trouble? J Cell Biol. 2002; 157:15–18.
33. Terada N, Hamazaki T, Oka M, Hoki M, Mastalerz DM, Nakano Y, Meyer EM, Morel L, Petersen BE, Scott EW. Bone marrow cells adopt the phenotype of other cells by spontaneous cell fusion. Nature 2002; 416:542–545.
34. Quesenberry PJ, Abedi M, Aliotta J, Colvin G, Demers D, Dooner M, Greer D, Hebert H, Menon MK, Pimentel J, et al. Stem cell plasticity: an overview. Blood Cells Mol Dis. 2004; 32:1–4.
35. Serakinci N, Keith WN. Therapeutic potential of adult stem cells. Eur J Cancer 2006; 42:1243–1246.
36. Till JE, Mc CE. A direct measurement of the radiation sensitivity of normal mouse bone marrow cells. Radiat Res. 1961; 14:213–222.
37. Pittenger MF, Mackay AM, Beck SC, Jaiswal RK, Douglas R, Mosca JD, Moorman MA, Simonetti DW, Craig S, Marshak DR. Multilineage potential of adult human mesenchymal stem cells. Science 1999; 284:143–147.
38. Wollert KC, Meyer GP, Lotz J, Ringes-Lichtenberg S, Lippolt P, Breidenbach C, Fichtner S, Korte T, Hornig B, Messinger D, et al. Intracoronary autologous bone-marrow cell transfer after myocardial infarction: the BOOST randomised controlled clinical trial. Lancet 2004; 364:141–148.
39. Quarto R, Mastrogiacomo M, Cancedda R, Kutepov SM, Mukhachev V, Lavroukov A, Kon E,Marcacci M. Repair of large bone defects with the use of autologous bone marrow stromal cells. N Engl J Med. 2001; 344:385–386.
40. Tateishi-Yuyama E, Matsubara H, Murohara T, Ikeda U, Shintani S, Masaki H, Amano K, Kishimoto Y, Yoshimoto K, Akashi H, et al. Therapeutic angiogenesis for patients with limb ischaemia by autologous transplantation of bone-marrow cells: a pilot study and a randomised controlled trial. Lancet 2002; 360:427–435.
41. Koc ON, Gerson SL, Cooper BW, Dyhouse SM, Haynesworth SE, Caplan AI, Lazarus HM. Rapid hematopoietic recovery after coinfusion of autologous-blood stem cells and culture-expanded marrow mesenchymal stem cells in advanced breast cancer patients receiving high-dose chemotherapy. J Clin Oncol. 2000; 18:307–316.
42. Takahashi K, Yamanaka S. Induction of pluripotent stem cells from mouse embryonic and adult fibroblast cultures by defined factors. Cell 2006; 126:663–676.
43. Welstead GG, Brambrink T, Jaenisch R. Generating iPS Cells from MEFS through forced expression of Sox-2, Oct-4, c-Myc, and Klf4. J Vis Exp. 2008; 14:734–737.
44. Huangfu D, Osafune K, Maehr R, Guo W, Eijkelenboom A, Chen S, Muhlestein W, Melton DA. Induction of pluripotent stem cells from primary human fibroblasts with only Oct4 and Sox2. Nat Biotechnol. 2008; 26:1269–1275.
45. Kim JB, Zaehres H, Wu G, Gentile L, Ko K, Sebastiano V, Arauzo-Bravo MJ, Ruau D, Han DW, Zenke M, et al. Pluripotent stem cells induced from adult neural stem cells by reprogramming with two factors. Nature 2008; 454:646–650.
46. Takahashi K, Tanabe K, Ohnuki M, Narita M, Ichisaka T, Tomoda K, Yamanaka S. Induction of pluripotent stem cells from adult human fibroblasts by defined factors. Cell 2007; 131: 861–872.
47. Nakagawa M, Koyanagi M, Tanabe K, Takahashi K, Ichisaka T, Aoi T, Okita K, Mochiduki Y, Takizawa N, Yamanaka S. Generation of induced pluripotent stem cells without Myc from mouse and human fibroblasts. Nat Biotechnol. 2008; 26:101–106.
48. Zuk PA, Zhu M, Mizuno H, Huang J, Futrell JW, Katz AJ, Benhaim P, Lorenz HP, Hedrick MH. Multilineage cells from human adipose tissue: implications for cell-based therapies. Tissue Eng. 2001; 7:211–228.
49. Shah NM, Groves AK, Anderson DJ. Alternative neural crest cell fates are instructively promoted by TGFbeta superfamily members. Cell 1996; 85:331–343.
50. Becker C, Laeufer T, Arikkat J, Jakse G. TGFbeta-1 and epithelial-mesenchymal interactions promote smooth muscle gene expression in bone marrow stromal cells: possible application in therapies for urological defects. Int J Artif Organs 2008; 31:951–959.

51. DiSandro MJ, Li Y, Baskin LS, Hayward S, Cunha G. Mesenchymal-epithelial interactions in bladder smooth muscle development: epithelial specificity. J Urol. 1998; 160:1040–1046, discussion 1079.
52. Bronner-Fraser M, Fraser SE. Cell lineage analysis reveals multipotency of some avian neural crest cells. Nature 1988; 335:161–164.
53. Oottamasathien S, Wang Y, Williams K, Franco OE, Wills ML, Thomas JC, Saba K, Sharif-Afshar AR, Makari JH, Bhowmick NA, et al. Directed differentiation of embryonic stem cells into bladder tissue. Dev Biol. 2007; 304:556–566.
54. Anumanthan G, Makari JH, Honea L, Thomas JC, Wills ML, Bhowmick NA, Adams MC, Hayward SW, Matusik RJ, Brock JW III, et al. Directed differentiation of bone marrow derived mesenchymal stem cells into bladder urothelium. J Urol. 2008; 180:1778–1783.
55. Chung SY, Krivorov NP, Rausei V, Thomas L, Frantzen M, Landsittel D, Kang YM, Chon CH, Ng CS, Fuchs GJ. Bladder reconstitution with bone marrow derived stem cells seeded on small intestinal submucosa improves morphological and molecular composition. J Urol. 2005; 174:353–359.
56. Shukla D, Box GN, Edwards RA, Tyson DR. Bone marrow stem cells for urologic tissue engineering. World J Urol. 2008; 26:341–349.
57. Lin F, Igarashi P. Searching for stem/progenitor cells in the adult mouse kidney. J Am Soc Nephrol. 2003; 14:3290–3292.
58. Poulsom R, Forbes SJ, Hodivala-Dilke K, Ryan E, Wyles S, Navaratnarasah S, Jeffery R, Hunt T, Alison M, Cook T, et al. Bone marrow contributes to renal parenchymal turnover and regeneration. J Pathol. 2001; 195:229–235.
59. Chen J, Park HC, Addabbo F, Ni J, Pelger E, Li H, Plotkin M, Goligorsky MS. Kidney-derived mesenchymal stem cells contribute to vasculogenesis, angiogenesis and endothelial repair. Kidney Int. 2008; 74:879–889.
60. Humes HD, Buffington DA, MacKay SM, Funke AJ, Weitzel WF. Replacement of renal function in uremic animals with a tissue-engineered kidney. Nat Biotechnol. 1999; 17:451–455.
61. Lanza RP, Chung HY, Yoo JJ, Wettstein PJ, Blackwell C, Borson N, Hofmeister E, Schuch G, Soker S, Moraes CT, et al. Generation of histocompatible tissues using nuclear transplantation. Nat Biotechnol. 2002; 20:689–696.
62. Minuth WW, Sorokin L, Schumacher K. Generation of renal tubules at the interface of an artificial interstitium. Cell Physiol Biochem. 2004; 14:387–394.
63. Brinster RL, Avarbock MR. Germline transmission of donor haplotype following spermatogonial transplantation. Proc Natl Acad Sci USA. 1994; 91:11303–11307.
64. Lo KC, Lei Z, Rao ChV, Beck J, Lamb DJ. De novo testosterone production in luteinizing hormone receptor knockout mice after transplantation of leydig stem cells. Endocrinology 2004; 145:4011–4015.
65. Toyooka Y, Tsunekawa N, Akasu R, Noce T. Embryonic stem cells can form germ cells in vitro. Proc Natl Acad Sci USA. 2003; 100:11457–11462.
66. Geijsen N, Horoschak M, Kim K, Gribnau J, Eggan K, Daley GQ. Derivation of embryonic germ cells and male gametes from embryonic stem cells. Nature 2004; 427:148–154.
67. Furuta A, Carr LK, Yoshimura N, Chancellor MB. Advances in the understanding of sress urinary incontinence and the promise of stem-cell therapy. Rev Urol. 2007; 9:106–112.
68. Cannon TW, Lee JY, Somogyi G, Pruchnic R, Smith CP, Huard J, Chancellor MB. Improved sphincter contractility after allogenic muscle-derived progenitor cell injection into the denervated rat urethra. Urology. 2003; 62:958–963.
69. Przyborski SA. Differentiation of human embryonic stem cells after transplantation in immune-deficient mice. Stem Cells 2005; 23:1242–1250.
70. Wognum AW, Eaves AC, Thomas TE. Identification and isolation of hematopoietic stem cells. Arch Med Res. 2003; 34:461–475.
71. Beyer Nardi N, da Silva Meirelles L. Mesenchymal stem cells: isolation, in vitro expansion and characterization. Handb Exp Pharmacol. 2006; 174:249–282.
72. Shibata D, Tavare S. Counting divisions in a human somatic cell tree: how, what and why? Cell Cycle 2006; 5:610–614.

73. Diamond DA, Caldamone AA. Endoscopic correction of vesicoureteral reflux in children using autologous chondrocytes: preliminary results. J Urol. 1999; 162:1185–1188.
74. Cho JH, Kim SH, Park KD, Jung MC, Yang WI, Han SW, Noh JY, Lee JW. Chondrogenic differentiation of human mesenchymal stem cells using a thermosensitive poly(N-isopropylacrylamide) and water-soluble chitosan copolymer. Biomaterials 2004; 25:5743–5751.
75. Keller G. Embryonic stem cell differentiation: emergence of a new era in biology and medicine. Genes Dev. 2005; 19:1129–1155.

Bio-synthetic Encapsulation Systems for Organ Engineering: Focus on Diabetes

Rylie A. Green, Penny J. Martens, Robert Nordon,
and Laura A. Poole-Warren

Abstract The development of a safe and effective bio-synthetic encapsulation system will potentially improve the treatment of diseases known to benefit from cell implantation therapies. This is of particular importance in the progression towards a permanent treatment for type 1 diabetes. Cell-based therapies for insulin-dependent diabetics ultimately aim to eliminate the need for exogenous insulin through the implantation of donor islet cells or, alternately, stem cells differentiated into glucose-sensitive islet-like cells. Encapsulation devices are required to protect implanted cells from destructive host immune factors. A successful design will be bio-synthetic, having well-defined permeability and providing appropriate biomolecules for physiological functionality and survival of the encapsulated cells.

Keywords Diabetes · Cell encapsulation · Stem Cells

1 Introduction

Regenerative medicine requires the replacement of damaged tissue while preserving normal physiological functions. Cellular therapy incorporating tissue engineering principles has the potential to be among the most effective therapies, due to recent advancements in technology allowing the control of specific interactions between bio-synthetic materials, cellular tissue, biological signalling factors and ligands [1].

Encapsulation is a technique proposed for providing a protective barrier for non-autologous cells implanted into patients. The primary functions of encapsulating materials are to prevent access of cellular and humoral immune factors, to allow diffusion of nutrients and waste and to support relevant therapeutic functions.

L.A. Poole-Warren (✉)
Graduate School of Biomedical Engineering, University of New South Wales, Sydney, NSW, Australia

G.M. Artmann et al. (eds.), *Stem Cell Engineering*,
DOI 10.1007/978-3-642-11865-4_16,

Examples where encapsulation is particularly appropriate include cell-based therapies that use allogeneic or xenogeneic cells such as bio-artificial pancreas and liver.

The control of in vivo behaviour in a time-dependent and space-dependent fashion is imperative to the development of bio-synthetic encapsulation systems. Hydrogels have been used as biofunctional vehicles for the introduction of these cell-based systems, and new techniques allow for the control of cell adhesion, proliferation, differentiation and other biological functions [1].

2 Cell-Based Therapies for Diabetes

Type 1 diabetes (T1D) or juvenile-onset diabetes is characterised by β-cell apoptosis in the islets of Langerhans, which comprise the endocrine portion of the pancreas. T1D is a multifactorial disease evolving from complex interactions between genetic and environmental risk factors, but is classified as an autoimmune disease due to strong evidence of T-lymphocyte up-regulation of inflammatory cytokines, responsible for β-cell destruction [2, 3]. The primary function of β cells is to maintain blood glucose levels through the production and release of insulin. Without insulin, cells are unable to adsorb glucose, resulting in hyperglycaemia and fatal ketoacidosis [2, 4]. Thus to survive, patients with T1D require insulin injections. Type 2 diabetes (T2D), unlike T1D, is a problem with the cells that respond to insulin rather than failure of the endocrine portion of the pancreas. It is a metabolic disorder characterised by high glucose, but without the fatal complication of ketoacidosis. It has an insidious onset in adulthood and is initially managed by diet and exercise. Oral hypoglycaemic agents or insulin injection is typically needed as the disease advances.

Maintenance of normoglycaemia is critical in delaying the onset and progression of secondary chronic complications including diabetic retinopathy, nephropathy, neuropathy and cardiovascular disease [3, 4]. Exogenous insulin and oral hypoglycaemics often fail to replicate the continuous regulation of insulin delivery performed by β cells in response blood glucose variations. Alternative strategies for re-establishing in situ blood glucose level maintenance have been the subject of intense research for more than 50 years [3].

The subset of diabetics who require insulin – all T1D and a minority of T2D patients – could potentially benefit from a cell-based therapy to replace insulin, though type 2 insulin-dependent diabetics may not benefit directly, since the metabolic disorder is unrelated to β-cell destruction. The prevalence of diabetes in the United States was estimated to be 7.8% in 2007 (around 24 million people). Approximately 14% of these patients were insulin-dependent [5]. The estimated economic cost of diabetes in the United States in 2007 was $174 billion, the majority of medical costs attributed to chronic complications and hospitalisation [6]. In the United Kingdom, the treatment of T1D accounts for approximately 5% of the national health budget [7].

2.1 Whole Organ Transplants

Both pancreas-only and combined pancreas–kidney transplants have been shown to normalise blood glucose levels and stabilise secondary complications. However, survival of whole organ grafts has been only 55% for pancreas-only transplants and 70% for combined pancreas–kidney transplants assessed at 5 years post-transplantation [8]. Transplant failure is caused by two primary mechanisms. In the initial period following surgery, organ recipients face a high risk of thrombosis which may lead to transplant removal and is responsible for significant morbidity and some mortality [9]. Over the subsequent years the need for immunosuppression to prevent graft rejection can lead to significant side effects including malignancies, peripheral oedema, anaemia, hypertension, hyperlipidaemia and renal complications including failure [3, 8, 10]. In some cases the negative side effects of immunosuppression outweigh the benefits of insulin independence and necessitate removal of the transplant [9, 10]. Despite the incidence of failure, the demand on donor organs is still significantly higher than supply, strongly indicating the need for alternate therapies.

Two cell-based alternatives to whole organ transplants are being explored as long-term and optimally permanent treatments for T1D. Substantial resources have been put into developing both cell-based gene therapy and cell transplantation therapy (Fig. 1).

2.2 Cell-Based Gene Therapy

Gene therapy attempts to prevent the effect of T-cell destruction of β cells by introducing an insulin-producing gene into non-β cells, which due to their differing cell type would evade the autoimmune attack [11]. An optimal β-cell replacement must be able to sense blood glucose levels, produce biologically active insulin and perform glucose-dependent regulation of insulin release. Experiments into insulin delivery through genetically modified non-islet tissues have been largely unsuccessful. The first attempts to produce an insulin-secreting non-β cell in vitro utilised several immortal cell lines including green monkey kidney (GMK) cells, NIH 3T3 fibroblasts and Chinese hamster ovarian (CHO) cells [11–14]. Primary transplants of liver hepatocytes, fibroblasts, pituitary cells and endocrine K-cells have also been transfected in an attempt to generate a glucose-sensing insulin-producing alternative to the β cell [11, 14–16]. A number of cell types were found to be capable of producing biologically active insulin, but most lacked the ability to sense blood glucose levels, resulting in uncontrolled insulin secretion [11, 16]. Attempts to produce cells with glucose-dependent transcription of the insulin gene were unable to reproduce the same complex and refined regulation of insulin production and release as β cells [16]. While gene therapy remains a potential treatment for diabetes, substantial improvements must be made in the area to make this therapy a clinically plausible option.

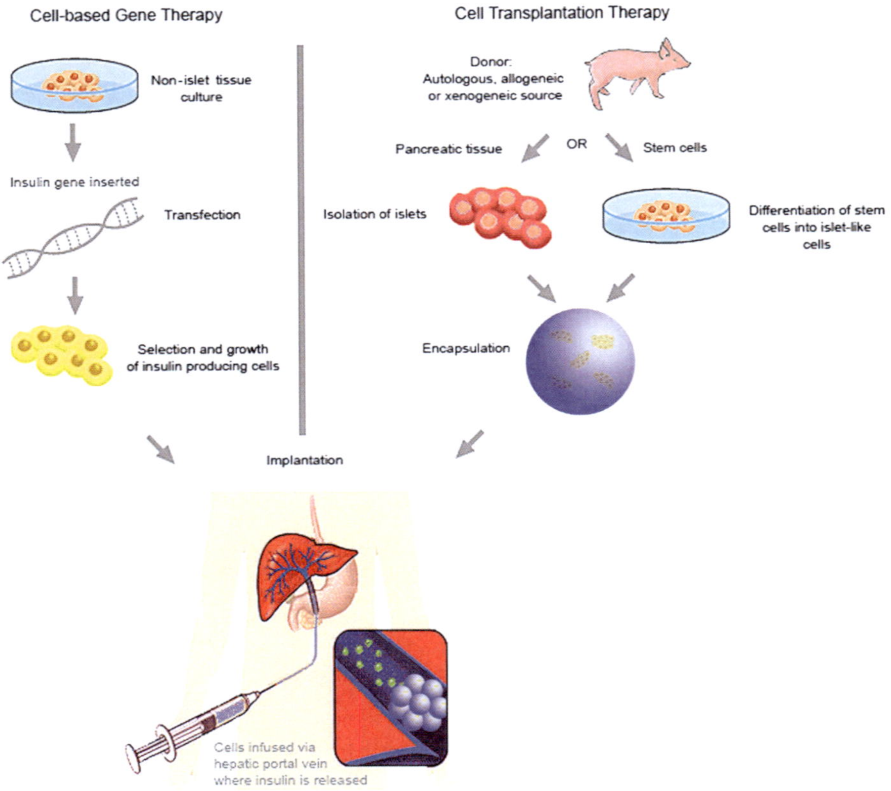

Fig. 1 Schematic depicting process for cell-based gene therapy vs. cell transplantation therapies

2.3 Cell Transplantation Therapy

Cell transplantation therapy has emerged as a promising approach for preventing the chronic complications of diabetes and while still restricted to use on severely ill patients, it has the potential to reduce the demand for transplantable organs [3]. Cell therapy attempts to restore insulin production, storage and release by introducing functional pancreatic cells from an alternate source. Clinical delivery needs to be patient-specific and at a scale that restores "functional β-cell mass" [17]. Given that insulin independence cannot be sustained beyond 2 years using the Edmonton protocol [18], it is likely that future protocols will require multiple sequential rounds of transplantation using a renewable β-cell source.

β-Cell or islet transplantation will, in principle, replace the damaged tissue while preserving normal physiological functions. Pancreatic cells are conventionally isolated from either xenogeneic or allogeneic cadavers. The most common method used to isolate islet cells was developed in 1986 and requires the slow digestion of the entire pancreas into fragments. Purified islets of Langerhans are quantitatively and qualitatively assessed to determine cell yields, purity, morphology, functionality and the presence of contaminating bacteria, prior to transplantation [19, 20].

Several organs have been investigated for the placement of isolated islet transplants, with animal models indicating that organs including the liver, thymus or subrenal capsule are possible locations. Most often the islets are transfused via the portal vein, where they are transported by the bloodstream into the liver [10]. Long-term intrahepatic presence and functionality has been confirmed for up to 5 years post-transplantation [21]. Transplantation into the liver has yielded suboptimal results in human patients, despite consistent positive results reported in small animal trials [22]. Investigations into the placement of cells within alternate organs attempt to increase the success of insulin independence through increased immunotolerance of transplants. Preliminary studies using the canine model have indicated that the thymus is a viable site for islet placement [22], and both xenografts and allografts applied to the rodent model demonstrate higher survival rates in the subrenal capsule than the liver [23, 24].

Regardless of the transplant site, widespread clinical application of pancreatic cell transplantation is still limited. Less than 30 cases of insulin independence have been documented from over 250 attempts between 1983 and 2004 [25]. Failure to achieve insulin independence is chiefly due to the deleterious side effects of immunosuppressive treatments necessary to prevent host rejection of transplanted cells. Current immunosuppressive regimens are capable of preventing islet failure for months to years, but the agents used in these treatments are expensive and may increase the risk for specific malignancies and opportunistic infections [10].

Many challenges still exist in developing a permanent treatment for diabetes, but cell therapies have the potential to control insulin levels in vivo minimising the risk of long-term complications associated with the disease. Current research is focused on preventing unwanted immune reactions with significant progress being made in the development of bio-synthetic encapsulation systems. A successful immunoisolation implant device will provide long-term insulin independence without the need for immune system suppression.

3 Cell Types and Sources

Establishing cell therapy as a conventional treatment for T1D requires widespread clinical assessment. This level of clinical use necessitates the use of inexhaustible numbers of cells, preferably obtained from a renewable source. Allogeneic and xenogeneic cells have emerged as a promising technology for the supply of transplantable β islets; however, clinical success has been limited by the need for host immune system protection through chronic immunosuppression [26, 27]. The discovery of pluripotent stem cells, as a source of precursor cells that can theoretically be differentiated into any desired cell type, has motivated research into the development of a potentially unlimited source of insulin-producing cells [3].

3.1 Xenogeneic Cell Sources

Xenogeneic cell sources have been investigated for islet and isolated β-cell transplantation since as early as 1893, when P.W. Williams implanted sheep pancreatic fragments into a 15-year-old boy [28]. Despite a century of research and technological advancements, animal trials have shown that non-encapsulated xenografts are generally destroyed by the host within a week [26, 29]. There are no documented clinically successful cases of xenogeneic islet transplantation without chronic immunosuppression [25], but non-clinical results obtained from animal studies indicate that cell encapsulation technology has the potential to overcome current limitations in xenograft application. A variety of polymeric and inorganic matrices and membranes have been investigated in this role with the most prevalent being alginate, alginate/poly-L-lysine/alginate composites (APA) and agarose. The desired properties for an optimal encapsulation system are discussed later in this chapter.

Experiments with animal models have indicated that long-term normoglycaemia can be obtained with encapsulated xenografts. Sample data of these animal models are listed in Table 1 and demonstrate that success relative to the animal lifespan has been obtained for both rodent and primate models.

Porcine islet cells have been suggested as a virtually unlimited supply of insulin-producing cells for transplantation [34]. Transgenic pigs that express human genes represent an immunologically superior xenograft which would lack xenoantigens and possibly allow for production of individually human leucocyte antigen (HLA)-matched donor animals [34].

Despite some positive results in animal models with encapsulated porcine islets, a number of efficacy and safety issues have prevented this treatment from being applied clinically. In the vast majority of xenograft transplant studies insulin dependence is reduced, but individual variation in the magnitude of response is substantial and insulin independence is rarely achieved [25, 26, 30, 33]. More importantly a number of safety concerns prevent porcine xenografts from being approved for clinical trials.

Table 1 Maintenance of normoglycaemia using xenogeneic islet sources for transplantation

Donor	Recipient	Average islet dose (IEQs/kg islets)	Encapsulation	Normogylcaemia (max. days)
Human	Balb/C mice	1,800	Alginate	270 [30]
SD rat	Balb/C mice	1,800	Alginate	268 [30]
Teleost fish	Balb/C mice	–[a]	Alginate	100 [31]
Porcine	Primate	37,000[b]	APA	804 [32]
Porcine	Primate	10,000	Alginate	>180[a] [33]

[a]Exact value not reported
[b]Total cell number distributed across multiple transplants

A number of reports document the leaking, degradation or failure of encapsulation devices [10, 25, 33, 34]. The resulting prolonged contact between donor xenogeneic tissue and the recipient raises the risk of human infection with animal retroviruses. Prior to clinical application safety protocols must be developed for the monitoring of xenogeneic transplant recipients for specific donor-associated viruses that may establish latent or persistent infection or have oncogenic potential.

3.2 Allogeneic Cell Sources

In 1972 Ballinger and Lacy demonstrated that pancreatic islets from allogeneic sources could be used to reverse the effects of chemically induced diabetes in the rat model [35]. Attempts to reproduce these results in large animal models and rodents with autoimmune diabetes have been largely unsuccessful, but allogeneic sources still remain the most common cell therapy used to treat severe chronic diabetes [19].

All clinical islet transplants where insulin independence has been achieved have been allografts; however, treatment has not been as successful as whole organ transplantation. Islet transplants have provided a number of patients with reduced exogenous insulin dependence and a prominent decrease in the occurrence of hypoglycaemic events [36], but prevention of immune-mediated destruction of transplants remains the greatest challenge in developing a long-term solution. As with xenografts, non-encapsulated islet allografts are generally destroyed within 2 weeks. Protection of transplants has been achieved through immunosuppression regimes and/or encapsulation devices.

Encapsulated allogeneic islets have been shown to provide long-term insulin independence in animal models, as detailed in Table 2, without the use of immunosuppression. Despite limited success in achieving long-term insulin independence in large animal and primate models, clinical trials suggest that the risk of complications resulting from chronic diabetes can be reduced through allogeneic islet transplants. A clinical trial pilot study commenced in 2006 has shown that alginate-encapsulated human allografts can reduce patient dependence on exogenous insulin and minimise hypoglycaemic episodes without the need for immunosuppression [37]. This study suggests that allogeneic islet encapsulation has the potential to provide the benefits of transplant therapy while removing the need for hazardous immunosuppression regimes.

While significant progress has been made towards cell transplant therapy, supply of allogeneic islets remains a factor in inhibiting widespread clinical application. Conventionally, allografts are obtained from cadavers and in addition to their limited supply, cells must be digested from whole organs. A possible solution reported by Matsumoto et al. is the use of live tissue, where a portion of pancreas is removed from a close relative donor, minimising the possibility of immune attack [20].

Table 2 Maintenance of normoglycaemia using allogeneic islet sources for transplantation

Donor	Recipient	Average islet dose (IEQs/kg islets)	Encapsulation	Normogylcaemia (max. days)
B6AF1 mice	NOD mice	950	Alginate	385[a] [38]
DA rat	Lewis rat	5,000	Alginate	91 [39]
Lewis rat	AO rat	3,850	APA	200 [40]
Canine	Canine	7,650	Agarose	49 [41]

[a]Natural lifespan of animal model precluded further investigation

Recent studies by Matsumoto et al. have shown that the tail section of the human pancreas is suitable for islet isolation, and living donor islet transplantation may be feasible using only this section of the pancreas [20]. Using this method a 27-year-old patient was infused with islets obtained from her 56-year-old mother and achieved insulin independence 22 days post-transplantation without immune suppression [42]. Researchers predict that the transplanted islets will be functional for approximately 5 years, after which the recipient will require another transplant [42]. Although living donor allografts have potential for successful treatment over a 5-year period, subsequent donations cannot be obtained from the original donor and hence this source is still limited.

3.3 Embryonic Stem Cells

Generation of pancreatic β cells from embryonic stem cells (ESCs) represents an appealing strategy that could potentially overcome the problems of lack of donor material and graft rejection. The mammalian body consists of around 200 distinct cell types, which all derive from the fertilised egg cell. The fertilised human egg divides and gives rise to the early embryo, which at the blastula stage contains a cluster of apparently totipotent cells from which pluripotent ESCs are derived [34]. Differentiation of ESCs into insulin-producing β cells requires in vitro development of ESCs along a defined pathway highlighted in Fig. 2.

In 2001 Soria et al. derived insulin-producing cells from mouse ESCs transfected with an insulin promoter [43]. The promoter selectively induces insulin-producing cells which produced normoglycaemia in chemically induced diabetic mice [43]. However, this procedure gives rise to proliferating cells, which are potentially malignant.

The first report of human ESCs being differentiated into β cells without the use of insulin promoter was reported in 2008 [27, 44]. In two separate studies differentiated human ESC-derived β cells successfully maintained normoglycaemia in chemically induced diabetic mice [27, 44]. Although these findings raise confidence about the future of human ESC-based cell therapy, several important issues remain in developing this technology for the treatment of diabetes.

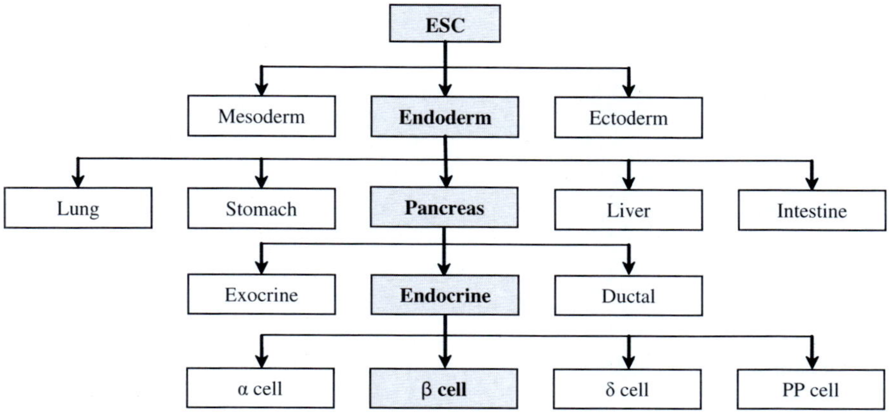

Fig. 2 Schematic representation of ESC pathway to insulin-producing β cells

Safety concerns regarding ESC technologies denote that this therapy remains experimental, and many questions must be answered prior to their incorporation into general clinical practice. In the Ricordi and Edlund study, the cumulative risk for tumours and tumourigenic cell changes was greater than 15% [27]. Kroon and colleagues stipulated the need for adequate purification of transplanted cells and development of improved maturation protocols [44]. In both studies the human ESC-derived cells were matured in vivo for several months before they presented with β cell-like maintenance of insulin synthesis and regulation. The implantation of immature pluripotent cells is not a clinical reality as it presents a serious risk for diabetic patients. While these studies confirm the potential of ESCs as a renewable source of pancreatic β cells, significant research is still required to make this source a clinical reality.

3.4 Adult Stem Cells

Recent advances in the in vitro manipulation of pancreatic stem cells suggest that this cell source is also capable of reducing patient insulin dependence and associated chronic complications. Insulin-secreting cells have been shown to develop from stem/progenitor cells isolated from a variety of tissues, such as bone marrow, liver and intestinal epithelium [45], but attempts to produce long-term insulin regulation using these cells have not been successful.

Pancreatic ductal epithelial cells isolated from pre-diabetic adult NOD mice provided promise when Ramiya et al. differentiated them into functioning islets containing α, β and δ cells [46]. Adult mouse pancreatic stem cells (PSCs) were identified and differentiated into β cells by Seaberg et al. in 2004 [47]. These PSCs when exposed to different growth factors or microenvironments give rise to islet-like cell clusters (ICCs) which temporarily express multiple endocrine hormones.

However, adult PSC-derived β cells acquired only an immature β-cell phenotype [45, 47].

PSCs are definitely a potential approach to islet cell replacement therapy, but only a few attempts have been made at the prospective isolation and differentiation of β cells [48]. Much more work is essential for full maturation of the in vitro growth of insulin-secreting cells. The identification of specific markers, gene expressions during the developmental stages of PSC-derived islets, and appropriate cocktails of growth factors or microenvironments essential for β-cell differentiation will represent a major breakthrough for the therapeutic intervention in the future. Additionally, design of encapsulation systems able to act as both barriers to components of the immune system and cell supporting matrices will be a key to producing more robust systems for bio-artificial pancreas replacement.

4 Design Requirements for Encapsulation Systems

Pancreatic islet encapsulation devices must provide permeability to factors required for physiological functionality including glucose, insulin and other metabolically active molecules imperative to β-cell survival. Concurrently, the device must protect the β cells from contacting cytotoxic cells, macrophages, antibodies and complement [25]. Apart from these immunoisolation properties, the device must allow cell survival and differentiated function of islet cells as well as maintain physical and mechanical properties for the required period.

Thus for encapsulation systems, the three fundamental design specifications that must be met to ensure the long-term safety and therapeutic efficacy of the device can be summarised as follows:

(1) the safety of the materials used, encompassing all of the physical and chemical aspects of the materials;
(2) appropriate barrier properties to allow long-term cell function;
(3) presence of appropriate matrix molecules for cell survival and function.

In addition, if encapsulated cells are to survive for an extended period of time, an adequate supply of blood is needed in the vicinity of the device for delivery of oxygen and nutrients and removal of waste products. These fundamental design requirements are similar for all tissue engineering devices in which cells need to be immunoisolated, although slight variations may be required for specific applications.

4.1 Physical and Chemical Material/Device Properties

Two main types of encapsulation devices have been defined based on the characteristics of the immunoisolation "compartment" and the implanted cell mass. Figure 3a, b shows a schematic of the key distinguishing characteristics of macro and

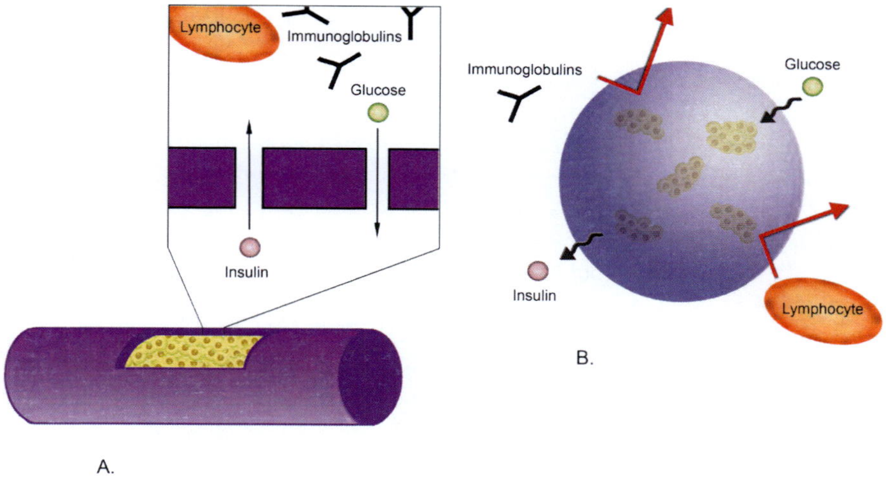

Fig. 3 Bio-synthetic encapsulation will allow transport of biofunctional molecules, while preventing contact with immune factors. The key distinguishing characteristics of (**a**) macroencapsulation and (**b**) microencapsulation are depicted

microencapsulation devices. Macroencapsulation refers to devices with semi-permeable membranes combined with large masses of cells within the device. Such systems include hollow fibre devices in which cells are maintained in one compartment with a semi-permeable membrane isolating the cells from the immune system, while theoretically allowing release of insulin. Early studies using hollow fibre systems noted that cell necrosis occurred and that this was likely due to poor diffusion within the aggregated cell mass in the fibres [49]. Tissue response to the materials used in hollow fibres also resulted in fibrosis around the fibres, further exacerbating the diffusion issue. Reducing the numbers of cells in aggregates within hollow fibres via microencapsulation proved to be effective in reducing this complication [49] and the microencapsulation approach has been a major focus of the research field in the recent past.

Microencapsulation, as represented in Fig. 3b, is generally based on the principle of encapsulating individual islets within microspheres ranging from 400 μm up to 1 mm [50, 51] in diameter. Most microencapsulation systems that have been studied are based on passive gels such as alginates which are designed to simply allow diffusion of oxygen, glucose, insulin and metabolic waste products while limiting diffusion of immunoglobulins and cytokines such as complement components.

Careful selection of materials used in fabricating encapsulation devices, as well as attention to the physical and mechanical properties of the device, is critical for assuring biocompatibility. Biocompatibility is defined as "the ability of a material to perform with an appropriate host response in a specific application" [52]. Therefore, in the case of islet encapsulation, it is critical for the application to maintain the barrier properties after implantation while not causing any toxic or abnormal responses in the encapsulated cells. Minimising fibrous encapsulation is important since fibrous capsules around the device can have detrimental effects on diffusion.

Thus materials should have low levels of leachable components and should present surface chemistry that causes minimal protein adhesion in order to minimise the inflammatory response invoked.

Natural and synthetic polymers, as well as a combination of the two have been studied for encapsulating islets. The research into natural materials has focused around the use of alginate, chitosan, cellulose and agarose, whereas a variety of synthetic polymers have been used including poly(ethylene glycol), polyacrylates and polystyrene sulphonic acid [53–55]. As mentioned earlier, alginate has been the natural polymer of choice for most microencapsulation systems. As a polysaccharide sourced from algae, the sugar composition of alternating macromers of guluronic and mannuronic acids can be somewhat variable. Some studies have suggested that alginate with a high ratio of mannuronic acid (MA) to guluronic acid (GA) is more likely to elicit an immune response and most studies tend to focus on the high guluronicate polymers [56]. With the development of alginates with better defined MA and GA contents, researchers have confirmed these findings that high GA polymers tend to be associated with higher survival of encapsulated cells [57]. This is likely due to a combination of the structure/pore size of the resulting gel and the tissue response to the material. The common practice of adding poly-L-lysine to various types of polymer capsules to improve mechanics and decrease pore size may have detrimental effects in vivo by increasing cellular adhesion and fibrosis thus limiting diffusion in and out of capsules [57].

4.2 Barrier Properties

In most cases encapsulated cells must be protected from the immune system to ensure their survival. Research into the use of stem cells for these applications has indicated that early-stage stem cells do not have a full complement of immune markers and therefore may not need to be separated from the immune system (as detailed in Sect. 3 above). However, immunoisolation of implanted cells is still considered to be a major design requirement to ensure the long-term survival of the device. Another consideration is the importance of protecting the host immune system from antigens produced by the transplanted cells. In some cases encapsulated cell implants fail as a result of antigen leakage out of the encapsulation device which causes destruction of islets through an intense pericapsular inflammatory response [58].

The vastly different molecular weights and sizes of active molecules of importance place stringent design requirements on the encapsulating material. Membranes need to allow ingress of oxygen and low molecular weight nutrients as well as permitting release of the insulin protein (~6,000 molecular weight) and waste products produced by the β cells. Preventing diffusion of immunoglobulins (IgG 150,000 molecular weight) and cytokines such as those produced by the complement system (C1q >400,000 molecular weight) is also critical for longer term survival of functional islets. It has been suggested that pore sizes of an ideal encapsulating material are of the order of 15 nm to protect cells from lower molecular weight cytokines [25]. Alginate systems are known to have a large range in their pore sizes, which is highly dependent on how the microcapsules are fabricated. The smaller pore sizes range from 5 to 20 nm [59, 60], whereas larger pores can range from the 100 nm to over 1 μm [61, 62].

To attain optimal immunoisolation, research efforts have focused on physically restricting molecular passage by controlling the pore size of the encapsulation membrane. It is a particularly difficult task to achieve optimum pore size since many cytokines, such as complement fragments and antigens shed by cells, have similar molecular size to various growth-promoting molecules. For example, complement fragments C3a and C5a have molecular weights close to small growth factors and insulin.

Permeability of encapsulation devices tends to be described in a number of ways. Semi-permeable hollow fibres typically quote a molecular weight cut-off that is determined via in vitro permeability studies of either one or two solute molecules or size exclusion chromatography [55, 63]. Solid scaffolds have true pores which are empty spaces in the structure and often these can be controlled to be within a very tight range. Hydrogels have "pores" in the angstrom size range and this is usually referred to as mesh size rather than pore size. This mesh size is the area between the polymeric chains and is often more flexible than pores as the polymer chains are essentially "dissolved" and can have small movement. While pores in solid scaffolds can be easily observed and characterised using standard optical techniques, the same is not true for measuring mesh size. Measurement of true mesh size of hydrogels is conducted in the hydrated state usually via diffusion or exclusion experiments to calculate/estimate an average mesh size.

Selection of an encapsulation material on the basis of its chemistry and physico-mechanical properties fulfils a large part of the design specifications for encapsulation of functional islet cells. This has been the basis for the large body of research conducted on passive microencapsulation systems. However, recent evidence has suggested that control of biocompatibility and pore size in isolation is not sufficient to ensure survival of implanted cells.

4.3 Biological Matrix Incorporation

All cells rely on both structural and soluble cues for their survival and for maintenance of differentiated function. Pancreatic β cells have exacting requirements in their surrounding matrix in order to maintain insulin gene expression [64, 65]. Currently, it is not clear whether these cells are capable of synthesising their own matrix, or instead rely on accessory cells such as capillary endothelial cells to produce vascular basement membrane which is thought to provide the ECM that β cells require [66]. Thus, the insoluble matrix presented to cells within a microencapsulation system may also be used to manipulate cell behaviour.

The simplest approach to enhancing biological functionality is via addition of adhesion peptides to enhance cell function [67]. While this approach is promising, use of intact extracellular matrix (ECM) molecules such as collagen IV is likely to be significantly more efficacious for cell adhesion than single peptides such as RGD [68]. Determining both the composition and presentation of the appropriate matrix molecules is necessary to generate the design parameters for optimal bio-synthetic encapsulation systems.

5 Challenges in Encapsulation of Stem Cells

Human embryonic stem cell (hESC) lines offer great promise as a renewable source of stem cells that can be differentiated into β cells or islet cell-like clusters [27, 69]. The most recent studies indicate that pancreatic endoderm derived from hESC generates glucose-responsive insulin-secreting cells in vivo [44]. Thus it may not be necessary or even desirable to fully differentiate β cells in vitro because in vitro differentiation protocols generate "polyhormonal" cells that are not able to regulate glucose.

Surprisingly few embryonic stem (ES) cell lines may be necessary to generate tissue with rudimentary histocompatibility [70]; however, little is known about the degree of disparity between donor and recipient tissue types that is acceptable for tolerance of ES cell-derived grafts [71]. The persistence of teratoma-forming cells in hESC-derived grafts will be a lingering safety concern, requiring development of highly sensitive and robust assays to estimate the frequency of teratoma-forming cells. Culture processes will need to virtually eliminate all teratoma-forming cells. Regulatory authorities may judge that even a single teratoma-forming cell within a graft ($1{:}10^9$ cells) poses a significant risk to the patient. Adult tissue mesenchymal cells have significant renewal capacity and are a safer stem cell source because they are not associated with teratoma formation. Mesenchymal-type cells, isolated postmortem from adult human islets, have been induced to undergo mesenchymal–epithelial transitions to form islet-like cell aggregates [72].

There are lessons to be learnt from early attempts to modulate the host immune system to induce graft tolerance. Experiments with animal models focused on transplantation into "immune-privileged" sites such as brain, testis or thymus. Graft survival was not related to physiological encapsulation, but local induction of immune tolerance by donor Fas ligand (FasL) expressing cells. Cytotoxic T cells reactive to allogeneic donor β cells were eliminated by co-transplantation with testis cells from the donor. Donor testis cells expressed FasL which triggered the apoptosis pathway of infiltrating cytotoxic T cells [73]. These observations led to an immune modulation strategy whereby gene therapy approaches were used to prolong islet survival by overexpression of death ligands such as FasL [74]. Despite encouraging early studies with an adenovirus that expressed FasL [75], mouse transgenic expression of FasL on β cells triggered early-onset diabetes [76].

Development of full tissue matching or autologous β-cell transplants would certainly eliminate the need for lifelong immunosuppression, but does not address the autoimmune aetiology of type I diabetes. Some sort of immunosuppressive therapy would still be required to prevent autoreactive memory T-cells triggering rejection of transplanted β cells. Thus it may only be possible to relax the requirement for immunosuppressant drugs if cells were physically encapsulated. As mentioned previously, the key challenges for development of encapsulation systems in the future are in identifying the appropriate composition and presentation of matrix for survival and function of β cells and in fabricating devices which incorporate this biological information while maintaining the desired barrier properties. Although there have been few reports on pancreatic matrix, most research suggests that various types of collagen, fibronectin and laminin are the most prevalent proteins [77]. In terms of cell adhesion and function, critical matrix components appear to be from the laminin family, with cell–matrix signalling being via the β-cell β1 integrin surface receptors [78]. However, there remains some controversy over the relative roles of laminin and collagen IV in promoting insulin gene expression and insulin production in various β-cell populations [65]. Although glycosaminoglycans (GAGs) are being increasingly recognised as key to cell–matrix signalling and essential factors in stimulating appropriate cell functions [79], their role in normal physiology of the pancreas has not been extensively studied and warrants further attention.

Fabrication challenges include development of methods for immobilising the biological matrix in such a way that it remains functional while also maintaining the barrier properties for both protection of the islets and release of insulin in response to glucose. This relies on developing approaches to controlling the mesh size of encapsulating hydrogels which can be applied to both the synthetic base material and the biological ligands incorporated.

6 Conclusions

There is clearly an unmet clinical need for more effective approaches to treating insulin-dependent diabetes. Recent advances in understanding stem cell biology have opened up this field with the promise of using adult stem cells or embryonic stem cells to replace the critical functions of the pancreas. Despite significant advances in developing alternate islet cell transplantation sources as well as safer immunosuppressive drugs, some form of physical encapsulation may still be required in the future to prevent alloimmune and autoimmune graft rejection. Immunoisolation technologies require development of biocompatible membrane materials which can be manufactured into appropriate geometries for the reciprocal transfer of nutrients and insulin. Future studies need to focus on understanding the microenvironment required for long-term survival of functional β-cell masses and to develop a technology that recapitulates this microenvironment within encapsulated devices.

References

1. Thanos CG, Emerich DF. On the use of hydrogels in cell encapsulation and tissue engineering systems. Recent Pat Drug Deliv Formul. 2008; 2(1):19–24.

2. Rother KI. Diabetes treatment – bridging the divide. N Engl J Med. 2007; 356(15):1499–1501.
3. Kabelitz D, et al. Toward cell-based therapy of type I diabetes. Trends Immunol. 2008; 29(2):68–74.
4. Alberti KGMM, Zimmet PZ. Definition, diagnosis and classification of diabetes mellitus and its complications. Geneva: Department of Noncommunicable Disease Surveillance, World Health Organisation; 1999.
5. Anonymous. National diabetes statistics, 2007. Bethesda: U.S. Department of Health and Human Services. National Institute of Health; 2007, p. 23.
6. American Diabetes Association. Economic costs of diabetes in the U.S. in 2007. Diabetes Care 2008; 31(3):596–615.
7. Pringle M, Houghton P. National service framework for diabetes: standards. London: Department of Health; 2001, p. 48.
8. Pavlakis M, Khwaja K. Pancreas and islet cell transplantation in diabetes. Curr Opin Endocrinol Diabetes Obes. 2007; 14(2):146–150.
9. Pavlakis M, Khwaja K. Transplantation for type 1 diabetes: whole organ pancreas and islet cells. Curr Diabetes Rep. 2007; 6(6):473–478.
10. Naftanel MA. Pancreatic islet transplantation. PLoS Med. 2004; 1(3):198–201.
11. Yoon J-W, Jun H-S. Recent advances in insulin gene therapy for type 1 diabetes. Trends Mol Med. 2002; 8(2):62–68.
12. Laub O, Rutter WJ. Expression of the human insulin gene and cDNA in a heterologous mammalian system. J Biol Chem. 1983; 258:6043–6050.
13. Iwata H, et al. Preparation of insulin-releasing Chinese Hamster ovary cell transfection of human insulin gene: its implantation into diabetic mice. In: Shalaby SW, et al., ed. Polymers of biological and biomedical significance. Washington: American Chemical Society; 1994.
14. Falqui L, et al. Production of mature human insulin by retrovirally engineered fibroblasts and primary rat hepatocytes. Exp Clin Endocrinol Diabetes 1996; 104:A37.
15. Bailey CJ, et al. Prospects for insulin delivery by ex-vivo somatic cell gene therapy. J Mol Med. 1999; 77:244–249.
16. Kofman AV, et al. Gene- and cell-based therapy of diabetes: between bench and bedside. Tsitologiia 2002; 44(12):1157–1177.
17. Henquin J. Pancreatic b-cell mass or b-cell function? That is the question! Diabetes Obes Metab. 2008; 10(suppl 4):1–4.
18. Shapiro AM, et al. International trial of the Edmonton protocol for islet transplantation. N Engl J Med. 2006; 355(13):1318–1330.
19. Ricordi C, Strom TB. Clinical islet transplantation: advances and immunological challenges. Nat Rev Immunol. 2004; 4:259–268.
20. Matsumoto S, et al. Efficacy of human islet isolation from the tail section of the pancreas for the possibility of living donor islet transplantation. Transplantation 2004; 78(6):839–843.
21. Ricordi C, et al. Detection of intrahepatic human islets following combined liver-islet allotransplantation. Pancreas 1992; 7(4):507–509.
22. Levy MM, et al. Intrathymic islet transplantation in the canine: I. Histological and functional evidence of autologous intrathymic islet engraftment and survival in pancreatectomized recipients. Transplantation 2002; 73(6):842–852.
23. Jaeger C, et al. Effect of transplantation site and culture pretreatment on islet xenograft survival (rat to mouse) in experimental diabetes without immunosuppression of the host. Acta Diabetol. 1994; 31(4):193–197.
24. Jaeger C, et al. Pancreatic islet xenografts at two different transplantation sites (renal subcapsular versus intraportal): comparison of graft survival and morphology. Exp Clin Endocrinol Diabetes 1995; 103(2):123–128.
25. Desai TA, et al. Nanoporous microsystems for islet cell replacement. Adv Drug Deliv Rev. 2004; 56(11):1661–1673.
26. Wilson JT, Chaikof EL. Challenges and emerging technologies in the immunoisolation of cells and tissues. Adv Drug Deliv Rev. 2008; 60(2):124–145.

27. Ricordi C, Edlund H. Toward a renewable source of pancreatic beta-cells. Nat Biotechnol. 2008; 26(4):397–398.
28. Williams PW. Notes on diabetes treated with extract and grafts of sheep's pancreas. BMJ 1894; 2:1303–1304.
29. Kobayashi T, et al. Indefinite islet protection from autoimmune destruction in nonobese diabetic mice by agarose microencapsulation without immunosuppression. Transplantation 2003; 75(5):619–625.
30. Schneider S, et al. Long-term graft function of adult rat and human islets encapsulated in novel alginate-based microcapsules after transplantation in immunocompetent diabetic mice. Diabetes 2005; 54(3):687–693.
31. Yang H, et al. Long-term function of fish islet xenografts in mice by alginate encapsulation. Transplantation 1997; 64(1):28–32.
32. Sun Y, et al. Normalization of diabetes in spontaneously diabetic cynomologus monkeys by xenografts of microencapsulated porcine islets without immunosuppression. J Clin Invest. 1996; 98(6):1417–1422.
33. Elliott RB, et al. Intraperitoneal alginate-encapsulated neonatal porcine islets in a placebo-controlled study with 16 diabetic cynomolgus primates. Transplant Proc. 2005; 37(8): 3505–3508.
34. Serup P, et al. Science, medicine, and the future: islet and stem cell transplantation for treating diabetes. BMJ 2001; 322(7277):29–32.
35. Ballinger WF, Lacy PE. Transplantation of intact pancreatic islets in rats. Surgery 1972; 72:175–186.
36. Ryan EA, et al. Five-year follow-up after clinical islet transplantation. Diabetes 2005; 54(7):2060–2070.
37. Calafiore R, et al. Microencapsulated pancreatic islet allografts into nonimmunosuppressed patients with type 1 diabetes: first two cases. Diabetes Care 2006; 29(1):137–138.
38. Duvivier-Kali VF, et al. Complete protection of islets against allorejection and autoimmunity by a simple barium-alginate membrane. Diabetes 2001; 50(8):1698–1705.
39. Omer A, et al. Long-term normoglycemia in rats receiving transplants with encapsulated islets. Transplantation 2005; 79(1):52–58.
40. de Vos P, et al. Long-term biocompatibility, chemistry, and function of microencapsulated pancreatic islets. Biomaterials 2003; 24(2):305–312.
41. Tashiro H, et al. Characterization and transplantation of agarose encapsulated canine islets of Langerhans. Ann Transplant 1997; 2:33–39.
42. Matsumoto S, et al. Insulin independence after living-donor distal pancreatectomy and islet allotransplantation. Lancet 2005; 365(9471):1642–1644.
43. Soria B, et al. Insulin-secreting cells derived from embryonic stem cells normalize glycemia in streptozotocin-induced diabetic mice. Diabetes 2001; 49:157–162.
44. Kroon E, et al. Pancreatic endoderm derived from human embryonic stem cells generates glucose-responsive insulin-secreting cells in vivo. Nat Biotechnol. 2008; 26(4): 443–452.
45. Suen PM, Leung PS. Pancreatic stem cells: a glimmer of hope for diabetes? J Pancreas. 2005; 6(5):422–424.
46. Ramiya VK, et al. Reversal of insulin-dependent diabetes using islets generated in vitro from pancreatic stem cells. Nat Med. 2000; 6(3):278–282.
47. Seaberg RM, et al. Clonal identification of multipotent precursors from adult mouse pancreas that generate neural and pancreatic lineages. Nat Biotechnol. 2004; 22:1115–1124.
48. Hori Y, et al. Enrichment of putative pancreatic progenitor cells from mice by sorting for prominin1 (CD133) and PDGFR{beta}. Stem Cells 2008; 26(11)2912–2920.
49. Lacy PE, et al. Maintenance of normoglycemia in diabetic mice by subcutaneous xenografts of encapsulated islets. Science 1991; 254(5039):1782–1784.
50. Maria-Engler SS, et al. Microencapsulation and tissue engineering as an alternative treatment of diabetes. Braz J Med Biol Res. 2001; 34(6):691–697.

51. Kizilel S, et al. The bioartificial pancreas: progress and challenges. Diabetes Technol Ther. 2005; 7(6):968–985.
52. Williams DF. The Williams dictionary of biomaterials. Liverpool University Press, 1999, ISBN 0-85323-921-5.
53. Silva AI, et al. An overview on the development of a bio-artificial pancreas as a treatment of insulin-dependent diabetes mellitus. Med Res Rev. 2006; 26(2):181–222.
54. Yang KC, et al. Chitosan/gelatin hydrogel as immunoisolative matrix for injectable bioartificial pancreas. Xenotransplantation 2008; 15(6):407–416.
55. Lacik I. Polymer chemistry in diabetes treatment by encapsulated islets of Langerhans: review to 2006. Aust J Chem. 2006; 59(8):508–524.
56. Kulseng B, et al. Transplantation of alginate microcapsules: generation of antibodies against alginates and encapsulated porcine islet-like cell clusters. Transplantation 1999; 67(7): 978–984.
57. King A, et al. The effect of capsule composition in the reversal of hyperglycemi in diabetic mice transplanted with microencapsulated allogeneic islets. Diabetes Technol Ther. 2003; 5:653–663.
58. Xu B-Y, et al. Rapid destruction of encapsulated islet xenografts by NOD Mice is CD4-dependent and facilitated by b-cells: innate immunity and autoimmunity do not play significant roles. Transplantation 2005; 80(3):402–409.
59. Dembczynski R, Jankowski T. Determination of pore diameter and molecular weight cut-off of hydrogel-membrane liquid-core capsules for immunoisolation. J Biomater Sci Polym Ed. 2001; 12(9):1051–1058.
60. Stewart WW, Swaisgood HE. Characterization of calcium alginate pore diameter by size-exclusion chromatography using protein standards. Enzyme Microb Technol. 1993; 15(11):922–927.
61. Partap S, et al. Preparation and characterisation of controlled porosity alginate hydrogels made via a simultaneous micelle templating and internal gelation process. J Mater Sci. 2007; 42(10):3502–3507.
62. Chan AW, et al. Kinetic controlled synthesis of pH-responsive network alginate. Biomacromolecules 2008; 9(9):2536–2545.
63. Brissová M, et al. Evaluation of microcapsule permeability via inverse size exclusion chromatography. Anal Biochem. 1996; 242(1):104–111.
64. Bosco D, et al. Importance of cell-matrix interactions in rat islet beta-cell secretion in vitro: role of alpha6beta1 integrin. Diabetes 2000; 49(2):233–243.
65. Kaido T, et al. Impact of defined matrix interactions on insulin production by cultured human {beta}-cells: effect on insulin content, secretion, and gene transcription. Diabetes 2006; 55(10):2723–2729.
66. Nikolova G, et al. The vascular basement membrane: a niche for insulin gene expression and [beta] cell proliferation. Dev Cell 2006; 10(3):397–405.
67. Weber LM, et al. The effects of cell-matrix interactions on encapsulated [beta]-cell function within hydrogels functionalized with matrix-derived adhesive peptides. Biomaterials 2007; 28(19):3004–3011.
68. Beck J, et al. Islet encapsulation: strategies to enhance islet cell functions. Tissue Eng. 2007; 13(3):589–599.
69. D'Amour KA, et al. Efficient differentiation of human embryonic stem cells to definitive endoderm. Nat Biotechnol. 2005; 23(12):1534–1541.
70. Taylor CJ, et al. Banking on human embryonic stem cells: estimating the number of donor cell lines needed for HLA matching. Lancet 2005; 366(9502):2019–2025.
71. Robertson NJ, et al. Embryonic stem cell-derived tissues are immunogenic but their inherent immune privilege promotes the induction of tolerance. Proc Natl Acad Sci USA. 2007; 104(52):20920–20925.
72. Gershengorn MC, et al. Epithelial-to-mesenchymal transition generates proliferative human islet precursor cells. Science 2004; 306(5705):2261–2264.

73. Takeda Y, et al. Protection of islet allografts transplanted together with Fas ligand expressing testicular allografts. Diabetologia 1998; 41(3):315–321.
74. Sanlioglu AD, et al. Molecular mechanisms of death ligand-mediated immune modulation: a gene therapy model to prolong islet survival in type 1 diabetes. J Cell Biochem. 2008; 104(3):710–720.
75. Jin Y, et al. Simultaneous stimulation of Fas-mediated apoptosis and blockade of costimulation prevent autoimmune diabetes in mice induced by multiple low-dose streptozotocin. Gene Ther. 2004; 11(12):982–991.
76. Petrovsky N, et al. The role of Fas ligand in beta cell destruction in autoimmune diabetes of NOD mice. Ann NY Acad Sci. 2002; 958:204–208.
77. Meyer T, et al. Extracellular matrix proteins in the porcine pancreas: a structural analysis for directed pancreatic islet isolation. Transplant Proc. 1998; 30(2):354–354.
78. Parnaud G, et al. Blockade of β1 integrin-laminin-5 interaction affects spreading and insulin secretion of rat β-cells attached on extracellular matrix. Diabetes 2006; 55(5):1413–1420.
79. Knox S, et al. Not all perlecans are created equal: interactions with fibroblast growth factor (FGF) 2 and FGF receptors. J Biol Chem. 2002; 277(17):14657–14665.

Stem Cell Engineering for Regeneration of Bone Tissue

Michael Gelinsky, Anja Lode, Anne Bernhardt, and Angela Rösen-Wolff

Abstract Due to their great potential, mesenchymal stem cells (MSC) are of special interest for regenerative therapies. One of the most investigated areas in this field is the regeneration of bone tissue using tissue engineering strategies. This chapter focuses on osteogenic differentiation and cultivation of human MSC in vitro and their utilisation for tissue engineering of bone tissue, discussing a number of protocols to induce and influence the osteogenic differentiation of MSC into the osteoblastic lineage in vitro.

According to the tissue engineering approach, 3D porous biomaterials (scaffolds) are seeded with autologous cells, followed by a period of in vitro cultivation which can be carried out under differentiation-inducing conditions and finally by implantation of the cell-matrix constructs into the defect. Here we describe as an example the in vitro cultivation and osteogenic differentiation of human MSC in porous scaffolds consisting of biomimetically mineralised collagen. This artificial extracellular bone matrix that resemble the nanocomposite made of fibrillar collagen type I and hydroxyapatite crystals in natural bone tissue was furthermore used to establish an in vitro model of the bone remodelling process: osteoblasts derived from human MSC and osteoclast-like cells which were also differentiated from their natural precursors were co-cultured on membranes of mineralised collagen to learn more about the cellular cross talk between both cell types.

The concept of in situ tissue engineering, a novel strategy to accelerate bone defect healing, aims at the utilization of natural chemotactic mechanisms to colonise scaffolds with MSC in vivo. Exemplarily, the chemoattraction of human MSC into porous 3D scaffolds of mineralised collagen using the stromal-derived factor 1α (SDF-1α) is demonstrated.

M. Gelinsky (✉)
The Max Bergmann Center of Biomaterials, Institute for Materials Science, Technische Universität Dresden, Dresden, Germany; DFG Research Center and Cluster of Excellence for Regenerative Therapies Dresden (CRTD),Technische Universität Dresden, Dresden, Germany
e-mail: michael.gelinsky@tu-dresden.de

G.M. Artmann et al. (eds.), *Stem Cell Engineering*,

DOI 10.1007/978-3-642-11865-4_17,

Keywords (Human) mesenchymal stem cells, (h)MSC · Osteogenic differentiation · Tissue engineering · In situ tissue engineering · Biomimetic collagen/hydroxyapatite composites (biomimetic mineralised collagen scaffolds) · Osteoclasts (osteoclast-like cells) · Co-culture · Bone remodelling · Chemoattraction · Stromal-derived factor 1α (SDF-1α)

1 Introduction

Isolation, characterisation and utilisation of mesenchymal stem cells (MSC) for various applications have been and are still extensively investigated over the last few decades ([43, 40] – and literature cited there). MSC have been first identified in bone marrow [25]. This rare cell population with stem cell characteristics, referred to as bone marrow-derived MSC or marrow stromal cells, has been widely characterised in a variety of species including human – but nevertheless distinct markers for the 'stemness' of these cells have not yet been identified [38]. Bone marrow-derived MSC are able to differentiate along the osteoblastic, chondrocytic, adipocytic and various other cell lineages [51, 15, 55, 2]. MSC can be easily enriched from bone marrow aspirates and expanded in vitro, allowing the therapeutic application of autologous cells that avoid complications caused by immunological rejection.

In the last years, MSC have also been purified from other sources like adipose tissue [72], umbilical cord blood [7, 8], placenta [34] and periosteum [50]. These MSC populations possess a proliferation capacity and multipotent differentiation potential which is comparable to that of bone marrow-derived MSC. Especially the stem cells isolated from adipose tissue are in the focus of many current investigations with the aim of their application for cell-based therapies like bone tissue regeneration. In comparison to bone marrow-derived MSC, a higher number of stem cells can be easily harvested from patients by a simple, less invasive method (liposuction). Autologous MSC isolated from adipose tissue have been successfully applied for the healing of widespread calvarial defects of a 7-year-old girl demonstrating their great potential in cases in which the amount of autologous MSC from bone marrow might be limited [44]. Regardless of the successful application of MSC from other sources, bone marrow is still the mostly used source in the field of hard tissue reconstruction [30, 21].

MSC play a pivotal role in bone regeneration and repair in vivo [61, 41, 22]. They form a reservoir of progenitor cells which can differentiate into cells of the osteoblastic lineage. In addition, MSC are involved in the regulation of the balance between the opposing actions of bone-forming (osteoblast) and bone-resorbing (osteoclast) cells [9], and they express bone morphogenetic proteins (BMPs) during the first phases of fracture repair which are known to play a highly significant role in the control of bone healing [54].

These functions of MSC in vivo and the difficulty in isolating vital autologous osteoblasts from bone tissue led to a growing interest in mesenchymal stem cells for regenerative therapies of bone defects and skeletal tissue engineering.

Non-expanded MSC – being part of the bone marrow – have already been used by surgeons since decades when transplanting autologous spongiosa or bone graft, e.g. from the iliac crest [64], for the therapy of critical size bone defects or problematic cases. Due to the disadvantages of the harvesting procedure of autologous bone (donor site morbidity, limited amount) compared to bone marrow aspiration, followed by MSC isolation, the latter attracts more and more attention. In addition, MSC are capable of self-replication which allows their extensive in vitro expansion [13] and therefore a remarkable cell number for therapeutic applications can be easily provided. Nevertheless, in most countries the use of even autologous MSC after in vitro expansion is complicated by unclear regulatory issues and therefore comprehensive clinical studies are still missing [48]. However, the great potential of MSC for regeneration of skeletal tissues has been demonstrated in numerous in vitro and animal studies and also in some first clinical trials [1, 16, 17, 29].

For the application of MSC in the field of hard tissue reconstruction, suitable carrier materials are necessary, mostly referred to as scaffolds. In principle, two therapeutic approaches are possible: when a biomaterial is combined with bone marrow or isolated MSC (expanded or non-expanded) directly before surgery, this procedure can be defined as in situ tissue engineering, because a differentiated tissue is only formed in vivo after implantation (the term 'in vivo tissue engineering' should not be used because it is contradictory). The second approach involves the combination of scaffolds with autologous MSC a few weeks before implantation. In the subsequent cultivation period, osteogenic differentiation can occur which results in the formation of new bone matrix in vitro. This approach is called just 'tissue engineering' in the original meaning of the word.

The present chapter focuses on in vitro differentiation and cultivation of human bone marrow-derived MSC and their utilisation for tissue engineering of bone tissue.

2 Osteogenic Differentiation of Human Mesenchymal Stem Cells

The utilisation of MSC for bone tissue engineering requires their differentiation into bone-forming osteoblasts. This differentiation is accompanied by an increase of expression and activity of the enzyme alkaline phosphatase (ALP, liver/bone/kidney isozyme), an enhanced expression and secretion of proteins of the extracellular matrix, e.g. collagen type I, osteocalcin and bone sialoprotein II (BSPII), and the mineralisation of the extracellular matrix to form calcified bone tissue. Hence, measurements of ALP activity, expression analyses of osteoblastic markers like ALP, osteocalcin and BSPII as well as qualitative or quantitative calcium assays determining mineral deposition are established methods which are used to investigate osteogenesis in vitro.

In vitro, differentiation of MSC can be directed towards the osteoblastic lineage by addition of soluble factors including chemical factors, growth factors and hormones to the cell culture medium. Table 1 gives an overview about compositions of such additives which are capable to induce differentiation of human MSC (hMSC)

Table 1 Soluble factors which are able to induce the differentiation of hMSC into osteoblasts in vitro

Osteogenic factors	Effect	References
Dexamethasone + β-glycerophosphate + L-ascorbic acid 2-phosphate or L-ascorbic acid	Increase of ALP activity, mineral deposition, no osteocalcin expression	Jaiswal et al. [37]
Parathyroid hormone (PTH) + 1,25-dihydroxyvitamin D_3	Increase of ALP activity, mineral deposition, osteocalcin expression	Sammons et al. [60]
Bone morphogenetic protein-6 (BMP-6) + β-glycerophosphate + L-ascorbic acid	Increase of ALP activity, mineral deposition, expression of osteoblast-related genes including osteocalcin and BSPII	Friedman et al. [26]
Hepatocyte growth factor (HGF) + 1,25-dihydroxyvitamin D_3 + β-glycerophosphate + L-ascorbic acid 2-phosphate	Increase of ALP activity, mineral deposition	D'Ippolito et al. [19]

into osteoblasts. Beyond it, there are a number of factors which are not sufficient to induce but able to promote osteogenesis of hMSC – those are summarised in Table 2.

Most commonly used to induce osteogenic differentiation of hMSC is a combination of the synthetic glycocorticoid dexamethasone, β-glycerophosphate and L-ascorbic acid 2-phosphate. Treatment of hMSC with this combination results in an increase of ALP activity as well as in mineral deposition; however, the expression of the bone matrix protein osteocalcin is repressed by dexamethasone [37].

Numerous investigations address the regulation of osteogenic differentiation, and insights in this multi-step event provide tools which can be used to direct and control osteogenic differentiation in vitro. Thus, many factors which are known to be involved in the complex regulation process in vivo are able to induce or promote osteogenic differentiation of hMSC also in vitro. One of these factors is the steroid hormone 1,25-dihydroxyvitamin D_3 (calcitriol), the active form of vitamin D. 1,25-Dihydroxyvitamin D_3 alone is not sufficient to induce, but support, osteogenic differentiation induced by dexamethasone/β-glycerophosphate/ascorbate (Table 2). The addition of 1,25-dihydroxyvitamin D_3 to these osteogenic supplements can compensate for the dexamethasone-mediated repression of osteocalcin synthesis and leads to a stronger increase of ALP activity [3, 37]. Moreover, 1,25-dihydroxyvitamin D_3 acts synergistically with parathyroid hormone (PTH) and hepatocyte growth factor (HGF) resulting in osteogenic induction of hMSC (Table 1).

Table 2 Soluble factors which are alone not sufficient to induce, but promote the differentiation of hMSC into osteoblasts in vitro

Osteogenic promoting factors	Effect	References
1,25-Dihydroxyvitamin D_3 + β-glycerophosphate + L-ascorbic acid 2-phosphate or L-ascorbic acid	Increase of ALP activity, no mineral deposition, osteocalcin but no BSPII expression	Beresford et al. [3] Liu et al. [45] D'Ippolito et al. [19]
1,25-Dihydroxyvitamin D_3 in combination with dexamethasone/β-glycerophosphate/L-ascorbic acid 2-phosphate or L-ascorbic acid	Stronger increase of ALP activity, osteocalcin expression	Beresford et al. [3] Jaiswal et al. [37]
Bone morphogenetic protein-7 (BMP-7) + β-glycerophosphate + L-ascorbic acid	Increase of ALP activity, no mineral deposition, expression of osteoblast-related genes including osteocalcin and BSPII	Friedman et al. [26]
BMP-7 in combination with dexamethasone/β-glycerophosphate/L-ascorbic acid 2-phosphate	Stronger increase of ALP activity	Tsiridis et al. [67]
Bone morphogenetic protein-2 (BMP-2) in combination with dexamethasone/β-glycerophosphate/L-ascorbic acid 2-phosphate	In some cases, stronger increase of ALP activity	Diefenderfer et al. [18] Schwartz et al. [63]
Transforming growth factor beta (TGF-β) + 1,25-dihydroxyvitamin D_3	Synergistically enhanced ALP activity	Liu et al. [45]
Basic fibroblast growth factor (bFGF) in combination with dexamethasone	Enhanced mineralisation	Pri-Chen et al. [56]
17-β estradiol in combination with dexamethasone/β-glycerophosphate/L-ascorbic acid 2-phosphate	Stronger increase of ALP activity, enhanced mineral deposition, osteocalcin expression	Hong et al. [32, 33]

Bone morphogenetic proteins (BMPs), the structurally related transforming growth factor beta (TGF-β) and all members of its superfamily [57], are also among the factors which are involved in the regulation of osteogenic differentiation. However, in contrast to the efficacy of BMPs in animal models, these growth factors are relatively ineffective inducers of osteogenesis in humans [18]. Among the 14 types of human BMP isozymes, only BMP-6 is sufficient to induce osteogenic differentiation of hMSC in vitro (Table 1). For BMP-2 and BMP-7, a promoting effect on hMSC osteogenesis has been demonstrated (Table 2), whereas BMP-4 has no significant influence on osteogenic differentiation of cultured hMSC [60]. TGF-β, that causes a chondroinduction and osteoinduction at early stages of differentiation in mice [23], enhances in combination with 1,25-dihydroxyvitamin D_3 synergistically the activity of ALP (Table 2). Treatment of hMSC with TGF-β alone results in an inhibition of ALP activity [45]. At late stages of the osteoblastic differentiation pathway, TGF-β acts as an inhibitor [58].

Basic fibroblast growth factor (bFGF) is a potent mitogen that exhibits stimulatory effects on bone tissue regeneration. Exogenous bFGF affects mainly the proliferation of hMSC [11, 49], but the expression of the osteogenic phenotype of dexamethasone-treated hMSC is also enhanced (Table 2). The estrogen 17-β estradiol is known to modulate multiple differentiations in bone, cartilage and adipose tissue and to enhance osteogenic differentiation of hMSC induced by dexamethasone/β-glycerophosphate/ascorbate (Table 2).

Besides soluble factors, there are further culture conditions which have a positive influence on osteogenesis of MSC. Thus, the cultivation of hMSC on substrates consisting of components of the extracellular matrix like collagen type I and vitronectin promotes their osteogenic differentiation [42]. In own work, we could demonstrate that hMSC showed a fourfold higher specific ALP activity when cultivated on membranes of mineralised collagen type I compared to those growing on standard polystyrene [4].

Another strategy to direct osteogenesis of MSC is the application of mechanical stimuli in the form of cyclic compression, stretch or fluid shear flow during cultivation in vitro (for review [31]). The effects of mechanical stimuli on differentiation of mesenchymal cells were already described in detail by Pauwels [53]. Jagodzinski and coworkers demonstrated by expression analyses of osteogenic markers that cyclic stretching of hMSC cultured on an elastic silicone dish is even a stronger differentiation factor than dexamethasone [35]. The perfusion culture is of high importance especially for the cultivation of MSC on 3D scaffolds. The better mass transport in the whole scaffold is reflected by higher cell numbers resulting from improved viability and proliferation as well as a homogenous distribution of the cells in the scaffold compared to a preferred colonisation of peripheral parts observed in static cultures [10]. Beyond it, a positive impact of the fluid flow in perfusion culture on osteogenic differentiation has been observed; however, the effect of cyclic compression of the cell-seeded scaffolds as osteogenic stimulus seems to be stronger [36].

3 Response of Osteogenically Induced hMSC to Biomimetic Mineralised Collagen Scaffolds

In the field of bone engineering, scaffold materials with properties close to those of natural bone, mainly consisting of collagen type I and the calcium phosphate phase hydroxyapatite, are highly advantageous [68]. Biomimetic collagen/hydroxyapatite composites, which are produced under nearly physiological conditions and therefore contain reconstituted collagen fibrils and hydroxyapatite as nanoscopical and for that reason resorbable mineral phase, are promising candidates for application in the field of bone regeneration [28]. Such materials can be seen as artificial extracellular matrices because they resemble the highly organised nanocomposite made of fibrillar collagen and hydroxyapatite crystals and therefore can be easily involved in the remodelling process (see Sect. 4). Based on a method developed by Bradt and coworkers in our laboratory [12] leading to such an artificial composite material ("mineralised collagen"), we have produced membrane-like 2D [14, 4] as well as porous 3D scaffolds [27].

The claim of the tissue engineering approach is to pre-seed porous 3D scaffolds with autologous cells, followed by a period of in vitro cultivation and finally by implantation of the cell–matrix constructs into the defect to induce formation of new bone tissue locally. To evaluate the suitability of porous scaffolds from mineralised collagen type I for bone engineering approaches, hMSC were seeded and cultivated on these scaffolds in vitro [5, 47]. The porous scaffolds of mineralised collagen can be seeded rather homogenously and with a high efficiency. Homogenous colonisation of 3D scaffolds with a minimal loss of cells is of critical importance for tissue engineering applications but can be – depending on the pore structure and geometry of the scaffolds – problematic. The interconnecting pores of the scaffolds from mineralised collagen with a mean pore diameter of 180 μm [27] facilitate an easy penetration of the cells which was demonstrated by staining of vital cells inside the scaffold (Fig. 1). The successful seeding of the inner parts of the mineralised collagen scaffolds could also be demonstrated by confocal laser scanning microscopy (cLSM, Fig. 2a) or scanning electron microscopy (SEM, Fig. 2b) of scaffold sections. Evaluation of seeding efficiency revealed that of 94% of all samples tested ($n = 60$), more than 90% of the cells became adherent on or in the porous matrices [46].

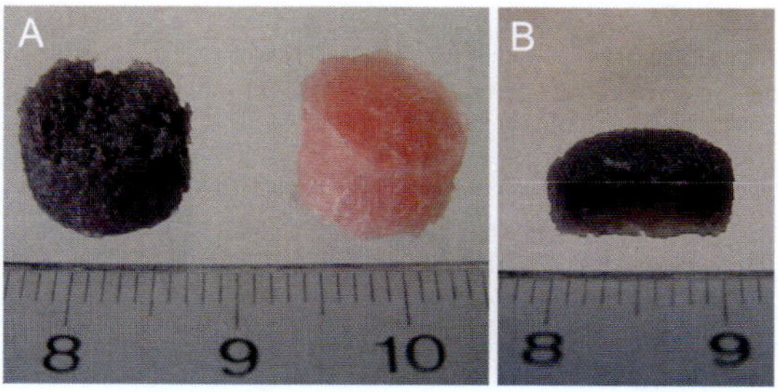

Fig. 1 MTT staining of cell-seeded porous 3D scaffolds from mineralised collagen after 14 days of culture. The intracellular conversion of the tetrazolium salt MTT to a *dark blue* insoluble formazan dye by living cells was used to visualise the distribution of hMSC within the scaffold. (**a**) Scaffold seeded with 2×10^5 hMSC after 14 days of cultivation compared to control scaffold without cells. (**b**) Longitudinal section of a cell-seeded, MTT-stained scaffold after 14 days of cultivation

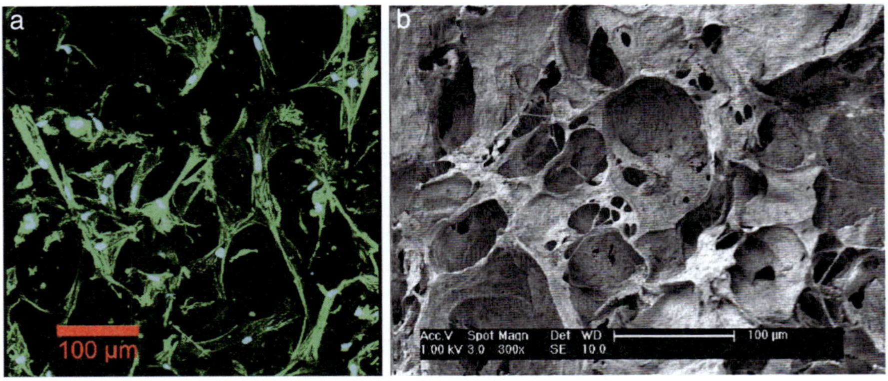

Fig. 2 Micrographs, showing cross sections of porous mineralised collagen 3D scaffolds seeded with hMSC and cultivated under osteogenic stimulation. (**a**) Confocal LSM image, taken after 18 days of culture. Actin cytoskeleton shows *green* and cell nuclei blue fluorescence (sample prepared by Dr. M. Kruse and image taken by Dr. Th. Hanke, TU Dresden). (**b**) SEM micrograph showing state after a cultivation period of 28 days. Both scale bars represent 100 μm

During cultivation of hMSC on porous scaffolds from mineralised collagen for 4 weeks, a 2.5–5 fold raise of DNA content was observed indicating that the material is suitable for in vitro expansion of cells. Expression of the osteogenic markers ALP and BSPII was found to be up-regulated for hMSC, which were induced to differentiate into the osteoblastic lineage by dexamethasone, β-glycerophosphate and ascorbic acid 2-phosphate, whereas non-induced hMSC on mineralised collagen scaffolds showed no increase in osteogenic marker expression [5]. Interestingly, a substantial influence of the seeding density was observed on proliferation and osteogenic differentiation of hMSC in mineralised collagen scaffolds. Cylindrical scaffolds with a diameter of 10 mm and a height of 4 mm were seeded with 5×10^4, 2×10^5 and 1×10^6 cells, respectively, and cultivated for up to 6 weeks in the presence of dexamethasone, β-glycerophosphate and ascorbic acid 2-phosphate. The experimental conditions have been described in detail elsewhere [47]. Highest proliferation rate and specific ALP activity of osteogenically induced hMSC were found for the cell–matrix constructs seeded with the lowest density (Fig. 3a, b).

The substantial influence of the seeding density on cellular functions observed in these experiments indicates the relevance of the seeding cell number for the generation of cell–matrix constructs in vitro. The usage of optimised seeding densities is not only advantageous for cell proliferation and differentiation but might also – as observed for the porous scaffolds from mineralised collagen – reduce the number of cells required for seeding and that in turn minimise the in vitro cell expansion steps. This is important for tissue engineering applications since extensive expansion of hMSC in vitro is not only expensive and time consuming but also lead to a progressive decrease of their proliferation capacity, loss of their multilineage differentiation potential and finally to senescence (for review [17]).

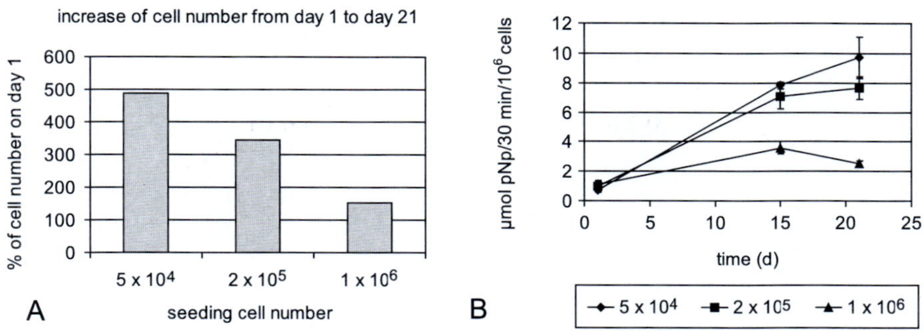

Fig. 3 (**a**) Influence of seeding density on proliferation and osteogenic differentiation of osteogenically induced hMSC cultured on 3D scaffolds from mineralised collagen over 21 days. To study proliferation, cell numbers were determined at different time points by measurement of the DNA content. (**b**) For investigation of osteogenic differentiation, ALP activity of osteogenically induced hMSC was determined and related to the cell number

4 Cocultivation of hMSC and Osteoclast-Like Cells on Synthetic Extracellular Matrices as In Vitro Model for Bone Remodelling

Bone remodelling requires the coordinated regulation of osteoblast and osteoclast activity. After detailed investigations of growth and osteogenic differentiation of hMSC on mineralised collagen scaffolds (see Sect. 3), additional work focused on osteoclast formation from progenitor cells and resorption of artificial mineralised collagen matrices in vitro. To simplify the experimental set-up, membranes of mineralised collagen were applied instead of porous 3D matrices. Osteoclast-like cells were differentiated from primary human monocytes (isolated from peripheral blood) under stimulation with macrophage colony-stimulating factor (MCSF) and receptor activator of NF-κ B ligand (RANKL). It was demonstrated that multinucleated and biologically active human osteoclast-like cells could be achieved and cultured on the membrane-like scaffolds [20]. A first model of bone remodelling on matrices of mineralised collagen was established by co-culture of osteoblast-like cells (murine cell line ST-2) and osteoclast-like cells, derived as described above. Cells from different species were chosen to be able to distinguish between both in PCR analyses. Osteoblast-like cells in co-culture on mineralised collagen showed osteogenic differentiation as confirmed by raised transcription levels of osteoblast-specific markers like ALP gene, osteocalcin gene and pro-collagen I gene – and osteoclast-like cells by up-regulated transcription of tartrate-resistant acidic phosphatase (TRAP), cathepsin K and chloride channel 7 (CLC 7) genes. In addition, active degradation of the scaffold material could be revealed by transmission electron microscopical (TEM) investigations [20].

In order to establish a model system which is more similar to the conditions in human bone, co-culture experiments with solely human cells were performed on

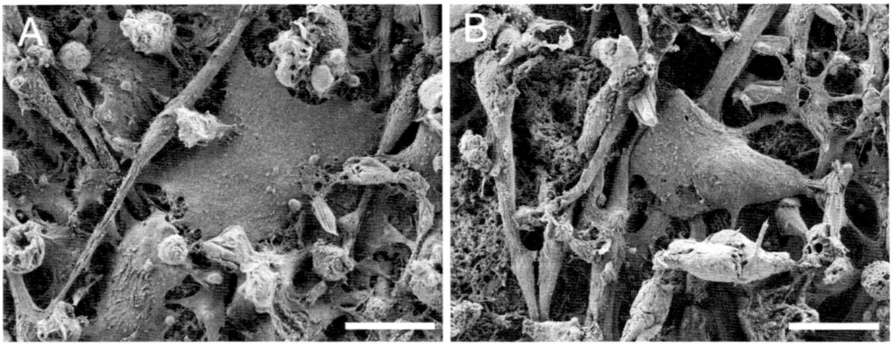

Fig. 4 SEM images of osteoclast-like cells, differentiated from human monocytes on membranes of mineralised collagen. (**a**) Single culture. (**b**) Indirect co-culture with osteogenically induced hMSC; magnification $\times 1000$, scale bar = 20 μm

membranes of mineralised collagen type I (Fig. 4). Human MSC, which were differentiated towards the osteogenic lineage, were cultured in the presence of human monocytes, differentiating into osteoclast-like cells. For this purpose, the cell culture medium was supplemented with the standard additives for osteogenic differentiation (dexamethasone, β-glycerophosphate and ascorbic acid 2-phosphate) as well as with MCSF and RANKL. For separate analysis of gene transcripts as well as specific enzyme activities and DNA content, both cell species were seeded on separate membranes and cultivated in transwell plates allowing exchange of soluble factors between both cell types. During this indirect cocultivation of osteogenically induced hMSC on membranes of mineralised collagen, cell number increased steadily over the whole cultivation period. Thereby the cell number in the single culture was lower compared to the indirect co-culture (Fig. 5a). Osteogenic differentiation of hMSC was analysed by quantification of specific ALP activity. We found a significantly

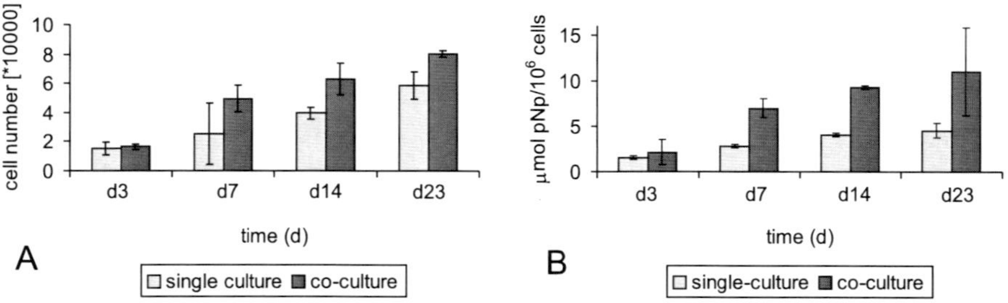

Fig. 5 (**a**) Proliferation of osteogenically induced hMSC on mineralised collagen membranes in single culture and indirect co-culture with human osteoclast-like cells. Cell number was determined by measurement of lactate dehydrogenase activity after cell lysis. (**b**) Specific ALP activity of osteogenically induced hMSC on mineralised collagen membranes in single culture and indirect co-culture with human osteoclast-like cells

higher amount of specific ALP activity for hMSC in the indirect co-culture compared to those in single culture (Fig. 5b) which is a strong hint for an active cell–cell communication between both cell types using soluble cytokines. Similar results were obtained by analysis of ALP gene expression (data not shown).

As confirmed by SEM investigations, the surface of the mineralised collagen membranes became rougher during cultivation of osteoclast-like cells, possibly due to partial solubilisation of the mineral phase which can occur when the pH of culture medium decreases. This altered surface of the mineralised collagen substrate may also have influenced proliferation and osteogenic differentiation of hMSC. Furthermore, upon resorption of the artificial extracellular matrix by osteoclast-like cells (again proven by TEM investigations), calcium ions may be released into the cell culture medium which could also stimulate osteogenic differentiation of hMSC.

This in vitro model of the bone remodelling processes consisting of functional osteoblasts and osteoclast-like cells, both derived from their natural precursors, and a scaffold consisting of biomimetically mineralised collagen, acting as an artificial extracellular bone matrix, might be useful for a better understanding of cellular cross talk between and mutual regulation of both cell types. An increased knowledge of these complex interactions should contribute to novel therapeutic approaches for systemic bone diseases like osteoporosis in which the balance between bone resorption and synthesis of new matrix is disturbed.

5 Chemoattraction of hMSC into Porous 3D Scaffolds: A Novel Possibility for Accelerated Bone Defect Healing

Whereas small bone defects or fractures of otherwise healthy bone can heal by themselves because of the high regenerative capacity of this tissue, large defects or those of predamaged bone (osteoporotic defects, state after radiotherapy, etc.) often do not. For these critical cases, the combination of biomaterials and autologous (stem) cells or implantation of in vitro generated tissue engineered bone would be highly beneficial. Unfortunately, just for such big constructs one has to face the initial lack of vascularisation: whereas even large cell-seeded scaffolds can be kept vital in vitro, using suitable perfusion cell culture systems, most of the cells will die shortly after implantation due to the slow ingrowth of blood capillaries [52, 39].

Actually therefore the new concept of in situ tissue engineering arouse interest which tries to utilise natural mechanisms of chemotaxis for colonisation of scaffold materials with MSC in vivo. The aim of this approach is to redundantise cell seeding of the scaffold ex vivo and so to overcome the problems concerning oxygen and nutrient supply directly after implantation. This might be achieved by attracting MSC into the defect site by implantation of a cell-free, but suitably functionalised scaffold material. MSC exhibit directed chemotaxis towards different compounds, e.g. fibronectin, platelet-derived growth factor (PDGF) and BMP-2 [24, 65]. Chemotactic interaction between stromal-derived factor 1α (SDF-1α) and cells expressing its receptor CXCR4 is well characterised.

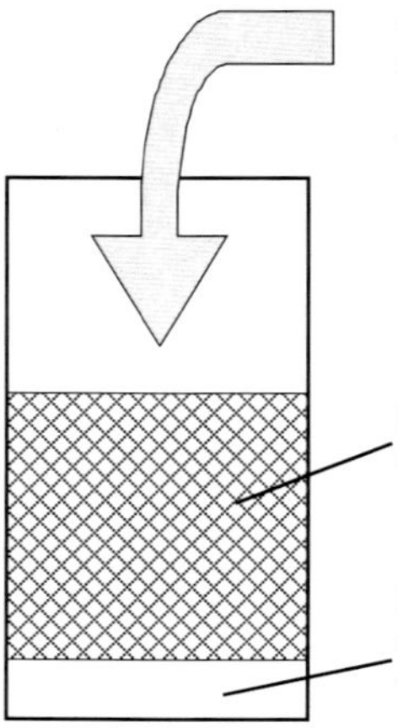

Fig. 6 Schematic diagram of migration assay using porous 3D scaffolds. A fixed scaffold enables the application of SDF-1α in the lower, basal compartment. MSC can be applied to the upper surface of the scaffold in order to examine their migration efficiency into the 3D scaffolds

Interestingly, hMSC also express the chemokine receptor CXCR4, but limited to a minor subpopulation, showing also SDF-1α directed chemotaxis [70, 6]. The attraction of circulating stem cells by SDF-1α also seems to play a crucial role in bone fracture repair [61, 69]. Schantz and coworkers attracted native MSC into small polycaprolactone scaffolds by infusion of SDF-1α using a micropump system [62]. However, colonisation and remodelling of scaffolds used for treatment of large bone defects would need an enlarged number of CXCR4 expressing MSC.

Therefore, in the authors' own experiments overexpression of the CXCR4 receptor in hMSC was induced by mRNA transfection in order to reinforce migration into porous 3D scaffolds made of mineralised collagen, already described [27], which release the chemokine SDF-1α [66].

To achieve overexpression of CXCR4 in hMSC, viral gene transfer had been used in several other studies [6, 71]. However, transient overexpression of the CXCR4 receptor could also be achieved by a novel mRNA transfection method which led to an SDF-1α-directed chemotaxis of hMSC [59]. The main advantage of mRNA transfection of MSC is its limited duration of overexpression due to intracellular degradation of transfected mRNA within 5 days reducing the risk of mutagenic events considerably.

In our own in vitro studies we showed that CXCR4 mRNA transfection could be used to enable hMSC to migrate actively into porous 3D scaffolds of mineralised collagen towards an SDF-1α gradient (Fig. 6).

This migration was clearly dependent on the presence of overexpression of CXCR4 and led to ingrowth of considerable amounts of cells into deeper regions of the scaffolds within 7 days (Fig. 7). No significant cell migration could be detected if no SDF-1α was added to the cell culture medium in the basal compartment [66].

In summary, the combined use of mRNA transfection of hMSC leads to transient overexpression of CXCR4 receptor, on the one hand, and release of SDF-1α

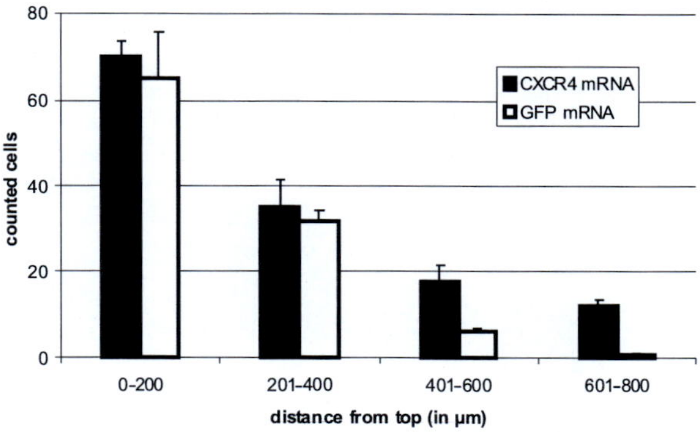

Fig. 7 Migration of CXCR4 mRNA transfected hMSC into porous 3D scaffolds of mineralised collagen. MSC had been transfected with CXCR4 mRNA (*black columns*) or GFP mRNA (*white columns*) 24 h before they were placed to the top of the scaffolds. The media at the bottom of the scaffold contained 100 ng/ml SDF-1α, leading to a concentration gradient inside of the porous matrix. After 5 days the scaffolds were fixed, cut into longitudinal slices and analysed after DAPI staining in order to count the number of hMSC which invaded the scaffold. The analyses were performed in triplicate

from biomimetic porous scaffolds made of mineralised collagen, on the other hand, denotes a promising strategy for clinical utilisation of in situ bone tissue engineering addressing the treatment of large bone defects or those with decreased self-healing capacity. Further studies will be directed to establish controlled release of SDF-1α from internal compartments of large 3D scaffolds which will make the constructs implantable.

6 Conclusions

Utilisation of autologous human mesenchymal stem cells of various sources for regenerative therapies and bone tissue engineering is an intensively investigated research area. The conditions for a stable osteogenic differentiation in vitro are known to a large extent, and several animal studies have shown that the combination of these cells with suitable scaffolds as carrier materials can be used to accelerate defect healing. However, concerning the application in humans, comprehensive clinical studies are still missing in which the use of autologous bone marrow is compared with that of isolated and expanded or non-expanded MSC. Novel approaches make use of chemoattraction of MSC towards identified chemokines like SDF-1α to overcome the problem of vascularisation of tissue-engineered constructs.

An increasing knowledge about the function of MSC in natural bone defect healing processes and the cross talk of osteoblasts, osteoclasts and the precursors of both will lead to improved therapeutic options, finally hopefully also for demanding cases like osteoporotic fractures.

Acknowledgements The studies described in this chapter were partly funded by the German Research Foundation (DFG), the Saxonian Ministry for Science and Arts (SMWK), the German Federal Ministry for Education and Research (BMBF) and the Center for Regenerative Therapies Dresden (CRTD). Human MSC used in our own investigations were kindly provided by the Medical Clinic I of Dresden University Hospital (Prof. Dr. M. Bornhäuser and coworkers). We want to thank several PhD students, coworkers and technicians for their contributions, namely Hagen Domaschke, Sebastian Thieme, Ortrud Zieschang, Katrin Navratiel and Utta Herzog.

References

1. Arthur A, Zannettino A, Gronthos S. The therapeutic applications of multipotent mesenchymal/stromal stem cells in skeletal tissue repair. J Cell Physiol. 2009; 218:237–245.
2. Baksh D, Song L, Tuan RS. Adult mesenchymal stem cells: characterization, differentiation, and application in cell and gene therapy. J Cell Mol Med. 2004; 8:301–316.
3. Beresford JN, Joyner CJ, Devlin C, Triffitt JT. The effects of dexamethasone and 1,25-dihydroxyvitamin D3 on osteogenic differentiation of human marrow stromal cells in vitro. Arch Oral Biol. 1994; 39:941–947.
4. Bernhardt A, Lode A, Boxberger S, Pompe W, Gelinsky M. Mineralised collagen – an artificial, extracellular bone matrix – improves osteogenic differentiation of bone marrow stromal cells. J Mater Sci Mater Med. 2008; 19:269–275.
5. Bernhardt A, Lode A, Mietrach C, Hempel U, Hanke T, Gelinsky M. In vitro osteogenic potential of human bone marrow stromal cells cultivated in porous scaffolds from mineralised collagen. J Biomed Mater Res A 2008; 90A:852–862.
6. Bhakta S, Hong P, Koc O. The surface adhesion molecule CXCR4 stimulates mesenchymal stem cell migration to stromal cell-derived factor-1 in vitro but does not decrease apoptosis under serum deprivation. Cardiovasc Revasc Med. 2006; 19–24.
7. Bieback K, Kern S, Klüter H, Eichler H. Critical parameters for the isolation of mesenchymal stem cells from umbilical cord blood. Stem Cells 2004; 22:625–634.
8. Bieback K, Klüter H. Mesenchymal stromal cells from umbilical cord blood. Curr Stem Cell Res Ther. 2007; 2:310–323.
9. Bielby R, Jones E, McGonagle D. The role of mesenchymal stem cells in maintenance and repair of bone. Injury 2007; 38(Suppl 1):26–32.
10. Bjerre L, Bünger CE, Kassem M, Mygind T. Flow perfusion culture of human mesenchymal stem cells on silicate-substituted tricalcium phosphate scaffolds. Biomaterials 2008; 29:2616–2627.
11. van den Bos C, Mosca JD, Winkles J, Kerrigan L, Burgess WH, Marshak DR. Human mesenchymal stem cells respond to fibroblast growth factors. Hum Cell 1997; 10:45–50.
12. Bradt JH, Mertig M, Teresiak A, Pompe W. Biomimetic mineralization of collagen by combined fibril assembly and calcium phosphate formation. Chem Mater. 1999; 11:2694–2701.
13. Bruder SP, Jaiswal N, Haynesworth SE. Growth kinetics, self-renewal, and the osteogenic potential of purified human mesenchymal stem cells during extensive subcultivation and following cryopreservation. J Cell Biochem. 1997; 64:278–294.
14. Burth R, Gelinsky M, Pompe W. Collagen-hydroxyapatite tapes – a new implant material. Tech Textile. 1999; 8:20–21.
15. Caplan AI. Mesenchymal stem cells. J Orthop Res. 1991; 9:641–650.
16. Devine SM. Mesenchymal stem cells: will they have a role in the clinic? J Cell Biochem Suppl. 2002; 38:73–79.
17. Derubeis AR, Cancedda R. Bone marrow stromal cells (BMSCs) in bone engineering: limitations and recent advances. Ann Biomed Eng. 2004; 32:160–165.
18. Diefenderfer DL, Osyczka AM, Reilly GC, Leboy PS. BMP responsiveness in human mesenchymal stem cells. Connect Tissue Res. 2003; 44(Suppl 1):305–311.

19. D'Ippolito G, Schiller PC, Perez-Stable C, Balkan W, Roos BA, Howard GA. Cooperative actions of hepatocyte growth factor and 1,25-dihydroxyvitamin D3 in osteoblastic differentiation of human vertebral bone marrow stromal cells. Bone 2002; 31:269–275.
20. Domaschke H, Gelinsky M, Burmeister B, Fleig R, Hanke T, Reinstorf A, Pompe W, Rösen-Wolff A. In vitro ossification and remodeling of mineralized collagen I scaffolds. Tissue Eng. 2006; 12, 949–958.
21. Drosse I, Volkmer E, Capanna R, De Biase P, Mutschler W, Schieker M. Tissue engineering for bone defect healing: an update on a multi-component approach. Injury 2008; 39S2:S9–S20.
22. Eghbali-Fatourechi GZ, Mödder UIL, Charatcharoenwitthaya N, Sanyal A, Undale AH, Clowes JA, Tarara JE, Khosla S. Characterization of circulating osteoblast lineage in humans. Bone 2007; 40:1370–1377. Erratum: Bone 2007; 41:741.
23. Erlebacher A, Filvaroff EH, Ye JQ, Derynck R. Osteoblastic responses to TGF-beta during bone remodeling. Mol Biol Cell 1998; 9:1903–1918.
24. Fiedler J, Roderer G, Gunther KP, Brenner RE. BMP-2, BMP-4, and PDGF-bb stimulate chemotactic migration of primary human mesenchymal progenitor cells. J Cell Biochem. 2002; 87:305–312.
25. Friedenstein AJ, Gorskaja JF, Kulagina NN. Fibroblast precursors in normal and irradiated mouse hematopoietic organs. Exp Hematol. 1976; 4:267–274.
26. Friedman MS, Long MW, Hankenson KD. Osteogenic differentiation of human mesenchymal stem cells is regulated by bone morphogenetic protein-6. J Cell Biochem. 2006; 98:538–554.
27. Gelinsky M, Welzel PB, Simon P, Bernhardt A, König U. Porous three dimensional scaffolds made of mineralised collagen: preparation and properties of a biomimetic nanocomposite material for tissue engineering of bone. Chem Eng J. 2008; 137:84–96.
28. Gelinsky M. Mineralised collagen as biomaterial and matrix for bone tissue engineering. In: Meyer U, Handschel J, Wiesmann HP, Meyer Th eds. Fundamentals of tissue engineering and regenerative medicine. Heidelberg and Berlin: Springer; 2009; chapter 36, 485–493.
29. Giordano A, Galderisi U, Marino IR. From the laboratory bench to the patient's bedside: an update on clinical trials with mesenchymal stem cells. J Cell Physiol. 2007; 211:27–35.
30. Handschel J, Wiesmann HP, Depprich R, Kübler NR, Meyer U. Cell-based bone reconstruction therapies – cell sources. Int J Oral Maxillofac Impl. 2006; 21:890–898.
31. Heng BC, Tong C, Stanton LW, Robson P, Olsen B. Strategies for directing the differentiation of stem cells into the osteogenic lineage in vitro. J Bone Miner Res. 2004; 19:1379–1394.
32. Hong L, Colpan A, Peptan IA. Modulations of 17-beta estradiol on osteogenic and adipogenic differentiations of human mesenchymal stem cells. Tissue Eng. 2006; 12:2747–2753.
33. Hong L, Colpan A, Peptan IA, Daw J, George A, Evans CA. 17-Beta estradiol enhances osteogenic and adipogenic differentiation of human adipose-derived stromal cells. Tissue Eng. 2007; 13:1197–1203.
34. Igura K, Zhang X, Takahashi K, Mitsuru A, Yamaguchi S, Takashi TA. Isolation and characterization of mesenchymal progenitor cells from chorionic villi of human placenta. Cytotherapy 2004; 6:543–553.
35. Jagodzinski M, Drescher M, Zeichen J, Hankemeier S, Krettek C, Bosch U, van Griensven M. Effects of cyclic longitudinal mechanical strain and dexamethasone on osteogenic differentiation of human bone marrow stromal cells. Eur Cells Mater. 2004; 7:35–41.
36. Jagodzinski M, Breitbart A, Wehmeier M, Hesse E, Haasper C, Krettek C, Zeichen J, Hankemeier S. Influence of perfusion and cyclic compression on proliferation and differentiation of bone marrow stromal cells in 3-dimensional culture. J Biomech. 2008; 41:1885–1891.
37. Jaiswal N, Haynesworth SE, Caplan AI, Bruder SP. Osteogenic differentiation of purified, culture-expanded human mesenchymal stem cells in vitro. J Cell Biochem. 1997; 64:295–312.
38. Jones E, McGonagle D. Human bone marrow mesenchymal stem cells in vivo. Rheumatology 2008; 47:126–131.

39. Kanczler JM, Oreffo RO. Osteogenesis and angiogenesis: the potential for engineering bone. Eur Cell Mater. 2008; 15:100–114.
40. Kauser K, Zeiher AM eds. Bone marrow-derived progenitors. Handbook of experimental pharmacology. Berlin and Heidelberg: Springer; 2007.
41. Koshla S, Eghbali-Fatourechi GZ. Circulating cells with osteogenic potential. Ann NY Acad Sci. 2006; 1068:489–497.
42. Kundu AK, Putnam AJ. Vitronectin and collagen I differentially regulate osteogenesis in mesenchymal stem cells. Biochem Biophys Res Commun. 2006; 347:347–357.
43. Lee KD. Applications of mesenchymal stem cells: an updated review. Chang Gung Med J. 2008; 31:228–236.
44. Lendeckel S, Jödicke A, Christophis P, Heidinger K, Wolff J, Fraser JK, Hedrick MH, Berthold L, Howaldt HP. Autologous stem cells (adipose) and fibrin glueused to treat widespread traumatic calvarial defects: case report. J Craniomaxillofac Surg. 2004; 32:370–377.
45. Liu P, Oyajobi BO, Russell RG, Scutt A. Regulation of osteogenic differentiation of human bone marrow stromal cells: interaction between transforming growth factor-beta and 1,25(OH)(2) vitamin D(3) In vitro. Calcif Tissue Int. 1999; 65:173–180.
46. Lode A, Bernhardt A, Boxberger S, Gelinsky M. Cultivation of mesenchymal stem cells on a three-dimensional artificial extracellular bone matrix. Proceedings of the International Conference "Biomaterials in Regenerative Medicine", Nadolny AJ, ed. Conference Proceedings and Monographs, Vienna Scientific Centre of the Polish Academy of Sciences, Vienna 2006, Vol. 6, 167–172.
47. Lode A, Bernhardt A, Gelinsky M. Cultivation of human bone marrow stromal cells on three-dimensional scaffolds of mineralized collagen: influence of seeding density on colonization, proliferation and osteogenic differentiation. J Tissue Eng Reg Med. 2008; 2:400–407.
48. Lucarelli E, Donati D, Cenacchi A, Fornasari PM. Bone reconstruction of large defects using bone marrow derived autologous stem cells. Transfus Apher Sci. 2004; 30:169–174.
49. Martin I, Muraglia A, Campanile G, Cancedda R, Quarto R. Fibroblast growth factor-2 supports *ex vivo* expansion and maintenance of osteogenic precursors from human bone marrow. Endocrinology 1997; 138:4456–4462.
50. Ng AM, Saim AB, Tan KK, Tan GH, Mokhtar SA, Rose IM, Othman F, Idrus RB. Comparison of bioengineered human bone constructs from four sources of osteogenic cells. J Orthop Sci. 2005; 10:192–199.
51. Owen M. Marrow stromal stem cells. J Cell Sci Suppl. 1988; 10:63–76.
52. Patel ZS, Mikos AG. Angiogenesis with biomaterial-based drug- and cell-delivery systems. J Biomater Sci Polym Ed. 2004; 15:701–726.
53. Pauwels F. Eine neue Theorie über den Einfluß mechanischer Reize auf die Differenzierung der Stützgewebe. Z Anat Entwicklungsgesch. 1960; 121:478–515 (in German).
54. Philips AM. Overview of the fracture healing cascade. Injury 2005; 36S:S5–S7.
55. Pittenger MF, Mackay AM, Beck SC, Jaiswal RK, Douglas R, Mosca JD, Moorman MA, Simonetti DW, Craig S, Marshak DR. Multilineage potential of adult human mesenchymal stem cells. Science 1999; 284:143–147.
56. Pri-Chen S, Pitaru S, Lokiec F, Savion N. Basic fibroblast growth factor enhances the growth and expression of the osteogenic phenotype of dexamethasone-treated human bone marrow-derived bone-like cells in culture. Bone 1998; 23:111–117.
57. Ramoshebi LN, Matsaba TN, Teare J, Renton L, Patton J, Ripamonti U. Tissue engineering: TGF-beta superfamily members and delivery systems in bone regeneration. Expert Rev Mol Med. 2002; 4:1–11.
58. Roelen BA, Dijke P. Controlling mesenchymal stem cell differentiation by TGF beta family members. J Orthop Sci. 2003; 8:740–748.
59. Ryser MF, Ugarte F, Thieme S, Bornhäuser M, Roesen-Wolff A, Brenner S. mRNA transfection of a CXCR4-GFP fusion protein – generated by a simple three step PCR protocol –

results in efficient migration of primary mesenchymal stem cells. Tissue Eng Part C 2008; 14:179–184.
60. Sammons J, Ahmed N, El-Sheemy M, Hassan HT. The role of BMP-6, IL-6, and BMP-4 in mesenchymal stem cell-dependent bone development: effects on osteoblastic differentiation induced by parathyroid hormone and vitamin D(3). Stem Cells Dev. 2004; 13:273–280.
61. Sata M, Tanaka K, Nagai R. Circulating osteoblast-lineage cells. N Engl J Med. 2005; 353:737–738.
62. Schantz JT, Chim H, Whiteman M. Cell guidance in tissue engineering: SDF-1 mediates site-directed homing of mesenchymal stem cells within three-dimensional polycaprolactone scaffolds. Tissue Eng. 2007; 13:2615–2624.
63. Schwartz Z, Simon BJ, Duran MA, Barabino G, Chaudhri R, Boyan BD. Pulsed electromagnetic fields enhance BMP-2 dependent osteoblastic differentiation of human mesenchymal stem cells. J Orthop Res. 2008; 26:1250–1255.
64. Sen MK, Miclau T. Autologous iliac crest bone graft: should it still be the gold standard for treating nonunsions? Injury 2007; 38S1:S75–S80.
65. Thibault MM, Hoemann CD, Buschmann MD. Fibronectin, vitronectin, and collagen I induced chemotaxis and haptotaxis of human and rabbit mesenchymal stem cells in a standardized transmembrane assay. Stem Cells Dev. 2007; 16:489–502.
66. Thieme S, Ryser M, Gentsch M, Brenner S, Stiehler M, Rölfing J, Gelinsky M, Rösen-Wolff A. Stromal cell derived factor-1α directed chemoattraction of transiently CXCR4 overexpressing mesenchymal stem cells into functionalised three-dimensional biomimetic scaffolds. Tissue Eng Part C 2009; 15:687–696.
67. Tsiridis E, Bhalla A, Ali Z, Gurav N, Heliotis M, Deb S, DiSilvio L. Enhancing the osteoinductive properties of hydroxyapatite by the addition of human mesenchymal stem cells, and recombinant human osteogenic protein-1 (BMP-7) in vitro. Injury 2006; 37(Suppl 3):S25–S32.
68. Wahl DA, Czernuszka JT. Collagen-hydroxyapatite composites for hard tissue repair. Eur Cell Mater. 2006; 11:43–56.
69. Ratajczak MZ, Zuba-Surma E, Kucia M, Reca R, Wojakowski W, Ratajczak J. The pleiotropic effects of the SDF-1-CXCR4 axis in organogenesis, regeneration and tumorigenesis. Leukemia 2006; 20:1915–1924.
70. Wynn RF, Hart CA, Corradi-Perini C, O'Neill L, Evans CA, Wraith JE, Fairbairn LJ, Bellantuono I. A small proportion of mesenchymal stem cells strongly expresses functionally active CXCR4 receptor capable of promoting migration to bone marrow. Blood 2004; 104:2643–2645.
71. Zhang D, Fan GC, Zhou X, Zhao T, Pasha Z, Xu M, Zhu Y, Ashraf M, Wang Y. Overexpression of CXCR4 on mesenchymal stem cells augments myoangiogenesis in the infarcted myocardium. J Mol Cell Cardiol. 2008; 44:281–292.
72. Zuk PA, Zhu M, Ashjian P, De Ugarte DA, Huang JI, Mizuno H, Alfonso ZC, Fraser JK, Benhaim P, Hedrick MH. Human adipose tissue is a source of multipotent stem cells. Mol Biol Cell 2002; 13:4279–4295.

Part IV
Techniques and Applications

Building, Preserving, and Applying Extracellular Culture Integrity Using New Cell Culture Methods and Surfaces

Thomas Brevig, Robin Wesselschmidt, and Masayuki Yamato

Abstract Cultivation of mammalian cells holds promise for generating cell and tissue models and transplants, but the shortcomings of current cell culture methods are limiting such use. In order to provide biological material that can represent cells and tissues of the body in drug development and regenerate organs and systems impaired by disease in the clinic, it is important to recognize that the extracellular integrity, the interphase between the cultureware surface and the cells, and the cell-to-cell contacts are an integral part of the cell culture. One example would be the deposition of type IV collagen, the main structural component of the basement membrane, by human endothelial cells, which can be manipulated by varying the cultureware surface and/or the level of oxidative stress in the culture. Even if culture protocols were optimized to make the cell environment more like the environment in the body, the integrity of the culture would be destroyed during cell harvesting, as matrix molecules are digested by enzymes used for dissociating the cells, and cannot be accumulated from one passage to another. The Thermo Scientific Nunc™ UpCell™ Surface enables dissociation of cells from the cultureware upon a simple change in temperature. Being slightly hydrophobic at 37°C, the surface allows cells to attach and grow, but hydrophilic when the temperature is reduced to below 32°C, the surface will bind water and swell, resulting in the release of adherent cells with their underlying matrix molecules. The preservation of matrix molecules enables the harvesting of contiguous cell sheets, which can be grafted onto another cell layer in vivo or in vitro without the use of fibrin glue or sutures. This chapter gives examples of how extracellular culture integrity can be built, preserved, and applied in cell analysis and tissue engineering.

Keywords Cell culture · Extracellular matrix · Oxidative stress · Cell surface proteins · Tissue engineering

---

T. Brevig (✉)
Research and Development, Thermo Fisher Scientific, Roskilde, Denmark
e-mail: thomas.brevig@thermofisher.com

G.M. Artmann et al. (eds.), *Stem Cell Engineering*,
DOI 10.1007/978-3-642-11865-4_18,

1 Introduction

Stem cells are the subject of ever-increasing interest and study due to their promise to inform our understanding of basic human development and transform the way we perceive and treat human disease, injury, and aging. In the clinic, hematopoietic stem cells are routinely used in disease treatment, while mesenchymal stem cells are being developed and tested to treat a variety of injuries and diseases as well. However, tissue-derived stem cells generally have limited differentiation capacity and limited growth in vitro. Interest in stem cell biology increased greatly in 1998 when Thomson and colleagues reported on the successful derivation of human embryonic stem cells [1]. Embryonic stem cells (ESCs) are pluripotent cells capable of producing cellular derivatives from all three germ layers while retaining their capacity for unlimited proliferation in vitro. The breakthrough finding that a small set of transcription factors is all that is required to induce the reprogramming of differentiated fibroblasts into pluripotent cells that are very much like ESCs has furthered interest in stem cells and their potential use in regenerative medicine regimes [2, 3]. These induced pluripotent stem cells (iPSCs) have the characteristics of human ESCs without requiring human embryos for their derivation, thus removing many of the objections over embryonic stem cell research. This technological breakthrough has already led to the isolation of iPSCs from patient populations, allowing scientists to follow some disease processes in the petri dish in order to elucidate disease pathways that are likely to lead to the development of novel treatments [4].

While the promising use and novel application of stem cells in the study and treatment of human diseases ignite the imagination, there remain many hurdles to overcome in order to make the culture and scale-up of stem cells efficient and routine. At this time, the study of stem cells is primarily a study in vitro, impeded by the shortcomings of the general methods for cultivation of cells. Cell culture processes need to reach a state where mammalian cells, including stem cells, can be grown using methods and materials that produce biological material that (1) represents cells and tissues in the body to an extent where widespread use in cellular analysis is warranted and (2) regenerates organs and systems impaired by disease in transplantation-based therapies.

Stem cells are grown in cultureware in an artificial milieu consisting of a mixture of defined and undefined components, many of which are of animal origin, such as bovine sera, extracellular matrix proteins, growth factors, and medium conditioned by other cells. In some cases, stem cells are grown in direct co-culture with other types of cells. The goal of these complex culture systems is to provide the stem cells with a microenvironment that allows growth and expansion in vitro while retaining their stem cell properties; this system can be thought of as an ex vivo stem cell niche.

This chapter will describe those interactions of a culture system that are related to cell-to-cultureware and cell-to-cell contacts, as well as the application of such preserved contacts in cell analysis and tissue engineering.

2 Cultureware Surfaces and Extracellular Culture Integrity

Most cell cultureware is made of polystyrene, an aromatic polymer that can be molded into colorless vessels with optical properties sufficient for general cell culture. Anchorage-dependent mammalian cells do not, however, attach well to molded polystyrene, unless the surface is modified, for example, by introducing oxygen- and/or nitrogen-containing functional groups in the hydrocarbon chains. When cells are seeded in culture, the surface of the cultureware binds proteins from the medium. Surface modification influences which medium-derived proteins are adsorbed, the degree of denaturation and clustering of these proteins on the surface [5], and, finally, how the adsorbed proteins are presented to the cells. Endothelial cells, for example, attach initially to serum fibronectin and vitronectin adsorbed to the cultureware surface from the medium [6, 7]. The surface of the cultureware may also be coated with proteins prior to the seeding of the cells. The interactions between the cells and the surface – with its adsorbed proteins – are a contributing factor in determining the fate of cells in culture.

The interphase between the cultureware surface and the cells, and the cell-to-cell contacts, referred to collectively herein as the *extracellular integrity* of the culture, is presumed to be an important product and property of a cell culture, one that can be exploited to generate cells and tissues that mimic the in vivo in assays and to repair tissues in disease. An underlying hypothesis would be that cells grown in culture can be stimulated to build an extracellular integrity resembling the environment the cells came from in the body, and that stem cells differentiated in culture can be stimulated to build an extracellular integrity resembling the habitat for the mature cell type in the body.

3 Building Culture Integrity

The interaction between the cells and the cultureware surface is dynamic and changes over the course of the cultivation. Whereas proteins from the medium adsorb momentarily to the cultureware surface, cells require minutes to hours to respond to the surface of the cultureware. Once the cells have settled on the surface of the cultureware and cell membrane proteins digested during trypsinization have been replenished, the cells recognize and bind to the adsorbed proteins. This is the basis for cell adhesion or cell attachment. Attached cells spread, migrate, and proliferate to a state where they cover the entire active surface of the cultureware (confluence). Even if there is no net increase in the number of cells at confluence, the culture is by no means static. Cells may, for example, change to perform new functions, form close connections with their neighbor cells, or secrete extracellular matrix molecules. Extracellular matrix molecules deposited under the cells form a subcellular matrix. In the body, the extracellular matrix is important for both the structure and the function of tissues. The basement membrane of endothelial cells, for example, is essential to the spatial organization of these cells at the luminal surface of heart and blood vessels, and during the growth and repair of the vasculature [8].

The subcellular deposition of extracellular matrix molecules by cells in culture can be manipulated by varying the culture conditions. Type IV collagen, which is the main structural component of basement membranes [9], is deposited by endothelial cells; the amount deposited depends not only on the duration of the culture but also on the concentration of O_2 in the culture atmosphere, the presence of antioxidants in the culture medium, and the surface of the cultureware [10]. In the latter study, pre-coating of oxidized polystyrene with fibronectin stimulated deposition of type IV collagen by human endothelial cells (Fig. 1). Type IV collagen deposition on cultureware with fibronectin was more pronounced when the endothelial cells were grown in an atmosphere of 5% O_2 and 5% CO_2 in N_2 compared to cells grown in the standard cell culture atmosphere of 5% CO_2 in air (20% O_2). Accumulation of type IV collagen under the cells was inhibited via oxidative stress,

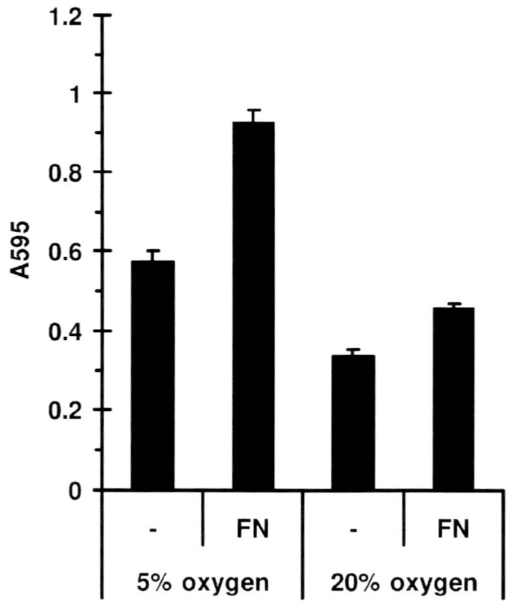

Fig. 1 Cultureware surface and culture–atmosphere oxygen concentration have an effect on the content of type IV collagen in endothelial cell subcellular matrix. Human umbilical vein endothelial cells were seeded at $6.0 \times 10^4/cm^2$ (near confluence) on oxidized polystyrene or on oxidized polystyrene with pre-adsorbed fibronectin (*FN*) and incubated for 4 days in either of two identical incubators (Forma Series II, model 3141; Thermo Fisher Scientific, OH, USA), but with different oxygen concentrations in the culture atmosphere (5% O_2 and 5% CO_2 in N_2 or 20% O_2 by using 5% CO_2 in air). The cells were then removed from the underlying matrix using a protocol that results in minimal digestion and loss of type IV collagen epitopes, and the content of type IV collagen was determined using an enzyme-linked immunoassay. Mean absorbance of 595-nm light (*A595*) and SD are given ($n = 5$). $P < 0.01$ in two-way ANOVA and $P < 0.01$ in all pairwise comparisons. Reproduced from Brevig et al. [10] with kind permission from *Artificial Organs*

because the inhibitory effect of 20% O_2 was antagonized by antioxidants, ascorbic acid (vitamin C; $\geq$0.01 mM) and ascorbic acid phosphate ($\geq$0.001 mM), and mimicked by prooxidant pyrogallol ($\geq$1.0 mM) and exogenous H_2O_2 ($\geq$0.1 mM). Whether oxidative stress reduces the amount of type IV collagen deposited under the cells by reducing the secretion of type IV collagen (gene expression or post-transcriptional events), by reducing incorporation of this protein into the matrix, by increasing degradation of the protein, or by a combination of these mechanisms remains to be determined. The lipid-soluble antioxidant tocopherol (vitamin E), which protects cells from oxidation of membrane lipids, did not stimulate type IV collagen accumulation under endothelial cells exposed to 20% O_2, indicating that oxidative stress inhibiting type IV collagen accumulation takes place in the aqueous compartment(s) of the cell culture. Ascorbic acid stimulates matrix synthesis not only by protecting cells from oxidative stress. It also increases the production of collagen, including type IV, by increasing the level of collagen mRNA, by being a cofactor for the enzymatic hydroxylation of prolyl residues in procollagen, and by stimulating procollagen secretion [11–13].

By avoiding oxidative stress in the culture, for example, by using a reduced-oxygen culture atmosphere or adding an antioxidant to the medium, and/or by pre-adsorbing certain biomolecules to the cultureware, stem cell biologists and tissue engineering scientists can manipulate and optimize the formation of the extracellular integrity of a cell culture.

4 Preserving Culture Integrity

One of the major problems with the traditional cell culture process is that the extracellular integrity of the culture is destroyed when adherent cells are harvested using the standard methods of enzymatic (trypsinization) and mechanical (scraping) dissociation from the cultureware. Enzymatic harvesting and mechanical harvesting disrupt cell-to-cell and cell-to-matrix contacts, as well as the subcellular matrix and the polarization of the culture. These harvesting methods bring the cell culture back to the starting point, that is, the state at the seeding of the cells, except, of course, that the cells may have proliferated, differentiated, or de-differentiated during the course of the cultivation. This cycling of cultivation and passaging does not result in the desired accumulation of extracellular integrity of the culture.

In order to employ the extracellular integrity of a cell culture in sub-cultures or cell analysis, a method for harvesting cells that does not destroy the extracellular integrity must be used. The Thermo Scientific Nunc™ UpCell™ Surface supports cultivation of adherent cells and enables cell harvesting without using enzymes or scraping. The surface responds to a change in temperature and releases cells with their underlying extracellular matrix upon a reduction in temperature to below 32°C (Fig. 2). The development and use of the temperature-responsive cell culture surface was pioneered by Okano and colleagues [14, 15] (videos available at www.nuncbrand.com/en/page.aspx?ID=11850).

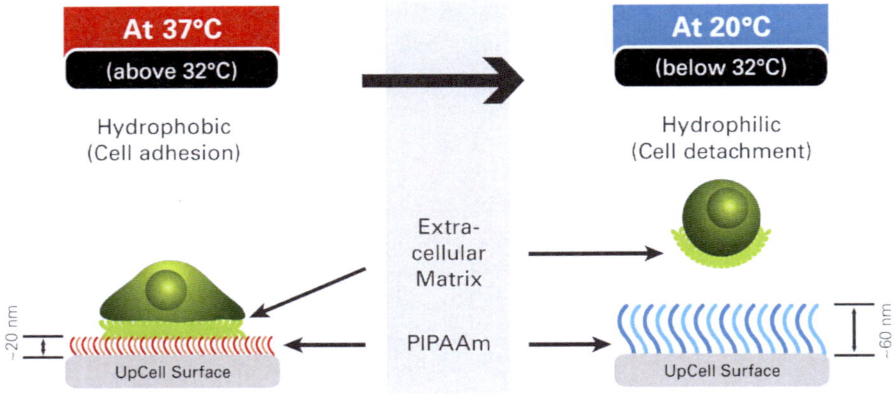

Fig. 2 The temperature-responsive Nunc UpCell Surface for non-enzymatic cell harvesting. UpCell cultureware has poly(*N*-isopropylacrylamide), or PIPAAm, covalently immobilized to the polystyrene surface. The PIPAAm layer is relatively hydrophobic at 37°C, allowing cells to attach and grow. When the temperature of the culture is reduced to below 32°C, the PIPAAm layer becomes very hydrophilic, binds water, and swells, resulting in the release of adherent cells with their underlying extracellular matrix from the cultureware

Cell-to-cell and cell-to-matrix contacts are mediated by proteins in the cell membrane. Cell surface proteins are also important for other interactions between the cell and the environment. Some cell surface proteins are involved in the response of a cell to growth factors and other soluble mediators, whereas other cell surface proteins are involved in the ion homeostasis of the cell. Finally, cell surface proteins are used as antigens or markers in cell analysis and enrichment procedures. Preservation of cell surface proteins in cell culture and harvesting is thus pivotal for maintaining the extracellular integrity of the culture.

Figure 3 shows the results of a flow cytometry analysis of the integrity of CD140a (a cell surface tyrosine kinase receptor for members of the platelet-derived growth factor family) on human bone marrow cells and preadipocytes harvested from the UpCell Surface by temperature reduction and from traditional cultureware (oxidized polystyrene) by trypsinization. Human bone marrow cells and preadipocytes harvested from the UpCell Surface by temperature reduction had preserved cell surface CD140a, whereas CD140a on cells harvested from traditional cultureware by a short trypsinization could barely be detected, demonstrating that using the UpCell Surface and temperature reduction preserves the integrity of cell surface proteins to a higher degree than using traditional cultureware and trypsinization.

Cultivation and harvesting of cells using the temperature-responsive UpCell Surface enable stem cell biologists and tissue engineering scientists to preserve the extracellular integrity of a cell culture for (1) cell analysis and purification, working with stem cell markers in the cell membrane and (2) cell transplantation, where intact cell surface receptors on donor cells may be needed for cells to promptly respond to cues in the graft environment.

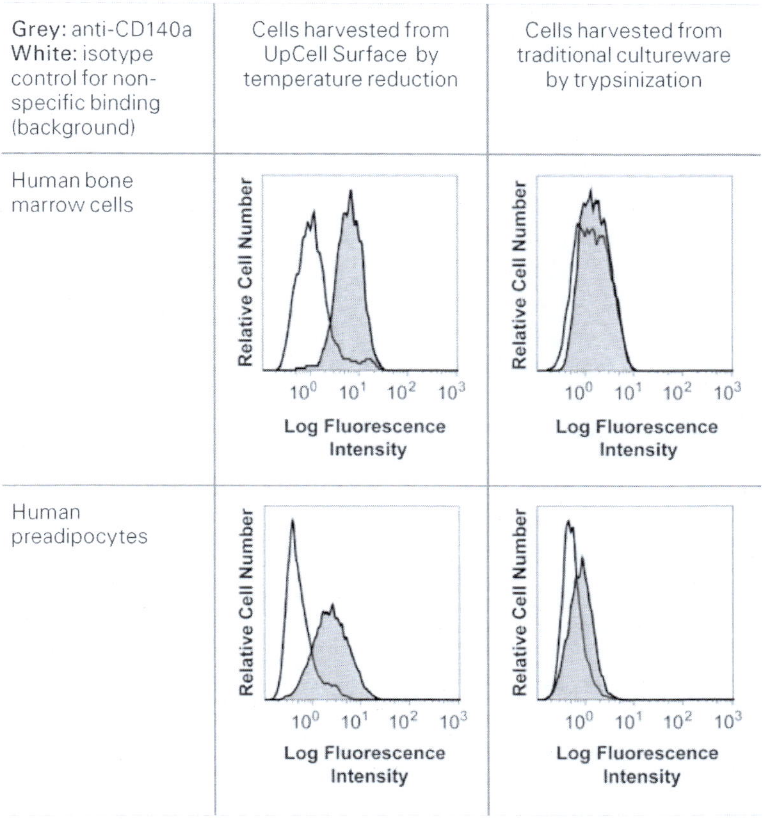

Fig. 3 Preservation of cell surface proteins during cell harvesting. Human bone marrow cells and human preadipocytes in Dulbecco's Modified Eagle's Medium supplemented with 10% fetal bovine serum were seeded at 6.8×10^3 cells/cm^2 on Nunc UpCell Surface (Thermo Fisher Scientific, Denmark) and on a traditional cultureware surface (oxidized polystyrene). Cells were incubated for 24 h at 37°C in a humidified atmosphere of 5% CO_2 in air and then harvested from the UpCell Surface using temperature reduction (20°C for 30 min) and from the oxidized polystyrene using a 3-min trypsinization. Cells were then washed twice by centrifugation, supernatants were discarded, and aliquots of 5.0×10^5 cells were incubated with 200 μl of phycoerythrin (PE)-conjugated mouse monoclonal antibody against human CD140a (5 μg/ml; BD Pharmingen, NJ, USA), shown as grey histograms, or PE-conjugated mouse IgG_{2a} isotype-control antibody (5 μg/ml; BD Pharmingen), shown as white histograms. After incubation at 4°C for 60 min, cells were washed with phosphate-buffered saline and analyzed by flow cytometry (FC500; Beckman Coulter, CA, USA)

5 Applying Culture Integrity

The Nunc UpCell Surface can be used not only for harvesting single cells with preserved surface proteins but also for harvesting contiguous sheets of cells from a confluent culture. The subcellular matrix of the cells is harvested, at least in part, together with the cells [16], and this protein layer provides the adhesive necessary for bonding the cell sheet to a graft site in the body or to another cell sheet in vitro [17]. The extracellular integrity of a cell culture can thus be preserved and applied to the engineering of 3D tissue models and transplants. In tissue engineering, tissue-like structures are typically prepared by seeding a cell suspension on a pre-fabricated scaffold. The scaffold materials are most often materials foreign to the body or from another species (xenogeneic), such as polylactic acid, polyglycolic acid, alginate, gelatin, and collagen. Problems often encountered when using scaffolds for tissue engineering include uneven cell distribution and difficulties in controlling the spatial distribution of different cell types. After transplantation, there can be a host of inflammatory reactions and fibrous tissue formation due to the exogenous scaffold material. The UpCell Surface enables the creation of tissue models and transplants held together only by normal cell junctions and the extracellular matrix deposited by the cells [18].

One example of how cell culture extracellular integrity can be applied is in the preparation of myocardial grafts. Shimizu and colleagues [19] stacked neonatal rat cardiomyocyte sheets harvested by temperature reduction. Sheets in these multi-layer constructs began to contract simultaneously, and alignment of connexin-43 was established between cells of apposing sheets. Four- to six-layer cardiomyocyte constructs transplanted into the subcutaneous tissue of athymic rats survived for 1 year with preservation of spontaneous and simultaneous contraction [20]. Size, conduction velocity, and contractile force of grafts increased in proportion to the growth of the host. Histological studies showed characteristic structures of heart tissue, including elongated cardiomyocytes, well-differentiated sarcomeres, and gap junctions within the grafts.

Monolayers of mesenchymal stem cells may also be useful for regenerating myocardium. Miyahara and colleagues [21] transplanted adipose tissue-derived mesenchymal stem cell sheets harvested by temperature reduction onto the scarred myocardium of rats that had undergone coronary artery ligation 4 weeks earlier. The engrafted sheet gradually grew to form a thick stratum that included newly formed vessels. It also reversed wall thinning in the scar area and improved cardiac function in rats with myocardial infarction. By contrast, transplantation of a fibroblast cell sheet did not result in any hemodynamic improvements.

Whereas preserving the extracellular integrity of cell cultures may enable the generation of myocardial grafts for heart repair in the future [22], preservation of extracellular integrity has already been used in cell sheet transplantation to reconstruct the human cornea, using autologous oral mucosal epithelial cells [39]b Moreover, endoscopic transplantation of cell sheets prepared from autologous oral mucosal epithelial cells has been demonstrated to enhance wound healing and prevent postoperative esophageal stenosis after endoscopic submucosal resections in dogs, inciting further studies on the effect of using cell cultures with preserved extracellular integrity in preventing postoperative stenosis and the need for balloon dilatations in individuals with resected esophageal cancer [23]. Cell cultures with preserved extracellular integrity might also be useful for sealing visceral pleural defects, a complication to pulmonary surgery, which requires the use of a material or graft that has the flexibility to respond to expansion and contraction during respiration. Kanzaki and colleagues [24] used autologous dermal fibroblast cell sheets harvested by temperature reduction to successfully seal defects in a pig visceral pleural injury model without the use of sutures or additional adhesive agents. Preserved extracellular cell culture integrity may also be applied to improve prostheses used in surgery. One example would be the functional epithelial lining of a tracheal prosthesis, which has been shown to be feasible, using tracheal epithelial cell sheets, in a rabbit model of a tracheal replacement [25]. Other examples of applications using harvesting of single cells and cell sheets by temperature reduction in order to preserve the extracellular culture integrity are provided in Table 1, along with references to the original publications. Building and preserving cell culture extracellular integrity thus has applications within cellular analysis, purification, and transplantation and in the engineering of tissues and bioprostheses.

Table 1 Examples of applications using single cells and cell sheets harvested by temperature reduction

Cell type	Application	References
Single cells		
Microglia (rat)	Analysis (detachment and function)	Nakajima et al. [26]
Kupffer cells (human)	Analysis (flow cytometry)	Alabraba et al. [27]
Basophilic cell line RBL-2H3 (rat)	Analysis (antigen-mediated degranulation measured by surface plasmon resonance)	Yanase et al. [28]
Monocytes and macrophages (human)	Activation and analysis (structural)	Gordon and Freedman [29]
Monocytes and macrophages (human)	Re-seeding/passaging	Collier et al. [30]
Cell sheet		
Aortic endothelial cells (cow)	Analysis (structural and matrix deposition)	Kushida et al. [16]
Keratinocytes (human)	Analysis (electron microscopy)	Yamato et al. [31]
Urothelial cells (human)	Analysis (electron microscopy)	Shiroyanagi et al. [32]
Retinal pigment epithelial cell line ARPE-19 (human)	Analysis (light microscopy)	Kubota et al. [33]
Kidney epithelial cells (human and dog)	Re-plating to traditional cultureware and analysis (electron and fluorescence microscopy)	Kushida et al. [34]
Lung cells (rat)	Re-plating to traditional cultureware and analysis (fluorescence microscopy)	Nandkumar et al. [35]
Smooth muscle cells and fibroblasts (human)	Re-plating to PIPAAm surface, analysis (functional), and transplantation	Hobo et al. [36]

Table 1 (continued)

Cell type	Application	References
Mesenchymal stem cells and skin fibroblasts (rat)	Analysis (structural and functional) and transplantation	Miyahara et al. [21]
Corneal stem cells (human and rabbit)	Analysis (structural) and transplantation	Nishida et al. [37]
Corneal endothelial cells (human)	Analysis (structural and functional) and transplantation	Sumide et al. [38]
Oral mucosal epithelial cells (human)	Analysis (structural) and transplantation	Nishida et al. [39]
Oral mucosal epithelial cells (dog)	Analysis (structural) and transplantation	Ohki et al. [23]
Tracheal epithelial cells (rabbit)	Transplantation	Kanzaki et al. [25]
Periodontal ligament cells (human)	Transplantation	Hasegawa et al. [40]
Tissue models by cell sheet engineering		
Aortic endothelial cells (human) and hepatocytes (rat)	Cultivation (3D co-culture) and analysis (structural)	Harimoto et al. [41]
Hepatocytes (mouse and human)	Analysis (structural and functional) and stacking during transplantation	Ohashi et al. [42]
Skeletal myoblast (dog)	Analysis (structural) and stacking during transplantation	Hata et al. [43]
Lung and skin fibroblasts (rat, rabbit, and pig)	Analysis (structural) and stacking during transplantation	Kanzaki et al. [44, 24]
Cardiomyocytes (rat)	Cultivation, analysis (structural and functional), and stacking before transplantation	Sekine et al. [45]; Shimizu et al. [19, 20] Sekiya et al. [46];

6 Summary

Methods to derive new stem cell lines of all types, coupled with technological advancements in high-throughput screening, genome-wide gene analysis, and proteomic approaches, are rapidly evolving. However, improved methods for the propagation and scale-up of stem cells are still badly needed. This chapter focused on interactions of cells and cultureware and demonstrated that cells grown in culture can be stimulated, by avoiding oxidative stress and/or by pre-adsorbing certain biomolecules to the cultureware, to build an extracellular integrity resembling the in situ environment. The extracellular integrity of a cell culture can be preserved and accumulated by harvesting cells and cell sheets without enzymatic digestion using the temperature-responsive Nunc UpCell Surface, enabling stem cell biologists to create intact cell and tissue material for analysis and transplantation.

References

1. Thomson JA, Itskovitz-Eldor J, Shapiro SS, Waknitz MA, Swiergiel JJ, Marshall VS, Jones JM. Embryonic stem cell lines derived from human blastocysts. Science 1998; 282: 1145–1147.
2. Takahashi K, Tanabe K, Ohnuki M, Narita M, Ichisaka T, Tomoda K, Yamanaka S. Induction of pluripotent stem cells from adult human fibroblasts by defined factors. Cell 2007; 131:1–12.
3. Yu J, Vodyanik MA, Smuga-Otta K, Antosiewics-Bourget J, Frane JL, Tian S, Nie J, Jonsdottir GA, Routti V, Stewart R, Slukvin II, Thomson JA. Induced pluripotent stem cell lines derived from human somatic cells. Science 2007; 318:1917–1920.
4. Ebert AD, Yu J, Rose FF, Mattis VB, Lorson CL, Thomson JA, Svendsen CN. Induced pluripotent stem cells from a spinal muscular atrophy patient. Nature 2009; 457:277–280.
5. Brevig T, Holst B, Ademovic Z, Rozlosnik N, Røhrmann JH, Larsen NB, Hansen OC, Kingshott P. The recognition of adsorbed and denatured proteins of different topographies by β2 integrins and effects on leukocyte adhesion and activation. Biomaterials 2005; 26:3039–3053.
6. Pompe T, Kobe F, Salchert K, Jørgensen B, Oswald J, Werner C. Fibronectin anchorage to polymer substrates controls the initial phase of endothelial cell adhesion. J Biomed Mater Res A 2003; 67:647–657.
7. Steele JG, Johnson G, McFarland C, Dalton BA, Gengenbach TR, Chatelier RC, Underwood PA, Griesser HJ. Roles of serum vitronectin and fibronectin in initial attachment of human vein endothelial cells and dermal fibroblasts on oxygen- and nitrogen-containing surfaces made by radiofrequency plasmas. J Biomater Sci Polym Ed. 1994; 6:511–532.
8. Davis GE, Senger DR. Extracellular matrix mediates a molecular balance between vascular morphogenesis and regression. Curr Opin Hematol. 2008; 15:197–203.
9. Yurchenco PD, Amenta PS, Patton BL. Basement membrane assembly, stability and activities observed through a developmental lens. Matrix Biol. 2004; 22:521–538.
10. Brevig T, Røhrmann JH, Riemann H. Oxygen reduces accumulation of type IV collagen in endothelial cell subcellular matrix via oxidative stress. Artif Organs 2006; 30:915–921.
11. Kitagawa Y. Role of vitamin C in basement membrane synthesis and cell differentiation. J Nutr Sci Vitaminol (Tokyo) 1992; Spec. No. 157–160.
12. Mahmoodian F, Peterkofsky B. Vitamin C deficiency in guinea pigs differentially affects the expression of type IV collagen, laminin, and elastin in blood vessels. J Nutr. 1999; 129:83–91.
13. Schwarz RI. Procollagen secretion meets the minimum requirements for the rate-controlling step in the ascorbate induction of procollagen synthesis. J Biol Chem. 1985; 260:3045–3049.
14. Okano T, Yamada N, Sakai H, Sakurai Y. A novel recovery system for cultured cells using plasma-treated polystyrene dishes grafted with poly(N-isopropylacrylamide). J Biomed Mater Res. 1993; 27:1243–1251.
15. Yamada N, Okano T, Sakai H, Karikusa F, Sawasaki Y, Sakurai Y. Thermo-responsive polymeric surfaces; control of attachment and detachment of cultured cells. Makromol Chem Rapid Commun. 1990; 11:571–576.
16. Kushida A, Yamato M, Konno C, Kikuchi A, Sakurai Y, Okano T. Decrease in culture temperature releases monolayer endothelial cell sheets together with deposited fibronectin matrix from temperature-responsive culture surfaces. J Biomed Mater Res. 1999; 45:355–362.
17. Yang J, Yamato M, Shimizu T, Sekine H, Ohashi K, Kanzaki M, Ohki T, Nishida K, Okano T. Reconstruction of functional tissues with cell sheet engineering. Biomaterials 2007; 28: 5033–5043.
18. Yang J, Yamato M, Kohno C, Nishimoto A, Sekine H, Fukai F, Okano T. Cell sheet engineering: recreating tissues without biodegradable scaffolds. Biomaterials 2005; 26:6415–6422.

19. Shimizu T, Yamato M, Isoi Y, Akutsu T, Setomaru T, Abe K, Kikuchi A, Umezu M, Okano T. Fabrication of pulsatile cardiac tissue grafts using a novel 3-dimensional cell sheet manipulation technique and temperature-responsive cell culture surfaces. Circ Res. 2002; 90:e40–e48.
20. Shimizu T, Sekine H, Isoi Y, Yamato M, Kikuchi A, Okano T. Long-term survival and growth of pulsatile myocardial tissue grafts engineered by the layering of cardiomyocyte sheets. Tissue Eng. 2006; 12:499–507.
21. Miyahara Y, Nagaya N, Kataoka M, Yanagawa B, Tanaka K, Hao H, Ishino K, Ishida H, Shimizu T, Kangawa K, Sano S, Okano T, Kitamura S, Mori H. Monolayered mesenchymal stem cells repair scarred myocardium after myocardial infarction. Nat Med. 2006; 12: 459–465.
22. Masuda S, Shimizu T, Yamato M, Okano T. Cell sheet engineering for heart tissue repair. Adv Drug Deliv Rev. 2008; 60:277–285.
23. Ohki T, Yamato M, Murakami D, Takagi R, Yang J, Namiki H, Okano T, Takasaki K. Treatment of oesophageal ulcerations using endoscopic transplantation of tissue-engineered autologous oral mucosal epithelial cell sheets in a canine model. Gut 2006; 55:1704–1710.
24. Kanzaki M, Yamato M, Yang J, Sekine H, Takagi R, Isaka T, Okano T, Onuki T. Functional closure of visceral pleural defects by autologous tissue engineered cell sheets. Eur J Cardiothorac Surg. 2008; 34:864–869.
25. Kanzaki M, Yamato M, Hatakeyama H, Kohno C, Yang J, Umemoto T, Kikuchi A, Okano T, Onuki T. Tissue engineered epithelial cell sheets for the creation of a bioartificial trachea. Tissue Eng. 2006; 12:1275–1283.
26. Nakajima K, Honda S, Nakamura Y, López-Redondo F, Kohsaka S, Yamato M, Kikuchi A, Okano T. Intact microglia are cultured and noninvasively harvested without pathological activation using a novel cultured cell recovery method. Biomaterials 2001; 22:1213–1223.
27. Alabraba EB, Curbishley SM, Lai WK, Wigmore SJ, Adams DH, Afford SC. A new approach to isolation and culture of human Kupffer cells. J Immunol Methods 2007; 326:139–144.
28. Yanase Y, Suzuki H, Tsutsui T, Uechi I, Hiragun T, Mihara S, Hide M. Living cell positioning on the surface of gold film for SPR analysis. Biosens Bioelectron. 2007; 23:562–567.
29. Gordon IO, Freedman RS. Defective antitumor function of monocyte-derived macrophages from epithelial ovarian cancer patients. Clin Cancer Res. 2006; 12:1515–1524.
30. Collier TO, Anderson JM, Kikuchi A, Okano T. Adhesion behavior of monocytes, macrophages, and foreign body giant cells on poly (N-isopropylacrylamide) temperature-responsive surfaces. J Biomed Mater Res. 2002; 59:136–143.
31. Yamato M, Utsumi M, Kushida A, Konno C, Kikuchi A, Okano T. Thermo-responsive culture dishes allow the intact harvest of multilayered keratinocyte sheets without dispase by reducing temperature. Tissue Eng. 2001; 7:473–480.
32. Shiroyanagi Y, Yamato M, Yamazaki Y, Toma H, Okano T. Transplantable urothelial cell sheets harvested noninvasively from temperature responsive culture surfaces by reducing temperature. Tissue Eng. 2003; 9:1005–1012.
33. Kubota A, Nishida K, Yamato M, Yang J, Kikuchi A, Okano T, Tano Y. Transplantable retinal pigment epithelial cell sheets for tissue engineering. Biomaterials 2006; 27:3639–3644.
34. Kushida A, Yamato M, Isoi Y, Kikuchi A, Okano T. A noninvasive transfer system for polarized renal tubule epithelial cell sheets using temperature-responsive culture dishes. Eur Cell Mater. 2005; 10:23–30.
35. Nandkumar MA, Yamato M, Kushida A, Konno C, Hirose M, Kikuchi A, Okano T. Two-dimensional cell sheet manipulation of heterotypically co-cultured lung cells utilizing temperature-responsive culture dishes results in long-term maintenance of differentiated epithelial cell functions. Biomaterials 2002; 23:1121–1130.
36. Hobo K, Shimizu T, Sekine H, Shin'oka T, Okano T, Kurosawa H. Therapeutic angiogenesis using tissue engineered human smooth muscle cell sheets. Arterioscler Thromb Vasc Biol. 2008; 28:637–643.
37. Nishida K, Yamato M, Hayashida Y, Watanabe K, Maeda N, Watanabe H, Yamamoto K, Nagai S, Kikuchi A, Tano Y, Okano T. Functional bioengineered corneal epithelial sheet grafts

from corneal stem cells expanded *ex vivo* on a temperature-responsive cell culture surface. Transplantation 2004a; 77:379–385.

38. Sumide T, Nishida K, Yamato M, Ide T, Hayashida Y, Watanabe K, Yang J, Kohno C, Kikuchi A, Maeda N, Watanabe H, Okano T, Tano Y. Functional human corneal endothelial cell sheets harvested from temperature-responsive culture surfaces. FASEB J. 2006; 20:392–394.
39. Nishida K, Yamato M, Hayashida Y, Watanabe K, Yamamoto K, Adachi E, Nagai S, Kikuchi A, Maeda N, Watanabe H, Okano T, Tano Y. Corneal reconstruction with tissue-engineered cell sheets composed of autologous oral mucosal epithelium. N Engl J Med. 2004b; 351:1187–1196.
40. Hasegawa M, Yamato M, Kikuchi A, Okano T, Ishikawa I. Human periodontal ligament cell sheets can regenerate periodontal ligament tissue in an athymic rat model. Tissue Eng. 2005; 11:469–478.
41. Harimoto M, Yamato M, Hirose M, Takahashi C, Isoi Y, Kikuchi A, Okano T. Novel approach for achieving double-layered cell sheets co-culture: overlaying endothelial cell sheets onto monolayer hepatocytes utilizing temperature-responsive culture dishes. J Biomed Mater Res. 2002; 62:464–470.
42. Ohashi K, Yokoyama T, Yamato M, Kuge H, Kanehiro H, Tsutsumi M, Amanuma T, Iwata H, Yang J, Okano T, Nakajima Y. Engineering functional two- and three-dimensional liver systems in vivo using hepatic tissue sheets. Nat Med. 2007; 13:880–885.
43. Hata H, Matsumiya G, Miyagawa S, Kondoh H, Kawaguchi N, Matsuura N, Shimizu T, Okano T, Matsuda H, Sawa Y. Grafted skeletal myoblast sheets attenuate myocardial remodeling in pacing-induced canine heart failure model. J Thorac Cardiovasc Surg. 2006; 132:918–924.
44. Kanzaki M, Yamato M, Yang J, Sekine H, Kohno C, Takagi R, Hatakeyama H, Isaka T, Okano T, Onuki T. Dynamic sealing of lung air leaks by the transplantation of tissue engineered cell sheets. Biomaterials 2007; 28:4294–4302.
45. Sekine H, Shimizu T, Yang J, Kobayashi E, Okano T. Pulsatile myocardial tubes fabricated with cell sheet engineering. Circulation 2006; 114 (Suppl.):I87–I93.
46. Sekiya S, Shimizu T, Yamato M, Kikuchi A, Okano T. Bioengineered cardiac cell sheet grafts have intrinsic angiogenic potential. Biochem Biophys Res Commun. 2006; 341:573–582.

Fabrication of Modified Extracellular Matrix for the Bone Marrow-Derived Mesenchymal Stem Cell Therapeutics

Hwal (Matthew) Suh

Abstract In regenerative medicine, there are several issues directly related to the production of cell-based therapeutics. As pluripotent embryonic or multipotent adult stem cells are committed and differentiated to specific tissue cells and proliferate in number, it is easily conjecturable that there may be internal and/or external cellular factors involved in such a biological event, and that internal factors depend on genetic characteristics while external cellular factors mainly exist within the extracellular matrix.

Keywords Extracellular Matrix (ECM) · Regenerative medicine

1 Issues in Cell Production for the Cell-Based Therapeutics

Several issues for cell-based therapy shall be considered in regenerative medicine (Table 1). From a gross view, the cellular therapeutics is developed through two stages: cell preparation and fabrication. The first stage is manipulating donor cells to have appropriate property to restore, replace, and/or regenerate target recipient tissue. Gene modification is frequently necessary to achieve genetic consistency of donor cells with recipient to escape from immunological rejection and to harmonize longevity except the case of using autologous cells and to correct autologous genes to treat congenital genetic disease. Recently the induced pluripotent stem/somatic (iPS) cells were introduced by reprogramming human adult somatic cells to have embryonic stem cell-like properties in various respects.

In any gene modification, the most hazardous problem is choice of the vector. Although many kinds of viral vectors have been used to deliver selected therapeutic genes into chromosomal DNA in the host cell through infection and demonstrated

H. (Matthew) Suh (✉)
Laboratory of Tissue Regenerative Medicine, Department of Medical Engineering, College of Medicine, Yonsei University, Seoul, Republic of Korea
e-mail: hwal@yuhs.ac

G.M. Artmann et al. (eds.), *Stem Cell Engineering*,
DOI 10.1007/978-3-642-11865-4_19,

Table 1 General issues of the regenerative medicine and their involved scientific tree

Scientific tree of the regenerative medicine	
Issue	Scientific tree linked to downstream
Regenerative medicine	– Clinical medicine – implantation – tissue restoration/replacement/regeneration – tissue – cell and extracellular matrix (ECM) hybridization
Cell	– Somatic cell – progenitor cell – adult stem cell – embryonic stem cell – blastocyst/iPS cell
Cell expansion	– Cell culture – cell differentiation/proliferation/ migration – signal transduction – integrins on cell surface/ligands in ECM
Cell integrity	– Cell-to-cell adhesion – intercellular adhesive proteins – cadherin (– α, β catenin)/desmin/connexin
Cell–ECM hybridization	– Cell-to-ECM adhesion – cell-adhesive protein – fibronectin/laminin/vitronectin – cell-adhesive peptide (RGD sequence)
Safety	– Safety assurance – toxicology/immunology – viability (– cell fabricated materials)/gene modification (– DNA/vectors)
Storage	– Cell/tissue reservation – cryopreservation – cryoprotectant – molecule transportation/ATP cycle

great transfection efficacy, there is no completely safe virus to use for gene transfection yet, because of unknown adventitious diseases and possible genetic mutation oriented from the nature of virus that continues replication cycle which may promote oncogenesis. Furthermore, difficulties in scale-up production, lack of appropriate cell selectivity, inadequate infectivity in vivo, susceptibility to neutralization by serum antibodies, and inflammatory responses are representative problems of the viral vectors.

Therefore, to avoid any disadvantage, non-viral vectors using physicochemical methods have been developed. Typical physical methods include injection or pressure-mediated direct insertion of a therapeutic gene into nucleus of the target cell and electroporation or particle-mediated gene transfer of hybridizing electrically (–)-charged DNA with (+)-charged, non-hazardous metal particles such as gold or tungsten to avoid repellence from (–)-charged cell membrane and provide an easy cell-membrane pathway for therapeutic DNA into cytoplasm for endocytosis. Some chemical vectors are also concerned on the escape from cell-membrane repellence by integrating therapeutic DNA with (+)-charged chemicals of liposome, calcium phosphates, or diethyl-amino-ethyl(DEAE)-dextran and polymers such as poly-L-lysine (PLL) and polyethyleneimine (PEI).

Non-viral vector is beneficial due to a simple and easy process in scale-up production with no side effect; moreover, it is possible to modify the cell surface. But low transfection efficiency, transient gene expression, low target specificity, and possible toxicity from the used chemicals are representative disadvantages in currently available technology. The efficiency of gene transfection by non-infectious, non-viral vectors depends on the so-called endosomal escape which permits detachment of delivered therapeutic DNAs from vectors taken up by endosomes and lysosomes in cytoplasm, and the DNAs can penetrate into the nucleus which triggers recombination of chromosome by delivery of DNA. For the therapeutic cell preparation stage, selection of appropriate DNA for the treatment purpose and development of the safe gene modification tools with ultimate efficacy of target cell's chromosome recombination without mutation are important issues.

Selection of appropriate therapeutic cell is the next step to produce practically applicable cells. Isolation of therapeutic cells from the gene-modified cells by discarding unfavorable and mutated cells can be supported by a cell-sorting technique, the so-called flow cytometry, in combination with specific antibodies for specific cell types; however, no perfect tool to detect mutated cells is available at present. In addition, immunofluorescent staining is applied to identify qualitative characteristics of cells.

Expansion of selected cells to deliver into a patient is usually carried out by cell culture technology. Suitable culture environment is controlled by temperature, humidity, media with nutrients, and cell-specific growth factors to promote fast differentiation and proliferation and mechanical coordination. Growth factors are definitely attractive substances engaged in signal transduction for specific cell-growth events. Nevertheless, as direct delivery of growth factors into a patient is not recommendable due to the risk of oncogenesis which has no method to be prevented at present, cells expanded with growth factors are also exposed to oncogenesis. Therefore, several culture methods that promote cell-growth potential without using growth factors are under investigation for clinics.

After cell expansion, heterogeneity of the cultured cells at various cell life cycles shall be also considered because cells with differentiation and proliferation capacity are needed only. Obtained cells without mutation and apoptic tendency are regarded as the therapeutic cells, respectively.

2 Issues in Extracellular Substances for Production of Cell-Based Therapeutics

The second stage to produce cellular therapeutics is the process of implanting the cells obtained at the first stage as a cell-based implantable substitute designed to post-operatively express characteristics of the cells within the body. Cellular therapeutics is in the form of either single cells or multilayered cellular structures mimicking tissues. Single cell suspension is usually designed to deliver individual therapeutic cells through local or systemic administration mainly for correction of genetic disorders, while multilayered cellular structure is an artificial tissue or organ for implantation manufactured by the so-called tissue engineering.

For single cell therapy, individual encapsulation by a membrane protecting antibody and/or complement attack while permitting inward supply of oxygen and nutritional matters and outward release of the cell-produced biological substance has been designed. Furthermore, specific signal receptor for homing the administered cells toward a target tissue through conduction is considered to be hybridized with the encapsulating membranes.

To produce multilayered cellular structures, tissue engineering technology has been introduced since the mid-1980s as an idea to overcome donor deficiency for transplantation by creating artificial tissues or organs made of somatic cells and structural scaffolds that simply support anatomical morphology through applying synthetic polymeric materials instead of structural proteins in the natural ECM.

Biodegradable polymers, for those post-operatively dissolved through hydrolysis by body fluid after implantation, had been the first consideration for hybridization with somatic cells as a structural support, or a scaffold in another term, at the early era. Biodegradable polymers, such as polylactic acid (PLA) and polyglycolic acid (PGA), are still widely used and have advantages in its easy manipulation at low cost; however, initial inflammatory reaction occurring from localized acidosis due to the extremely low pH often limits viability of both delivered and neighbor recipient's cells after implantation.

To increase biocompatibility of synthetic polymers and to provide favorable environment for the integrated therapeutic cells, various chemical modification methods have been introduced. For example, when hydrophilic polyethyleneglycol (PEG) is grafted upon the polylactic-glycolic acids (PLGA), it provides compensation to neutral condition from the local acidic environment produced by PLGA during degradation.

But every trial to mimic natural structural protein by using synthetic biomaterials has demonstrated unavoidable limitations since synthetics are not natural. Each cell type lives in each specific extracellular environment that consists of not only structural proteins but also cell–cell and cell–ECM integrating proteins,

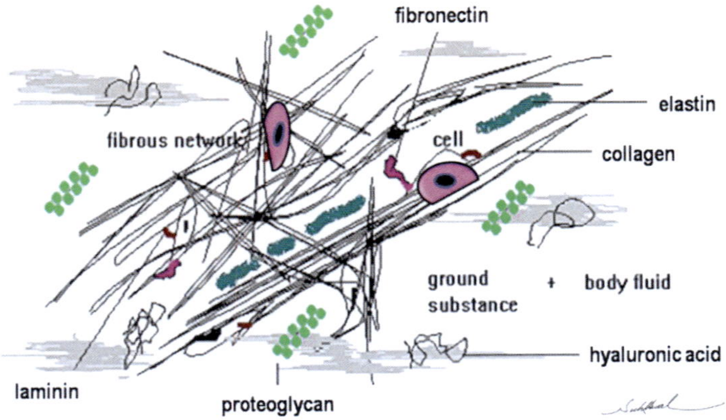

Fig. 1 General composition of the extracellular matrix

cytokines and signal-transmitting substances, and metabolic ground substances (Fig. 1). Therefore, many researchers are concerning hybridization of the therapeutic cells with ECM components to control cellular functions.

3 Characteristics of ECM Components

Every tissue consists of cell and extracellular matrix (ECM) including structural proteins, glycoproteins, cell-adhesive proteins, cytokines, ground substances and body fluid, and stem cells. Ideally, hybridized cells with ECM may stay with natural niche until domiciliation at the target tissue, but unfortunately, ECM is not completely reproducible by presently available methods, and some component has been purified and applied in numerous biological technologies.

Collagen is a representative structural protein that exists in any animal and has been considered as the first choice implantable biomaterial. Various collagen-fabricating methods have been already introduced and applied to clinics (Fig. 2).

The most attractive characteristic of collagen is the ability to maintain its harvested, anatomical and biological conditions of the donor tissue (e.g., cryopreserved cornea, bone chip and skin, cross-linked porcine aortic heart valve, and others in the tissue bank systems) for transplantation or biological characteristics while it avoids the cell-derived immune reactions in forms of decellularized collagenous biomaterials (e.g., demineralized bone, acellular amniotic, and small intestine submucosal membranes) when implanted after hybridization with the donor cells

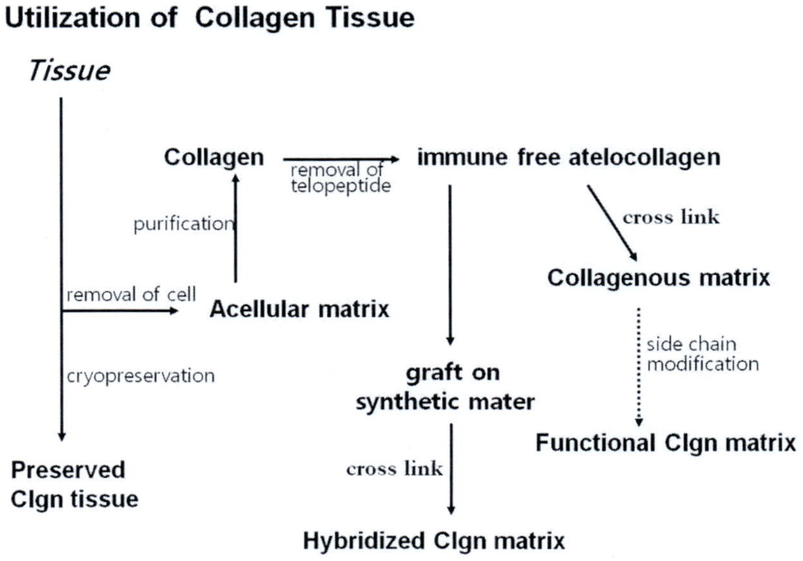

Fig. 2 Fabrication of collagen for various purposes

as ECM, containing every natural component and providing complete biological niche.

In the choice of ECM for fabrication with cells, resource must be considered. Ideal acellular matrix for hybridization with therapeutic cell is from autologous tissue with the same anatomical structure of target lesion. For example, autologous hepatocytes hybridized with a decellularized autologous liver fragment can be regarded as a replica of the autologous liver, but the problem is that the autologous target ECM is already involved in diseases or defects from the treatment target tissue. Using allogeneic acellular matrix is also as practical as using an allogeneic tissue to transplant. In addition, selected xenogeneic acellular matrices have been considered for the cell hybridizing ECM.

Currently, in xenogeneic resources, porcine is expansively being considered since its histological structure and size of organs are similar to human's and it is safer than bovine from the life-threatening risks such as transmissible spongious encephalopathy like human Creutzfeldt–Jakob's disease, although undefined and unexpected side effects may still exist. As culturing the human chondrocyte on a porcine acellular knee cartilage produces a reliable artificial cartilage implant, the undifferentiated mesenchymal stem cells (MSCs) hybridized with acellular matrix of the target tissue may provide favorable niche to lead to committing differentiation toward the designed target tissue cells.

Though acellular matrices provide natural niche to cells, immune reaction oriented from xenogeneic resources is controversial except tissues with less distributed blood vessels such as cartilage and cornea. Therefore, modifying the purified ECM treated to demonstrate less immune reaction is practically preferable.

Collagen is a representative structural ECM usually existing in forms of fibrous matrix, and the telopeptides at both C- and N-terminals of the molecule express immunological characteristics. Hence, atelocollagen molecules produced by removing telopeptides are recommended as implantable biomaterials. Application of the atelocollagen molecule is relatively convenient to be hybridized with cells for encapsulation or tissue construction with significant cell-adhesive characteristics along the abundant arginine–glycine–aspartic acid (RGD) sequences in the molecular peptides.

Gelatin is a denatured collagen single chain that has already been used in pharmaceuticals for sustained drug delivery system and is applied to produce cell encapsulating gel. Multiple peptide chains including mammalian atelocollagen molecules are dissociated at temperatures over about 40°C and become gelatin which consists of single peptide strands. Though gelatin does not reassemble to form fibrous matrix as temperature decreases, the physical property is controllable from sol to jelly-like gel, according to the purpose with maintaining the cell-adhesive RGD sequences in the strands.

It is possible to reassemble atelocollagen molecules to fibrous matrix through cross-linking the molecules by induction of intermolecular binding among carboxyl and amine groups using chemicals such as glutaraldehydes and/or random amine–amine bonds with physical methods as γ-ray irradiation, but even a cross-linked

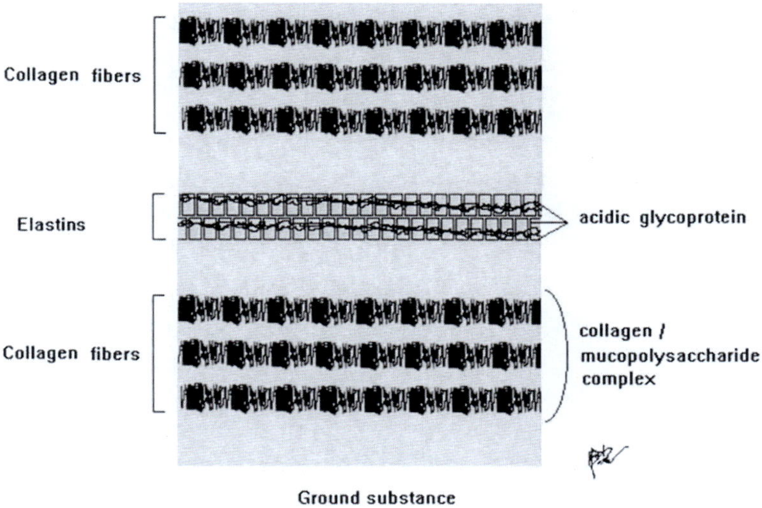

Fig. 3 Diagram of connective tissue. Elastin exists between collagen fiber bundles and provides elasticity to connective tissue

atelocollagen matrix is rather brittle because there is no linear molecule–molecule binding between the telopeptide-removed atelocollagen molecules.

Another main structural ECM component is elastin which provides elasticity of tissue as exists between collagen fibers containing acidic glycogens in connective tissues (Fig. 3). By hybridizing elastin with atelocollagen molecules, a flexible structural membrane can be specifically used in fabricating cells for the soft tissues.

Hyaluronic acid (HA) molecule is another important substance in ECM. This highly hydrophilic polysaccharide consists of repeating two uronic acids without binding to core proteins as other polysaccharides, such as glycoproteins, and demonstrates acceptable viscosity with high water content as in synovial fluids or vitreous humor in nature and maintains tissue homeostasis. HA mediates cell signaling by interacting with cells through cell surface receptors such as CD44. The non-immunogenicity of HA is a practically attractive characteristic which permits to be applied in cell-based therapy.

Several attempts for applying hyaluronic acids with collagen to either cell therapy or multicellular implants are introduced, and cartilage replacement by hyaluronic acid-encapsulated chondrocytes or hyaluronic acid membrane for epithelial cell culture is one of them.

Delivered cells must be integrated with other cells and ECM after implantation as a part of recipient's tissue. Cell–cell adhesion is governed by cadherin depending on Ca^+, and cell–ECM adhesion is via Mn^{2+}- and Mg^{2+}-dependent integrin, cell-adhesive proteins, and ligands on fibrous ECM network. In addition to the cell-adhesive RGD sequences in structural ECM, additive fibronectin, laminin, or vitronectin increases binding intensity and stability to hybridize cells with structural proteins.

Growth factors play important roles in specific differentiation and proliferation of cells via signal transduction from ECM; therefore, cell carrier of structural proteins containing specific growth factors induces stem cells to be committed and differentiated toward specific tissue cells, through growth factors may cause production of various cell-mixed teratoma or induction of oncogenesis at present.

3.1 Application of Acellular Matrix

3.1.1 For Cartilage

To produce articular cartilage, an aseptically harvested porcine cartilage fragment was placed in solution consisting of 1% tert-octylphenyl-polyoxyethylene with 0.02% EDTA in phosphate-buffered saline without Ca^{++} and Mg^{++} for 24 h, together with RNase A (20 μg/ml) and DNase (0.2 mg/ml) under continuous shaking at 4°C. Obtained acellular cartilage was washed with PBS several times to remove residual substances and was stored in PBS at 4°C prior to application (Fig. 4).

Human bone marrow-derived MSCs were seeded at a density of 2×10^5/matrix in serum-free media and allowed to adhere for 4 h and cultured in acellular matrix derived from bovine articular cartilage in a high-glucose DMEM with 10% fetal bovine serum and 1% (v/v) penicillin/streptomycin at 37°C in a humidified atmosphere of 5% CO_2 incubator (Fig. 5).

After 20 days, Alcian blue staining for glycosaminoglycan specific to chondrogenesis was performed to define chondrogenic differentiation of the cultured human MSCs on acellular matrix along with confocal microscopic observation after DAPI and Texas red stains. The cells were completely differentiated and proliferated and became artificial cartilage chondrocytes (Fig. 6).

Fig. 4 Light microscopic view of acellular cartilage matrix stained by Mason's trichrome

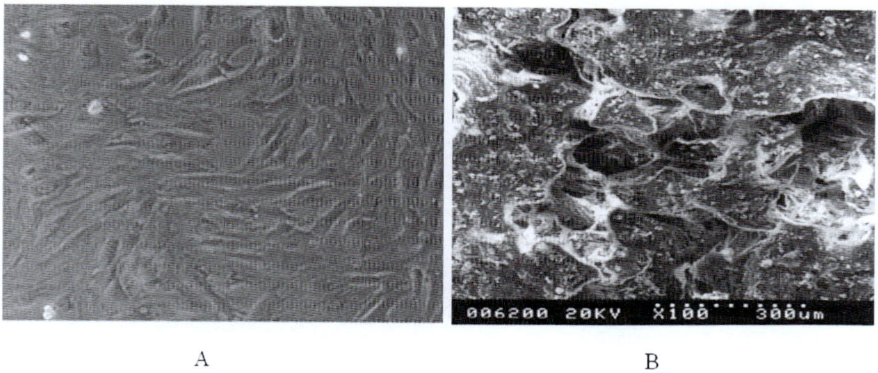

Fig. 5 Cultured human MSCs (**a**), cultured on a porcine acellular cartilage matrix for 20 days (**b**)

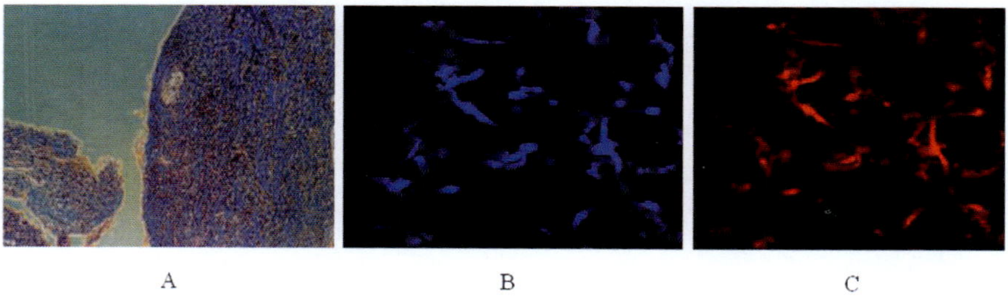

Fig. 6 MSCs cultured on acellular matrix stained by Alcian blue (**a**), and MSCs demonstrating chondrocytic characterisistics by DAPI (**b**) and Texas red (**c**) stains via a confocal microscopy

Human MSCs hybridized with porcine cartilage acellular matrix were implanted into an artificially created rectangular defect with 7 × 3 mm on knee joint of a rabbit as a xenogeneic cartilageous tissue, and the harvested specimen after 12 weeks demonstrated a favorably replaced cartilage (Fig. 7). There was no rejective immune reaction because cartilage is a less vascularized tissue.

3.1.2 For Esophagus

Esophagus provides peristalsis and is often involved in cancer. Therefore, a stent insertion has been recommended to maintain luminal diameter against the obstruction oriented from grown mucosal layer. But the inserted stent is likely to fail because of recurred obstruction following post-operative mucosal growth through lattices of stent-nets. Several tubular synthetic polymeric (e.g., polyurethane, silicone) implants have been applied, but providing appropriate compliance for the

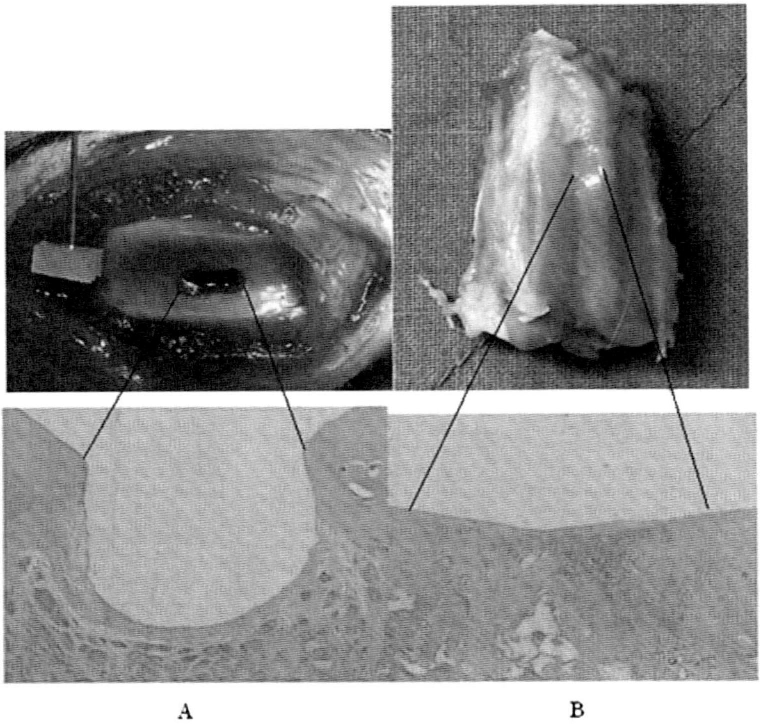

Fig. 7 A human MSC cultured on porcine acellular matrix for 20 days was implanted into a rabbit knee joint (a) and observed after 12 weeks (b). Histological investigation revealed a restored cartilage by chondrocytes differentiated from human MSCs without any immune and inflammatory reaction

peristaltic wave remained as a problem to the biomechanical property of used materials.

Small intestine is a typical tissue for obtaining many blood vessels for nutritional agents of digestive products – absorption and expanded nerve innervations for simultaneous muscle contraction and segmentation which are applicable mechanical characteristics to the esophageal peristalsis.

Submucosal layer of porcine intestine was detached from porcine small intestine, and an acellular small intestine submucosa (SIS) was produced by applying 1% trypsin and 0.02% EDTA mixed solution for 4 h and was lyophilized at –40°C (Fig. 8).

Human MSCs were co-cultured with autologous oral mucosal cells in MEM with 5% fetal bovine serum and 1% v/v penicillin/streptomycin at 37°C in a humidified atmosphere of 5% CO_2 incubator for 7 days to commit toward mucosal fibroblasts. The cultured cells and smooth muscle cells were transported to fabricated porcine SIS, and culture in a MEM without any serum for 14 days was required to produce a

Fig. 8 A freeze-dried tubular porcine SIS

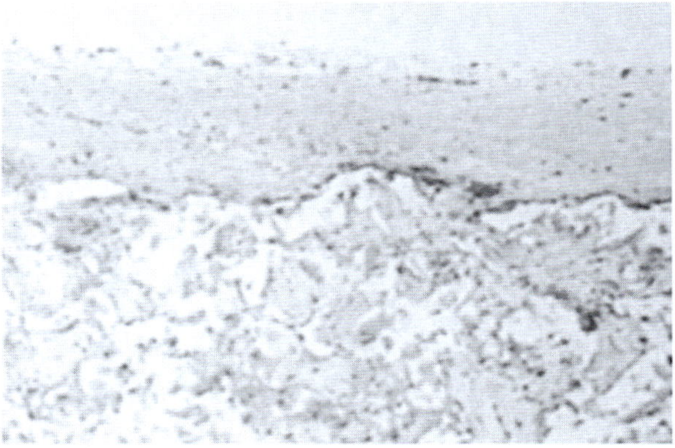

Fig. 9 Histological view of an artificial esophagus by the hMSC-derived cell hybridized SIS

complete artificial esophagus made of SIS hybridized with hMSC-oriented mucosal epithelium (Fig. 9).

Though it was possible to produce esophagus replica in vitro, control of muscle cells to form circumferential alignment to the long axis for providing compliance against the deglutition was a problem remained to be solved.

3.2 Application of ECM

3.2.1 Collagen Encapsulation of MSC for Cell Therapy

Cell suspension is usually designed to deliver individual cells toward target tissue via injection, and encapsulation of cells by alginate is a well-known method permitting selective mass transportation with simultaneously protecting cell from host antibody. However, alginate-encapsulated cells do not proliferate since the material is bioinert.

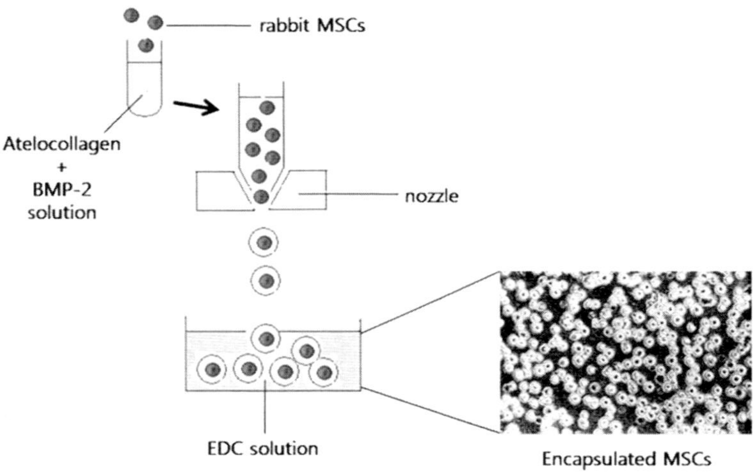

Fig. 10 Diagram of encapsulating cells by collagen solution

As the denaturated collagen molecules, gelatin, have long been used in pharmaceutical industries as capsules for drugs, atelocollagen is applicable to produce cell microcapsules.

One percent atelocollagen in 0.5-N acetic acid solution with controlled pH at 7.4 by NaOH solution was heated at 42°C for 24 h, and recombinant human bone morphogenetic protein (BMP)-2 was mixed with the solution and stored at 25°C. Rabbit bone marrow-derived MSCs were placed in the solution that was placed in a chamber and dropped into cool 1-ethyl-3(3-dimethylaminopropyl) carbodiimide (EDC) solution, with stirring at 5°C, and following centrifugation at $500\times g$ produced a precipitate of microcapsules containing single cell (Fig. 10).

The atelocollagen capsules containing BMP-2 in membrane and coating MSCs were intramuscularly injected into the rabbit's dorsum and observed for 3 weeks to find an ectopic osteogenetic event oriented from MSC differentiation toward bony cells. In histological investigation, osteogenesis with osteocytes and osteoblast-like cells, which might have been differentiated from MSC committed by the hybridized BMP-2 within the encapsulating membrane, was observed (Fig. 11).

This result suggests that encapsulating MSC with collagen hybridized with specific tissue cell growth factors may play a role as a niche committing stem cells to differentiate toward specific tissue cells.

3.2.2 Collagen Matrix for Multicellular Structural Implants

In nature, tissue formation is achieved by selective cell–cell or cell–ECM adhesions. Cell–cell adhesion is a calcium-dependent event through cadherins on cell surface, and cell–ECM adhesion depends on manganese or magnesium and occurs through integrins on cell surface binding to ligands on structural proteins and cell-adhesive proteins in ECM. Arginine–glycine–aspartic acid (RGD) sequence plays

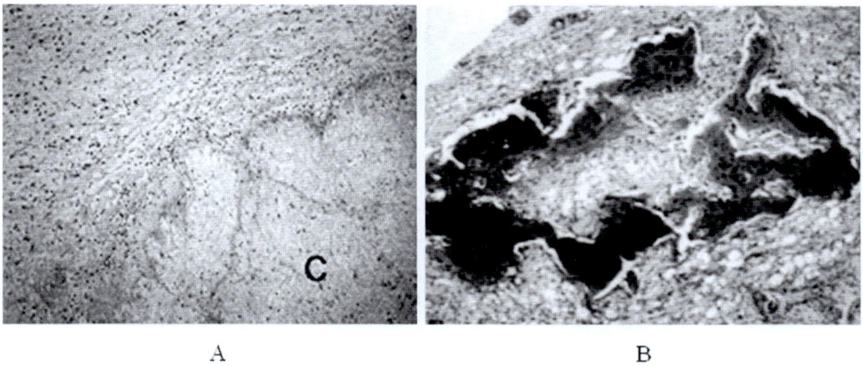

Fig. 11 Histological view of an ectopic osteogenesis by collagen-hrBMP encapsulated rabbit MSCs after 3 days (**a**) and 21 days (**b**). "C" in (**a**) indicates the collagen-rhBMP-2 microcapsules containing MSC

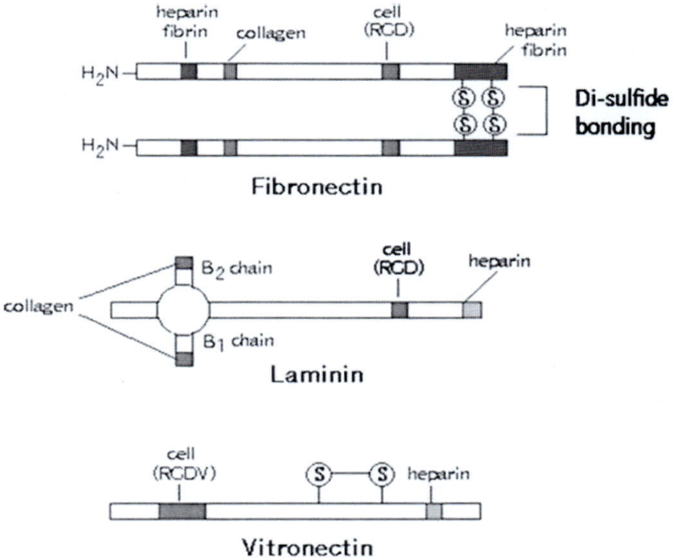

Fig. 12 Adhesive ligands on the cell-adhesive protein molecules

the main role as the cell-adhesive ligand in both structural proteins as in collagen molecule and cell-adhesive proteins as in fibronectin, laminin, vitronectin (Figs. 12 and 13). Though structural proteins support tissue structure and provide general ligands to which integrins bind, cell-adhesive proteins also bind with selective cells, and integrins transmit cell signals controlling cell viability and functions.

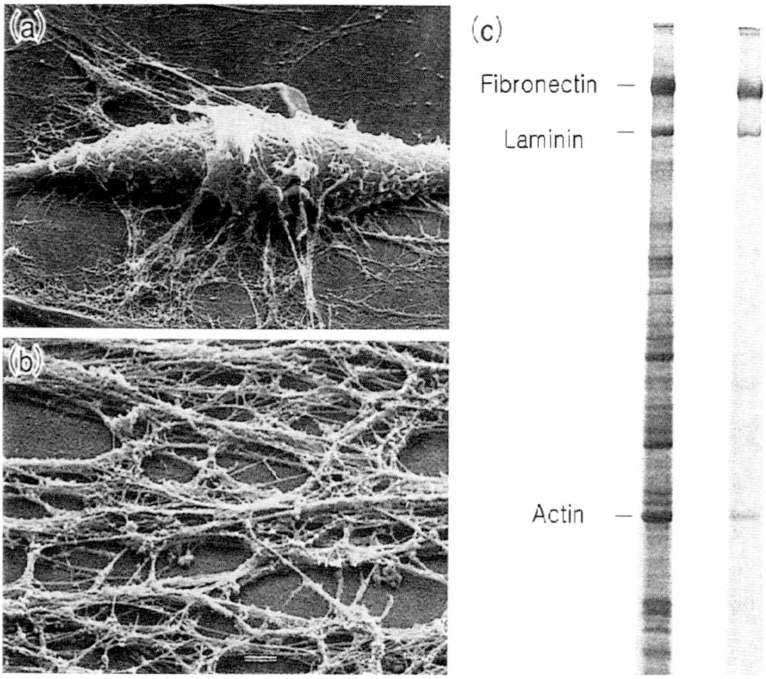

Fig. 13 Cell-adhesive proteins observed in the acellular matrix produced by a fibroblast culture. (**a**) A cultured fibroblast, (**b**) acellular matrix after treatment by surfactant, and (**c**) SDS-PAGE analysis

Ligament plays an important role in joint movements. Therefore, leaving a ruptured ligament often leads to limited mobility as atrophy occurs since mechanical stress stops delivering vital signals. As ligament has an outstanding elasticity, an ECM matrix composed of collagen and elastin was designed. Type I porcine origin atelocollagen and elastin nanofibrous bundle were produced by electrospinning and cross-linked by EDC. Seven-percent solution of bovine type I atelocollagen and elastin (8:2) was loaded into a syringe, positive output lead of a high-voltage supply set to 20 kV was attached to a blunt needle, and target mandrel was rotated at 3,500 rpm. Fabricated nanofiber was cross-linked by 1-ethyl-(3-3-dimethylaminopropyl)carbodiimide hydrochloride (EDC) solution. Fibronectin molecules were hybridized with the produced collagen–elastin nanofibers. Human mesenchymal stem cells (hMSC) were cultured on the aligned collagen–elastin nanofibers and cyclically strained to promote cell differentiation toward ligament array. Human MSCs were cultured on collagen–elastin nanofibers using a spinner bioreactor at 60 rpm for 8 h, and nanofibers matrix with seeded hMSCs was strained at 0.75 Hz in a CO_2 incubator for 6, 12, 18, or 24 h at 1.5 Hz of frequency (Figs. 14 and 15).

To investigate the effect of mechanical strain on differentiation of hMSC toward ligament cell, real-time polymerase chain reaction (RT-PCR) was performed, and

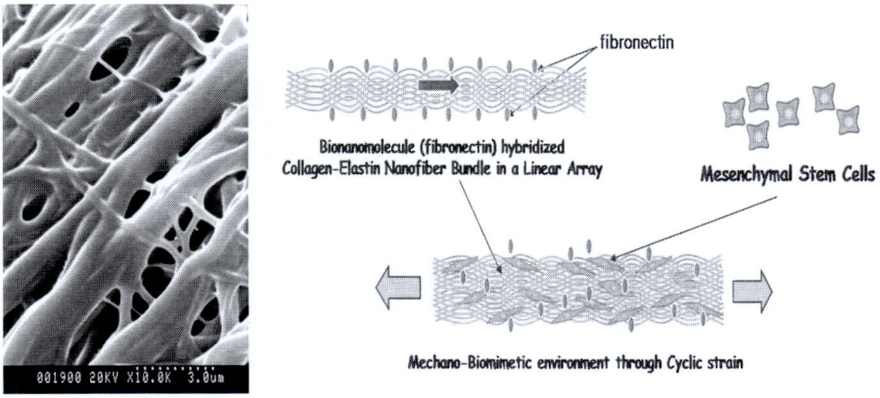

Fig. 14 Structure of a fibronectin-hybridized collagen–elastin nanofiber on which MSCs were cultured to produce a ligament. Delivering a cyclic strain aligns cells along the biomechanical stress direction

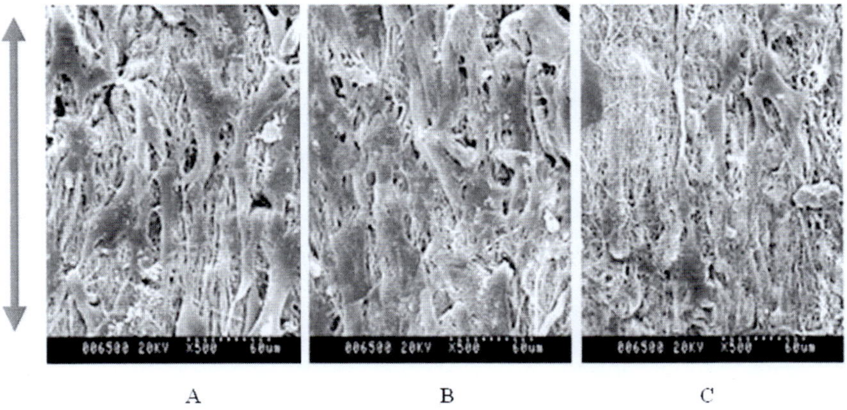

Fig. 15 SEM images of hMSC in collagen–elastin nanofibers after 24 h static culture (**a**), applying 6 h strained and 18 h static culture (**b**), and applying 12 h strained and 12 h static culture (**c**). The *arrow* beside (**a**) indicates the strain direction

the result demonstrated that mRNA expression of the type I and III collagen increased distinctively as the cyclic tensile stress time increased (Fig. 16).

After culture for 7 days in a spinner bioreactor, the hMSC-cultured collagen–elastin membrane was implanted in a resected (length 5 mm) rabbit's Achilles tendon ligament. On the 90th post-operative day, the resected tendon ligament was completely regenerated by the implanted hMSC hybridized membrane (Fig. 17).

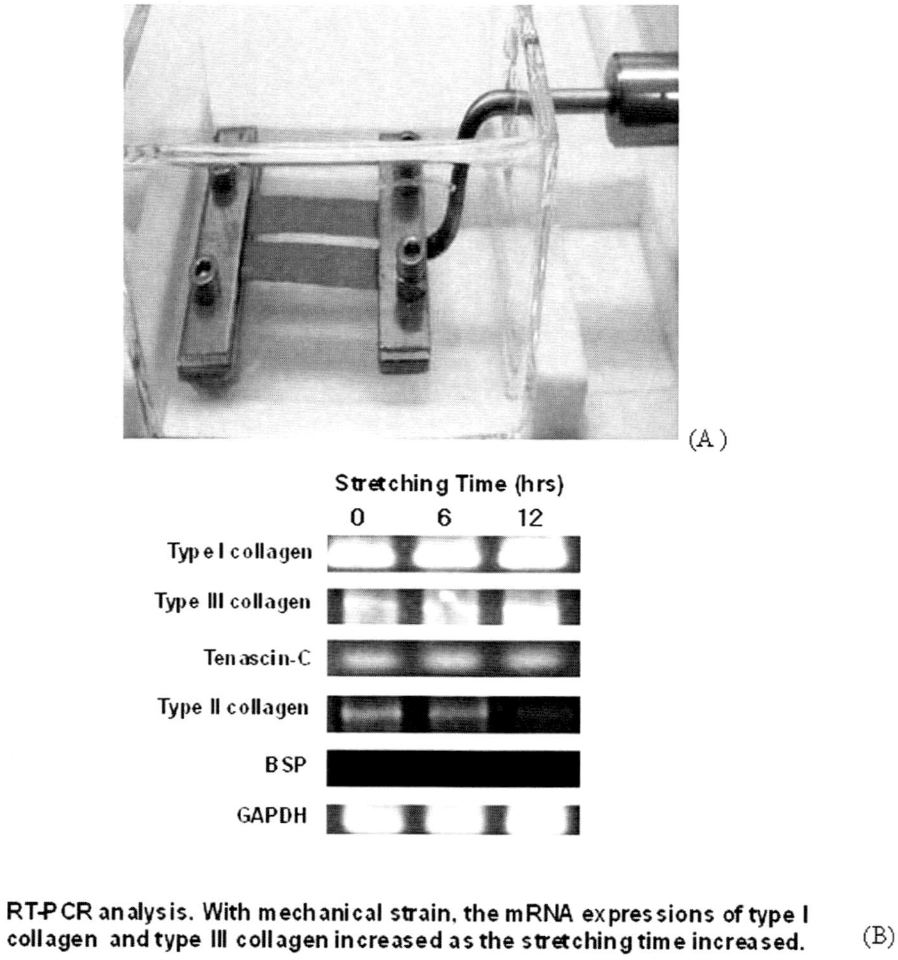

Fig. 16 Cyclic strain chamber (**a**) and the result of RT-PCR (**b**)

4 Summary

Though it has been possible to restore anatomical defects or deformities as biocompatible non-vital substances (e.g., metals, ceramics, synthetic or natural polymers) have been introduced for implantation, replacement of diseased tissues or organs has long depended on transplantation. Then the concept of engineered tissue, which is an artificial tissue consisting of adult cells hybridized with synthetic biomaterials to overcome limitation of donors, has been developed since the mid-1980s.

Meanwhile, invention of tools for gene transfection in the 1970s opened a new era of cell manipulation which modify biological functional characteristics of individual cells and has led to the so-called cell therapy purposing treatment of diseases from

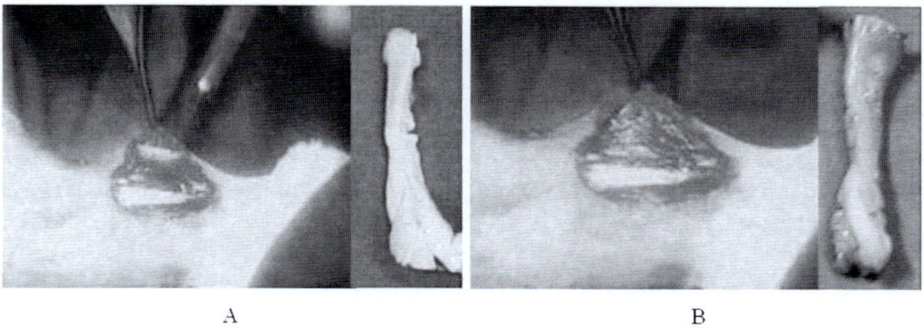

Fig. 17 In situ experiment of the cultured hMSC hybridized collagen–elastin nanofibrous membrane containing fibronectin. Rabbit's Achilles tendon ligament was resected (**a**). After post-operative 90 days, the membrane-implanted lesion was completely regenerated (**b**)

genetic disorders by delivering cells with corrected gene. The complete decoding of the human genomes in 1999 has greatly contributed to the understanding and control of genetic events. After discovery of pluripotent and/or multipotent stem cells in the mid-1990s, gene replication has made it possible to reprogram adult cells to have even pluripotency as embryonic stem cell in 2008.

To apply stem cells in clinics, establishment of safe tools is required to avoid oncogenesis since pluripotent or multipotent cell may be differentiated toward either normal or cancer cells as known from evidence that a simple expansion of stem cell produces teratoma. Therefore, development of correct cell signal transduction for appropriate commitment of stem cells toward specific tissue cells as in the developmental stage is important. If such a commitment is controllable, control of differentiating cell number will be required to prevent malignant tumor formation due to expansive overgrowth of the delivered cells. In nature, commitment to specific tissue cells and formation of appropriate anatomical volume of specific tissue are managed through signal transductions by the host niche which is a part of ECM.

Therefore, providing perfectly mimicked ECM for delivered stem cells will support a successful approach to the regeneration of tissue.

References

1. Suh H. Collagen fabrication for the cell-based implants in regenerative medicine. In: Bioengineering in cell and tissue research. Berlin: Springer; 2008, pp. 159–192.
2. Suh H, Okazaki M, Kimura M. The biological response of the apatite-collagen composite. In: Apatite. Vol. 1. Tokyo: Takayama Press; 1992, pp. 293–298.
3. Li TZ, Shin S-H, Cho HH, Kim JH, Suh H. Growth factor-free cultured rat bone derived mesenchymal stem cells towards hepatic progenitor cell differentiation. Biotechnol Bioprocess Eng. 2008;13:511–523.
4. Kim JH, Park S-N, Suh H. Generation of insulin-producing human mesenchymal stem cells using recombinant adeno-associate virus. Yonsei Med J. 2007;48:109–119.

5. Cha JM, Park S-N, Noh SH, Suh H. Time-dependent modulation of alignment and differentiation of smooth muscle cells seeded on a porous substrate undergoing cyclic mechanical strain. Artif Organs 2006;30:250–258.
6. Choi YS, Park S-N, Suh H. Adipose tissue engineering using mesenchymal stem cells attached to injectable PLGA spheres. Biomaterials 2005;26:5855–5863.
7. Kim H, Kim HW, Suh H. Sustained release of ascorbate-2-phosphate and dexamethasone from porous PLGA scaffolds for bone tissue engineering using mesenchymal stem cells. Biomaterials 2003;24:4671–4679.
8. Suh H, Song MJ, Park YN. Behaviors of isolated rat oval cells in porous collagen scaffold. Tissue Eng. 2003;9:411–420.
9. Kim H, Lee J-H, Suh H. Interaction of mesenchymal stem cells and osteoblasts for in vitro osteogenesis. Yonsei Med J. 2003;44:187–197.
10. Park S-N, Lee HJ, Lee KH, Suh H. Biological characterization of EDC-crosslinked collagen-hyaluronic acid matrix in dermal tissue restoration. Biomaterials 2003;24:1631–1641.
11. Park S-N, Park J-C, Kim HO, Song MJ, Suh H. Characterization of porous collagen/hyaluronic acid scaffold modified by 1-ethyl-3(3-dimethylaminopropyl)carbodiimide cross-linking. Biomaterials 2002;23:1205–1212.
12. Suh H, Han D-W, Park J-C, Lee DH, Lee WS, Han CD. A bone replaceable artificial bone substitute: osteoinduction by combining with bone inducing agent. Artif Organs 2001;25: 459–466.
13. Suh H. Tissue restoration, tissue engineering and regenerative medicine. Yonsei Med J. 2000;41:681–684.
14. Suh H. Treatment of collagen for tissue regenerative scaffold. Biomed Eng Appl Basis Commun. 1999;11:167–173.
15. Suh H. Recent advances in biomaterials. Yonsei Med J. 1998;39:87–96.
16. Suh H. Fundamental concepts for tissue engineering. Biomater Res. 1998;2:1–7.
17. Suh H, Lee C. Biodegradable ceramic-collagen composite implanted in rabbit tibiae. ASAIO J. 1995;41:652–656.

Neural Stem Cells: From Cell Fate and Metabolic Monitoring Toward Clinical Applications

Jan Pruszak, Máté Döbrössy, Jochen Kieninger, Kuppusamy Aravindalochanan, Gerald A. Urban, and Guido Nikkhah

Abstract Stem cells are characterized by their ability to self-renew and to generate a diverse range of physiological cell types (Singec et al., Annu Rev Med. 2007; 58:313–328). In mammals, stem cells found in the mature organism (somatic stem cells) generate progeny of a specific cell lineage or tissue type (multipotency). For example, a hematopoietic stem cell gives rise to the cell lineages found in bone marrow and blood, while a central nervous system (CNS) neural stem cell generates neurons, oligodendrocytes, and astroglia (Murry and Keller, Cell. 2008; 132:661–680). At an earlier developmental stage, embryonic stem (ES) cells obtained from the inner cell mass of blastocyst-stage embryos exhibit limitless capacity to self-renew and can give rise to any cell type of the organism, thereby defining pluripotency (Murry and Keller, Cell. 2008; 132:661–680; Singec et al. Annu Rev Med. 2007; 58:313–328). The term "totipotent" is reserved for cells that can also give rise to the trophoblast and extraembryonic tissue in vivo, such as the fertilized egg (zygote). These designations exemplify how differentiation toward a specific mature cell phenotype is accompanied by an increasingly limited spectrum of potential descendant cell types and by a diminished proliferative capacity (Yeo et al., Hum Reprod. 2008; 23:67–73) (Fig. 1). In addition, recent advances in reprogramming of adult somatic cell types into pluripotent cells (Takahashi et al., Cell. 2007; 131:861–872) are aimed at controlling the regulatory mechanisms that govern stemness and pluripotency, which may soon enable even more refined modulation of cell fate (Yeo et al., Hum Reprod. 2008; 23:67–73; Pruszak and Isacson, Development and engineering of dopamine neurons. Austin, TX: Landis Bioscience; 2008; Jaenisch and Young, Cell. 2008; 132:567–582). To achieve translation of stem cell biology into clinical applications, somatic, embryonic, and reprogrammed stem cell sources alike are presently being investigated.

J. Pruszak (✉)
Laboratory of Molecular Neurosurgery, Department of Stereotactic and Functional Neurosurgery, Albert Ludwigs University Freiburg, Freiburg, Germany; Center for Neuroregeneration Research, Harvard Medical School, McLean Hospital, Belmont, MA, USA

G.M. Artmann et al. (eds.), *Stem Cell Engineering*,
DOI 10.1007/978-3-642-11865-4_20,

Despite the fast-paced progress of stem cell research as a field, many aspects of cell development in the dish are still not fully understood. To ensure appropriate patterning, signaling parameters for cell lineage specification need to be identified, and generating the phenotype of interest requires close monitoring and controlled modulation of the microenvironmental conditions in the dish.

We summarize universal principles, advancements, and ongoing challenges in deriving and characterizing therapeutic cell types from pluripotent and multipotent stem cells for clinical and scientific biomedical scenarios. Specifically, this overview illustrates the paradigm of neural stem cell differentiation and the application of microphysiometer systems to monitor and control conditions fostering the generation, maturation, and survival of specific neural cell populations aimed at treating neurological disease.

Keywords Cell culture monitoring · Micro physiometry · Micro technology

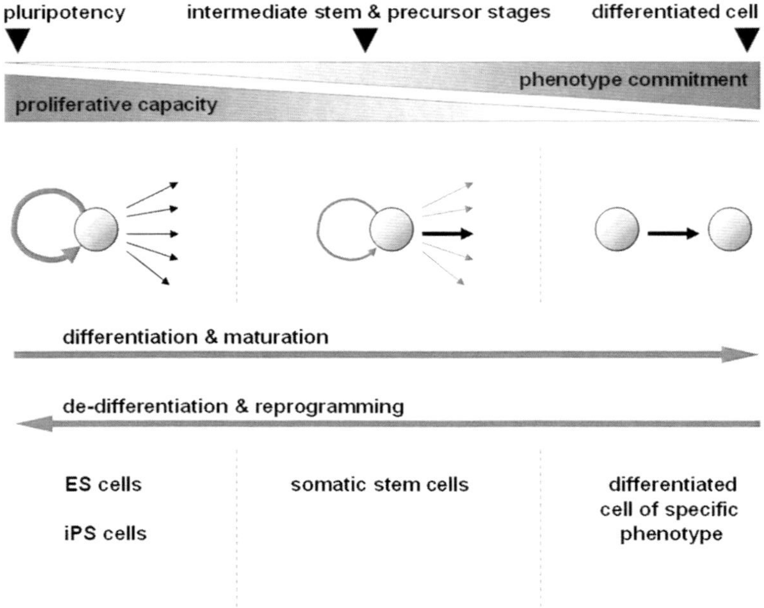

Fig. 1 Schematic of stem cell differentiation. Representing a rather primordial source, embryonic stem (ES) cells are pluripotent and exhibit limitless self-renewal. True pluripotent cells show little to no phenotypic commitment and can differentiate into all major tissue types. Over the course of development, i.e., upon differentiation in vitro, proliferative capacity usually decreases while phenotypic restriction increases. An appropriately patterned precursor cell will only give rise to its corresponding descendant lineage. However, recent reprogramming efforts can reverse a differentiated somatic cell back to an earlier immature pluripotent stage, rendering it with pluripotency. This enables scenarios such as the generation of autologous neuronal cells from a patient's skin cell

1 Stem Cell Differentiation and Fate Specification

A viable cell source for regenerative medicine, scientific study, and pharmacological testing must provide ample amounts of medical grade, standardized, physiologically differentiated specific cell populations in a controlled manner. How can stem cell sources be exploited to generate biomedically relevant functional cell types in high purity?

Pluripotency and unwanted growth. It is paramount to maintain an awareness of safety issues when cell therapeutic application is the eventual goal. Pluripotent cells exhibit and will maintain self-renewal as long as kept under defined conditions in vitro. Differentiation is accompanied by a marked downregulation of immaturity markers [1, 2] including the transcription factors Oct4, Sox2, Nanog; the surface antigens Tra-1-81, Tra-1-60, SSEA3, SSEA4; and the enzymatic marker alkaline phosphatase. However, proliferative immature stem cells remaining as "contaminants" in biomedical cell preparations can lead to formation of tumors. Importantly, even after eliminating any residual pluripotent cells from the cell suspension prior to grafting, tumors can develop from proliferative precursor cells (non-teratoma type) [3]. It should be noted that an authentic pluripotent cell source (such as the classic embryonic stem cell) is dissimilar from a cancer cell in that it responds to differentiation cues and will therefore eventually differentiate. In contrast, alternative stem cell sources such as immortalized neural cell lines containing genetic constructs of inducible oncogenes may bear a higher risk of developing malignancies. Insertion of suicide genes [4] may be a viable option as a "safety net." All cell sources require critical evaluation before translation to clinical applications. As cells change over time in vitro, particularly after clonal expansion, it is necessary to regularly monitor chromosomal integrity and stability through karyotyping [5, 6]. Key strategies to combat the manifold barriers to rapid clinical translation include the optimization and monitoring of in vitro differentiation conditions, gene engineering, and cell purification.

Patterning, growth factors, and culture conditions. When differentiating stem cells toward a phenotype of interest, it seems most sensible to induce genes in a sequence mimicking the chronology of normal development. Given the complexity of cell–cell signaling and gene regulatory transcriptional cascades, the design of in vitro differentiation protocols has to take into account the orchestrated interaction of factors governing physiological cell fate specification in vivo during embryo development [7]. Toward that goal, protocols for generating dopaminergic neurons from human pluripotent stem cells made use of successive inductive cues active during normal development, such as Noggin, Sonic hedgehog, and GDNF, to yield functional neural cell types [8–10]. Such work provided the proof of principle that phenotypes of clinical relevance could be generated from mouse, as well as human pluripotent stem cells and grant therapeutic benefit [3, 10]. Yet further optimization is required to obtain biomedical cell preparations from stem cells.

Cell development in vitro. Existing differentiation procedures of pluripotent stem cells or neural stem cells augment certain cellular subtypes, but are insufficient to coordinate the birth and development of cells to the same degree seen in normal ontogenesis, in part because we do not yet fully understand and control the precise cues provided by complimentary cell populations and by the microenvironment. Plating density, handling of the cells, and cell–cell interactions can interfere with the administered media supplements and substrates [11]. As a result, only a minor fraction of the differentiated cells may represent an authentic equivalent of the anticipated cell type. Moreover, cells at varying stages of maturation and lineage are present in culture, causing a cellular heterogeneity that impedes experimental and clinical utility [2, 7, 12, 13]. Such problems have been observed in the differentiation of hematopoietic cell types from human ES cells, which failed to functionally reconstitute the hematopoietic system in vivo, despite mimicking marker expression in vitro [14]. Similarly, dopaminergic neurons generated from embryonic stem cells do not yet represent a direct copy of in vivo-grown midbrain cell types [15, 10]. Ongoing research efforts will increase our understanding of the transcriptional regulatory cascades and epigenetic conditions of phenotype control, such that we can better induce relevant neural subtypes from stem cells. Throughout, it is critical to identify stringent criteria that define whether a cell product is appropriate for cell therapy:

Characterization of functional phenotypes. To monitor the differentiation of stem cells, a combination of genetic, morphological, immunophenotypic, and functional criteria is used. What characterizes the neural cell phenotypes generated in vitro for therapeutic studies? For instance, precursors to dopaminergic neurons should express ventral midbrain region-specific markers analogous to normal development, regardless of whether they are derived from pluripotent or neural stem cell sources

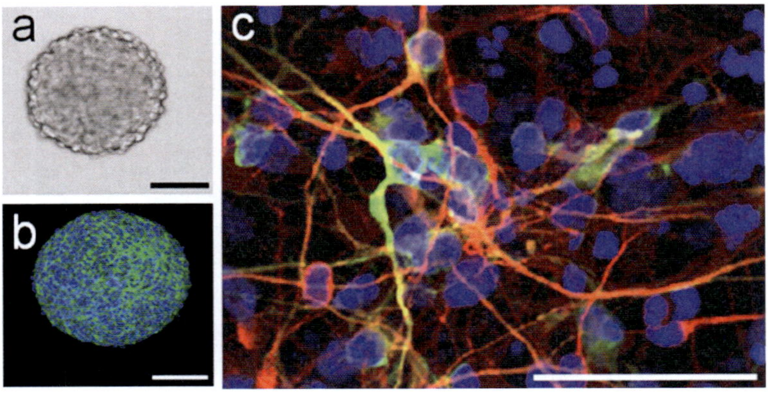

Fig. 2 Differentiation of human neural stem cells. Neural stem cells (the somatic stem cells of the nervous system) can be propagated as neurosphere cultures applying mitogens such as epidermal growth factor (EGF) and basic fibroblast growth factor (bFGF) for continued expansion. Phase-contrast image of a neurosphere (**a**) derived from human ventral midbrain after long-term expansion (scale bar = 100 μm). Immunostaining shows that neurospheres contain large numbers of nestin-positive cells (**b**) (*green* – nestin, *blue* – DAPI; scale bar 100 μm). For appropriate differentiation, here into dopaminergic neurons (**c**), a cocktail of patterning factors including brain-derived neurotrophic factor is applied (*green* – tyrosine hydroxylase, *red* – β-III-tubulin, *blue* – DAPI; scale bar = 50 μm). Modified from [27]

(Fig. 2). As mentioned earlier, physiologically appropriate timing during application of extrinsic factors and supplements, as well as the influence of cell–cell interactions in the dish has to be taken into account. Tight control of marker expression over time, as detected by immunocytochemistry or mRNA expression analyses, ensures that the cascade of events mimics the normal in vivo situation during development. In the case of dopamine neurons, such midbrain-region markers monitored at multiple times during differentiation in vitro include the various stage-characteristic transcription factors FoxA2, Nurr1, Pitx3, Lmx1a, Lmx1b, En1/2, among others. Ideally, any stem cell derivative should express such region-typical markers at the corresponding stages of in vitro maturation, in order to ensure that the blueprint of a fetal DA neuron has been matched as closely as possible. As described in greater detail below, microtechnological measuring devices can determine functionally relevant parameters such as neurotransmitter release and electrophysiological properties. Notably, differentiation studies of stem cells in vitro and after transplantation can serve as a valuable tool to further increase our understanding of development, using human cells as a novel model organism that complements our established animal models of genetics and embryology.

Gene engineering. Our growing insight into the genetic regulatory cascades governing neural phenotype can be directly applied to increase the fraction of cells of therapeutic relevance in the dish. One way to enhance dopamine neuron development from human embryonic stem cells utilized a combined approach, expressing two dopamine neuron-characteristic factors, Pitx3 and Nurr1 together [16]. Another strategy identified and exploited Lmx1a to successfully increase midbrain dopamine neuronal specification [17]. It should be noted that forced gene expression requires close monitoring to avoid generation of non-physiological cell types. For example, using a tetracycline (tet)-inducible mouse ES cell line, Nurr1 expression at the nestin-positive precursor cell stage induced gene expression of dopaminergic markers such as AADC, DAT, and AHD2 even in non-neuronal cells [18], while other transcription factors relevant for physiological midbrain development such as Otx2, En-1, GBX2, Pitx3, and Lmx1b remained unaffected by the tet-induced Nurr1 expression. Similar experiments will continue to elucidate the function of transcriptional interactions underlying cell fate decisions. On the other hand, this work illustrates that forced expression of key modulators can perturb the complex regulatory networks governing cell phenotype and requires careful fine-tuning when intended for therapeutic purposes.

Cell sorting and purification. Our increased insight into cell specification also opens possibilities to harness population-specific genes as fluorescence-labeled markers. For example, Sox1-GFP characterizes neural precursors and Pitx3-GFP or TH-GFP serve as markers for dopamine neurons, which has proven useful for enriching cellular subtypes by cell sorting. Fluorescence-activated cell sorting (FACS) optimized for selection of fragile neural cells [2] has been successfully applied to eliminate tumor-generating proliferative cells from ES cell-derived [19, 20, 2] as well as iPS-cell-derived [21] neural cell populations. Furthermore, genetic fluorescent markers such as Tau-GFP [22], Pitx3-GFP [19], and also dye labeling [23] have been used to purify specific rat and mouse neuronal cell types. Finally, a Synapsin promoter-driven GFP was harnessed to successfully isolate differentiated human neurons from a diverse cell population derived from embryonic stem cells [2]. Importantly, FACS is routinely used in hematological cell transplantation to generate clinical-grade cell preparations of high purity [24] by means of a broad array of surface antigen combinations to define specific subsets of the hematopoietic lineage. For more widespread applications in other tissue types, markers must be identified that define developmental maturity and lineage specification according cluster of differentiation (CD) antigens. This will enable the isolation of specific subpopulations by FACS or by immunomagnetic cell separation ("MACS"). Alternative cell selection procedures using microfluidics and optical switches may open additional options for clinical-grade cell selection in the near future, exemplifying how technological advances synergize with increased biological insight. For reasons of safety and reproducibility, cell selection procedures are likely a prerequisite to future applications in a clinical setting.

2 Microphysiometer Systems

The ultimate positive readout of biomedical effectiveness of a particular cell preparation will be the restoration of function after transplantation into the patient. So far, proof-of-principle studies demonstrating functionality in appropriate pre-clinical animal models, such as the toxic-lesion primate models of Parkinson's disease, are still required. To ensure consistent development of future algorithms for monitoring cell differentiation, more cost-effective and practical surrogate measures of phenotype specification and cellular functionality are needed for outcome prediction and quality control. There is therefore a need for functional assays that can predict clinical safety and efficacy [25]. As outlined earlier, gene expression profiles can be defined that match the genetic fingerprint of in vitro-generated sources to a validated gold standard. A routine clinical readout can then be employed, for example, one based on gene arrays. Similarly, physiological assays that measure metabolites, electrical responsiveness, and other functional parameters can be devised as chip-based assays and microsensor systems. Specifically, pericellular parameter fluctuations in the dish due to endogenously secreted factors, physical factors, and plating density [7, 25] can be monitored and potentially modulated using microsystem technology. Microphysiometer systems usually monitor several physiological parameters (e.g., oxygen, pH, glucose, lactate, nitric oxide, neurotransmitters) simultaneously. Most of these parameters require measuring on a microenvironmental level, pericellularly, as they significantly vary within a cell culture system due to diffusion gradients (such as oxygen) or due to decomposition (e.g., nitric oxide, which has a very small half-life in the presence of oxygen). Both in vivo and in vitro, there is a demand for application-specific small, integrated sensors, and microsystems technology allows a cost-effective fabrication of such devices. Currently, microphysiometer systems are usually developed for in vitro applications. Which metabolic factors are both relevant for stem cell differentiation and accessible to detection by micro sensors?

Metabolic parameters. With respect to monitoring devices, neural stem cell cultures can be considered as being equivalent to other adherent cell cultures in general, such as cancer cell lines, as long as there are no toxicity effects and the cells proliferate and differentiate in the presence of sensors in the same way as without. Regular monitoring of small molecules (oxygen, protons, nitric oxide) as well as biomolecules (glucose, lactate, neurotransmitters) is important to maintain a stable microatmosphere and can help to obtain more reproducible and comparable results from cell culture experiments. Additionally, it yields new insight into regulatory mechanisms of the cells and of the microenvironmental conditions pertinent to their development. In this section we focus on the extracellular measurement of these parameters in vitro.

Microphysiometry principles. Physiologically relevant molecules as listed earlier can be classified into two groups. Those of the first group have *both* a source and a sink inside the cell culture system. Oxygen, for example, is consumed due to respiration of the cells and can therefore be regarded as a sink. The atmosphere surrounding the cell culture is defined as an infinite source. Hydrogen ions excreted by the cells are dissipated by the buffer system of the medium, which thereby acts as sink, as long as the buffer capacity is not exceeded. Nitric oxide (NO) released from the cells has its sink in the reaction with oxygen to nitrogen dioxide (NO_2). All of these parameters can show non-monotone sensor readings over time, as the flux stemming from sink and source can be independent from one another. The second group of molecules has *either* a sink *or* a source inside the cell culture system. This is illustrated by glucose which is depleted from the medium is through cellular metabolism (sink only). Concurrently, the lactate concentration is increasing (source only system). The common characteristic of this second group of molecules is the monotone behavior of the sensor readings over time, therefore demanding a broader sensitivity range for long-term experiments (days to weeks).

Oxygen. The most prominent molecule in cell physiology is oxygen, which is consumed by cellular respiration. In neural cell culture, lowered levels of environmental oxygen have, for example, been demonstrated to promote dopaminergic differentiation [26, 27]. By measuring the pericellular oxygen concentration, two types of information can be obtained: first, the absolute oxygen concentration, which influences the cells' aerobic or anaerobic metabolic pathways; second, assuming an undisturbed cell culture and therefore a diffusion gradient along the medium height, the overall cellular respiration rate can be calculated [28]. There are multiple oxygen measurement systems described in the literature. Usually, oxygen is measured by electrochemical reduction at platinum [29, 30] or palladium electrodes [31]. An advantage of this approach is that the measurement signal ideally has a defined zero-point. The major drawback is the immanent consumption of the analyte, which can be minimized, but never completely avoided. An example of such a sensor system for in vitro applications is depicted in Fig. 3. Alternative approaches utilize fluorescent dyes that display quenched emission spectra in the presence of oxygen [32]. While an advantage of such optical sensors is the absence of analyte consumption, they usually do not have a defined zero-point, requiring at least a two-point calibration. If the application of such sensors is intended for long-term experiments,

Fig. 3 Micro sensor systems. Microsensor systems can be used to determine metabolites, electrical responsiveness, and other functional parameters. Most of the physiological parameters of relevance, such as oxygen, pH, glucose, lactate, nitric oxide, neurotransmitters, require measuring on a microenvironmental level, due to significant diffusion gradients within a cell culture system. Using microphysiometry, pericellular fluctuations of these physiological parameters due to endogenously secreted factors, physical factors, or plating density can be monitored. The images display a sensing cell culture flask (SCCF) comprising four amperometric oxygen sensors, enlarged on the *right panel* [29]

methods have to be devised to compensate for the inherent drift: for example, the application of special measurement protocols for electrochemical sensors [29] or the usage of phase-sensitive measurement methods in an optical system [33].

Measuring pH. Stability of pH is a requirement for routine stem cell culture, exemplified by the colorimetric indicator phenol red in standard tissue culture media. In a conventional cell culture medium the buffer system is usually strong (e.g., the bicarbonate system). Notably, even potent buffers cannot completely compensate for acidification of the medium due to cellular respiration. Still, significant change in pH can only be observed during long-term experiments (days to weeks) or under extremely dense versus sparse plating conditions. Nevertheless, several reports [31, 34] describe high acidification rates over relatively short periods of time, which is achieved by the usage of a weakly buffered cell culture medium. To measure pH, ion-selective field effect transistors (ISFETs) are frequently used as sensors. These pH-ISFETs have a non-metallic, pH-dependent gate exposed to the measurement solution. One inherent disadvantage of pH-ISFETS is their drift, restricting their usage to short-time applications or relative measurements. An alternative path is the potentiometric pH measurement with metal oxides, a method exhibiting a significantly lower drift rate as well as improved performance in long-term experiments. For the metal oxide electrodes mainly iridium oxide films are used [35].

Glucose metabolism and lactate. Driven by the need for glucose measurements in the blood of diabetic patients, numerous efforts have been made to devise glucose sensors. The electrochemical approach is the most promising one for long-term monitoring of glucose levels in vivo as well as in vitro. Glucose-specific enzymes such as glucose oxidase are immobilized on electrodes or embedded in hydrogels on top of the electrodes [36, 37]. Glucose concentration is measured indirectly by the oxidation of hydrogen peroxide at a platinum electrode. Lactate can be measured in an analogous way, by choosing appropriate enzymes and membranes [37, 38].

Optically, glucose exhibits a characteristic electromagnetic absorption spectrum. From the amount of light absorbed in the near infrared region glucose levels can be deduced [39]. Nevertheless, because of scattering of light inside tissue, the in vivo situation limits the access to accurate absolute measurements. An alternative optical sensor approach uses the glucose-binding protein of *Escherichia coli*, to which a fluorophore is conjugated [40]. For in vivo as well as cell culture applications all described methods can be combined with microdialysis, in order to measure glucose concentrations without the need of implantation or integration of the sensors [41, 42].

Nitric oxide (NO). Nitric oxide's manifold physiological roles are exemplified by its acting as a neural transmitter in the central and peripheral nervous systems [43]. NO is a free radical and its half-life in the human body is below 6 s. Hence, quantifying NO represents a challenging task. A number of detection methods have been proposed depending on nitric oxide's unique physical and chemical properties. Chemiluminescence is a technique based on the reaction of gaseous NO with ozone to form nitrogen dioxide and thereby emitting photons. A red-sensitive photomultiplier tube captures these photons, enabling quantification. Of greater relevance for cell culture, dissolved NO in biological medium can be measured by purging tissue fluid with helium or nitrogen, which remains to be a commonly applied method [44]. Another sophisticated, albeit cumbersome, method is electron spin resonance spectroscopy which utilizes the spin-trapping of NO's free electron [45]. An electrochemical approach toward a NO sensor was introduced by Shibuki et al. [46]. Later, Malinski et al. developed a microsensor by modifying the sensor surface with electropolymerized nickel porphyrin and Nafion membranes [47]. The latter supports the selective detection of nitric oxide, which opens new ways to quantify NO in an elegant and simple manner. Many groups continue to work on devising NO sensors that enable sensitive quantification of NO without interference by other molecules. Their main focus is put on enhancing the selectivity for NO by exploring different electrodes, selective membranes, and improved measurement techniques [48–50].

Neurotransmitters. In the generation of specific therapeutic neuronal subtypes, functionality can be assessed by monitoring the concentration of synthesized neurotransmitters such as acetylcholine, dopamine, norepinephrine, epinephrine, serotonin, histamine, γ-aminobutyric acid, glycine, and glutamate. In order to evaluate the function of intrinsic or of transplanted neurons, microdialysis can be exploited. Here, a probe is introduced into the brain tissues that contains a semipermeable membrane allowing tissue fluid to diffuse inside the probe [51]. The concentration of the neurotransmitters can then be externally analyzed by liquid chromatography, electrochemistry, and fluorophore methodologies. Additional functional tests utilize brain imaging. Positron emission tomography (PET) is a spectrometric, non-invasive technique by which chemical agents are administered that emit positrons dependent on the presence of neurotransmitters [52], enabling the indirect quantification of neurotransmitters. Functional magnetic resonance imaging (fMRI) represents another spectroscopic principle of measuring neurotransmitters based on hydrogen atomic spins [53]. Electrochemical techniques are also well suited for in vivo real-time experiments. Different neurotransmitter molecules such as serotonin, dopamine, and other catecholamines can be measured by these techniques due to different electrode materials and electrochemical measurement methodologies: amperometry, chronoamperometry, differential pulse voltammetry, and fast-scan voltammetry [54–56].

Microtechnological aspects. The emerging use of microtechnology in sensor fabrication has a strong impact on the development of novel microphysiometry systems. Mainly two approaches in microtechnology are used: First, there is the classic top-down approach, where conventional dip-in sensors benefit from improved fabrication methods by miniaturization to achieve appropriate dimensions. Second and more relevant is the thin-film technology, adapted from the semiconductor fabrication field. To obtain pericellular sensor readings the cells should either settle directly on a sensor chip or miniaturized dip-in sensors should be positioned very close to the cells. For example, by microfabrication, biochips comprising an array of metal microelectrodes embedded on a glass substrate can be devised (multi-electrode arrays, MEA), which are used to stimulate and record electrical activity of excitable biological cells, such as neurons or myocytes, from multiple sites in parallel. A critical point of the sensor chips is the chip surface, which should either allow the cells to adhere to or allow the usage of cell culture-specific coatings. In contrast, dip-in sensors avoid this issue, as the cells do not have to settle onto the sensor surface [57]. But the effort to achieve an exact and reliable positioning of the sensor probes in the near vicinity of the cells is higher than with sensor chips. For sensor chips, directly settled by cells, silicon is often used as substrate [31, 58]. These sensor chips have the significant disadvantage that they are all optical opaque and therefore prohibit the usage of inverted microscopy for observation and appraisal of the cells. To overcome that, transparent glass chips can be used [29], which allows optical inspection. By using a transparent conductive material for the connection lines (e.g., indium tin oxide) nearly the whole chip surface (expect sensor points comprising metallic electrodes) could be accessed by inverted microscopy. The main focus on recent research is in the field of improved combination of electrochemical sensor principles with thin-film technology. Taken together, microtechnology systems can be helpful in the verification and monitoring of physiological phenotypes through biochemical and neurophysiological in vitro and in vivo assays. Such refinement of protocols further approximates the differentiation of stem cells to normal development with the goal to create authentic, functional, therapeutically relevant neural subpopulations.

3 Toward Clinical Applications

Principles of cell therapy. The option of generating cells in vitro to replace tissue lost due to degenerative processes, injury, and other lesions has opened an exciting, rapidly developing field of biomedical research and applications [59]. A universal model for the novel disciplines of cell therapy and regenerative medicine entails the following sequence: (1) An initially intact organ or tissue system develops a more or less circumscribed defect (or becomes defective during development in utero).

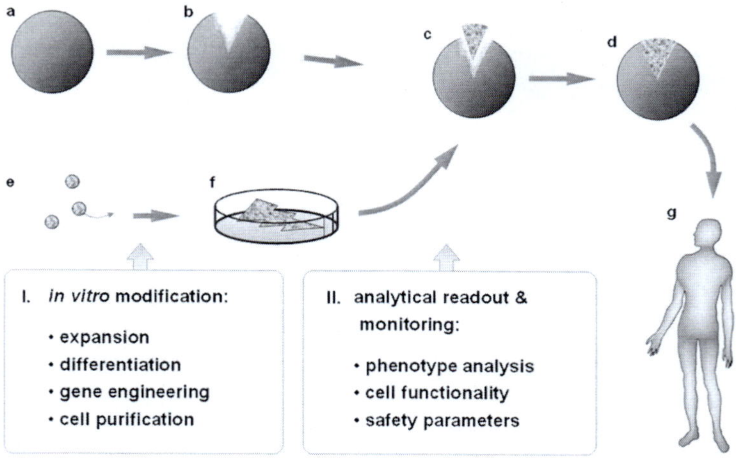

Fig. 4 Cell therapeutic approaches in regenerative medicine. An intact organ or organ system (**a**) experiences a defect, i.e., cell loss (**b**), due to aging, trauma, and/or genetic disposition. Strategies for tissue repair (**c**) replace specific cell populations with therapeutic cell preparations derived from stem cells in order to achieve functional integration into host tissue (**d**). Cell sources (**e**) to be considered include somatic, embryonic, and induced pluripotent stem cells, which are differentiated and modified in vitro (**f**, see *Box I*). Detailed characterization using a panel of assays, including microsystem technology-based functional monitoring, and safety controls are included in clinical-grade cell derivation protocols to provide a safe and functional therapeutic product (see *Box II*). Ultimately, assessment by clinical, physiological, behavioral readouts (**g**) will be used to evaluate restoration of function in the patient

(2) Cells are used to replace the damaged tissue, specifically replenishing the lost or damaged cell type. (3) The grafted cells integrate adequately within the host tissue and restore function (Fig. 4). Cell transplantation therapy has to prove superior to existing therapies and must be safe. Current studies address these paradigms from a variety of angles, utilizing concepts and methodologies from discipline as diverse as developmental biology, stem cell science, cancer biology, transplantation immunology, and tissue engineering. There are a multitude of impending clinical applications, among the most studied currently being focal damages to islet cells (diabetes mellitus) [60, 61], cardiomyocytes (ischemic cardiac necrosis, heart attack) [62], and the loss of dopamine neurons (Parkinson's disease) [63, 64, 10]. This synopsis focuses on the development of cell therapeutic options for diseases of the nervous system (for review, see [65–67]). With respect to the nervous system, disease entities ranging from retinal macular degeneration [68], spinal cord injury [69], stroke, Alzheimer's disease, Huntington's disease, ALS [70], and Parkinson's disease [67, 64] are currently under investigation.

Recruitment of endogenous stem cells. Some organisms, such as planarian flatworms, can regenerate the entire organism from a piece of tissue. And while, for example, the newt (*Notophthalmus viridescent*) shows the ability to regenerate its dopaminergic system after toxic injury [71], Parkinson patients cannot replenish the lost dopaminergic neurons from endogenous cell sources. The adult brain in mammals has comparably poor regenerative capacity. In principle, the presence of somatic stem cells in various tissues offers the opportunity to actively recruit these cells for repair. Neurogenesis in the adult provides a case in point: most brain regions are not permissive for neurogenesis under normal in vivo conditions and neurogenesis is restricted to the olfactory bulb and dentate gyrus of the hippocampus[72]. Under pathological conditions both neurogenic and non-neurogenic regions can promote the recruitment of new neurons [73]. The limited supply of adult NSCs and restricted location combined with nonpermissive microenvironments in vivo represent significant impediments, and it remains to be seen whether stimulation of endogenous somatic stem cells in the brain or other regions can eventually be clinically exploited for brain repair.

Transplantation in experimental models. Clinical trials of cell replacement therapy for PD and HD patients are based on more than 20 years of experimental research. Scientific interest looking into transplantation date back to the work of Elizabeth Dunn in the early part of the twentieth century who, despite the rudimentary nature of her work, highlighted several principal factors that are still recognized today to be critical in producing successful grafts such as the age and identity of the donor tissue and the method and the region of delivery of the graft [74]. After many decades of hiatus, studies into cell transplantation were revived in the early 1970s when Olson and colleagues performed intraocular grafts of neural tissue into animals [75]. However, research into intracerebral grafts was started in the late 1970s by Stenevi, Björklund, and Svendgaard, with their work on catecholamine grafts into rodent forebrain [76]. Equally important, animal models of neurodegenerative diseases started appearing in the 1970s: Ungerstedt developed the 6-hydroxydopamine-based striatal dopamine depleting lesion as a rodent model for PD [77], while Coyle and Schwarcz lesioned striatal projection neurons by injecting excitotoxins into the striatum as a way of simulating HD pathology [78].

The experimental cell transplantation field developed concurrently on several levels. The first – and still on-going – wave of research activity following Björklund and colleagues' work was dedicated to provide the proof of principle that neuronal grafting can function in animal models. Key areas of research can be divided into questions relating to (i) cell preparation and make up; (ii) cell survival and anatomical integration; (iii) cell delivery; and (iv) graft-mediated behavioral and functional recovery.

The viability of cell replacement therapy has been tested in the context of several model diseases, such as stroke, acute brain injury, and multiple sclerosis, but the most comprehensive evidence is associated with models of PD and HD [65]. There is overwhelming data demonstrating that, in models of these neurodegenerative diseases, neural tissue can survive transplantation into the adult mammalian brain, integrate and connect with the host nervous system, and influence the host's behavior. From the earliest studies, prior to the availability of embryonic stem cells or neuronal progenitor cells, it has been self-evident to use primary embryonic neural tissue – from the developmental stage just prior to the appearance of the required phenotype – as these cells are inherently more plastic and robust. Furthermore, the area of dissection for the grafted tissue did not generate much debate either as in both PD and HD models the tissue is prepared from the developing anatomical region that gives rise to the affected area later on, that is, the ventral mesencephalon (substantia nigra) and the ganglionic eminence (striatum), respectively. However, whereas in the case of HD model the grafts are homotopic (striatal tissue into the striatum), models of PD are based on ectopic grafts (nigral tissue into the striatum), as research into homotopic "bridge grafts" suggest that the adult environment receiving the embryonic tissue is not permissive for the grafts to make contact with the distal striatum where the lost dopamine needs to be replaced [79]. Cell preparation and delivery methods have been generally standardized across the research field as well. Apart from a few exceptions, the transplants are introduced into the brain in the form of dissociated single cell suspension as it was observed, in contrast to tissue pieces, that this leads to improved survival partly due to better vascularization [80]. The transplants have been predominantly allografts, but with improved tracking methods, the availability of green-fluorescent protein donors, species-specific markers, and more reliable immunosuppression methods, xenografts have allowed to explore key issues [81]. There is little doubt across the field that using cells of the right age, grafts can survive – although maximizing survival is an acute issue in the PD model – and integrate anatomically with the host by developing appropriate efferent and afferent connections [82, 83]. Delivery of the cells to the desired areas relies on the precision that only stereotactic neurosurgery methods can provide, and further refinements have been achieved through microtransplantation procedures that reduce the surgery-related tissue trauma [84]. Furthermore, beyond the grafting methodologies, data suggest that the environmental conditions and training experience of the animal can affect the plasticity of transplanted cells, that is, the cells remain accessible for manipulation even following grafting [85]. This last point is crucial on several grounds. First, both in the PD and in the HD models, primary grafts have shown to confer functional benefits in motor, sensorimotor, and more cognitive tasks [86–89]. However, except in the case of the simplest motor tasks, the graft-mediated recovery tends to be partial: improved performance compared to the lesioned animals, but not as good as the control animals. This suggests that grafting methodologies can still be improved. Second, the observation that training and environmental manipulation can influence graft development and recovery has implications in the preparation of post-transplantation clinical programs.

Using primary embryonic cells in animal models are crucial to establish the limits of cell replacement therapy and to achieve the "gold standard": the levels of anatomical reconstruction and functional recovery one needs to achieve with alternative cell sources. In the clinic, the ethical and logistic issues surrounding the use of aborted human tissue seriously limit the widespread application of this treatment. Research into stem cells, both pluripotent embryonic and lineage-restricted adult neural stem cells, offers the way toward identifying suitable alternative cell sources for transplantation [90].

Clinical transplantation. While stimulation of intrinsic neurogenesis, and obviously also neuroprotective strategies continue to be explored, the application of exogenous cells has already yielded clinical proof-of-principle results [64, 91–93]. Midbrain tissue containing dopamine neurons obtained from aborted fetuses was grafted into PD patients, and in some but not all cases could restore function. The grafted dopaminergic neurons could survive in the brain tissue, as assessed by functional imaging and postmortem immunohistochemical analysis, and some cases showed significant clinical improvements as measured by different clinical rating scales and self-judgment of the patients. Further scientific and clinical progress in this field will depend on multidisciplinary efforts from basic and clinical scientists. Among the current primary goals of neuroregeneration research are the development of unlimited, appropriate neural cell resources, the optimization of the implantation methods, and gaining a better understanding of graft–host interactions. Stem cell therapies must be compared to conventional therapies and be guided by the medical commitment to first do no harm. Ethical concern and the limited supply of tissue have posed an obstacle to fetal midbrain-derived grafts, and while stem cells are also not free of ethical concern, some types can represent a potentially unlimited cell source. One advantage of using neurons derived from expandable multipotent or pluripotent cells is the large number of cells that can be generated in vitro. For example, in the aforementioned clinical trials of cellular transplantation therapy for Parkinson's disease, about $1.1–1.3 \times 10^6$ cells were obtained per human fetal ventral midbrain dissected [92]. In comparison, using neurons differentiated from a single embryonic stem cell colony, more than 3×10^7 cells can be harvested at late stage of differentiation from one 100-mm Petri dish, an amount of cells that also would allow for the inclusion of cell selection steps prior to transplantation. Still, a variety of cell types are currently being explored to provide the therapeutic phenotypes for the various disease entities. Regardless of the source, the specificity and safety of the therapeutic product generated must be ascertained prior to use in clinical cell replacement applications. Cell transplantation approaches must minimize confounding factors such as biological variability (batch-to-batch differences of cell preparations), contamination with unwanted cell types or with microorganisms, and also minimize immunological adverse reactions by donor–recipient alloantigen matching. Importantly, while immunological factors have to be considered in the development of cell therapeutic paradigms [94, 95], the CNS is deemed an immune-privileged site [96]. As described in detail earlier, the absence of potential tumor-initiating remaining pluripotent and multipotent stem cells in therapeutic cell preparations has to be ensured. Dynamic changes in surface antigen expression can be exploited for the experimental separation of key cell populations derived from developing human stem cells [2], and further identification of such biomarkers will enable the selection of not only hematopoietic [14] but also neural [2], mesenchymal [97], and other lineages according to surface antigen codes. In addition to their direct therapeutic use for tissue repair and regenerative medicine, stem cell derivatives may contribute to the treatment of human diseases in other ways and are being explored for a number of ex vivo applications.

Standardized cell assays and novel disease models. Stem cell-derived neural cells can be considered a useful tool for cell therapeutic, basic developmental, and pharmacological studies alike. Stem cell technology offers the opportunity to develop in vitro assays based on specific well delineated cell populations, e.g., for large-scale drug screening and testing and toxicity studies. Currently under investigation are novel cellular models of human disease to study aspects of complex pathogenetic interrelations using reprogrammed, induced pluripotent stem (iPS) cells derived from patients' somatic cells (Parkinson's disease, ALS). Testing drugs or factors on such patient-derived cell populations will enable "disease modeling," enhance our understanding of human cellular neurobiology, and eventually lead to its translation to the treatment of diseases. The original isolation of iPS cells from mice utilized retrovirus-mediated transduction of oncogenes and drug resistance selection for Fbx15, Oct4, or Nanog activation [98]. More recently, iPS cell technology enabled the derivation of patient-specific pluripotent cell lines, and thereby opens the door for speculations on future customized therapy of diseases with autologous cells, potentially circumventing the potential immunological issues. In this cell therapeutic context, critical proof-of-principle studies recently showed that mouse iPS cell-derived cells were able to restore function in mouse models of sickle cell anemia [99] and after differentiation into dopaminergic neurons in a rat model of Parkinson's disease [21]. It must be noted that while iPS technology revolutionized the field, such reprogramming approaches face a number of challenges before introducing patient-specific transplantation therapy. Introduction of oncogenes such as c-myc or Oct4 could further enhance the risk of tumor formation, adding to the baseline risk inherent in the application of pluripotent cell sources. Moreover, using retroviral vectors can be accompanied by insertional mutations, again potentially increasing tumor risks. Strategies to circumvent these and similar issues are currently underway, such as eliminating c-myc [100] or using adenovirus instead of retrovirus [101] to introduce reprogramming factors. Our increased understanding of molecular networks underlying stemness and pluripotency and of the mechanisms of reprogramming is prone to eventually be biomedically fruitful [102]. The successful translation to the clinical level of this or other stem cell resources will depend on technologies to generate, characterize, store, and administer such cell-based products in a reliable, safe, and also cost-effective manner [103, 104].

4 Summary

Stem cells represent a powerful source for cell therapy and scientific studies. Authentic human cell populations generated in vitro can be modulated for and utilized in drug discovery paradigms, genetic assays modeling human diseases, and in detailed developmental studies. Our growing understanding of stemness and pluripotency is likely to redefine current concepts of differentiation and cell phenotype and may yield clinical applications in the not-too-distant future. As outlined here, biomedical stem cell research represents a multidisciplinary and multilayered challenge, and parallel developments in technology and biology will synergize to promote its translation to clinical applications.

References

1. Ginis I, Luo Y, Miura T, Thies S, Brandenberger R, Gerecht-Nir S, Amit M, Hoke A, Carpenter MK, Itskovitz-Eldor J, Rao MS. Differences between human and mouse embryonic stem cells. Dev Biol. 2004; 269:360–380.
2. Pruszak J, Sonntag KC, Aung MH, Sanchez-Pernaute R, Isacson O. Markers and methods for cell sorting of human embryonic stem cell-derived neural cell populations. Stem Cells 2007; 25:2257–2268.
3. Roy NS, Cleren C, Singh SK, Yang L, Beal MF, Goldman SA. Functional engraftment of human ES cell-derived dopaminergic neurons enriched by coculture with telomerase-immortalized midbrain astrocytes. Nat Med. 2006; 12:1259–1268.
4. Fareed MU, Moolten FL. Suicide gene transduction sensitizes murine embryonic and human mesenchymal stem cells to ablation on demand – a fail-safe protection against cellular misbehavior. Gene Ther. 2002; 9:955–962.
5. Imreh MP, Gertow K, Cedervall J, Unger C, Holmberg K, Szoke K, Csoregh L, Fried G, Dilber S, Blennow E, Ahrlund-Richter L. In vitro culture conditions favoring selection of chromosomal abnormalities in human ES cells. J Cell Biochem. 2006; 99:508–516.
6. Hedlund E, Pruszak J, Ferree A, Vinuela A, Hong S, Isacson O, Kim KS. Selection of embryonic stem cell-derived enhanced green fluorescent protein-positive dopamine neurons using the tyrosine hydroxylase promoter is confounded by reporter gene expression in immature cell populations. Stem Cells 2007; 25:1126–1135.
7. Pruszak J, Isacson. O. Molecular and cellular determinants for generating ES-cell derived dopamine neurons for cell therapy. In: Pasterkamp R, Smidt M, Burbach J, eds. Development and engineering of dopamine neurons. Austin: Landis Bioscience; 2008.
8. Perrier AL, Tabar V, Barberi T, Rubio ME, Bruses J, Topf N, Harrison NL, Studer L. Derivation of midbrain dopamine neurons from human embryonic stem cells. Proc Natl Acad Sci USA. 2004; 101:12543–12548.
9. Yan Y, Yang D, Zarnowska ED, Du Z, Werbel B, Valliere C, Pearce RA, Thomson JA, Zhang SC. Directed differentiation of dopaminergic neuronal subtypes from human embryonic stem cells. Stem Cells 2005; 23:781–790.
10. Sonntag KC, Pruszak J, Yoshizaki T, van Arensbergen J, Sanchez-Pernaute R, Isacson O. Enhanced yield of neuroepithelial precursors and midbrain-like dopaminergic neurons from human embryonic stem cells using the bone morphogenic protein antagonist noggin. Stem Cells 2007; 25:411–418.
11. Vunjak-Novakovic G. Patterning stem cell differentiation. Cell Stem Cell 2008; 3:362–363.
12. Carson CT, Aigner S, Gage FH. Stem cells: the good, bad and barely in control. Nat Med. 2006; 12:1237–1238.
13. Stewart MH, Bosse M, Chadwick K, Menendez P, Bendall SC, Bhatia M. Clonal isolation of hESCs reveals heterogeneity within the pluripotent stem cell compartment. Nat Methods 2006; 3:807–815.
14. Murry CE, Keller G. Differentiation of embryonic stem cells to clinically relevant populations: lessons from embryonic development. Cell 2008; 132:661–680.
15. Yurek DM, Fletcher-Turner A. Comparison of embryonic stem cell-derived dopamine neuron grafts and fetal ventral mesencephalic tissue grafts: morphology and function. Cell Transplant. 2004; 13:295–306.
16. Martinat C, Bacci JJ, Leete T, Kim J, Vanti WB, Newman AH, Cha JH, Gether U, Wang H, Abeliovich A. Cooperative transcription activation by Nurr1 and Pitx3 induces embryonic stem cell maturation to the midbrain dopamine neuron phenotype. Proc Natl Acad Sci USA. 2006; 103:2874–2879.
17. Andersson E, Tryggvason U, Deng Q, Friling S, Alekseenko Z, Robert B, Perlmann T, Ericson J. Identification of intrinsic determinants of midbrain dopamine neurons. Cell 2006; 124:393–405.
18. Sonntag KC, Simantov R, Kim KS, Isacson O. Temporally induced Nurr1 can induce a non-neuronal dopaminergic cell type in embryonic stem cell differentiation. Eur J Neurosci. 2004; 19:1141–1152.

19. Hedlund E, Pruszak J, Lardaro T, Ludwig W, Vinuela A, Kim KS, Isacson O. Embryonic stem cell-derived Pitx3-enhanced green fluorescent protein midbrain dopamine neurons survive enrichment by fluorescence-activated cell sorting and function in an animal model of Parkinson's disease. Stem Cells 2008; 26:1526–1536.
20. Chung S, Shin BS, Hedlund E, Pruszak J, Ferree A, Kang UJ, Isacson O, Kim KS. Genetic selection of sox1GFP-expressing neural precursors removes residual tumorigenic pluripotent stem cells and attenuates tumor formation after transplantation. J Neurochem. 2006; 97:1467–1480.
21. Wernig M, Zhao JP, Pruszak J, Hedlund E, Fu D, Soldner F, Broccoli V, Constantine-Paton M, Isacson O, Jaenisch R. Neurons derived from reprogrammed fibroblasts functionally integrate into the fetal brain and improve symptoms of rats with Parkinson's disease. Proc Natl Acad Sci USA. 2008; 105:5856–5861.
22. Wernig M, Tucker KL, Gornik V, Schneiders A, Buschwald R, Wiestler OD, Barde YA, Brustle O. Tau EGFP embryonic stem cells: an efficient tool for neuronal lineage selection and transplantation. J Neurosci Res. 2002; 69:918–924.
23. Ozdinler PH, Macklis JD. IGF-I specifically enhances axon outgrowth of corticospinal motor neurons. Nat Neurosci. 2006; 9:1371–1381.
24. Herzenberg LA, Parks D, Sahaf B, Perez O, Roederer M, Herzenberg LA. The history and future of the fluorescence activated cell sorter and flow cytometry: a view from Stanford. Clin Chem. 2002; 48:1819–1827.
25. Kirouac DC, Zandstra PW. The systematic production of cells for cell therapies. Cell Stem Cell 2008; 3:369–381.
26. Studer L, Csete M, Lee SH, Kabbani N, Walikonis J, Wold B, McKay R. Enhanced proliferation, survival, and dopaminergic differentiation of CNS precursors in lowered oxygen. J Neurosci. 2000; 20:7377–7383.
27. Maciaczyk J, Singec I, Maciaczyk D, Nikkhah G. Combined use of BDNF, ascorbic acid, low oxygen, and prolonged differentiation time generates tyrosine hydroxylase-expressing neurons after long-term in vitro expansion of human fetal midbrain precursor cells. Exp Neurol. 2008; 213:354–362.
28. Metzen E, Wolff M, Fandrey J, Jelkmann W. Pericellular PO_2 and O_2 consumption in monolayer cell cultures. Respir Physiol. 1995; 100:101–106.
29. Kieninger J, Dannenberg A, Aravindalochanan K, Jobst G, Pettersen E, Urban G. Amperometric oxygen sensor array with novel chronoamperometric protocols for hypoxic tumor cell cultures. Solid-State Sensors, Actuators and Microsystems Conference, TRANSDUCERS 2007, Lyon, France, 2007, pp. 1907–1910.
30. Wittkampf GC, Chemnitius K, Cammann MR, Mokwa W. Silicon thin film sensor for measurement of dissolved oxygen. Sens Actuators B Chemi. 1997; 43:40–44.
31. Brischwein M, Motrescu ER, Cabala E, Otto AM, Grothe H, Wolf B. Functional cellular assays with multiparametric silicon sensor chips. Lab Chip 2003; 3:234–240.
32. Wolfbeis OS, Leiner MJP, Posch HE. A new sensing material for optical oxygen measurement, with the indicator embedded in an aqueous phase. Microchimica Acta 1986; 90:359–366.
33. Lippitsch ME, Pusterhofer J, Leiner MJP, Wolfbeis OS. Fibre-optic oxygen sensor with the fluorescence decay time as the information carrier. Analytica Chimica Acta 1988; 205:1–6.
34. Henning T, Brischwein M, Baumann W, Ehret R, Freund I, Kammerer R, Lehmann M, Schwinde A, Wolf B. Approach to a multiparametric sensor-chip-based tumor chemosensitivity assay. Anti-Cancer Drugs 2001; 12:21–32.
35. Marzouk SAM. Improved electrodeposited iridium oxide pH sensor fabricated on etched titanium substrates. Anal Chem. 2003; 75:1258–1266.
36. Gough DA, Lucisano JY, Tse PH. Two-dimensional enzyme electrode sensor for glucose. Anal Chem. 1985; 57:2351–2357.
37. Jobst G, Moser I, Svasek P, Varahram M, Trajanoski Z, Wach P, Kotanko P, Skrabal F, Urban GA. Mass producible miniaturized flow through a device with a biosensor array. Sens Actuators B 1997; 43:121–125.

38. Jobst G, Moser I, Varahram M, Svasek P, Aschauer E, Trajanoski Z, Wach P, Kotanko P, Skrabal F, Urban G. Thin-film microbiosensors for glucose-lactate monitoring. Anal Chem. 1996; 68:3173–3179.
39. Heise HM. Non-invasive monitoring of metabolites using near infrared spectroscopy. Horm Metab Res. 1996; 10:527–534.
40. Marvin J, Hellinga H. Engineering biosensors by introducing fluorescent allosteric signal transducers: construction of a novel glucose sensor. J Am Chem Soc. 1998; 120:7–11.
41. Ge X, Rao G, Tolosa L. On the possibility of real-time monitoring of glucose in cell culture by microdialysis using a fluorescent glucose binding protein sensor. Biotechnol Progr. 2008; 24:691–697.
42. Summers LK, Clark ML, Humphreys SM, Bugler J, Frayn KN. The use of microdialysis to monitor rapid changes in glucose concentration. Horm Metab Res. 1999; 7:424–428.
43. Murphy S, Simmons ML, Agullo L, Garcia A, Feinstein DL, Galea E, Reis DJ, Minc-Golomb D, Schwartz JP. Synthesis of nitric oxide in CNS glial cells. Trends Neurosci. 1993; 16:323–328.
44. Hobbs AJ, Fukuto JM, Ignarro LJ. Formation of free nitric oxide from L-arginine by nitric oxide synthase: direct enhancement of generation by superoxide dismutase. Pharmacology 1994; 91:10992–10996.
45. Singh RJ, Hogg N, McHaourab HS, Kalyanaraman B. Physical and chemical interactions between nitric oxide and nitroxides. Biochim Biophys Acta 1994; 1201:437–441.
46. Shibuki K. An electrochemical microprobe for detecting nitric oxide release in brain tissue. Neurosci Res. 1990; 1:69–76.
47. Malinski T, Taha Z. Nitric oxide release from a single cell measured in situ by a porphyrinic-based microsensor. Nature 1992; 358:676–678.
48. Aravindalochanan K, Kieninger J, Jobst G, Igel G, Urban GA. Micro electrode array for sensing nitric oxide in hypoxic cell culture environment. Sens Proc. 2007; 1:111–116.
49. Bedioui F, Trevin S, Devynck J. Chemically modified microelectrodes designed for the electrochemical determination of nitric oxide in biological systems. Electroanalysis 1996; 8:1085–1091.
50. Oni J, Pailleret A, Isik S, Diab N, Radtke I, Blöchl A, Jackson M, Bedioui F, Schuhmann W. Functionalised electrode array for the detection of nitric oxide released by endothelial cells using different NO-sensing chemistries. Anal Bioanal Chem. 2004; 378:1594–1600.
51. Watson CJ, Venton BJ, Kennedy RT. In vivo measurements of neurotransmitters by microdialysis sampling. Anal Chem. 2006; 78:1391–1399.
52. Christian BT, Narayanan TK, Shi B, Mukherjee J. Quantitation of striatal and extrastriatal D-2 dopamine receptors using PET imaging of [18F]fallypride in nonhuman primates. Synapse 2000; 38:71–79.
53. Knutson B, Fong GW, Bennett SM, Adams CM, Hommerb D. A region of mesial prefrontal cortex tracks monetarily rewarding outcomes: characterization with rapid event-related fMRI. NeuroImage 2003; 18:263–272.
54. Crespi F, Paret J, Keane PE, Morre M. An improved differential pulse voltammetry technique allows the simultaneous analysis of dopaminergic and serotonergic activities in vivo with a single carbon-fibre electrode. Neurosci Lett. 1984; 52:159–164.
55. Hoffman AF, Gerhardt GA. Differences in pharmacological properties of dopamine release between the substantia nigra and striatum: an in vivo electrochemical study. J Pharmacol Exp Ther. 1999; 289:455–463.
56. Leszczyszyn DJ, Jankowski JA, Viveros OH, Diliberto EJ, Near JA, Wightman RM. Secretion of catecholamines from individual adrenal medullary chromaffin cells. J Neurochem. 1991; 56:1855–1863.
57. Pettersen EO, Larsen LH, Ramsing NB, Ebbesen P. Pericellular oxygen depletion during ordinary tissue culturing, measured with oxygen microsensors. Cell Prolif. 2005; 38: 257–267.
58. Lehmann M, Baumann W, Brischwein M, Gahle HJ, Freund I, Ehret R, Drechsler S, Palzer H, Kleintges M, Sieben U, Wolf B. Simultaneous measurement of cellular respiration and acidification with a single CMOS ISFET. Biosens Bioelectron. 2001; 16:195–203.

59. Keller G. Embryonic stem cell differentiation: emergence of a new era in biology and medicine. Genes Dev. 2005; 19:1129–1155.
60. D'Amour KA, Bang AG, Eliazer S, Kelly OG, Agulnick AD, Smart NG, Moorman MA, Kroon E, Carpenter MK, Baetge EE. Production of pancreatic hormone-expressing endocrine cells from human embryonic stem cells. Nat Biotechnol. 2006; 24:1392–1401.
61. Zhou Q, Brown J, Kanarek A, Rajagopal J, Melton DA. In vivo reprogramming of adult pancreatic exocrine cells to beta-cells. Nature 2008; 455:627–632.
62. Laflamme MA, Chen KY, Naumova AV, Muskheli V, Fugate JA, Dupras SK, Reinecke H, Xu C, Hassanipour M, Police S, O'Sullivan C, Collins L, Chen Y, Minami E, Gill EA, Ueno S, Yuan C, Gold J, Murry CE. Cardiomyocytes derived from human embryonic stem cells in pro-survival factors enhance function of infarcted rat hearts. Nat Biotechnol. 2007; 25:1015–1024.
63. Nikkhah G, Rosenthal C, Falkenstein G, Roedter A, Papazoglou A, Brandis A. Microtransplantation of dopaminergic cell suspensions: further characterisation and optimisation of grafting parameters. Cell Transplant. 2009; 18:119–133.
64. Mendez I, Vinuela A, Astradsson A, Mukhida K, Hallett P, Robertson H, Tierney T, Holness R, Dagher A, Trojanowski JQ, Isacson O. Dopamine neurons implanted into people with Parkinson's disease survive without pathology for 14 years. Nat Med. 2008; 14:507–509.
65. Bjorklund A, Lindvall O. Cell replacement therapies for central nervous system disorders. Nat Neurosci. 2000; 3:537–544.
66. Lindvall O, Kokaia Z. Stem cells for the treatment of neurological disorders. Nature 2006; 441:1094–1096.
67. Astradsson A, Cooper O, Vinuela A, Isacson O. Recent advances in cell-based therapy for Parkinson disease. Neurosurg Focus 2008; 24:E6.
68. Vugler A, Carr AJ, Lawrence J, Chen LL, Burrell K, Wright A, Lundh P, Semo M, Ahmado A, Gias C, da Cruz L, Moore H, Andrews P, Walsh J, Coffey P. Elucidating the phenomenon of HESC-derived RPE: anatomy of cell genesis, expansion and retinal transplantation. Exp Neurol. 2008; 214:347–361.
69. Coutts M, Keirstead HS. Stem cells for the treatment of spinal cord injury. Exp Neurol. 2008; 209:368–377.
70. Dimos JT, Rodolfa KT, Niakan KK, Weisenthal LM, Mitsumoto H, Chung W, Croft GF, Saphier G, Leibel R, Goland R, Wichterle H, Henderson CE, Eggan K. Induced pluripotent stem cells generated from patients with ALS can be differentiated into motor neurons. Science 2008; 321:1218–1221.
71. Parish CL, Beljajeva A, Arenas E, Simon A. Midbrain dopaminergic neurogenesis and behavioural recovery in a salamander lesion-induced regeneration model. Development 2007; 134:2881–2887.
72. Lie DC, Song H, Colamarino SA, Ming GL, Gage FH. Neurogenesis in the adult brain: new strategies for central nervous system diseases. Annu Rev Pharmacol Toxicol. 2004; 44:399–421.
73. Snyder EY, Yoon C, Flax JD, Macklis JD. Multipotent neural precursors can differentiate toward replacement of neurons undergoing targeted apoptotic degeneration in adult mouse neocortex. Proc Natl Acad Sci USA. 1997; 94:11663–11668.
74. Dunn EH. Primary and secondary findings in a series of attempts to transplant cerebral cortex in the albino rat. J Comp Neurol. 1917; 27:565–582.
75. Olson L, Seiger A. Brain tissue transplanted to the anterior chamber of the eye. 1. Fluorescence histochemistry of immature catecholamine and 5-hydroxytryptamine neurons reinnervating the rat iris. Z Zellforsch Mikrosk Anat. 1972; 135:175–194.
76. Stenevi U, Bjorklund A, Svendgaard NA. Transplantation of central and peripheral monoamine neurons to the adult rat brain: techniques and conditions for survival. Brain Res. 1976; 114:1–20.
77. Ungerstedt U. Postsynaptic supersensitivity after 6-hydroxy-dopamine induced degeneration of the nigro-striatal dopamine system. Acta Physiol Scand Suppl. 1971; 367:69–93.

78. Coyle JT, Schwarcz R. Lesion of striatal neurons with kainic acid provides a model for Huntington's chorea. Nature 1976; 263:244–246.
79. Bentlage C, Nikkhah G, Cunningham MG, Bjorklund A. Reformation of the nigrostriatal pathway by fetal dopaminergic micrografts into the substantia nigra is critically dependent on the age of the host. Exp Neurol. 1999; 159:177–190.
80. Bjorklund A, Schmidt RH, Stenevi U. Functional reinnervation of the neostriatum in the adult rat by use of intraparenchymal grafting of dissociated cell suspensions from the substantia nigra. Cell Tissue Res. 1980; 212:39–45.
81. Kelly CM, Precious SV, Penketh R, Amso N, Dunnett SB, Rosser AE. Striatal graft projections are influenced by donor cell type and not the immunogenic background. Brain 2007; 130:1317–1329.
82. Bolam JP, Freund TF, Bjorklund A, Dunnett SB, Smith AD. Synaptic input and local output of dopaminergic neurons in grafts that functionally reinnervate the host neostriatum. Exp Brain Res. 1987; 68:131–146.
83. Clarke DJ, Dunnett SB. Synaptic relationships between cortical and dopaminergic inputs and intrinsic GABAergic systems within intrastriatal striatal grafts. J Chem Neuroanat. 1993; 6:147–158.
84. Nikkhah G, Cunningham MG, Jodicke A, Knappe U, Bjorklund A. Improved graft survival and striatal reinnervation by microtransplantation of fetal nigral cell suspensions in the rat Parkinson model. Brain Res. 1994; 633:133–143.
85. Dobrossy MD, Dunnett SB. Optimising plasticity: environmental and training associated factors in transplant-mediated brain repair. Rev Neurosci. 2005; 16:1–21.
86. Nikkhah G, Falkenstein G, Rosenthal C. Restorative plasticity of dopamine neuronal transplants depends on the degree of hemispheric dominance. J Neurosci. 2001; 21:6252–6263.
87. Klein A, Metz GA, Papazoglou A, Nikkhah G. Differential effects on forelimb grasping behavior induced by fetal dopaminergic grafts in hemiparkinsonian rats. Neurobiol Dis. 2007; 27:24–35.
88. Brasted PJ, Watts C, Robbins TW, Dunnett SB. Associative plasticity in striatal transplants. Proc Natl Acad Sci USA. 1999; 96:10524–10529.
89. Dobrossy MD, Dunnett SB. The corridor task: striatal lesion effects and graft-mediated recovery in a model of Huntington's disease. Behav Brain Res. 2007; 179:326–330.
90. Singec I, Jandial R, Crain A, Nikkhah G, Snyder EY. The leading edge of stem cell therapeutics. Annu Rev Med. 2007; 58:313–328.
91. Kordower JH, Chu Y, Hauser RA, Freeman TB, Olanow CW. Lewy body-like pathology in long-term embryonic nigral transplants in Parkinson's disease. Nat Med. 2008; 14:504–506.
92. Mendez I, Sanchez-Pernaute R, Cooper O, Vinuela A, Ferrari D, Bjorklund L, Dagher A, Isacson O. Cell type analysis of functional fetal dopamine cell suspension transplants in the striatum and substantia nigra of patients with Parkinson's disease. Brain 2005; 128: 1498–1510.
93. Li JY, Englund E, Holton JL, Soulet D, Hagell P, Lees AJ, Lashley T, Quinn NP, Rehncrona S, Bjorklund A, Widner H, Revesz T, Lindvall O, Brundin P. Lewy bodies in grafted neurons in subjects with Parkinson's disease suggest host-to-graft disease propagation. Nat Med. 2008; 14:501–503.
94. Brunlid G, Pruszak J, Holmes B, Isacson O, Sonntag KC. Immature and neurally differentiated mouse embryonic stem cells do not express a functional Fas/Fas ligand system. Stem Cells 2007; 25:2551–2558.
95. Chen Z, Palmer TD. Cellular repair of CNS disorders: an immunological perspective. Hum Mol Genet. 2008; 17:R84–R92.
96. Ferguson TA, Green DR, Griffith TS. Cell death and immune privilege. Int Rev Immunol. 2002; 21:153–172.
97. Barberi T, Willis LM, Socci ND, Studer L. Derivation of multipotent mesenchymal precursors from human embryonic stem cells. PLoS Med. 2005; 2:e161.

98. Takahashi K, Tanabe K, Ohnuki M, Narita M, Ichisaka T, Tomoda K, Yamanaka S. Induction of pluripotent stem cells from adult human fibroblasts by defined factors. Cell 2007; 131:861–872.
99. Hanna J, Wernig M, Markoulaki S, Sun CW, Meissner A, Cassady JP, Beard C, Brambrink T, Wu LC, Townes TM, Jaenisch R. Treatment of sickle cell anemia mouse model with iPS cells generated from autologous skin. Science 2007; 318:1920–1923.
100. Nakagawa M, Koyanagi M, Tanabe K, Takahashi K, Ichisaka T, Aoi T, Okita K, Mochiduki Y, Takizawa N, Yamanaka S. Generation of induced pluripotent stem cells without Myc from mouse and human fibroblasts. Nat Biotechnol. 2008; 26:101–106.
101. Stadtfeld M, Nagaya M, Utikal J, Weir G, Hochedlinger K. Induced pluripotent stem cells generated without viral integration. Science 2008; 322:945–949.
102. Jaenisch R, Young R. Stem cells, the molecular circuitry of pluripotency and nuclear reprogramming. Cell 2008; 132:567–582.
103. Halme DG, Kessler DA. FDA regulation of stem-cell-based therapies. N Engl J Med. 2006; 355:1730–1735.
104. Burger SR. Current regulatory issues in cell and tissue therapy. Cytotherapy 2003; 5: 289–298.

Adult Stem Cells in Drug Discovery

Stefan Golz, Andreas Geerts, and Andreas Wilmen

Abstract Significant advancements in stem cell research have provided important information on stem cell biology and offer great promise for developing novel successful stem cell-based medical treatments. Further investigations appear to be necessary to translate the basic knowledge into clinical therapeutic applications in humans. The identification of specific biomarkers to each type of adult stem/progenitor cells relative to their more committed and mature progeny is important for the characterization of their specific physiological functions.

Keywords Drug discovery · Screening · Target identification · Target validation · Cancer · Cardiovascular diseases

1 Drug Discovery

The ability to divide and to differentiate makes a cell a stem cell. Adult stem cells are usually organ specific. They can differentiate into the cell types of an organ they belong to. The ability to repair or better rebuild a damaged tissue makes them promising tools for a lot of different diseases. They can show their versatility by injecting into a damaged region or by growing tissues outside the body to create replacement parts.

In drug discovery the adult stem cells could be used in screening after differentiation into a defined cell type. It will be very helpful to produce exactly cell type and number of cells needed. This will be a major advantage but there is a lot of research to do before. But it is also helpful to support the body by making use of its own stem cells. In case of damaged tissues the recruitment of stem cells can be increased by mimicking chemokines. On the other hand, some tumors use the stem cells to build new vessels to improve their oxygen supply. In these cases, interrupting recruitment

S. Golz (✉)
Bayer Schering Pharma AG, Target Discovery, Wuppertal, Germany
e-mail: stefar.Golz@Bayerhealthcare.com

G.M. Artmann et al. (eds.), *Stem Cell Engineering*,
DOI 10.1007/978-3-642-11865-4_21,

or differentiation of stem cells can be helpful to stop tumor growing. Indeed there are some experiments that use the ability of tumors to attract stem cells against the tumor. Genetically engineered stem cells are used like a Trojan horse to mark or to destroy tumor cells.

The modern drug research process has reversed the classical pharmacological strategy [10]. Today, research programs are initiated based on biological evidence suggesting a particular gene or a gene product to be a meaningful target for small-molecule drugs useful for therapy. This process was established to identify target-specific modulators lacking several side effects. Also, it allows setting up a linear drug discovery process starting from target identification to finally delivering molecules for clinical development. One central element is lead discovery through high-throughput screening (HTS) of comprehensive corporate compound collections. Pharma research-in most organizations-is organized in discrete phases building together a "value chain" along which discovery programs process to finally drug candidated for clinical testing.

This pipeline is fueled by targets suggested from external or in-house generated data suggesting a gene or a gene product to be disease relevant. Today a large arsenal of technologies are available for researchers in industry and academia to generate data in support of a functional link between a given gene and a disease state. Pharmaceutical companies typically have established specialized groups dedicated to target biology.

The modulation of identified targets could affect different organ or cell functions, whereas diverse cell types are addressed. One type of cell is adult stem cell, whose impact on different diseases increased. Adult stem cells are undifferentiated cells found throughout the body after embryonic development that divide into replenish dying cells and regenerate damaged tissues. Also known as somatic stem cells, they can be found in children, as well as adults. Today hematopoietic stem cells are the preferred class of adult stem cells in pharmaceutical research. Other classes of adult stem cells are olfactory, neural, endothelial, mammary, or testicular stem cells. In addition the recruitment of endothelial progenitor cells are in the focus of cancer research.

2 Target Identification

Generation of a hypothetical disease linkage is frequently followed by a series of experiments designed to further validate the initial hypothesis. In many cases, high-throughput technologies like real-time PCR using TaqMan detection systems (see respective chapter) have been implemented to systematically screen for genes with interesting expression patterns or significant regulation in relevant disease state. As a result, most companies have established comprehensive bioinformatics databases containing molecular biology and expression data. Such data are further supplemented by information derived from generic and population studies and transgenic animals. The introduction of RNAi technology [1] marks another technological advancement employed for target identification and validation in vitro and in laboratory animals. Following the sequencing and publication of the human

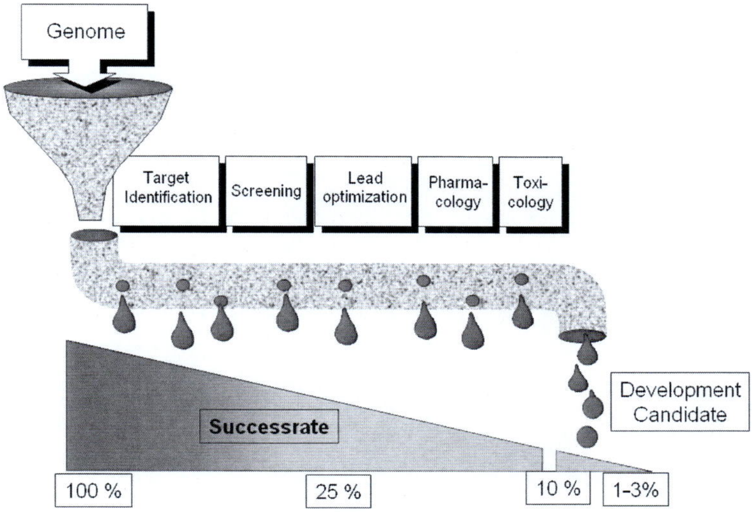

Fig. 1 Schematic overview of success rates of drug discovery process (Bayer Schering Pharma)

genome in 2001, the catalogue of potential target candidates is publically available. However, our understanding of gene function and their potential relevance for the treatment of diseases still remains obscure.

Following a technical assessment of the target "druggability" [13], the probability to identify small-molecule modulators and the technical feasibility of target-specific assays are developed to probe the corporate compound collection for meaningful leads.

A major characteristic of the "research pipeline" is leakiness (Fig. 1). All steps are characterized by their intrinsic attrition rates resulting in a very low overall probability of success. The frequently cited "productivity gap" in pharmaceutical research and the widely observed drought in pharma's pipelines challenge the current drug discovery paradigms. Clearly, the lack of meaningful leads in late stages of discovery or development cannot be compensated by increasing the number of targets introduced into the pipeline. Increasing the number of poorly validated targets is likely to further fuel failures in proofing the therapeutic concept, thus leading to a "vicious cycle". This general challenge to the "target approach" of drug discovery has to be reflected in novel pharma strategies trying to balance innovation around established and novel target mechanisms.

2.1 Microarray Analysis

Microarray-based expression analysis enables researchers to simultaneously monitor genome-wide expression profiles. This global view helps scientists understand biological mechanisms of complex diseases and processes, and identify and validate new drug targets.

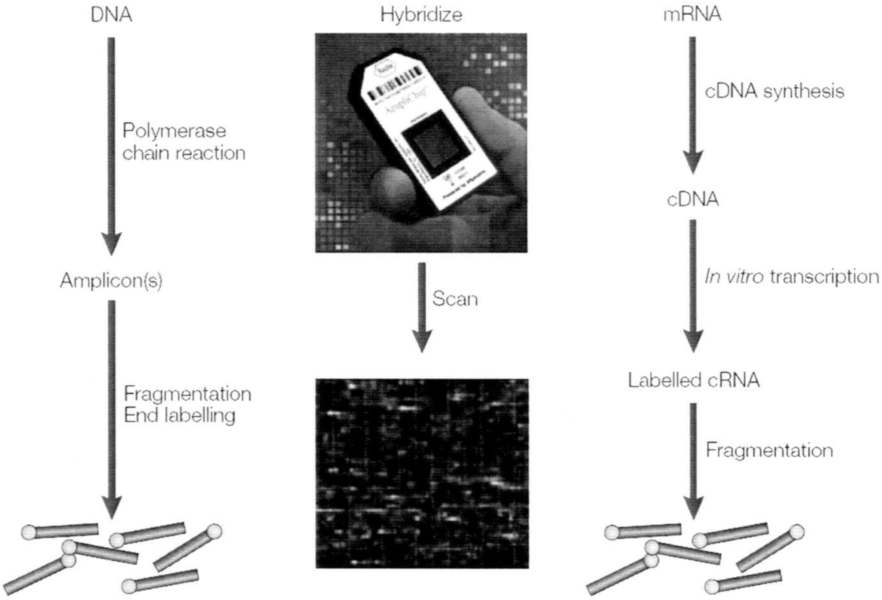

Fig. 2 Workflow of microarray-based expression analysis (Koch [11])

GeneChip microarrays consist of small DNA fragments, chemically synthesized at specific locations on a coated quartz surface. Extracting and labeling nucleic acids from experimental samples, and then hybridizing those prepared samples to the array, enables a wide range of applications on a whole-genome scale – including gene- and exon-level expression analysis, novel transcript discovery, genotyping, and resequencing (Fig. 2).

2.2 TaqMan Analysis

TaqMan is a recently developed technique in which the release of a fluorescent reporter dye from a hybridization probe (see Fig. 3) in real time during a polymerase chain reaction (PCR) is proportional to the accumulation of the PCR product. Quantification is based on the early, linear part of the reaction, and by determining the threshold cycle (CT), at which fluorescence above background is first detected (Fig. 3).

The TaqMan technology [9] can be used to compare differences between expression profiles of normal tissue and diseased tissue. Expression profiling has been used in identifying genes, which are up- or downregulated in a variety of diseases. An interesting application of expression profiling is temporal monitoring of changes in gene expression during disease progression and drug treatment or in patients versus healthy individuals. The premise in this approach is that changes in the pattern of gene expression in response to physiological or environmental stimuli (e.g., drugs)

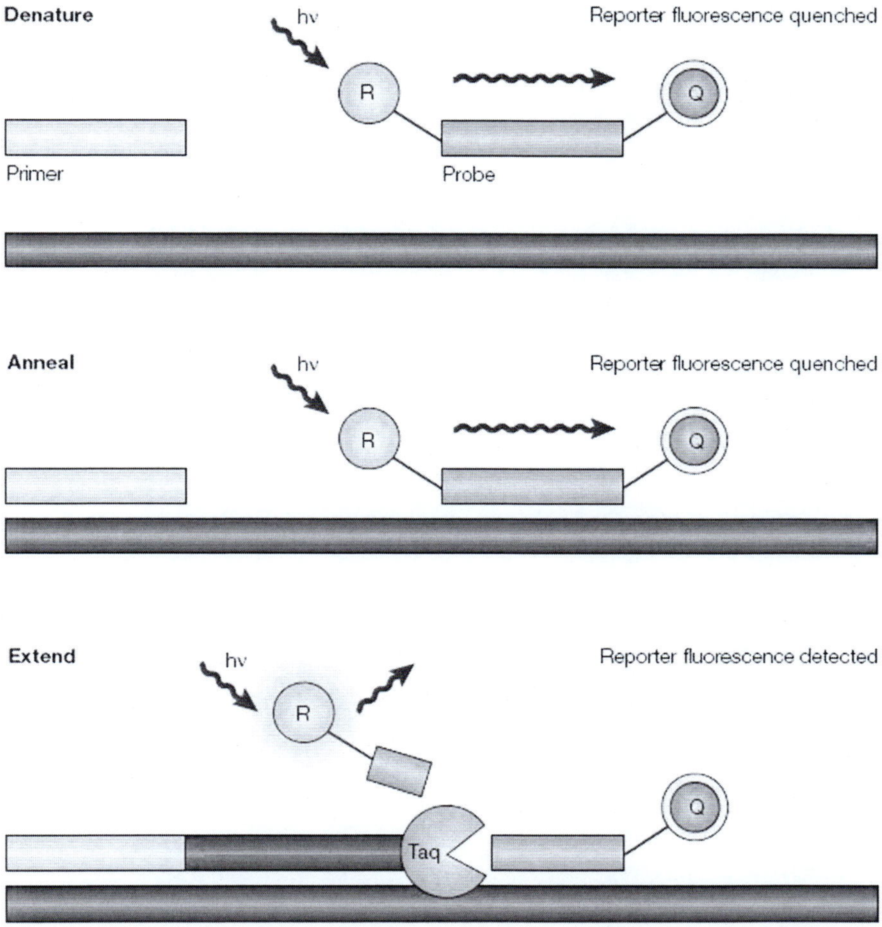

Fig. 3 Technological description of real-time, PCR-based TaqMan expression analysis (Taq, Polymerase; Q, quencher; R, fluorescent dye; $h\nu$, excitation) (Koch [11])

may serve as indirect clues to disease-causing genes or drug targets. Moreover, the effects of drugs with established efficacy on global gene expression patterns may provide a guidepost, or a genetic signature, against which a new drug candidate can be compared.

3 Target Validation

Target validation involves proving that DNA, RNA, or a protein molecule is directly involved in a disease process and can be a suitable target for development of a new therapeutic drug. Target validation is the functional study of the target that has been identified in genomic or proteomic investigations. The inhibition of the protein's function in in vitro or in vivo models followed by the investigation of the resulting phenotype is the focus of target validation.

The main strategies involved in target validation are gene knockout studies, knockdown by antisense RNA or RNA interference, and direct inhibition of the protein by small molecules, peptides, antibodies, aptamers, or any other class of inhibitors. Functional validation is required to show that the inhibition of the target protein results in a desirable biological end point in terms of disease progression. A target is validated if its inhibition reverts a disease phenotype in a model system that is as close to the disease as possible.

3.1 In Vitro Target Validation

3.1.1 RNAi Technology

RNA interference (RNAi) is a selective method of silencing protein expression at the post-transcriptional level. Double-stranded RNA specific to the gene to be silenced is introduced into the cell, where it is processed into short single-stranded RNA fragments. Antisense strands, complementary to the fragments, bind to the target RNA and prompt enzymes to disable it. The effect is to destroy all the target messenger RNAs, effectively halting the production of the protein [2].

3.1.2 Cell Biology

The scope of cell based target validation is the demonstration that a molecular target is involved in a disease process, whereas its modulation can result in a therapeutic effect. Various cell based technologies are regularly used: cell proliferation, apoptosis, cell cycle arrest, chemotaxis, invasion, angiogenesis, tumorigenicity, cytokine release and many more. The cell based target validation approach often builds the bridge between the molecular mechanism and the observed *in vivo* effects.

3.2 In Vivo Target Validation

One of the most important tests for a potential drug is an assessment of its role in disease in an animal model. In vivo target validation using gene knockouts, in which genes are deleted or disrupted to halt their expression, is a powerful method of predicting drug action. This kind of target validation is based on the assumption that knocking out the gene for the potential target has the same effect as administering a highly specific inhibitor of the target protein in vivo. Furthermore, target-specific side effects can be discovered before time and money is invested in drug design. Drugs can also be tested for toxicity and their potential therapeutic activity against the target more easily than in mammals.

In vivo testing involves whole organisms. It assesses both pharmacology and biological efficacy in parallel. Animal models have specific characteristics that mimic human diseases. The technologies for the creation of transgenic animals, where certain genes are deleted, modulated, or added, have progressed tremendously in the last decade. As a consequence, the predictive power of animal models for human disease and pharmacology is improving. Pharmaceutical companies have long used model organisms in preclinical efficacy and safety studies. With the emerging knowledge of whole genomes, researchers are now increasingly seeking animal models not only of specific diseases but also of their particular underlying pathways to broaden assays from pharmacology to include the mechanism of action. Disease models are usually incomplete models of pathology or mechanism, and their utility in drug screening is limited by the validity of the pathway in human disease pathogenesis.

3.2.1 Knockout

Knockout of genes that are essential in development is very often lethal. And the embryonic phenotype does not always reflect the function of the gene of interest within the adult organism. Therefore, techniques that allow controlling time and degree of gene silencing in vivo are of outstanding interest.

RNA interference through expression of short hairpin (sh)RNAs provides an efficient approach for gene function analysis in mouse genetics. Several systems are nowadays available, which allow the temporal control of ubiquitous shRNA expression in mice. Depending on the dose of the inductor, the knockdown efficiency reaches up to significant levels.

These systems facilitate a rapid performance of in vivo gene function studies, providing a generally applicable tool for reverse mouse genetics research. Moreover, it permits the investigation of the effect of gene knockdown after the onset of a chronic or an acute disease. This aspect is of particular interest for pharmaceutical drug target validation, as inducible gene silencing in mouse models of human disease can provide an ideal surrogate for the treatment of patients with antagonistic drugs. Even multiple dose regimes can be tested in preclinical studies by applying several rounds of induction. Furthermore, reversible knockdown of drug targets can help to distinguish target-dependent from target-independent effects of drug action.

3.2.2 Animal Models

An area of research that today generates great optimism is the use of stem cells for treatment of human diseases. Pluripotency and the capacity for continuous self-renewal make these cells an attractive donor source for cell replacement strategies.

Two different approaches are commonly used in in vivo animal models: the mobilization and activation of endogenous stem cells and the exogenous transplantation of genetically modified stem cells.

Hematopoietic stem cell transplantation is a treatment option, for example, for hematological malignancies. Current mobilization regimes frequently result in inadequate numbers of HSCs for transplant; therefore alternative methods of mobilization are required. The chemokine receptor CXCR4 and the ligand SDF-1 are integrally involved in HSC homing and mobilization. In animal models, disruption of the SDF-1/CXCR4 axis by CXCR4 antagonists has shown to improve HSC mobilization. Increase in stem cells by mobilizing principles can be easily detected by analyzing overall blood parameters. For example, antagonizing CXCR4 results in a significant increase in the white blood cell (WBC) count of up to nearly threefold in the plasma, peaking at around 6 h postdosing and returning to normal within 24 h.

Using adequate animal models, regimens for an optimal endogenous mobilization of stem cells can be defined. Clinical studies have to show whether these results can be translated to human beings.

In animal models, transplanted bone marrow cells, which have been genetically modified, e.g., by transfection of the green fluorescent protein (GFP), have been subsequently detected in certain tissues like skeletal and cardiac muscles, vascular endothelium, liver, lung, gut and skin epithelia, pancreatic beta cell islets, renal glomeruli, neural tissue, and tumors (Fig. 4). Thereby, the GFP labeling allows an easy and convenient detection of these exogenously transplanted cells in the animal. Even the xenobiotic transplantation of human-derived pluripotent stem cells in immunosuppressed mice is possible.

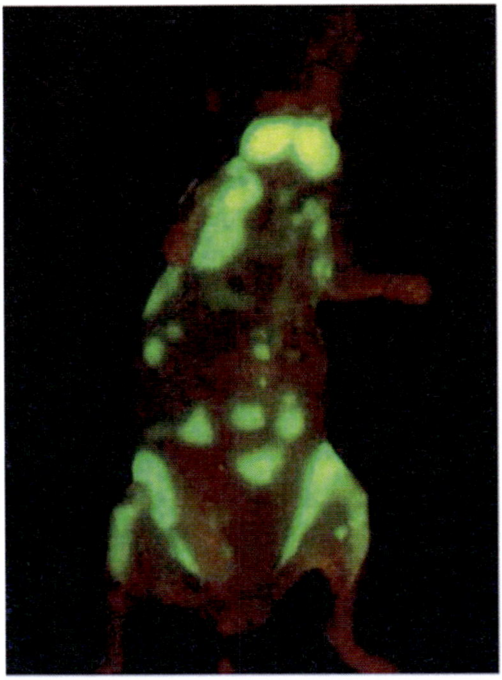

Fig. 4 In vivo fluorescent imaging of a mouse lymphoma following reconstitution with stem cells infected with a GFP-tagged p53 shRNA Michael T. Hemann (http://mit.edu/biology/www/facultyareas/facresearch/hemann.html)

3.3 Screening

Lead discovery in the pharmaceutical industry today still depends largely on experimental screening of compound collections. To this end, the industry has invested heavily in expanding their compound files and established appropriate screening capabilities to handle large numbers of compounds within a reasonable period of time. "High-throughput screening" (HTS) started roughly one decade ago with the introduction of laboratory automation to handle the different assay steps typically performed in microtiter plates. HTS technologies during the last decade have witnessed remarkable developments. Assay technologies have advanced to provide a large variety of various cell-based and biochemical test formats for a large spectrum of disease-relevant target classes [3]. In parallel, further miniaturization of assay volumes and parallelization of processing have further increased the test throughput. At Bayer Healthcare, HTS is performed entirely in 1,536-well plates with assay volumes between 5 and 10 μl. This assay carrier together with fully automated robotic systems allows for testing in excess of 200,000 compounds per day.

The ultra-high throughput is required to fully exploit Bayer Healthcare compound file of >2.5 million compounds (compound library; Fig. 5), which are routinely screened in each project. The comprehensive substance collection together with sophisticated screening technologies has resulted in a clear advantage in lead discovery especially for poorly druggable targets with a poor track record in the past. In addition, size and composition of the library with compounds largely proprietary to the company facilitate the identification of leads, providing a straightforward path

Fig. 5 Fully automated chemical compound repository (Bayer Schering Pharma)

toward establishing intellectual property on the composition of matter. Generally, the requirement to generate solid IP around drug candidates is a major driving force for discovery of novel lead structures. The productivity of HTS has been questioned recently and the poor druggability of many molecular targets persuaded in research programs together with a poor quality of compounds in screening libraries has been identified as a major drawback responsible for low success rates. Improvements in the design of screening libraries aimed at providing diverse collections of compounds with physicochemical properties expected from "drug-like", orally available molecules and increasing awareness of druggability aspects in target selection are consequences of the above-mentioned debate.

3.4 Lead Structure Optimization

Lead optimization compares the properties of various lead compounds and provides information to help select the compound or compounds with the greatest potential to be developed into safe and effective medicines. Often during this same stage of development, lead prioritization studies are conducted in in vivo and in vitro studies to compare various lead compounds and how they are metabolized and affect the body.

Medicinal chemistry combines empirical knowledge from the structure–function relationships of known drugs with rational designs optimizing the physicochemical properties of drug molecules. For example, medicinal chemists improve drug efficacy, particularly with respect to stability and bioavailability, by developing mechanism-based pro-drugs.

4 Adult Stem Cells in Hematological Diseases

Hematopoietic stem cells (HSCs) are stem cells that give rise to all the blood cell types including myeloid (monocytes and macrophages, neutrophils, basophils, eosinophils, erythrocytes, megakaryocytes/platelets, and dendritic cells) and lymphoid lineages (T cells, B cells, and NK cells). The hematopoietic tissue contains cells with long-term and short-term regeneration capacities and committed multipotent, oligopotent, and unipotent progenitors.

HSCs are found in the bone marrow of adults, which includes femurs, hip, ribs, sternum, and other bones. Cells can be obtained directly by removal from the hip using a needle and syringe or from the blood following pretreatment with cytokines, such as G-CSF (granulocyte colony-stimulating factors), that induce cells to be released from the bone marrow compartment. HSCs have a higher potential than other immature blood cells to pass the bone marrow barrier and thus may travel in the blood from the bone marrow in different organs and vessels.

HSCs can be mobilized by pharmacological treatment and migrate in different organs and tissues. This ability allows and enables the identification of target genes whose modulation could be used to treat diverse diseases by stem cell activation.

5 Adult Stem Cells in Cancer

Research accomplished during the last few years on embryonic, fetal, amniotic, umbilical cord blood, and adult stem cells has constituted a revolution in cancer therapies by providing the possibility of generating multiple therapeutically useful cell types. In clinical practice, HSCs homing in the BM compartment may be used after their isolation from BM or peripheral blood (PB) from patients or healthy donors. BM-derived HSCs may be collected from BM aspirate or by apheresis after their mobilization in PB by using diverse mobilizing agents such as granulocyte colony-stimulating factor, granulocyte–macrophage colony-stimulating factor, or synthetic chemical compounds as chemokine receptor modulators [4].

Chemokines are highly basic small secreted proteins consisting on average of 70–125 amino acids which mediate their effects through to G protein-coupled receptors (GPCRs). Over the past few years, chemokines are the focus of anti-cancer drug discovery, which elucidated their crucial role at all stages of neoplastic transformation and progression. Thus, tumor cells extensively express functional chemokine receptors, which can sustain proliferation, angiogenesis, survival, and promote organ-specific localization of distant cancer metastases.

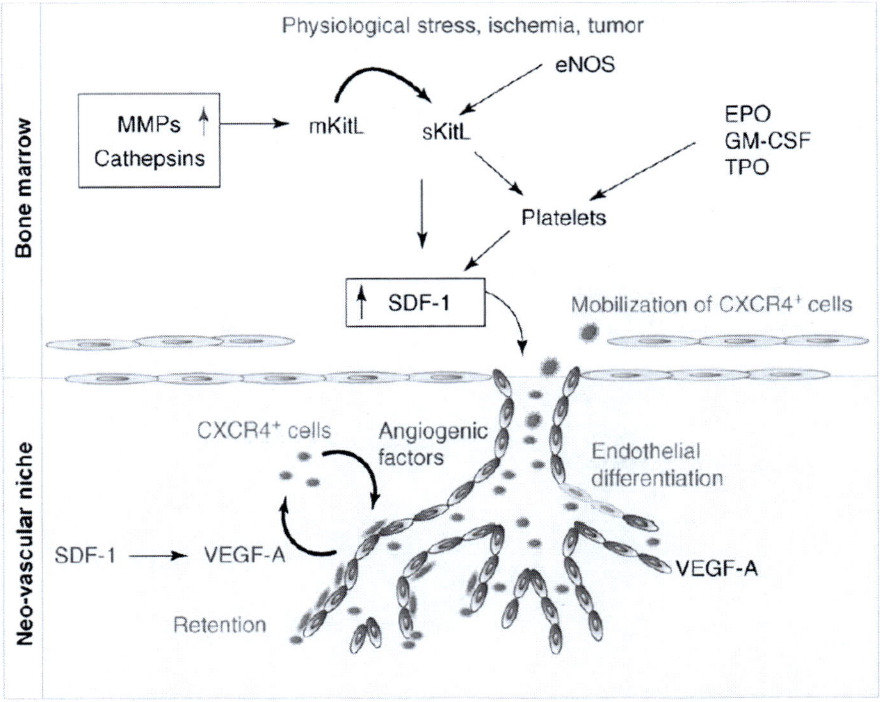

Fig. 6 Involvement of bone marrow-derived stem cells in neoangiogenesis (Petit et al. [12])

The chemokine receptor CXCR4 possesses multiple fundamental functions in both normal and pathologic physiology (Fig. 6). CXCR4 is a GPCR that transduces signals of its endogenous ligand the chemokine CXCL12 (stromal cell-derived factor-1, SDF-1). The interaction between CXCL12 and CXCR4 plays a critical role in the migration of progenitors during cardiovascular, hematopoietic, and central nervous system development. This interaction is also known to be involved in several intractable disease processes, including HIV infection, cancer cell metastasis, leukemia cell progression, rheumatoid arthritis, asthma, pulmonary fibrosis, and angiogenesis [5].

6 Adult Stem Cells in Cardiovascular Diseases

Cardiovascular disease is an umbrella term for a variety of different diseases affecting the heart and vessels. As of 2007, it is the leading cause of death in most countries. Cardiovascular diseases cover a variety of different diseases such as myocardial infarction, heart failure, and vascular diseases such as PAOD.

6.1 Myocardial Infarction

Bone marrow-derived circulating endothelial progenitor cells (EPCs) contribute to regeneration of infracted myocardium by enhancement of neovascularization and putative paracrine effects such as secretion of pro-angiogenic factors (Fig. 7). Patients with coronary artery disease (CAD) are known to have reduced numbers of ex vivo-cultivated EPCs. Furthermore, the functional activities of progenitor cells in CAD patients are compromised and might thereby limit the potential for neovascularization processes within the myocardium after AMI [6].

The granulocyte colony-stimulating factor (G-CSF) is well established to mobilize hematopoietic stem cells (HSCs) from the bone marrow in patients with hematologic disorders undergoing radiochemotherapy and stem cell transplantation. Thereby, G-CSF activates progenitor cell-releasing factors such as neutrophil elastases and matrix metalloproteinase to release the stem cells from the bone marrow.

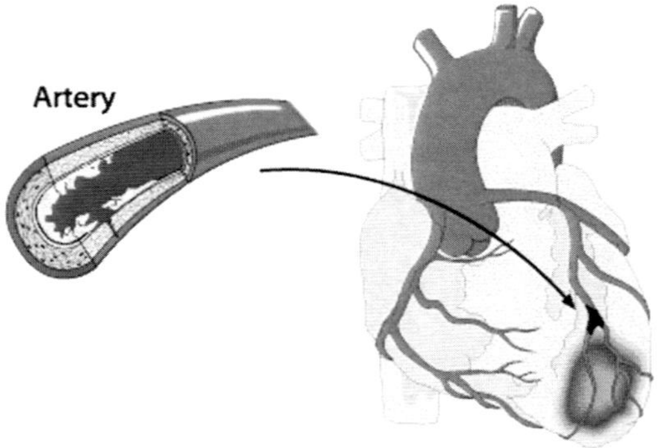

Fig. 7 Myocardial infarction (http://www.heartpoint.com/mimore.html)

6.2 Heart Failure

Heart failure remains the leading cause of morbidity and mortality worldwide. Despite that advances in medical treatment and interventional procedures have reduced the mortality rate in patients with coronary artery disease, the number of patients with refractory myocardial ischemia and congestive heart failure is rapidly increasing. Experimental studies have demonstrated that bone marrow (BM) contains adult stem cells that can induce neovascularization and improve heart function in ischemic myocardium [7]. Recent insights into the understanding of the mechanisms involved in proliferation, recruitment, mobilization, and incorporation of BM-derived stem cells into the myocardium and blood vessels have prompted the development of cellular transplantation therapy for heart diseases refractory to conventional therapy. Initial preliminary clinical studies indicated potential clinical benefit of BM therapy in patients with acute myocardial infarction and chronic myocardial ischemia (Fig. 8).

Cytokine therapy has been well proven to be an effective method for mobilization of BM stem cells into the peripheral circulation. In patients with chronic stable angina, cytokine therapy with granulocyte–macrophage colony-stimulating factor (GM-CSF) or G-CSF has also been investigated as a sole or an adjunctive therapy. Importantly, current therapies do not address the central pathophysiology underlying the development of heart failure, namely the loss of functional cardiomyocytes. Novel therapeutic strategies are aimed at repairing and replacing cardiomyocytes using stem cells.

6.3 PAOD

Peripheral vascular diseases are defined as vascular diseases in which arterial and/or venous flow is reduced resulting in an imbalance between blood supply

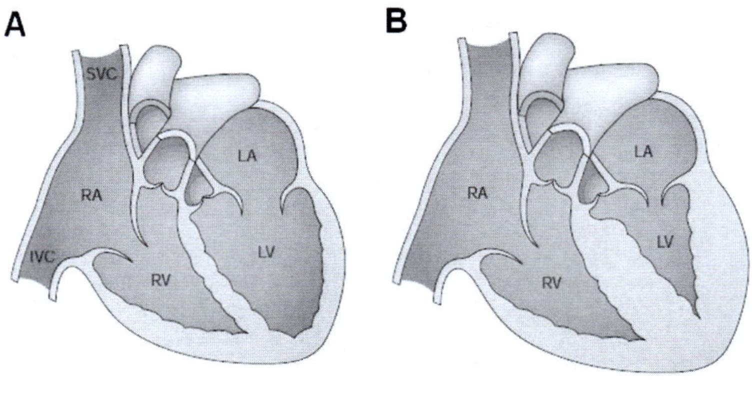

Fig. 8 Cardiomyopathy A: normal heart, B: cardiomyopathy[14]

and tissue oxygen demand. It includes chronic peripheral arterial occlusive disease (PAOD), acute arterial thrombosis and embolism, inflammatory vascular disorders, Raynaud's phenomenon, and venous disorders. Furthermore atherosclerosis, in which the vessel wall is remodeled, compromising the lumen of the vessel, is often an accompanying disease. The atherosclerotic remodeling process involves accumulation of cells, both smooth muscle cells and monocyte/macrophage inflammatory cells, in the intima of the vessel wall.

These cells take up lipid, likely from the circulation, to form a mature atherosclerotic lesion. Although the formation of these lesions is a chronic process, occurring over decades of an adult human life, the majority of the morbidity associated with atherosclerosis occurs when a lesion ruptures, releasing thrombogenic debris that rapidly occludes the artery. When such an acute event occurs in the coronary artery, myocardial infarction can ensue and in the worst case can result in death.

The formation of the atherosclerotic lesion can be considered to occur in five overlapping stages such as migration, lipid accumulation, recruitment of inflammatory cells, proliferation of vascular smooth muscle cells, and extracellular matrix deposition.

Circulating endothelial progenitor cells (EPCs) are thought to contribute to angiogenesis following vascular injury, stimulating interest in their ability to mediate therapeutic angiogenesis. However, the number of EPCs in the blood and tissue is low and can be increased by mobilization of EPCs from bone marrow. Mobilized EPCs displayed a greater proliferative capacity than did EPCs isolated from baseline blood. The number of angiogenic cells CACs and EPCs in the blood at baseline is low, limiting their delivery to sites of ischemia and subsequent stimulation of angiogenesis. There is evidence that treatment with certain cytokines induces angiogenic cell mobilization from the bone marrow into the blood, potentially overcoming this limitation. In particular, granulocyte colony-stimulating factor (G-CSF) has been shown to mobilize EPCs in various animal models and humans. Based on these observations, several clinical trials of G-CSF-induced EPC mobilization following acute myocardial infarction, cancer, and hematology have been performed. An alternative approach to increase angiogenic cell delivery to sites of ischemia is the direct injection of enriched cell populations of EPCs or CACs into the ischemic area.

In animal studies it was shown that the inhibition of CXCR4 leads to a rapid mobilization of angiogenic cells, providing a promising strategy to stimulate therapeutic angiogenesis following acute vascular injury [8] or PAOD.

7 Adult Stem Cells in Drug Discovery

Significant advancements in stem cell research have provided important information on stem cell biology and offer great promise for developing novel successful stem cell-based medical treatments. Further investigations appear to be necessary to translate the basic knowledge into clinical therapeutic applications in humans.

The identification of specific biomarkers to each type of adult stem/progenitor cells relative to their more committed and mature progeny is important for the characterization of their specific physiological functions [4].

References

1. Fjose A, Ellingsen S, Wargelius A, Seo HC. RNA interference: mechanisms and applications. Biotechnol Annu Rev. 2001; 7:31–57.
2. Smith C, et al. Hitting the target. Nature 2003; 422:341–347.
3. Walters WP, Namchuk M. Designing screens: how to make your hits a hit. Nat Rev Drug Discov. 2003; 2:259–266.
4. Mimeault M, et al. Stem cells: a revolution in therapeutics—recent advances in stem cell biology and their therapeutic applications in regenerative medicine and cancer therapies. Clin Pharmacol Ther. 2007 Sep; 82(3):252–264.
5. Lavrovsky Y, et al. CXCR4 receptor as a promising target for oncolytic drugs. Mini Rev Med Chem. 2008; 2008(8):1075–1087.
6. Honold J, et al. Effects of granulocyte colony stimulating factor on functional activities of endothelial progenitor cells in patients with chronic ischemic heart disease. Arterioscler Thromb Vasc Biol. 2006; 26:2238–2243.
7. Tse HF, et al. Therapeutic angiogenesis with bone marrow-derived stem cells. J Cardiovasc Pharmacol Ther. 2007; 12(2):89–97.
8. Shepherd RM, et al. Angiogenic cells can be rapidly mobilized and efficiently harvested from the blood following treatment with AMD3100. Blood 2006 Dec 1; 108(12):3662–3667.
9. Bustin SA. Absolute quantification of mRNA using real-time reverse transcription polymerase chain reaction assays. J Mol Endocrinol. 2000; 25(2):169–193.
10. Golz S, Hüser J. Discovery of a new drug – from target identification to ultra-high-throughput screening. Clin Lab. 2007; 53(1–2):77–79.
11. Koch W. Technology platforms for pharmacogenomic diagnostic assays. Nat Rev Drug Discov. 2004 Sep; 3:749–761.
12. Petit I, Jin D, Rafii S. The SDF-1-CXCR4 signaling pathway: a molecular hub modulating neo-angiogenesis. Trends Immunol. 2007; 28(7):299–307.
13. Hopkins AL, Groom CR. The druggable genome. Nat Rev Drug Discov. 2002 Sep; 1(9):727–730.
14. Towbin J.A. and Bowles N.E. The failing heart. Nature. Jan 10; 4151(6868):227–233.

Embryonic Stem Cells as a Tool for Drug Screening and Toxicity Testing

Bernd Denecke and Silke Schwengberg

Abstract Embryonic stem cells (ESCs) are recognized by the general public mainly as a possible source of cellular replacement therapy and transplantation. However, the usefulness of ESCs and ESC-derived tissue cell types for cell-based in vitro test systems in drug screening and toxicity was already recognized early after the discovery of ESCs in the late 1980s.

In the present chapter, the employment of ESCs in cell-based test systems is described in the light of state-of-the-art methods in the field of drug discovery, and the advantages and disadvantages of the applications are discussed. The possible use of ESCs in the drug development process is illustrated by three examples: (1) the use of ESCs for basic research in early developmental biology, (2) the embryonic stem cell test for embryotoxicity, and (3) the use of ESC-derived cardiomyocytes for detection of cardiac toxicity. Potential applications of human ESCs, induced pluripotent stem cells (IPS cells), and personalized medicine are described in the last part.

Keywords Cardiac toxicity · Cell-based test system · Developmental biology · Drug discovery · Embryotoxicity · iPS · Personalized medicine

1 Conventional Methods in Drug Screening and Toxicity Testing

Today, the development of a drug starting from a new chemical entity (NCE) usually follows a structured, tiered approach. Although variations may be applied by different drug-developing companies, e.g. in the realization of the different steps, registration of drugs requires a certain amount of information that has to be provided by the company filing the application. In order to get the desired information, different types of studies will be conducted, which can be mainly divided into non-clinical or preclinical and clinical studies.

B. Denecke (✉)
IZKF "BIOMAT.", RWTH Aachen, Germany
e-mail: bernd.denecke@rwth-aachen.de

G.M. Artmann et al. (eds.), *Stem Cell Engineering*,
DOI 10.1007/978-3-642-11865-4_22, © Springer-Verlag Berlin Heidelberg 2011

To perform non-clinical studies for efficacy, metabolization, and toxicity of a given drug, a researcher can apply different in vivo and in vitro methods. Most likely, both types of studies will be used in a tiered approach, starting from simple biochemical assays, moving to more sophisticated in vitro test systems, and ending up with in vivo studies using different animal species [1, 2].

The functional testing of the efficacy of an NCE and the prediction of unwanted side effects may require different testing strategies. Whereas signalling effects of a certain molecule can be studied in isolated pathway assay systems or using HTS suitable cell lines, complex animal data have to be analysed in order to predict misbehaviour of a given drug candidate in humans. Even more, while functional development focusing on cellular behaviour is analysed with most modern techniques, toxicological analyses is confined to relatively old test strategies.

1.1 In Vivo Methods

Traditionally, a holistic, physiological in vivo approach (the use of a living, either conscious or anaesthetized, animal given the drug,) was used to identify the desired action of a new drug as well as unwanted side effects within the drug development process [3]. Until the dawn of molecular biology, most diseases could not be broken down to a singular cause. Therefore, drug screening using the whole organism or at least isolated organs was the only option to identify new drugs.

With this attempt, determination of the mode of action of an NCE (and even of an already known drug) is difficult, because effects could not be observed directly, neither at the target of the drug nor at the structural, cellular or molecular "cause" of the disease. Nevertheless, the approach was successful in identifying medication for many common diseases, like cardiovascular, gastrointestinal, metabolic, and psychiatric disorders [3]. In case that a disease itself is manifested only at the level of a whole organism, the holistic approach may be even advantageous over mechanistic molecular attempts.

Even after the introduction of high-throughput and high-content in vitro assays in the drug development process, in vivo testing is still needed to answer key questions in pharmacokinetics, metabolization, and toxicity.

For toxicity studies, in vivo testing has the advantage that negative effects can be identified even without knowing the exact mode of action of the tested compound. Also, chronic exposure or multigeneration effects can hardly been mimicked using classical cell-based in vitro test systems.[1] Despite the use in general safety studies, PK/PD (pharmacokinetics/pharmacodynamics) testing as well as ADME toxicity (absorption, distribution, metabolization, and excretion) demands in vivo testing to determine dosage protocols and safety margins for clinical studies in humans [4].

[1] It has to be kept in mind that the results of in vivo toxicity have to be linked to other data, like route of exposure, environmental concentrations, and the mode of action of a compound, to correctly predict the risk for human beings.

In vivo testing is also still mandatory for most regulatory demands. Today, 73 out of 86 OECD (Organisation for Economic Co-operation and Development) guidelines for the testing of chemicals for health effects require the use of animals or isolated organs (http://www.oecd.org/document/40/0,3343,en_2649_34377_37051368_1_1_1_1,00.html). The US Department of Health and Human Services, Food and Drug Administration (FDA) in its "Common Technical Document for the Registration of Pharmaceuticals for Human Use" demands for various toxicity studies for drug registration (http://www.fda.gov/cder/guidance/#International%20Conference%20on%20Harmonisation%20-%20Safety). For registration, in vitro methods are mainly accepted only as first step in a tiered approach, e.g. in vitro genotoxicity using non-mammalian or mammalian cells, followed by subsequent in vivo studies.

Beside its successful employment during the last decades of drug discovery and toxicity testing, the holistic in vivo approach has some drawbacks:

- The whole organism represents a "black box", and compounds are tested in a blind manner. Therefore, identification of the target(s) and the mode of action of a given drug exerting an effect in the whole organism is difficult.
- Inter-species differences tend to be higher at the level of the whole organism then at the molecular, biochemical, or cellular level.
- Results of in vivo toxicity studies are often not correlated with the human situation in terms of dosage, effect levels, and affected organs [5].
- Due to the high costs of large animal breeding, most in vivo studies are performed with inbred, genetically homogeneous mice or rats. Although the number of animals needed per study is lower due to low variation, results can be highly strain specific [6].[2]
- There is growing ethical concern about animal experiments and a high demand for alternative methods, e.g. for the European directive for Registration, Evaluation, and Assessment of Chemicals (REACh) of the European Union [7].

1.2 In Vitro Methods

To investigate the mode of action of drugs in more detail and to identify the cause of a disease as well as possible targets for a cure, scientists in drug development employ either ex vivo (an isolated organ or a piece of tissue) or in vitro (primary cells isolated from tissues or organs or cell lines) models and test systems.

Ex vivo models have already been used for decades, with the isolated frog leg for experiments in neurophysiology being well-known even to undergraduate students. For determination of efficacy and toxicity, the "isolated perfused mammalian

[2] Schardein described in his book *Chemically Induced Birth Defects* the effect of the well-known teratogen thalidomide on different strains of laboratory animals: see Schardein [105].

heart preparation" established by Oscar Langendorff in 1898 [8] for effects on cardiac physiology, as well as isolated intestinal mucosa strips for determination of intestinal absorption and secretion using the Ussing chamber developed by Hans Ussing in the 1950s [9], is widely used.

The employment of isolated cells and the generation of cell lines were first described already in the early twentieth century and became a standard method during the 1950s (http://www.ncbi.nlm.nih.gov/books/bv.fcgi?db=Books&rid=mboc4.table.1516). Nowadays, the use of in vitro cell culture techniques is common in all kinds of laboratories, from basic molecular biology over physiology up to drug development and toxicology. In most experiments, cell lines may be applied, because they are well described and easy to obtain. The work with primary cells isolated from organs and tissues (of any desired species) requires more technical effort, better training, and last but not least the use of animals or humans as source of the biological material to work on.

In drug development, in vitro test systems are widely used during the first steps of preclinical studies as well as for basic research on the underlying mechanism of diseases and the identification of drug targets. Today, two different approaches can be identified (with numerous intermediate states that may be useful for certain types of studies):

(1) High-throughput screening (HTS), that is, the screening of large compound libraries with up to several hundreds of thousands of entities with a simple type of cell-based assay.
(2) High-content screening (HCS), that is, the screening of a small number of "hits" that came from HTS in a cell-based test system that is build to investigate simultaneously several different endpoints [10].

Besides the obvious advantages of in vivo assays at certain steps within the drug development process, rising ethical concerns on animal testing demand for alternatives. In Europe as well as in the United States, two authorities, namely ECVAM (European Center for the Validation of Alternative Methods, www.ecvam.jrc.eu) and ICCVAM (Interagency Co-ordinating Committee on the Validation of Alternative Methods, http://iccvam.niehs.nih.gov), are assigned to develop, validate, and promote in vitro tests as alternatives to animal testing. During the last centuries, more than 25 in vitro tests have gained scientific validity, many of those developed in consortia of academia, industry, and governmental institutions. These strategies ensure that tests are developed that fulfil the needs of (pharmaceutical) industry as well as of regulatory authorities.

1.2.1 Limitations of State-of-the-Art In Vitro Systems

Although in vitro tests are widely used in basic research as well as within the drug development process, the cell lines and primary cells currently used for those assays exhibit several disadvantages.

Continuous, immortal cell lines are derived either from tumour material or by transduction with viruses, e.g. the Epstein–Barr virus (EBV). Although these cell lines are easy to obtain and will proliferate indefinitely, they rather exhibit the normal physiology of their non-transformed counterpart. Usually, these cells exhibit a dedifferentiated phenotype and express higher levels of telomerase, which allows for indefinite proliferation [11]. Tumour cells lack the normal growth inhibition displayed by most tissue cells that are fully differentiated and not intended to proliferate anymore. Moreover, certain biochemical pathways may be regulated differently in tumour cells compared to normal tissue cells [12]. Therefore, it is at least questionable to what extent results obtained using such cell lines can be linked to desired or unwanted effects in any human being receiving the drug.

A problem widely overlooked when working with cell lines is that subclones are generated easily when different protocols are used to maintain the cells.[3] Moreover, recent studies by the German Collection of Microorganisms and Cell Cultures (Deutsche Sammlung von Mikroorganismen und Zellkulturen, DSMZ) revealed that many cell lines are accidentally contaminated with other cell lines cultured in the same laboratory (http://www.dsmz.de/human_and_animal_cell_lines/main.php?contentleft_id=21) [13]).

Primary cells are mainly isolated from animal species, although certain human cell types are now even commercially available (e.g. http://www.tno.co.jp/Pharma/Humanprimarycells.pdf; http://www.promocell.com). Primary cells usually should display all physiological features of their in vivo counterparts, but may be affected by the isolation procedure. Differences in the procedure itself can be a cause for large variation and poor standardization of the quality and responsiveness of the material used in different laboratories. Even within a certain laboratory, the quality of primary cells depends on the "quality" of the animal used for preparation, as well as on the technical skills of the operator.

In contrast to cell lines, most cell types isolated from tissue will not proliferate in culture, and many of them can be cultured only for a few days before their physiological relevance starts to decline. Therefore, only a limited time frame for the conduction of experiments based on these cell types can be utilized. Moreover, most primary cell types are not available in large quantities or can be hold as a frozen stock for future experiments, which requires careful planning and logistics prior to experimentation.

2 Embryonic Stem Cells as a Tool for In Vitro Test Systems

The limitations described above make the development of a standardized, easy-to-handle, and convenient cell system desirable that combines the advantages of tumour cells (indefinite proliferation, high standardization) with those of primary

[3] From the author's own experience, the generation of a subclone of the cell line T84 was achieved within 3 months (starting from the same material) only by (a) different passaging intervals and (b) not ensuring that all cells are removed from the plates after trypsinization for passaging.

cells (physiological relevance compared to the in vivo situation). Embryonic stem cells (ESCs), in their undifferentiated state, possess high proliferative capacity. The development of directed differentiation protocols for different cell types during the last years [14–16] allows for standardized production of cells that to a large extent exhibit the normal physiological behaviour of the corresponding primary cell. Therefore, ESCs may represent an almost ideal cell source for in vitro test systems for drug development and toxicity testing.

2.1 Definition of ESCs

During the entire lifetime of a multicellular organism, adult stem cells are present. They are found in organs, in the bone marrow, and also in the cord blood. The task of these cells in adult organisms is to not only act as a repair system for the body, replenishing specialized cells, but also maintain the normal turnover of regenerative organs, such as blood, skin, or intestinal tissues (For review see [17]). By the differentiation process in general, the cells loss the ability to proliferate. Adult stem cells are able to differentiate only into a limited number of cell types. Normally they can differentiate into the different cell types of one tissue. This is the reason why they are called multipotent.

In contrast to multipotent adult stem cells, ESCs can only be isolated very early in the development [18, 19]. ESCs are derived from the inner cell mass of an early stage embryo known as blastocyst. Human embryos reach the blastocyst stage 4–5 days postfertilization, at which time they consist of 50–150 cells. Murine embryos reach the blastocyst stage a little bit earlier at day 3.5 postfertilization.

ESCs are able to differentiate into all derivatives of the three primary germ layers [20]: ectoderm (e.g. nervous system, epidermis, and the outer part of integument), endoderm (e.g. liver, gastrointestinal tract, and lungs), and mesoderm (e.g. cardiovascular system, bones, urogenital tract, blood, and muscles). These include each of the more than 220 cell types in the adult body. However, ESCs lack the potential to contribute to extra-embryonic tissue, such as the placenta, which is the reason why ESCs are called pluripotent and not totipotent. When given no stimuli for differentiation (i.e. when grown in vitro), ESCs maintain pluripotency throughout multiple cell divisions [21].

Besides the basic definition of ESCs as self-replicating cells derived from the embryo that can differentiate into almost all of the cells of the body, a more precise definition for research purposes was necessary. Therefore, specific criteria that help scientists to better define ESCs were proposed. Among others, Smith has offered a list of essential characteristics that define ESCs more precisely [22]:

Embryonic stem cells

- are derived from the inner cell mass/epiblast of the blastocyst;
- are capable of undergoing an unlimited number of symmetrical divisions without differentiating (long-term self-renewal);

- exhibit and maintain a stable, full (diploid), normal set of chromosomes (karyotype);
- can give rise to differentiated cell types that are derived from all three primary germ layers of the embryo (endoderm, mesoderm, and ectoderm);
- are capable of integrating into all foetal tissues during development;[4]
- are capable of colonizing the germ line and giving rise to egg or sperm cells (not shown for human ESCs);
- are clonogenic, that is, a single ESC can give rise to a colony of genetically identical cells, or clones, which have the same properties as the original cell;
- express the transcription factor Oct-4, which then activates or inhibits a host of target genes and maintains ESCs in a proliferative, non-differentiating state;

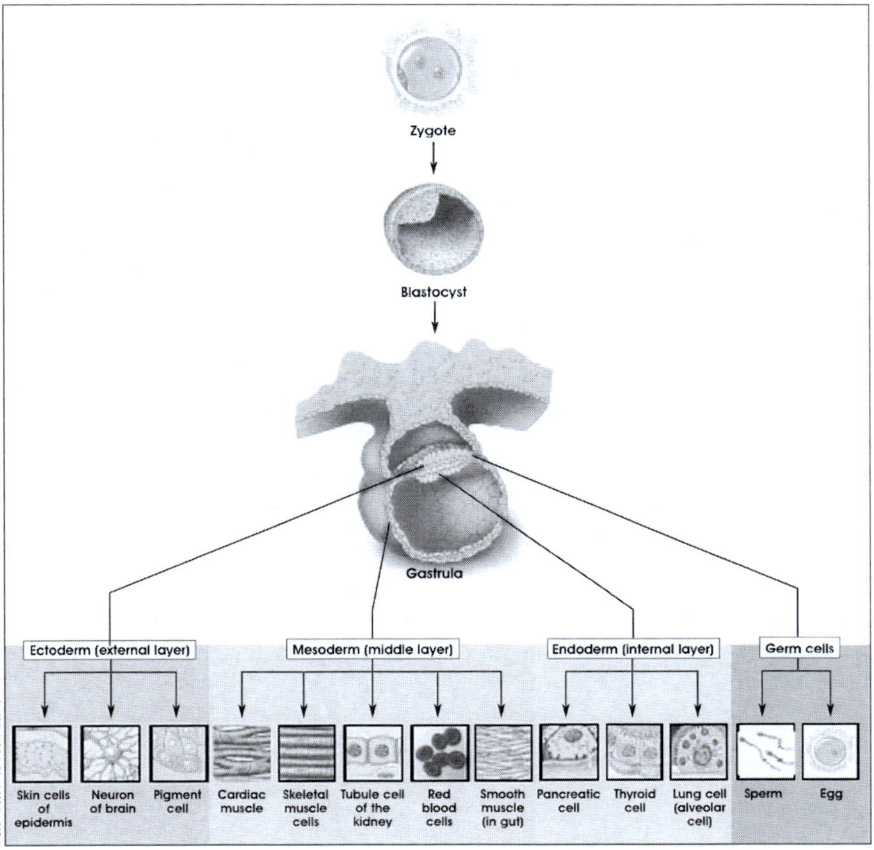

Fig. 1 Organs derived from each germ layer. Image from: *Stem Cells: Scientific Progress and Future Research Directions*. Department of Health and Human Services. June 2001

[4] Mouse ESCs maintained in culture for long periods can still generate any tissue when they are reintroduced into an embryo to generate a chimeric animal. Not shown for human ESCs.

- can be induced to continue proliferating or to differentiate;
- lack the G1 checkpoint in the cell cycle;[5]
- do not show X inactivation. In every somatic cell of a female mammal, one of the two X chromosomes becomes permanently inactivated. X inactivation does not occur in undifferentiated ESCs.

Another new strategy for classifying stem cells and supporting the idea that pluripotency and self-renewal are under tight control by specific molecular networks was recently published by Mueller et al. [23]. They uncovered a protein–protein network (PluriNet) that is shared by the pluripotent cells (ESCs, embryonal carcinomas, and induced pluripotent cells). By analysis of published data, they could show that the PluriNet seems to be a common characteristic of pluripotent cells, including mouse ESCs, induced pluripotent cells (IPS cells), and human oocytes.

ESCs were first derived from mouse embryos in 1981 by Evans and Kaufman [18] and independently by Martin [19]. The successful cultivation of human embryonic stem cell lines was first reported in November 1998 by Thomson et al., who first developed a technique to isolate and grow ESCs when derived from blastocysts [24].

The time lag of about 17 years to extend the mouse results to humans may be due to the fact that human embryonic stem cell lines are derived from embryos generated by in vitro fertilization (IVF). Culture media for human embryos produced by IVF were sub-optimal. Thus, it was difficult to culture single-cell fertilized embryos long enough to obtain healthy blastocysts, a precondition for the derivation of embryonic stem cell lines. Additionally, experience made with mouse ESCs could not be completely applied to the derivation of human ESCs. Based on the experience gained by the isolation of non-human primate embryonic stem cell lines [25, 26] together with improvements in culture medium for human IVF-produced embryos, finally the isolation and the cultivation of human embryonic stem cell lines were enabled [24].

2.2 Cultivation of ESCs

To maintain ESCs at their full developmental potential, optimal growth conditions should be provided in culture. Suppression of differentiation programms is an essential requirement for maintenance of embryonic stem cell pluripotency. Inappropriate culture conditions entail progressively genetic lesions that will compromise the differentiation potential of ESCs.

Classically, mouse ESCs are cultivated on inactivated mouse embryonic fibroblasts (feeder cells) in the presence of FCS (foetal calf serum; lots have to be tested for stem cell cultivation). Spontaneous differentiation is inhibited by leukaemia inhibitory factor (LIF), which is one of the active components produced by feeder cells [27]. Meanwhile it is possible to substitute the feeder cells with purified LIF.

[5] ESCs are most of the time in the S phase of the cell cycle, during which they synthesize DNA. Unlike differentiated somatic cells, ESCs do not require any external stimulus to initiate DNA replication.

However, most laboratories continue to use feeder cells to culture murine ESCs and in many cases they additionally supplement the medium with LIF. In serum-free cultures, LIF alone is not sufficient to prevent the murine ESCs from differentiation. Ying et al. [28] reported for the first time that the expression of inhibitor of differentiation (Id) proteins, the inhibition of extracellular receptor kinase (ERK), and the inhibition of p38 mitogen-activated protein kinases (MAPKs), all induced by BMP, as well as STAT3 induced by LIF [29], are sufficient to keep murine ESCs undifferentiated and support self-renewal in a serum-independent manner. Mouse ESCs survive and propagate well with only leukaemia inhibitory factor, bone morphogenetic protein, transferrin, and gelatine [30].

In contrast, self-renewal of human ESCs is not enabled by LIF in the presence of serum due to the lack of the activated LIF/STAT3 pathway in these cells. Furthermore, the addition of BMPs leads to differentiation of human ESCs [31–34]. Basic fibroblast growth factor enables the growth of human ESCs in the presence of fibroblasts using a serum replacement [35]. As shown by Xu et al., it is also possible to use a feeder cell-free human embryonic stem cell culture system in which Matrigel or laminin can be used as protein matrix and basic fibroblast growth factor-containing medium previously conditioned by co-culture with fibroblasts is applied [36, 37]. Defined culture conditions for human ESCs are also described by Ludwig

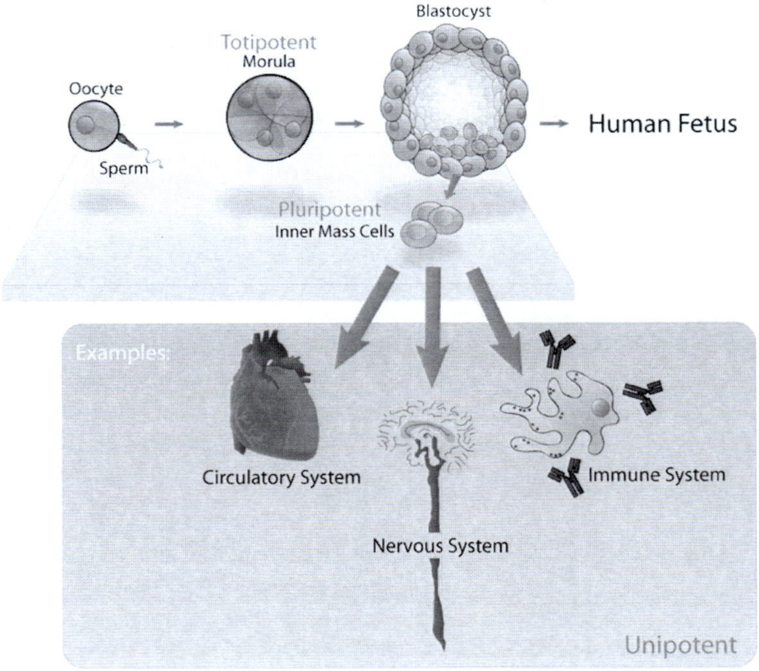

Fig. 2 Pluripotent, embryonic stem cells originate as inner mass cells within a blastocyst. The stem cells can become any tissue in the body, excluding a placenta. Only the morula cells are totipotent, able to become all tissues and a placenta. Image from http://schools-wikipedia.org/images/735/73539.png.htm (Original work by Mike Jones for Wikipedia)

et al. [38]. They established feeder-independent culture conditions using protein components solely derived from recombinant sources or purified from human material. This defined TeSR1 medium contained among others basic fibroblast growth factor, lithium chloride, γ-aminobutyric acid, pipecolic acid, and tumour growth factor beta. A complete list of the formulation of the TeSR1 medium is given in the supplement of the paper [38]. Another quite different defined medium to maintain human ESCs in an undifferentiated state was described by Lu et al. [39]. This human embryonic stem cell cocktail (HESCO) includes basic fibroblast growth factor, Wingless-type MMTV integration site family – member 3A, April (a proliferation-inducing ligand)/BAFF (B-cell-activating factor of the tumour necrosis factor family), albumin, cholesterol, insulin, transferrin, and fibronectin as coating material. Both, TeSR1 and HESCO share reagents like albumin, transferrin, insulin, and basic fibroblast growth factor, a fact that suggests that these factors are probably crucial for the self-renewal of human ESCs.

One prerequisite to use ESCs as a resource in regenerative medicine but also for drug screening and development studies is the elimination of products from foreign species in the culture medium and a clear definition of the components contained in the medium. Although differences exist between murine and human ESCs regarding culture conditions, it seems that cultivation of both kinds of stem cells is enabled under defined conditions, which is useful for drug development as well as for basic research in developmental biology.

2.3 Present Use of ESC

Already in 1992, Wobus et al. [40] described the possible use of ESCs and ESC-derived cells in toxicology and pharmacology. During the last years, many possible applications of ESCs in drug development were described in the literature (for review [41, 42]).

In the following section, the application of ESCs in basic research of early embryonic development as well as in two different test systems is described. The first test system (the embryonic stem cell test for embryotoxicity, EST) is based on the differentiation process of ESCs itself, whereas the second test system (ESC-derived cardiomyocytes for detection of cardiac toxicity) is based on a tissue-specific cell type generated from differentiated ESCs. It has to be noted that the tests described are chosen personally by the authors to represent two different possible approaches for the employment of ESCs in the drug development process. Many more tests are already under development or even employed in research institutes and the laboratories of the pharmaceutical industry.

2.3.1 Example 1: Use of ESCs for Basic Research of Early Developmental Biology

Due to the size and the accessibility of mammalian and in particular human embryos, it is very difficult to study the early processes of mammalian (human) embryogenesis, i.e. germ layer formation.

The circumstance that ESCs can generate all three germ layers (ectoderm, mesoderm, and endoderm) in vitro favoured this cell type as a useful tool to study the biology of early development. Particularly for the study of human development, the availability of cells from human embryos is limited. The ability of nearly unlimited propagation and hence the possibility to genetically modify ESCs is an additional advantage. By genetic modifications (overexpression or inactivation) of specific factors, cell lines can be created that allow the interpretation of the function of the modified gene in terms of early development. By withdrawal of factors preventing differentiation (e.g. LIF for murine stem cells), ESCs can aggregate to embryoid bodies (EBs), a structure that has visceral endoderm in the outer layer and other lineages derived from primitive ectoderm inside this three-dimensional structure.

Also, the formation of the proamniotic cavity is traced in EBs allowing, e.g. the study of the egg-cylinder formation. In EBs, the differentiation to nearly all lineages including mesoderm and neuronal lineages as well as precursors of germ cells can be induced [43, 44]. Despite the far less organization of the structure of an EB in comparison to an embryo in vivo, the time course of the expression of several important molecules in EBs is comparable to that observed in embryos. For

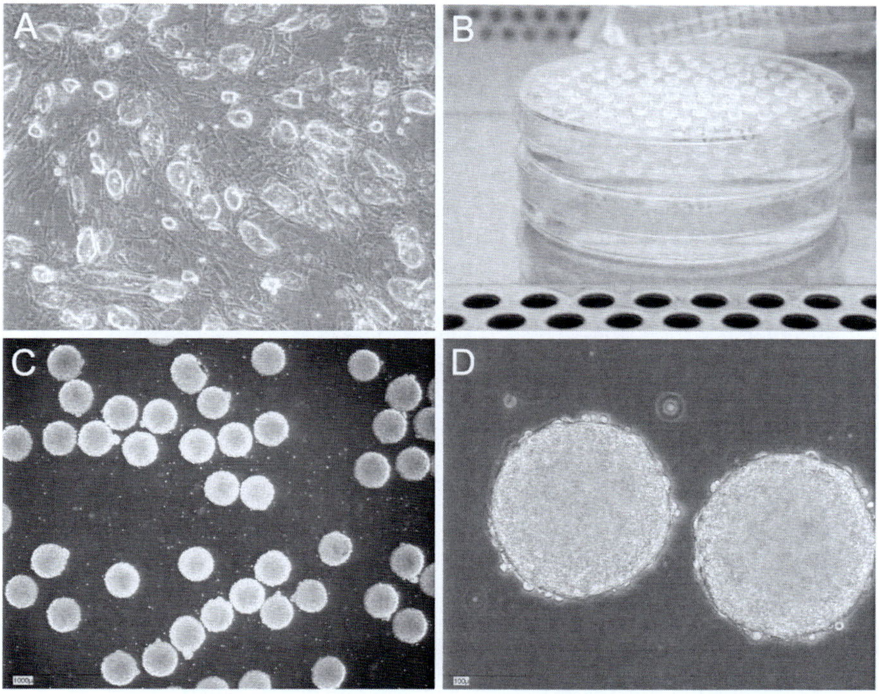

Fig. 3 Murine ESCs cultivated (**a**) on feeder cells, (**b**) as hanging drops with 400 cells per 20 μl, (**c**) embryoid bodies in suspension, and (**d**) embryoid bodies of (**c**) in a higher magnification. B. Denecke, unpublished.

these reasons, the culture of EBs has become a standard method for induction of differentiation in mouse as well as in human ESCs.

To understand the cell biology of embryogenesis, specific embryonic cell types can be isolated from EBs. Since this method supports the differentiation of many different cell lineages also including unwanted cell types, desired cells have to be purified, e.g. using antibodies against lineage-specific markers. Meanwhile a series of markers for distinct lineages and stages are defined [45, 46]. In combination with purification procedures, the EB method provides a convenient tool to isolate several cell types for further biological analyses. Additionally, ESCs enable the analysis of very early processes of embryonic differentiation including the early specification of the germ layers. In in vitro assays, concentration gradients responsible for specifying cells in a position (concentration)-dependent manner could be assessed without disturbance of the system by unwanted factors which could functionally interfere.

In all these procedures the expression level of one marker alone is not sufficient for an unambiguous definition of a definitive cell type.

Since large numbers of cells can be generated from ESC differentiation cultures, this offers the possibility to use highly paralleled technologies generating high-content data for further investigations. By gene expression profiling in a genome-wide manner (e.g. gene expression arrays), a complete discovery of genes involved in the specific embryonic processes is enabled. In this way it is possible to draw a map that displays the relationship among various intermediates isolated from embryonic stem cell differentiation cultures. In Fig. 4, expression profiles of murine mesenchymal stem cells, murine mesenchymal stem cells treated with AZA/TSA, undifferentiated murine ESCs (line R1), R1 cells differentiated into macrophages, and murine macrophages derived from the bone marrow are compared. It can be seen that murine mesenchymal stem cells and undifferentiated R1 cells are very different in the expression profile of the selected genes. There was no noticeable change in the expression profile of the mesenchymal stem cells after 5-aza-2′-deoxycytidine (AZA) and trichostatin A (TSA) treatment, two well-characterized pharmacologic inhibitors of DNA methylation and histone deacetylation. On the other hand, the expression pattern of the undifferentiated R1 cells shifts to the expression pattern of bone marrow macrophages by induction of macrophage-specific differentiation. This is also an indication that macrophages derived from ESCs are very similar to those isolated from the bone marrow. This kind of parallel analyses is more reliable than the analysis of single factors. Microarray expression analyses are also useful for the detection of novel genes involved in the earliest stage of germ layer formation.

2.3.2 Example 2: The Embryonic Stem Cell Test (EST) for Embryotoxicity

All compounds that are released into the environment may interact with the complex process of embryonic development and therefore may damage the developing organism. To prevent such risks, developmental toxicity screening is mandatory for new compounds prior to registration. Up to now, test strategies are based mainly on

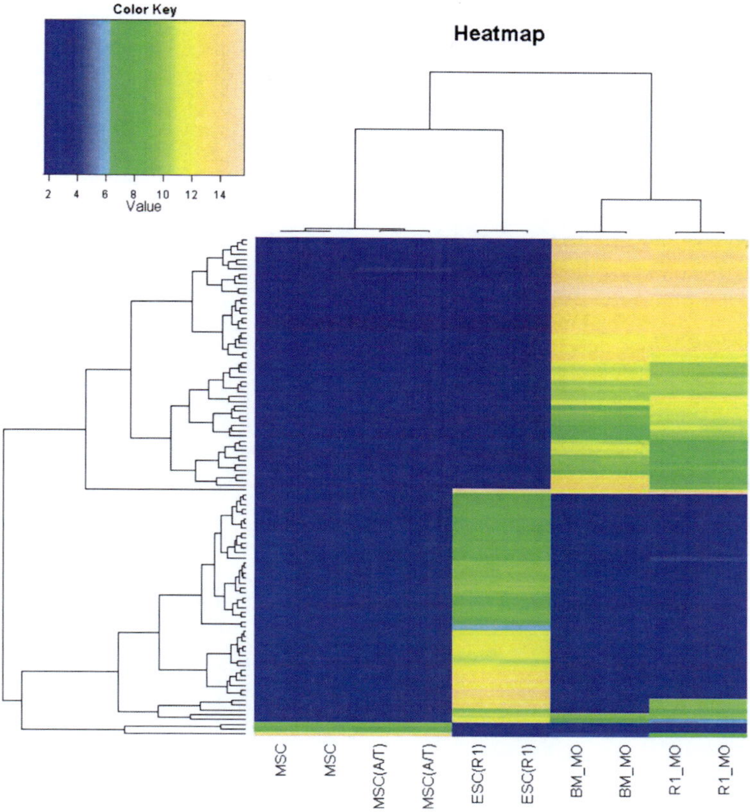

Fig. 4 Heat map of expression values of some genes from Affymetrix GeneChip experiments with mouse embryonic stem cells (ESC (R1)), mesenchymal stem cells (MSC), macrophages derived from ESCs (R1_MO), bone marrow macrophages (BM_MO), and AZA/TSA-treated mesenchymal stem cells (MSC (A/T)). The data were clustered based on Euclidean distances. Denecke, Szesny, and Gan, unpublished data

animal testing (e.g. according to OECD guideline 414 or 425), which is time and cost consuming and often a bottleneck within compound development.

During the last years, in vitro test systems have been examined to provide an alternative to the current animal testing. The embryonic stem cell test for detection of embryotoxic and/or teratogenic effects of test compounds was first described by Spielmann in 1997 [47]. The test is based on the ability of ESCs to differentiate in vitro into various tissues of all three germ layers, with cardiac tissue already described by Doetschman in 1985 [48]. Exposure of ESCs to embryotoxic and teratogenic compounds during this process can interfere with correct differentiation and will alter the resulting amount of a given tissue, e.g. cardiac tissue. Since the heart is the first organ developed in the embryo, cardiac differentiation was chosen as a robust biomarker for embryotoxic and/or teratogenic effects.

The aim of the test is to detect compounds that either exhibit toxic effects on differentiating ESCs (as a marker for embryolethality or growth retardation) or influence the delicate interaction of processes involved in differentiation of ESCs (as a marker for malformation), which then also influence cardiac differentiation. The EST includes a test for differentiation of ESCs as well as tests for cytotoxicity of the compound on ESCs and on a fibroblast cell line (INVITOX protocol 113, ECVAM Database Service on Alternative Methods to Animal experimentation, http://ecvam-dbalm.jrc.cec.eu.int). Data derived from three different tests are then calculated and interpreted using a prediction model which allows to discriminate between non-embryotoxic, low-embryotoxic, and high-embryotoxic compounds [49]. A flowchart of the EST is shown in Fig. 5.

	Test for Differentiation	**Test for Cytotoxicity**
Prior to exp	Culture of undifferentiated ES cells	Culture of undifferentiated ES cells and 3T3 fibroblasts (maternal cell type)
Day 1	Generation of ESC aggregates ("embryoid bodies"; EBs) by Hanging Drop technique, plus test compounds or solvent	Seeding of cells in 96-well plates, addition of test compounds or solvent after 3h
Day 3	Culture of EBs in suspension, plus test compounds or solvent	Change of Media, addition of fresh test compounds or solvent
Day 7	Plating of individual EBs in 24 well plates, plus test compounds or solvent	Change of Media, addition of fresh test compounds or solvent
Day 10	Microscopic observation and counting of beating areas	Detection of cytotoxicity by an MTT or XTT test (conversion of MTT or XTT by metabolic active cells to Formazan, measurement of absorption)

Fig. 5 Flow charts for "test for differentiation" and "test for cytotoxicity" are shown

For the EST originally developed by Spielmann et al. [47], the differentiation towards cardiac tissue is monitored by microscopic observation and counting of so-called beating areas, comprising spontaneous beating cardiomyocyte clusters within the differentiating ES cell aggregates [47, 50, 49]. Since this method is time consuming, bias sensitive, and requires highly trained personnel, other methods to detect cardiac cells have been tested during the last years: the use of cardiac-specific expression of the green fluorescent protein (GFP) to either detect cardiomyocytes by flow cytometry [51] or quantify the GFP in cell lysates by fluorescence spectroscopy [52] and the detection of molecular markers specific for cardiac differentiation [53]. Molecular markers are also used to detect effects of embryotoxic compounds on other cell types, e.g. neuronal cells or chondrocytes [54]. The usefulness of this approach is investigated at the moment within the ReProTect consortium (www.reprotect.eu) within the 6th EU framework program [55].

The EST was validated by the European Center for the Validation of Alternative Methods (ECVAM) in a formal validation study. For the validation study, 20 different compounds [56] with known embryotoxic and/or teratogenic behaviour in vivo were tested in four different laboratories. The results of the validation study were published in 2004 [57]. The predictive capacity of the test is shown in Table 1:

Table 1 Predictive capacity of the EST in the ECVAM validation study using 20 compounds extracted from [57]

	Laboratory I	Laboratory J	Laboratory K	Laboratory L
Specificity	71	50	86	64
Sensitivity	88	85	92	81
Negative predictive value	77	64	86	64
Positive predictive value	85	76	92	81
Accuracy	83	73	90	75

Although the results obtained with the compounds used within the validation study are very promising, the test did not get regulatory approval so far [58].[6] However, the EST is a very useful tool to prioritize compounds according to their embryotoxic behaviour within the development process of a new drug or chemical [59]. The test will help to identify possible risks for patients and consumers and therefore minimize the risk for compound attrition at late stages of their development, or even after they are already put to market.

Since human and mouse embryos differ significantly, particularly in the formation, structure, and function of the foetal membranes and placenta, and the formation of an embryonic disc instead of an egg cylinder [60–62], the use of human ESCs should be considered for the future. An advantage of such test will be that human-specific development not observed in the mouse model is also covered [5].

2.3.3 Example 3: Use of ES Cell-Derived Cardiomyocytes for Detection of Cardiac Toxicity

The most common reason for the delay of regulatory approval of new drugs or the withdrawal of drugs from the market are cardiac side effects [63]. Therefore, cardiac safety testing is a major issue for the pharmaceutical industry as well as regulatory authorities.

Cardiac toxicity is defined here as either an unwanted effect on the physiology of cardiac cells [e.g. induction of Torsade de Pointes (TdP) tachycardia by delay of ventricular repolarization] or a cytotoxic effect specifically affecting cardiomyocytes, but not necessarily other cell types, like fibroblasts. For testing of cardiac safety, several different in vivo, ex vivo, and in vitro tests are used. In vivo studies using different species (e.g. rats, rabbits, guinea pigs, dogs, pigs, and monkeys) are to be performed before a potential new drug is first applied in phase I clinical studies (FDA guideline, above, ICH guideline S7B). The advantages of

[6] One reason for this may be the relatively low number of compounds tested in the validation study. It seems to be necessary to increase these numbers in order to judge for the suitability of the test not only within the drug development process but also for additional applications, like the testing of chemicals within the REACh approach of the European Union [106]. At the moment, seven additional compounds are tested with the EST within the ReProTect consortium, but results are not published yet.

in vivo experiments using conscious or anaesthetized animals are that adverse effects of the NCEs on different physiological properties of the cardiovascular systems can be readily identified [64], and even long-term effects can be investigated [65–67].

As for many in vivo models, the disadvantages are the low throughput, high costs, long timelines, and possible ethical problems.

Several ex vivo models are belonging to the "gold standard" of current testing strategies for cardiac side effects: the Langendorff perfused heart (mainly from guinea pig or rabbit) [68–70] and isolated tissues like Purkinje fibres or papillary muscle strips [71] are widely used. The disadvantages of these models are the very laborious and time-consuming preparation that needs highly skilled personnel, low standardization, and a low throughput. Moreover, the explants can be used only for a short time period, and problems arising during preparation may affect the outcome of the studies.

In vitro experiments to assess cardiac toxicity are performed either with primary cardiomyocytes or with cell lines. Primary cardiomyocytes are of perfect physiological relevance, because all necessary elements of a cardiomyocyte (e.g. ion channels, contractile sub-cellular structures) can be studied in their native environment [72]. Nevertheless, the model also has some disadvantages: Since fully differentiated cardiomyocytes are not proliferating, the number of cells that can be used is limited. As for the ex vivo models, the preparation of cardiomyocytes is time consuming and laborious, and needs highly skilled personnel. Adult cardiomyocytes are electrically coupled via gap junctions; therefore, disruption of cell-to-cell contacts during the preparation procedure may alter the physiology of the cells. Moreover, cardiomyocytes represent only one fraction of all cells comprising the heart, and preparations of cardiomyocytes are therefore always contaminated with other cell types (e.g. fibroblasts), which may rapidly overgrow the non-proliferating cardiomyocytes [73].

Besides primary cardiomyocytes, different cell lines are also applied in cardiac safety testing. These cell lines are derived by either immortalization of cardiac cells of different origin or genetic modification of non-cardiac cell lines, e.g. HEK or CHO. Cell lines derived from cardiac cells, like HL-1 [73] or H9c2 [74, 75], have been widely used to explore cardiac cytotoxicity [76, 77] or the negative effects of oxidative stress [78]. Non-cardiac cell lines expressing heterologous ion channels are useful to study compound effects on a single type of ion channel. They allow for high-throughput applications like automated patch clamp analysis [79, 80]. The advantages of cell lines are that they are easy to obtain and to maintain, and can be prepared in ready-to-use frozen stocks. The major disadvantage of those cell lines is that they lack the physiological components of normal cardiomyocytes. The reliability of the data generated is therefore at least questionable.

ESC-derived cardiomyocytes combine the physiological relevance of primary cardiac cells with the high level of standardization, convenience, and the availability of large quantities known from cell lines. Differentiation of cardiac cells from ESCs was already described by Doetschman et al. in 1985 [48]. The first pharmacological experiment that employed ESCs was the investigation of the chronotropic effects of cardiovascular drugs on ESC-derived cardiomyocytes [81]; however, these cardiomyocytes were not purified.

In order to obtain 100% pure ESC-derived cardiomyocytes, ESCs were genetically modified such that a bicistronic vector was inserted where the cardiac-specific α-myosin heavy chain promoter drives the expression of both a puromycin resistance gene and green fluorescent protein (aGFP) cassette [82]. As soon as the first beating cardiomyocytes appeared around day 7 of differentiation, aGFP as well as the puromycin resistance gene is expressed. To kill all non-cardiac cells, puromycin is added to the cultures at 10 μg/ml for 3 days, resulting in clusters of bright green beating cardiomyocytes (Fig. 6). The clusters can be enzymatically dissociated, and the resulting single cells can be frozen and thawed retaining the morphological as well as the physiological phenotype [83] (Fig. 6).

Using this technique, highly standardized ESC-derived cardiomyocytes can be supplied as frozen stocks in large quantities. That allows to use them routinely for electrophysiological experiments [84, 85] as well as for determination of cardiac cytotoxicity [86]. To this end, compounds are tested in standard multiwell formats (e.g. 96-well plates) in various concentrations on ESC-derived cardiomyocytes and on a non-cardiac, non-proliferating reference cell type, e.g. mitotically inactivated fibroblasts. This effect is demonstrated in Fig. 7, where anti-neoplastic compounds from the class of anthracyclines showed a much higher effect on cardiomyocytes than on inactivated mouse fibroblasts (MEFs). Cytotoxicity can be detected by different read-outs, e.g. the neutral red uptake test (NRU test) or the MTT test as described above. This strategy allows to distinguish plain general cytotoxic compounds from those that exhibit a specific cytotoxic effect on cardiac cells. Even more, detailed investigation on the mechanism of action of a given compound is possible, e.g. to distinguish necrotic from apoptotic events by determination of either LDH release or caspase activity (e.g. Promega ApoOne and CytoOne Assay, www.promega.com).

The data demonstrate that ESC-derived cardiomyocytes provide a highly relevant and robust system that accurately detects cardiac-specific cytotoxicity. Since they

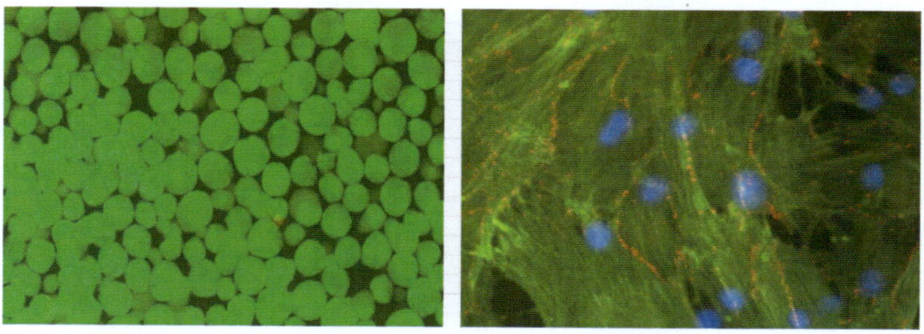

Fig. 6 ESC-derived cardiomyocytes expressing GFP. *Left panel*: EBs at day 3 after addition of puromycin. Note that EBs are composed mainly of *bright green* cardiomyocytes. *Right panel*: Microphotograph of GFP-expressing cardiomyocytes co-stained for connexin-43 expression (*red dots*) and nuclei (DAPI, *blue*)

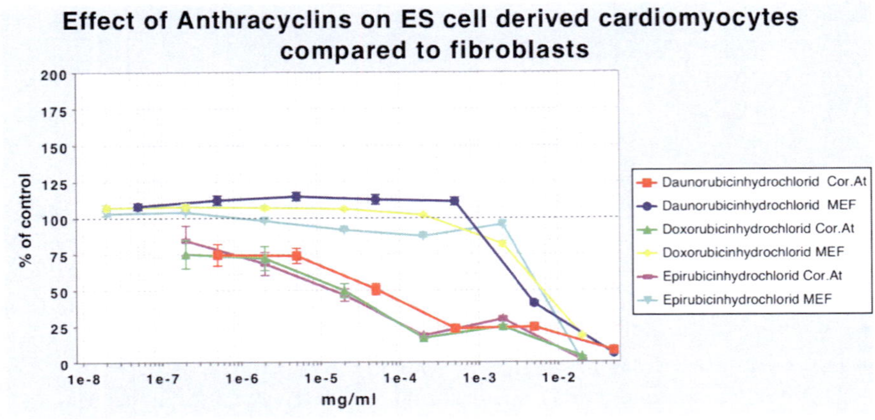

Fig. 7 Different toxic effects of anthracyclines on ESC-derived cardiomyocytes (Cor.At) and mouse embryonic fibroblasts (MEFs) [77]. After thawing, cells were cultured for 3 days prior to compound addition and incubated with compounds at different concentrations or DMSO as vehicle control for 48 h. Cytotoxicity was determined using the neutral red uptake test

can be used for medium- to high-throughput screenings, they may be employed at early stages of drug development. Moreover, the cells are also suitable for more advanced assay systems, e.g. label-free detection systems like impedance measurements as published recently [87].

3 Outlook: Potential Advanced In Vitro Models

3.1 Use of Human ESCs

As described above, most preclinical test systems, in vivo as well as in vitro, are based on animal sources. Results obtained with these methods are the basis for starting doses and safety margins for the first clinical studies. Nevertheless, sometimes the results obtained with animal systems are highly species specific. For this reason, it cannot be excluded that the effect of a given drug, e.g. in the murine system will not be seen in humans. Moreover, toxic effects may be overlooked or overestimated in the animal system and may not be transferable to humans. This may result either in the withdrawal of potent drugs from further development or in severe problems with unwanted side effects in the clinical studies. For these reasons, the use of human cell systems for use in preclinical studies is desirable. Today, sources of human cells are artificially immortalized (e.g. by viral transduction) cell lines or primary cells isolated from human tissue, which are difficult to provide in sufficient quantity. Moreover, primary cells vary from batch to batch in quality, they often lose their physiological features during the cultivation process, and have limited proliferative capacity. Thus, most of the tests using human cells are based on cell lines, which do not necessarily represent normal cell physiology [88].

The use of human ESCs can overcome these limitations. Because ESCs can be continuously propagated in vitro and manipulated to generate many differentiated cell types (in principle, all cell types of the body), they could provide a valuable source of human cells. However, before human ESCs are feasible as a potential source for drug screening as well as for toxicity tests, it is necessary to generate cells of sufficient quality and quantity in a standardized manner. One prerequisite for this is the establishment of cultivation systems under fully defined conditions without xenogeneic contaminants in the substrate or the media (see also Sect. 2.2). Only the more recently produced human ESC lines are free of xenogeneic contaminants and should be chosen for test systems to be developed. To obtain comparable results, standardized systems for culture conditions must be established that are capable of delivering human ESCs at a scale required for drug discovery. Up to now, such technologies are successfully developed only for mouse ESCs, and a one-to-one transfer of these technologies to human ESCs seems to be not possible [89].

The development and the optimization of specific differentiation protocols for human ESCs are in progress. It is assumed that in the near future, human cells in adequate amounts with properties of primary cells will be available. In the future, thousands of compounds could be quickly tested in a high-throughput manner on a wide assortment of cell types derived from human ESCs, making drug discovery and toxicity testing more efficient and cost effective. Under these aspects, there is a particular interest in the differentiation into cardiomyocytes, specific neuronal cells, and hepatocytes. Also, low molecular weight agents for the control of differentiation could be discovered using human ESCs. Christie et al. have developed two synthetic molecules (EC23 and EC19), which can be used to differentiate stem cells [90]. The effectiveness of both factors on four types of stem cells was evaluated. EC23 was found to be particularly effective in producing neurons, which can be used in laboratory testing of drugs for brain disorders such as Alzheimer's disease and Parkinson's disease. EC19 was found to be particularly effective at producing epithelial cells. These results will help to elucidate the problems of how ESCs may be reprogrammed to become different tissue types. One advantage of synthetic molecules is the high degree of standardization, ensuring more reliable and robust results. This will not only accelerate the use of human ESCs in drug development but also potentially reduce the numbers of animals used in such research. Human-specific toxicity could be examined very early during the drug development. Drugs showing human-specific toxicity could be removed from further examinations and in this way a lot of money could be saved.

Due to the lack of efficient human-specific test systems for investigation of drug metabolism, compounds causing liver toxicity or exhibiting pharmacokinetic problems depending on the metabolism of the drug are often recognized not until late phases of drug development. Moreover, very often the metabolic products of drugs are the real cause for unwanted, and partial fatal, side effects. To expand the informative value of in vitro experiments on drug metabolism, a test system using functional human hepatocytes would be helpful. In such a system, pharmacokinetic issues of new drugs, which are among others mainly dependent on the liver-specific metabolism, could be analysed. For this kind of analyses, the maintenance of cell integrity and function in vitro is required. Hepatocytes must express functional active enzymes involved in drug-metabolic and transport processes to keep metabolic activity intact. These features are in general not given in established human liver cell lines like HepG2 [91]. These cells, if any, express the enzymes in a very low amount, which makes them useless for such kind of assays. Hepatocytes differentiated from human ESCs in a reproducible manner offer an attractive approach to fulfil the multiple and complex functions of the liver with respect to synthesis, metabolism, and detoxification of substances. To realize such a functional active system, specific requirements are necessary for the cultivation of hepatocytes [92].

By the use of human ESCs with different genetic background, drugs can be analysed in cell culture in preclinical studies depending on the genetic background or the ethnic origin. These cells can be derived either by establishing human embryonic stem cell lines from embryos with genetic disorders [93], which can be detected during the practice of preimplantation genetic diagnosis, or by using induced pluripotent stem cells (IPS, see below) derived from donors that have a disease-correlated defect. These new cell lines may provide an excellent in vitro model for studies on the effects of genetic mutations on cell proliferation and differentiation. In this way, a natural human cell-based in vitro disease model is available for testing the effects of drugs in a native system. In contrast to an animal model system, all interacting molecules are from human origin. Species-specific differences regarding the interactions of specific factors different from man can be eliminated.

3.2 Generation of Personalized Induced Pluripotent Cells (IPS Cells)

IPS cells are genetically reprogrammed cells from a patient that exhibit the properties of pluripotent stem cells. Generation of mouse IPS cells was first described by Takahashi and Yamanaka [94], who have identified four embryonic transcription factors (Oct4, Sox2, c-myc, and Klf4) whose overexpression restores pluripotency to adult somatic cells. This work was confirmed by two other groups [95, 96], demonstrating that pluripotent stem cells can be directly generated from adult fibroblast cultures. The cells produced by Yamanaka as well as the other laboratories displayed all key features of ESCs, including the ability to form chimeric mice and contribute to the germ line.

In November 2007, Yamanaka as well as Thomson [97, 98] announced a similar breakthrough with human skin cells that were transformed into batches of cells that look and act like ESCs. This may enable the generation of patient-specific IPS cell lines that could potentially be used for cellular replacement therapies as well as for screening of pharmaceuticals that are most suitable for the individual patient in terms of efficacy and safety. In addition, IPS cell lines from patients with a variety of genetic diseases can be produced providing invaluable models to study those diseases.

While this work is already a huge accomplishment for science, there is still much work to be done before this technology can be used for the treatment of diseases. For example, the genes used to reprogram the skin cells into ES-like cells were added by the use of retroviruses that can cause mutations and lead to the risk of possible cancers, although recent research by Yamanaka's group has made advances in avoiding this particular problem [99]. Future work is aimed at attempting to reprogram without permanent genetic manipulation of the cells with viruses. This could be accomplished by either small molecules or other methodologies to express these reprogramming genes.

However, a first indication that IPS cell technology can in rapid succession lead to new cures is its use by the group of Jaenisch to cure mice of sickle cell anaemia [100].

3.3 Personalized Medicine

A huge challenge for the pharmaceutical industry is the establishment of "personalized medicine", i.e. the medical intervention is customized, taken into account the phenotypic and genotypic properties of the individual.

In principle, all tools to analyse drug effects in a personalized manner have already been invented: genome-wide analysis tools (e.g. gene expression arrays) and the ability to generate personalized, cell-based assays by using induced pluripotent stem cells (IPS cells) from the patient (see above) [101, 102].

Effects of individual differences in the treatment of patients could be recently exemplified by an HIV-1-infected patient with an acute myeloid leukaemia (AML). Based on the knowledge that homozygosity for the naturally occurring delta 32 deletion in the HIV co-receptor CCR5 confers resistance to HIV-1 infection, the patient underwent an allogeneic stem cell transplant selected not only for the appropriate HLA-identical donor but also for the CCR5–delta 32 mutation which has a frequency of about 1% [103, 104]. By the targeted selection of a donor with the CCR5–delta 32 mutation, the patient appears to have been cured of AIDS 20 months after receiving bone marrow transplant normally used to fight leukaemia. Although it cannot be excluded that this case is only a fluke, it may inspire a greater interest to utilize this mechanism (inaccessibility of CCR5 for HIV-1) as a therapeutic target.

This example shows that depending on the individual genetic state, patients respond differently to exogenous compounds. To enable drug discovery and development for individual patients, specific biological profiles must be generated. Based on protein expression profiles, combinations of therapeutic agents could be developed that enable more effective treatment of diseases with less adverse reactions. In the future, genome-sequencing data of every patient may provide clinicians with extensive data, enabling a better overall disease management strategy. In this context, the adaptation of microarrays as a tool for routine diagnostic testing in disease and for drug applications is a big challenge. The combination of different techniques like genomic arrays and patient-specific IPS cells as well as data integration from multiple sources (transcriptome, proteome, and genome) will give the clinician a thorough understanding of the biological mechanism of various diseases. Functional testing at the individual cellular level provided by IPS cells will probably replace pharmacological screening based on isolated molecular targets. A significant increase in the use of complex cellular models based on patient-specific IPS cells for efficacy screening as well as toxicology studies will occur. In the first step, molecularly targeted drugs will probably be designed to responsive patient populations before the vision of individually personalized medicine will be really realized. Ultimately, not a few drugs will be designed for the masses, but masses of drugs will be developed targeted to people with a specific "biological fingerprint", either individual patients or groups of (phenotypically or genetically) related patients.

4 Conclusion

ESCs provide a highly standardized and robust, but nevertheless, flexible source for physiologically relevant cell systems for drug development and toxicity testing. Several protocols for the differentiation of specific cell types from ESCs as well as for test systems using either ESCs or ESC-derived cell types have already been described. Up to now, mouse ESCs are mainly used for the test systems described above. To overcome species-specific differences, the use of human ESCs or IPS cells is favourable. Although a lot of promising possibilities for the use of human ESCs or IPS cells in preclinical studies in drug development or toxicity could be allowed, these novel systems have to be integrated in the regulatory guidelines until they are widely accepted beyond their optional application in non-regulated parts of the drug development process.

References

1. Caldwell GW, Ritchie DM, Masucci JA, Hageman W, Yan Z. The new pre-preclinical paradigm: compound optimization in early and late phase drug discovery. Curr Top Med Chem. 2001; 1:353–366.
2. The Academy of Medical Sciences. Safer medicines report. Pre-clinical toxicology working group report; 2005. http://www.acmedsci.ac.uk/p103.html
3. Lindsay MA. Finding new drug targets in the 21st century. Drug Discov Today. 2005; 10:1683–1687.
4. Guttendorf RJ. The emerging role of A.D.M.E. in optimizing drug discovery and design. http://www.netsci.org/Science/Special/feature06.html
5. Brent RL. Utilization of animal studies to determine the effects and human risks of environmental toxicants (drugs, chemicals, and physical agents). Pediatrics. 2004; 113:984–995.
6. Guerassimov A, Hoshino Y, Takubo Y, Turcotte A, Yamamoto M, Ghezzo H, Triantafillopoulos A, Whittaker K, Hoidal JR, Cosio MG. The development of emphysema in cigarette smoke-exposed mice is strain dependent. Am J Respir Crit Care Med. 2004; 170:974–980.
7. Pedersen F, de Bruijn J, Munn S, van Leeuwen K. Assessment of additional testing needs under REACH. Ispra: Publication of the European Commission, Directorate General JRC, Joint Research Centre, Institute for Health and Consumer Protection; 2003.

8. Skrzypiec-Spring M, Grotthus B, Szelag A, Schulz R. Isolated heart perfusion according to Langendorff—still viable in the new millennium. J Pharmacol Toxicol Methods. 2007; 55:113–126.
9. van de Kerkhof EG, de Graaf IA, Groothuis GM. In vitro methods to study intestinal drug metabolism. Curr Drug Metab. 2007; 8:658–675.
10. Haney SA, LaPan P, Pan J, Zhang J. High-content screening moves to the front of the line. Drug Discov Today. 2006; 11:889–894.
11. Park JK, Kim BH, Han YS, Park IK. The effect of telomerase expression on the escape from M2 crisis in virus-transformed human retinal pigment epithelial cells. Exp Mol Med. 2002; 34:107–113.
12. Poste G, Greig R. On the genesis and regulation of cellular heterogeneity in malignant tumors. Invasion Metastasis. 1982; 2:137–176.
13. MacLeod RAF, Nagel S, Scherr M, Schneider B, Dirks WG, Uphoff CC, Quentmeier H, Drexler HG. Human leukemia and lymphoma cell lines as models and resources. Curr Med Chem. 2008; 15:339–359.
14. Sachinidis A, Fleischmann BK, Kolossov E, Wartenberg M, Sauer H, Hescheler J. Cardiac specific differentiation of mouse embryonic stem cells. Cardiovasc Res. 2003; 58:278–291.
15. Dang L, Tropepe V. Neural induction and neural stem cell development. Regen Med. 2006; 1:635–652.
16. Spence JR, Wells JM. Translational embryology: using embryonic principles to generate pancreatic endocrine cells from embryonic stem cells. Dev Dyn. 2007; 236:3218–3227.
17. Valtieri M, Sorrentino A. The mesenchymal stromal cell contribution to homeostasis. J Cell Physiol. 2008; 217:296–300.
18. Evans M, Kaufman M. Establishment in culture of pluripotential cells from mouse embryos. Nature. 1981; 292:154–156.
19. Martin G. Isolation of a pluri-potent cell line from early mouse embryos cultured in medium conditioned by teratocarcinoma stem cells. Proc Natl Acad Sci USA. 1981; 78(12):7634–7638.
20. Bradley A, Evans M, Kaufman MH, Robertson E. Formation of germ-line chimaeras from embryo-derived teratocarcinoma cell lines. Nature. 1984; 309:255–256.
21. Boyer LA, Mathur D, Jaenisch R. Molecular control of pluripotency. Curr Opin Genet Dev. 2006; 16(5):455–462.
22. Smith AG. Embryo-derived stem cells: of mice and men. Annu Rev Cell Dev Biol. 2001; 17:435–462.
23. Mueller FJ, KostkaKostka D, Laurent L, Ulitsky I, Williams R, Lu C, Rao MS, Shamir R, Schwartz PH, Schmidt NO, Loring JF. Regulatory networks define phenotypic classes of human stem cell lines. Nature. 2008; 455:401–405.
24. Thomson JA, Itskovitz-Eldor J, Shapiro SS, Waknitz MA, Swiergiel JJ, Marshall VS, Jones JM. Embryonic stem cell lines derived from human blastocysts. Science. 1998; 282:1145–1147.
25. Thomson JA, Kalishman J, Golos TG, Durning M, Harris CP, Becker RA, Hearn JP. Isolation of a primate embryonic stem cell line. Proc Natl Acad Sci USA. 1995; 92:7844–7848.
26. Thomson JA, Kalishman J, Golos TG, Durning M, Harris CP, Hearn JP. Pluripotent cell lines derived from common marmoset (*Callithrix jacchus*) blastocysts. Biol Reprod. 1996; 55:254–259.
27. Williams RL, Hilton DJ, Pease S, Willson TA, Stewart CL, Gearing DP, Wagner EF, Metcalf D, Nicola NA, Gough NM. Myeloid leukaemia inhibitory factor maintains the developmental potential of embryonic stem cells. Nature. 1988; 336:684–687.
28. Ying QL, Nichols J, Chambers I, Smith A. BMP induction of Id proteins suppresses differentiation and sustained embryonic stem cell self-renewal in collaboration with STAT3. Cell. 2003; 115:281–292.
29. Cartwright P, McLean C, Sheppard A, Rivett D, Jones K, Dalton S. LIF/STAT3 controls ES cell self-renewal and pluripotency by a Myc-dependent mechanism. Development. 2005; 132:885–896.

30. Hyslop LA, Armstrong L, Stojkovic M, Lako M. Human embryonic stem cells: biology and clinical implications. Expert Rev Mol Med. 2005; 7:1–21.
31. Dahéron L, Optiz SL, Zaehres H, Lensch MW, Andrews PW, Itskovitz-Eldor J, Daley GQ. LIF/STAT3 signaling fails to maintain self-renewal of human embryonic stem cells. Stem Cells. 2004; 22:770–778.
32. Humphrey RK, Beattie GM, Lopez AD, Bucay N, King CC, Firpo MT, Rose-John S, Hayek A. Maintenance of pluripotency in human embryonic stem cells is STAT3 independent. Stem Cells. 2004; 22:522–530.
33. Pera MF, Andrade J, Houssami S, Reubinoff B, Trounson A, Stanley EG, Ward-van Oostwaard D, Mummery C. Regulation of human embryonic stem cell differentiation by BMP-2 and its antagonist noggin. J Cell Sci. 2004; 117:1269–1280.
34. Xu RH, Chen X, Li DS, Li R, Addicks GC, Glennon C, Zwaka TP, Thomson JA. BMP4 initiates human embryonic stem cell differentiation to trophoblast. Nat Biotechnol. 2002; 20:1261–1264.
35. Amit M, Carpenter MK, Inokuma MS, Chiu CP, Harris CP, Waknitz MA, Itskovitz-Eldor J, Thomson JA. Clonally derived human embryonic stem cell lines maintain pluripotency and proliferative potential for prolonged periods of culture. Dev Biol. 2000; 227:271–278.
36. Xu C, Inokuma MS, Denham J, Golds K, Kundu P, Gold JD, Carpenter MK. Feeder-free growth of undifferentiated human embryonic stem cells. Nat Biotechnol. 2001; 19:971–974.
37. Amit M, Shariki K, Margulets V, Itskovitz-Eldor J. Feeder layer- and serum-free culture of human embryonic stem cells. Biol Reprod. 2004; 70:837–845.
38. Ludwig TE, Levenstein ME, Jones JM, Berggren WT, Mitchen ER, Frane JL, Crandall LJ, Daigh CA, Conard KR, Piekarczyk MS, Llanas RA, Thomson JA. Derivation of human embryonic stem cells in defined conditions. Nat Biotechnol. 2006; 24:185–187.
39. Lu J, Hou R, Booth CJ, Yang SH, Snyder M. Defined culture conditions of human embryonic stem cells. Proc Natl Acad Sci USA. 2006; 103:5688–5693.
40. Wobus AM, Hescheler J. Development of an in vitro cardiomyocytes cell model for embryotoxicological and pharmacological studies. ALTEX. 1992; 9:29–42.
41. Davila JC, Cezar GG, Thiede M, Strom S, Miki T, Trosko J. Use and application of stem cells in toxicology. Toxicol Sci. 2004; 79:214–223.
42. McNeish J. Embryonic stem cells in drug discovery. Nat Rev Drug Discov. 2004; 3:70–80.
43. Hübner K, Fuhrmann G, Christenson LK, Kehler J, Reinbold R, De La Fuente R, Wood J, Strauss JF III, Boiani M, Schöler HR. Derivation of oocytes from mouse embryonic stem cells. Science. 2003; 300:1251–1256.
44. Geijsen N, Horoschak M, Kim K, Gribnau J, Eggan K, Daley GQ. Derivation of embryonic germ cells and male gametes from embryonic stem cells. Nature. 2003; 427: 148–154.
45. Nishikawa SI, Nishikawa S, Hirashima M, Matsuyoshi N, Kodama H. Progressive lineage analysis by cell sorting and culture identifies FLK1+VE-cadherin+ cells at a diverging point of endothelial and hemopoietic lineages. Development. 1998; 125:1747–1757.
46. Ng ES, Azzola L, Sourris K, Robb L, Stanley EG, Elefanty AG. The primitive streak gene Mixl1 is required for efficient haematopoiesis and BMP4-induced ventral mesoderm patterning in differentiating ES cells. Development. 2005; 132:873–884.
47. Spielmann H, Pohl I, Döring B, Liebsch M, Moldenhauer F. The embryonic stem cell test, an in vitro embryotoxicity test using permanent mouse cell lines: 3T3 fibroblasts and embryonic stem cells. In Vitro Toxicol. 1997; 10:119–127.
48. Doetschman TC, Eistetter H, Katz M, Schmidt W, Kemler R. The in vitro development of blastocyst-derived embryonic stem cell lines: formation of visceral yolk sac, blood islands and myocardium. J Embryol Exp Morphol. 1985; 87:27–45.
49. Genschow E, Spielmann H, Scholz G, Seiler A, Brown N, Piersma A, Brady M, Clemann N, Huuskonen H, Paillard F, Bremer S, Becker K. The ECVAM international validation study

on in vitro embryotoxicity tests: results of the definitive phase and evaluation of prediction models. European Centre for the Validation of Alternative Methods. ATLA. 2002; 30: 151–176.

50. Scholz G, Pohl I, Genschow E, Klemm M, Spielmann H. Embryotoxicity screening using embryonic stem cells in vitro: correlation to in vivo teratogenicity. Cells Tissues Organs. 1999; 165:203–211.
51. Paparella M, Kolossov E, Fleischmann BK, Hescheler J, Bremer S. The use of quantitative image analysis in the assessment of in vitro embryotoxicity endpoints based on a novel embryonic stem cell clone with endoderm-related GFP expression. Toxicol In Vitro. 2002; 16:589–597.
52. Schwengberg S, Bohlen H, Kleinsasser N, Kehe K, Seiss M, Walther UI, Hickel R, Reichl FX. In vitro embryotoxicity assessment with dental restorative materials. J Dent. 2005; 33:49–55.
53. Pellizzer C, Adler S, Corvi R, Hartung T, Bremer S. Monitoring of teratogenic effects in vitro by analysing a selected gene expression pattern. Toxicol In Vitro. 2004; 18:325–335.
54. zur Nieden NI, Kempka G, Ahr HJ. Molecular multiple endpoint embryonic stem cell test—a possible approach to test for the teratogenic potential of compounds. Toxicol Appl Pharmacol. 2004; 194:257–269.
55. Hareng L, Pellizzer C, Bremer S, Schwarz M, Hartung T. The integrated project ReProTect: a novel approach in reproductive toxicity hazard assessment. Reprod Toxicol. 2005; 20: 441–452.
56. Brown NA. Selection of test chemicals for the ECVAM international validation study on in vitro embryotoxicity tests. European Centre for the Validation of Alternative Methods. ATLA. 2002; 30:177–198.
57. Genschow E, Spielmann H, Scholz G, Pohl I, Seiler A, Clemann N, Bremer S, Becker K. Validation of the embryonic stem cell test in the international ECVAM validation study on three in vitro embryotoxicity tests. ATLA. 2004; 32:209–244.
58. Bremer S, Hartung T. The use of embryonic stem cells for regulatory developmental toxicity testing in vitro—the current status of test development. Curr Pharm Des. 2004; 10: 2733–2747.
59. Paquette JA, Kumpf SW, Streck RD, Thomson JJ, Chapin RE, Stedman DB. Assessment of the Embryonic Stem Cell Test and application and use in the pharmaceutical industry. Birth Defects Res B Dev Reprod Toxicol. 2008; 83:104–111.
60. Castellucci M, Scheper M, Scheffen I, Celona A, Kaufmann P. The development of the human placental villous tree. Anat Embryol (Berlin) 1990; 18:117–128.
61. Luckett WP. The development of primordial and definitive amniotic cavities in early Rhesus monkey and human embryos. Am J Anat. 1975; 144:149–167.
62. Luckett WP. Origin and differentiation of the yolk sac and extra-embryonic mesoderm in presomite human and rhesus monkey embryos. Am J Anat. 1978; 152:59–97.
63. Gussak I, Litwin J, Kleiman R, Grisanti S, Morganroth J. Drug-induced cardiac toxicity: emphasising the role of electrocardiography in clinical research and drug development. J Electrocardiol. 2004; 37:19–24.
64. Lawrence CL, Pollard CE, Hammond TG, Valentin JP. Nonclinical proarrhythmia models: predicting Torsades de Pointes. J Pharmacol Toxicol Methods. 2005; 52:46–59.
65. Moe GW, Marin-Garcia J, Konig A, Goldenthal M, Lu X, Feng Q. In vivo TNF-alpha inhibition ameliorates cardiac mitochondrial dysfunction, oxidative stress, and apoptosis in experimental heart failure. Am J Physiol Heart Circ Physiol. 2004; 287:1813–1820.
66. Kubota T, McTiernan CF, Frye CS, Slawson SE, Lemster BH, Koretsky AP, Demetris AJ, Feldman AM. Dilated cardiomyopathy in transgenic mice with cardiac-specific overexpression of tumor necrosis factor-alpha. Circ Res. 1997; 81:627–635.
67. Recchia FA, McConnell PI, Bernstein RD, Vogel TR, Xu X, Hintze TH. Reduced nitric oxide production and altered myocardial metabolism during the decompensation of pacing-induced heart failure in the conscious dog. Circ Res. 1998; 83:969–979.

68. Valentin JP, Hoffmann P, De Clerck F, Hammond TG, Hondeghem L. Review of the predictive value of the Langendorff heart model (Screenit system) in assessing the proarrhythmic potential of drugs. J Pharmacol Toxicol Methods. 2004; 49:171–181.
69. Testai L, Breschi MC, Martinotti E, Calderone V. QT prolongation in guinea pigs for preliminary screening of torsadogenicity of drugs and drug-candidates. II. J Appl Toxicol. 2007; 27:270–275.
70. Heavner JE. Cardiac toxicity of local anesthetics in the intact isolated heart model: a review. Reg Anesth Pain Med. 2002; 27:545–555.
71. Arrigoni C, Crivori P. Assessment of QT liabilities in drug development. Cell Biol Toxicol. 2007; 23:1–13.
72. Davie C, Pierre-Valentin J, Pollard C, Standen N, Mitcheson J, Alexander P, Thong B. Comparative pharmacology of guinea pig cardiac myocyte and cloned hERG (I(Kr)) channel. J Cardiovasc Electrophysiol. 2004; 15:1302–1309.
73. White SM, Constantin PE, Claycomb WC. Cardiac physiology at the cellular level: use of cultured HL-1 cardiomyocytes for studies of cardiac muscle cell structure and function. Am J Physiol Heart Circ Physiol. 2004; 286:823–829.
74. Kimes BW, Brandt BL. Properties of a clonal muscle cell line from rat heart. Exp Cell Res. 1976; 98:367–381.
75. Hescheler J, Meyer R, Plant S, Krautwurst D, Rosenthal W, Schultz G. Morphological, biochemical, and electrophysiological characterization of a clonal cell (H9c2) line from rat heart. Circ Res. 1991; 69:1476–1486.
76. Li K, Sung RY, Huang WZ, Yang M, Pong NH, Lee SM, Chan WY, Zhao H, To MY, Fok TF, Li CK, Wong YO, Ng PC. Thrombopoietin protects against in vitro and in vivo cardiotoxicity induced by doxorubicin. Circulation. 2006; 113:2211–2220.
77. Menna P, Salvatorelli E, Minotti G. Doxorubicin degradation in cardiomyocytes. J Pharmacol Exp Ther. 2007; 322:408–419.
78. Hofgaard JP, Sigurdardottir KS, Treiman M. Protection by 6-aminonicotinamide against oxidative stress in cardiac cells. Pharmacol Res. 2006; 54:303–310.
79. Priest BT, Swensen AM, McManus OB. Automated electrophysiology in drug discovery. Curr Pharm Des. 2007; 13:2325–2337.
80. Pehl U, Leisgen C, Gampe K, Guenther E. Automated higher-throughput compound screening on ion channel targets based on the *Xenopus laevis* oocyte expression system. Assay Drug Dev Technol. 2004; 2:515–524.
81. Wobus AM, Wallukat G, Hescheler J. Pluripotent mouse embryonic stem cells are able to differentiate into cardiomyocytes expressing chronotropic responses to adrenergic and cholinergic agents and $Ca2^{+}$ channel blockers. Differentiation. 1991; 48:173–182.
82. Kolossov E, Bostani T, Roell W, Breitbach M, Pillekamp F, Nygren JM, Sasse P, Rubenchik O, Fries JW, Wenzel D, Geisen C, Xia Y, Lu Z, Duan Y, Kettenhofen R, Jovinge S, Bloch W, Bohlen H, Welz A, Hescheler J, Jacobsen SE, Fleischmann BK. Engraftment of engineered ES cell-derived cardiomyocytes but not BM cells restores contractile function to the infarcted myocardium. J Exp Med. 2006; 203:2315–2327.
83. Kettenhofen R, Bohlen H. Preclinical assessment of cardiac toxicity. Drug Discov Today. 2008; 13:702–707.
84. Kolossov E, Lu Z, Drobinskaya I, Gassanov N, Duan Y, Sauer H, Manzke O, Bloch W, Bohlen H, Hescheler J, Fleischmann BK. Identification and characterization of embryonic stem cell-derived pacemaker and atrial cardiomyocytes. FASEB J. 2005; 19:577–579.
85. Hescheler J, Halbach M, Egert U, Lu ZJ, Bohlen H, Fleischmann BK, Reppel M. Determination of electrical properties of ES cell-derived cardiomyocytes using MEAs. J Electrocardiol. 2004; 37:110–116.
86. Schwengberg S, Kettenhofen R, Clarke E, Bohlen H. Use of embryonic stem cell derived cardiomyocytes for determination of cardiac cytotoxicity. Safety Pharmacology Society Annual Meeting, Madison, WI; 2008.
87. Abassi Y, Bohlen H. The application of cell-based impedance technology in drug discovery. SBS Symposium Europe, Dresden, Germany; 2008.

88. Thedinga E, Ullrich A, Drechsler S, Niendorf R, Kob A, Runge D, Keuer A, Freund I, Lehmann M, Ehret R. In vitro system for the prediction of hepatotoxic effects in primary hepatocytes. ALTEX. 2007; 24:22–34.
89. Thomson H. Bioprocessing of embryonic stem cells for drug discovery. Trends Biotechnol. 2007; 25:224–230.
90. Christie VB, Barnard JH, Batsanov AS, Bridgens CE, Cartmell EB, Collings JC, Maltman DJ, Redfern CPF, Marder TB, Przyborski S, Whiting A. Synthesis and evaluation of synthetic retinoid derivatives as inducers of stem cell differentiation. Org Biomol Chem. 2008; 19:3497–3507.
91. Aden DP, Fogel A, Plotkin S, Damjanov I, Knowles BB. Controlled synthesis of HBsAg in a differentiated human liver carcinoma-derived cell line. Nature. 1979; 282: 615–616.
92. Zeilinger K, Holland G, Sauer IM, Efimova E, Kardassis D, Obermayer N, Liu M, Neuhaus P, Gerlach JC. Time course of primary liver cell reorganization in three-dimensional high-density bioreactors for extracorporeal liver support: an immunohistochemical and ultrastructural study. Tissue Eng. 2004; 10:1113–1124.
93. Verlinsky Y, Strelchenko N, Kukharenko V, Rechitsky S, Verlinsky O, Galat V, Kuliev A. Human embryonic stem cell lines with genetic disorders. Reprod Biomed Online. 2005; 10:105–110.
94. Takahashi K, Yamanaka S. Induction of pluripotent stem cells from mouse embryonic and adult fibroblast cultures by defined factors. Cell. 2006; 126:663–676.
95. Wernig M, Meissner A, Foreman R, Brambrink T, Ku M, Hochedlinger K, Bernstein BE, Jaenisch R. In vitro reprogramming of fibroblasts into a pluripotent ES-cell-like state. Nature. 2007; 448:318–324.
96. Maherali N, Sridharan R, Xie W, Utikal J, Eminli S, Arnold K, Stadtfeld M, Yachechko R, Tchieu J, Jaenisch R, Plath K, Hochedlinger K. Directly reprogrammed fibroblasts show global epigenetic remodeling and widespread tissue contribution. Cell Stem Cell. 2007; 1:55–70.
97. Yu J, Vodyanik MA, Smuga-Otto K, Antosiewicz-Bourget J, Frane JL, Tian S, Nie J, Jonsdottir GA, Ruotti V, Stewart R, Slukvin II, Thomson JA. Induced pluripotent stem cell lines derived from human somatic cells. Science. 2007; 318:1917–1920.
98. Takahashi K, Tanabe K, Ohnuki M, Narita M, Ichisaka T, Tomoda K, Yamanaka S. Induction of pluripotent stem cells from adult human fibroblasts by defined factors. Cell. 2007; 31: 861–872.
99. Okita K, Nakagawa M, Hyenjong H, Ichisaka T, Yamanaka S. Generation of mouse induced pluripotent stem cells without viral vectors. Science. 2008; 322:949–953.
100. Hanna J, Wernig M, Markoulaki S, Sun CW, Meissner A, Cassady JP, Beard C, Brambrink T, Wu LC, Townes TM, Jaenisch R. Treatment of sickle cell anemia mouse model with iPS cells generated from autologous skin. Science. 2007; 21:1920–1923.
101. Maherali N, Ahfeldt T, Rigamonti A, Utikal J, Cowan C, Hochedlinger K. A high-efficiency system for the generation and study of human induced pluripotent stem cells. Cell Stem Cell. 2008; 3:340–345.
102. Park IH, Lerou PH, Zhao R, Huo H, Daley GQ. Generation of human-induced pluripotent stem cells. Nat Protoc. 2008; 3:1180–1186.
103. Perez EE, Wang J, Miller JC, Jouvenot Y, Kim KA, Liu O, Wang N, Lee G, Bartsevich VV, Lee YL, Guschin DY, Rupniewski I, Waite AJ, Carpenito C, Carroll RG, Orange JS, Urnov FD, Rebar EJ, Ando D, Gregory PD, Riley JL, Holmes MC, June CH. Establishment of HIV-1 resistance in $CD4^+$ T cells by genome editing using zinc-finger nucleases. Nat Biotechnol. 2008; 26:808–816.
104. Hütter G, Nowak D, Mossner M, Ganepola S, Allers K, Schneider T, Hofmann J, Blau IW, Hofmann WK, Thiel E. Treatment of HIV-1 infection by allogeneic CCR5-delta32/delta32 stem cell transplantation: a promising approach. Poster, 15th Conference on Retroviruses and Opportunistic Infections, Boston, MA, February 3–6, 2008.

105. Schardein JL Chemically induced birth defects. 3rd edition New York NY: Marcel Dekker; 2000.
106. Van der Jagt K, Munn S, Tørsløv J, de Bruijn J. Alternative approaches can reduce the use of test animals under REACH. Addendum to the report "Assessment of additional testing needs under REACH. Effects of (Q)SARs, risk based testing and voluntary industry initiatives". JRC Report EUR 21405 EN, Ispra, Italy; 2004, 25pp.

Embryonic Stem Cells: A Biological Tool to Translate the Mechanisms of Heart Development

Omonigho A. Aisagbonhi and Antonis K. Hatzopoulos

Abstract Heart organogenesis is sensitive to both genetic and environmental factors and heart malformations account for the majority of birth defects. The development of the mammalian heart is a complex morphogenetic process that ultimately results in a contractile, four-chambered organ that is formed from progenitor cells of mesodermal and neural crest origins. Studies in animal models with primitive hearts such as drosophila, zebrafish, and xenopus have provided important insights into the way the heart begins to form and the molecular cues that guide the early stages of cardiac morphogenesis. Much of what we know about the development of the heart in higher vertebrates comes mostly from embryological investigations in chick and mouse. In addition, genetic studies in zebrafish and mouse, including loss- and gain-of-function approaches, have been especially valuable for gaining information about the role of individual genes in heart development. However, the various animal models also have a number of limitations regarding lack of genetic tools (chick, xenopus), inaccessibility to observation and embryological manipulation (mouse), or unsuitability for high-throughput screens to identify pharmacologic compounds to treat cardiac defects. Mouse embryonic stem (ES) cells give rise to a wide variety of organ-specific cell types, offering an accessible and relevant system to investigate how cell lineages emerge and grow. The ES cell model is particularly pertinent for studying the molecular mechanisms regulating the specification and differentiation of cardiovascular progenitor cells, because these cells appear relatively early during embryonic development and ES cell differentiation. This chapter reviews a selective number of studies that have applied the in vitro differentiation of embryonic stem cells toward understanding the regulatory steps of cardiac development.

Keywords Embryonic stem cells · Cardiomyocyte differentiation · Cardiac development · Signaling pathways

---

A.K. Hatzopoulos (✉)
Division of Cardiovascular Medicine, Department of Cell and Developmental Biology, Department of Medicine, Vanderbilt University Medical Centre, Nashville, TN, USA
e-mail: antonis.hatzopoulos@vanderbilt.edu

G.M. Artmann et al. (eds.), *Stem Cell Engineering*,
DOI 10.1007/978-3-642-11865-4_23,

1 Heart Development: An Overview of Morphogenesis

The development of the vertebrate heart begins shortly after gastrulation and proceeds through the neonatal stages [1]. Heart development is a well-orchestrated process that follows a complex sequence of morphogenetic movements, incorporates a variety of cell types, and requires the precise execution of a number of genetic programs [2, 3]. During gastrulation, mesodermal cells destined to form the heart delaminate from the primitive ectoderm in the anterior region of the primitive streak [4]. As the rest of the embryo grows, the cardiac cells wrap around the endocardial cells, the two populations forming the original two-layered heart tube. The early cardiac cells turn on the expression of sarcomeric proteins like cardiac troponin and α-myosin heavy chain and begin to contract so that circulation commences as soon as the embryonic vascular loop is closed [5].

The rapidly growing heart tube then loops rightward, bringing the atria anterior and dorsal to the ventricles [6]. As the primary heart tube loops, cells from the surrounding pharyngeal mesoderm (also known as the secondary heart field) migrate into and incorporate in the growing heart tube [7, 8]. These cells from the secondary heart field eventually contribute to the population of cells that form the outflow tract, atria, and right ventricle [8].

Two additional populations also migrate and become incorporated into the developing heart. First, cells from the splanchnic mesothelium, adjacent to the heart, form a "cauliflower-like" structure that attaches to the heart and spreads around it. The migrating cells eventually cover the entire heart surface, giving rise to the external layer of the epicardium. Epicardial cells further differentiate to produce the cellular components of the coronary blood vessels, interstitial fibroblasts, and also cardiomyocytes [9–11]. Second, migrating neural crest cells invade the outflow tract and contribute to the aorticopulmonary septum that separates the aorta and pulmonary arteries [12]. Neural crest cells also give rise to the neuronal network of the heart [13].

After looping, the heart still functions as a linear tube lacking septa and valves to separate its chambers. The process of chamber separation and valve formation begins with the expansion of the cardiac jelly, which separates the interior endocardial from the outer myocardial layer, at the junctions between the atria and the ventricles (AV) and at the outflow tract (OFT), to form cardiac cushions. The superior and inferior AV cushions fuse to form the septum intermedium. At this time, septa begin growing within the atria and ventricles to separate the chambers into left and right halves. The atrial and ventricular septa grow to fuse with the septum intermedium, completing the process of chamber separation [14]. Elongation and remodeling of protrusions from the AV cushions give rise to the mitral and tricuspid valve leaflets [15]. The aortic and pulmonary valves are formed upon remodeling and septation of the OFT cushions by invading neural crest cells [12, 16].

2 Model Organisms to Study Heart Development

Most of our current understanding of heart development comes from studies in model organisms such as drosophila, zebrafish, xenopus, chick, and mouse. The complex morphogenetic movements of heart formation have been mostly studied in the chick and to a lesser extent in xenopus. In both species, embryos are accessible to visualize cell movements as well as to do transplantation experiments to identify areas that contribute to heart induction like the anterior endoderm. However, both organisms lack comprehensive genetic tools for mutational analysis to identify genes critical for heart formation. This task was taken primarily on organisms such as drosophila, zebrafish, and mouse, which have relatively small genome sizes and are amenable to gene manipulations and mutational screens. In addition, there are a number of available sophisticated genetic tools for relatively rapid cloning of mutated genes.

2.1 Drosophila

The drosophila heart is a linear tube called the dorsal vessel. Formation of the dorsal vessel resembles the early stages of vertebrate heart development – from the specification of mesodermal heart precursor cells to the formation of the linear heart tube. The similarities between vertebrates and drosophila in early heart tube formation and the conserved genetic networks governing this process serve as the basis for using drosophila to study heart development.

A seminal example of a gene discovered to be important in drosophila dorsal vessel formation, and subsequently found to be of importance in vertebrate heart development, is the drosophila homeobox transcription factor, *tinman*. Flies with mutations in *tinman* do not form dorsal vessels [17]. This discovery led to the identification of *Nkx2.5*, the *tinman* gene orthologue in other species. Mutations in the *Nkx2.5* gene in humans lead to a number of congenital heart disorders [18, 19].

Although drosophila continues to serve as an important model organism for identifying genes involved in early heart development, the simple structure of the linear dorsal vessel limits its usefulness as a model for obtaining a complete understanding of the mechanisms behind mammalian heart development. This is illustrated by the finding that unlike the situation in drosophila, the heart forms in *Nkx2.5* gene knockout mice, although looping morphogenesis is arrested [20].

2.2 Zebrafish

The attributes of the zebrafish embryo that make it a useful model for cardiovascular development include a closed circulatory system and a cardiac cycle that is similar to human physiology, an external development that is easily observed by microscopy, and an ability to obtain oxygen and nutrients by diffusion so that severe cardiac defects remain sub-lethal throughout embryogenesis. In addition to these embryological features, zebrafish has a well-annotated genome and a host of advanced genetic tools.

Forward genetic approaches have identified more than 35 zebrafish genes associated with defects in cardiovascular development [21]. Phenotypic and molecular analyses of zebrafish mutations have provided new insights in heart development and human congenital diseases. For example, the mutation *heartstrings* (*hst*) leads to abnormalities in cardiac differentiation and fin development as a result of a mutation in the transcription factor *Tbx5* [22]. Interestingly, in humans, mutations in *Tbx5* are associated with Holt–Oram syndrome, which is characterized by cardiac septation defects and limb abnormalities [23]. However, the zebrafish heart is more primitive than the heart of higher vertebrates having only two chambers, so not all aspects of heart development can be elucidated in this lower vertebrate system.

2.3 Mouse

The mouse has an anatomically similar heart to human and is amenable to genetic manipulation, including tissue-specific gain-of-function of genes of interest, targeted loss-of-function of genes by homologous recombination, conditional deletions, and lineage tracing by Cre–loxP strategies. In addition, embryological techniques like chimera generation and tetraploid embryo complementation have made the mouse a highly desirable model to understand the molecular mechanisms regulating every stage of mammalian heart development [24].

However, the small size of early-stage mouse embryos makes it challenging to access and study the initial stages of heart development and has thereby limited the utility of the mouse embryo for the study of early heart development events like cardiac precursor specification and lineage commitment. Gene manipulation techniques are also hard to control in a precise spatiotemporal manner and it is difficult to study their effects at the single cell level. Moreover, gain- and loss-of-function studies are expensive and time consuming, raising the need for an alternative system for systematic studies of gene function in the differentiation of cardiac cells.

3 Lingering Questions About Heart Development

Animal models of cardiac development have provided invaluable insight into the morphological and molecular events that occur during heart development, but as more effective therapies for congenital and degenerative heart diseases need to be developed, a more comprehensive understanding of the molecular mechanisms controlling the distinct morphological stages is necessary. Knowledge of the molecular machineries guiding mammalian cardiac precursor specification and lineage commitment would be especially useful for the expansion of cardiac cells that can potentially be used to replace damaged tissue after myocardial infarction or other degenerative heart diseases. The small size of the early-stage embryos of commonly used animal models makes it challenging to isolate and study early pre-cardiac and cardiac cells in order to better understand how cardiac development is initiated and regulated.

Embryonic stem (ES) cells are able to differentiate into a broad variety of organ-specific cell types in vitro including all cardiovascular lineages [25, 26]. Moreover, ES cells self-renew and thus practically expand indefinitely in culture offering a tool to investigate how lineages become specified and how they differentiate. This is particularly true for cardiovascular cells, because these cells appear early during both embryonic development and ES cell differentiation [27–29]. The in vitro differentiation of pluripotent stem cells into cardiomyocytes and other heart cells can thus be applied toward gaining information about the molecular events surrounding the earliest stages of mammalian heart development. The rest of this chapter summarizes representative examples of the use of ES cell differentiation as a model to understand the development of cardiac cells.

4 The In Vitro Differentiation of Embryonic Stem Cells into Cardiomyocytes Recapitulates the Early Stages of In Vivo Embryonic Heart Development

Mouse embryonic stem cells are isolated from the inner cell mass of E3.5 blastocyst-stage embryos [30, 31]. ES cells are pluripotent, forming teratocarcinomas when injected into adult mice [30]. When injected in blastocysts that are then implanted, ES cells are able to give rise to chimeric animals and contribute to all somatic lineages and germ cells [32].

ES cells can be expanded in culture and can differentiate into a large number of different cell types in vitro [33, 34], including cardiac cells [35, 36]. To synchronize their differentiation program in vitro, ES cells are first dissociated in single-cell suspension and then placed in hanging drops [37, 38]. The procedure allows cultivation of a defined number of ES cells in a single drop leading to formation of embryoid bodies (EBs) of homogeneous size, which is necessary for synchronized differentiation. After EBs form, they can be maintained in suspension in bacterial petri dishes or allowed to attach and spread on tissue culture plates.

Although EBs do not build a functional heart in culture, morphological and molecular analyses suggest that the cardiomyocytic differentiation program that becomes initiated can be compared to events that take place during heart development in the embryo. Like with embryonic heart development, cardiomyocyte differentiation in ES cells progresses through distinct stages. First, analogous to in vivo specification of cardiac precursors and the cardiac crescent stage which, respectively, occur on days 6.5 and 7.75 of mouse embryonic development [8], day 3.25 EBs are specified to give rise to the cardiac lineage and begin to express

Nkx and Gata transcription factors by day 4.25 [39]. Second, similar to the spontaneous contraction of the heart tube which occurs on mouse embryonic day 8 [40], in vitro specified cardiac cells begin beating on day 6 [38]. And, third, ES cell-derived cardiomyocytes display myofibrillar arrangements and electrophysiological properties that are consistent with atrial, ventricular, and conduction system myocytes by day 10 of their differentiation [37]. Therefore, the ES cell model can be applied to interrogate the molecular and cellular mechanisms that regulate the emergence and differentiation of the various cardiovascular lineages that are challenging to address in vivo. A number of selected examples are discussed below.

5 ES Cell Differentiation as a Model for Cardiac Fate Mapping

Classical lineage tracing experiments were performed in the chick embryo. The ease of performing surgical grafts in the developing chick made it the model of choice for fate mapping using isotopic and heterotopic chick-quail grafts [41, 42]. Complementary studies labeling specific regions with dyes or reporter gene constructs were also performed. These experiments showed that prospective myocardial cells originate from the anterior region of the primitive streak [4, 41] and prospective epicardial cells from the proepicardium [9, 43]. However, grafting studies in the avian embryos could not determine the origin of endocardial cells or determine whether cells in particular regions were fully committed to one lineage or had the potential to generate cells of other cardiac lineages. For example, it is not clear if cells from the anterior region of the primitive streak could become endothelial or smooth muscle cells, in addition to becoming cardiomyocytes [44].

Although the mouse embryo is not easily amenable to microsurgeries, the Cre–loxP strategy has been used to trace the fate of mammalian cardiac precursors. Interestingly, mouse genetic fate mapping added another level of complexity to the question of the developmental origin, fate, and potential of cardiac lineages. One mouse study, for example, showed that cardiac precursor cells expressing *Nkx2.5* give rise to endocardial cells in addition to previously described myocardial cells [45], thereby suggesting that the different cardiac lineages – endocardial, myocardial, and cardiac smooth muscle cells – may arise from a multipotent progenitor rather than from distinct region-restricted precursors as implied by the fate mapping studies from the chick embryo. However, a caveat of the Cre–loxP system for fate mapping is that a promoter can be turned on or off in different cell types so that labeled cells that express the gene may not necessarily be part of the same lineage. ES cells offer the possibility to generate early mesodermal cells in large numbers, isolate specific subpopulations, and assess their differentiation potential. To this end, three seminal studies applied the ES cell differentiation model to tackle the question of cardiac lineage fate and potential.

In the first study, Kattman and colleagues [46] isolated cells from very early-stage EBs at day 3.25, the time that mesodermal cells become specified but have not yet differentiated to particular cell types. The dissociated cells were separated into three groups based on their expression of the genes *T-brachyury* and *Flk-1 (VEGFR-2)*, two early mesodermal markers, as follows: $Bry^{neg}/Flk\text{-}1^{neg}$, $Bry^{pos}/Flk\text{-}1^{neg}$, and $Bry^{pos}/Flk\text{-}1^{pos}$. The different subpopulations were then allowed to differentiate and their fate was analyzed using a panel of cell-specific markers. The results show that contracting heart cells arose from the Bry^{pos}/Flk^{neg} population. Upon studying the differentiation potential of the Bry^{pos}/Flk^{neg} group in more detail, the investigators found that these cells generated a second Flk^{pos} population after 24 h in culture. Further analysis revealed that the second Flk^{pos} population was the source of the contracting myocardial cells and, moreover, also gave rise to endothelial and smooth muscle cells. Based on these observations, it appears that myocardial, endothelial, and cardiac smooth muscle cells arise from a common $Flk\text{-}1^{pos}$ progenitor. These results have been recently extended to human ES cells [29].

In another study, Wu and colleagues [47] applied the ES cell differentiation model to determine the potential of $Nkx2.5^{pos}$ progenitor cells. The researchers sorted cells from day 5 EBs, a time point at which cardiac precursors have been specified but not yet differentiated, into two groups: $Nkx2.5^{pos}$ and $Nkx2.5^{neg}$ cells. The differentiation potential of the cells was then analyzed. It was observed that the $Nkx2.5^{pos}$ group matured into myocardial and smooth muscle cells, therefore confirming that myocardial and smooth muscle cells share a common $Nkx2.5^{pos}$ progenitor.

The third study by Moretti and colleagues [48] traced the fate of Isl-1-positive progenitors. Preceding mouse embryo studies had shown that *Isl-1* gene expression marks the secondary heart field and is important for the development of the secondary heart field-derived heart segments – right ventricle, atria, and outflow tract [49]. However, it was unclear whether Isl-1-expressing progenitors gave rise to all the cardiac lineages or only to myocardial cells. Morreti and coworkers isolated Isl-1-positive cells from early-stage EBs and followed their differentiation fate and potential. They observed that the Isl-1-positive cells gave rise to subsets of Nkx2.5- and Flk-1-expressing cells that further differentiated into endothelial, myocardial, and smooth muscle cells. As a result, the researchers concluded that *Isl-1* expression denotes a common multipotential cardiac lineage progenitor.

These three studies showed that the ES cell differentiation model can be used to clarify questions regarding the origin, fate, and potential of cardiac lineages. Furthermore, they point to a clonal model of cardiac lineage derivation whereby a multipotential progenitor becomes sequentially committed into the various cardiac lineages.

6 Embryonic Stem Cell Differentiation to Elucidate the Molecular Cues Guiding the Specification of Mammalian Cardiac Progenitor Cells

At the cellular level, cardiogenesis involves the specification, proliferation, migration, and terminal differentiation of cardiac progenitor cells. Specification, the process through which naïve cells are inducted into the cardiac lineage, is the first stage of heart development. Mouse genetic fate maps show that cardiac progenitors in the anterior primitive streak express *Mesp-1* [50] and *Flk-1* [51]. By the time the cardiac progenitors form a crescent in the pre-cardiac mesoderm, they express the genes *Nkx2.5* [52]; *GATA-4,-5,-6* [53–55]; and *Tbx5* [56]. FGF-10 and Isl-1 expression by medially located secondary heart field progenitors can also be detected at this stage [48, 57]. Terminal differentiation into contractile myocytes occurs at the heart tube stage and is associated with the expression of proteins that include cardiac troponin (cTn), alpha myosin heavy chain (α-MHC), and atrial and ventricular myosin light chain 2 (MLC-2a and MLC-2v, respectively) [35–38, 58].

Tissue transplantation experiments performed in amphibian, avian, and mouse embryo explants at various stages of cardiac development have revealed the importance of the primitive endoderm (hypoblast in chick and anterior visceral endoderm in mouse) and definitive anterior lateral endoderm for cardiogenesis. The prevailing consensus is that primitive endoderm-derived signals induce cardiogenesis by acting on the pre-gastrula epiblast [59], while the anterior lateral endoderm facilitates cardiogenesis in the pre-cardiac mesoderm and can induce cardiogenesis in non-cardiogenic posterior blood-forming mesoderm [60, 61].

Hypoblast-secreted activin, nodal, and TGFβ, along with definitive endoderm-derived fibroblast growth factors (FGFs), bone morphogenetic proteins (BMPs), and Wnt antagonist proteins, have been implicated in the cardiogenic process. The roles these signals play in cardiac precursor specification, migration, proliferation, and terminal differentiation as obtained from drosophila, zebrafish, amphibian, and avian models are further reviewed below. In addition, new insights into the cardiogenic function and mechanism of action of these signals gained from the embryonic stem cell in vitro differentiation model are highlighted.

6.1 Activin, TGFβ1, and Nodal Secreted by the Primitive Endoderm Guide Epiblast Cells into the Mesendodermal Fate

Expression of the proteins Activin [62], TGFβ1 [63], and Nodal [64] has been detected in the primitive endoderm. Although Activin and TGFβ1 can substitute for hypoblast in inducing cardiogenesis in posterior epiblast explants [65], these proteins are unlikely to be the main cardiac mesoderm specification factors because in vivo heart development is only minimally affected upon their deletions; Activin-null mice have no defects in mesoderm formation [62] while TGFβ1-null mice primarily have immunological, but not cardiac defects [66–68].

On the other hand, Nodal-deficient mice die in utero with failure to form mesendoderm [64], though there is an initial overproliferation of ectoderm [69]. Mice lacking the Nodal receptor ALK4 also fail to form mesendoderm [70]. In light of the extreme phenotype of Nodal-deficient mice, the cardiogenic inductive abilities of Activin and TGFβ1 in epiblast explants are likely attributed to engaging Smad2 and Smad3 in the same manner as Nodal. Further evidence of the importance of Smad2, downstream of Nodal, Activin, or TGFβ1, to the formation of mesendoderm is presented with the embryonic stem cell differentiation model. Suppression of Smad2 in early-stage embryonic stem cells, before the formation of EBs, inhibits mesendoderm formation, but promotes the formation of neuroectoderm [71]. Taken together, these data suggest that primitive endoderm-derived factors, most significantly Nodal, signaling through Smad2/3 induce cardiogenesis by guiding epiblast cells into the mesendodermal fate.

6.2 BMPs and Wnts Play Stage-Specific Roles in Cardiogenesis

BMPs and canonical Wnt pathway inhibitors secreted by anterior lateral endoderm promote cardiogenesis in mesoderm explants [72, 73]. The importance of Wnt inhibition in inducing the cardiac fate is supported by experiments which showed that Cre-mediated deletion of the canonical Wnt signal transducer, *β-catenin*, from the mouse embryonic endoderm resulted in the formation of multiple hearts [74]. However, the global mouse knockout of *β-catenin* interrupted gastrulation, causing a failure in the development of all three germ layers and early embryonic lethality [75]. The likely explanation for these two different *β-catenin* mutant phenotypes is that canonical Wnt signaling activity is necessary for epiblast proliferation and the initial formation of cells of the three germ layers, including cardiac progenitor cells, but upon development of mesendoderm, canonical Wnt signaling inhibition favors the formation of cardiac mesoderm over endoderm and hematopoietic mesoderm.

Despite the fact that inhibition of canonical Wnt pathway promotes cardiogenesis in avian and amphibian pre-cardiac mesoderm explants, post-gastrulation inactivation of wg in drosophila results in the selective loss of cardiac precursors [76]. These confounding and opposite effects of Wnt inhibition were recently explained with the ES cell in vitro differentiation model. Naito and coworkers [77] showed that canonical Wnt signaling played biphasic roles in heart development; activation of canonical Wnt signaling in early EBs promoted cardiomyogenesis while activation of canonical Wnt signaling in late-stage EBs inhibited cardiomyocyte generation. These results suggest that canonical Wnt signaling promotes the expansion of cardiac precursors, but inhibits the terminal differentiation of specified cardiac cells. In support of this model of canonical Wnt action, a constitutively active mutant of *β-catenin* was shown to support the proliferation of Nkx2.5- and Isl-1-expressing cardiac progenitors [78]. The knockout of *β-catenin* in Isl-1-expressing cells resulted in hypomorphic right ventricle and OFT defects due to a significant decrease in cell proliferation and an increase in cell death [79]. In addition, Qyang and colleagues [80] observed that Wnt3a promoted the expansion but inhibited the terminal differentiation of ES cell-derived Isl-1-positive cardiac progenitors. Along with promoting the expansion of specified cardiac progenitors, canonical Wnt/β-catenin signaling appears to drive pre-cardiac cells into the secondary heart field lineage; Klaus and coworkers [81] observed that ablation of *β-catenin* in Mesp1-expressing cells caused a reduction in Isl-1 expression by secondary heart field progenitors and blocked the formation of second heart field-derived structures. Furthermore, the *Isl-1* promoter has been found to contain β-catenin-responsive LEF1 binding sites [79].

Taken together, these data support a model whereby canonical Wnt signaling acts within narrow developmental windows to play stage-specific roles in cardiogenesis. At initial stages, canonical Wnt signaling promotes the expansion of epiblast cells to support gastrulation, but inhibits the induction of mesendodermal cells into the cardiac lineage. Later on, canonical Wnt signaling activity promotes the expansion of Mesp-1-expressing cardiac precursors and drives specified cardiac progenitors into the secondary heart field lineage but then inhibits their terminal differentiation into contractile cardiomyocytes.

Thus, the initially seemingly contradictory effects of canonical Wnt inhibition in amphibian and avian explants in comparison to drosophila embryos can be explained by the times Wnt signaling was interrupted in these models. In the drosophila embryos, loss of cardiac precursors was observed because Wnt signaling was interrupted at a time in which Wnt activity is required for the expansion of cardiac progenitors, while cardiogenesis was observed in avian and chick explants because Wnt signaling was interrupted after progenitors had expanded and Wnt inhibition promoted their terminal differentiation into contractile myocytes and other cardiac cells.

Similarly, BMPs play stage-specific roles in heart development, and the ES cell in vitro differentiation model was useful in elucidating these roles. The importance of BMPs to cardiogenesis was initially discovered in drosophila where it was shown that decapentaplegic (DPP), the homologue of BMP, functions upstream of the transcription factor *tinman* [82]. A PCR screen for DPP-like molecules in anterolateral endoderm revealed the presence of BMP-2 [83]. When combined with FGF-4, BMP-2 is able to mimic anterolateral endoderm in inducing cardiogenesis in explants of non-cardiogenic posterior mesoderm [83]. Furthermore, the ectopic hearts formed upon ablation of *β-catenin* in embryonic mouse endoderm were associated with ectopic BMP-2 expression [74]. However, BMP-2 alone or in combination with FGF-4 cannot mimic hypoblast and Activin in inducing cardiogenesis in epiblast explants [65]. These data suggest that Activin/TGFβ/Nodal induce mesendoderm formation while BMPs drive mesendodermal cells into the cardiac lineage.

Although the mouse knockout of the type 1 BMP receptor revealed that BMP signaling is required for epiblast proliferation in order to generate all three germ layers during gastrulation [84], BMPs appear to have a negative effect on mesendoderm formation because treating early-stage ES cells with the BMP inhibitor noggin resulted in a twofold increase in the number of cells expressing the mesendodermal marker *T-brachyury* along with a sixfold increase in the level of T-brachyury expressed per cell [85].

Upon inducing mesendodermal cells to enter the cardiac lineage, BMP signaling appears to subsequently direct the generation of the primary heart field precursors; when the type 1 BMP receptor was ablated in Mesp1-expressing cells, the expression of the primary heart field markers Tbx5 and eHand could not be detected and primary heart field-derived structures failed to form [81]. Upon specification of the pre-cardiac mesoderm and the primary heart field, the continued presence of active BMP signals is required for the proliferation and terminal differentiation of cardiac progenitors. Barron and colleagues [86] observed that though a transient exposure to BMP was sufficient to specify cardiogenesis in a modest percentage of explants from non-cardiogenic posterior mesoderm, the continued presence of BMP supported cardiogenesis in 100% of the explants. In addition, noggin-mediated inhibition of BMP signaling in explants from specified pre-cardiac mesoderm resulted in a loss of Nkx2.5-expressing cardiac progenitors [87], and the knockout of noggin resulted in mice with thicker myocardium due to increased cell proliferation [88]. In summary, BMP signals play stage-specific roles in cardiogenesis initially promoting the proliferation of epiblast cells before gastrulation, then inhibiting the development of mesendoderm, but specifying mesendodermal cells into the cardiac lineage and then promoting the proliferation and terminal differentiation of specified cardiac progenitors.

6.3 FGFs Potentiate the Actions of Wnt and BMP Signals During Cardiogenesis

The presence of FGF-4 is required for BMP-2 to induce cardiogenesis in posterior mesoderm explants [83, 86]. FGF-8 is also able to induce ectopic cardiogenesis, but this only occurs in BMP-expressing regions [89]. In addition, the concomitant increase in FGF-3, -10, -16, and -20 was observed in Isl-1-expressing cardiac progenitors that had proliferated in response to the activation of the canonical Wnt signaling pathway [90]. These data indicate that FGFs cooperate with BMPs and Wnts in regulating cardiogenesis.

6.4 Cardiac Crescent Expressed Transcription Factors Do Not Induce the Cardiac Specification During Mammalian Heart Development

Altogether, drosophila, zebrafish, xenopus, chick, mouse, and the in vitro ES cell differentiation models have aided in identifying Nodal, BMPs, and Wnts as signals that control cardiac fate decisions and elucidated the timely roles these signals play in maintaining the balance between cell proliferation and lineage commitment/fate restriction at the various stages of heart development (Fig. 1). However, the genetic switch/transcription factor that triggers specification into the cardiac fate downstream of Wnt and BMP signals remains to be clearly identified.

Based on being expressed during the cardiac crescent stage of heart development, the transcription factors *Nkx2.5*, *GATA-4*, and *Isl-1* have been proposed as cardiac specification factors. *Nkx2.5* was suspected to be a cardiac specification factor because of studies in drosophila that showed that *tinman*, the orthologue of mammalian *Nkx2.5*, is a marker of early cardiac progenitors required for dorsal vessel formation [17]. However, targeted deletions of *Nkx2.5* in mice did not affect cardiac specification as heart development proceeded up to the contracting linear

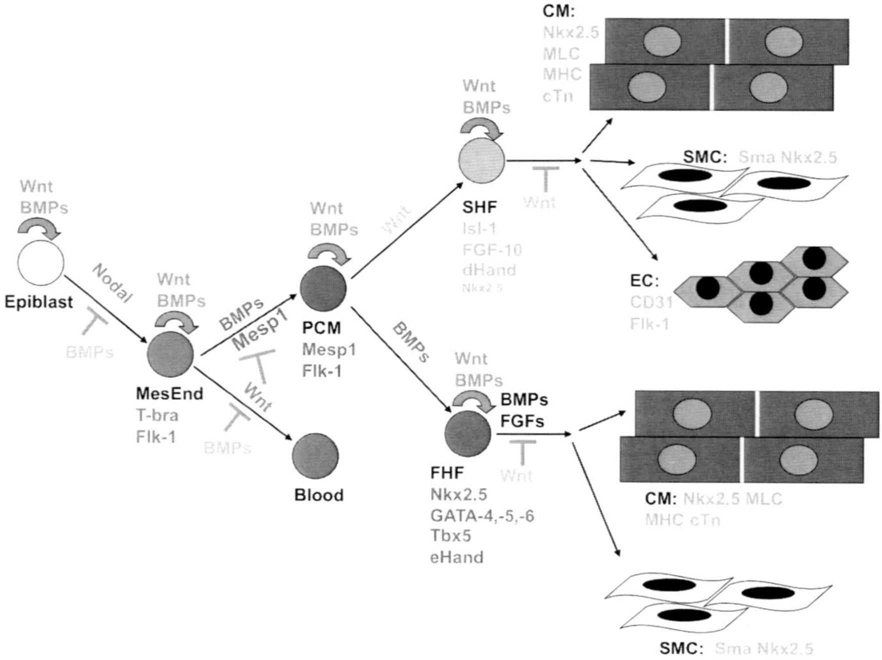

Fig. 1 Nodal, Wnt, and BMP signals direct cell fate decisions as a common multipotential cardiovascular progenitor becomes sequentially committed into the cardiac lineage. PCM – precardiac mesoderm; FHF – primary heart field progenitors; SHF – secondary heart field progenitors; CM – cardiomyocytes; SMC – smooth muscle cells; EC – endothelial cells

heart tube stage [20]. Moreover, Nkx2.5-null embryonic stem cells were able to differentiate into contracting cardiomyocytes in vitro [20]. Thus, *Nkx2.5* does not appear to be the mammalian cardiac specification factor.

GATA-4,-5,-6 are another set of transcription factors expressed in the cardiac crescent and initially presumed to be involved in cardiac fate induction. Like with *Nkx2.5*, the drosophila GATA homologue, *pannier*, is a cardiac identity gene [91]. However, cardiac specification occurs normally in mice with null mutations in GATA proteins. *GATA-4*-null mice develop cardia bifida [92] and *GATA-4*-null embryonic stem cells are able to differentiate into cardiomyocytes in vitro and in chimeric embryos [93]. *GATA-5*-null mice do not have problems with heart formation [94], and though *GATA-6*-null mice die soon after implantation [95], this has been attributed to its function in the development of extra-embryonic endoderm and not heart development; when tetraploid embryo complementation was used to circumvent the extra-embryonic need for GATA6, null embryos formed hearts [96]. The possibility of GATA factors playing redundant roles in cardiac specification was recently tested. Tetraploid embryo complementation was used to generate mice from GATA4/6 double-null embryonic stem cells; the embryos formed neither hearts nor contractile myocytes; however, the expression of early cardiac mesodermal markers *Mesp1* and *Mesp2* and crescent-associated cardiac progenitor markers *Nkx2.5* and *Isl-1* could be detected [97]. This suggests that GATA factors are important for cardiac progenitors to migrate and differentiate into contractile myocytes, but they might not be the initial cardiac specification factors.

Expression of the LIM homeodomain transcription factor *Isl-1* can also be detected at the cardiac crescent stage of heart development. Though the secondary heart field-derived structures fail to form in Isl-1-null mice, heart formation proceeds up to the contracting linear heart tube stage [49]. In addition, because the Isl-1-null mutation generated by Cai and colleagues [49] was a deletion in the LIM homeodomain, the researchers were able to track Isl-1 expression in mutant mice. They observed that Isl-1-positive cells were formed in the mutant embryos but were progressively lost from E8.5 to E10 due to decreased proliferation and increased apoptosis, suggesting that *Isl-1* is not required for the initial induction of the Isl-1-expressing secondary heart field lineage but is important for their survival and proliferation.

6.5 Mesp1 May Be the Cardiac Specification Factor

Evidence from mouse genetic mutations and ES cell differentiation in vitro indicates that cardiac crescent-associated factors do not specify mammalian cardiac fate. *Mesp1*, on the other hand, is currently the earliest gene known to be expressed by cardiac precursors [98]. Lineage tracing with Mesp1-cre showed that Mesp1-expressing precursors give rise to all the cells of the heart, irrespective of whether they are primary or secondary heart field derived [50, 98]. Though specification occurred in *Mesp1* knockout mice [50], this was later attributed to the compensatory action of *Mesp2*. Mesp1/2 double knockouts failed to generate cardiac mesoderm though primitive streak and mesendoderm were generated as detected by *T-brachyury* expression [99]. Because the observed defects in the formation of cardiac mesoderm could alternatively be interpreted as migration rather than specification defects, the Mesp1/2 double knockout was insufficient to conclude that Mesp1 is the cardiac specification transcription factor. Recent experiments in ES cells do, however, provide evidence that Mesp1 is likely the cardiac specification factor.

In the first study, David and colleagues [100] overexpressed human Mesp1 in mouse ES cells and allowed the cells differentiate in vitro. They observed a surge in beating areas and sprouting vascular structures. The increase in beating areas was associated with higher expression levels of cardiac contractile proteins along with induction of cardiac crescent-associated factors *Nkx2.5* and *GATA-4*. Analysis of earlier stages of ES cell differentiation revealed an increase in the expression of Flk-1 but no change in the levels of T-brachyury. Furthermore, injection of Mesp1-overexpressing plasmid DNA into two-cell stage xenopus embryos induced ectopic contracting cardiac tissue [100]. Their results suggest that Mesp1 is able to induce cardiac mesoderm formation.

In the second study, Lindsley and coworkers [101] generated ES cell lines with doxycycline-inducible expression of primitive streak and early mesoderm-associated transcription factors. They found that only Mesp1 was able to induce a Wnt-independent increase in Flk-1-positive cells. The increase in Flk-1-expressing cells was not attributed to increased cell proliferation, suggesting instead that Mesp1 recruited cells into the Flk-1 lineage. Detailed analysis of mesodermal cell lineages showed that Mesp1 promoted the formation of pre-cardiac mesoderm – as determined by analysis for the expression of early cardiac genes *Nkx2.5*, *Isl-1*, *GATA-4*, and other transcription factors. However, gain of Mesp1 function did not support the formation of hematopoietic cells or induce expression of T-brachyury and other markers of early mesendoderm. Furthermore, Mesp1 overexpression was associated with an increase in endothelial and smooth muscle cells. These results show that Mesp1 is able to direct cells to enter the three lineages of cardiac, endothelial, and smooth muscle cells.

The last study by Bondue and colleagues [102] showed that *Mesp1* acts as a key cardiac regulatory transcription factor. Similar to the findings outlined above, the transient expression of Mesp1 in differentiating ES cells enhanced the formation of cardiovascular progenitor cells, which resulted in a marked increase in the number of terminally differentiated cardiomyocytes, smooth muscle cells, and endothelial cells. Chromatin immunoprecipitation (ChIP) analysis showed that many of the key cardiac transcription factors upregulated by Mesp1 contain Mesp1-binding sites in their promoter regions. In addition, the researchers observed that Mesp1 represses the expression of key genes regulating pluripotency (*Oct4, Nanog, Sox2*), as well as early mesoderm (*T-brachyury*) and endoderm (*Sox17, Foxa2*) cell fates. Interestingly, repressed genes also contained Mesp1-binding sites in their regulatory regions, suggesting that Mesp1 may simultaneously induce cardiovascular cells and suppress other lineages.

Altogether, these studies demonstrate that the in vitro differentiation of the ES cells is a useful model for understanding the molecular cues regulating cardiac specification and other stages of early heart development. Furthermore, they provide strong support for the possibility that Mesp1 is the cardiac specification factor. However, it remains to be determined if Mesp1 or Mesp1/2 double knockout ES cells are able to generate cardiac progenitors and differentiate into cardiovascular cells in vitro.

7 Limitations of the Cultured Embryonic Stem Cells as a Model for Heart Development

The in vitro differentiation of pluripotent stem cells into cardiac cells can be used to study the molecular mechanisms governing the early stages of mammalian heart development and indeed has been used to gain new insights into the mechanisms of cardiac specification and lineage commitment and the roles of Wnt and BMP signals in these early events of embryonic development. However, ES cells do not generate a contractile, functional, four-chambered heart. As a result, the ES cell differentiation model may have limited use to study complex temporospatial events that occur during heart morphogenesis such as cardiac looping, left–right axis patterning, and septation.

8 Conclusions and Future Directions

The regulated in vitro differentiation of ES cells into cardiac cells has emerged as a compelling model to study the molecular and cellular mechanisms of mammalian heart development. The ES cell model has already yielded important new information about genes and signaling pathways that control the specification, lineage commitment, and maturation of cardiovascular cells. This knowledge, besides helping us to tease out the molecular interactions that take place during heart development, can be applied to generate new drug therapies for congenital and acquired heart diseases as well as to develop novel strategies for the expansion of cardiac progenitors for regenerative transplantation after myocardial infarction or other degenerative heart diseases using small molecular compounds [103]. The in vitro differentiation of human pluripotent stem cells into cardiomyocytes also offers the possibility of using the stem cell model to assess the cardiotoxicity of newly developed drugs, potentially decreasing the risks associated with drug safety studies performed in animals and ultimately human volunteers. Finally, there is a lot of excitement about the generation of inducible pluripotent stem (iPS) cells with ES cell characteristics from adult fibroblasts or epithelial cells [104, 105]. Information gained in ES cells would be instrumental in the quest to use autologous iPS cells for cardiac repair and regeneration or to utilize iPS cells as model systems to understand the molecular and cellular mechanisms of human cardiomyopathies.

Acknowledgements This work was supported by an American Heart Association Fellowship to OAA and NIH grants HL083958 and HL087403 to AKH.

References

1. Snarr BS, Kern CB, Wessels A. Origin and fate of cardiac mesenchyme. Dev Dyn. 2008 Oct; 237(10):2804–2819.
2. Brand T. Heart development: molecular insights into cardiac specification and early morphogenesis. Dev Biol. 2003; 258:1–19.
3. Olson EN. Gene regulatory networks in the evolution and development of the heart. Science 2006; 313(5795):1922–1927.
4. Martisen BJ. Reference guide to the stages of chick heart embryology. Dev Dyn. 2005; 233:1217–1237.
5. van den Berg G, Moorman AF. Concepts of cardiac development in retrospect. Pediatr. Cardiol. 2009; 30(5): 580–587.
6. Manner J. The anatomy of cardiac looping: a step towards the understanding of the morphogenesis of several forms of congenital cardiac malformations. Clin Anat. 2009; 1:21–35.
7. Waldo K, Kumiski D, Wallis K, Stadt H, Hutson M, Platt D, Kirby ML. Conotruncal myocardium arises from a secondary heart field. Development 2001; 128:3179–3188.

8. Buckingham M, Meilhac S, Zaffran S. Building the mammalian heart from two sources of myocardial cells. Nat Rev Genet. 2005; 6:826–835.
9. Ratajska A, Czarnowska E, Ciszek B. Embryonic development of the proepicardium and coronary vessels. Int J Dev Biol. 2008; 52:229–236.
10. Cai CL, Martin JC, Sun Y, et al. A myocardial lineage derives from Tbx18 epicardial cells. Nature 2008; 454(7200):104–108.
11. Zhou B, Pu T. More than a cover: epicardium as a novel source of cardiac progenitor cells. Regen Med. 2008; 5:633–635.
12. Waldo K, Miyagawa-Tomita S, Kumiski D, Kirby ML. Cardiac neural crest cells provide new insight into the septation of the cardiac outflow tract: aortic sac to ventricular septal closure. Dev Biol. 1998; 196:129–144.
13. Brown CB, Baldwin HS. Neural crest contribution to the cardiovascular system. Adv Exp Med Biol. 2006; 589:134–154.
14. Person A, Klewer S, Runyan R. Cell biology of cardiac cushion development. Int Rev Cytol. 2005; 243:287–324.
15. Armstrong EJ, Bishoff J. Heart valve development: endothelial cell signaling and differentiation. Circ Res. 2004; 95:459–470.
16. Kirby ML, Gale TF, Stewart DE. Neural crest cells contribute to aorticopulmonary septation. Science 1983; 220:1059–1061.
17. Bodmer R. The gene tinman is required for specification of the heart and visceral muscles in drosophila. Development 1993; 118:719–729.
18. Schott J, Benson D, Basson C, et al. Congenital heart disease caused by mutations in the transcription factor Nkx2.5. Science 1998; 281:108–111.
19. Pashmforoush M, Lu JT, Chen H, et al. Nkx2-5 pathways and congenital heart disease: loss of ventricular myocyte lineage specification leads to progressive cardiomyopathy and complete heart block. Cell 2004; 117:373–386.
20. Lyons I, Parsons L, Hartley L, Li R, Andrews J, Robb L, Harvey R. Myogenic and morphogenetic defects in the heart tubes of murine embryos lacking the homeobox gene Nkx2.5. Genes Dev. 1995; 9:1654–1666.
21. Schoenbeck JJ, Yelon D. Illuminating cardiac development: advances in imaging add new dimensions to the utility of zebrafish genetics. Semin Cell Dev Biol. 2007; 1:27–35.
22. Garrity D, Childs S, Fishman M. The heartstrings mutation in zebrafish causes heart/fin Tbx5 deficiency syndrome. Development 2002; 129:4635–4645.
23. Mori A, Bruneau B. TBX5 mutations and congenital heart disease: Holt-Oram syndrome revealed. Curr Opin Cardiol. 2004; 19:211–215.
24. Moon A. Mouse models of congenital cardiovascular disease. Curr Top Dev Biol. 2008; 84:171–248.
25. Doetschman TC, Eistetter H, Katz M, Schmidt W, Kemler R. The in vitro development of blastocyst-derived embryonic stem cell lines: formation of visceral yolk sac, blood islands and myocardium. J Embryol Exp Morphol. 1985; 87:27–45.
26. Murry CE, Keller G. Differentiation of embryonic stem cells to clinically relevant populations: lessons from embryonic development. Cell 2008; 132:661–680.
27. Kehat I, Kenyagin-Karsenti D, Snir M, et al. Human embryonic stem cells can differentiate into myocytes with structural and functional properties of cardiomyocytes. J Clin Invest. 2001; 108:407–414.
28. Xu C, Police S, Rao N, Carpenter MK. Characterization and enrichment of cardiomyocytes derived from human embryonic stem cells. Circ Res. 2002; 91:501–508.
29. Yang L, Soonpaa MH, Adler ED, et al. Human cardiovascular progenitor cells develop from a KDR+ embryonic stem-cell-derived population. Nature 2008; 453:524–528.
30. Evans MJ, Kaufman MH. Establishment in culture of pluripotential cells from mouse embryos. Nature 1981; 292:154–156.
31. Martin GR. Isolation of a pluripotent cell line from early mouse embryos cultured in medium conditioned by teratocarcinoma stem cells. Proc Natl Acad Sci USA. 1981; 78:7634–7638.

32. Roberston E, Bradley A, Kuehn M, Evans M. Germline transmission of genes introduced into cultured pluripotent stem cells. Nature 1986; 323(6087):445–448.
33. O'Shea KS. Embryonic stem cell models of development. Anat Rec New Anat. 1999; 257:32–41.
34. Keller G. Embryonic stem cell differentiation: emergence of a new era in biology and medicine. Genes Dev. 2005; 19:1129–1155.
35. Sachinidis A, Fleischmann B, Kolossov E, Wartenberg M, Sauer H, Hescheler J. Cardiac specific differentiation of mouse embryonic stem cells. Cardiovasc Res. 2003; 58:278–291.
36. Gerecht-Nir S, Fishman B, Itskovitz-Eldor J. Cardiovascular potential of embryonic stem cells. Anat Rec A 2004; 276A:58–65.
37. Hescheler J, Fleishmann B, Lentinis S, Maltsev V, Rohwedel J, Wobus A, Addicks K. Embryonic stem cells: a model to study structural and functional properties in cardiomyogenesis. Circ Res. 1997; 36:149–162.
38. Boheler KR, Czyz J, Tweedie D, Yang H, Anisimov S, Wobus A. Differentiation of pluripotent embryonic stem cells into cardiomyocytes. Circ Res. 2002; 91:189–201.
39. Kattman S, Adler E, Keller G. Specification of multipotential cardiovascular progenitor cells during embryonic stem cell differentiation and embryonic development. Trends Cardiovasc Med. 2007; 17:240–246.
40. Cohen ED, Tian Y, Morrisey EE. Wnt signaling: an essential regulator of cardiovascular differentiation, morphogenesis and progenitor self-renewal. Development 2008; 135:789–798.
41. Garcia-Martinez V, Schoenwolf G. Primitive streak origin of the cardiovascular system in avian embryos. Dev Biol. 1993; 159:706–719.
42. LeDouarin N, Dieterlen-Lièvre F, Creuzet S, Teillet MA. Quail-chick transplantations. Methods Cell Biol. 2008; 87:19–58.
43. Viragh S, Gittenberger-de Grot A, Poelmann R, Kalman F. Early development of quail heart epicardium and associated vascular and glandular structures. Anat Embryol (Berlin) 1993; 188(4):381–393.
44. Eisenberg C, Bader D. QCE-6: a clonal cell line with cardiac myogenic and endothelial cell potentials. Dev Biol. 1995; 167:469–481.
45. Stanley EG, Biben C, Elefanty A, Barnett L, Koentgen F, Robb L, Harvey R. Efficient cre-mediated deletion in cardiac progenitor cells conferred by a 3'UTR-Ires-Cre allele of the homeobox gene Nkx2.5. Int J Dev Biol. 2002; 46:431–439.
46. Kattman S, Huber T, Keller G. Multipotent Flk-1+ cardiovascular progenitor cells give rise to the cardiomyocyte, endothelial and vascular smooth muscle lineages. Dev Cell 2006; 11:723–732.
47. Wu S, Fujiwara Y, Cibulsky S, Clapham D, Lien C, Schultheiss T, Orkin S. Developmental origin of a bipotential myocardial and smooth muscle cell precursor in the mammalian heart. Cell 2006; 127:1–14.
48. Moretti A, Caron L, Nakano A, et al. Multipotent embryonic Isl1+ progenitor cells lead to cardiac, smooth muscle and endothelial cell diversification. Cell 2006; 127:1–15.
49. Cai C, Liang X, Shi Y, Chu P, Pfaff S, Chen J, Evans S. Isl1 identifies a cardiac progenitor that proliferated prior to differentiation and contributes a majority of cells to the heart. Dev Cell 2003; 5:877–889.
50. Saga Y, Miyagawa-Tomita S, Takagi A, Kitajima S, Miyazaki J. Mesp1 is expressed in the heart precursor cells and required for the formation of a single heart tube. Development 1999; 126:3437–3447.
51. Ema M, Takahashi S, Rossant J. Deletion of the selection cassette, but not *cis*-acting elements, in targeted Flk1-lacZ allele reveals Flk1 expression in multipotent mesodermal progenitors. Blood 2006; 107:111–117.
52. Lints T, Parsons L, Hartley L, Lyons I, Harvey R. Nkx2.5 a novel murine homeobox gene expressed in early heart cells and their myogenic descendants. Development 1993; 109:419–431.

53. Heikinheimo M, Scandrett J, Wilson DB. Localization of transcription factor GATA-4 to regions of the mouse embryo involved in cardiac development. Dev Biol. 1994; 164:361–373.
54. Morrisey EE, Ip HS, Tang Z, et al. GATA-5: a transcriptional activator expressed in a novel temporally and spatially-restricted pattern during embryonic development. Dev Biol. 1997; 183:21–36.
55. Morrisey EE, Ip HS, Lu MM, et al. GATA-6: a zinc finger transcription factor that is expressed in multiple lineages derived from lateral mesoderm. Dev Biol. 1996; 177:309–322.
56. Bruneau BG, Logan M, Davis N, et al. Chamber-specific cardiac expression of Tbx5 and heart defects in Holt Oram syndrome. Dev Biol. 1999; 211:100–108.
57. Kelly R, Brown N, Buckingham M. The arterial pole of the mouse heart forms from Fgf10-expressing cells in pharyngeal mesoderm. Dev Cell 2001; 1:435–440.
58. Lim S, Woodroofe N, Limanski LF. An analysis of contractile proteins in developing chick heart by SDS polyacrylamide gel electrophoresis and electron microscopy. J Embryol Exp Morphol. 1983; 77:1–14.
59. Yatskievych T, Ladd A, Antin P. Induction of cardiac myogenesis in avian pregastrula epiblast: the role of the hypoblast and activin. Development 1997; 124:2561–2570.
60. Schultheiss T, Xydas S, Lassar A. Induction of avian cardiac myogenesis by anterior endoderm. Development 1995; 121:4203–4214.
61. Sugi Y, Lough J. Activin-A and FGF-2 mimic the inductive effects of anterior endoderm on terminal cardiac myogenesis in vitro. Dev Biol. 1995; 169:567–574.
62. Mitrani E, Ziv T, Thomsen G, Shimoni Y, Melton DA, Bril A. Activin can induce the formation of axial structures and is expressed in the hypoblast of the chick. Cell 1990; 63(3):495–501.
63. Sanders EJ, Prasad S, Hu N. The involvement of TGF beta 1 in early avian development: gastrulation and chondrogenesis. Anat Embryol (Berlin) 1993; 187(6):573–581.
64. Conlon FL, Lyons KM, Takaesu N, Barth KS, Kispert A, Hermann B, Robertson EJ. A primary requirement for nodal in the formation and maintenance of the primitive streak in the mouse. Development 1994; 120(7):1919–1928.
65. Ladd A, Yatskievych T, Antin P. Regulation of avian cardiac myogenesis by activin/TGFβ and bone morphogenetic proteins. Dev Biol. 1998; 204:407–419.
66. Matzuk MM, Kumar TR, Vassalli A, Bickenbach JR, Roop DR, Jaenisch R, Bradley A. Functional analysis of activins during mammalian development. Nature 1995; 374(6520):354–356.
67. Kulkarni AB, Karlsson S. Transforming growth factor-beta 1 knockout mice. A mutation in one cytokine causes a dramatic inflammatory disease. Am J Pathol. 1993; 143:3–9.
68. Christ M, McCartney-Francis N, Kulkarni AB, et al. Immune dysregulation in TGF-beta 1-deficient mice. J Immunol. 1994; 153:1936–1946.
69. Zhou X, Sasaki H, Lowe L, Hogan BL, Kuehn MR. Nodal is a novel TGF-beta-like gene expressed in the mouse node during gastrulation. Nature 1993; 361:543–547.
70. Gu Z, Nomura M, Simpson B, Lei H, Feijen A, van den Eijnden-van Raaj J, Donahoe PK, Li E. The type 1 activin receptor ActRIB is required for egg cylinder organization and gastrulation in the mouse. Genes Dev. 1998; 6:844–857.
71. Kitamura R, Takahashi T, Nakajima N, Isodono K, Asada S, Ueno H, Ueyama T, Yoshikawa T, Matsubara H, Oh H. Stage-specific role of endogenous Smad2 activation in cardiomyogenesis of embryonic stem cells. Circ Res. 2007; 101:78–87.
72. Lough J, Sugi Y. Endoderm and heart development. Dev Dyn. 2000; 217:327–342.
73. Marvin M, Di Rocco G, Gardiner A, Bush S, Sive H, Lassar A. Inhibition of wnt activity induces heart formation from posterior mesoderm. Genes Dev. 2001; 15:316–327.
74. Lickert H, Kutsch S, Kanzler B, Tamai Y, Taketo M, Kemler R. Formation of multiple hearts in mice following deletion of beta-catenin in the embryonic endoderm. Dev Cell. 2002; 3:171–181.

75. Haegel H, Larue L, Ohsugi M, Fedorov L, Herrenknecht K, Kemler R. Lack of beta-catenin affects mouse development at gastrulation. Development 1995; 121:3529–3537.
76. Wu X, Golden K, Bodmer R. Heart development in drosophila requires the segment polarity gene wingless. Dev Biol. 1995; 169:619–628.
77. Naito A, Shiojima I, Akazawa H, Hidaka K, Morisaki T, Kikuchi A, Komuro I. Developmental stage-specific biphasic roles of wnt/β-catenin signaling in cardiomyogenesis and hematopoiesis. PNAS 2006; 103:19812–19817.
78. Kwon C, Arnold J, Hsiao EC, Taketo MM, Conklin BR, Srivastava D. Canonical wnt signaling is a positive regulator of mammalian cardiac precursors. PNAS 2007; 104:10894–10899.
79. Lin L, Cui L, Zhou W, Dufort D, et al. β-catenin directly regulates Islet1 expression in cardiovascular progenitors and is required for multiple aspects of cardiogenesis. PNAS 2007; 104:9313–9318.
80. Qyang Y, Puig S, Chiravuri M, et al. The renewal and differentiation of Isl+ cardiovascular progenitors are controlled by a wnt/β-catenin pathway. Cell Stem Cell 2007; 1:1–15.
81. Klaus A, Saga Y, Taketo M, Tzahor E, Birchmeier W. Distinct roles of wnt/β-catenin and Bmp signaling during early cardiogenesis. PNAS 2007; 104:18531–18536.
82. Frasch M. Induction of visceral and cardiac mesoderm by ectodermal Dpp in the early drosophila embryo. Nature 1995; 374:464–467.
83. Lough J, Barron M, Brogley M, Sugi Y, Bolender D, Zhu X. Combined BMP-2 and FGF-4 but neither factor alone, induces non-precardiac embryonic mesoderm. Dev Biol. 1996; 178:198–202.
84. Mishina Y, Suzuki A, Ueno N, Behringer R. Bmpr encodes a type 1 bone morphogenetic protein receptor that is essential for gastrulation during mouse embryogenesis. Genes Dev. 1995; 9:3027–3037.
85. Yuasa S, Itabashi Y, Koshimizu U, et al. Transient inhibition of BMP signaling by Noggin induces cardiomyocyte differentiation of mouse embryonic stem cells. Nat Biotechnol. 2005; 23:607–611.
86. Barron N, Gao M, Lough J. Requirement for BMP and FGF signaling during cardiogenic induction in non-precardiac mesoderm is specific, transient and cooperative. Dev Dyn. 2000; 218:383–393.
87. Schlange T, Andree B, Arnold H, Brand T. BMP2 is required for early heart development during a distinct time period. Mech Dev. 2000; 91:259–270.
88. Choi M, Stottmann R, Yang Y, Meyers EN, Klingensmith J. The bone morphogenetic protein antagonist Noggin regulates mammalian cardiac morphogenesis. Circ Res. 2007; 100:220–228.
89. Alsan B, Schultheiss T. Regulation of action cardiogenesis by Fgf8 signaling. Development 2002; 129:1935–1943.
90. Cohen ED, Wang Z, Lepore J, Lu M, Taketo M, Epstein D, Morrisey EE. Wnt/β-catenin signaling promotes expansion of Isl-1-positive cardiac progenitor cells through regulation of FGF signaling. JCI 2007; 117:1794–1804.
91. Gajewski K, Fossett N, Molkentin J, Kim Y, Choi C, Schulz R. The zinc finger proteins pannier and Gata4 function as cardiogenic factors in drosophila. Development 1999; 126:5679–5688.
92. Kuo C, Morrisey E, Anandappa R, Sigrist K, Lu M, Parmacek M, Soudais C, Leiden J. GATA 4 transcription factor is required for ventral morphogenesis and heart tube formation. Genes Dev. 1997; 11:1048–1060.
93. Narita N, Bielinska M, Wilson D. Wild-type endoderm abrogates GATA-4 deficiency in the mouse. Dev Biol. 1997; 189:270–274.
94. Molkentin J, Tymitz K, Richardson J, Olson E. Abnormalities of the genitourinary tract in female mice lacking GATA5. Mol Cell Biol. 2000; 20:5256–5260.
95. Morrisey EE, Tang Z, Sigrist K, Lu M, Jiang F, Ip H, Parmacek M. Gata6 regulates HNF4 and is required for differentiation of visceral endoderm in the mouse embryo. Genes Dev. 1998; 12:3579–3590.

96. Zhao R, Watt AJ, Li J, Luebke-Wheeler J, Morrisey EE, Duncan SA. GATA6 is essential for embryonic development of the liver but is dispensable for early heart formation. Mol Cell Biol. 2005; 25:2622–2631.
97. Zhao R, Watt A, Battle M, Li J, Bondow B, Duncan S. Loss of both GATA4 and GATA6 blocks cardiac myocyte differentiation and results in acardia in mice. Dev Biol. 2008; 317:614–619.
98. Saga Y, Kitajima S, Miyagawa-Tomita S. Mesp1 expression is the earliest sign of cardiovascular development. Trends Cardiovasc Med. 2000; 10:345–352.
99. Kitajima S, Takagi A, Inuoue T, Saga Y. Mesp1 and Mesp2 are essential for the development of cardiac mesoderm. Development 2000; 127:3215–3226.
100. David R, Brenner C, Stieber J, et al. Mesp1 drives vertebrate cardiovascular differentiation through Dkk-1 mediated blockade of wnt signaling. Nat Cell Biol. 2008; 10:338–345.
101. Lindsley R, Gill J, Murphy T, et al. Mesp1 coordinately regulates cardiovascular fate restriction and epithelial-mesenchymal transition in differentiating ESCs. Cell Stem Cell 2008; 3:55–68.
102. Bondue A, Lapouge G, Paulissen C, Semeraro C, Iacovino M, Kyba M, Blanpain C. Mesp1 acts as a master regulator of multipotent cardiovascular progenitor specification. Cell Stem Cell 2008; 3:69–84.
103. Hao J, Daleo MA, Murphy CK, et al. Dorsomorphin, a selective small molecule inhibitor of the BMP signaling, promotes cardiomyogenesis in embryonic stem cells. PLoS One 2008; 3:e2904.
104. Takahashi K, Tanabe K, Ohnuki M, et al. Induction of pluripotent stem cells from adult human fibroblasts by defined factors. Cell 2007; 131:861–872.
105. Okita K, Ichisaka T, Yamanaka S. Generation of germ-line competent induced pluripotent stem cells. Nature 2007; 448:313–317.

Index

Note: The letters 'f' and 't' following the locators refer to figures and tables respectively

G.M. Artmann et al. (eds.), *Stem Cell Engineering*,
DOI 10.1007/978-3-642-11865-4, © Springer-Verlag Berlin Heidelberg 2011

C

D

H